D0086348

BICENTENNIAL 1807 WILEY 2007 BICENTENNIAL

THE WILEY BICENTENNIAL–KNOWLEDGE FOR GENERATIONS

Each generation has its unique needs and aspirations. When Charles Wiley first opened his small printing shop in lower Manhattan in 1807, it was a generation of boundless potential searching for an identity. And we were there, helping to define a new American literary tradition. Over half a century later, in the midst of the Second Industrial Revolution, it was a generation focused on building the future. Once again, we were there, supplying the critical scientific, technical, and engineering knowledge that helped frame the world. Throughout the 20th Century, and into the new millennium, nations began to reach out beyond their own borders and a new international community was born. Wiley was there, expanding its operations around the world to enable a global exchange of ideas, opinions, and know-how.

For 200 years, Wiley has been an integral part of each generation's journey, enabling the flow of information and understanding necessary to meet their needs and fulfill their aspirations. Today, bold new technologies are changing the way we live and learn. Wiley will be there, providing you the must-have knowledge you need to imagine new worlds, new possibilities, and new opportunities.

Generations come and go, but you can always count on Wiley to provide you the knowledge you need, when and where you need it!

William J. Pesce — Peter Booth Wiley

WILLIAM J. PESCE
PRESIDENT AND CHIEF EXECUTIVE OFFICER

PETER BOOTH WILEY
CHAIRMAN OF THE BOARD

FUNDAMENTALS OF MOLECULAR

VIROLOGY

NICHOLAS H. ACHESON

Department of Microbiology and Immunology
McGill University

BICENTENNIAL 1807 WILEY 2007 BICENTENNIAL

JOHN WILEY & SONS

SENIOR EDITOR	Kevin Witt
PRODUCTION MANAGER	Kelly Tavares
PRODUCTION EDITOR	Courtney Leshko
EXECUTIVE MARKETING MANAGER	Clay Stone
SENIOR DESIGNER	Madelyn Lesure
SENIOR ILLUSTRATION EDITOR	Anna Melhorn
PHOTO RESEARCHER	Teri Stratford
COVER PHOTO	Courtesy of Dr. Jean-Yves Sgro, University of Wisconsin-Madison.

This book was set in 10/12 Janson by Matrix Publishing Services, printed and bound by Courier. The cover was printed by Phoenix Color.

This book is printed on acid free paper. ∞

Copyright @ 2007 John Wiley & Sons, Inc. All rights reserved. No part of this publication may be reproduced, stored in a retrieval system or transmitted in any form or by any means, electronic, mechanical, photocopying, recording, scanning or otherwise, except as permitted under Sections 107 or 108 of the 1976 United States Copyright Act, without either the prior written permission of the Publisher, or authorization through payment of the appropriate per-copy fee to the Copyright Clearance Center, Inc. 222 Rosewood Drive, Danvers, MA 01923, website www.copyright.com. Requests to the Publisher for permission should be addressed to the Permissions Department, John Wiley & Sons, Inc., 111 River Street, Hoboken, NJ 07030-5774, (201)748-6011, fax (201)748-6008, website http://www.wiley.com/go/permissions.

To order books or for customer service please, call 1-800-CALL WILEY (225-5945).

ISBN-13 978- 0-471-35151-1
ISBN-10 0-471-35151-2

Printed in the United States of America

10 9 8 7 6 5 4 3 2 1

I dedicate this book to four mentors whose enthusiasm for virology stimulated my interest when I was a student, and who encouraged me to follow my own path.

Johns Hopkins III
James D. Watson
Igor Tamm
Purnell Choppin

Preface

This book is written for students who are learning about viruses for the first time in a university course at the undergraduate or graduate level. As the title implies, it concentrates on the molecular mechanisms of virus replication, and on the interactions between viruses and the cells in which they replicate. The book approaches learning about virology by presenting a set of chapters each of which covers a specific virus family, using one or two well-studied viruses as examples. Each chapter is designed to tell a story about the virus under discussion, and to portray the "personality" of that virus, with the idea that this will help students to learn about and remember each virus.

This organizational scheme has been used in a number of successful virology textbooks, including Salvador Luria's classic 1953 book, *General Virology*. Luria was one of the founding members of the "phage group", a loose coalition of physicists, biologists and chemists who chose during the 1940's to study bacteriophages in order to understand the molecular basis of life, and who invented the field of molecular biology. Their approach was to study how the proteins and nucleic acids of viruses interact with cellular molecules and organelles, transforming the cell into a factory that can produce many new progeny virus particles. Their underlying hope, largely achieved, was to use viruses as a tool to help understand how cells work.

The amount of knowledge that has accumulated about viruses has expanded enormously in recent years, as in many other areas of biomedical sciences. *Fields Virology* has become the classic reference book for knowledge about human and animal viruses during the past 20 years; that book is also organized in chapters that cover specific virus families. My own teaching experience and conversations with numerous colleagues convinced me that there is a real need for a concise, up-to-date textbook organized around the concept of virus families and designed specifically for teaching.

The problem was to make such a book accessible for beginning students but not to oversimplify the material. My approach was to ask a number of prominent virology researchers and teachers to write chapters on viruses that they knew well, using a set of criteria that I provided. I then edited and sometimes rewrote these chapters into a common style, and in many cases I redesigned or created the illustrations. No individual can possibly write knowledgeably about the large spectrum of viruses that a virology course should cover, so a collaborative approach was necessary. However, a textbook that is an effective learning tool must have a coherent organization and a clear and consistent style of writing and illustration. My job has been to craft the original chapters that I received into what I hope are readable and easily understood units.

The emphasis of this textbook is on virus replication strategies; it is directed towards university students studying microbiology and the biomedical sciences. It does not go deeply into pathogenesis, epidemiology, disease symptoms, or host responses to virus infection. However, some information and stories about medical and historical aspects of virology are included, particularly in introductory sections of each chapter. Students who understand what diseases are caused by particular viruses, and the importance of these diseases in human history, may be motivated to learn more about those viruses.

How To Use This Book

This textbook is designed to be used in a modular fashion. No course would be expected to use all the chapters in the book, nor necessarily in same order in which they appear. The organization of the book gives wide latitude to course coordinators to make their own choices of which virus groups will be covered. Chapters are designed to accompany a 50-minute lecture on the subject, or in some cases, two or three such lectures. Each chapter should be able to be read in 30-60 minutes, including examination of figures and tables. Lecturers might want to supplement material given in the text with experimental methods or results, which are not covered because of lack of space.

The book is organized into nine sections and 34 chapters. Four introductory chapters in section I cover the virus life cycle, virus structure, virus classification, and the entry of viruses into animal cells. The chapters in section II cover four well-studied bacteriophages. These are included because bacteriophages are among the best known viruses, and because much of our knowledge of molecular biology and virology began with their study. Furthermore, bacteriophages are the source of many tools commonly used in modern molecular and cell biology laboratories.

Sections III through VII cover viruses that primarily infect animals, with emphasis on viruses that infect humans. The division into sections is based on the nature of the virus genome and virus replication strategies: DNA viruses (sections III and IV), positive-strand

RNA viruses (section V), negative strand and double-stranded RNA viruses (section VI), and viruses that use a reverse transcriptase (section VII). Within a section, smaller and simpler viruses are discussed first, then larger and more complex viruses. In this way, concepts that are learned about simpler viruses can be applied when more complex viruses are encountered.

Section VIII covers small infectious entities that are not viruses: viroids, which are virus-like nucleic acids that replicate but code for no proteins; and prions, which are infectious proteins that contain no detectable nucleic acid. Section IX finishes the book by reviewing some important applications in virology: interferons, antiviral chemotherapy, virus vectors, and vaccines.

Each chapter begins with an outline. For chapters that cover virus families, these outlines are "thumbnail sketches" that contain some basic information about virion structure, genome organization, replication strategies, diseases caused, and distinctive characteristics shared by viruses in that family. These outlines are designed to serve as study aids that will help students understand and remember common features of the viruses they study.

Subheadings within each chapter are explanatory phrases, telling the reader what will be discussed in the next several paragraphs. These subheadings (collected in the Table of Contents) can also be read separately to provide an overview of the material presented in the chapter, and to follow the steps of the virus replication cycle. Figures concentrate on individual well-studied steps in virus replication. Most figures are designed to be simple and easily understood while reading the accompanying text, rather than comprehensive descriptions of the entire replication cycle. Figure legends are kept to a minimum.

Specialized terms that may be unfamiliar to students are presented in bold type at their first appearance in the chapter. These terms are collected at the end of each chapter as a review aid, and definitions are given in a combined glossary at the end of the book. Many chapters have text boxes that cover intriguing applications or recent developments in research. A short list of references to reviews and recent articles is available at the end of each chapter. This is meant to provide an entry into the literature for students; no references are cited within the chapter texts.

Key Features of This Book

- A concise, up-to-date textbook designed for university-level virology courses for students in biomedical sciences and microbiology
- Written in a simple and clear style for students with a background in cell and molecular biology
- Explains replication mechanisms of viruses representing many of the major virus families
- Each chapter is designed to tell a story about a specific virus family and to portray the "personality" of the virus covered
- Chapter introductions give historical background and information about viral diseases
- Includes study aids such as thumbnail sketches of each virus group, informative chapter subheadings, text boxes outlining recent research and applications, and a comprehensive glossary with definitions of numerous terms
- Many two-color figures complement the text and illustrate virus structure, genome organization and individual steps in virus replication
- Introductory section provides basic information about virus replication, structure, classification and entry into cells
- A section on bacteriophages covers four of the best-known phages: single-stranded RNA phages, phiX174, T7 and lambda
- Five sections containing 20 chapters cover a wide variety of animal and insect viruses, with emphasis on viruses that cause human disease
- Includes chapters that cover important human pathogens such as smallpox virus, measles virus, poliovirus, herpes viruses, human immunodeficiency virus, hepatitis B virus, Ebola virus, SARS coronavirus, West Nile virus and others
- A separate chapter on viroids: small infectious nucleic acids that code for no proteins but cause important plant diseases
- A separate chapter on prions: infectious proteins that cause mad cow disease and Creutzfeld-Jacob disease in humans
- A concluding section with four chapters on interferons, antiviral chemotherapy, virus vectors, and vaccines

Acknowledgments

This textbook is an outgrowth of an undergraduate science course in virology taught by myself and colleagues in the Department of Microbiology and Immunology at McGill University during the past 25 years. I am grateful to Professors Dal Briedis, Mike DuBow, and John Hassell, with whom I collaborated in designing and offering this course. Their high academic standards and constant effort ensured its success. Among other colleagues who contributed significantly to this course during recent years are Alan Cochrane, Matthias Gotte, John Hiscott, Arnim Pause, and Mark Wainberg.

David Harris, then acquisitions editor at John Wiley and Sons, enthusiastically endorsed and welcomed my book project when I first proposed it. During its gestation, I have been ably helped by a succession of editors at Wiley: Joe Hefta, Keri Witman, Patrick Fitzgerald, and finally Kevin Witt.

A number of university and college teachers of virology reviewed the concept of the book, or parts of the manuscript at various stages, and offered helpful suggestions and comments. On behalf of the 40 people who contributed chapters to this book, I would like to thank the following reviewers:

Karen Beemon, *Johns Hopkins University*
Martha Brown, *University of Toronto*
Craig E. Cameron, *Pennsylvania State University*
Howard Ceri, *University of Calgary*
Jeffrey DeStefano, *University of Maryland, College Park*
Rebecca Ferrell, *Metropolitan State College of Denver*
Sidney Grossberg, *Medical College of Wisconsin*
Tarek Hamouda, *University of Michigan Medical Center*
Richard W. Hardy, *Indiana University*
Richard Kuhn, *Purdue University*
Alexander CK Lai, *Oklahoma State University*
Michael Roner, *The University of Texas at Arlington*
David A. Sanders, *Purdue University*
Jeff Sands, *Lehigh University*
Miroslav Sarac, *Our Lady of the Lake College*
Ann M. Sheehy, *College of the Holy Cross*
Suresh Subramani, *University of California, San Diego*
Carol St. Angelo, *Hofstra University*
Milton Taylor, *Indiana University*
Michael N. Teng, *Pennsylvania State University*
Darlene Walro, *Walsh University*
Jeannine Williams, *College of Marin*

Preliminary versions of a number of chapters from this book were made available to students taking our virology course at McGill from 2001 through 2004, and many of those students gave precious feedback that has improved the book. Furthermore, a number of chapters were read and reviewed in detail by the following McGill undergradaute students, who contributed insightful comments and suggestions: Jonathan Bertram, Yasmin D'Souza, Eric Fox, Caroline Lambert, Kathryn Leccese, Edward Lee, Alex Singer, Brian Smilovici, and Claire Trottier. Claire Trottier helped organize these student reviews.

Thanks to the following McGill University undergraduate students who worked with me on chapter summaries, permissions and editing in the final phases of preparation of the book: Meoin Hagege, Jennifer LeHuquet, Melany Piette, and Pooja Raut. Thanks also to Joan Longo and Mei Lee of the Department office for their help.

Work on this book began during a sabbatical year I spent in the laboratory of Steve Harrison at Harvard University. Thanks to McGill for approving my sabbatical leave, and to Steve and the members of the Harrison and Wiley laboratories for their stimulation and support.

Finally, this book is the result of an enjoyable and fruitful collaboration between myself and 39 other virologists from around the world who contributed chapters. Their expertise, energy and enthusiasm made this book possible. Thank you, all.

Brief Contents

I. INTRODUCTION

1. Introduction to Virology 1
Nicholas H. Acheson, McGill University
2. Virus Structure 17
Stephen C. Harrison, Harvard University
3. Virus Classification: The World of Viruses 30
Nicholas H. Acheson, McGill University
4. Virus Entry 43
Ari Helenius, Swiss Federal Institute of Technology Zurich

II. BACTERIOPHAGES

5. Single-Stranded RNA Bacteriophages 53
Jan van Duin, University of Leiden
6. Bacteriophage ϕ-X 174 63
Bentley Fane, University of Arizona
7. Bacteriophage T7 71
William C. Summers, Yale University
8. Bacteriophage Lambda: A Piñata of Paradigms 79
Michael Feiss, University of Iowa

III. SMALL DNA VIRUSES

9. Parvoviruses 89
Peter Beard, Swiss Institute for Experimental Cancer Research
10. Polyomaviruses 98
Nicholas H. Acheson, McGill University
11. Papillomaviruses 114
Greg Matlashewski, McGill University

IV. LARGER DNA VIRUSES

12. Adenoviruses 123
Phillip Branton, McGill University
Richard C. Marcellus, Geminix Biotech Inc.
13. Herpes Simplex Virus 134
Bernard Roizman, University of Chicago
Gabriella Campadelli-Fiume, University of Bologna
14. Baculoviruses 147
Eric Carstens, Queen's University
15. Poxviruses 157
Richard C. Condit, University of Florida

V. POSITIVE-STRAND RNA VIRUSES

16. Picornaviruses 169
Bert L. Semler, University of California, Irvine
17. Flaviviruses 182
Richard Kuhn, Purdue University
18. Togaviruses 191
Milton Schlesinger, Washington University in St. Louis
Sondra Schlesinger, Washington University in St. Louis
19. Coronaviruses 201
Mark Denison, Vanderbilt University
Sadie Coberley, Vanderbilt University

VI. NEGATIVE-STRAND AND DOUBLE-STRANDED RNA VIRUSES

20. Paramyxoviruses and Rhabdoviruses 214
Nicholas H. Acheson, McGill University
Daniel Kolakofsky, University of Geneva
Christopher Richardson, University of Toronto
21. Filoviruses 226
Hans-Dieter Klenk, University of Marburg
Heinz Feldmann, Public Health Agency of Canada, Winnipeg
22. Bunyaviruses 238
Richard M. Elliott, University of St. Andrews
23. Orthomyxoviruses 248
Dalius J. Briedis, McGill University
24. Reoviruses 262
Terence S.Dermody, Vanderbilt University
James D. Chappell, Vanderbilt University

VII. VIRUSES THAT USE A REVERSE TRANSCRIPTASE

25. Retroviruses 272
Alan Cochrane, University of Toronto
26. Human Immunodeficiency Virus Type 1 284
Alan Cochrane, University of Toronto
27. Human T-cell Leukemia Virus Type 1 294
John Hiscott, McGill University
Ya'el Mamane, McGill University
28. Hepadnaviruses 303
Christopher Richardson, University of Toronto
Robert G. Garces, University of Toronto

VIII. VIROIDS AND PRIONS

29. Viroids and Hepatitis Delta Virus 315
Jean-Pierre Perreault, Université de Sherbrooke
Martin Pelchat, University of Ottawa
30. Prions 323
Dalius J. Briedis, McGill University

IX. ANTIVIRAL AGENTS, VECTORS, AND VACCINES

31. Interferons 333
John Hiscott, McGill University
Pierre Genin, McGill University
32. Antiviral Chemotherapy 346
Donald M. Coen, Harvard University
33. Eukaryotic Virus Vectors 358
Rénald Gilbert, NRC Biotechnology Research Institute, Montreal
Bernard Massie, NRC Biotechnology Research Institute, Montreal
34. Viral Vaccines 370
Brian Ward, McGill University

Contents

I. INTRODUCTION

1. Introduction to Virology 1

THE NATURE OF VIRUSES 2
Viruses consist of a nucleic acid genome packaged in a protein coat 2
Viruses are dependent on living cells for their replication 2
Virus particles break down and release their genomes inside the cell 2
Virus genomes are either RNA or DNA, but not both 2

WHY STUDY VIRUSES? 3
Viruses are important disease-causing agents 3
Viruses can infect all forms of life 3
Viruses are the most abundant form of life on Earth 4
The study of viruses has led to numerous discoveries in molecular and cell biology 4

A BRIEF HISTORY OF VIROLOGY: THE STUDY OF VIRUSES 5
The scientific study of viruses is very recent 5
Viruses were first distinguished from other microorganisms by filtration 5
The crystallization of tobacco mosaic virus challenged conventional notions about genes and the nature of living organisms 5
The "phage group" stimulated studies of bacteriophages and helped found the field of molecular biology 7
Study of tumor viruses led to discoveries in molecular biology and understanding of the nature of cancer 7

DETECTION AND TITRATION OF VIRUSES 8
Most viruses were first detected and studied by infection of intact organisms 8
The plaque assay arose from work with bacteriophages 8
Eukaryotic cells cultured in vitro have been adapted for plaque assays 8
Hemagglutination is a convenient and rapid assay for many viruses 9
Virus particles can be seen and counted by electron microscopy 10
The ratio of physical virus particles to infectious particles can be much greater than 1 10

THE VIRUS REPLICATION CYCLE: AN OVERVIEW 10
The single-cycle virus replication experiment 10
An example of a virus replication cycle: mouse polyomavirus 11
Analysis of viral macromolecules reveals the detailed pathways of virus replication 12

STEPS IN THE VIRUS REPLICATION CYCLE 12
1. Virions bind to receptors on the cell surface 12
2. The virion (or the viral genome) enters the cell 12
3. Early viral genes are expressed: the Baltimore classification of viruses 13
The six groups in the Baltimore classification system 13
4. Early viral proteins direct replication of viral genomes 14
5. Late messenger RNAs are made from newly-replicated genomes 15
6. Late viral proteins package viral genomes and assemble virions 15
7. Progeny virions are released from the host cell 15

2. Virus Structure 17

The molecular structure of virus particles 17
How virus structure is studied: viruses come in a variety of sizes and shapes 18
Small viruses come in simple, symmetrical packages 18
Many virus capsids have icosahedral symmetry 18
Some examples of virions with icosahedral symmetry 21
The concept of quasi equivalence 21
How many subunits can be accomodated on the capsid surface? 22
Other structures, large and small, display icosahedral symmetry 23
Many virus capsids are organized as helical tubes 23
Larger viruses come in more complex packages 24
Specific packaging signals direct incorporation of viral genomes into virions 25
Core proteins may accompany the viral genome inside the capsid 25
Scaffolding proteins help in virion assembly but are not incorporated into the mature virion 25
Viral envelopes are made from lipid bilayer membranes 26
Viral glycoproteins are inserted into the lipid membrane to form the envelope 26
Budding is driven by interactions between viral proteins 27
Assembly and disassembly of virions: the importance of an irreversible step 27

3. Virus Classification: The World of Viruses 30

VIRUS CLASSIFICATION 30
Many viruses, infecting virtually all known life forms, have been discovered 30
Virus classification is based on molecular architecture, genetic relatedness, and host organism 31

Viruses are grouped into species, genera, and families 31
Distinct naming conventions and classification schemes have developed in different domains of virology 32

MAJOR VIRUS GROUPS 32
Study of the major groups of viruses leads to understanding of shared characteristics and replication pathways 32
Viruses with single-stranded DNA genomes are small and have few genes 33
Viruses with double-stranded DNA genomes include the largest known viruses 34
Most plant viruses and many viruses of vertebrates have positive-strand RNA genomes 35
All viruses with negative-strand RNA genomes have helical nucleocapsids and some have fragmented genomes 36
Viruses with double-stranded RNA genomes have fragmented genomes packaged in capsids with icosahedral symmetry 37
Viruses with a reverse transcription step in their replication cycle can have either RNA or DNA genomes 38
Satellite viruses and satellite nucleic acids require a helper virus to replicate 39
Viroids do not code for proteins, but replicate independently of other viruses 39

THE EVOLUTIONARY ORIGIN OF VIRUSES 40
The first steps in the development of life on earth: the RNA world 40
Viroids and RNA viruses may have originated in the RNA world 40
The transition to the DNA-based world 40
Small and medium-sized DNA viruses could have arisen as independently-replicating genetic elements in cells 41
Large DNA viruses could have evolved from cells that became obligatory intracellular parasites 41

4. Virus Entry 43

How do virions get into eukaryotic cells? 43
Enveloped and nonenveloped viruses have distinct penetration strategies 44
Some viruses can pass directly from cell to cell 44
A variety of cell surface proteins can serve as specific virus receptors 45
Receptors interact with viral glycoproteins, surface protrusions, or 'canyons' in the surface of the virion 45
Many viruses enter the cell via receptor-mediated endocytosis 45
Passage from endosomes to the cytosol is often triggered by low pH 47
Membrane fusion is mediated by specific viral "fusion proteins" 47
Fusion proteins undergo major conformational changes that lead to membrane fusion 48
Nonenveloped viruses penetrate by membrane lysis or pore formation 49
Virions and capsids are transported within the cell in vesicles or on microtubules 49
Import of viral genomes into the nucleus 50
The many ways in which viral genomes are uncoated and released 50

5. Single-Stranded RNA Bacteriophages 53

The discovery of RNA phages stimulated research into messenger RNA function and RNA replication 53
RNA phages are among the simplest known organisms 54
Two genera of RNA phages have subtle differences 54
RNA phages bind to the F-pilus and use it to insert their RNA into the cell 55
Phage RNA is translated and replicated in a regulated fashion 55
RNA secondary structure controls translation of lysis and replicase genes 56
Ribosomes translating the coat gene disrupt secondary structure, allowing replicase translation 57
Ribosomes terminating coat translation can reinitiate at the lysis gene start site 57
Replication versus translation: competition for the same RNA template 58
Genome replication requires four host cell proteins plus the replicase 58
A host ribosomal protein directs polymerase to the coat start site 59
Polymerase skips the first A residue but adds a terminal A to the minus strand copy 59
Synthesis of plus strands is less complex and more efficient than that of minus strands 59
The start site for synthesis of maturation protein is normally inaccessible to ribosomes 61
Synthesis of maturation protein is controlled by delayed RNA folding 61
Assembly and release of virions 62

6. Bacteriophage φX 174 63

φX174: a tiny virus with a big impact 63
Overlapping reading frames allow efficient use of a small genome 64
φX174 binds to glucose residues in lipopolysaccharide on the cell surface 65
φX174 delivers its genome into the cell through spikes on the capsid surface 66
Stage I DNA replication generates double-stranded replicative form DNA 66
Gene expression is controlled by the strength of promoters and transcriptional terminators 66
Replicative form DNAs are amplified via a rolling circle mechanism 67
Summary of viral DNA replication mechanisms 67
Procapsids are assembled by the use of scaffolding proteins 67

Scaffolding proteins have a flexible structure 68
Single-stranded genomes are packaged into procapsids as they are synthesized 68
Role of the J protein in DNA packaging 69
Cell lysis caused by E protein leads to release of phage 69
Did all icosahedral ssDNA virus families evolve from a common ancestor? 69

7. Bacteriophage T7 71

T7: a model phage for DNA replication, transcription, and RNA processing 71
T7 genes are organized into three groups based on transcription and gene function 72
Entry of T7 DNA into the cytoplasm is powered by transcription 73
Transcription of class II and III genes requires a novel T7-coded RNA polymerase 73
Class II genes code for enzymes involved in T7 DNA replication 74
T7 RNAs are cleaved by host cell ribonuclease III to smaller, stable mRNAs 74
Regulation of class III gene expression 74
DNA replication starts at a unique internal origin and is primed by T7 RNA polymerase 75
Large DNA concatemers are formed during replication 76
Concatemers are processed and packaged into preformed proheads 76
Special features of the T7 family of phages 76

8. Bacteriophage Lambda: A Piñata of Paradigms 79

In the beginning . . . 80
Uptake of λ DNA depends on cellular proteins involved in sugar transport 80
The λ lytic transcription program is controlled by termination and antitermination of RNA synthesis at specific sites on the genome 81
The CI repressor blocks expression of the lytic program by regulating three nearby promoters: P_L, P_R, and P_{RM} 82
Cleavage of CI repressor in cells with damaged DNA leads to prophage induction 83
The Cro repressor suppresses CI synthesis and regulates early gene transcription 83
Making the decision: Go lytic or lysogenize? 83
A quick review 85
Breaking and entering: The insertion of λ DNA into the bacterial chromosome 85
The great escape: The liberation of λ DNA from the bacterial chromosome 86
Int synthesis is controlled by retroregulation 86
λ DNA Replication is directed by O and P, but carried out by host cell proteins 87
Assembly of λ heads involves chaperone and scaffolding proteins 87
DNA is inserted into preformed proheads by an ATP-dependent mechanism 87
Host cell lysis 88

9. Parvoviruses 89

Parvoviruses have very small virions and a linear, single-stranded DNA genome 89
Parvoviruses replicate in cells that are going through the cell cycle 90
Discovery of mammalian parvoviruses 90
Parvoviruses have one of the simplest known virion structures 91
Parvoviruses have very few genes 91
Single-stranded parvovirus DNAs have unusual terminal structures 92
Uncoating of parvovirus virions takes place in the nucleus and is cell-specific 92
DNA replication begins by extension of the 3′ end of the terminal hairpin 93
The DNA "end replication" problem 93
Steps in DNA replication 93
Non-structural proteins are multifunctional 95
Adenovirus functions that help AAV replication 96
In the absence of helper virus, AAV DNA can integrate into the cell genome 96
Parvovirus pathogenesis: the example of B19 virus 96

10. Polyomaviruses 98

Mouse polyomavirus was discovered as a tumor-producing infectious agent 98
Simian virus 40 was found as a contaminant of Salk poliovirus vaccine 99
Polyomaviruses are models for studying DNA virus replication and tumorigenesis 99
Polyomavirus capsids are constructed from pentamers of the major capsid protein 99
The circular DNA genome is packaged with cellular histones 100
Circular DNA becomes supercoiled upon removal of histones 100
Supercoiled DNA can be separated from relaxed or linear DNA molecules 101
Polyomavirus genes are organized in two divergent transcription units 102
Virions enter cells in caveolae and are transported to the nucleus 102
The viral minichromosome is transcribed by cellular RNA polymerase II 103
Four early mRNAs are made by differential splicing of a common transcript 104
T antigens share common N-terminal sequences but have different C-terminal sequences 105
T antigens bring resting cells into the DNA synthesis (S) phase of the cell cycle 105

Small T antigen inhibits protein phosphatase 2A and induces cell cycling 105
Middle T antigen stimulates protein tyrosine kinases that signal cell proliferation and division 105
Large T antigen activates or suppresses transcription of cellular genes by binding to a number of important cellular regulatory proteins 107
Large T antigen hexamers bind to the origin of DNA replication and locally unwind the two DNA strands 108
Large T antigen hexamers assemble cellular DNA synthesis machinery to initiate viral DNA replication 110
High levels of late transcripts are made after DNA replication begins 111
Three late mRNAs are made by alternative splicing 112
How do polyomaviruses transform cells in vitro and cause tumors *in vivo*? 112

11. Papillomaviruses 114

Papillomaviruses cause warts and other skin and mucosal lesions 114
Oncogenic human papillomaviruses are a major cause of genital tract cancers 115
Papillomaviruses are not easily grown in cell culture 115
Papillomavirus genomes are circular, double-stranded DNA 116
The infectious cycle follows differentiation of epithelial cells 116
Viral mRNAs are made from two promoters and two polyadenylation signals 117
Viral E1 and E2 proteins bind to the replication origin and direct initiation of DNA replication 118
Viral E7 protein interacts with cell cycle regulatory proteins, particularly Rb 118
Viral E6 protein controls the level of cellular p53 protein 120
Synergism between E6 and E7 and the predisposition to cancer 121
Cells transformed by papillomaviruses express E6 and E7 gene products from integrated viral DNA 121
Future prospects for diagnosis and treatment of diseases caused by papillomaviruses 121

12. Adenoviruses 123

Adenoviruses cause respiratory and enteric infections in humans 124
Adenoviruses can be oncogenic, but not in humans 124
Virions have icosahedral symmetry and are studded with knobbed fibers 124
Fibers make contact with cellular receptor proteins to initiate infection 124
Expression of adenovirus genes is controlled at the level of transcription 126
E1A proteins are the kingpins of the adenovirus growth cycle 127
E1A proteins bind to the retinoblastoma protein and activate E2F, a cellular transcription factor 127
E1A proteins also activate other cellular transcription factors 128
E1A proteins indirectly induce apoptosis by activation of cellular p53 protein 128
E1B proteins suppress E1A-induced apoptosis, allowing viral replication to proceed 128
The preterminal protein primes DNA synthesis carried out by viral DNA polymerase 129
Single-stranded DNA is circularized via the inverted terminal repeat 130
The major late promoter is activated after DNA replication begins 131
Five different poly(A) sites and alternative splicing generate multiple late mRNAs 131
The tripartite leader ensures efficient transport of late mRNAs to the cytoplasm 132
The tripartite leader directs efficient translation of late adenovirus proteins 132
Adenoviruses kill cells by apoptosis, aiding virus release 132
Cell transformation and oncogenesis by human adenoviruses 132

13. Herpes Simplex Virus 134

Herpesviruses are important human pathogens 135
Most herpesviruses can establish latent infections 135
Herpes simplex virus genomes contain both unique and repeated sequence elements 135
Nomenclature of herpes simplex virus genes and proteins 137
The icosahedral capsid is enclosed in an envelope along with tegument proteins 137
Entry by fusion is mediated by envelope glycoproteins and may occur at the plasma membrane or in endosomes 138
Viral genes are sequentially expressed during the replication cycle 138
Tegument proteins interact with cellular machinery to activate viral gene expression and to degrade cellular messenger RNAs 139
Immediate early (α) genes regulate expression of other herpesvirus genes 140
β gene products set the stage for viral DNA replication 140
Herpesvirus begins with bidirectional DNA replication 141
Rolling-circle replication subsequently produces multimeric concatemers of viral DNA 141
DNA replication leads to activation of γ_1 and γ_2 genes 142
Viral nucleocapsids are assembled on a scaffold in the nucleus 143
Envelopment and egress: three possible routes 143
Many viral genes are involved in blocking host responses to infection 143
The establishment and maintenance of virus latency 145
Latency-associated transcripts include stable introns 145

14. Baculoviruses 147

Insect viruses were first discovered as pathogens of silkworms 148
Baculoviruses are used for pest control and to express eukaryotic proteins 148
Baculoviruses produce two kinds of particles: "budded" and "occlusion-derived" virions 149
Baculoviruses have large, circular DNA genomes and encode many proteins 150
Insects are infected by ingesting occlusion bodies; infection spreads within the insect via budded virions 151
Viral proteins are expressed in a timed cascade regulated at the transcription level 152
Immediate early gene products control expression of early genes 152
Early gene products regulate DNA replication, late transcription, and apoptosis 153
Late genes are transcribed by a novel virus-coded RNA polymerase 153
Baculoviruses are widely used to express foreign proteins 156

15. Poxviruses 157

Smallpox was a debilitating and fatal worldwide disease 158
Variolation led to vaccination, which has eradicated smallpox worldwide 158
Poxviruses remain a subject of intense research interest 159
Linear vaccinia virus genomes have covalently sealed hairpin ends and lack introns 159
Two forms of vaccinia virions have different roles in spreading infection 160
Poxviruses replicate in the cytoplasm 162
Poxvirus genes are expressed in a regulated transcriptional cascade controlled by viral transcription factors 162
Virus-coded enzymes packaged in the core carry out early RNA synthesis and processing 162
Enzymes that direct DNA replication are encoded by early mRNAs 163
Poxviruses produce large concatemeric DNA molecules that are resolved into monomers 164
Postreplicative mRNAs have 5′ end poly(A) extensions and 3′ end heterogeneity 164
Mature virions are formed within virus "factories" 165
Extracellular virions are extruded through the plasma membrane by actin tails 166
Poxviruses make several proteins that target host immune defenses 167

16. Picornaviruses 169

Picornaviruses cause a variety of human and animal diseases including poliomyelitis and the common cold 170
Poliovirus: a model picornavirus for vaccine development and studies of replication 170
Picornavirus virions bind to cellular receptors via depressions or loop regions on their surface 171
Genome RNA may pass through pores formed in cell membranes by capsid proteins 171
Translation initiates on picornavirus RNAs by a novel internal ribosome entry mechanism 172
Essential features of picornavirus IRES elements 173
Interaction of picornavirus IRES elements with host cell proteins 175
Picornavirus proteins are made as a single precursor polyprotein that is autocatalytically cleaved by viral proteinases 176
Picornaviruses make a variety of proteinases that cleave the polyprotein and some cellular proteins 176
Replication of picornavirus RNAs is initiated in a multiprotein complex bound to proliferated cellular vesicles 176
RNA synthesis is primed by VPg covalently bound to uridine residues 177
Virion assembly involves cleavage of VP0 to VP2 plus VP4 178
Inhibition of host cell macromolecular functions 179

17. Flaviviruses 181

Flaviviruses cause several important human diseases 182
Yellow fever is a devastating human disease transmitted by mosquitoes 182
A live, attenuated yellow fever virus vaccine is available and widely used 183
Hepatitis C virus: a recently discovered member of the Flaviviridae 183
The flavivirus virion contains an icosahedral nucleocapsid wrapped in a tightly fitted envelope 183
Flavivirus E protein directs both binding to receptors and membrane fusion 184
Flaviviruses enter the cell by pH-dependent fusion 185
Flavivirus genome organization resembles that of picornaviruses 185
The polyprotein is processed by both viral and cellular proteinases 186
Nonstructural proteins organize protein processing, viral RNA replication, and capping 187
Flavivirus RNA synthesis is carried out on membranes in the cytoplasm 188
Virus assembly also takes place at intracellular membranes 189

18. Togaviruses 191

Most togaviruses are arthropod borne, transmitted between vertebrate hosts by mosquitoes 192
Togavirus virions contain a nucleocapsid with icosahedral symmetry wrapped in an envelope of the same symmetry 192
Togaviruses enter cells by low pH-induced fusion inside endosome vesicles 193

Nonstructural proteins are made as a polyprotein that is cleaved by a viral protease 193
Partly-cleaved nonstructural proteins catalyze synthesis of full-length antigenome RNA 194
Replication and transcription: synthesis of genome and subgenomic RNAs 196
Structural proteins are cleaved during translation and directed to different cellular locations 196
Assembly of virions and egress at the plasma membrane 197
Effects of mutations in viral proteins on cytopathic effects and on pathogenesis 198
Alphaviruses have been modified to serve as vectors for the expression of heterologous proteins 199
Alphavirus vectors have multiple potential uses 199

19. Coronaviruses 201

Coronaviruses cause common colds in humans and important veterinary diseases 202
A newly emerged coronavirus caused a worldwide epidemic of severe acute respiratory syndrome (SARS) 202
The SARS coronavirus may have passed from animals to humans via direct contact 202
Coronaviruses have large, single-stranded, positive sense RNA genomes 203
Coronaviruses fall into three groups based on genome sequences 203
Coronaviruses have enveloped virions containing helical nucleocapsids 204
Coronavirus virions contain multiple envelope proteins 204
Coronavirus spike proteins bind to a variety cellular receptors 205
The virus envelope fuses with the plasma membrane or an endosomal membrane 206
The replicase gene is translated from genome RNA into a polyprotein that is processed by viral proteinases 206
RNA polymerase, RNA helicase, and RNA modifying enzymes are coded by the replicase gene 207
Replication complexes are associated with cytoplasmic membranes 207
Genome replication proceeds via a full-length negative-strand intermediate 208
Transcription produces a nested set of subgenomic mRNAs 208
Subgenomic mRNAs are most likely transcribed from subgenomic negative-sense RNA templates 208
The alternative model of discontinuous transcription of antigenome RNA is unlikely to be correct 209
Assembly of virions takes place at intracellular membrane structures 211
Adaptability of Coronaviruses 212

20. Paramyxoviruses and Rhabdoviruses 214

The mononegaviruses: a group of related negative-strand RNA viruses 215
Rabies is a fatal human encephalitis caused by a rhabdovirus 215
Measles is a serious childhood disease caused by a paramyxovirus 215
Paramyxovirus and rhabdovirus virions have distinct morphologies 216
Viral envelope proteins are responsible for receptor binding and fusion with cellular membranes 217
Genome RNA is contained within helical nucleocapsids 218
Paramyxoviruses enter the cell by fusion with the plasma membrane at neutral pH 218
Gene order is conserved among different paramyxoviruses and rhabdoviruses 219
Viral messenger RNAs are synthesized by an RNA polymerase packaged in the virion 220
Viral RNA polymerase initiates transcription exclusively at the 3' end of the viral genome 220
The promoter for plus-strand RNA synthesis consists of two sequence elements separated by one turn of the ribonucleoprotein helix 220
mRNAs are synthesized sequentially from the 3' to the 5' end of the genome RNA 222
The P/C/V gene codes for several proteins by using alternative translational starts and by mRNA "editing" 223
Functions of P, C and V proteins 224
N protein levels control the switch from transcription to genome replication 224
Virions are assembled at the plasma membrane 224

21. Filoviruses 226

Marburg and Ebola viruses: sporadically emerging viruses that cause severe, often fatal disease 227
Filoviruses are related to paramyxoviruses and rhabdoviruses 228
Filoviruses cause hemorrhagic fever 228
Filovirus genomes contain seven genes in a conserved order 228
Filovirus transcription, replication and assembly 230
Cloned cDNA copies of viral mRNAs and viral genome RNA are used for study of filoviruses 230
Multi-plasmid transfection systems allow recovery of infectious filoviruses 230
Filovirus glycoprotein mediates both receptor-binding and entry by fusion 231
Ebola virus uses RNA editing to make two glycoproteins from the same gene 232
Does the secreted glycoprotein play a role in virus pathogenesis? 232
Minor nucleocapsid proteinVP30 activates viral mRNA synthesis in Ebola virus 233
Matrix protein VP40 directs budding and formation of filamentous particles 234
Most filovirus outbreaks have occurred in equatorial Africa 234

Filovirus infections are transmitted to humans from an unknown animal origin 235
Spread of filovirus infections among humans is limited to close contacts 235
Pathogenesis of filovirus infections 236
Clinical features of infection 236

22. Bunyaviruses 238

Most bunyaviruses are transmitted by arthropod vectors, including mosquitoes and ticks 239
Some bunyaviruses cause severe hemorrhagic fever, respiratory disease, or encephalitis 240
Bunyaviruses encapsidate a segmented RNA genome in a simple enveloped particle 240
Bunyavirus protein coding strategies: negative-strand and ambisense RNAs 240
L RNA codes for viral RNA polymerase 241
M RNA codes for virion envelope glycoproteins 242
S RNA codes for nucleocapsid protein and a nonstructural protein 243
After attachment via the virion glycoproteins, bunyaviruses enter the cell by endocytosis 243
Bunyavirus mRNA synthesis is primed by the capped 5? ends of cellular mRNAs 243
Coupled translation and transcription may prevent premature termination of mRNAs 244
Genome replication begins once sufficient N protein is made 244
Virus assembly takes place at Golgi membranes 245
Evolutionary potential of bunyaviruses via genome reassortment 246

23. Orthomyxoviruses 248

Influenza viruses cause serious acute disease in humans, and occasional pandemics 249
Influenza virus infections of the respiratory tract can lead to secondary bacterial infections 249
Orthomyxoviruses are negative-strand RNA viruses with segmented genomes 249
Eight influenza virus genome segments code for a total of ten different viral proteins 251
Hemagglutinin protein binds to cell receptors and mediates fusion of the envelope with the endosomal membrane 252
M_2 is an ion channel that facilitates release of nucleocapsids from the virion 252
Nucleocapsids enter the nucleus, where mRNA synthesis and RNA replication occur 253
Capped 5′ ends of cellular pre-messenger RNAs are used as primers for synthesis of viral mRNAs 254
Viral mRNAs terminate in poly(A) tails generated by "stuttering" transcription 255
Two influenza A mRNAs undergo alternative splicing in the nucleus 255
Genome replication begins when newly synthesized NP protein enters the nucleus 255
Nucleocapsids are exported from the nucleus in a complex with matrix protein and NS2 256
The NS1 protein interferes with polyadenylation of cellular mRNAs 256
NS1 also inhibits activation of PKR, an important antiviral pathway induced by interferon 257
Viral envelope proteins assemble in the plasma membrane and direct budding of virions 257
Neuraminidase cleaves sialic acid, the cellular receptor that binds to HA 257
Influenza virus strains vary in both transmissibility and pathogenicity 257
Genetic variability generates new virus strains that can cause pandemics 258
The 1918 pandemic influenza A virus was probably not a reassortant virus 258
Genome sequences from some previous influenza A virus strains confirm the antigenic shift hypothesis 258
Highly pathogenic influenza A strains in poultry farms could lead to a new pandemic 259

24. Reoviruses 261

Reoviruses were the first double-stranded RNA viruses discovered 262
Some members of the Reoviridae are important pathogens 262
Reoviridae have segmented genomes made of double-stranded RNA 262
Reovirus virions contain concentric layers of capsid proteins 263
The attachment protein binds to one or two cellular receptors 265
During entry, the outer capsid is stripped from virions and the core is released into the cytoplasm 265
Enzymes in the viral core synthesize and cap messenger RNAs 266
Translation of reovirus mRNAs is regulated 267
Interferon and PKR: effects on viral and cellular protein synthesis 267
Synthesis of progeny double-stranded genomes occurs within subviral particles 268
Reoviruses induce apoptosis via activation of transcription factor NF-κB 269
Studies of reovirus pathogenesis in mice 270

25. Retroviruses 272

Retroviruses have a unique replication cycle based on reverse transcription and integration of their genomes 273
Viral proteins derived from the gag, pol and env genes are incorporated in virions 273
Retroviruses enter cells by the fusion pathway 274

Viral RNA is converted into a double-stranded DNA copy by reverse transcription 275
A copy of proviral DNA is integrated into the cellular genome at a random site 277
Sequence elements in the long terminal repeats direct transcription and polyadenylation by host cell enzymes 277
Differential splicing generates multiple mRNAs 279
The Gag/Pol polyprotein is made by suppression of termination and use of alternative reading frames 279
Virions mature into infectious particles after budding from the plasma membrane 280
Acute transforming retroviruses express mutated forms of cellular growth signalling proteins 281
Retroviruses lacking oncogenes can transform cells by insertion of proviral DNA near a proto-oncogene 282

26. Human Immunodeficiency Virus Type 1 284

Human immunodeficiency virus type 1 (HIV-1) and acquired immunodeficiency syndrome 285
HIV-1 infection leads to a progressive loss of cellular immunity and increased susceptibility to opportunistic infections 285
HIV-1 is a complex retrovirus 287
HIV-1 targets cells of the immune system by recognizing CD4 antigen and chemokine receptors 287
Virus mutants arise rapidly because of errors generated during reverse transcription 288
Unlike other retroviruses, HIV-1 directs transport of proviral DNA into the cell nucleus 288
Latent infection complicates the elimination of HIV-1 289
The Tat protein increases HIV-1 transcription by stimulating elongation by RNA polymerase II 289
The Rev protein mediates cytoplasmic transport of viral mRNAs that code for HIV-1 structural proteins 290
Together, the Tat and Rev proteins strongly upregulate viral protein expression 291
The Vif protein increases virion infectivity by counteracting a cellular deoxycytidine deaminase 291
The Vpr protein enables the preintegration complex to be transported to the nucleus 291
The Vpu protein enhances release of progeny virions from infected cells 291
The Nef protein is an important mediator of pathogenesis 292

27. Human T-Cell Leukemia Virus Type 1 294

Discovery of the first human retrovirus 294
Like lentiviruses, HTLV-1 codes for regulatory proteins by producing doubly-spliced mRNAs 295
HTLV-1 Rex regulates polyadenylation, splicing, and nuclear export of viral RNAs 296
HTLV-1 Tax regulates transcription of viral and cellular genes 297
Cell transformation by HTLV-I is mediated by Tax 299
The interleukin-2†autocrine loop stimulates T-cell proliferation 299
Activation of the Jak-Stat pathway by $p12^{I}$ mimics interleukin-2 stimulation 299
Cell Cycle Progression: $p16^{INK4A}$ and cyclin-dependent kinases 300
The mitotic spindle checkpoint and MAD1 300
Downregulation of p53 activity by Tax allows T cell proliferation 301
Diseases caused by HTLV-1 develop slowly and can be severe 301
Coinfection by HIV-1 and HTLV-1 is an emerging problem 302
Antiviral therapy of disease caused by HTLV-1 has not met with great success 302

28. Hepadnaviruses 303

At least seven distinct viruses cause human hepatitis 304
The discovery of hepatitis B virus 304
Dane particles are infectious virions; abundant noninfectious particles lack nucleocapsids 305
The viral genome is a circular, partly single-stranded DNA with overlapping reading frames 305
Nucleocapsids enter the cytoplasm via fusion and are transported to the nucleus 305
Transcription of viral DNA gives rise to several mRNAs and a pregenome RNA 307
The roles of hepatitis B virus proteins 308
The pregenome RNA is packaged by interaction with polymerase and core proteins 309
Genome replication occurs via reverse transcription of pregenome RNA 309
Virions are formed by budding in the endoplasmic reticulum 311
Hepatitis B virus can cause chronic or acute hepatitis, cirrhosis, and liver cancer 312
Hepatits B virus is transmitted by blood transfusions, contaminated needles, and unprotected sex 313
A recombinant vaccine is available 313
Antiviral drug treatment has real but limited success 313

29. Viroids and Hepatitis Delta Virus 315

Viroids are small, circular RNAs that do not encode proteins 316
Group A and group B viroids have distinct properties 316
Viroids replicate via linear multimeric RNA intermediates 317
Three enzymatic activities are needed for viroid replication 317

How do viroids cause disease? 318
Interaction of viroid RNA with cellular RNAs or proteins may disrupt cell metabolism 318
Plant satellite RNAs resemble viroids but are encapsidated 320
Hepatitis delta virus is a human viroid-like satellite virus 320
Hepatitis delta virus may use two different cellular RNA polymerases to replicate 320
RNA editing generates two forms of hepatitis delta antigen 322
Conclusion: viroids may be a link to the ancient RNA world 322

30. Prions 323

Prions are proteins that cause fatal brain diseases 323
Prion diseases were first detected in domestic ruminants 324
Human prion diseases can be either inherited or transmitted 324
The infectious agent of prion diseases contains protein but no detectable nucleic acid 325
PrP^{Sc} is encoded by a host cell gene 325
Differences between PrP^{C} and PrP^{Sc} 326
The prion hypothesis: formation of infectious and pathogenic prions from normal PrP^{C} 326
Is the prion hypothesis correct? 327
Pathology and diagnosis of prion diseases 330
Genetics of prion diseases 330
Prion diseases are not usually transmitted among different species 330
Strain variation and crossing of the species barrier 331
The nature of the prion infectious agent 331

31. Interferons 333

Virus-infected cells secrete interferons, which protect nearby cells against virus infection 334
Interferons are a first line of host defense against viruses but therapeutic use has been limited 334
Interferons α, β, and γ are made by different cells and have distinct functions 335
Transcription of interferon genes is activated by virus infection or double-stranded RNA 335
Transcriptional activation occurs by binding of transcription factors to interferon gene enhancers 335
Interferon signal transduction is carried out via the Jak-Stat pathway 337
Antiviral activities induced by interferon 338
Interferons have diverse effects on the immune system 340
The adaptive immune system 340
Interferons stimulate antigen processing and presentation 342
Interferon and the development of CD4-positive helper T-cells 342
The role of interferon in macrophage activation and cellular immunity 343
Effects of interferons on antibody production 343
Interferons regulate cell growth and apoptosis 343
Viruses have developed numerous strategies to evade the interferon response 343
Conclusion: interferons are a first line of defense against virus infection 344

32. Antiviral Chemotherapy 346

The discovery and widespread use of antiviral compounds began only recently 346
Importance of antiviral drugs for basic science 347
How are antiviral drugs obtained? 347
Targeting drugs to specific steps of virus infection 347
Capsid-binding drugs prevent attachment and entry of virions 348
Amantadine blocks ion channels and inhibits uncoating of influenza virions 349
Nucleoside analogues target viral DNA polymerases 349
Acyclovir is selectively phosphorylated by herpesvirus thymidine kinases 350
Acyclovir is preferentially incorporated by herpesvirus DNA polymerases 351
Cytomegalovirus encodes a protein kinase that phosphorylates ganciclovir 351
HIV-1 reverse transcriptase preferentially incorporates azidothymidine into DNA, leading to chain termination 354
Nonnucleoside inhibitors selectively target viral replication enzymes 354
Protease inhibitors can interfere with virus assembly and maturation 354
Ritonavir: a successful protease inhibitor of HIV-1 that was developed by rational methods 354
Neuraminidase inhibitors suppress release and spread of influenza virus 355
Antiviral chemotherapy shows promise for the future 357

33. Eukaryotic Virus Vectors 358

Many viruses can be engineered to deliver and express specific genes 358
Virus vectors are used to produce high levels of specific proteins in cultured cells 359
Gene therapy is an expanding application of virus vectors 360
Virus vectors are produced by transfection of cells with plasmids containing deleted genomes 360
Virus vectors are engineered to produce optimal levels of gene products 361

ADENOVIRUS VECTORS 362
Adenovirus vectors are widely used in studies of gene transfer and antitumor therapy 362
Replication-defective adenovirus vectors are propagated in complementing cell lines 362
Replication-competent adenovirus vectors are useful tools in antitumor therapy 363
Advantages and limitations of adenovirus vectors 363
RETROVIRUS VECTORS 364
Retrovirus vectors incorporate transgenes into the cell chromosome 364
Packaging cell lines express retrovirus enzymatic and structural proteins 364
Strategies for controlling transgene transcription 364
Lentivirus vectors are used for gene delivery to nondividing cells 365
Production of lentivirus vectors requires additional cis-acting sequences 366
Applications of retrovirus vectors: treatment of blood disorders 367
Advantages and limitations of retrovirus vectors 367
ADENO-ASSOCIATED VIRUS VECTORS 367
Adeno-associated virus vectors can insert transgenes into a specific chromosomal locus 367
Production of AAV vectors usually requires a helper virus 367
Applications of adeno-associated virus vectors: treatment of hemophilia 368
Advantages and limitations of AAV vectors 368

34. Viral Vaccines 370
A brief history of viral vaccines 371
Early vaccine technology was crude but effective 372
Embryonated chicken eggs and cell culture played major roles in recent vaccine development 373
Major categories of viral vaccines 373
Advantages and drawbacks of vaccine types 374
How do viral vaccines work? 375
The role of the immune system in fighting viral infections 375
Some vaccines target mainly antibody production; others target the cellular immune response 376
New approaches to vaccine development show great promise 376
The changing vaccine paradigm 378
Vaccine-associated adverse events 378
Ethical issues in the use of viral vaccines 380

Introduction to Virology

THE NATURE OF VIRUSES

Viruses consist of:

- *A nucleic acid genome (either DNA or RNA)*
- *A protein coat (capsid) that encloses the genome*
- *In some cases, a lipid membrane (envelope)*

The infectious virus particle is called a virion.

Virus particles are very small: between 20 and 500 nanometers (nm) in diameter.

Viruses are obligatory intracellular parasites.

Viruses multiply inside cells by expressing and replicating their genomes.

Viruses need the following machinery provided by cells:

- *Enzyme systems that synthesize amino acids, nucleotides, carbohydrates, and lipids*
- *Enzyme systems that generate useable chemical energy in the form of ATP*
- *Ribosomes, tRNAs, and enzymes used in protein synthesis*
- *Membranes that concentrate cellular macromolecules, small molecules, and ions*

WHY STUDY VIRUSES?

Viruses are important disease-causing agents.

Probably all different forms of life can be infected by viruses.

Viruses are important players in the regulation of the Earth's ecology.

Viruses can be engineered to prevent and cure disease.

Study of viruses reveals basic mechanisms of gene expression, cell physiology, and intracellular signaling pathways.

A BRIEF HISTORY OF VIROLOGY

Viruses were first distinguished from other microorganisms by their small size and their ability to pass through fine filters that retain bacteria.

Viruses can be crystallized: They lie on the edge between chemical compounds and life.

Study of bacteriophages by the "phage group" led to understanding of the nature of genes and helped found molecular biology.

In vitro culture of eukaryotic cells led to rapid advances in the study of viruses and in vaccine production.

Study of tumor viruses led to the discovery of viral and cellular oncogenes.

DETECTION AND MEASUREMENT OF VIRUSES

The plaque assay is widely used to measure virus infectivity.

Hemagglutination is a cheap and rapid method for detection of virus particles.

Virus particles can be seen and counted by electron microscopy.

The ratio of physical particles to infectious particles is greater than 1 for many viruses.

VIRUS REPLICATION CYCLE

1. The virion binds to cell surface receptors.

2. The virion or viral genome enters the cell; the viral genome is uncoated.
3. Early viral genes are expressed.
4. Early viral proteins direct replication of the viral genome.
5. Late viral genes are expressed from newly replicated viral genomes.
6. Late viral proteins package genomes and assemble progeny virus particles.
7. Virions are released from the host cell.

THE NATURE OF VIRUSES

Viruses consist of a nucleic acid genome packaged in a protein coat

Viruses are the smallest and simplest forms of life on Earth. They consist of a set of nucleic acid genes enclosed in a protein coat or **capsid**, which in some cases is surrounded by a lipid membrane or **envelope** (Figure 1.1). The viral genome encodes proteins that enable it to replicate and to be transmitted from one cell to another, or from one organism to another. The infectious virus particle is called a **virion**.

Viruses are dependent on living cells for their replication

Viruses can replicate only within living cells. Another way of saying this is that viruses are **obligatory intracellular parasites**. Viruses depend on cells for their replication because they lack the following basic elements required for growth and replication, which are present in all living cells:

- Enzyme systems that produce the basic chemical building blocks of life: nucleotides, amino acids, carbohydrates, and lipids
- Enzyme systems that generate useable chemical energy, usually in the form of adenosine triphosphate (ATP), by photosynthesis or by metabolism of sugars and other small molecules
- Ribosomes, transfer RNAs, and the associated enzymatic machinery that directs protein synthesis
- Membranes that localize and concentrate in a defined space these cellular macromolecules, the small organic molecules involved in growth and metabolism, and specific inorganic ions

Virus particles break down and release their genomes inside the cell

Viruses are not the only obligatory intracellular parasites known. A number of small unicellular organisms including chlamydiae and rickettsiae, certain other bacterial species, and some protozoa can multiply only inside other host cells. However, viruses replicate by a pathway that is very different from the mode of replication of these other intracellular parasites. Virus replication begins with at least partial disintegration of the virus particle, and the release (**uncoating**) of the viral genome within the cell. Once released, the viral genome can be used as a template for synthesis of messenger RNAs and viral proteins,

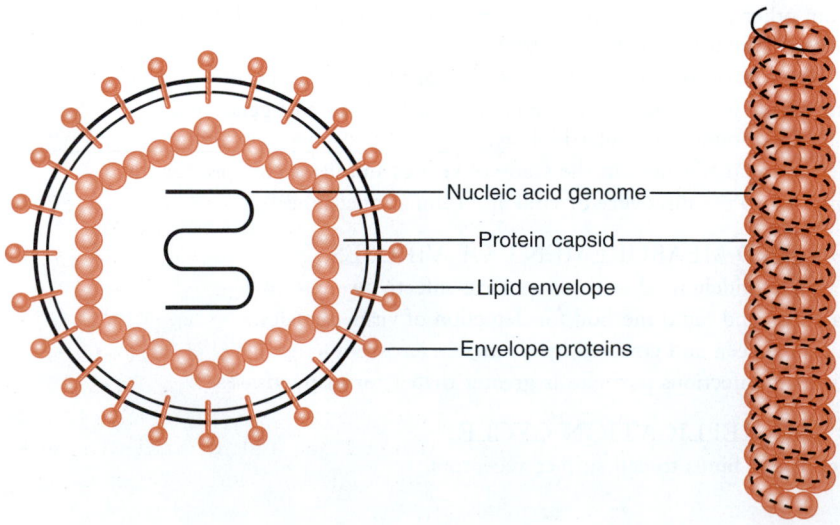

FIGURE 1.1 Schematic diagram of virus particles. Illustrated are the two most common capsid morphologies: a roughly spherical shell (left) and a tubular rod (right). Some virus particles have an external envelope (left) and some do not (right). Nucleic acid genomes are shown as black curved lines, capsid proteins as orange spheres, and envelope proteins as orange knobbed spikes.

using the enzyme systems, energy, ribosomes, and molecular building blocks that are present in the cell. Newly synthesized viral proteins then direct replication of the viral genome; other viral proteins encapsidate the newly-replicated genomes to form progeny virus particles.

In contrast, unicellular organisms that replicate inside other cells invariably remain intact and retain their genomes within their own cellular membranes. They replicate not by disintegration and reconstitution, but by growth and division into daughter cells. Such cellular parasites always contain their own ribosomes and protein synthetic machinery, and their genes code for enzymes that direct many of the basic metabolic pathways.

In summary, viruses in their simplest form contain a nucleic acid genome, packaged in a protein coat. To replicate, a virus must transport its genome into a host cell, where the genome codes for viral proteins, is replicated and is packaged. The host cell provides the virus with all of the other biological molecules required for its reproduction.

Virus genomes are either RNA or DNA, but not both

There are many different viruses in the world, and probably all organisms on Earth can be infected by at least one virus. Viruses can have a variety of distinct morphologies, genome and particle sizes, and mechanisms of replication. The smallest known viruses are 20 nm in diameter; their genomes contain fewer than 2000 nucleotides, and they code for as few as 2 proteins. The largest known viruses are some 500 nm in diameter; their genomes are as large as 1.2 million nucleotides, and they can code for over 1200 proteins. An overview of the variety of known viruses is given in Chapter 3.

All viruses contain genomes made of one and only one type of nucleic acid. Depending on the virus, the genome can be either RNA or DNA, and it can be either single-stranded or double-stranded. Some viral genomes are circular and others are linear.

In fact, viruses are the only known forms of life that can have RNA genomes. All cellular organisms (that is, all organisms except viruses) store the information required to sustain life, to grow, and to reproduce exclusively in DNA molecules, and all cellular RNA molecules are transcribed from DNA sequences. RNA-containing viruses are therefore unique, and they face two related problems as a result of their RNA genomes: (1) how to synthesize messenger RNAs from an RNA template, and (2) how to replicate their genome RNA. Most RNA viruses encode their own RNA-dependent RNA polymerases to carry out both these functions.

WHY STUDY VIRUSES?

Viruses are important disease-causing agents

As living organisms arose and evolved during the past 4 billion years on Earth, they were probably always accompanied by viruses that could replicate within cells and pass from cell to cell. Many of these viruses interfere with normal cellular processes and cause disease. Some of the most feared, widespread, and devastating human diseases are caused by viruses (see Table 1.1). These include smallpox, influenza, poliomyelitis, yellow fever, measles, and AIDS (acquired immunodeficiency syndrome). Viruses are responsible for many cases of human encephalitis, meningitis, pneumonia, hepatitis, and cervical cancer, as well as warts and the common cold. Viruses causing respiratory infections, gastroenteritis, and diarrhea in young children lead to millions of deaths each year in less-developed countries.

A number of newly emerging human diseases are caused by viruses. In addition to the worldwide AIDS epidemic that started in the early 1980s, there have been localized outbreaks in Africa of the highly fatal Marburg and Ebola hemorrhagic fevers during the past 30 years, a short-lived epidemic in southern Asia and Canada of severe acute respiratory syndrome (SARS) in 2003, and spread of acute and chronic hepatitis via both hepatitis B and C viruses. An invasion of North America by West Nile virus, transmitted by mosquitoes, began in 1999 and fortunately has only caused disease and death in a limited number of victims. There are fears that a new deadly pandemic of human influenza could occur if a recently emerged, highly pathogenic strain of avian influenza virus mutates to a form that is easily transmitted among humans.

Viruses can infect all forms of life

Viruses also infect animals, plants, and insects of importance to humans. Outbreaks of virus diseases in domesticated animals can lead to destruction of thousands or millions of animals to avoid even more widespread epidemics. These diseases include avian influenza; foot-and-mouth disease of cattle; infectious gastroenteritis and bronchitis in pigs, cattle, and chickens; sheep lung tumors caused by a retrovirus; canine distemper; and feline immunodeficiency disease. Virus diseases affecting domesticated plants such as potatoes, tomatoes, tobacco, coconut trees, and citrus trees are common and widespread. Insect viruses that kill silkworms, used for centuries in Asia and Europe to produce silk, have plagued that industry over the ages. Viruses can also infect and kill bacteria, archae, algae, fungi, and protozoa.

TABLE 1.1 Some human diseases caused by viruses

Disease	Virus	Family
Acquired immunodeficiency syndrome (AIDS)	HIV-1	Retrovirus
Cervical carcinoma	Human papillomavirus types 16, 18, 31	Papillomavirus
Chickenpox	Varicella virus	Herpesvirus
"Cold sores"	Herpes simplex virus type 1	Herpesvirus
Common cold	Adenoviruses	Adenovirus
	Coronaviruses	Coronavirus
	Rhinoviruses	Picornavirus
Diarrhea	Norwalk virus	Calicivirus
	Rotaviruses	Reovirus
Genital herpes	Herpes simplex virus type 2	Herpesvirus
Hemorrhagic fevers	Dengue virus	Flavivirus
	Ebola and Marburg viruses	Filovirus
	Lassa fever virus	Arenavirus
Hepatitis	Hepatitis A virus	Picornavirus
	Hepatitis B virus	Hepadnavirus
	Hepatitis C virus	Flavivirus
Influenza	Influenza A and B virus	Othomyxovirus
Measles	Measles virus	Paramyxovirus
Mononucleosis	Epstein-Barr virus	Herpesvirus
	Cytomegalovirus	Herpesvirus
Mumps	Mumps virus	Paramyxovirus
Poliomyelitis	Poliovirus types 1, 2, and 3	Picornavirus
Rabies encephalitis	Rabies virus	Rhabdovirus
Severe acute respiratory syndrome (SARS)	SARS coronavirus	Coronavirus
Smallpox	Variola virus	Poxvirus
Warts	Human papillomavirus types 1, 2, 4	Papillomavirus
Yellow fever	Yellow fever virus	Flavivirus

Viruses are the most abundant form of life on Earth

Recent studies of soil and seawater have revealed that bacterial viruses, also called bacteriophages, are much more numerous than previously imagined. There are 10–50 million bacteriophages on average per ml of seawater, and even more in many soils. Given the enormous volume of the oceans, scientists have calculated that there may be as many as 10^{31} bacteriophages in the world. This is about 10-fold greater than the estimated number of bacteria. In terms of mass, this many phages would weigh about 100 million tons, or the equivalent of 1 million blue whales (the largest animal on Earth). More astonishing, these 10^{31} phages, if lined up head-to-tail, would stretch some 200 million light years into space—that is, far into the universe beyond many of our known neighboring galaxies (see text box on page 5)!

More important is the ecological role played by bacteriophages and viruses that infect unicellular eukaryotic organisms such as algae. From 95 to 98% of the biomass in the oceans is microbial (the remaining 2–5% being made up of all other forms of life, including fish, marine invertebrates, marine mammals, birds, and plants), and roughly half of the oxygen in the Earth's atmosphere is generated by photosynthetic activity of marine microbes. It has been estimated that 20% of the microbes in the Earth's oceans are destroyed each day by virus infections. Therefore, these viruses play a major role in the carbon and oxygen cycles that regulate our atmosphere and help feed the world's population.

The study of viruses has led to numerous discoveries in molecular and cell biology

Because viruses replicate within cells but express a limited number of viral genes, they are ideal tools for understanding the biology of cellular processes. The intensive study of bacteriophages led to discovery of some of the fundamental principles of molecular biology and genetics. Research on animal, insect, and plant viruses has shed light on the functioning of these organisms,

Phages lined up through the universe

Scientists estimate that there are approximately 10^{31} tailed bacteriophages on Earth. Each phage measures approximately 200 nm (0.2 μm) in length from top of head to base of tail. Aligned head to tail, these phages would therefore cover the following distance:

- $10^{31} \times 0.2\ \mu m = 0.2 \times 10^{25}$ meters $= 2 \times 10^{24}$ meters = **2×10^{21} kilometers**.

Because 1 light year (distance traveled by light in one year) = 10^{13} kilometers,

- 2×10^{21} kilometers $= 2 \times 10^{21}/10^{13}$ light years = **2×10^{8} light years** (200 million light years).

Note that our Milky Way galaxy measures approximately 150 light years edge to edge, and the furthest visible galaxies in the universe are approximately 10×10^{9} light years distant.

their diseases, and their molecular mechanisms of replication, cell division, and signaling pathways. For example:

- Study of gene expression in small DNA viruses led to the identification of promoters for eukaryotic RNA polymerases.
- Research on the replication of bacteriophage and animal virus DNAs laid the foundations for understanding the enzymes involved in cellular DNA replication.
- RNA splicing in eukaryotic cells was first discovered by studying messenger RNAs of DNA viruses.
- Study of cancer-producing viruses led to the isolation of numerous cellular oncogenes and the understanding that cancer is caused by their mutation or unregulated expression.

Given this track record, the study of viruses will undoubtedly continue to shed light on many important aspects of cell and molecular biology.

A BRIEF HISTORY OF VIROLOGY: THE STUDY OF VIRUSES

The scientific study of viruses is very recent

Although viruses were probably present among the first forms of life and have evolved over several billion years, humans only began to understand the nature of viruses near the end of the nineteenth century (see Table 1.2). It had been appreciated for some time that infectious diseases were transmitted by air, water, food, or close contact with sick individuals. Many diseases were considered to be caused by mysterious elements in fluids termed *virus* ("poison" in Latin), but the distinction between what are now called viruses and cellular microorganisms was not clear. Scientists had begun to use light microscopes to discover and describe fungi and bacteria in the first half of the nineteenth century. Louis Pasteur and Robert Koch firmly established the science of bacteriology in the latter part of that century by isolating and growing a variety of bacteria, some of which were shown to cause disease (e.g., tuberculosis). Even though effective vaccines against smallpox (Edward Jenner, 1796) and rabies (Louis Pasteur, 1885) were developed, there was no understanding of the nature of these disease agents, which we now know to be viruses.

Viruses were first distinguished from other microorganisms by filtration

In the last decade of the nineteenth century, Russian scientist Dimitrii Ivanovski and Dutch scientist Martinus Beijerinck independently showed that the agent that causes tobacco mosaic disease could pass through fine earth or porcelain filters, which retain bacteria. Shortly afterward, similar experiments were carried out on agents that cause foot-and-mouth disease in cattle, and yellow fever in humans. These landmark experiments established that certain infectious agents are much smaller than bacteria, and they were called **filterable viruses**. For some time, it was not clear whether viruses were a form of soluble small molecules ("infectious living fluid"), or alternatively a particle like bacteria but too small to be retained by these filters.

Using filtration as a diagnostic tool, numerous viruses infecting humans, other vertebrate animals, plants, insects, and bacteria were described during the first half of the twentieth century. A tumor virus, Rous Sarcoma virus, was isolated from sarcomas of chickens in 1911, and was only many years later recognized as a representative of the important retrovirus family, of which human immunodeficiency virus (HIV) is a member. Scientists working in England and France discovered in 1915–1917 that bacterial cultures could be lysed by filterable agents, the first known bacterial viruses. Vaccines against the important human pathogens responsible for influenza and yellow fever were developed during the 1930s.

The crystallization of tobacco mosaic virus challenged conventional notions about genes and the nature of living organisms

Wendell Stanley found in the mid-1930s that highly purified tobacco mosaic virus could form crystals. This discovery shook the scientific world, because it placed viruses at the edge between living organisms and sim-

TABLE 1.2 Some milestones in virology research

Discovery	Date	Scientists	Nobel prize awarded
Smallpox vaccine	1798	Edward Jenner	
Rabies vaccine	1885	Louis Pasteur	
Filterable viruses:			
Tobacco mosaic virus	1892	Dimitrii Ivanovski	
	1898	Martinus Beijerinck	
Foot and mouth disease (cattle)	1898	Friedrich Loeffler and Paul Frosch	
Yellow fever (humans: transmitted by mosquitoes)	1900	Carlos Finlay and Walter Reed	
Discovery of Rous Sarcoma virus	1911	Peyton Rous	1966
Discovery of bacteriophages and the plaque assay	1915	Frederick Twort	
	1917	Felix d'Herelle	
Vaccine against yellow fever	1930s	Max Theiler	1951
Crystallization of tobacco mosaic virus	1935	Wendell Stanley and John Northrup	1946
Studies with bacteriophages	1940s	Max Delbruck and Salvador Luria	1969
Growth of poliovirus in cultured cells	1949	John Enders, Frederick Robbins, and Thomas Weller	1954
Bacteriophage lambda and lysogeny	1950s	Andre Lwoff	1965
Bacteriophage genes are DNA	1952	Alfred Hershey and Martha Chase	1969
Discovery of interferon	1957	Alick Isaacs and Jean Lindenmann	
Poliovirus vaccines:			
killed	1955	Jonas Salk	
live	1960	Albert Sabin	
Studies on polyomavirus: a tumor virus	1960s	Renato Dulbecco	1975
Kuru is caused by an infectious agent	1965	D. Carleton Gajdusek	1976
Discovery of hepatitis B virus	1968	Baruch Blumberg	1976
Reverse transcriptase in retroviruses	1971	Howard Temin and David Baltimore	1975
Virus vectors and genetic engineering	1970s	Paul Berg	1980
Cellular oncogene in a retrovirus	1976	Michael Bishop and Harold Varmus	1989
RNA splicing in adenovirus	1977	Phillip Sharp and Richard Roberts	1993
Prions: infectious proteins	1975–1990	Stanley Prusiner	1997

ple chemical compounds like sodium chloride. It posed the question, Are viruses living or inanimate? We now know that viruses are inanimate when their genomes are packaged in virions, but they share many attributes of life, including the ability to mutate, evolve, and reproduce themselves, when they enter cells that can support their replication.

Studies by Stanley and others showed that viruses contain both protein and nucleic acids. At that time, most scientists believed that genes were made of proteins, not nucleic acids, because only proteins were believed sufficiently complex to encode genetic information. The development of the electron microscope in the late 1930s allowed scientists for the first time to ac-

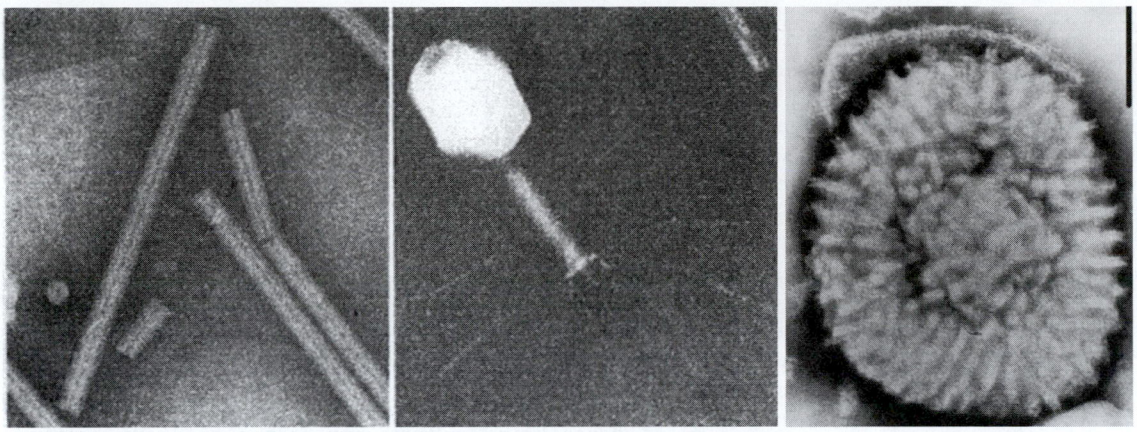

(a) (b) (c)

FIGURE 1.2 Electron micrographs of some representative virus particles. (a) Tobacco mosaic virus. (b) Bacteriophage T4. (c) Vaccinia virus.

tually see viruses—tobacco mosaic virus (a long rod-shaped virus), bacteriophages with their polygonal heads and tubular tails, and vaccinia virus, one of the largest animal viruses (Figure 1.2).

The "phage group" stimulated studies of bacteriophages and helped establish the field of molecular biology

During the late 1930s and early 1940s, a group of scientists led by German physicist Max Delbruck, American biologist Emory Ellis, and Italian biologist Salvador Luria decided that the study of bacterial viruses (bacteriophages) could lead to understanding of the basic processes of life at a molecular level. They reasoned that bacteriophages show heritable traits, and therefore must contain and express genes as do all other organisms. Because bacteriophages are small and simple and can be propagated easily in bacterial cultures, they would be a fertile terrain for scientific discovery. These scientists formed an informal network called the "phage group," which stimulated studies of bacteriophages and their host bacteria by numerous physicists, chemists, and biologists. These studies led to the isolation and analysis of phage genomes, the mapping of phage and bacterial genes by genetic crosses, and the elucidation of phage replication cycles. The phage group helped to found the field of molecular biology, which developed rapidly during the 1950s and 1960s.

One outcome of the phage group's work was the demonstration, by Alfred Hershey and Martha Chase in 1952, that the DNA of a bacteriophage is injected into the host cell and the protein coat remains outside the cell. This strongly backed up the chemical and enzymatic data published eight years earlier by Oswald Avery, Maclyn McCarty, and Colin MacLeod showing that a bacterial gene was made of DNA. Thus studies of a bacteriophage were important in proving the chemical nature of genes. Hershey and Chase's paper was followed a year later by the proposal by James Watson (a student of Luria) and Francis Crick of the double-helical model of DNA, which galvanized thinking and research throughout biology, but particularly in virology and molecular biology.

Study of tumor viruses led to discoveries in molecular biology and understanding of the nature of cancer

Virus research underwent an explosive development in the second half of the twentieth century that led to discovery of many new viruses and basic concepts in cell and molecular virology. Among the most important

Bacteriophages can be used as targeted antibiotics against bacterial diseases

Bacteriophages may be called upon to play an important role in the fight against bacterial diseases in humans and animals. Because bacteriophages can attack and kill specific bacteria, they have long been considered as possible alternatives to antibiotics in treating disease. Bacteriophages specific to a variety of pathogenic bacteria have been isolated and characterized. There is a long history of their medical use, particularly in the Republic of Georgia, in an institute co-founded by one of the discoverers of bacteriophages, Felix d'Herelle. However, their use as antibacterial agents in humans is not accepted in most countries; there are significant unsolved problems, not the least of which is the induction of immune reactions to the bacteriophage. Future research and development may well reveal situations in which their use will be able to control otherwise runaway infections by bacteria that have developed resistance to many commonly used antibiotics.

Virus vectors can replace defective genes, serve as vaccines, and combat cancer

Advances in molecular cloning have allowed the construction of numerous virus vectors (see Chapter 33). These are viruses in which some or all viral genes are removed by genetic engineering and are replaced by foreign genes. Because viruses can efficiently target specific cell types and express their genes at high levels, they can be used as vectors for expression of a variety of genes. Introduction of virus vectors into host cells or organisms can correct genetic diseases in which specific gene products are missing or defective. Vectors can also be used as vaccines that generate immune responses against a variety of unrelated pathogens. In a new twist, virus vectors have recently been used to combat cancers by expressing proteins that specifically kill tumor cells.

were the discovery and intensive study of DNA tumor viruses (polyomaviruses, papillomaviruses, and adenoviruses; see Chapters 10–12) and RNA tumor viruses (retroviruses; see Chapters 25–27). Research on DNA tumor viruses led to the discovery of viral **oncogenes**, whose protein products (tumor antigens) interact with numerous cell signaling pathways to stimulate cell growth and division.

Research on RNA tumor viruses led to the discovery of **reverse transcriptase**, an enzyme that can make a DNA copy of an RNA molecule, upsetting the one-way central dogma that "DNA makes RNA makes proteins." Numerous cellular oncogenes were discovered incorporated into retrovirus genomes. These oncogenes are normal cellular regulatory genes whose mutation and/or overexpression can lead to the development of cancer; their protein products are involved in a variety of cellular signaling pathways. The study of viral and cellular oncogenes has led to major advances in the detection and treatment of cancer.

DETECTION AND TITRATION OF VIRUSES

Most viruses were first detected and studied by infection of intact organisms

Many viruses cause disease in the host organism, and this is how scientists and medical doctors usually become aware of their existence. The original methods for the study of such viruses relied on inoculation of animals or plants with filtered extracts from infected individuals, and their observation for the effects of virus infection. However, this is expensive and time-consuming work, and is not ethically acceptable when applied to humans. Experimental laboratory animals such as suckling mice, in which many animal viruses are able to replicate, were adopted for use because they are relatively easy and inexpensive to raise. A number of animal viruses have also been adapted to grow in embryonated chicken eggs, which are readily available from farms; this reduced both the expense and the time of virus assays.

The plaque assay arose from work with bacteriophages

Bacterial viruses can be easily studied by inoculating bacteria grown in tubes or on Petri dishes at the lab bench. Intact bacteria diffract visible light and therefore dense liquid cultures appear cloudy. Many bacteriophages lyse their host cell, and this **lysis** causes a loss in light diffraction leading to clearing of the bacterial culture; "clear lysis" serves as an indicator of phage replication. A simpler and more quantitative application of cell lysis is to spread bacteria on the surface of nutrient agar in a Petri dish and to apply dilutions of a phage suspension. Wherever a phage binds to a bacterial cell and replicates, that cell releases the progeny phage particles, which are then taken up by neighboring cells and further replicated. After several such cycles all the cells in a circular area surrounding the original infected cell are lysed. The lysis area can be seen as a clear "plaque" against the cloudy background of the uninfected cells, which grow in multiple layers on the surface of the agar in the Petri dish. Use of this **plaque assay** (Figure 1.3) allows scientists to count the number of infectious virus particles in a suspension with a high degree of precision and reproducibility. The chance observation of such plaques played an important role in the discovery of bacteriophages.

Eukaryotic cells cultured *in vitro* have been adapted for plaque assays

In vitro culture of human, animal, insect, and plant cells was achieved in the mid-twentieth century, and allowed more convenient and cheaper growth and titration of many viruses. In the case of animal viruses, cells of many different tissues (especially from embryos or newborn animals) can be induced to grow in a monolayer on a glass or plastic surface underneath liquid media. Cultured cells also facilitated production of numerous viral vaccines (see Chapter 34), starting with the vaccines against poliovirus that were developed in the 1950s.

Plaque assays were subsequently developed for animal, plant, and insect viruses using cells cultured *in vitro*. When cells growing in a monolayer in liquid medium are infected, the progeny virus is released into the medium and can travel to distant sites, infecting other cells. To restrict diffusion of the progeny virus, the infected cells are overlaid with agar containing nutrient medium. In this way, released virus can infect only

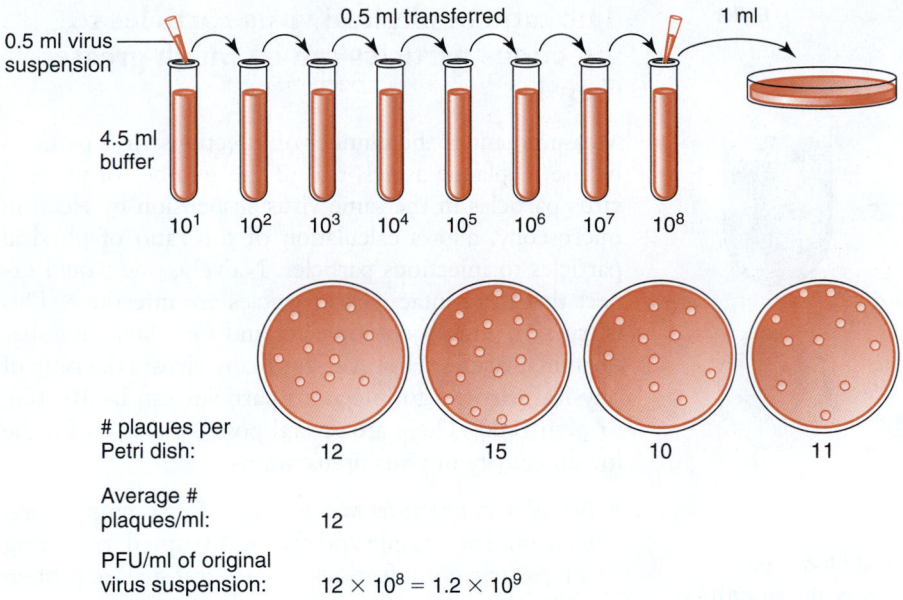

FIGURE 1.3 **Plaque assay: an example.** A virus suspension was subjected to 10-fold serial dilutions by adding 0.5 ml of the original suspension to 4.5 ml of a dilution buffer. After mixing, 0.5 ml of that dilution was added to 4.5 ml of fresh dilution buffer and mixed; these steps were repeated a total of eight times. Each successive tube contains a 10-fold dilution of the contents of the previous tube. The eighth tube therefore is diluted by a factor of 10^8 compared with the original virus suspension. 1-ml aliquots from this 10^8-fold dilution were applied to four different Petri dishes of susceptible cells and plaques (bottom) were allowed to develop. Plaque-forming units (PFU) per ml are calculated as shown.

nearby cells, forming a local area of dead cells or a plaque. Because cell monolayers are too thin to diffract light well, plaques in cultured animal or insect cells are usually visualized by staining the cells. When cells die and/or lyse, they do not stain well; therefore plaques are seen as clear, unstained circular areas on the background of the stained cell monolayer. Virus present in individual plaques can be isolated by sampling with a needle or Pasteur pipet, allowing the "cloning" of progeny virus derived from a single virion that initiated the infection leading to the plaque.

When a plaque assay is used to measure the infectious titer of a virus suspension, the results are usually expressed as **plaque-forming units (PFU)** per ml of suspension. To determine the titer, the number of plaques on a plate is multiplied by the factor by which the original virus suspension was diluted before an aliquot was applied to the plate. For an example, see Figure 1.3.

Hemagglutination is a convenient and rapid assay for many viruses

A number of animal viruses bind to **sialic acid** residues or other carboyhydrates on cell surface proteins and lipids. Red blood cells have carbohydrate-containing receptors on their surface, and have the advantage of being visible because of their color. Furthermore, they can be isolated easily from the blood of a variety of animals, are sturdy during manipulation, and can be stored for days or weeks. This makes red blood cells an ideal substrate for assaying viruses. Virus particles have multiple copies of receptor-binding proteins on their surface, and red blood cells contain many copies of surface receptors. Binding between an excess of virus particles and an aliquot of red blood cells forms an interlaced network of cells, held together by virus particles that form bridges between adjacent cells. These "agglutinated" red blood cells, when allowed to settle, form a light pink hemispherical shell in the bottom of a tube or well. In contrast, individual red blood cells slide to the bottom of the tube and form a compact, dark red pellet (Figure 1.4). This is the basis of **hemagglutination assays** for viruses.

Virus suspensions are diluted, usually in twofold steps, and the dilutions are added to aliquots of red blood cells in a buffer and mixed in tubes or multiple-well plates. After allowing the cells to settle, the tubes or plates are examined; the highest dilution that will agglutinate the aliquot of cells is considered to have one hemagglutinating unit (HAU) of virus. Such assays are sensitive to pH, temperature, and buffer composition, and some viruses will agglutinate only cells of a particular mammalian or avian species. Because

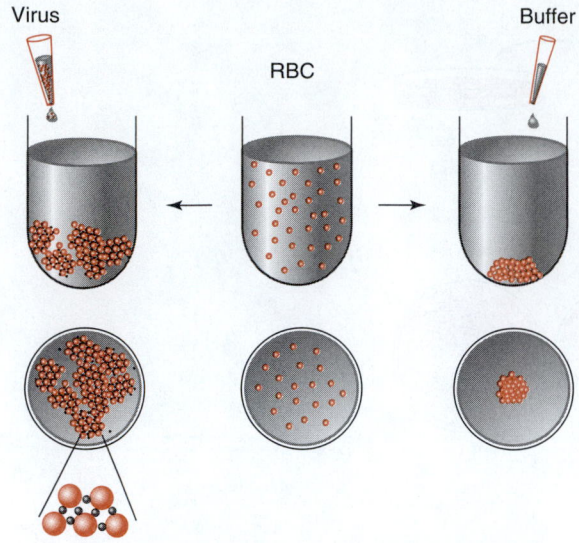

FIGURE 1.4 **Hemagglutination assay.** Red blood cells (RBC; small orange circles in central tube) are mixed with virus (small gray spheres), or with buffer, and are allowed to settle. Individual red blood cells settle to form a compact pellet in the bottom of the tube (right), but when agglutinated, form a thin shell on sides and bottom of tube (left). The lower set of images show what is seen when tubes are viewed from below. An enlargement at left shows red blood cells bound together by virus particles.

many (approximately 10^5) red blood cells are used in each tube, one hemagglutinating unit represents 10^5 or more virus particles. Therefore, hemagglutination assays are much less sensitive than plaque assays, but they are rapid and cheap. They can also be used to detect antibodies that bind to viral surface antigens, because addition of such antibodies will inhibit hemagglutination.

Virus particles can be seen and counted by electron microscopy

A variety of staining or shadowing methods can be used to detect virus particles by electron microscopy. One of the simplest methods is to mix a virus suspension with an electron-dense stain, usually phosphotungstate or uranyl acetate, and to spread the mixture on a grid for examination in the electron microscope. The "negative stain" tends to form electron-dense pools around virus particles; the virus particle excludes the stain and therefore shows up in negative contrast, and much fine surface detail can be seen (Figure 1.2). Measured aliquots of dilutions of virus suspensions can be applied to grids, and the number of virus particles in a given area can be counted. Standard suspensions of tiny latex spheres or other small uniform objects are often added to the virus suspension to help establish absolute numbers of virus particles per unit volume.

The ratio of physical virus particles to infectious particles can be much greater than 1

Measurement of the number of infectious virus particles by use of plaque assays, and of the number of physical virus particles in the same virus suspension by electron microscopy, allows calculation of the ratio of physical particles to infectious particles. Naively, one would expect that most intact virus particles are infectious. This is true for some bacteriophages and for a small number of animal viruses. However, for many viruses the ratio of physical particles to infectious particles can be 10, 100, or even 1000! There are several possible reasons for the low infectivity of virus preparations:

- *Not all virus particles may be intact.* For example, viral envelopes are fragile and can be disrupted, rendering the particle noninfectious. Some viral surface protein molecules can be denatured and therefore unable to bind to the cell receptor. However, virions contain numerous copies of receptor-binding proteins on their surface, so the loss or denaturation of a few protein molecules should not lead to loss of infectivity.
- *Some virus particles may contain defective genomes.* Mutations, including deletions in viral genes, occur frequently during genome replication, and such defective genomes are often incorporated into virus particles. In extreme cases, 90% or more of a virus preparation can contain particles with defective genomes.
- *"Empty" capsids that contain no viral genome can be made in large numbers.* Some viruses can form capsids in the absence of the viral genome. Others incorporate cellular DNA or RNA instead of the viral genome into the capsid. However, many viruses have specific packaging signals that ensure incorporation only of viral nucleic acid into virions.
- *Cells have antiviral defense mechanisms.* Many virus preparations consist of fully intact virions that contain infectious genomes. However, cells have a variety of defense mechanisms that can interfere with many steps in virus infection. Therefore, even though a cell takes up an intact and potentially infectious virion, it may not produce any progeny virus. This can be a major cause of the high ratio of physical particles to infectious particles of some viruses.

THE VIRUS REPLICATION CYCLE: AN OVERVIEW

The single-cycle virus replication experiment

Use of the plaque assay enables the quantitative study of the kinetics of virus replication. To understand the time course of events taking place during the replication cycle, scientists usually study cultures containing thousands to millions of infected cells, because only then can sufficient viral nucleic acid or proteins be isolated and analyzed. All cells must be infected simultaneously, and with some luck the events of the virus replication cycle will be synchronized so that similar steps will be taking place at the same time in all cells. To ensure simultaneous infection, cell cultures are infected with a sufficient number of virus particles such that each cell receives at least one infectious particle.

These considerations led to the concept of **multiplicity of infection (m.o.i.)**. This is defined as the number of infectious virus particles added per susceptible cell. M.o.i.'s of 10 to 100 plaque-forming units per cell are often used in studies of bacterial or animal viruses. In such a manner, nearly all cells in a culture are infected simultaneously and there remain very few uninfected cells. In practice, there are limitations in this method for distinguishing the time course of the various steps in a viral growth cycle; different steps in the cycle invariably overlap somewhat. However, the single-cycle approach does simplify the study of virus replication and is nearly universally used. Some bacteriophages complete their replication cycles in as little as 20 minutes; some animal viruses can take several days to complete one replication cycle. Certain viruses do not always undergo a productive growth cycle, but instead lodge one or more "silent" copies of their genome in the host cell until conditions are appropriate for a lytic cycle ("lysogeny" or "latent infection"; see Chapters 8, 9, and 13).

An example of a virus replication cycle: mouse polyomavirus

An idealized example of a single-cycle growth curve for mouse polyomavirus (see Chapter 10), a small DNA virus that replicates in cultured baby mouse kidney cells, is shown in Figure 1.5. Each 6-cm diameter Petri dish contains approximately 5 million cells, and 0.5 ml of a suspension containing 10^8 PFU/ml of polyomavirus is used to infect the cells (m.o.i. = 10 PFU/cell). After 1 hour, fresh medium is added to each culture and the cells are incubated at 37°C in a humidified chamber with a 5% CO_2 atmosphere. At intervals, samples of the infected cells or of the medium are harvested and analyzed for infectious virus, viral mRNA, viral proteins, or viral DNA. The results are expressed as PFU/cell for virus (logarithmic scale on left) or as a percent of the maximum amount of the viral macromolecule (linear scale on right).

During the first hour or two after adding virus, most of the infecting virions are taken up into the cells and thus disappear from the medium. This leads to an initial loss in the titer of virus detected in the medium. Eventually many of the virus particles taken up into the cells are uncoated, rendering intracellular virus noninfectious, and thus the titer of infectious intacellular virus also drops. This phase has been called the "latent phase" of infection because the infecting virus has disappeared and no new progeny virus has yet been made.

Some time later (in this case, starting at 15–20 hours after infection), new progeny virus begins to appear. Polyomavirus particles are assembled in the cell nucleus and are not efficiently released from the cell until after

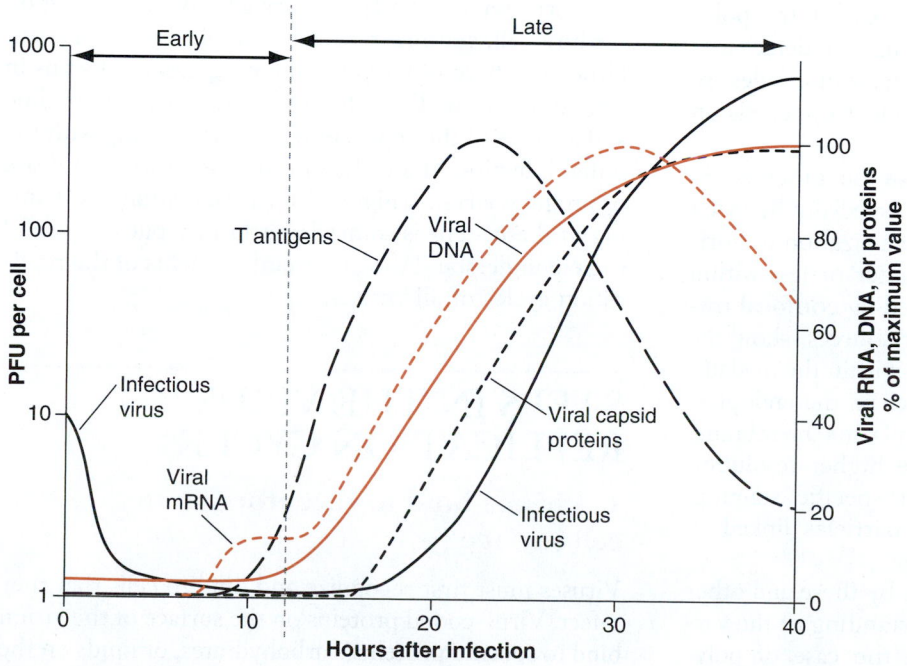

FIGURE 1.5 Replication cycle of mouse polyomavirus. The time course of appearance of infectious virus particles, viral mRNA, DNA, and proteins during a typical replication cycle of mouse polyomavirus in baby mouse kidney epithelial cells.

cell death. This means that most progeny virus can be detected only by lysing the cells after harvesting them. A much lower virus titer is detected in the medium surrounding the cells (not shown). This is true of many nonenveloped viruses that replicate and assemble in the nucleus or in the cytoplasm of eukaryotic cells, and of most bacteriophages. In contrast, some enveloped viruses form virions at the cell surface, where they acquire a lipid envelope, and therefore formation of virus particles can be simultaneous with release of the progeny virion into the medium. There is often a long period during which progeny virus is being produced, but at some point virus replication stops, often because the host cell dies. At this point no further virus is made, but release can take some time as cells slowly disintegrate after death.

Analysis of viral macromolecules reveals the detailed pathways of virus replication

This analysis of the virus growth cycle is very elementary and gives only an overview of the beginning and the end of steps in virus replication. Many important events of the virus replication cycle take place during the latent period. These steps are studied by extracting from the infected cell the macromolecules that constitute viral genes, messenger RNAs, and proteins. Their interactions with cellular macromolecules and organelles can be analyzed, and their size, number per cell, and rates of synthesis and turnover can be studied. This is often done by using radioactive compounds that are incorporated into viral or cellular DNA, RNA, protein, lipid, or carbohydrates. Antibodies against viral or cellular proteins can be used to detect specific molecules with fluorescent dyes or other assays. Molecular hybridization with oligonucleotide probes, or the polymerase chain reaction (PCR), can be used to detect specific viral DNA or RNA molecules. Macromolecules are often separated from each other by electrophoresis on polyacrylamide or agarose gels.

Staining of macromolecules can be observed in fixed or living cells by microscopy, especially by using computerized methods that allow visualization of horizontal layers of thickness 1 micrometer or less within individual cells. These views, revealed by **confocal microscopes** using focused laser light sources, show the localization of viral macromolecules within the nucleus, or in association with organelles such as the endoplasmic reticulum, lysosomes, or the plasma membrane. Other localization methods use the higher-resolution electron microscope coupled with specific staining, for example, using colloidal gold particles linked to antibodies.

Study of virus replication cycles by these and other methods has led to a general understanding of the various steps in virus replication. In the case of polyomavirus, Figure 1.5 shows that small amounts of "early" viral messenger RNAs are made beginning about 8–10 hours after infection, and that early viral proteins, known as T antigens, begin to appear shortly afterward. This is followed by the beginning of viral DNA replication at 12–15 hours after infection. Once viral DNA replication has begun, much larger amounts of viral messenger RNAs can be detected, and these "late" messenger RNAs are copied from a different set of viral genes. Mature progeny virions begin to appear by 18–20 hours after infection, and virus titers increase slowly for the next 24 hours, by which time most of the infected mouse cells are dead.

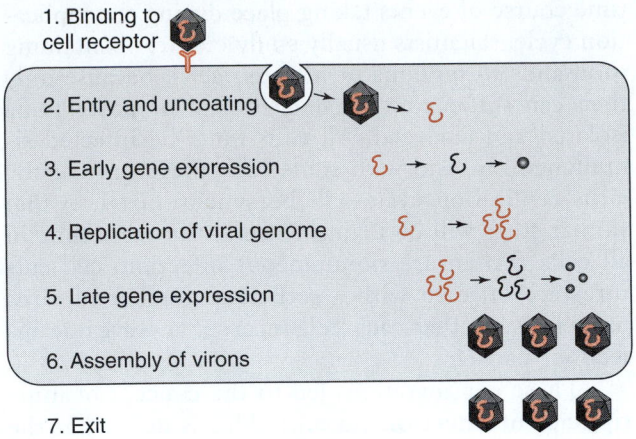

FIGURE 1.6 Steps in virus replication cycle. A schematic diagram showing a simplified virus replication cycle in seven steps, as described in text. The host cell is shown as a rounded box. Capsids are represented as gray hexagons, genomes as orange curved lines, messenger RNAs as gray lines, viral proteins as gray curved circles.

Virus replication cycles vary greatly, depending on the host cell, genome type, and complexity of the virus. However, there is a logic to the progression of steps in virus replication. The following section gives an outline and analysis of the replication pathway starting with the initial infection of a cell and progressing to the release of progeny virions (Figure 1.6). Designation and naming of these steps is somewhat arbitrary but it is useful to help understand the important elements of the replication cycles of all viruses.

STEPS IN THE VIRUS REPLICATION CYCLE

1. Virions bind to receptors on the cell surface

Viruses must first recognize and bind to cells that they infect. Virus-coded proteins on the surface of the virion bind to specific proteins, carbohydrates, or lipids on the cell surface. Structural surface proteins of many viruses bind to carbohydrate residues found in surface glycoproteins and glycolipids that are widely distributed on many cell types. Other viruses bind to cell surface proteins that are found only on specific cell types, and are often species-specific, thus limiting the **tropism** of the virus to a particular tissue and organism. Many viruses first bind to a relatively nonspecific (primary) receptor such as a carbohydrate, and subsequently bind to a specific cell surface protein that serves as a secondary receptor.

2. The virion (or the viral genome) enters the cell

Once they have located an appropriate cell, bacteriophages and plant viruses are presented with the problem of passing through a rigid cell wall, as well as the outer cell membrane(s). Many bacteriophages have specialized tails that drill holes in the cell wall and membranes, and serve as a conduit for passage of the DNA genome through the hole into the cell. Plant viruses often penetrate as a result of damage to the cell wall caused by abrasion or a wound caused by an insect.

Some enveloped animal viruses fuse their lipid envelopes directly with the plasma membrane of the cell, releasing the viral capsid and genome into the cell. Many other animal viruses are taken up into the cytoplasm in vesicles formed at the plasma membrane. These vesicles then release the virion or its genome into the cytoplasm, and in some cases the genome is transported to the nucleus. Release can happen by disintegration of the vesicle membrane or by fusion of the viral envelope with the vesicle membrane. In cases where the capsid is released into the cell, a variety of pathways leads to disintegration of the capsid and release of the genome, called uncoating. Entry of animal viruses into cells is discussed in detail in Chapter 4.

3. Early viral genes are expressed: the Baltimore classification of viruses

Once in the appropriate compartment of the cell, the viral genome must direct expression of "early" proteins that will enable genome replication. The molecular pathways that lead to synthesis of early viral proteins from viral messenger RNAs depend on the chemical nature and strandedness of the genome. David Baltimore first recognized that all viruses can be divided into six groups based on the pathways leading to mRNA and protein synthesis; this has come to be known as the *Baltimore classification system* (see Figure 1.7). One advantage of this system is that it draws our attention to the distinct kinds of RNA polymerases needed by viruses in each group, and indicates whether these enzymes are available in the cell or must be provided by the virus.

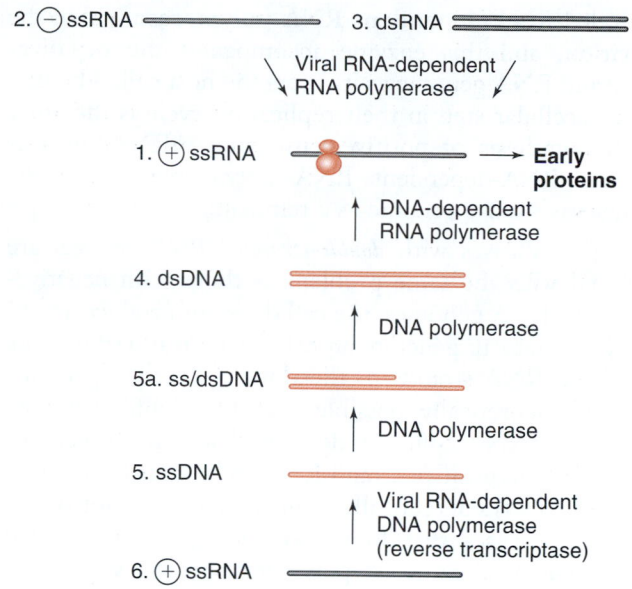

FIGURE 1.7 **Baltimore classification of viruses.** Six categories of viruses are distinguished based on the pathway leading from the viral genome to early messenger RNAs. Numbering of groups is arbitrary. RNA and DNA polymerases that carry out different steps are shown. DNAs are shown as orange lines and RNAs as black lines. ss: single-stranded; ds: double stranded; +: positive-strand RNA; −: negative-strand RNA. Messenger RNA is shown being translated by a ribosome.

Viruses with single-stranded RNA genomes fall into two categories. Some viruses package the "sense" RNA strand, which contains coding regions that can be directly translated into viral proteins. By convention, this is labeled the **positive** (or **plus**) **strand**. Other viruses package the "antisense" strand, which cannot be translated because it does not contain meaningful coding regions; only its complementary copy codes for viral proteins. This is labeled the **negative** (or **minus**) **strand**. Viruses in these two groups differ fundamentally in how their replication cycle begins.

The six groups in the Baltimore classification system

1. Viruses with *positive-strand RNA genomes* bring their messenger RNA directly into the cell in the form of the genome. This RNA can bind to ribosomes and is translated into viral proteins as the first step in virus replication. These proteins will then direct replication of the viral genome.

2. Viruses with *negative-strand RNA genomes* must first synthesize complementary positive-strand copies of the genome, which serve as viral messenger RNAs. However, the cell does not produce an enzyme able to carry out this step, so the virus itself must supply this enzyme. Negative-strand RNA viruses package a virus-coded **RNA-dependent RNA polymerase** within the virion, and this enzyme accompanies the negative-strand RNA genome as it enters the host cell. The first intracellular step in their replication cycle is therefore the synthesis of positive-sense viral mRNAs by this viral RNA-dependent RNA polymerase, using the negative-sense genome as a template.

3. Viruses with *double-stranded RNA genomes* are faced with the same problem as those with negative-strand RNA genomes: the cell does not produce an enzyme that can generate an mRNA by transcribing one of the RNA strands in the double-stranded genome. It is theoretically possible that the double-stranded genome RNA could be denatured within the cell, allowing the positive-strand RNA to be translated; however, this is energetically unlikely and does not occur. These viruses therefore also package a virus-coded RNA-dependent RNA polymerase in the virion and bring this enzyme into the host cell along with the genome. This RNA polymerase can specifically recognize the double-stranded RNA genome and transcribe its negative strand into positive-strand messenger RNAs.

4. Viruses with *double-stranded DNA genomes* must first transcribe their genomes into messenger RNAs, using a **DNA-dependent RNA polymerase**. Cellular messenger RNAs are synthesized in exactly the same way, and therefore all cells contain such enzymes. Most DNA viruses use a host cell RNA polymerase for this purpose, but some of the larger DNA viruses bring their own RNA polymerase into the cell along with their genome, or make an additional RNA polymerase later in the cycle to transcribe a subset of their genes.

5. Viruses with *single-stranded DNA genomes* must first convert their DNA into a double-stranded form, as there are no known enzymes that will transcribe single-stranded DNA directly into RNA. This conversion is carried out by a cellular DNA polymerase. The resulting double-stranded DNA is then transcribed by a cellular RNA polymerase to generate viral mRNAs. Note that either of the complementary DNA strands can be packaged in virions. Most viruses preferentially package only one of the two DNA strands, but some viruses package both strands, making two kinds of virions that are identical except for the polarity of their single-stranded DNA genomes. In all cases the single-stranded DNA is converted into a double-stranded form before being transcribed, so the polarity of the packaged strand is of little importance.

6. Retroviruses have a *single-stranded positive-sense RNA genome*, but they follow a pathway different from other positive-strand RNA viruses. Retroviruses package a virus-coded *reverse transcriptase* (an **RNA-dependent DNA polymerase**) within the virion, and use this enzyme to produce a double-stranded DNA copy of their RNA genome after they enter the cell. Once the double-stranded DNA is made (and integrated into the cellular genome: see Chapter 25 for details), it is transcribed into viral messenger RNAs by host cell RNA polymerase II, as for most DNA viruses.

5a. There remains a maverick category of viruses in the Baltimore classification system, of which the best known representatives are the **hepadnaviruses** (see Chapter 28). These viruses package a circular DNA genome that is partly double-stranded, with a single-stranded gap of variable length on one strand. Hepadnaviruses can be seen as a special case of the category of viruses that have single-stranded DNA genomes (#5 above), because like those viruses, their genomes are made fully double-stranded by cellular DNA polymerases after entering the cell. This double-stranded DNA is subsequently transcribed into messenger RNAs by cellular RNA polymerase, as for other DNA viruses. However, like retroviruses and unlike other DNA viruses, hepadnaviruses code for a reverse transcriptase. In the case of the hepadnaviruses, this reverse transcriptase is used to synthesize the partially single-stranded DNA genome that is packaged in virions, starting from an RNA intermediate made from the fully double-stranded DNA genome. See Chapter 28 for a full description of how this occurs.

4. Early viral proteins direct replication of viral genomes

Once early viral proteins are made, they act to promote replication of the viral genome. All RNA viruses (except retroviruses) must synthesize an RNA-dependent RNA polymerase to replicate their genomes, as this enzyme is not present in host cells. A number of other early proteins, as well as a variety of cellular proteins, help form RNA replication complexes in the cell. These replication complexes are often associated with cellular membranes.

The early proteins of many DNA viruses induce formation of a number of cellular enzymes that are involved in the synthesis of DNA and of its building blocks, deoxyribonucleoside triphosphates. This is often achieved by interaction of early viral proteins with cellular signaling pathways that affect the cell cycle and direct the cell to enter the DNA synthesis (S) phase. Small DNA viruses usually use host cell DNA polymerases to replicate their genomes; larger DNA viruses often code for their own DNA polymerases as well as other enzymes involved in DNA replication.

Hundreds to tens of thousands of copies of the viral genome are made in each cell. These progeny genomes can be used as templates for synthesis of more viral messenger RNAs or for further genome replication. At this stage, the cell has become a factory for the expression and replication of viral genomes.

5. Late messenger RNAs are made from newly replicated genomes

Many viruses synthesize a distinct set of "late" messenger RNAs after genome replication has begun. Because of genome replication, there can be many templates for late messenger RNA synthesis, and these mRNAs may therefore be abundant. The mechanisms that control the switch between early and late messenger RNA synthesis have been extensively studied as paradigms for the control of gene expression in a variety of viruses and their host cells.

6. Late viral proteins package viral genomes and assemble virions

The most abundant late proteins are structural proteins used in the packaging of viral genomes and the assembly of the capsid. The simplest virus capsids consist of one protein that forms either a closed shell or a helical tube within which the virus genome is packaged (Figure 1.1). Larger and more complex viruses may have numerous capsid proteins, and some viruses make **scaffolding proteins** that are involved in virion assembly but are subsequently discarded and do not form part of the mature virion. Many bacteriophages have both a polygonal head that contains the genome and a tubular tail involved in attachment to the cell and delivery of the genome. Enveloped viruses code for glycoproteins that are inserted into lipid membranes and that direct formation of the viral envelope around the capsid by a process called **budding**. For a detailed discussion of virion structure and assembly, see Chapter 2.

7. Progeny virions are released from the host cell

Once formed, virions leave the cell to find and infect new host cells and reinitiate the replication cycle. For viruses that infect unicellular organisms, this usually involves death and lysis of the host cell, and many viruses code for specialized late proteins that lead to cell death. However, some bacteriophages are formed by extrusion from the cell membrane and do not kill their host cells.

For viruses that infect multicellular organisms, the problem becomes more complex. Viruses can spread from one cell to another within an organism, or they can move from one organism to another; virions that follow these two pathways are not necessarily identical. For example, many viruses of higher plants spread throughout a plant by passage through intracellular channels (**plasmodesmata**), using specific viral proteins to enhance passage of unencapsidated viral genomes. However, spread from plant-to-plant is often carried out by insects that feed on plant juices, ingesting virions and reinjecting them into other plants.

Some insect viruses spread as individual virions from cell to cell within the host organism, but wrap one or more virions in a specialized protein coat that is released when the insect dies. This protein coat protects virions in the environment and is ingested by other insects, in which it dissolves and releases the infectious virions to start another cycle (see Chapter 14).

Release of animal viruses can occur by a number of pathways. Some viruses accumulate within host cells and are released only on cell death. Certain viral proteins function to retard cell death, prolonging the period of virus replication, and other viral proteins can kill the host cell, leading to virus release. Many enveloped viruses are assembled by budding at the plasma membrane, and can be released with little or no effect on the host cell. Others bud at internal cell membranes or form their envelopes *de novo*; many of these viruses use cellular transport vesicles to reach the cell membrane and to exit the cell.

SUGGESTED READING

A number of virology textbooks are useful as sources of more detailed information about viruses and their diseases, or as condensed overviews of the field of virology. Following is a selection of some of the most useful textbooks presently available, with brief descriptions of their content and approach.

1. Cann, A.J. (2005). Principles of molecular virology, 4th Ed. Paperback, 340 pp. Academic Press, San Diego, CA.

A condensed textbook directed toward undergraduate or graduate virology students. Organized around topics (Particles, Genomes, Replication, Expression, Infection, Pathogenesis) rather than virus groups.

2. Fauquet, C.M., Mayo, M.A., Maniloff, J., Desselberger, U., and Ball, L.A., Eds. (2005). Virus taxonomy: VIIIth report of the international committee on taxonomy of viruses. Hardcover, 1259 pp. Elsevier Academic Press, San Diego, CA.

An essential reference book that describes present understanding of the taxonomy of all known viruses. Articles present basic information about each virus group and lists of known virus species.

3. Flint, S.J., Enquist, L.W., Racaniello, V.R., and Skalka, A.M. (2004). Principles of virology: Molecular biology, pathogenesis, and control of animal viruses, 2nd Ed. Hardcover, 918 pp. ASM Press, Washington D.C.

An elaborately illustrated, up-to-date textbook written by expert authors. Many full-color figures; useful appendices covering virus replication cycles and virus diseases. Oriented toward students with a high knowledge level. The section on molecular biology is organized to cover individual steps in virus replication cycle, using examples from a number of viruses as illustrations. Extensive section on virus pathogenesis and antiviral treatment.

4. Knipe, D.M. and Howley, P.M., Eds.; Griffin, D.E., Lamb, R.A., Martin, M.A., Roizman, B., and Straus, S.E.,

Assoc. Eds. (2001). Fields Virology, 4th Ed. Hardcover, 3087 pp (two volumes). Lippincott Williams and Wilkins, Philadelphia, PA.

Has become the "bible" of human medical virology; an in-depth textbook with extensive articles on most groups of human viruses and their pathogenesis, written by experts in the field. More useful as a reference book than a textbook for a virology course.

5. Knipe, D.M. and Howley, P.M., Eds.; Griffin, D.E., Lamb, R.A., Martin, M.A., Roizman, B., and Straus, S.E., Assoc. Eds. (2001). Fundamental virology, 4th Ed. Hardcover, 1395 pp. Lippincott Williams and Wilkins, Philadelphia, PA.

Selected chapters from *Fields Virology*, covering basic aspects of molecular biology of most groups of viruses that infect humans as well as some other organisms.

6. Levine, A.J. (1991). Viruses. Hardcover, 240 pp. Scientific American Library.

A book written for knowledgeable laypeople who want to learn about viruses. Includes an extensive section on the history of virology research.

7. Sompayrac, L. (2002). How pathogenic viruses work. Paperback, 115 pp. Jones and Bartlett Publishers, Sudbury, MA.

A short review of pathogenic viruses, written in a chatty and accessible style for students or educated laypeople.

8. Strauss, J.H., and Strauss, E.G. (2002). Viruses and human disease. Hardcover, 383 pp. Academic Press, San Diego, CA.

Covers most human viruses, broken down by nature of genome and virus family, from the standpoints of molecular biology, pathogenesis, and epidemiology. Up-to-date coverage, well illustrated in full color.

KEY TERMS

Budding
Capsid
Confocal microscope
DNA-dependent RNA polymerase
Envelope
Filterable virus
Hemagglutination assay
Hepadnavirus
Lysis
Multiplicity of infection (m.o.i.)
Negative (or minus) strand
Obligatory intracellular parasite
Oncogene
Plaque assay
Plaque-forming units (PFU)
Plasmodesmata
Positive (or plus) strand
Retrovirus
Reverse transcriptase
RNA-dependent DNA polymerase
RNA-dependent RNA polymerase
Scaffolding proteins
Sialic acid
Tropism
Uncoating
Virion

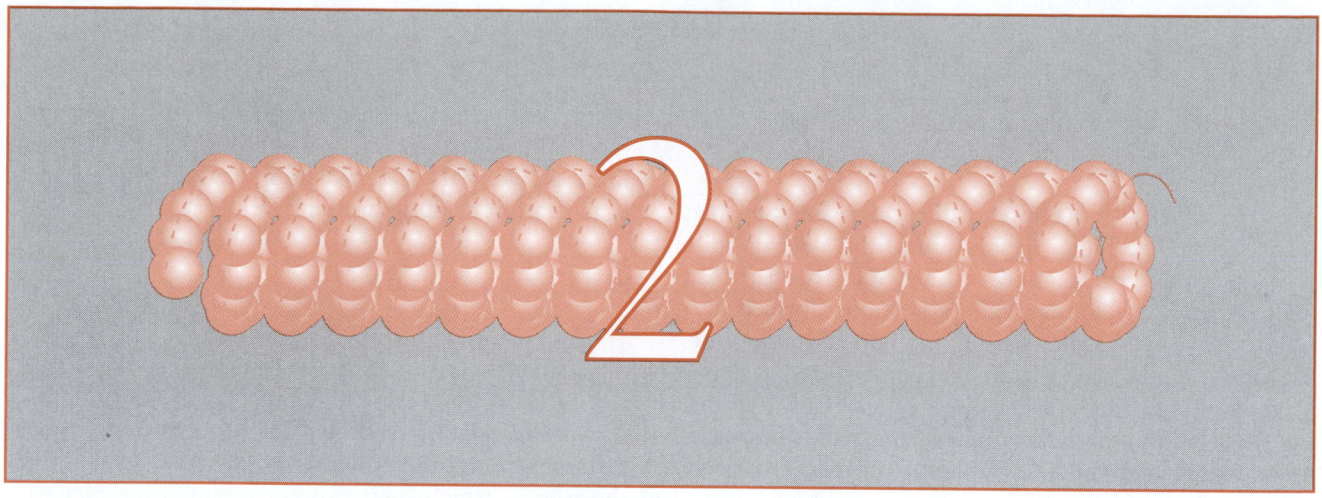

Virus Structure

CHARACTERISTICS OF CAPSIDS

Virus capsids are assembled from many copies of a small number of viral proteins: principle of genetic economy.
Capsids made from multiple identical subunits have regular, symmetrical shapes.
Symmetrical capsids are strong, stable, and can self-assemble.
Most capsids are constructed as either:

- *Closed spherical shells with icosahedral symmetry or*
- *Elongated tubes with helical symmetry*

Most icosahedral capsids contain 60 × T protein subunits.
T is called the triangulation number; $T = h^2 + hk + k^2$, where h and k are integers.
The polypeptide chain of many capsid proteins is folded as a jelly-roll β-barrel.
Some viruses include internal or core proteins bound to the nucleic acid.

VIRUS ENVELOPES

Many viruses enclose the nucleocapsid in a lipid membrane, called an envelope.
Virus envelopes contain viral glycoproteins inserted into the lipid membrane via a transmembrane anchor.
Most virus envelopes are formed by budding from preexisting cellular membranes.
Some virions contain a layer of matrix protein between the nucleocapsid and the envelope.

VIRION ASSEMBLY

Many capsids self-assemble, but others use scaffolding proteins to assemble virions and package genomes.
Specific packaging signals on virus genomes are recognized and bound by viral proteins.
Virion proteins often undergo proteolytic cleavage during or after assembly, preparing the virion for uptake by target cells.

The molecular structure of virus particles

Virus particles are molecular structures that package virus genomes in infected cells and transmit them to new host cells. The word **virion** is used to denote a *complete, infectious virus particle*. The structures of virions have evolved to carry out the task of transmitting the RNA or DNA genome of a virus from one cell to another and from one host organism to another. Virions must be correctly assembled, escape the cell in which they are made, withstand the extracellular environment, attach to and enter another cell, and release the viral genome. These different functions determine the structural characteristics of virions.

What do we mean by the term *structure*? In everyday life, there can be packages with different structural characteristics. A cardboard box is rigid, symmetrical, and fixed in volume and shape; a plastic bag is flexible, variable in volume and shape, and roughly spherical when stuffed full. At the molecular level, virions can be composed of rigid, symmetrical containers called **capsids** (derived from the Latin *capsa* = box). When the viral nucleic acid genome is packaged within the capsid, the resulting structure is called a **nucleocapsid**. Some viruses enclose their nucleocapsids in lipid-bilayer membranes called **envelopes**; these virions are more flexible and can have a variety of shapes. Virions without envelopes are sometimes called "naked" virions.

How virus structure is studied: viruses come in a variety of sizes and shapes

Most virus particles are too small to be visible by light microscopy. The higher resolution of the electron microscope reveals the general outlines of a virus particle, and shows the bumps and projections formed by oligomers of viral structural proteins on the surface of the virion (Figure 2.1). Virions can be visualized by **negative staining**, which reveals the structure of the virion by the exclusion of electron-absorbing molecules such as uranyl acetate (Figures 2.1a–g). Alternatively, proteins, nucleic acids, and lipids are revealed by **positive staining** in plasticized thin sections of infected cells (e.g., human immunodeficiency virus; Figure 2.1j). However, both of these methods can distort or blur structural details of virions.

In **cryoelectron microscopy**, samples are flash-frozen in liquid nitrogen and kept at that temperature in the microscope. This method improves preservation of structure, but image contrast is low. Data from images of many different virions are combined by computer programs to build an overall averaged image of the structure at a higher resolution than possible by direct imaging of individual particles (e.g., Semliki Forest virus; Figure 2.1h). Studies of the hepatitis B virus core particle by cryoelectron microscopy revealed even the α-helical secondary-structure elements in the protein coat.

When the particle can be crystallized, as is the case for many naked virions, a very detailed model can be obtained from **x-ray diffraction** data (e.g., simian virus 40; Figure 2.1i). Because x-rays have a much shorter wavelength than electrons, the resolution of structures determined by x-ray diffraction is much greater; in the most favorable cases, this allows precise determination of the position of every atom in the virion. However, x-rays cannot be focused, and therefore direct imaging of the virion is not possible; instead, diffraction patterns are captured and analyzed. High-intensity x-rays are produced in giant **synchrotrons** and the diffraction data are analyzed by sophisticated computer programs to yield the three-dimensional structure of the virion.

In contrast to many highly symmetrical naked virions, some enveloped viruses such as influenza virus (Figure 2.1f) have virions with variable shape and size. The lipid bilayer of the envelope ensures impermeability to ions or nucleases that might damage the viral genome within the capsid that lies inside the envelope. The viral proteins that are inserted into this membrane do not need to form a perfectly tight barrier as in a capsid, and therefore are not usually organized in a symmetrical fashion. Because of their structural variability, the virions of most enveloped viruses cannot form crystals, and therefore high-resolution structures cannot be obtained by x-ray diffraction. Nonetheless, individual viral membrane proteins, such as the influenza virus hemagglutinin and neuraminidase, can be purified and crystallized, allowing the determination of the precise atomic organization of these proteins.

Small viruses come in simple, symmetrical packages

The simplest virus particles assemble spontaneously from their protein subunits and nucleic acid genome; this is called **self-assembly**. The information that determines their three-dimensional structures is built into the way in which their protein subunits interact with each other and with the packaged RNA or DNA. A major constraint on the evolution of designs for virus particles is **genetic economy**. Consider a 5000-nucleotide single-stranded RNA or DNA genome, which can code for proteins totalling about 1500 amino acid residues (1 amino acid per 3 nucleotides). If highly condensed, the genome will occupy a spherical cavity about 18 nm in diameter. A protein shell 3 nm thick surrounding such a cavity will have about 30,000 amino acid residues, or a total molecular weight of about 3×10^6 Daltons. It is clearly impossible for the viral genome to code for proteins of that size. It follows that the protein shell must be made of a large number of identical protein subunits, in order not to exhaust the coding capacity of the genome.

Genetic economy leads to a requirement for symmetry of capsid structures. The large number of identical protein subunits will not form a tight and enclosed package if they are linked by random contacts. Instead, the interactions between neighboring proteins must be well-defined and specific. If all protein subunits on the surface of a capsid are identical, then the contacts will be identical (or nearly so), generating a symmetrical object. The rigorous definition of symmetry involves operations in space, such as rotations, that bring a symmetrical object into self-coincidence. The symmetry of the object is defined by the collection of all such operations that apply to it. A symmetry axis that involves rotation by 180 degrees is called a **twofold axis**; one that involves rotation by 120 degrees is called a threefold axis (Figure 2.2a); and so forth. Note the distinction between shape and symmetry: the *shape* of an object refers to the geometry of its outline, whereas its *symmetry* refers to the rotational and translational operations that describe it.

Many virus capsids have icosahedral symmetry

Geometry dictates that closed shells made of identical subunits can have one of only three symmetries—tetrahedral (a tetrahedron is a polygon with 4 identical triangular faces), cubic (a cube has 6 square faces), and

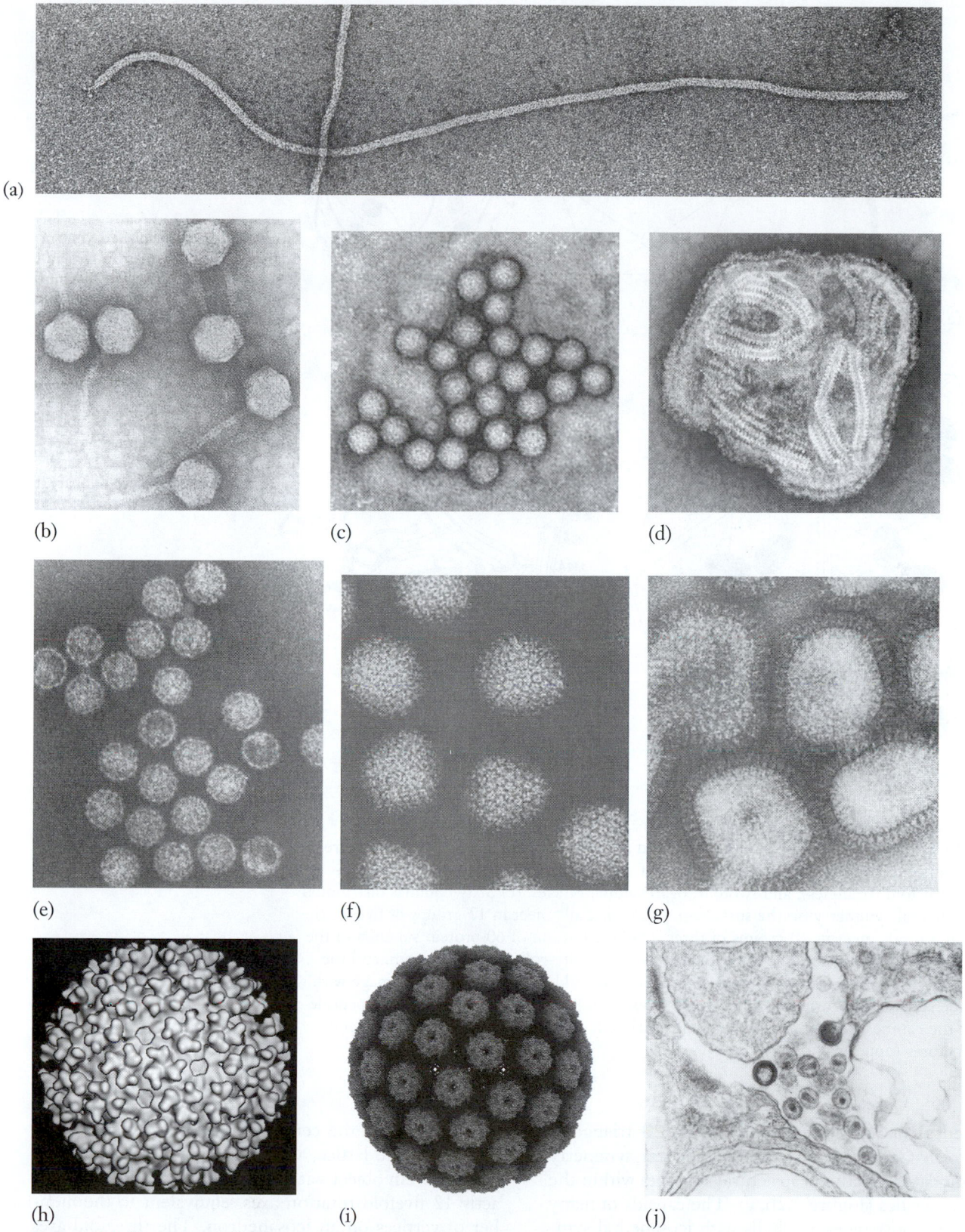

FIGURE 2.1 Structures of some typical virus particles. (a) Bacteriophage fd. (b) Bacteriophage lambda. (c) Hepatitis A virus (a picornavirus). (d) Human parainfluenza virus (a paramyxovirus; in this image, the helical nucleocapsid is revealed by penetration of stain inside the particle). (e) Human papillomavirus. (f) Adenovirus. (g) Influenza virus (an orthomyxovirus). (h) Semliki Forest virus (a togavirus). (i) Simian virus 40 (a polyomavirus). (j) Human immunodeficiency virus (a retrovirus). Parts a–g are electron micrographs of negatively stained virions, part h is a computer-generated image derived from cryoelectron microscopy, part i is a computer-generated image derived from x-ray diffraction, and part j is an electron micrograph of a thin section of infected cells. Scales are different for each image.

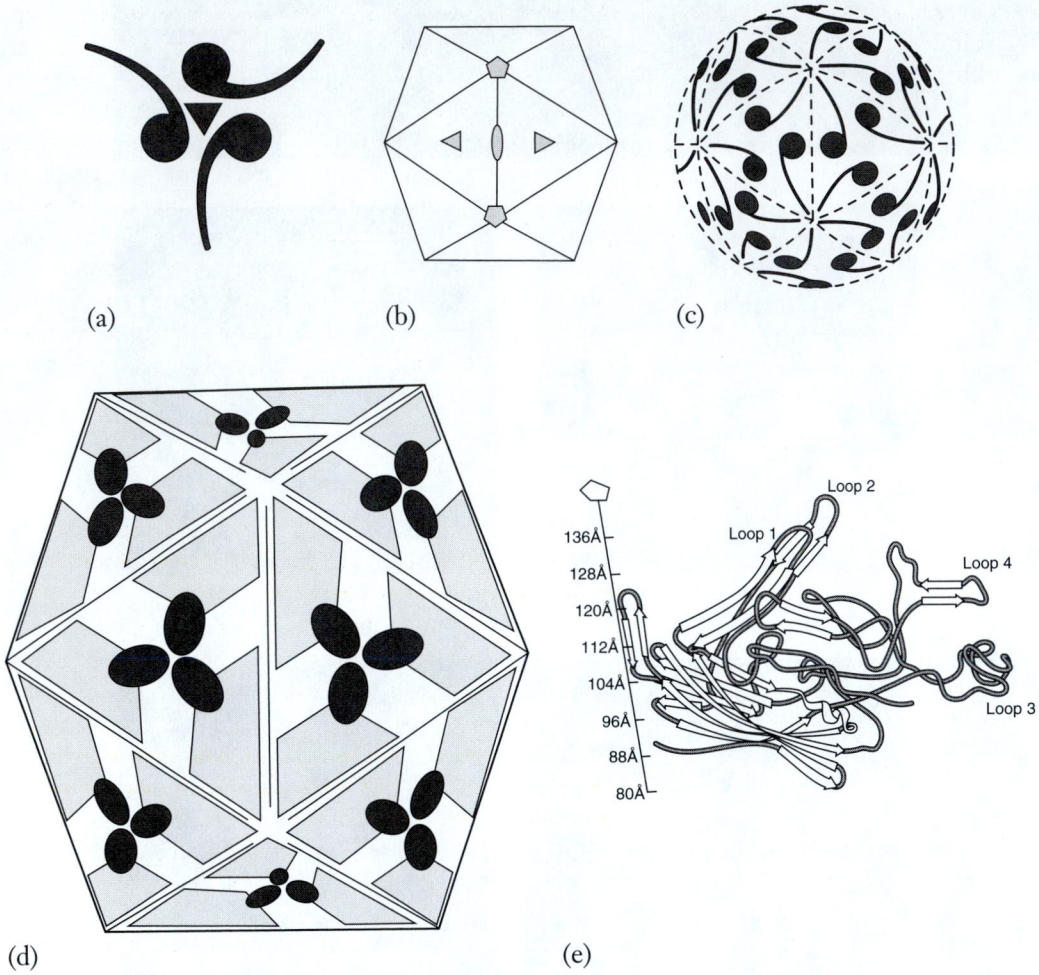

FIGURE 2.2 **Elements of icosahedral symmetry.** (a) Three (asymmetrical) commas, arranged with threefold symmetry. The threefold axis passes perpendicularly to the figure through the center of the triangle. (b) Outline of an icosahedron, showing fivefold (pentagons), threefold (triangles), and twofold (oval) symmetry axes. (c) Sixty commas distributed with icosahedral symmetry on the surface of a sphere. Tails meet in 12 groups of five on the surface, and heads meet in 20 groups of three. (d) Distribution of 60 protein subunits in the shell of a parvovirus capsid. The trapezoids represent a jelly-roll β-barrel domain, and the ovals represent loops that form contacts about the threefold axes. Compare this image with that of the sphere in C, oriented in the same manner. (e) Diagram showing the polypeptide chain of the canine parvovirus capsid protein, with jelly-roll β-barrel (lower left half) and loop structures (upper right half).

icosahedral (a regular icosahedron has 20 triangular faces; from Greek *eikosi* = 20). Icosahedral symmetry accommodates exactly 60 structural subunits within the shell it describes (Figure 2.2b, c). The capsids of many viruses are constructed as shells with icosahedral symmetry; tetrahedral and cubic symmetries have not been observed.

An object with icosahedral symmetry possesses fivefold, threefold, and twofold rotation axes. Figures 2.2b and c show side-by-side representations of an icosahedron and a sphere onto which 60 commas are distributed in a **surface lattice** with icosahedral symmetry. The fivefold rotation axes (pentagons on the icosahedron) pass from the center of the sphere through the vertices of the lattice, where the tails of five commas converge. An object with icosahedral symmetry has exactly 12 fivefold rotation axes, equivalent to the number of vertices on an icosahedron. The threefold axes (triangles) pass through the centers of the spherical triangles outlined by the dashed lines on the image (see also Figure 2.2a). Objects with icosahedral symmetry have 20 threefold rotation axes, the number of triangular faces on an icosahedron. The twofold axes (oval) pass through the midpoints of the edges of those triangles, where the heads of two commas lie opposite each other. Objects with icosahedral symmetry have 30 twofold rotation axes. Note that the sphere shown in Figure 2.2c has icosahedral *symmetry*, but it does not have the *shape* of an icosahedron (Figure 2.2b).

Some examples of virions with icosahedral symmetry

Parvoviruses are among the simplest viruses, with a 5 Kb single-stranded DNA genome. Their capsids contain 60 copies of a protein of about 520 amino acid residues. The parvovirus capsid protein therefore uses up about one-third of the coding capacity of the viral genome. Parvovirus capsid protein subunits are arranged on the surface of the capsid with icosahedral symmetry (Figure 2.2d; compare with Figures 2.2b and c). This is the simplest type of capsid that can be constructed with icosahedral symmetry. The folded structure of the parvovirus capsid protein is based on a domain known as a **jelly-roll β-barrel** (Figure 2.2e). The polypeptide chain is folded into a series of β strands that interact with each other to form a rigid structure. Arm-like extensions from the β strands of the subunits tie together the mature virion by interacting with neighboring subunits. The same fold is found in many other icosahedral virus capsids.

Poliovirus (Figure 2.3) illustrates one way in which viruses have evolved to produce larger icosahedral shells and thus to package larger genomes. The poliovirus capsid has an internal cavity about 19 nm in diameter, which contains the single-stranded RNA genome of about 8 Kb. The genome encodes three structural protein subunits, known as VP1, VP2, and VP3, which are processed by cleavage from a single precursor polypeptide. Again, about one-third of the genome is used to specify the virion structural proteins. Like parvoviruses, the three poliovirus capsid proteins are all folded as a jelly-roll β-barrel. The capsids of poliovirus and other members of the picornavirus family therefore contain three sets of 60 similarly folded protein subunits, for a total of 180 subunits.

The concept of quasi equivalence

A more economical way to build a capsid with 180 subunits is to use 180 identical protein subunits, rather than 60 copies each of three different homologous proteins. However, these subunits will not all be in identical positions on the surface of the capsid, and therefore they must be allowed to interact differently, corresponding to their symmetrically distinct environments. Capsids containing certain multiples of 60 subunits permit all the subunits to interact with their neighbors in *nearly* equivalent ways. This concept is illustrated in Figure 2.4a, which can be compared with Figure 2.2c. All 180

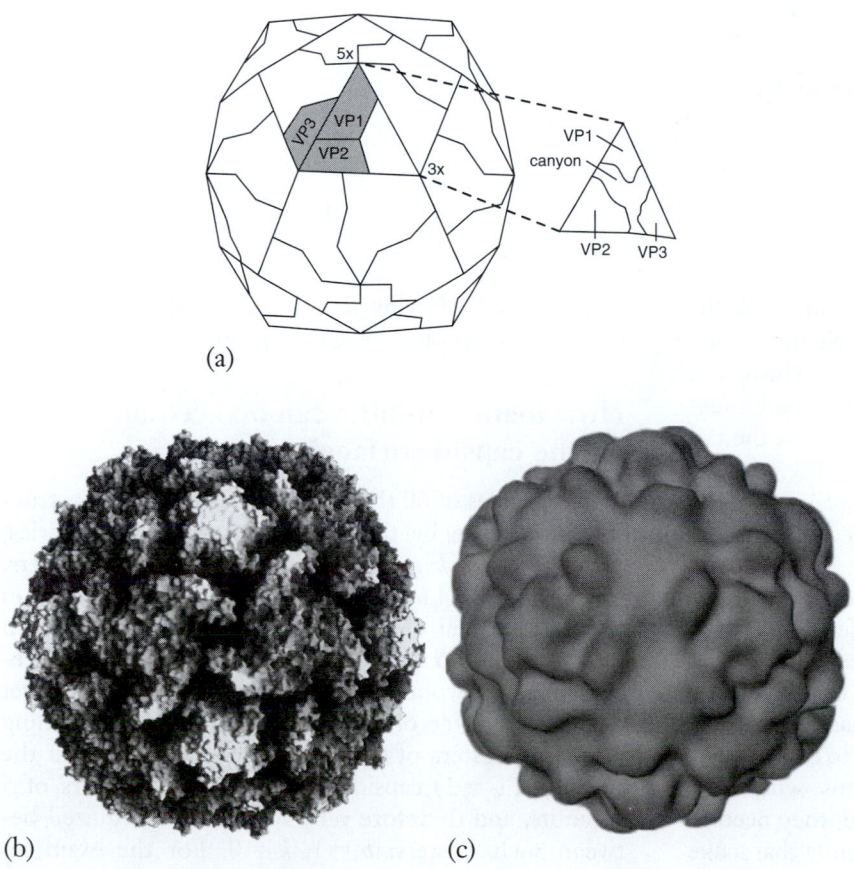

FIGURE 2.3 Picornavirus capsids. (a) Packing of subunits, showing structural proteins VP1, VP2, and VP3 on the surface of the virion (VP4 is in the interior). (b) Model of poliovirus type I based on x-ray crystallography. (c) Model based on cryoelectron microscopy.

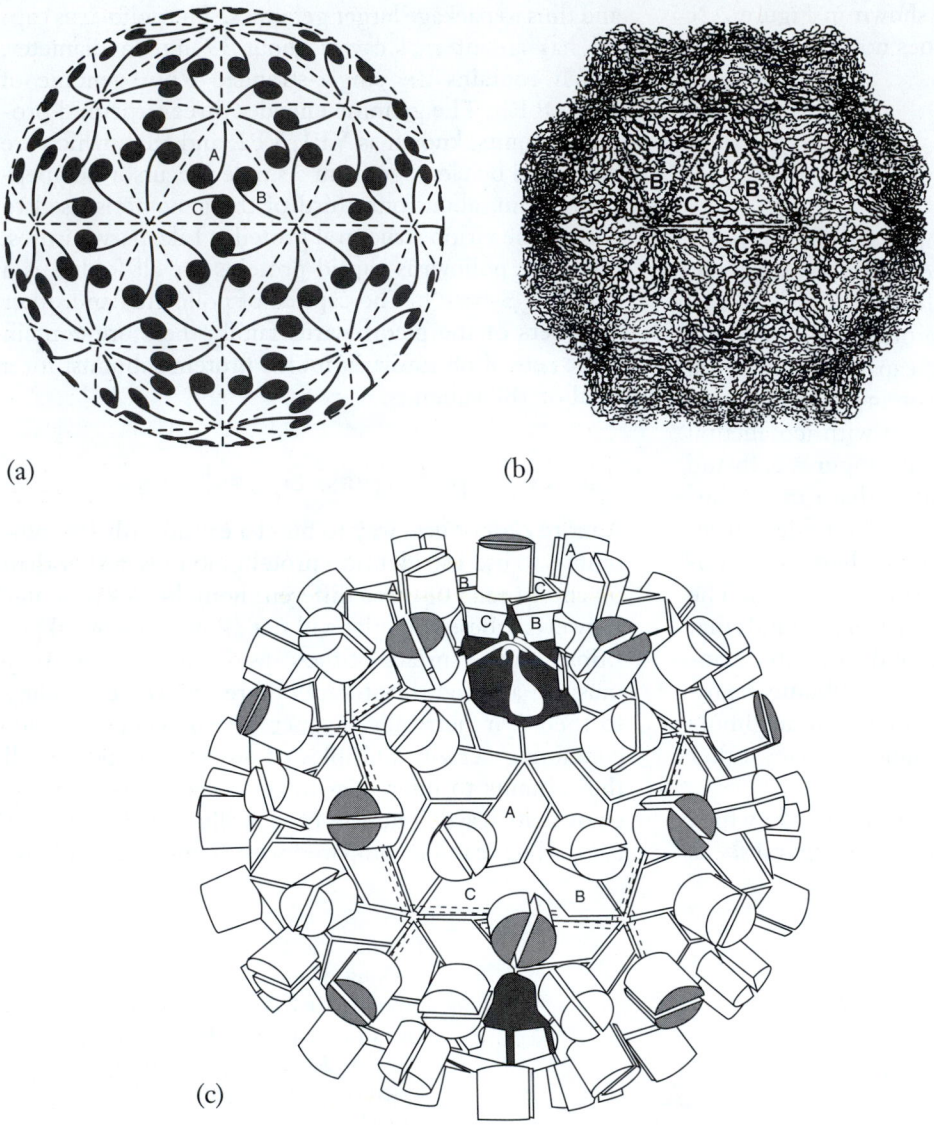

FIGURE 2.4 More complex capsid protein arrangements. (a) Quasi-equivalent arrangement of 180 commas in a T = 3 icosahedral surface lattice on a sphere. Compare with Figure 2.2b, a T = 1 arrangement. (b) Capsid of Norwalk virus, a human pathogen with a T = 3 structure. In this computer-drawn image, the outlines of the polypeptide chains are shown. (c) Tomato bushy stunt virus, a plant virus with a T = 3 capsid structure. The capsid is formed by 180 identical subunits, shown as 60 groups of three subunits labeled A, B, and C. The 90 cylindrical protrusions are constructed from portions of two adjacent subunits.

commas in Figure 2.4a have head-to-head, neck-to-neck, and tail-to-tail interactions with their neighbors. Sixty of them have tail-to-tail interactions in clusters of five; for example, look at the 5 commas whose tails meet at a point immediately above the equator, near the tail of the comma marked "A." The remaining 120 commas have tail-to-tail interactions in clusters of six; for example, look at the clusters on the equator including the commas marked "B" and "C."

Provided that the proteins can evolve to tolerate the distortions required by these slightly different contacts, they can use the same domains to make stable contacts at all 180 positions on the capsid surface. In other words, if the capsid corresponds to one of these **quasi-equivalent** designs, the protein subunits do not need to evolve three alternative binding sites for interactions with each other, as in the case of poliovirus; instead, they need to have built-in *flexibility*. The protein subunits that make up the capsids of many simple viruses exhibit this property. Two examples are shown in Figures 2.4b and c.

How many subunits can be accomodated on the capsid surface?

The multiples of 60 that lead to quasi-equivalent structures are given by the **triangulation number** series, $T = h^2 + hk + k^2$, where h and k are any two integers, including zero. The total number of protein subunits in a capsid is equal to $60 \times T$. The integers h and k can be considered to represent the number of "jumps" required along two axes on the surface of a capsid to get from one cluster of 5 subunits to the next, touching down on clusters of 6 subunits along the way. On the simplest (T = 1) capsids, there are only clusters of 5 subunits, and therefore only one jump is required between such clusters: $h = 1$, $k = 0$. For the examples shown in Figures 2.4a and b, it takes two single jumps in two different directions to get from the fivefold axis above the equator to the fivefold axis below the equator. Therefore $h = 1$ and $k = 1$, and $T = 1^2 + 1^2 + 1^2 = 3$. Examples of T = 1, 3, 4, 7, 13, 16, and 25 are known. Capsids with larger T numbers require additional "glue" proteins to stabilize the interactions between subunits.

An extension of the concept of quasi-equivalence is seen in the diagrams of a bacteriophage head and the capsid of human immunodeficiency virus (HIV-1), shown in Figure 2.5. These diagrams do not show symbols for individual subunits, like the commas in Figures 2.2 and 2.4. Instead, they show the lattice that connects positions of threefold symmetry on the surface of the capsid. The lattice can be filled with subunits in sets of six within the hexagons and sets of five within the pentagons. The curved, closed ends of these elongated capsid structures have icosahedral symmetry, and contain clusters of five subunits. These closed ends are separated by sets of six subunits that form the elongated part of the capsid. The same protein can form both parts of the capsid, via quasi-equivalent interactions that allow sets of either five or six subunits to form.

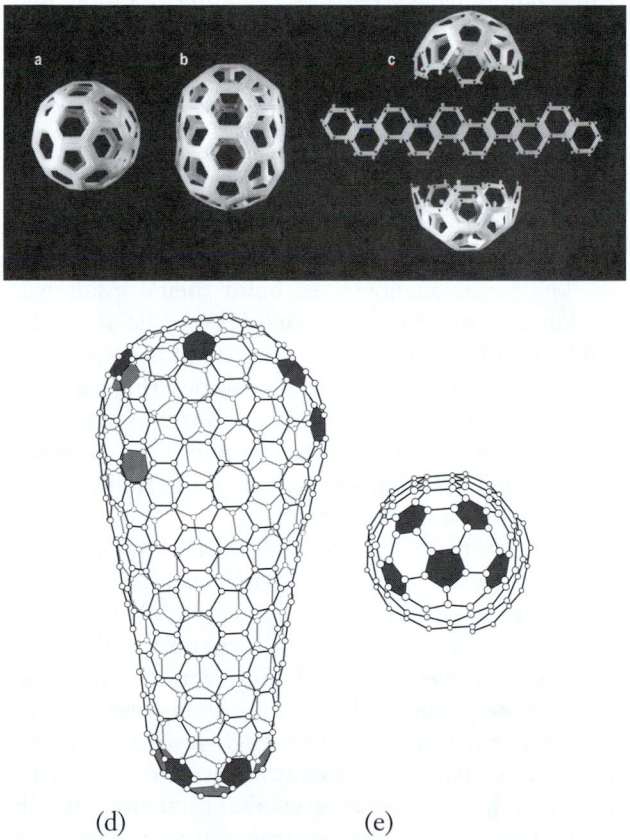

FIGURE 2.5 Variations on icosahedral symmetry. Top: A schematic diagram showing how elongated phage phi 29 capsids (b) can be constructed from a T = 3 lattice (a) by adding a band of 10 hexagons around the equator of the structure (c). Bottom: The mature capsid of human immunodeficiency virus type 1 (d,e). This cone-like structure can be represented as containing slightly more than half of a T = 3 structure at its top, and slightly less at its bottom, with the connecting lattice being made of hexagons.

Other structures, large and small, display icosahedral symmetry

The principles of construction of closed shells using identical subunits are universal. The architect Buckminster Fuller exploited these principles to construct "geodesic domes" of various sizes from steel bars clustered in groups of 5 or 6 on their surface; a good example is the dome built for Expo '67 in Montreal (Figure 2.6a). Many soccer balls are made from 12 pentagonal and 20 hexagonal subunits (Figure 2.6b). A molecule in the shape of a stable closed shell composed of 60 carbon atoms was recently discovered and named **buckminsterfullerene**. This molecule has icosahedral symmetry; the 60 carbon atoms lie at the vertices of 12 pentamers and 20 hexamers on the surface of the shell (Figure 2.6c; compare with Figure 2.5a). These objects range in size from tens of meters (geodesic domes) to 0.7 nm in diameter (C_{60}), and all share unusual strength, stability, and simplicity of construction based on their use of identical subunits joined in a symmetrical pattern.

Many virus capsids are organized as helical tubes

An alternative to a closed shell for packaging viral genomes is an elongated tube. The same arguments for genetic economy, simplicity, and strength apply to these structures: They must be built from identical subunits and are therefore symmetric. In this case, the symmetry is **helical**. Helical structures extend along a single helix axis. The symmetry of a helical assembly of subunits is defined by two parameters: u, the number of subunits per turn, and p, the displacement along the helix axis between one subunit and the next. P, the pitch of the helix, is the distance along the helix axis that corresponds to exactly one turn of the helix; $P = u \times p$. Figure 2.7 shows a diagram of the tobacco mosaic virus particle, in which many copies of a single protein subunit bind to the viral RNA and form a tube with helical symmetry. Here u is 16.33 (there are 49 subunits in three turns of the helix), p is 0.14 nm, and P is 2.3 nm.

In tobacco mosaic virus, as well as in the nucleocapsids of negative-strand RNA viruses like measles or influenza, the RNA winds along a groove that follows the helical path of the protein subunits. Each subunit binds to a fixed number of nucleotides in the RNA: 3 in the case of tobacco mosaic virus; 6 in the case of paramyxoviruses. By contrast, the protein coat of the filamentous bacteriophage fd forms a sleeve that sur-

(a)

(b)

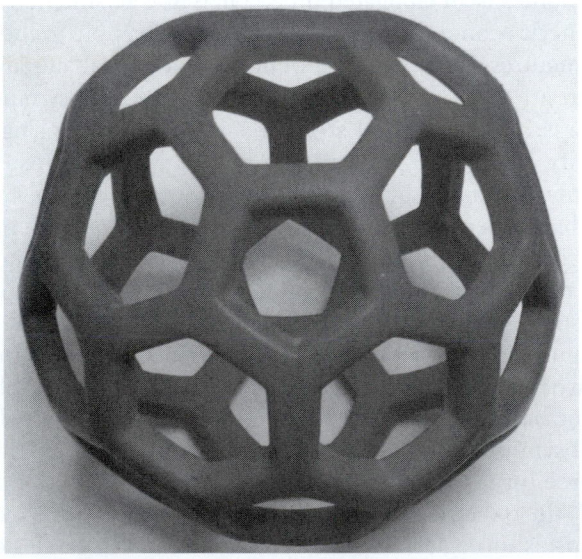

(c)

FIGURE 2.6 Structures that share symmetry elements with icosahedral viruses.
(a) The geodesic dome, designed by Buckminster Fuller, erected for Expo '67 in Montreal. (b) A soccer ball. (c) C_{60}, known as buckminsterfullerene.

rounds the circular, single-stranded DNA, without a specific interaction between each subunit and a particular number of nucleotides in the DNA.

Larger viruses come in more complex packages

Quasi-equivalent icosahedral designs are neat and economical, but accurate assembly of larger particles is hard to achieve based solely on this principle. A protein subunit must take on the appropriate configuration when it enters a growing icosahedral capsid, and the more choices it has available, the more likely it is to make a mistake—which could result in a fatal flaw in the particle. The structures of larger and more complex virions therefore depart in one way or another from the simple designs described above. The polyomaviruses (Figure 2.8a) and papillomaviruses have capsids that assemble from 72 pentamers of the major capsid protein, VP1. These pentamers lie at the points of a T = 7 icosahedral lattice, but 60 of the 72 lattice points have six nearest neighbors rather than five. Therefore, the pentamers at these positions have to fit into a "sixfold hole" on the capsid surface, even though they have fivefold symmetry. Apparently, this is easier to accomplish than forming pentamers for the fivefold positions and hexamers for the sixfold positions. A pattern of interlocking arms from the C-termini of the capsid subunits holds the structure together (Figure 2.8b).

The larger adenoviruses build their capsids with two kinds of structural subunits. Hexons lie at the 240 sixfold positions of a T = 25 icosahedral lattice, and pentons lie at the 12 fivefold positions (Figure 2.8c). The hexon is actually constructed as a trimer of the adenovirus hexon protein, but each protein subunit contains two very similar subdomains, so that the assembled hexon has a pseudo-sixfold character.

Both of these examples show that defined *subassemblies* of protein subunits confer significant advantage. Thus, the major capsid protein of polyomaviruses forms pentamers as a stable substructure, and mechanisms have evolved for specific interactions with either five or six neighbors. Likewise, the adenovirus hexon is a defined subassembly. The high-resolution crystal structures of polyomavirus capsid proteins and of the adenovirus hexon protein reveal that these proteins, like the parvovirus and picornavirus capsid proteins, are folded as jelly-roll β-barrels.

Virions of a number of bacteriophages contain both elements with icosahedral symmetry (the capsid) and elements with helical symmetry (the tail). Phage tails are specialized structures that allow binding to cellular receptors and help to introduce the viral genome by piercing the bacterial cell wall and membranes. These particles elegantly illustrate the above principles of con-

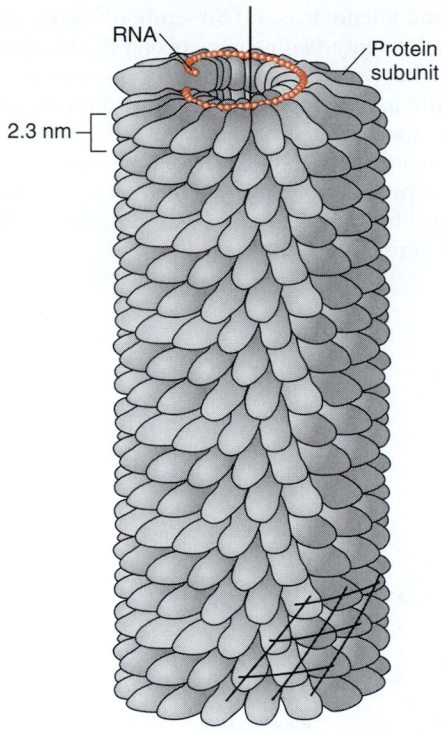

FIGURE 2.7 **The helical nucleocapsid of tobacco mosaic virus.** At the top is a cutaway view to reveal the RNA, protected within a groove, in the interior of the capsid. Each protein subunit binds to 3 nucleotides of the single-stranded viral RNA. Each turn of the helix is separated by 2.3 nm.

struction of molecular packages from a limited number of proteins.

Specific packaging signals direct incorporation of viral genomes into virions

The viral genome has a unique nucleotide sequence and is present in one or a few segments per virion. By definition, therefore, packaged genomes are not organized with the same symmetry as the multiple protein subunits in the capsid. A variety of mechanisms have evolved to ensure specific genome packaging. Small icosahedral positive-strand RNA viruses, including some plant viruses and the togaviruses, have a positively charged N-terminal arm on their capsid proteins. This arm can interact specifically with a packaging sequence on the genomic RNA, or nonspecifically with other segments of the RNA. The specific interaction often involves an RNA sequence element containing a stem-loop structure (the folded **packaging sequence**). During capsid assembly, one or a few capsid protein subunits participate in this specific contact. The majority of the capsid subunits bind to RNA nonspecifically, perhaps just through neutralization of the negative charge on the phosphate groups. Recognition of the packaging sequence can serve to initiate capsid assembly. The encapsidated RNA is highly condensed, with much of its stem-loop secondary structure retained during assembly.

In helical capsids, such as tobacco mosaic virus (Figure 2.7) or the nucleocapsids of many negative-strand RNA viruses, the RNA extends along a groove in the helical array of protein subunits. The RNA packaging sequence fits particularly well into a short length of this groove, giving the RNA genome an assembly advantage over any cellular RNAs that might be present during packaging.

More complex viruses have more elaborate packaging processes. The DNA of herpesviruses and of many bacteriophages is inserted into a preformed precursor shell by an ATP-requiring spooling mechanism (see Chapters 6, 8, and 13). This machinery recognizes specific packaging sequences on viral DNA, and often involves cleavage of the DNA genome from a larger replicative intermediate.

Core proteins may accompany the viral genome inside the capsid

Some DNA viruses package not just the viral genome but a nucleoprotein complex. **Core proteins** condense viral DNA by neutralizing the negative charges of the phosphate groups on DNA. Polyomaviruses and papillomaviruses use cellular histones to condense their DNA genomes, which therefore resemble cellular chromatin and are called "minichromosomes." Perhaps because of this similarity, those viruses can also package degraded fragments of host cell chromatin that can arise during virus infection; such virus particles are called **pseudovirions.** Other DNA viruses such as adenoviruses and herpesviruses code for their own histone-like core proteins. Some viruses incorporate cellular polyamines such as spermidine into their capsids to neutralize the charge of phosphate groups and condense the nucleic acid genome.

Scaffolding proteins help in virion assembly but are not incorporated into the mature virion

Even though self-assembly is an inherent characteristic of virion structural proteins, many viruses code for other proteins that take part in capsid assembly but are eventually discarded and do not remain part of the mature virion. Because of their specialized function, these are called **scaffolding proteins**. Scaffolding proteins help to "dock" and stabilize subassemblies of viral structural proteins during the process of virion assembly. They can also help in the packaging of viral genomes into pre-

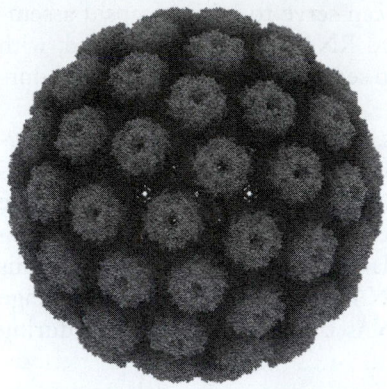

(a)

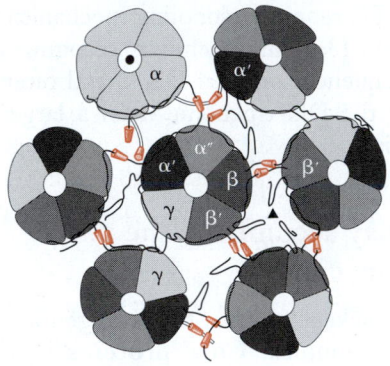

(b)

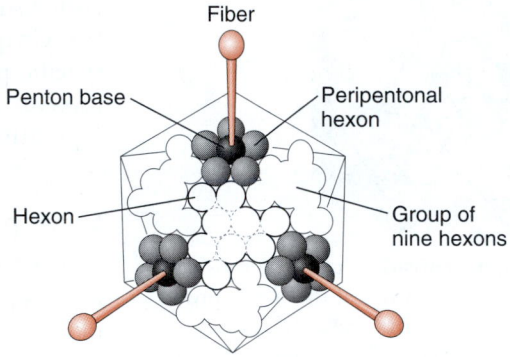

(c)

FIGURE 2.8 Simian virus 40 and adenovirus. (a) Structure of the capsid of simian virus 40 (a polyomavirus). Seventy-two pentamers of VP1 are centered on lattice points of a T = 7 icosahedral surface lattice. Only 12 of these pentamers are located at centers of fivefold symmetry; the other 60 pentamers are each surrounded by 6 neighboring pentamers. (b) The arrangement of VP1 subunits in a 6-coordinated group. C-terminal arms of VP1 link adjacent pentamers together. (c) Structure of the adenovirus capsid. Hexons surrounding the penton base, and hexons in groups of 9, are formed from trimers of the hexon protein. Other proteins (not shown) link together hexons and pentons.

assembled provirions. Examples can be found in Chapters 6, 8, and 13.

Viral envelopes are made from lipid bilayer membranes

Most viral envelopes are acquired at a cellular membrane by a process of **budding** (Figure 2.9). During budding, the viral nucleocapsid is wrapped in a membrane into which viral glycoproteins have been inserted. Viral envelopes contain a lipid bilayer, whose lipid content reflects the composition of the cellular membrane from which the envelope was constructed. Viruses that bud outward from the cell surface, like influenza virus, contain phospholipids and cholesterol in the proportions characteristic of the plasma membrane. Flaviviruses, which bud into the endoplasmic reticulum, have much lower amounts of cholesterol, as does the endoplasmic reticulum membrane itself.

The lipid bilayer provides a uniform permeability barrier, so that enveloped viruses do not need to assemble as perfectly as nonenveloped particles. Some of the smallest enveloped viruses, such as the togaviruses (Figure 2.1h), have virions with essentially perfect icosahedral symmetry. Each viral glycoprotein molecule in the envelope contacts a capsid protein molecule, with the result that the viral glycoproteins are distributed on the envelope with the same symmetry as the proteins in the underlying capsid. However, virions of larger enveloped viruses are often somewhat irregular in shape and their envelopes have no obvious symmetry. Influenza virus (Figure 2.1f), which contains helical nucleocapsids, can adopt a variety of shapes and sizes from round to filamentous, and one virion may contain many copies of the segmented influenza virus genome.

Viral glycoproteins are inserted into the lipid membrane to form the envelope

Most of the proteins found in viral envelopes have a relatively large glycosylated external domain (**ectodomain**), a hydrophobic **transmembrane anchor** of about 20 amino acid residues, and a short internal tail. The tail is sometimes described as "cytoplasmic," since it faces the cytoplasm of the cell before the virus buds. The 20 nonpolar amino acids of the transmembrane anchor form an alpha helix that neatly spans the 3 nm thick hydrophobic part of a lipid bilayer membrane. Glycosylation of the ectodomain ensures that the external surface of virus particles is hydrated, reducing protein/protein interactions that could aggregate the virions.

These proteins are synthesized on ribosomes bound to the endoplasmic reticulum and are inserted into the cell membrane by the standard route for cell-surface integral membrane proteins. Modifications such as glycosylation of the ectodomain and **palmitoylation** of cysteines near the inner margin of the lipid bilayer occur in the endoplasmic reticulum and in the Golgi complex. Most viral envelope proteins are **type I integral membrane proteins**, with the N-terminus facing outward and the transmembrane anchor nearest the C-terminus of the polypeptide. However, some are **type II integral membrane proteins** of the opposite polarity, with the C-terminus facing outward and the anchor near the inward-facing N-terminus. Type I membrane proteins have a **signal sequence** that is cleaved from their N-terminus by a peptidase as they are inserted in the endoplasmic reticulum during synthesis. In type II proteins, the transmembrane anchor itself serves as the signal for membrane insertion.

Budding is driven by interactions between viral proteins

Viral envelope proteins inserted into a cell membrane tend to aggregate in local two-dimensional pools from which most of the cellular membrane proteins are excluded. This is possible because membrane proteins can move laterally in the fluid membrane environment, and viral glycoproteins establish close interactions with each other. Regions of the membrane containing such pools of envelope proteins will eventually form the viral envelope by a process of budding.

Budding involves curvature of the membrane as it is wrapped around the viral nucleocapsids. There are several different patterns of budding. Icosahedral togavirus nucleocapsids assembled in the cytoplasm approach the plasma membrane and bind to the cytoplasmic tails of viral envelope proteins (Figure 2.9a). The capsid effectively wraps itself in the membrane, and progressive formation of contacts between the surface of the capsid and the cytoplasmic tails of the viral glycoproteins drives pinching-off of the bilayer by membrane fusion.

Many enveloped viruses with helical nucleocapsids have an internal **matrix protein** that mediates the association between the viral glycoproteins inserted into the plasma membrane and the nucleocapsids in the cytoplasm. Matrix proteins located at the cytoplasmic face of the membrane bind to the nucleocapsids and cooperate with the glycoprotein layer to drive the budding process. This is shown in Figure 2.9b for influenza virus.

A third variation on budding is carried out by some retroviruses, such as HIV-1. Instead of being assembled in the cytoplasm and migrating to the membrane, the nucleocapsid is assembled from the Gag precursor protein directly at the membrane during budding (Figure 2.1g; see Chapter 25). The matrix and nucleocapsid proteins are generated by cleavage of the Gag precursor during and after virion formation.

Budding is driven primarily by the interactions between envelope glycoproteins and matrix or nucleocapsid proteins. However, some viruses can generate empty envelopes that bud in the absence of nucleocapsids; in this case, the envelope glycoproteins themselves drive budding. In contrast, some retroviruses that have lost their envelope glycoprotein genes can form bald particles that contain nucleocapsids wrapped in membranes lacking envelope proteins. In these cases, the layer of Gag proteins at the inner surface of the plasma membrane interacts sufficiently strongly with the lipid bilayer (in many cases through an N-terminal covalently bound fatty acid, **myristate**) that its assembly into a nucleocapsid can drive budding. In many cases, viruses need to recruit cellular proteins to complete the final pinching-off of the lipid bilayer and release a completed particle.

Assembly and disassembly of virions: the importance of an irreversible step

Virus particles are more than passive packages. As illustrated by many of the examples above, they are vehicles for delivering a payload (the viral genome) to a host cell according to a programmed set of instructions (see Chapter 4). Assembly and disassembly are not simply reverse processes. Proteolytic cleavage is the most common way in which an assembled structure is primed for the delivery program. For example, autolytic cleavage of the capsid protein VP0 in the picornaviruses creates a small internal peptide, VP4, which emerges and helps mediate penetration when receptor binding forces the virus particle to expand or dissociate (Chapter 16). Cleavage of fusion proteins such as influenza virus hemagglutinin or human immunodeficiency virus gp160 primes them for a dramatic, fusion-inducing conformational rearrangement triggered by low pH (hemagglutinin; Chapter 23) or by receptor and co-receptor binding (gp160; Chapter 26).

In the nonenveloped reoviruses, a *protector protein* called $\sigma 3$ is cleaved to expose another protein, $\sigma 1$, that is responsible for penetration of the virus core into the cell. Similarly, the enveloped flavirus PrM protein is cleaved to reveal the E protein, responsible for fusion with the plasma membrane of the target cell. "Jack-in-

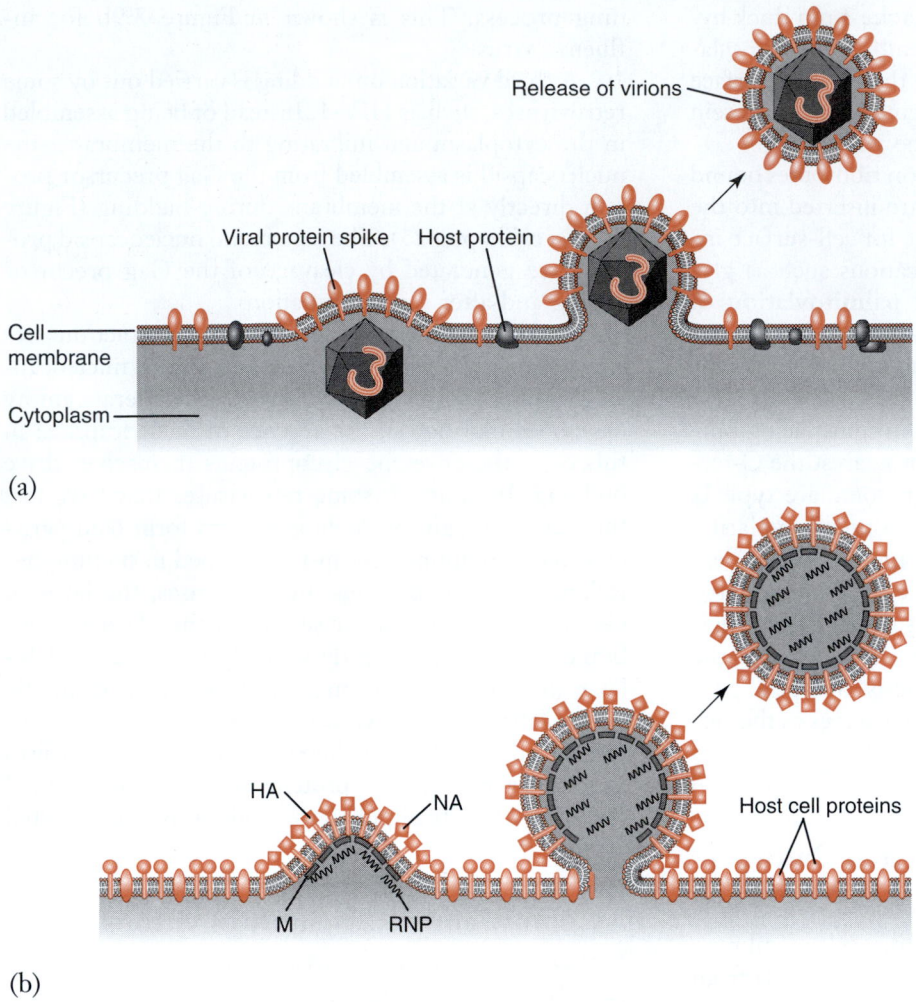

FIGURE 2.9 **Release of enveloped virions from the cell surface by budding.** (a) The icosahedral nucleocapsid of Semliki Forest virus interacts directly with the cytoplasmic tail of virus envelope glycoproteins during budding. (b) The matrix (M) protein of influenza virus interacts both with the cytoplasmic tail of surface glycoproteins (HA and NA) and with helical nucleocapsids (RNP) during budding.

the box" and "spring-loading" have been used as metaphors for the metastable state created by this proteolytic event. Some triggering step is then needed to unleash the spring, and this step usually involves interaction with a specific receptor on the cell surface. The metaphor is particularly apt in the case of herpesviruses or certain DNA bacteriophages, in which DNA is coiled in a spring-like fashion into the virion by the action of an energy-requiring protein spooling machine. Release of a latch that seals the particle leads to release of the genome. This occurs at the cell surface for bacteriophages, but only at the nuclear pore for herpesviruses. The DNA of other bacteriophages (Chapter 7, bacteriophage T7) is not spring-loaded sufficiently to drive complete injection into the cell; rather, it is actively spooled into the cell during transcription.

Plant viruses, which are injected into host cells by their insect vectors, do not require a spring-loading mechanism. They appear to undergo a reversible structural transition when exposed to the cytoplasm, allowing ribosomes or initiation factors access to their RNA. The ribosomal translocation process then drives uncoating. A related process may mediate uncoating of some positive-strand RNA animal viruses.

SUGGESTED READING

1. Harrison, S.C. (2005). Viral fusion proteins. *Advances in Virus Research*, In press.

2. Leiman, P.G., Kanamaru, S., Mesyanzhinov, V.V., Arisaka, F., and Rossmann, M.G. (2003). Structure and morphogenesis of bacteriophage T4. *Cellular and Molecular Life Sciences* **60,** 2356–2370.

3. Tang, L. and Johnson, J.E. (2002). Structural biology of viruses by the combination of electron cryomicroscopy and x-ray crystallography. *Biochemistry* **41,** 11517–11524.

4. http://mmtsb.scripps.edu/viper/. An extremely useful web site with many virus structures determined by x-ray crystallography or cryoelectron microscopy.

KEY TERMS

Budding
Buckminsterfullerene
Capsid
Core proteins
Cryoelectron microscopy
Ectodomain
Envelope
Genetic economy
Helical symmetry
Icosahedral symmetry
Jelly-roll β-barrel
Matrix protein
Myristate
Negative staining
Nucleocapsid
Packaging sequence
Palmitoylation
Positive staining
Pseudovirion
Quasi-equivalent
Scaffolding proteins
Self-assembly
Signal sequence
Surface lattice
Synchrotron
Transmembrane anchor
Triangulation number
Twofold axis
Type I integral membrane protein
Type II intregal membrane protein
Virion
X-ray diffraction

Virus Classification: The World of Viruses

VIRUS CLASSIFICATION

Virus are classified into related groups based on:

- *Genome composition (DNA or RNA, single- or double-stranded)*
- *Genome topology (linear or circular; single or multiple segments)*
- *Capsid symmetry*
- *Presence or absence of an envelope*
- *Genetic relatedness (nucleotide and amino acid sequence similarity)*
- *Mechanisms for expressing messenger RNAs and replication of genomes*

Host organisms

THE MAJOR VIRUS GROUPS

Viruses are classified into species, genera, families, and (in some cases) orders, within the following categories:

- *Viruses with single-stranded DNA genomes*
- *Viruses with double-stranded DNA genomes*
- *Viruses with positive-strand RNA genomes*
- *Viruses with negative-strand RNA genomes*
- *Viruses with double-stranded RNA genomes*
- *Viruses that use a reverse transcriptase*
- *Satellite viruses, satellite nucleic acids, and viroids*

THE EVOLUTIONARY ORIGIN OF VIRUSES

RNA viruses may be relics of the prebiotic RNA world.

Small DNA viruses could have evolved from escaped, self-replicating fragments of cellular genome DNA.

Some large DNA viruses could have evolved from degenerate intracellular parasites.

VIRUS CLASSIFICATION

Many viruses, infecting virtually all known life forms, have been discovered

Several thousand different viruses have been isolated and characterized during the approximately 100 years since the science of virology was established. These viruses infect numerous organisms including bacteria, fungi, algae, higher plants, invertebrates, and vertebrates. Several hundred viruses have been studied in some detail. To make sense out of this large collection of viruses, scientists have attempted to classify viruses into related groups. Classification of organisms that are genetically and evolutionarily related to each other is useful because detailed knowledge about one organism can often be generalized and applied to related organisms. The classification of viruses into **species** and into a hierarchy of related groups of species (**genera**, **families**, and **orders**) helps virologists to compare viruses and to analyze and understand many aspects of virus structure, replication pathways, interactions with host cells, and disease transmission. The science of classification is called **taxonomy**. This chapter presents an overview of the different virus groups that have been recognized and named.

Virus classification is based on molecular architecture, genetic relatedness, and host organism

A rational classification of viruses must take into account the nature of the viral genome and the structure of the virus particle, because these are fundamental, inherited characteristics of viruses. The following molecular criteria are used to classify viruses:

- Type of nucleic acid genome (DNA or RNA)
- Strandedness of nucleic acid (single- or double-stranded)
- Topology of nucleic acid (linear or circular)
- Symmetry of capsid (icosahedral, helical, or no symmetry)
- Presence or absence of an envelope

Genetic relatedness among viruses can be determined by comparison of viral genome sequences and amino acid sequences of viral proteins. Other characteristics such as the order of genes on viral genomes, the mechanism of viral messenger RNA synthesis, the presence of viral genes coding for DNA or RNA polymerases, and the use of reverse transcriptase during replication, are also useful criteria for virus classification.

Many viruses have been associated with their host organisms during millions to billions of years of evolution. Viruses must be able to enter and replicate in their host organism, and must then be released and transmitted to another individual. Viruses therefore adapt to the host organisms in which they replicate, and evolve as these organisms evolve. Some examples of these adaptations are:

- Specific mechanisms for transmission between individual cells or organisms (via direct contact, diffusion through liquid media, dispersion via the circulatory system, respiratory droplets, stool or urine contamination of food or water, insect vectors, etc.)
- Entry pathways into cells (binding to specific cell surface receptors, passage through cell walls, entry by membrane fusion, uptake via vesicles, etc.)
- Interaction with cellular RNA or DNA polymerases, RNA processing enzymes, protein synthesizing systems, membrane systems, cellular signaling proteins, and other cellular proteins used by the virus to express and replicate its genes

As a result of these adaptations, viruses that are genetically and evolutionarily related tend to infect related organisms. Virus taxonomy takes into account the ability of viruses to infect different categories of organisms. For this purpose, it is convenient to divide all living organisms into the following five broad categories:

1. Vertebrates (including humans)
2. Invertebrates
3. Plants
4. Lower eukaryotes (fungi, protozoa, algae)
5. Prokaryotes (bacteria and archaea)

Viruses that are evolutionarily and genetically related tend to infect organisms within only one of the above five categories. However, this is not always the case. For example, a number of viruses of vertebrates are transmitted from one animal to another by bloodsucking invertebrate arthropods (mosquitoes, ticks, sandflies), in which they can also replicate. Many plant viruses are also transmitted by insects. This means that such viruses can evolve and spread in both vertebrates (or plants) and invertebrates. As a result, a number of virus families include viruses that can infect organisms in two or more of the above categories.

Viruses are grouped into species, genera, and families

This chapter outlines information about the major virus groups presently identified, with particular emphasis on viruses that infect bacteriophages and vertebrates, the subject of most of the remainder of this book. This description of virus groups is not meant to be exhaustive; the periodical reports on virus taxonomy published by the International Committee on Virus Taxonomy (ICTV) are the definitive source on known viruses and their classification. Much of the information in this chapter is derived from the ICTV's 8th Report (2005). Following their rules, names of virus families are italicized and end in the Latin suffix *-viridae*, and names of genera end in the suffix *-virus*. However, in this book, virus family names are usually referred to in their English version for sake of simplicity (e.g., "picornaviruses" instead of "*Picornaviridae*").

Different virus isolates are generally considered to be members of a virus *species* if they share a high degree of nucleic acid sequence identity and most of their proteins have highly similar amino acid sequences and common antigenic properties. Members of a species usually infect a limited number of organisms or specific target cells and tissues. This implies that they are in genetic contact with each other and are therefore members of a common genetic **lineage**.

Virus species are grouped into *genera* by virtue of shared characteristics such as genome organization and size, virion structure, and replication strategies. Although they are related by evolution, species that are members of a common genus may show significant divergence in nucleotide and amino acid sequences. Different members of a genus may infect different organisms or different tissues in the same organism, and therefore may not usually come in contact with each other.

Genera are in turn grouped into virus *families*. Members of a virus family share overall genome organization, virion structure, and replication mechanisms. Because of this, they are presumed to be evolutionarily related. However, different members of a virus family may vary significantly in virion size and genome length, and may have a variety of unique genes not shared by other family members. Because members of a virus family may have evolved separately over considerable periods of time in distinct host organisms, they often share only limited nucleotide or amino acid sequence homology in a small number of genes. Presently the ICTV recognizes 73 virus families and 24 genera that are not yet assigned to families.

Distinct naming conventions and classification schemes have developed in different domains of virology

Research on viruses has historically been carried out by scientists who study viruses that infect only one category of organism: for example; bacteria, fungi, algae, plants, insects, fish, domesticated animals, or humans. As a result, distinct domains such as bacterial virology, plant virology, veterinary virology, and medical virology have arisen. Although there has been some convergence among these fields of study, their different histories have led to adoption of distinct conventions for naming viruses and for virus classification. For example, bacteriophages are given names that include the name of the host bacterial genus and an arbitrary number or Greek letter (e.g., Enterobacteria phage T4, Bacillus phage SP01, Enterobacteria phage Mu). In practice, these official names are often shortened to "bacteriophage" or "phage" followed by the number.

Naming of plant viruses usually makes reference to the plant species in which the virus was first isolated and to the disease symptoms caused by the virus (e.g., tobacco mosaic virus, tomato bushy stunt virus, maize streak virus). Many insect viruses are named using the Latin name of the insect species from which the virus was isolated, followed by the virus genus name (e.g., *Aedes aegypti* densovirus, *Autographica californica* multiple nucleopolyhedrovirus, *Chironomus luridus* entomopoxvirus). A variety of naming conventions are used for viruses of vertebrates. These include the species of origin (e.g., bovine papillomavirus, simian virus 40, porcine parvovirus), the locality where the virus was first isolated (e.g., Semliki Forest virus, Ebola virus, Norwalk virus), or the disease caused by the virus (e.g., measles virus, influenza A virus, hepatitis B virus).

Because most virology research has been carried out in the fields of bacterial, plant, and vertebrate viruses, the majority of viruses that have been discovered infect organisms in one of these groups. There undoubtedly remain many viruses yet to be discovered, particularly in poorly studied organisms such as archaea, protists, small invertebrates, and marine organisms.

Scientists who classify viruses that infect distinct groups of organisms tend to use different criteria. For example, bacteriophages are currently classified into families based on the nature of their genomes and on their capsid and tail morphology. Bacteriophages can mutate rapidly because of their short growth cycles, and they can readily exchange genes with each other by recombination or by integration into (and excision from) the genomes of their host bacteria. This means that bacteriophages tend to be genetic mosaics, and determining evolutionary or genetic relatedness among different phages is difficult if not impossible. Plant viruses are presently grouped into 18 different families, but a large number of genera (most of which contain viruses with positive-strand RNA genomes) have not yet been assigned to families. Viruses of vertebrates and invertebrates are more evenly represented in all genome categories; they are currently grouped into 28 distinct families and only 3 unassigned genera.

MAJOR VIRUS GROUPS

Study of the major groups of viruses leads to understanding of shared characteristics and replication pathways

For convenience and simplicity, many of the major known virus groups (families and unassigned genera) are listed below in seven separate tables, based on the nature of their genomes and their replication strategy. Within each table, these lists are divided into sections based on host organism, starting with the simplest (prokaryotes) and ending with the most complex (vertebrates). Within each section, virus groups are listed in order of increasing genome length and therefore complexity. Virus groups that are covered in detail in this book are highlighted in orange.

The attentive reader will note that some interesting observations can be made about viruses grouped in these different tables. For example: (1) viruses with single-stranded DNA genomes tend to be very small; (2) most plant viruses have positive-strand RNA genomes; (3) many fungal viruses have double-stranded RNA genomes; (4) most bacteriophages are tailed and have double-stranded DNA genomes; and (5) viruses with very large double-stranded DNA genomes (400 to 1200 Kb) are limited to free-living unicellular algae and protozoa. Understanding these and other conclusions that are derived from our knowledge of virus classification will be an interesting future challenge for virologists.

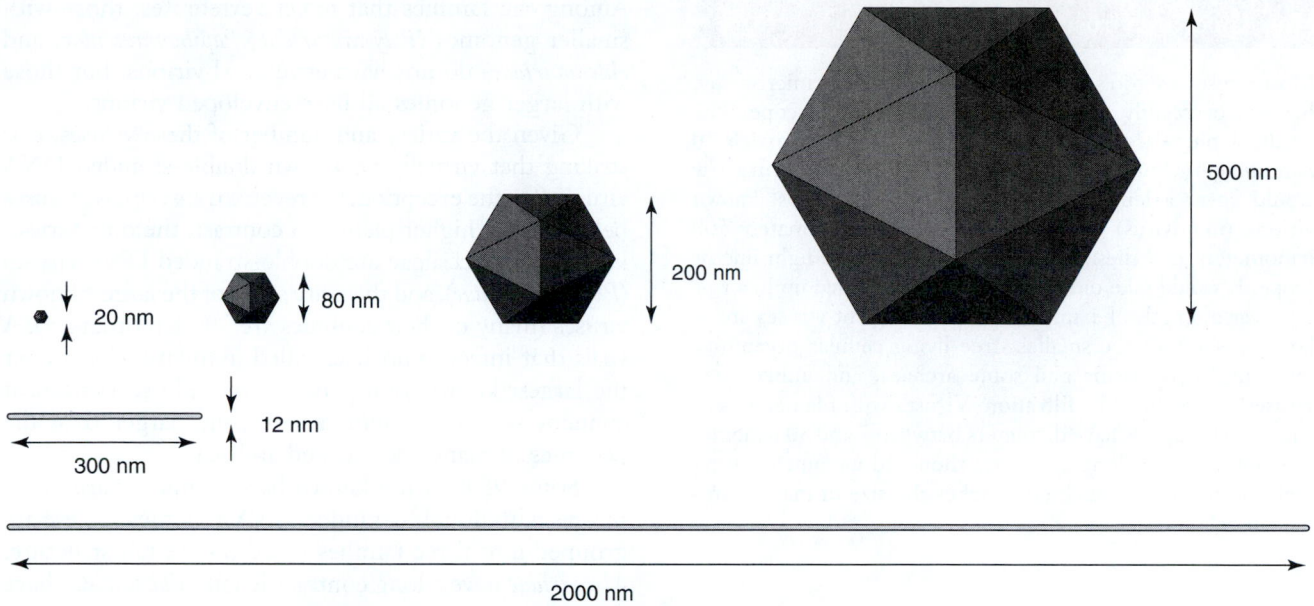

FIGURE 3.1 Size variation of viral capsids. Shown are examples of the smallest and largest known capsids with icosahedral symmetry, and two intermediate-size capsids. From left to right: a parvovirus (beak and feather disease virus), diameter 20 nm; adenovirus, diameter 80 nm; a phycodnavirus (Paramecium bursaria Chlorella virus 1), diameter 200 nm; and mimivirus, diameter 500 nm. The ratios of the volumes of these capsids is 1:64:1000:15,000. Below are examples of helical capsids: tobacco mosaic virus, 12 × 300 nm; and pseudomonas phage Pf1, 7 × 2000 nm.

Viruses with single-stranded DNA genomes are small and have few genes

Given the great variations in size and shape of known viruses (Figure 3.1), it is striking that viruses with single-stranded DNA genomes (Table 3.1) lie within a very narrow size range. Their genomes are among the smallest of all viral genomes, ranging from less than 2 Kb up to 9 Kb. The genomes of five of the six families are circular, and those of the sixth family, *Parvoviridae*, have hairpin ends. These characteristics may reflect the fragility of single-stranded DNA compared with double-stranded DNA; small size and circularity or hairpin ends ensure that the genome DNA cannot be easily attacked by nucleases.

None of these viruses has a lipid envelope, and viruses in five out of the six families have icosahedral capsids. It is tempting to speculate that most viruses with single-stranded DNA genomes may have evolved from a common ancestor. There are presently no known viruses with single-stranded DNA genomes that infect archaea or lower eukaryotes.

TABLE 3.1 Viruses with single-stranded DNA genomes

Virus group	Hosts	Example	Genome segments	Genome size (Kb) and topology		Capsid symmetry	Envelope
Microviridae	Bacteria	Bacteriophage phiX-174	1	4–6	Circular	Icosahedral	No
Inoviridae	Bacteria	Bacteriophage M13	1	5–9	Circular	Helical	No
Geminiviridae	Plants	Maize streak virus	1–2	3–6	Circular	Icosahedral	No
Nanoviridae	Plants	Banana bunchy top virus	6–8	6–9	Circular	Icosahedral	No
Circoviridae	Vertebrates	Beak and feather disease virus	1	2–4	Circular	Icosahedral	No
Parvoviridae	Vertebrates & invertebrates	Adeno-associated virus	1	4–6	Linear	Icosahedral	No

How small are viruses?

Most viruses are too small to be detected by light microscopy, but can be readily seen with the electron microscope. The smallest spherical virions have a diameter of approximately 20 nanometers (Figure 3.1); 50,000 such virions lined up in a row would cover a length of 1 millimeter. The largest known virions (mimivirus) have a diameter of approximately 500 nanometers, and these can be barely seen with a light microscope. It would take only 2000 such virions lined up in a row to cover a length of 1 millimeter. These giant viruses are as large as some of the smallest free-living cellular organisms, including mycoplasma and some archaea, and unlike most viruses, are retained by filtration. Viruses with filamentous or rod-shaped capsids have diameters between 7 and 30 nanometers but can be as long as several thousand nanometers (several micrometers), which approaches the size of many mammalian cells.

Viruses with double-stranded DNA genomes include the largest known viruses

Viruses with double-stranded DNA genomes (Table 3.2) show a wide variation in genome size (between 5 and 1180 Kb; Figure 3.2). Without exception, all of these viruses have unfragmented genomes; that is, all viral genes are contained in a single DNA molecule. Some of these DNA genomes are circular; others are linear. The large genome size of some of these viruses is probably a result of the superior qualities of double-stranded DNA as a gene storage facility compared with other forms of nucleic acids: increased resistance to degradation, a low error rate during DNA replication, and the ability to recognize and repair mutations. Among the families that infect vertebrates, those with smaller genomes (*Polyomaviridae*, *Papillomaviridae*, and *Adenoviridae*) do not have enveloped virions, but those with larger genomes all have enveloped virions.

Given the variety and number of these viruses, it is striking that virtually no known double-stranded DNA viruses (for the exception, see reverse transcriptase viruses below) infect higher plants. In contrast, the only viruses known to infect algae are double-stranded DNA viruses (*Phycodnaviridae*), and these are some of the largest known viruses (many of their genomes are 300 Kb or larger). A virus that infects amoebae, called mimivirus, has by far the largest known viral genome; recently sequenced, it contains nearly 1.2 million base pairs, larger than the genomes of many bacteria and archaea.

Some 95% of the known bacteriophages are tailed phages with double-stranded DNA genomes. These are grouped into three families based on the tail structure: *Myoviridae* have a long contractile tail, *Siphoviridae* have

The great disparity of viral genome size and complexity

Viral genomes (Figure 3.2) range from slightly less than 2000 nucleotides (coding for 2–3 proteins) up to nearly 1.2 million nucleotides (coding for over 1200 proteins). In comparison, the smallest cellular genomes contain approximately 400,000 nucleotides, and the genomes of mammals contain approximately 3 billion nucleotides. Viral genomes are very compact and economical; most nucleotide sequences are used for protein coding, and a good yardstick is one protein coded per 1000 nucleotides. In contrast, genomes of complex plants and animals contain mostly noncoding sequences; for example, the human genome codes for about 30,000 proteins, or one protein per 100,000 nucleotides.

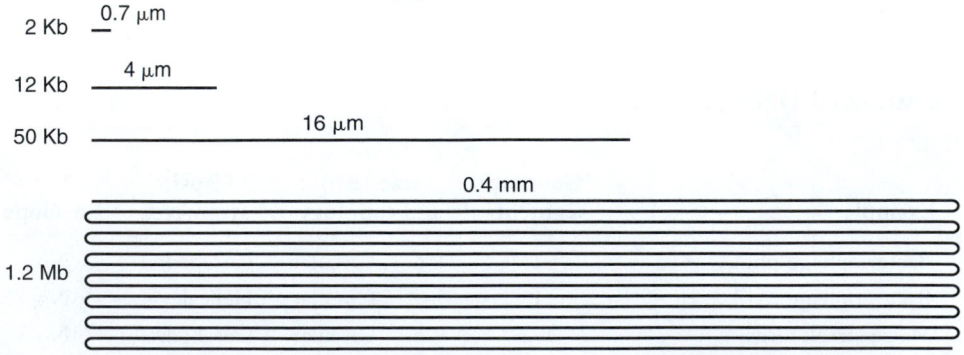

FIGURE 3.2 Size variation of viral genomes. Lines are drawn in proportion to genome lengths. Genome sizes in Kb (thousands of nucleotides) are shown on left, and genome lengths in micrometers (μm) or mm are shown above each line. Top: Circoviruses have genomes of less than 2 Kb; many positive-strand RNA viruses have genomes of approximately 12 Kb; medium-size phages such as lambda have genomes of 50 Kb; the largest known viral genome, that of mimivirus, is 1.2 Mb long.

TABLE 3.2 Viruses with double-stranded DNA genomes

Virus group	Hosts	Example	Genome segments	Genome size (Kb) and topology		Capsid symmetry	Envelope
Corticoviridae	Bacteria	Bacteriophage PM2	1	10	Circular	Icosahedral	Internal
Plasmaviridae	Mycoplasma	Bacteriophage L2	1	12	Circular	Asymmetric	Yes
Tectiviridae	Bacteria	Bacteriophage PRD1	1	15	Linear	Icosahedral	Internal
Podoviridae	Bacteria	Bacteriophage T7	1	16–70	Linear	Icosahedral, tailed	No
Siphoviridae	Bacteria & archaea	Bacteriophage lambda	1	22–121	Linear	Icosahedral, tailed	No
Myoviridae	Bacteria & archaea	Bacteriophage T4	1	34–497	Linear	Icosahedral, tailed	No
Phycodnaviridae	Algae	Paramecium bursaria Chlorella virus 1	1	100–560	Linear or circular	Icosahedral	No
Mimivirus	Protozoa (amoeba)	Acanthamoeba polyphaga mimivirus	1	1180	Linear	Icosahedral	Internal
Baculoviridae	Invertebrates	*A. californica* multiple nucleopolyhedrovirus	1	80–180	Circular	Helical	Yes
Polyomaviridae	Vertebrates	Mouse polyomavirus	1	5	Circular	Icosahedral	No
Papillomaviridae	Vertebrates	Human papillomavirus 16	1	8	Circular	Icosahedral	No
Adenoviridae	Vertebrates	Human adenovirus 5	1	25–45	Linear	Icosahedral	No
Asfarviridae	Vertebrates	African swine fever virus	1	170–190	Linear	Icosahedral	Yes
Herpesviridae	Vertebrates	Herpes simplex virus 1	1	125–240	Linear	Icosahedral	Yes
Iridoviridae	Vertebrates & invertebrates	Frog virus 3	1	135–303	Linear	Icosahedral	Yes
Poxviridae	Vertebrates & invertebrates	Variola (smallpox) virus	1	130–375	Linear	Complex	Yes

a long noncontractile tail, and *Podoviridae* have a short noncontractile tail. Within each of these families there is a wide range of genome sizes, gene organization, and replication mechanisms. Because it is based primarily on virion morphology, this classification system is somewhat arbitrary. However, the members of individual genera within each family share significant biological characteristics.

Virions of three additional families of double-stranded DNA phages contain lipid membranes either outside or inside their protein capsids. Each of these families includes only one or a small number of different phage isolates. A number of phages that infect archaea have been recently described; these all contain double-stranded DNA genomes and their virions have distinct and unusual morphologies. They are not included in these tables as their classification is still only rudimentary.

Most plant viruses and many viruses of vertebrates have positive-strand RNA genomes

Members of eight different virus families with positive-strand RNA genomes infect vertebrates, and positive-strand RNA viruses classified into 22 different families or unassigned genera infect plants (Table 3.3). The majority of known plant viruses have positive-strand RNA genomes. There is only one family of bacteriophages with positive-strand RNA genomes; no archaea phages with RNA genomes have yet been discovered. All known RNA viruses have linear genomes. However, some very small satellite viruses and viroids (see below) have circular RNAs.

Most of these viruses have genomes in the size range of 5 to 15 Kb; an exception is the *Coronaviridae*, whose genomes are as long as 31 Kb (the largest known RNA

TABLE 3.3 Viruses with positive-strand RNA genomes

Virus group	Hosts	Example	Genome segments	Genome size (Kb) and topology		Capsid symmetry	Envelope
Leviviridae	Bacteria	Bacteriophage Qβ	1	3–4	Linear	Icosahedral	No
Marnaviridae	Algae	Heterosigma akawisho RNA virus	1	9	Linear	Icosahedral	No
Barnaviridae	Fungi	Mushroom bacilliform virus	1	4	Linear	Icosahedral	No
(3 families and 8 genera)	Plants	Tomato bushy stunt virus	1–3	4–15	Linear	Icosahedral	No
(6 families and 5 genera)	Plants	Tobacco mosaic virus	1–5	6–19	Linear	Helical	No
(2 families and 1 genus)	Invertebrates	Cricket paralysis virus	1–2	6–10	Linear	Icosahedral	No
Roniviridae	Invertebrates	Gill-associated virus	1	26	Linear	Helical	Yes
Nodaviridae	Vertebrates & invertebrates	Flock house virus	2	4–5	Linear	Icosahedral	No
Astroviridae	Vertebrates	Human astrovirus	1	6–7	Linear	Icosahedral	No
Picornaviridae	Vertebrates	Poliovirus	1	7–9	Linear	Icosahedral	No
Caliciviridae	Vertebrates	Norwalk virus	1	7–8	Linear	Icosahedral	No
Flaviviridae	Vertebrates & invertebrates	Yellow fever virus	1	10–12	Linear	Icosahedral	Yes
Togaviridae	Vertebrates & invertebrates	Semliki Forest virus	1	10–12	Linear	Icosahedral	Yes
Arteriviridae	Vertebrates	Equine arteritis virus	1	13–16	Linear	Icosahedral	Yes
Coronaviridae	Vertebrates	SARS coronavirus	1	28–31	Linear	Helical	Yes

virus genomes). As with single-stranded DNA genomes, single-stranded RNA genomes are susceptible to degradation by nucleases or by divalent cation attack, and this may be a factor that limits their size. Most viruses with RNA genomes smaller than 10 Kb have naked (unenveloped) capsids, while the capsids of viruses with larger genomes (*Flaviviridae*, *Togaviridae*, *Arteriviridae*, *Coronaviridae*, and *Roniviridae*) are enveloped; envelopes may afford additional protection to the viral genome. Viruses with the largest RNA genomes (*Coronaviridae*, *Roniviridae*), as well as many plant viruses, have helical nucleocapsids. The close association between capsid proteins and the single-stranded RNA in helical nucleocapsids may afford additional protection against degradation, particularly inside the cell.

All viruses with negative-strand RNA genomes have helical nucleocapsids and some have fragmented genomes

Some of the most widespread and/or deadly human diseases (measles, influenza, rabies, Ebola hemorrhagic fever, Lassa fever) are caused by viruses with negative-strand RNA genomes (Table 3.4). All known viruses in this group have helical nucleocapsids. This may be related to their mechanism of replication; the genome RNA must first be transcribed within the cell to generate complementary, positive-strand messenger RNAs, and the helical nucleocapsid may provide stability and structure to the genome during this process. All negative-strand RNA viruses that infect vertebrates are enveloped.

Viruses in this group have genomes with a limited size range of 9–19 Kb. Members of four families (*Bornaviridae*, *Paramyxoviridae*, *Rhabdoviridae*, and *Filoviridae*) share several distinctive characteristics: a single genome segment with a common gene order, transcribed in series from a single promoter at the 3′ end of the genome. These four families have been grouped into the order *Mononegavirales* (for "single segment, negative-strand RNA viruses"). All other families in this group have fragmented genomes (as many as eight genome segments). In most cases, each negative-strand genome segment codes for a single

TABLE 3.4 Viruses with negative-strand RNA genomes

Virus group	Hosts	Example	Genome segments	Genome size (Kb) and topology		Capsid symmetry	Envelope
Ophiovirus	Plants	Citrus psorosis virus	3–4	11–12	Linear	Helical	No
Tenuivirus	Plants & invertebrates	Rice stripe virus	4–6	17–18	Linear	Helical	No
Bornaviridae	Vertebrates	Borna disease virus	1	9	Linear	Helical	Yes
Arenaviridae	Vertebrates	Lassa fever virus	2	11	Linear	Helical	Yes
Orthomyxoviridae	Vertebrates & invertebrates	Influenza A virus	6–8	10–15	Linear	Helical	Yes
Rhabdoviridae	Vertebrates, invertebrates, & plants	Rabies virus	1	11–15	Linear	Helical	Yes
Paramyxoviridae	Vertebrates	Measles virus	1	13–18	Linear	Helical	Yes
Filoviridae	Vertebrates	Ebola virus	1	19	Linear	Helical	Yes
Bunyaviridae	Vertebrates, invertebrates, & plants	California encephalitis virus	3	11–19	Linear	Helical	Yes

messenger RNA and one or two viral proteins. Viruses with fragmented genomes face the problem of ensuring that at least one copy of each genome segment is packaged into the virion. However, in some plant viruses, the different genome segments are separately packaged into distinct virions. In those cases, successful infection requires that two or more distinct virions, each with a different genome segment, enter the same cell.

Viruses with double-stranded RNA genomes have fragmented genomes and capsids with icosahedral symmetry

Most viruses with double-stranded RNA genomes (Table 3.5) contain multiple genome segments, each of which codes for a single messenger RNA and usually a single viral protein. All of these viruses have capsids with icosahedral symmetry, in contrast to the helical nucleocapsids of viruses with negative-strand RNA genomes. This may be related to their mechanism of transcription. The capsid or a subviral particle remains intact in the cell while messenger RNAs are transcribed from each genome segment by viral RNA polymerase molecules packaged with the genome RNAs; the resulting messenger RNA is extruded into the cytoplasm. The capsid therefore provides a structure that positions each RNA polymerase molecule and the various genome segments, and this allows reiterated transcription of each segment and directional transfer of the mRNAs to the cytoplasm.

Different members of the *Reoviridae* family can infect vertebrates, invertebrates, plants, or fungi; this is one of the few virus families (along with the negative-strand *Rhabdoviridae* and *Bunyaviridae*) whose members can replicate in such a wide range of host organisms. Viruses with negative-strand or double-stranded RNA genomes import their RNA polymerases into the host cell, and the capsids serve as tiny intracellular machines for synthesizing viral messenger RNAs. Because of this life style, and because they make their own RNA replication machinery, these viruses may be less dependent on the specific cellular environment where they replicate, giving them freer range to adapt to a wide variety of organisms.

The length of viral and cellular genomes

One turn of the DNA double helix contains 10 base pairs and is 3.4 nanometers in length. This means that a double-stranded viral DNA genome of 2000 nucleotides measures approximately 700 nanometers (0.7 micrometers, μm) from end-to-end (Figure 3.2). A viral genome containing 1.2 million base pairs is 600 times longer, or 420 micrometers (0.4 mm). Thus, just three molecules of the DNA genome of mimivirus, the largest known virus, stretched out end-to-end, would cover a length of more than 1 millimeter. In comparison, the total length of the genome contained within the 23 pairs of chromosomes in a single human cell, 3 billion base pairs, would be approximately 1 meter if stretched out end-to-end. Because there are approximately 10^{13} cells in a single human being, the aggregate length of all of the DNA in a single individual is 10^{13} meters (10^{10} kilometers), or 50 times the distance between the Earth and the sun!

TABLE 3.5 Viruses with double-stranded RNA genomes

Virus group	Hosts	Example	Genome segments	Genome size (Kb) and topology		Capsid symmetry	Envelope
Cystoviridae	Bacteria	Bacteriophage phi6	3	13	Linear	Icosahedral	Yes
Totiviridae	Fungi & protozoa	Saccharomyces cerevisiae virus L-A	1	5–7	Linear	Icosahedral	No
Chrysoviridae	Fungi	Penicillium chrysogenum virus	4	13	Linear	Icosahedral	No
Partitiviridae	Plants & fungi	Penicillium stoloniferum virus S	2	3–4	Linear	Icosahedral	No
Birnaviridae	Vertebrates & invertebrates	Infectious bursal disease virus	2	5–6	Linear	Icosahedral	No
Reoviridae	Vertebrates, invertebrates, plants, & fungi	Rotavirus A	10–12	19–31	Linear	Icosahedral	No

Viruses with a reverse transcription step in their replication cycle can have either RNA or DNA genomes

Three virus families are grouped separately from other viruses because they have a reverse transcriptase that makes a DNA copy of an RNA during their replication cycle (Table 3.6). Therefore, unlike all other viruses, genome sequences can occur as both DNA and RNA at different stages of replication. *Retroviridae* package the RNA copy of the genome, while *Hepadnaviridae* and *Caulimoviridae* package the DNA copy of the genome. Viruses that use reverse transcriptase package this enzyme in the virion, as viruses with negative-strand or double-stranded RNA genomes package their respective RNA polymerases in the virion.

All retroviruses package two identical copies of the viral genome. This is unique to retroviruses, and it may be important in the reverse transcription step that makes a full-length DNA copy of the genome within the capsid once it has entered the cell (see Chapter 25). During reverse transcription, the growing DNA chain is transposed ("jumps") from one end of the RNA template to the other end; chain transfer can alternatively occur to the second packaged copy of the RNA genome. Having a second copy may be a backup mechanism in case one of the RNA molecules is damaged. Retroviruses are found only in vertebrates, but a variety of retrovirus-like elements, some of which make virus-like particles, have been found in plants, invertebrates, and fungi.

Hepadnaviruses have an unusual DNA genome: it is circular and mostly double-stranded, with a single-stranded region of variable length. This results from interruption of synthesis of the second DNA strand by the viral reverse transcriptase (DNA polymerase) during virion maturation (see Chapter 28). Caulimoviruses have a circular double-stranded DNA genome that contains strand breaks at specific sites. Viruses in these two families carry out the reverse transcription step at the end of the replication cycle and package the resulting DNA molecule in the virion, whereas retroviruses carry out the reverse transcription step at the beginning of the replication cycle.

TABLE 3.6 Viruses with a reverse transcriptase

Virus group	Hosts	Example	Genome segments	Genome size (Kb) and topology		Capsid symmetry	Envelope
Caulimoviridae	Plants	Cauliflower mosaic virus	1	7–8	Circular ds DNA	Icosahedral	No
Hepadnaviridae	Vertebrates	Hepatitis B virus	1	3–4	Circular ss/ds DNA	Icosahedral	Yes
Retroviridae	Vertebrates	Human immunodeficiency virus 1	1 (dimer)	7–13	Linear + RNA	Icosahedral	Yes

Satellite viruses and satellite nucleic acids require a helper virus to replicate

A number of very small virus-like genomes replicate only in association with other viruses. These genomes can replicate only if present in a cell that is simultaneously infected with a **helper virus**; therefore they have been called **satellite viruses** and **satellite nucleic acids** (Table 3.7). Genomes of satellite viruses encode their own capsid proteins, while genomes of satellite nucleic acids either contain no coding regions or encode only noncapsid proteins; their genomes are encapsidated by structural proteins of the helper virus. Satellite nucleic acids are presently classified based on the nature, length, and topology of their genomes.

Most known satellite viruses and nucleic acids replicate in plants, but an insect satellite virus (chronic bee paralysis virus), several double-stranded RNA satellite nucleic acids of fungi and protozoa, and a human satellite virus (hepatitis delta virus) are known. The genomes of all known satellite viruses are less than 2 Kb in length, and those that contain no protein-coding regions are less than 700 nucleotides in length. Strictly speaking, adeno-associated viruses, which are single-stranded DNA viruses whose replication depends on coinfection with adenoviruses or herpesviruses, should be included in this category, but those viruses are closely related to numerous other parvoviruses that replicate independently of helper viruses.

Viroids do not code for proteins, but replicate independently of other viruses

Viroids (Table 3.7) are small, virus-like RNAs that are similar to satellite nucleic acids, but viroids are able to replicate and cause disease in a variety of plants in the absence of helper viruses. Viroid genomes are single-stranded, circular RNAs from 250 to 400 nucleotides long that do not code for any proteins and are not encapsidated. Viroids use cellular DNA-dependent RNA polymerases to replicate their RNA genomes. These genomes have a high degree of internal base-pairing and therefore resemble double-stranded RNAs over much of their length.

Some viroids, as well as hepatitis delta virus RNA and the small, circular single-stranded RNA satellites, have the ability to self-cleave and self-ligate. These enzymatic activities play important roles in viroid replication, during which RNA oligomers are synthesized. These oligomers can cleave themselves at specific sites to generate linear monomers, which then self-ligate into circular monomers. These enzymatic activities link viroids and circular RNA satellites to the ancient RNA world (see below), in which it is postulated that RNAs were able to replicate themselves in the absence of proteins by serving as enzymes (**ribozymes**).

TABLE 3.7 Satellite viruses, satellite nucleic acids, and viroids

Virus group	Hosts	Example	Genome	Genome segments	Genome size (Kb) and topology		Function of encoded protein
Satellite viruses	Plants & invertebrates	Tobacco necrosis satellite virus	ss RNA	1–3	1–3	Linear	Capsid protein
Single-stranded DNA satellites	Plants	Tomato leaf curl virus satellite DNA	ss DNA	1	0.7–1.3	Circular	Unknown function
Double-stranded RNA satellites	Fungi & protozoa	M Satellites of Saccharomyces cerevisiae L-A virus	ds RNA	1	0.5–1.8	Linear	Toxin
Circular single-stranded RNA satellites	Plants	Tobacco ringspot virus satellite RNA	ss RNA	1	0.35	Circular	No encoded proteins
Small linear single-stranded RNA satellites	Plants	Cucumber mosaic virus satellite RNA	ss RNA	1	0.7	Linear	No encoded proteins
Large single-stranded RNA satellites	Plants	Bamboo mosaic virus satellite RNA	ss RNA	1	0.8–1.5	Linear	Replication
Deltavirus	Humans	Hepatitis delta virus	ss RNA	1	1.7	Circular	Replication and packaging
Viroids	Plants	Avocado sunblotch viroid	ss RNA	1	0.25–0.4	Circular	No encoded proteins

THE EVOLUTIONARY ORIGIN OF VIRUSES

Because viruses need living cells for their replication, they must have evolved hand-in-hand with their hosts as these organisms evolved. But how and when did viruses first appear? Because viruses leave no fossil record, we can only speculate on their first appearance based on what we know about their genomes, their proteins, and their replication strategies.

The first steps in the development of life on Earth: the RNA world

Within perhaps 300 million years after the formation of the Earth, once the oceans had formed and cooled, macromolecules may have arisen by spontaneous polymerization of simple organic molecules such as ribonucleotides and amino acids. Some primitive RNAs may have by chance contained nucleotide sequences that enabled them to bind to other ribonucleotides and form covalent bonds. This would enable such RNAs to extend themselves and, by using base-pairing, replicate themselves in the absence of any protein-based enzymes. These RNAs may have represented the first self-replicating macromolecules in the **RNA world**.

As these RNAs developed and evolved, some of them may have acquired the property of binding covalently to a limited variety of amino acids, a role presently carried out by transfer RNAs and amino acid–activating enzymes. Simple polypeptides could have been assembled by such adaptor RNAs on templates of self-replicating RNAs, and those polypeptides that helped to protect the RNAs and increase their rate of replication would have enhanced selection and further multiplication of their template RNAs. Thus a self-replicating RNA linked to a primitive protein-synthesizing system developed. This could have eventually led to the selection of RNAs that coded for a polypeptide with RNA polymerase activity, increasing the rate and efficiency of RNA replication beyond that of a purely RNA-based enzymatic activity.

At some point, these RNAs and their associated polypeptides may have become localized and concentrated within lipid droplets. Polypeptides coded by these RNA genomes could be inserted into the lipid membranes and control the passage of small molecules and ions, forming the first primitive cells. These primitive cells would have contained small RNA genomes that coded for only a few polypeptides, whose main functions were to replicate the genome RNA and to provide for the entry of substrate nucleotides and amino acids.

Viroids and RNA viruses may have originated in the RNA world

Viroids and small single-stranded RNA satellites share some unique characteristics that indicate that they may be relics of the RNA world that survive to this day. These small, circular RNAs are able to catalyze some steps (cleavage and ligation) in their replication in the absence of proteins, but depend on RNA polymerases coded by cellular genes or the genes of helper viruses. They could be descended from self-replicating RNA molecules that arose in the RNA world.

Viruses with RNA genomes may also have arisen from primitive RNA-based life forms. They are the only known present-day biological entities with RNA genomes. To replicate, these viruses must encode their own RNA-dependent RNA polymerases, because such enzymes are not made by their host organisms. These enzymes may well have been inherited from the RNA world.

The transition to the DNA-based world

At some point during the evolution of the RNA world, polypeptides produced by self-replicating RNAs could have acquired the abilities to catalyze (1) the production of deoxyribonucleotides from ribonucleotides, and (2) their polymerization into DNA copies of the RNA templates, by acting as a primitive reverse transcriptase. In this way, the presently known DNA world could have emerged. Because DNA molecules are more stable than RNAs, they would have a tendency to accumulate even before they could be replicated. At some point, one of the polypeptide products made by a self-replicating RNA, perhaps a mutation of the RNA-dependent RNA polymerase or the reverse transcriptase, could have acquired the ability to act as a primitive DNA-dependent RNA polymerase. This alllowed the DNA molecule to take over the function of storing the information for protein synthesis, while the RNA copies could continue to serve as messengers and amino-acid binding adaptors for templating protein synthesis. A final development would have been the generation of a polypeptide with DNA polymerase activity, enabling the replication of the DNA genomes. The fundamental properties of cellular life as we know it would have then been established. Self-replicating RNAs could have coexisted for some time alongside of the DNA genomes in these primitive cells.

The steps postulated to have taken place in the evolution from simple self-replicating RNAs to a DNA-based cellular life imply the existence of nucleic acid polymerases of four types: (1) RNA-dependent RNA polymerases, to help replicate RNAs in the RNA world; (2) RNA-dependent DNA polymerases (reverse transcriptases), to transfer the information in these

RNAs to DNA and begin the DNA world; (3) DNA-dependent RNA polymerases, to express the information in the DNA via messenger RNAs; and (4) DNA-dependent DNA polymerases, to replicate the DNA genomes. But as DNA-based cells evolved, they would no longer need the RNA-dependent RNA polymerase or reverse transcriptase activities to survive, and they therefore could have lost genes that code for these enzymes.

Small and medium-sized DNA viruses could have arisen as independently replicating genetic elements in cells

A number of small DNA viruses replicate their genomes and express their messenger RNAs by using host cell DNA and RNA polymerases. The limited number of viral genes code for (1) virion structural proteins, (2) proteins that stimulate production of the cellular DNA replication machinery, and (3) proteins that bind to the viral DNA and assemble the cellular DNA replication machinery (for examples, see bacteriophage PhiX 174, parvoviruses, polyomaviruses, and papillomaviruses, in Chapters 6, 9, 10, and 11). It is possible that these small DNA viruses originated from fragments of cellular DNA that broke away from the cell genome and were able to replicate independently within the cell as plasmids. Such DNA fragments must have contained a nucleotide sequence that was recognized by the cellular DNA replication machinery as a replication origin.

As it evolved, this independently-replicating DNA could have acquired a gene coding for a protein that bound to the DNA, providing a package for its extracellular delivery. Such a DNA would therefore have become a primitive virus, able to jump from one cell to the next. As it evolved, it would have undergone selection for the production of a stronger capsid that could better package and protect the genome DNA.

Capsid proteins that recognize and bind to receptors on the surface of susceptible cells would have arisen. As the virus replicated and evolved, a virus-coded protein that specifically binds to its origin of replication and attracts the cellular DNA replication machinery would ensure preferential replication of its DNA. Some of the larger DNA viruses could have evolved from these small viruses through recombination and gene duplication events.

Large DNA viruses could have evolved from cellular forms that became obligatory intracellular parasites

An alternative origin has been suggested for some of the largest DNA viruses. These viruses contain hundreds of genes that supply many functions usually provided by cellular genomes. It is possible that they began as independently replicating cells, which at some point parasitized other cell types and became obligatory intracellular parasites. With time, some of the progeny of these intracellular parasites would lose genes that were no longer needed, and other genes that can package the viral genome and transmit it to susceptible cells could have evolved.

This argument makes most sense for the poxviruses (Chapter 15), which package within the virion core a large number of enzymes needed to express and process viral early messenger RNAs. These RNAs are synthesized within the cores after they enter the host cell, and are transported through the core membrane into the cytoplasm. This strongly resembles the mechanisms by which cellular RNAs are expressed, modified, and transported from the cell nucleus to the cytoplasm. After the production of poxvirus early mRNAs, the poxvirus core membrane is broken down, allowing viral DNA replication to be directed by early viral proteins. This resembles the breakdown of the nuclear membrane during mitosis. At the end of the virus life cycle, lipid membranes reform around the progeny DNA and associated proteins to make the virus core, just as nuclei are reformed by enclosure of chromatin within a lipid membrane after mitosis.

An interesting twist on this hypothesis has been proposed. Instead of being derived from a degenerate intracellular parasite, a large DNA virus could have been the precursor to the nucleus in the evolution of the first eukaryotic cells. This argument suggests that a large, enveloped DNA-containing virus that had evolved to code for most of the functions required for its replication could have established itself persistently in a prokaryotic cell. Eventually, the resident viral genes, localized within an internal membrane system, could have taken over the functions of the prokaryotic genome, which would have subsequently been lost. In this way, a nucleus would be established, producing the first eukaryotic cell.

Because there are very few hard data on which to base theories of the origin of viruses, the above ideas are really only informed speculation. However, it is striking that (1) many viruses have nucleic acid genomes not found within any other life form (single-stranded DNA, positive- or negative-strand RNA, double-stranded RNA), (2) some viruses code for replication proteins not normally found in other organisms (RNA-dependent RNA polymerases, reverse transcriptase), and (3) the smallest viruses are almost totally dependent on their host cells for replication machinery, while the largest viruses require only substrates, energy, and protein synthesis machinery. The great variation among virus genomes and replication strategies suggests that there may have been multiple origins of viruses during the evolution of life.

SUGGESTED READING

1. Domingo, E., Webster, R.G., and Holland, J.J. Eds. (1999). Origin and evolution of viruses. Hardcover, 499 pp. Academic Press, San Diego, CA.

2. Fauquet, C.M., Mayo, M.A., Maniloff, J., Desselberger, U., and Ball, L.A. Eds. (2005). Virus taxonomy: VIIIth report of the international committee on taxonomy of viruses. Hardcover, 1259 pp. Elsevier Academic Press, San Diego, CA.

3. Villareal, L.P. (2005). Viruses and the evolution of life. Hardcover, 395 pp. ASM press, Washington, DC.

KEY TERMS

Family
Genus (plural, genera)
Helper virus
Lineage
Order
Ribozyme
RNA world
Satellite nucleic acid
Satellite virus
Species
Taxonomy
Viroid

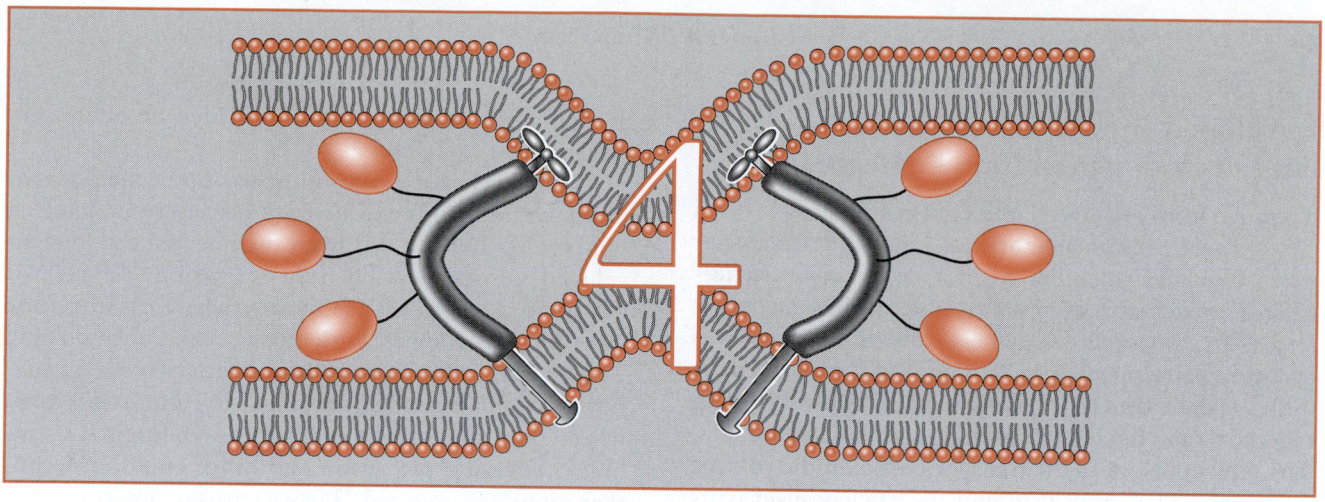

Virus Entry

STEPS IN VIRUS ENTRY

Virions bind to specific receptors on the cell surface.
Virions enter the cell either in specialized vesicles or by direct penetration through the plasma membrane.
Virions or nucleocapsids are transported within the cell to sites of transcription and replication (cytoplasm or nucleus).
Viral genomes are released into the cell by passage through the vesicle membrane or by disintegration of capsids within the cell (uncoating).

PENETRATION THROUGH CELLULAR MEMBRANES

Membrane penetration is mediated by conformational changes of virion proteins.
These changes are triggered by

- *interaction of viral surface proteins with cellular receptors in membranes, or*
- *exposure to low pH inside endosomes.*

Enveloped viruses penetrate by fusion of the virus envelope with either

- *the plasma membrane, or*
- *membranes of intracellular vesicles.*

Non-enveloped viruses enter the cytosol either by

- *forming transmembrane channels through which the viral nucleocapsid or genome can pass, or*
- *inducing lysis of endocytic vesicles.*

INTRACELLULAR TRANSPORT

Virions and virus nucleocapsids are transported within the cell along microtubules using the cellular transport motor dynein.
Viruses that replicate in the nucleus transport their genomes into the nucleus via the nuclear pore complex.

How do virions get into eukaryotic cells?

Virus particles have a single mission: to mediate the transport of the viral genome from an infected cell to an uninfected cell, thus allowing the spread of infection from cell to cell and from organism to organism. The transfer process begins with assembly of virions in the infected cell, followed by their release, transmission within the body or in the environment to uninfected cells, and finally, entry into a new host cell.

This chapter focuses on the final stage of the journey for viruses of eukaryotes: entry of the virion into a host cell, its transport to the proper intracellular location, and disassembly of the virion to release the viral genome. This complex process involves several steps and exploits a variety of cellular activities ranging from endocytosis to nuclear import. In many ways, viruses work like Trojan horses; they can only infect a cell that provides extensive assistance during entry. Viruses follow the molecular rules and regulations that control events inside the cell, but manipulate these pathways to serve their own ends. In this way, viruses can take advantage of the cell while remaining virtually invisible to the cell's antiviral defenses. Bacteriophages have evolved distinct mechanisms for passage of their genomes through the cell wall and membranes of bacteria, and some of these are described in detail in Chapters 5 through 8.

1 line short

Enveloped and nonenveloped viruses have distinct penetration strategies

To move from cell to cell, the viral genome has to cross two or more cellular membranes. This is not trivial, because biological membranes provide effective barriers against movement of proteins and nucleic acids, let alone large macromolecular assemblies such as virus particles. Enveloped viruses circumvent this problem by using **vesicle transport** and coupled **fusion** ("coming together") and **fission** ("breaking apart") of membranes. The viral envelope is the transport vessel in the passage from one cell to the next, and the nucleocapsid is its cargo (Figure 4.1). The viral envelope forms around the nucleocapsid and other internal components such as matrix proteins during **budding** at a cellular membrane. Often budding takes place at the (external) plasma membrane, but some viruses use the membranes of the nucleus, the endoplasmic reticulum, the intermediate compartment, endosomes, or the Golgi complex. The enveloped virion is then released into the extracellular space directly, or after transport through the secretory pathway.

Delivery of the nucleocapsid cargo into the **cytoplasm** of the target cell occurs when the viral envelope fuses with a cellular membrane (Figure 4.1, right). Fusion can take place either at the plasma membrane or at the membranes of early or late **endosomes**. Endosomes are membranous organelles in which the virion is enclosed after undergoing **receptor-mediated endocytosis** (see below). The membrane fusion pathway has the advantage that viral macromolecules do not need to pass directly across the hydrophobic interior of any cellular membrane. In fact, in a topological sense, the capsid never leaves the cytoplasmic compartment. Given the simplicity and elegance of this strategy, it is not surprising that the entry of enveloped viruses such as parainfluenzaviruses or togaviruses is often very efficient.

Nonenveloped viruses must use other mechanisms to penetrate membrane barriers because they have no lipid membrane. To exit from the infected cell, nonenveloped viruses typically rely on rupture of cellular membranes. Cell death is usually necessary to release virus particles trapped in the nucleus or the cytoplasm. These viruses enter host cells in a variety of ways. Adenoviruses, for example, are taken up into endosomes, where they proceed to rupture the endosomal membrane, releasing the virion and other endosomal contents into the **cytosol**. Picornaviruses, on the other hand, undergo conformational changes during binding to surface receptors or after entering endosomes, so that viral proteins form membrane channels that allow the genome direct access to the cytosol.

Some viruses can pass directly from cell to cell

In a few cases, infection can spread from one cell to the next without formation of a complete virus particle. Measles and some other enveloped viruses induce fusion of infected cells with neighboring uninfected cells. This results in the formation of multinucleated cells called **syncytia**. The same viral proteins that serve as fusion agents in the virion envelope are responsible for cell–cell fusion when they are incorporated into the plasma membrane of the infected cell. Transmission of infection in this way is obviously limited to cells located next to each other; spread of infection between organisms cannot occur by this mechanism unless there is exchange of cells.

In plants, the cell wall prevents diffusion of virus particles between cells. Instead, infection spreads directly through special channels, called **plasmodesmata**, that

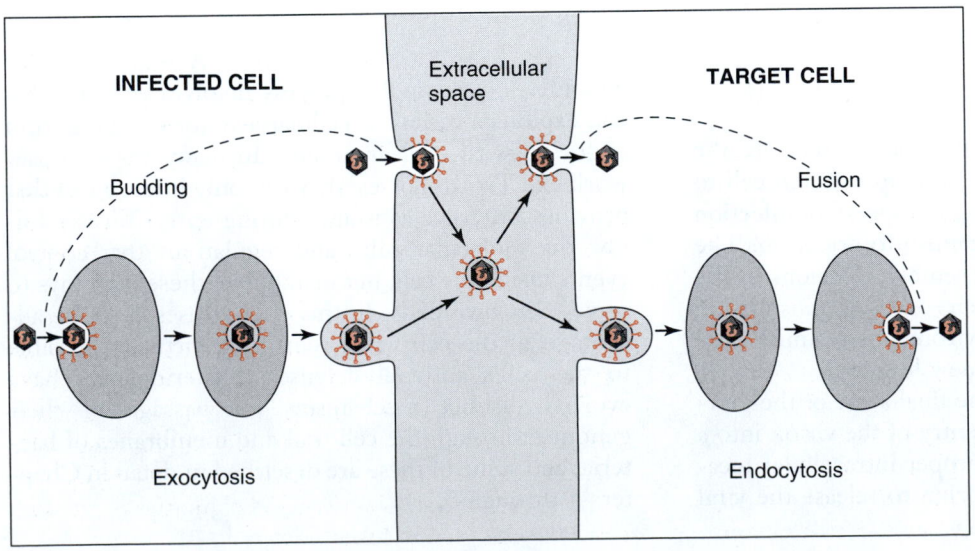

FIGURE 4.1 Exit and entry of enveloped viruses occurs by vesicle transport. Left: formation of a simple enveloped virus particle by budding at the plasma membrane (above) or at an internal cellular membrane (below), followed by release via exocytosis. Center: virions in the extracellular space. Right: uptake by fusion at the plasma membrane, or within a vesicle after endocytosis, releases the viral nucleocapsid into the cytosol.

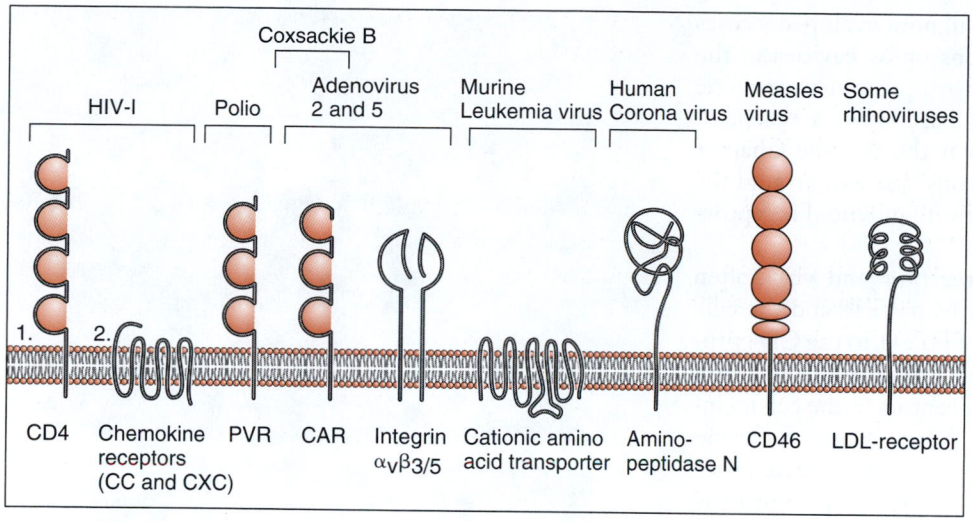

FIGURE 4.2 **Cell surface glycoproteins used as virus receptors.** Schematic views of several known virus receptors, shown protruding from the plasma membrane at the cell surface. Members of the immunoglobulin family have typical domains shown as orange spheres. PVR = poliovirus receptor; CAR = Coxsackie virus and adenovirus receptor; LDL = low density lipoprotein.

connect the cytosolic compartments of neighboring cells. These channels traverse the cell wall and provide a conduit for transport of viral genomes and other macromolecules. In some cases, only the viral genome is transported in plasmodesmata; in other cases, assembled virions, or nucleocapsids are transported. For some viruses, the plasmodesmata are too narrow to allow passage; virus-encoded "movement proteins" are expressed that cause expansion of plasmodesmata into wider channels.

A variety of cell surface proteins can serve as specific virus receptors

The first step in infection of a target cell is binding of the virion to the cell surface. Many different molecules located in the plasma membrane can serve as receptors for viruses (Figure 4.2). These molecules have their own cellular functions, and are only fortuitously exploited by viruses; therefore they tend to be highly conserved, and are unlikely to undergo mutations to avoid virus uptake. Since virus binding is a prerequisite for infection, the presence or absence of an appropriate receptor is one of the most important factors determining which species and which cell types can be infected by a given virus.

Cell surface proteins exploited by viruses as receptors include members of the **immunoglobulin family** (CD4, PVR, CAR, CD46); the low-density lipoprotein (LDL) receptor family; a variety of multimembrane-spanning transport proteins; **integrins**; and others. Some viruses bind to specific carbohydrate groups present on **glycoproteins**, **proteoglycans**, and **glycolipids**. For example, influenza viruses and paramyxoviruses bind to **sialic acid** residues present on glycoproteins and glycolipids. While this allows them to attach to virtually any cell, binding may also occur to cells or membrane fragments that cannot support productive infection. To avoid such nonproductive binding, these viruses carry **neuraminidases** in their envelopes. Whenever necessary, these "receptor-destroying enzymes" release the virions by cleaving the sialic acid residues to which they are bound.

A number of viruses require more than one type of receptor (Figure 4.2: HIV-1 and adenovirus). The primary receptor is usually sufficient for attachment while the secondary receptor (or "co-receptor") is needed to trigger endocytosis or penetration. Binding often occurs sequentially because the binding sites for the co-receptor are exposed only after interaction of the virion with the primary receptor. For example, the envelope glycoproteins of HIV-1 and other lentiviruses first bind to CD4, a glycoprotein of the immunoglobulin family. Only after CD4 binding do these viruses engage the chemokine receptors CCR4 or CXCR5, which allow penetration by membrane fusion (see Figure 4.2 and Chapter 26).

Receptors interact with viral glycoproteins, surface protrusions, or "canyons" in the surface of the virion

Binding of viruses to their receptors occurs via specific binding sites exposed on the surface of the virion. Because each virus particle carries many such sites, there are usually multiple receptor–virus contacts. Attachment can therefore be exceptionally tight even though the affinity of individual interactions may be rather low. Enveloped viruses have specific surface spike glycoproteins that serve as the binding sites for receptors. Some viruses devote a glycoprotein exclusively to attachment. More often, the same glycoproteins mediate both attachment and membrane fusion. For example, the HA1 domain of the trimeric influenza virus hemagglutinin (HA) protein contains the sialic acid binding sites with which the virus binds to the cell surface. The HA2 domain of the protein serves independently as an acid-activated membrane fusion factor.

The receptor-binding sites in nonenveloped viruses are formed either by protrusions or by cavities in the capsid surface. Adenoviruses have a prominent trimeric fiber attached to each penton base, with a receptor-binding site located in a knob at the tip (see Chapter 12). In contrast, the deep "canyons" present around the pentons in picornaviruses constitute indented receptor-binding sites (see Chapter 16).

The interaction between receptor and virus often leads to changes in both the virus particle and the cell. As already mentioned, binding of HIV-1 to the CD4 protein triggers a change in the HIV-1 glycoprotein trimers that allows them to bind to co-receptors in the cell membrane. This in turn triggers the fusion-active conformation in the spikes and leads to penetration. Attachment of the adenovirus fibers to the primary receptor protein, CAR, triggers fiber dissociation. This exposes specific amino acid sequences in the penton base proteins that are recognized by binding sites in integrins. Interaction with integrins is needed for internalization by endocytosis and subsequent penetration of the adenovirus virion into the cytosol. Attachment of adenovirus to the $\alpha_v\beta_3$ integrins also activates a signaling cascade inside the cell that temporarily increases endocytic activity and prepares the cell for the incoming virus.

Many viruses enter the cell via receptor-mediated endocytosis

Virus binding to receptors is often followed by receptor-mediated endocytosis, a process that occurs most frequently via **clathrin-coated pits** and vesicles (Figure 4.3). Instead of fusing with or penetrating through the plasma membrane, the virus allows itself to be carried as cargo inside an endocytic vesicle into the endocytic pathway. This vesicle ferries the virus particle through the cortical network of cytoskeletal fibers, which could otherwise constitute a major obstacle, and transports it away from the cell surface and toward the cell nucleus. A few minutes after leaving the cell surface, virions are delivered, like other substances endocytosed by clathrin-coated vesicles, into so-called **early endosomes**. After internalization, the virus particle is sepa-

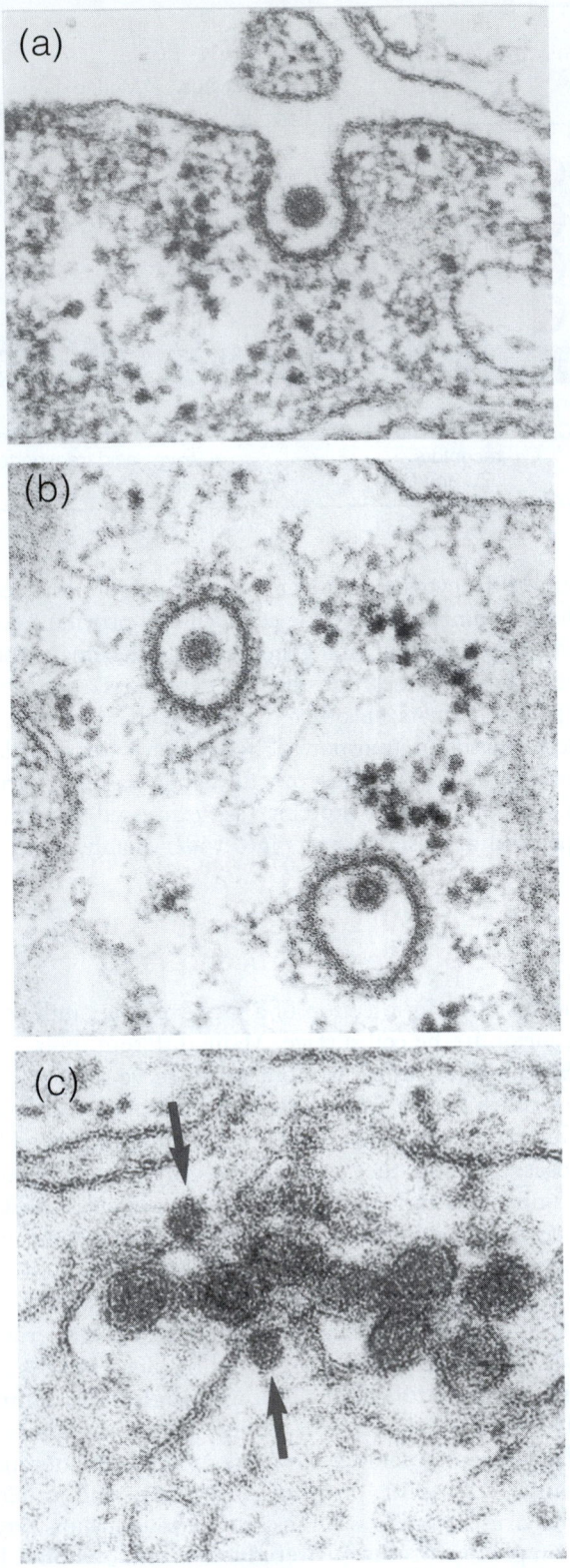

FIGURE 4.3 **Receptor-mediated endocytosis and acid-activated penetration of Semliki Forest virus.** (a.) Virus particles bound to the cell surface via their spike glycoproteins are trapped in clathrin-coated pits. The clathrin coat is seen in these thin-section electron micrographs as a densely stained layer on the cytosolic side of the plasma membrane. (b.) The coated vesicles detach from the plasma membrane and the clathrin coat dissociates. (c.) The vesicle fuses with early endosomes, from which the nucleocapsid is extruded into the cytosol (see arrows) upon fusion of the viral envelope with the endosomal membrane at a pH below 6.2.

rated from the cytosol by the barrier of the endosomal membrane, which it can overcome by the same strategies that other viruses use at the plasma membrane.

Early endosomes are complex organelles composed of tubular and bulbous membrane elements, found in the peripheral cytoplasm. They serve the cell as a sorting and recycling station for incoming fluid, **ligands**, receptors, and membrane proteins. Early endosomes are responsible for dissociation of incoming ligand–receptor complexes, for recycling of empty receptors to the cell surface, and for dispatching non-recycled material to late endosomes and **lysosomes** to be degraded. Early endosomes are characterized by a mildly acidic pH (6.3–5.5) maintained by membrane-associated vacuolar-type ATPases that pump protons from the cytosol into the lumen. Incoming molecules are exposed to a successively decreasing level of pH as they proceed from early to late endosomes, and ultimately to lysosomes.

Passage from endosomes to the cytosol is often triggered by low pH

Endosomes are a major site of virus penetration to the cytosol. The trigger for penetration in this compartment is usually low pH, which induces a conformational change in the spike glycoproteins of enveloped viruses or in the capsids of nonenveloped viruses. Low pH signals the virus that it has reached an intracellular location and that the penetration reaction should now begin. Virus particles attached to the plasma membrane can be artificially activated to penetrate prematurely simply by lowering the pH in the medium bathing the cells, showing that low pH is all that is needed as a cue for penetration. Most viruses adjust the timing of their penetration by carefully titrating the activation threshold to a pH in the 6.5–5.3 range. Depending on the threshold pH, conversion occurs 5–30 min after endocytic uptake, before the virus can be delivered to lysosomes.

Table 4.1 lists viruses known to enter through an endosomal, acid-triggered pathway, and viruses that use pH-independent entry. pH-independent penetration is triggered by interactions between the virus and its specific receptor. Consequently, many of these viruses can penetrate directly through the plasma membrane, where they first encounter their receptors. Some viruses (e.g., adenoviruses) need both a receptor-induced and a low pH–induced conformational change to penetrate successfully.

Not all viruses that are internalized are carried into the cytoplasm by clathrin-coated vesicles. For example, simian virus 40, a nonenveloped polyomavirus, is internalized via **caveolae**. These are small, flask-shaped invaginations in the plasma membrane enriched in cholesterol and sphingolipids, and in a cellular protein called **caveolin**. Caveolae and the caveolar pathway are thought to play a role in signaling and in the uptake of cholesterol and selected ligands. They are also responsible for transport across endothelial cells in blood vessels. Virions of simian virus 40 are brought by caveolae to the smooth endoplasmic reticulum, where penetration to the cytosol takes place.

Membrane fusion is mediated by specific viral "fusion proteins"

Penetration of enveloped viruses involves a membrane fusion event mediated by viral glycoproteins. Viral fusion proteins are generally **type I membrane glycoproteins** most of whose mass lies external to the viral membrane, and they are usually organized as oligomers. Because of their unusual function and their central role in infection, many fusion proteins have been characterized in great detail.

Like other viral glycoproteins, fusion proteins are synthesized, folded, and assembled into oligomers in the endoplasmic reticulum of infected cells. To become fusion-active, they generally must undergo several maturation steps. In transit from the endoplasmic reticu-

TABLE 4.1 Acid-activated and pH-independent viruses

Acid-activated penetration	pH-independent penetration
Orthomyxoviruses (influenza A and B viruses)	Most retroviruses (HIV-1)
Togaviruses (Semliki Forest and Sindbis viruses)	Herpesviruses (HSV-1)
Flaviruses (tick-borne encephalitis virus)	Paramyxoviruses (Sendai, measles)
Rhabdoviruses (rabies, vesicular stomatitis virus)	Some coronaviruses
Filoviruses (Ebola virus)	Some picornaviruses (poliovirus)
Bunyaviruses (Uukuniemi virus)	Reoviruses
Adenoviruses (types 2 and 5)	
Some picornaviruses (rhinovirus type 2)	
Some retroviruses	

lum through the Golgi complex to the plasma membrane, fusion proteins may be cleaved by cellular proteases. An example is the influenza virus hemagglutinin protein (HA), which is cleaved into a peripheral N-terminal fragment, HA1, and a C-terminal transmembrane fragment, HA2 (Figure 4.4). This cleavage renders the HA conformationally metastable, like a loaded spring; it responds to exposure to low pH with a major conformational change. In the case of HA and some other viral spike glycoproteins, the activating cleavage occurs just to the N-terminal side of a hydrophobic sequence of about 10–12 amino acid residues. In the cleaved protein, this conserved sequence, the **fusion peptide**, often constitutes the N-terminus of the transmembrane subunit, and plays a central role in the fusion reaction.

Fusion proteins undergo major conformational changes that lead to membrane fusion

When activated by low pH or by receptor binding, fusion proteins undergo a series of changes that lead to a dramatic alteration in protein architecture, properties, and membrane association. These have been studied in most detail for the trimeric influenza virus hemagglutinin protein (Figure 4.4). Immediately after exposure to low pH, the fusion peptides in the HA2 subunits are exposed and insert into the outer bilayer leaflet of the target membrane (second panel, Figure 4.4). The HA1 subunits remain attached but no longer form a trimeric domain at the top of the oligomer. As a result, the viral HA2 protein is now attached both to the viral membrane (via the transmembrane sequence) and to the target membrane (via the fusion peptide).

After further conformational changes, a rosette-like cluster of HA trimers is formed in the plane of the membrane (third panel). The HA trimers then fold back on themselves like a jacknife, forcing the two membranes closer together in the middle of the cluster. When the hydration shell that normally prevents fusion is eliminated in a small local spot between them, the two bilayers proceed to fuse (fourth panel). The outermost bilayer leaflets fuse first (this is called **hemifusion**), followed by the inner bilayer halves. A narrow aqueous channel is formed between the internal cavities of the virus and the cytosol (bottom panel). After flickering for a short while, the fusion channel remains permanently open and starts to expand. The nucleocapsids (not shown in the figure) are now exposed, and can move from the virion into the cytosol. Altogether, the fusion process is rapid and efficient.

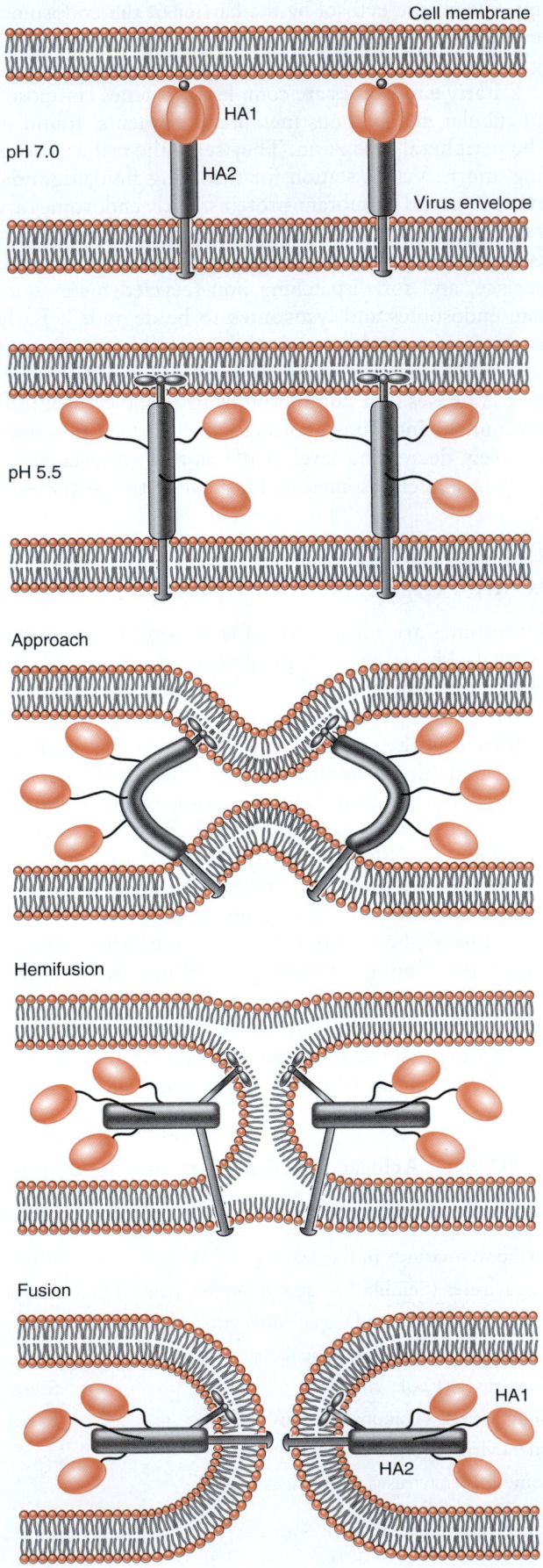

FIGURE 4.4 **A model for influenza virus hemagglutinin-mediated membrane fusion.** Steps from initial binding at pH7 (top) to fusion of virus envelope with endosomal membrane (bottom).

Although the step-by-step process described above for fusion is best characterized for the influenza A virus hemagglutinin protein, it is still only partially understood. A similar mechanism probably occurs with several other enveloped viruses that have **class I fusion proteins**, including retroviruses, paramyxoviruses, and filoviruses. The three-dimensional structure of fragments of the glycoproteins of these viruses reveals a trimeric spike dominated by long central alpha-helical **coiled coils**. These coiled coils can adopt two distinct conformations, which correspond to the form present in virions and the post-activation jacknife form.

Other viruses have **class II fusion proteins** with distinctly different designs. These include togaviruses, flaviviruses, rhabdoviruses, and hepadnaviruses. For some of these families the mechanism of fusion is starting to become known. The fusion-activating changes involve major alterations at the level of interactions between protein subunits (quaternary structure), accompanied by only minor protein conformational changes. Fusion peptides are not located at the N-terminus, and alpha-helices do not play a central role in these class II fusion proteins.

Nonenveloped viruses penetrate by membrane lysis or pore formation

The proteins of nonenveloped viruses also can undergo major conformational changes in response to receptor binding or low pH in endosomes. As a result, these proteins either induce the formation of pores in the cellular target membrane through which the genome can escape, or they actually lyse the membrane barrier.

Adenoviruses are the best-studied example of viruses that penetrate by acid-activated, virus-dependent endosome lysis (see Figure 4.5c). The virion is subjected to a series of changes prior to its activation in the endosome. The fibers dissociate, the penton base proteins bind to arginine-glycine-aspartate sequences in the $\alpha_v\beta_{3/5}$ integrin (the co-receptor), and when the virion is exposed to low pH it converts to the lytic form. As a result, the endosome membrane ruptures, and the virion together with other contents of the endosome is released into the cytosol.

Other nonenveloped viruses pass through cellular membrane barriers without apparent lysis or major membrane perturbations. Picornaviruses such as poliovirus enter at the plasma membrane by a pH-independent mechanism. Binding to the receptor triggers an initial conformational change that loosens up the capsid wall and makes the particles hydrophobic. These so-called "A particles" then interact hydrophobically with the cellular membrane and somehow induce the formation of a channel through which the viral RNA can slip into the cytosol, leaving an empty capsid behind. (see Figure 16.3) What is left is an RNA-free particle, the "B-particle." Rhinoviruses release their RNA by a similar mechanism, but it is induced by low pH within endosomes.

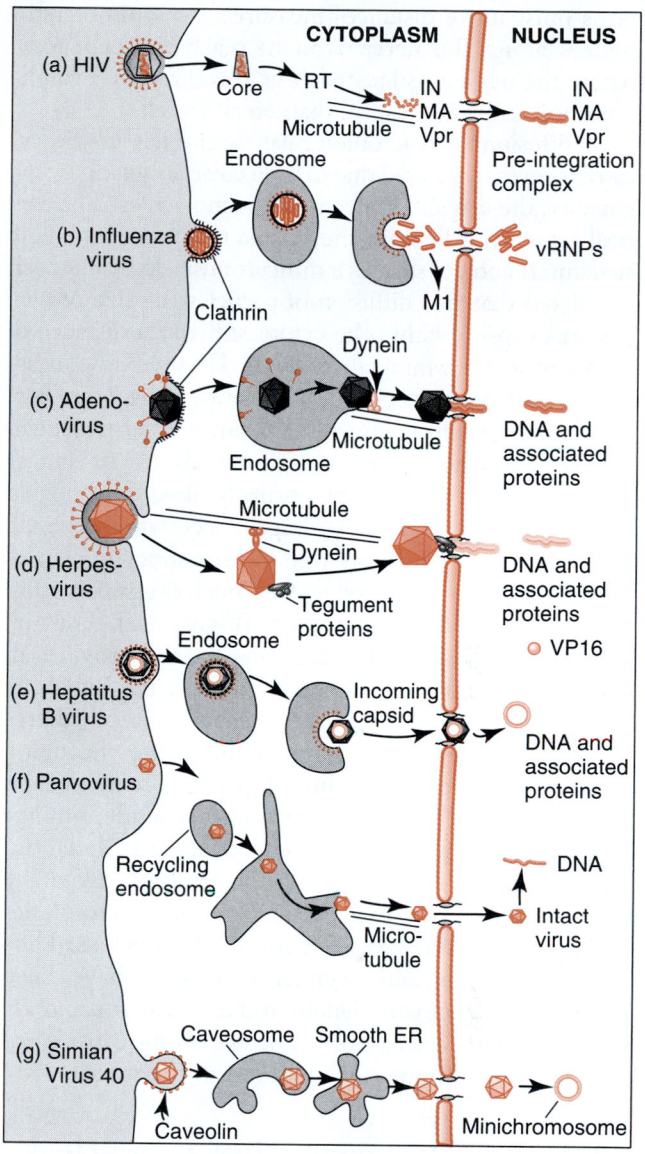

FIGURE 4.5 Entry and intracellular transport of viruses that replicate in the nucleus. Distinct entry and transport pathways of seven viruses are shown schematically. The extracellular space is on the left, the cytoplasm is in the center, and the nucleus is on the right. The double nuclear membrane is shown traversed by nuclear pores.

Virions and capsids are transported within the cell in vesicles or on microtubules

For most viruses, reaching the cytosol is not the end of the journey. They still have to find the proper site for replication, either in the nucleus or in some defined location in the cytoplasm. In extreme cases, virus particles or capsids may still have to travel extremely long distances inside the cell. For example, upon entering a peripheral neuron, herpes simplex virus type 1 or rabies virus must move distances measured in centimeters or meters along the nerve axon to reach the cell body, where the nucleus is located. The travel distance can be a million times the capsid diameter!

Diffusion of macromolecular assemblies in the cytosol is severely limited due to molecular crowding by organelles, the cytoskeleton, and macromolecules. The cortical region just beneath the plasma membrane is highly structured and constitutes a difficult obstacle. It has been calculated that free diffusion of particles the size of herpesvirus capsids within the cytoplasm over a distance of 10 micrometers would take 2 hours. However, the measured rate of movement during retrograde transport of herpesvirus capsids in neurons is 400 times faster than that. Not surprisingly, a variety of viruses make use of cellular transport systems that are normally used for moving membrane vesicles and other large objects within the cell.

Virions can be transported inside endocytic vesicles (prior to penetration), or as free nucleocapsids within the cytoplasm (after penetration) (Figure 4.5). The former strategy has the advantage that the cell provides all the systems and signals needed to move the incoming vesicle in the direction of the nucleus. The cell therefore grants free passage for its lethal cargo through a maze of obstacles within the cytoplasm. The molecular motors used include the microtubule-dependent motor **dynein** and an adaptor complex called **dynactin**. Dynein moves particles along microtubules from the cell periphery in the direction of the microtubule-organizing center, located close to the nucleus. Herpesvirus capsids and adenovirus capsids have been shown to interact with dynein in the cytoplasm, and use it for retrograde transport. The incoming capsids apparently contain proteins that bind to the motors.

Import of viral genomes into the nucleus

Many animal and plant viruses, particularly DNA-containing viruses, use the nucleus as their site of replication. There viruses profit from the presence of the cellular machinery for transcription, RNA processing, RNA export, and DNA replication. Alternatively, viruses can establish a latent state in the nucleus, or integrate their genes into host chromosomes. While dependence on the host cell nucleus carries with it many advantages, it also poses a problem: The incoming virus must deliver its genome and necessary viral proteins to the nucleus. To accomplish this, most viruses piggyback on the machinery that cells utilize for protein and nucleic acid trafficking. The mechanisms and strategies by which incoming viruses enter the nucleus differ from one family to the next. A number of viruses interact with nuclear targeting receptors such as **importins α** and **β** (also called **karyopherins**), and their genomes enter the nucleus through the nuclear pore complexes.

Dissociation of the nuclear membrane during mitosis. Most retroviruses (with the exception of lentiviruses) have no means of entering the intact, interphase nucleus. Therefore, their capsids ("preintegration complexes"), which contain proviral DNA made by reverse transcription as well as several viral proteins, must wait in the cytoplasm until the cell divides. During cell division the nuclear envelope disintegrates, allowing the preintegration complexes to gain access to the nucleus when the nuclear envelope reforms. The proviral DNA is therefore able to integrate into the cellular genome in the reformed nucleus. Consequently, infection is only possible in cells that undergo mitosis. This explains why terminally differentiated (nondividing) cells in the body are not usually targets for retrovirus infection.

Import after partial disassembly in the cytoplasm. Lentiviruses like HIV-1, in contrast to other retroviruses, direct their preintegration complexes to enter the nucleus via the nuclear pores. The matrix and Vpr proteins within these complexes presumably interact with importins (Figure 4.5a). Each of the eight distinct genome RNAs of influenza virus is individually packaged into ribonucleoprotein complexes small enough to pass through the nuclear pores. After release of these complexes in the cytoplasm by fusion with the endosome membrane, they interact with the importins, and are rapidly transported to and through the nuclear pore complexes (Figure 4.5b).

Import after disassembly at the nuclear pore. Adenoviruses and herpesviruses have large isometric capsids that are too bulky to pass through the nuclear pores. These capsids bind to fibers that extend from the cytoplasmic face of the pore, and the capsids are subsequently disassembled, allowing the viral DNA and associated proteins to enter through the nuclear pore (Figure 4.5c, d). Adenovirus capsids disintegrate in the process, but intact, empty herpesvirus capsids remain bound to the pore fibers after the viral DNA has entered the nucleus.

Transport of intact virions through the nuclear pore complex. Only a few viruses have capsids small enough to penetrate through the nuclear pore complexes without dissociation or deformation. Electron microscopy shows that hepatitis B virus capsids (diameter of 34 nm) and simian virus 40 capsids (diameter of 48 nm) can pass intact through the nuclear pores. Parvoviruses (diameter of 20 nm) are also thought to enter the nucleus in this fashion (Figure 4.5 e-g).

The many ways in which viral genomes are uncoated and released

The ultimate goal of virus entry is to release the viral genome within the host cell in a form able to be translated, transcribed, or replicated. Depending on the virus, this "uncoating" step may occur simultaneously with the first step of entry, or it may only come after a complex series of membrane fusion and transport steps as outlined above. The mechanism of uncoating of only a few viruses is understood in any detail.

The nonenveloped picornaviruses release their genome RNA directly into the cytosol, leaving the viral capsid at the plasma membrane or within an endosomal vesicle. Therefore virus entry and uncoating are the same event for this virus family.

At the opposite extreme, uncoating of the DNA genomes of hepatitis B virus, parvoviruses, and polyomaviruses takes place only after the capsids have been transported into the nucleus through the nuclear pores. Uncoating of herpesvirus and adenovirus genomes takes place at the nuclear pore, as the capsids are dissassembled there. In the case of adenovirus, a viral cysteine protease is activated by entry into the reducing cytosolic environment. This protease cleaves some capsid proteins, releasing the DNA from its attachment to the inside surface of the capsid shell. The DNA genomes of most of these viruses are not released as naked DNAs, but are complexed with cellular or viral proteins that compact the DNA and neutralize its charge. For example, polyomaviruses release their circular DNA as a minichromosome, covered with nucleosomes made from cellular histones.

In the case of togaviruses, uncoating of the viral RNA occurs only minutes after low pH–induced fusion of the viral envelope with an endosomal membrane. Ribosomal 60S subunits serve as cellular uncoating factors. They have affinity for viral capsid proteins, and by binding to the capsid they help to release the viral RNA, which is then ready for translation.

Influenza A viruses encode a protein (M2) that forms membrane channels allowing free passage of H^+ ions at low pH (see Figure 32.3). Small amounts of this protein are incorporated into the virion envelope. As a result, when virions are present in acidifying endosomes, the interior of the particle is also exposed to a low pH, and this releases the individual nucleoproteins from their association with each other and with the matrix protein (M1). Upon fusion of the viral envelope with the endosomal membrane, the individual genome segments are released freely into the cytosol. However, like other negative-strand RNA viruses, influenza virus genomes remain associated with nucleocapsid proteins at all times.

Another variation is seen for reoviruses, which are nonenveloped double-stranded RNA viruses. They enter the cytosol by a pH-independent mechanism through the plasma membrane or an endosomal membrane, during which an outer layer of capsid proteins is removed, but the inner capsid remains intact. Therefore uncoating is incomplete, and viral double-stranded RNA is never released to the cytosol. Synthesis and capping of viral mRNAs occurs within these capsids, and the mRNAs are extruded through channels in the "turrets" that lie at the axes of fivefold symmetry (see Chapter 24).

How can virus entry be prevented?

Since viruses are pathogens and cause major medical and veterinary problems, every step in the replication cycle must be viewed as a potential target for inhibition by antiviral drugs. The most direct means of preventing virus entry is to intercept the virus with neutralizing antibodies before it reaches the cell. This is the basis of vaccination (Chapter 34), still by far the most successful antiviral strategy available. An alternative method is to flood the extracellular space with truncated, soluble versions of the cellular receptor protein for a particular virus. These soluble receptors can bind to virions, thus preventing their interaction with the natural receptor on the cell surface. Conversely, proteins or antibodies that bind to the receptor on the cell surface can be used. Such agents work either by occupying the receptor sites, or by inducing receptor internalization and down-regulation.

Entry can be blocked by interfering with cellular processes needed for virus internalization or penetration. This can be achieved by treatment with agents that interfere with the formation of clathrin-coated vesicles or block transport via the nuclear pore complexes. Compounds that prevent acidification of endosomes are often used as tools to study entry of acid-activated viruses. They include **lysosomotropic agents** such as ammonium chloride, chloroquine, and other organic weak bases that allow protons to escape from acidic compartments. **Carboxylic ionophors**, such as monensin and nigericin, are also effective in breaking down proton gradients, as are inhibitors of the vacuolar proton ATPase such as **bafilomycin A1**. To prevent membrane fusion more directly, peptides have been designed that bind to conformational intermediates of viral fusion proteins and inhibit their function. Inhibitors of influenza virus neuraminidase have recently reached the market and are being used as drugs against flu infections.

Finally, there are agents that inhibit uncoating. These include **amantadine**, a small molecular weight molecule that specifically blocks the M2 proton channel of influenza A virus and thus prevents acidification of the virus interior. "WIN compounds" (named for the company that originally synthesized them) inhibit human rhinoviruses by binding to a pocket under the receptor-binding canyon. They stabilize the virus capsid and prevent uncoating.

SUGGESTED READING

1. Kielian, M. (1995). Membrane fusion and the alphavirus life cycle. *Advances in Virus Research* **45,** 113–151.

2. Klasse, P.J., Bron, R., and Marsh, M. (1998). Mechanisms of enveloped virus entry into animal cells. *Advanced Drug Delivery Reviews* **34,** 65–91.

3. Rossmann, M.G., Bella, J., Kolatkar, P.R., He, Y., Wimmer, E., Kuhn, R.J., and Baker, T.S. (2000). Cell recognition and entry by rhino- and enteroviruses. *Virology* **269,** 239–247.

4. Smith, A.E. and Helenius, A. (2004). How viruses enter animal cells. *Science* **304,** 237–242.

5. Weissenhorn, W., Dessen, A., Calder, L.J., Harrison, S.C., Skehel, J.J., and Wiley, D.C. (1999). Structural basis for membrane fusion by enveloped viruses. *Molecular Membrane Biology* **16,** 3–9.

6. Whittaker, G., Kann, M., and Helenius, A. (2000). Viral entry into the nucleus. *Annual Review of Cell and Developmental Biology* **16,** 627–651.

KEY TERMS

Amantadine
Bafilomycin A1
Budding
Carboxylic ionophors
Caveolae
Caveolin
Class I fusion proteins
Class II fusion proteins
Clathrin-Coated pits
Coiled-coil
Cytoplasm
Cytosol
Dynactin
Dynein
Early endosome
Endosome
Fission (Membrane)
Fusion (Membrane)
Fusion peptide
Glycolipid
Glycoprotein
Hemifusion
Immunoglobulin family
Importin
Integrins
Karyopherin
Ligand
Lysosome
Lysosomotropic agent
Neuraminidase
Plasmodesmata
Proteoglycan
Receptor-mediated endocytosis
Sialic acid
Syncytium (plural: synctia)
Type I transmembrane protein
Vesicle transport

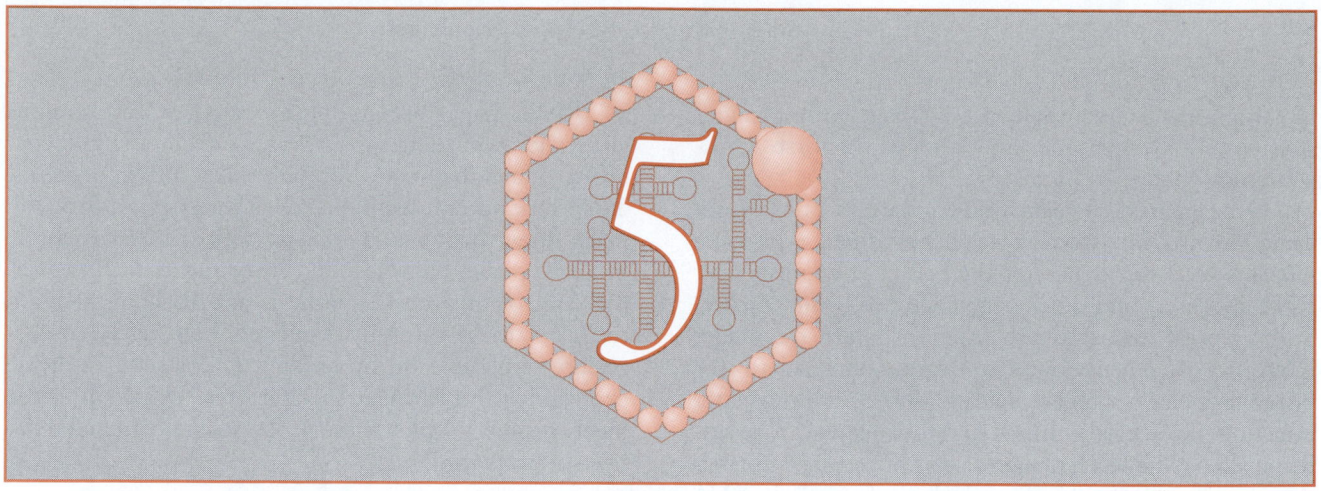

Single-Stranded RNA Bacteriophages

Leviviridae

From Latin *levis* (light), referring to small size of phage

VIRION

Naked icosahedral capsid made from 180 copies of coat protein (T = 3)
Diameter 26 nm
One copy of maturation protein bound to RNA genome

GENOME

Linear ss RNA, positive sense, 4 Kb
High degree of internal base-pairing

GENES AND PROTEINS

Four virus-coded proteins:

- *Coat protein*
- *Maturation protein*
- *Lysis protein*
- *RNA replicase*

VIRUSES AND HOSTS

Two genera:

- *Levivirus (Examples: f2, MS2, R17)*
- *Allolevivirus (Example: Q_β)*

Found in a variety of gram-negative bacteria

DISTINCTIVE CHARACTERISTICS

One of the smallest and simplest viruses known.
Binds to F-pili projecting from surface of F+ (male) *E. coli.*
Translation of genes is controlled via RNA secondary structures at ribosome binding sites.
RNA polymerase contains virus-coded RNA replicase plus three cellular proteins.

The discovery of RNA phages stimulated research into messenger RNA function and RNA replication

The RNA coliphages were discovered by Tim Loeb and Norton Zinder in 1961 as the result of a search for phages whose replication depends on *E. coli* F-pili, used for bacterial conjugation. Loeb and Zinder plated filtered samples of raw New York City sewage on *E. coli* strains and screened for phages that would produce plaques on male (F+) but not female (F−) bacteria. The first isolate, named f1, made turbid plaques but the second isolate, named f2, made clear plaques; Loeb and Zinder decided to concentrate their work on phage f2.

f1 turned out to be a filamentous phage with a single-stranded DNA genome, and f2 was the first RNA-containing phage to be discovered.

Over the next 20 years, study of the RNA phages became a growth industry. These phages came on the scene just as the first evidence for the existence of messenger RNA appeared, and the genetic code was being cracked. They represented a superb source of pure messenger RNA that could be produced in large amounts: up to 10^{13} phage particles per ml are made within a few hours after infection of bacterial cultures, and the phages can be easily purified. The RNA phages also attracted attention because scientists were intrigued about how their RNA genomes were replicated, since cells normally make RNA only from a DNA template. Many scientists who were to become dominant figures in molecular biology in the latter part of the twentieth century got their first taste of research by working with RNA phages.

RNA phages are among the simplest known organisms

RNA phages occur wherever *E. coli* lives, for example, in the intestinal tract of humans and other animals. Studies have shown that sewage samples worldwide contain from 10^2 up to 10^7 RNA phage particles per ml. However, for humans RNA phages are harmless creatures. In fact, RNA phages share the same habitat and the same inactivation characteristics in sewage treatment processes as numerous pathogenic enteroviruses (polioviruses, hepatitis A virus, etc.). However, the RNA phages are much easier to detect by a rapid plaque assay on male *E. coli* hosts. For this reason RNA phages are used as index organisms to reveal the potential presence of pathogenic enteroviruses in drinking water.

RNA phages are among the smallest autonomous viruses known; their capsid is made from a single coat protein, and they code for only three other proteins. Their task upon infecting a cell is very straightforward: make their proteins, replicate their RNA, assemble progeny phage particles, and leave the cell. However, the control of protein synthesis and RNA replication turns out to be carefully regulated, both by RNA secondary structure and by binding of coat and replicase proteins to the RNA at specific sites. Unravelling the mechanisms by which these processes are controlled has been a fascinating puzzle.

Two genera of RNA phages have subtle differences

RNA phages have been placed into two genera on the basis of genetic maps, serological cross-reactions, and RNA size. Members of the *Levivirus* genus (also called supergroup A) have RNAs approximately 3500 nt in length, and code for four proteins: maturation (or A) protein, coat protein, lysis protein, and the RNA replicase, encoded in that order proceeding from the 5′ end of the genome (Figure 5.1). The open reading frame for the lysis protein overlaps with those of the coat and replicase proteins. MS2 phage will be used to discuss translational control in this chapter. In contrast, members of the *Allolevivirus* genus (supergroup B) have a longer genome of approximately 4200 nucleotides. The difference comes mostly from a region on the genome used to code for a protein called readthrough protein or protein A1.

Readthrough in phage Q_β arises when the UGA stop codon at nt 1742 that terminates the normal coat protein gene is occasionally read as a tryptophan codon (UGG). This occurs with a probability of about 6%. A

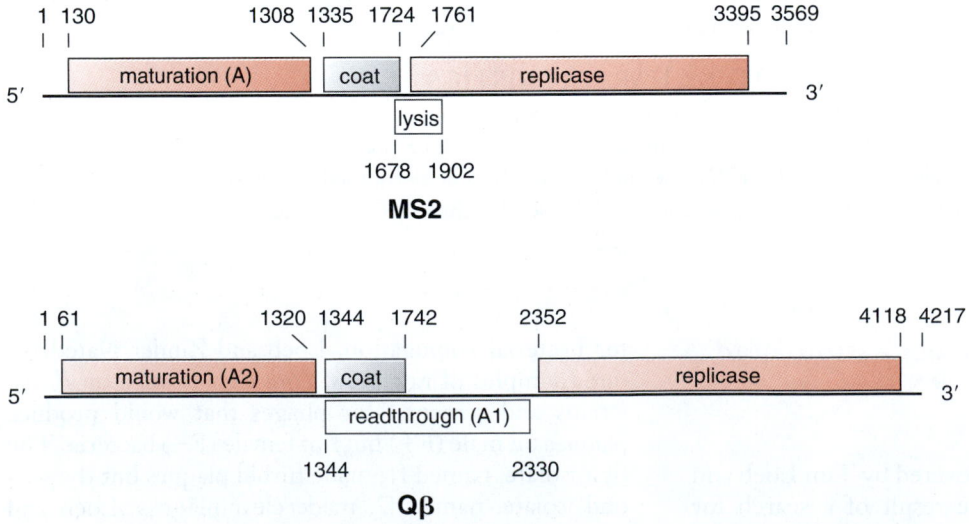

FIGURE 5.1 Genetic maps of phage MS2 and Q_β RNAs. First and last nucleotides are shown for all open reading frames (boxes).

small number of copies of the readthrough protein are incorporated into the capsid; their function is unknown. These phages have no distinct lysis protein; lysis is carried out by the maturation (A2) protein. Phage Q_β is the most intensively studied allolevivirus. The RNA replicase of Q_β can be easily purified, while the replicases of leviviruses are much more difficult to work with, and so most of the work on RNA replication has been done with Q_β.

RNA phages bind to the F-pilus and use it to insert their RNA into the cell

RNA phage capsids contain 180 copies of the coat protein arranged in a T = 3 icosahedral shell that encloses the RNA (Figure 5.2). A single copy of the maturation protein is bound to the encapsidated RNA at sites near both its 5′ and 3′ ends. The encapsidated RNA is resistant to ribonuclease treatment. However, RNA in virions made with defective or missing maturation protein is sensitive to RNAse degradation.

RNA phages do not have the specialized tail assemblies that are used by many DNA phages to inject their genomes. These phages instead subvert the bacterium's **F-pili** (singular: pilus) to maneuver their single-stranded RNA genomes into the cell. F-pili are filamentous structures of *E. coli* and other bacteria that enable F+ bacteria to transfer a partial (single-stranded) copy of their chromosome to F− cells, which do not possess pili. F-pili are made from a single protein polymerized into a long (1 to 2 μm), ribbon-like structure that protrudes from the cell. Virions attach to the side of the pilus via their maturation protein (Figure 5.3). Upon contact with the pilus, the maturation protein is cleaved into two fragments. This releases the RNA from the virion, and it becomes sensitive to RNAse degradation. How the RNA then gets inside the cell is not known. A likely possibility is that the pilus with the attached RNA retracts into the cell, dragging the RNA and the cleaved maturation protein along with it.

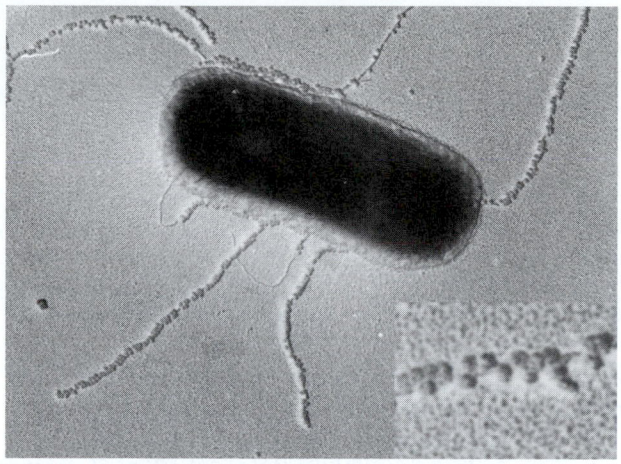

FIGURE 5.3 **Binding of RNA phage to F-pili.** Electron micrograph of a male *E. coli* cell showing numerous MS2 phages attached to F-pili (enlarged in inset). Normally, only one or a few phages bind and initiate infection.

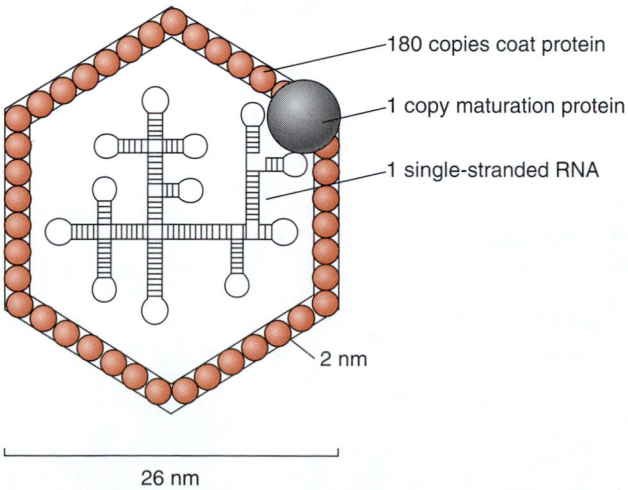

FIGURE 5.2 **Diagram of an MS2 virion.** Note that the RNA is highly structured. Alloleviviruses have in addition about 12 copies of the readthrough protein in their capsid.

Phage RNA is translated and replicated in a regulated fashion

Once inside the cell, the phage RNA begins to function as a messenger RNA for the synthesis of phage proteins. This turns out to be a highly regulated process, for two reasons. First, different amounts of each protein are needed. For every 180 copies of the coat protein, the phage needs only 1 copy of the maturation protein to construct virions, and a few copies of the lysis protein to leave the cell. Also, since a single replicase protein can make multiple copies of RNA, fewer replicase proteins are needed than coat proteins.

Second, replication and translation of the same RNA molecule can lead to problems. Replication starts at the 3′ end of the RNA and proceeds toward the 5′ end. Translation, on the other hand, moves in the opposite direction on the RNA. If the phage had not made the proper arrangements, the replicase and the translating ribosome would meet somewhere on the RNA and sit facing each other forever (Figure 5.4)! Thus a molecule of phage RNA that begins to be translated must not be allowed to begin replication, and vice-versa.

To regulate translation, the access of ribosomes to the AUG start codons of phage genes is strongly restricted. In fact, only the coat gene translational start site (nt 1335 on MS2 RNA) is able to bind directly to ribosomes. Experiments have shown that if the start codon of the coat gene is deleted, preventing ribosomes

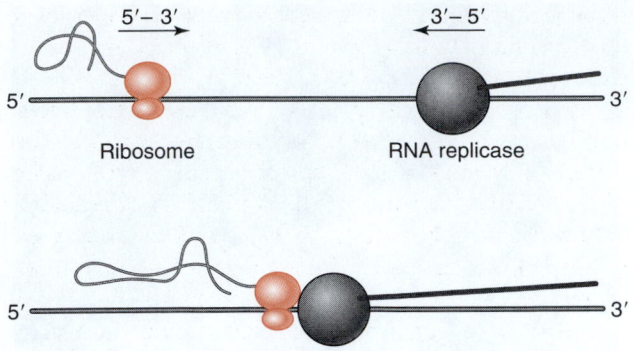

FIGURE 5.4 Collision between ribosome and RNA polymerase. Shown is an encounter between a ribosome translating an RNA in the 5′ to 3′ direction, and an RNA polymerase making a copy of the RNA in the 3′ to 5′ direction.

from translating this gene, there is no synthesis of either the lysis or the replicase protein.

RNA secondary structure controls translation of lysis and replicase genes

How is the access of ribosomes to the start sites of the lysis and replicase genes made dependent on translation of the coat gene? The general answer is: by changing RNA secondary structures. To understand this it is useful to consider the forces that drive the binding of ribosomes to translational start regions on prokaryotic messenger RNAs. There are at least three contributors to this binding energy: (1) base complementarity between 16S ribosomal RNA and the **Shine-Dalgarno sequence** just upstream of the start codon on the messenger RNA; (2) interaction of the anticodon on the initiator fmet-tRNA with the AUG start codon on the messenger RNA; and (3) binding of ribosomal protein S1 to pyrimidine-rich sequences frequently found upstream of the Shine-Dalgarno sequence.

If a strong secondary structure in the messenger RNA prevents interaction with one of the above components, ribosomes will not efficiently bind and therefore protein synthesis will not begin at that site. Secondary structure in RNA consists of regions of self-complementary base sequence that form stem-loop structures. RNA phage genomes contain extensive regions of secondary structure. In fact, the beginning of the lysis and replicase genes lie within RNA secondary structures that are too stable to allow ribosome binding (Figure 5.5), and this is the reason that they are not translated on their own.

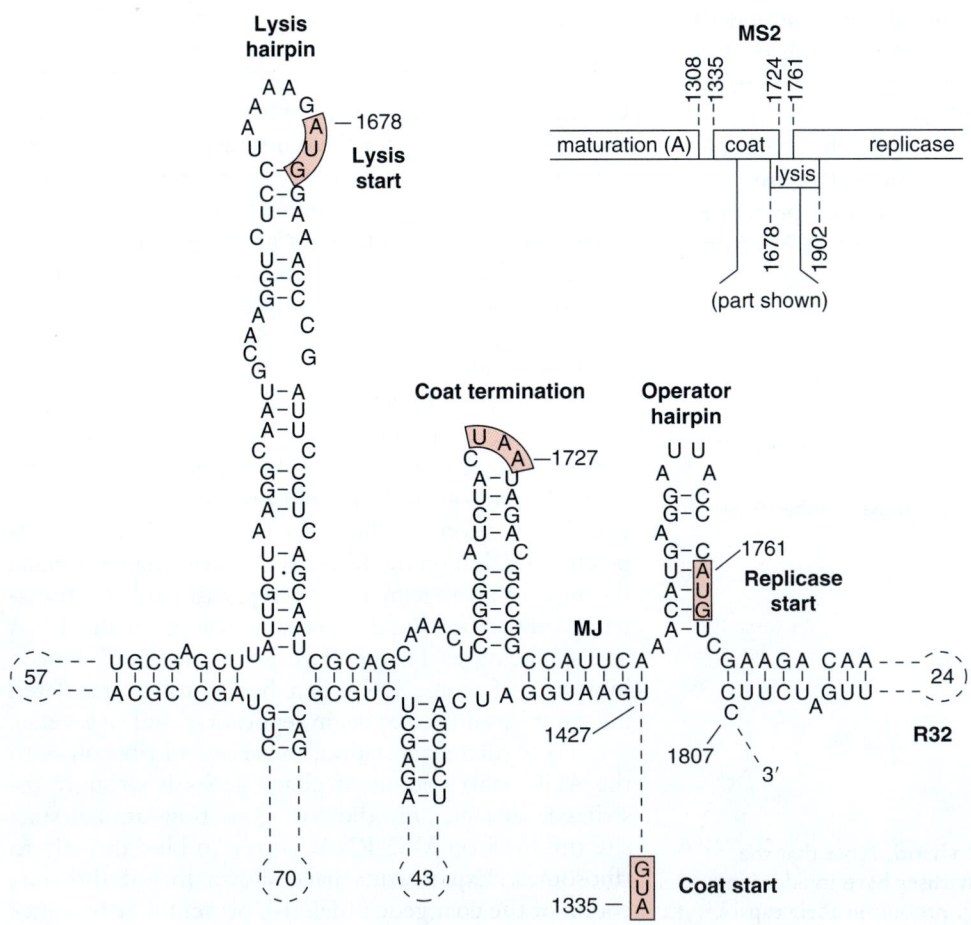

FIGURE 5.5 Translational control by secondary structure. Secondary structure of MS2 RNA between nts 1427 and 1807. Start and stop codons are shaded in orange. Regions irrelevant for translational controls are shown as dashed lines; numbers in dashed hairpins show how many nucleotides have been omitted.

Ribosomes translating the coat gene disrupt secondary structure, allowing replicase translation

Ribosomes contact the messenger RNA over a stretch of about 20 nt upstream and 15 nt downstream from the initiator AUG. Figure 5.5 shows that three regions of secondary structure near the replicase start site could contribute to impeding the ribosome from binding to that site. These are stem MJ, the "operator hairpin" containing the AUG itself, and stem R32. Mutagenesis experiments showed that when base-pairing at stem MJ is abolished by the introduction of mismatches or by deleting the sequence 1427–1434, there is a large increase in replicase synthesis, which is independent of translation of the coat gene. Apparently, the remaining operator and R32 hairpin structures together are not strong enough to block the entry of ribosomes.

The coupling between translation of the coat and replicase genes works like this. Every time a ribosome reads the coat gene, it disrupts base-pairing at stem MJ as it passes through that part of the coat gene open reading frame (Figure 5.6). Once this happens, other ribosomes can bind to the replicase start site and initiate translation there, even though this site is some 340 nt downstream from the position where the ribosome is translating the coat protein gene. Further support for this model comes from the finding that introduction of translational stop codons into the coat protein gene upstream of nt 1427 inactivates translation of the replicase gene (the ribosome never gets to stem MJ). However, stop codons placed beyond nt 1434 allow replicase synthesis to proceed at wild type levels.

There is a second level of control of the translation of replicase. When the concentration of coat protein becomes sufficiently high in the cell, dimers are formed. These dimers bind to the operator hairpin, precluding further translational starts of the replicase gene by excluding ribosome binding. This protein-RNA interaction is believed to lead to capsid formation by serving as a nucleation site for the further addition of coat dimers.

Ribosomes terminating coat translation can reinitiate at the lysis gene start site

Independent access of ribosomes to the lysis gene start site (nt 1678) is precluded by the "lysis hairpin" (Figure 5.5). In the absence of coat gene translation, the lysis gene is not expressed. Mutations that destabilize the lysis hairpin by introducing base mismatches allow lysis gene translation independent of the coat gene. However, activation of the lysis gene is not directly coupled to opening of this hairpin by translation of the overlapping coat gene, as with the replicase gene. Instead, initiation of translation at the lysis gene depends on termination of translation at the end of the coat gene, 50 nt downstream (Figure 5.5). If this UAA stop codon and the UAG stop codon that immediately follows it are mutated so that termination does not take place until the next UGA codon (25 nt farther along the RNA),

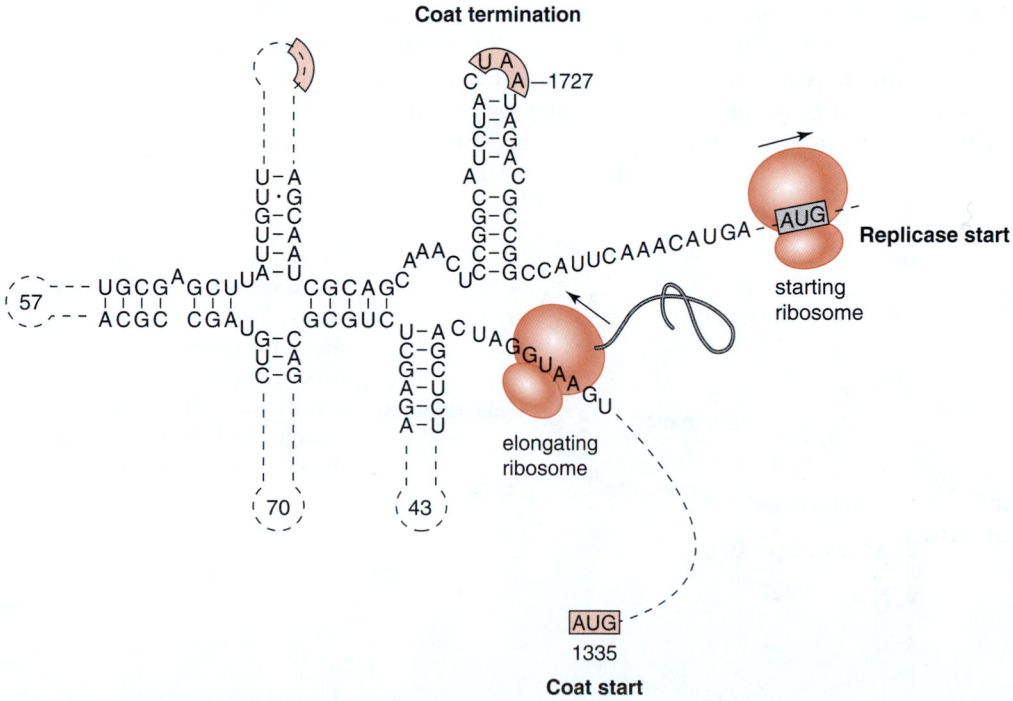

FIGURE 5.6 **Activation of replicase translation.** A ribosome translating the coat gene disrupts the MJ interaction, thus opening up the replicase start site for ribosome binding.

there is no expression of the lysis gene. On the other hand, if stop codons in the coat gene are introduced by mutagenesis between nt 1678 and 1725 or even somewhat upstream of the lysis protein start codon, lysis protein is made. In fact, the closer the coat gene stop codon is to the lysis start codon, the more lysis protein is made.

These results suggest that a ribosome that has reached the coat stop codon and released the coat protein does not immediately leave the mRNA, but makes a short random walk along the RNA in either direction. If a start codon is encountered, the ribosome may reinitiate translation (Figure 5.7); if not, it will eventually detach from the mRNA. From the proportion of lysis to coat protein synthesized by phage MS2, the probability for such successful reinitiation is only 5%.

This "scanning" model was further tested by introducing an additional start codon a short distance downstream from the authentic lysis start. The new start codon is used in place of the authentic start codon, as it is closest to the termination codon and thus forms a barrier that stops ribosomes from reaching the authentic start site. This shows that ribosomes look for the lysis start codon from the 3′ direction.

However, when a termination codon for coat gene translation was engineered upstream of the lysis start codon, the authentic lysis initiation site was again used, because it is closer to the termination site. This shows that ribosomes that have recently terminated can search for nearby start codons in both the 5′ and 3′ directions. *In vitro* experiments have shown that on short messenger RNAs containing only 5 codons, the ribosome shuffles back and forth between the start and stop codons, making multiple copies of a pentapeptide without ever leaving the template. This is a rarely used mechanism of initiation of protein synthesis. Normally when ribosomes come to the end of a coding region they are released from the messenger RNA if no reinitiation sites are nearby.

Replication versus translation: competition for the same RNA template

Once replicase is made, it can begin to generate new copies of phage RNA. The plus-strand genome is first copied into a complementary minus-strand RNA. This RNA is in turn used as a template to produce more plus-strand RNAs. Although they are fully complementary, plus and minus strands do not anneal to form a double-stranded RNA. Such annealing is inhibited by the high degree of internal secondary structure found in each single-stranded RNA.

How does the replicase copy the genome RNA starting at its 3′ end while ribosomes are translating the same RNA in the opposite direction (Figure 5.4)? The switch from translation to replication works as follows. Replicase binds to the RNA at two internal positions, one of which is the start of the coat gene. As a result there is competition between replicase and ribosomes for this site. If the replicase arrives first, there will be no new translation, allowing replicase to copy the RNA with no oncoming traffic. On the other hand, if the ribosome binds first to the coat protein gene, replicase will not be able to bind to its template, and therefore the ribosome is free to complete its voyage unhindered.

Genome replication requires four host cell proteins plus the replicase

How the replicase actually binds to the coat protein start site is fascinating. When scientists isolated the RNA polymerase activity from infected cells, they found to their surprise that it is a complex made of four proteins, only one of which (the replicase protein) is coded by the phage genome. The three other proteins in the complex are cellular proteins: the S1 protein of the small ribosomal subunit, and two translational elongation factors called EF-Tu and EF-Ts (Table 5.1). The active polymerase

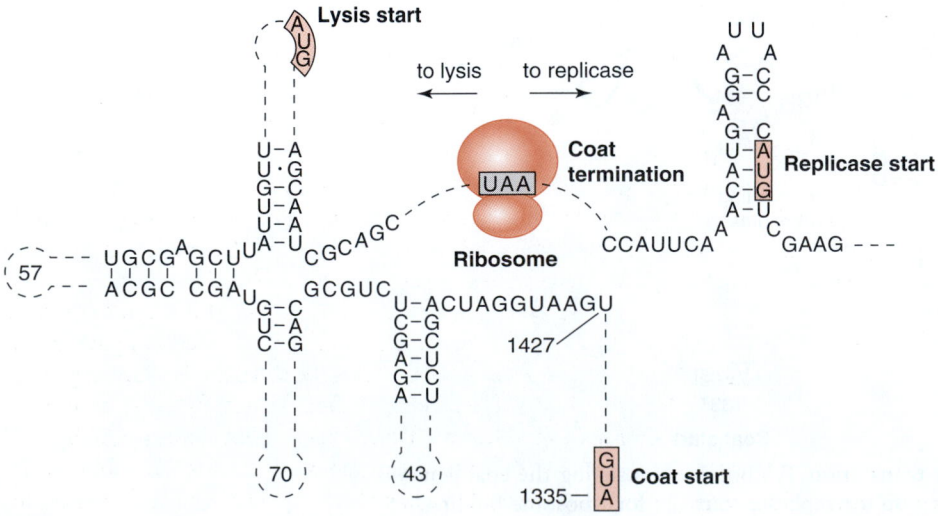

FIGURE 5.7 Termination of translation followed by reinitiation. A ribosome translating the coat gene arrives at its stop codon, releases coat protein, and carries out a short random walk during which it may encounter the lysis or replicase start sites.

TABLE 5.1 Proteins needed for Q_β RNA replicase activity

Protein	Coded by genome of	*Required for synthesis of:* Minus strand	Plus strand	Function
Replicase	Q_β	Yes	Yes	RNA-dependent RNA polymerase
EF-Tu	*E. coli*	Yes	Yes	Initiation of RNA synthesis by bound GTP
EF-Ts	*E. coli*	Yes	Yes	
S1	*E. coli*	Yes	No	Binding to plus strand RNA; inhibition of ribosome binding to coat start site
Host Factor	*E. coli*	Yes	No	Unwinding of 3′ terminal hairpin on plus strand RNA

complex contains one copy of each of the four proteins (we will refer to the active enzyme as *polymerase* while reserving the term *replicase* for the phage-coded subunit).

The three host cell proteins are crucial factors in the cellular translational apparatus. The S1 protein is instrumental in binding messenger RNA sequences to the ribosome during initiation of translation in bacteria. EF-Tu positions charged tRNAs on the ribosome during elongation of growing peptide chains. This activity involves binding of GTP and its hydrolysis to GDP. EF-Ts recycles the resulting EF-Tu/GDP to EF-Tu/GTP; an intermediate in this process is a complex of EF-Tu and EF-Ts.

A host ribosomal protein directs polymerase to the coat start site

It turns out that binding of the polymerase to the coat protein start site is facilitated by the S1 protein, which carries out the same function for the ribosome when it is binding to the coat protein start site. Thus the competition between ribosomes and replicase for the coat protein start site is in fact mediated by the same (cellular) protein!

Once the polymerase complex binds, assuring absence of ribosomes on the RNA it is about to copy, it looks for the 3′ end of the RNA. This requires a fifth protein, also a host cell protein called Host Factor. Host Factor is a poly(A) binding protein that, in the uninfected cell, stimulates translation of the mRNA of a transcriptional sigma factor. This protein, active as a hexamer, does not associate with the replicase but acts directly on the phage RNA and is thought to expose the 3′ terminal residues (CCA_{OH}), which are base-paired to nearby G residues. If two more Cs are added at the 3′ end of phage RNA ($CCCCA_{OH}$) the CCA_{OH} sequence will stick out. This mutant is no longer dependent on Host Factor for its replication.

Polymerase skips the first A residue but adds a terminal A to the minus strand copy

Surprisingly, polymerase initiates RNA synthesis not at the 3′ terminal A, but at the second nucleotide, a C. Thus the 5′ end of the new minus strand starts with a G residue. Here is where EF-Tu may act. Since EF-Tu has a GTP binding site, it could bring GTP to the replicase to initiate RNA synthesis. Therefore, EF-Tu would be acting as a kind of primer for the initiation of minus-strand RNA synthesis. The role of EF-Ts remains obscure.

Once synthesis has begun, it can continue in the absence of S1, EF-Tu, and EF-Ts, since the replicase protein has RNA polymerase activity by itself (but the complex usually stays intact). When the polymerase reaches the end of the plus-strand template, it completes the minus strand by copying the G's at the 5′ end of the plus strand into C's. Then the replicase adds on a single, untemplated A residue—so the 3′ end of the minus strand looks like the initial 3′ end of the plus strand, C's with a terminal A.

Synthesis of plus strands is less complex and more efficient than that of minus strands

The next step is to copy the minus-strand RNA into plus strands. This does not require either S1 or the Host Factor. The polymerase is not competing with ribosomes for binding to the minus-strand RNA, explaining the lack of requirement for S1. The Host Factor is not needed because the 3′ end of the minus-strand RNA is apparently not hidden in a stem structure and is therefore available as a template. Perhaps for all these reasons, plus-strand synthesis is more efficient than minus-strand synthesis, and therefore more plus strands are made. This makes sense, for the phage encapsidates only plus-strand RNA, and only this RNA can be translated to make more phage proteins.

A final question is why the Q_β plus strand conceals its 3′ end even though replication needs unpaired nucleotides to initiate. The best guess is that any protruding nucleotides will be a target for host exonucleases (tRNA, which does have a protruding CCA_{OH} 3′ end, is continuously repaired by tRNA nucleotidyl transferase). One has also to bear in mind that the plus

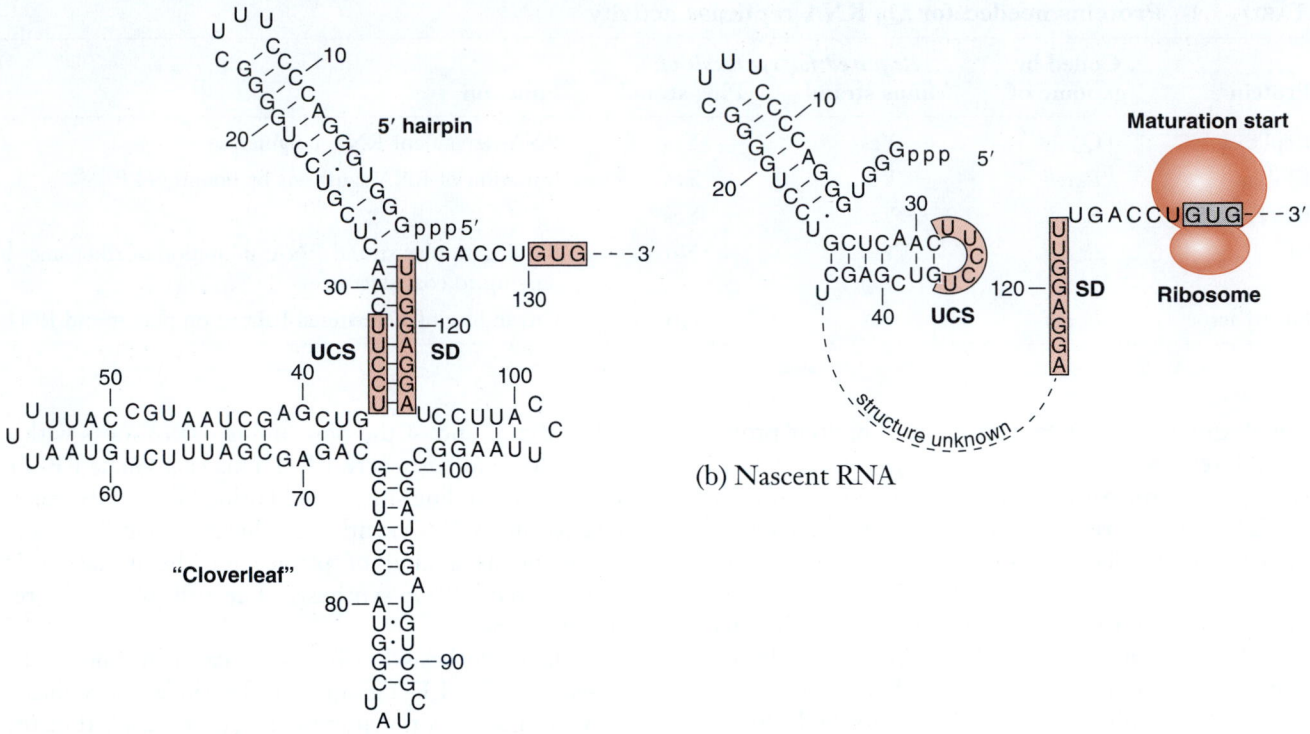

FIGURE 5.8 Translation of maturation gene on nascent RNA. (a) Equilibrium folding of the 5′ untranslated region of MS2 RNA. The secondary structure formed between the upstream complementary sequence (UCS) and the Shine-Dalgarno sequence (SD) does not allow ribosome binding. (b) Folding intermediate of the same region on nascent RNA, allowing translation of the maturation gene. The loop of the folding intermediate (kinetic trap) is shaded in orange.

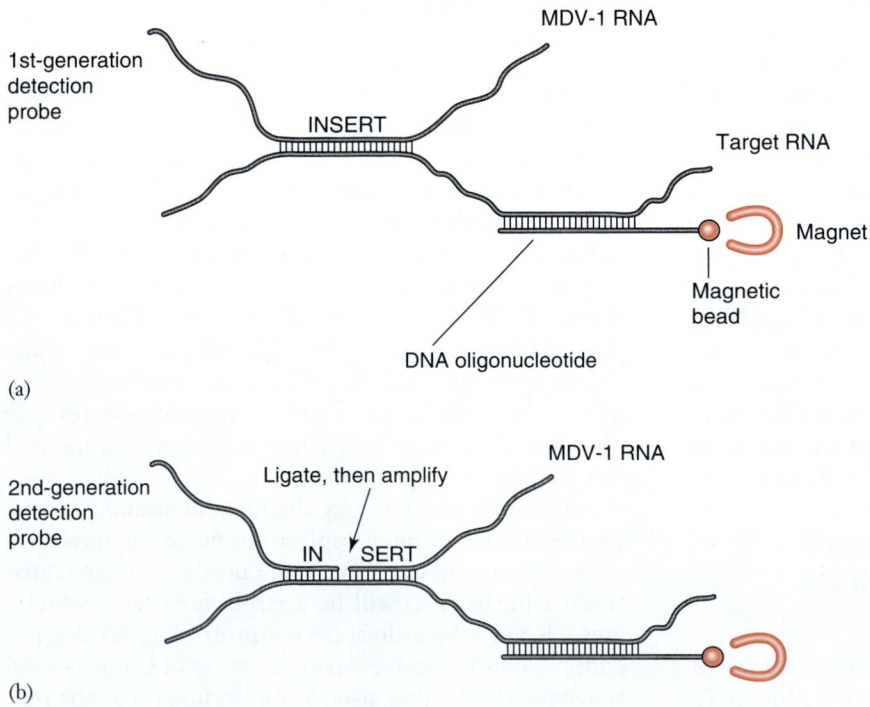

FIGURE 5.9 Detection of minute amounts of target RNAs using Q_β replicase. See text for details.

strand needs extra protection because it is the infectious agent. A single molecule must survive during infection long enough to be translated and then replicated at least once. The minus strand, which does have protruding C's, may occasionally perish. It can always be replaced by a new transcript.

The start site for synthesis of maturation protein is normally inaccessible to ribosomes

So far we have ignored translation of the maturation protein, whose coding region lies near the 5′ end of the phage RNA. Figure 5.8a shows the secondary structure of the 5′ untranslated region of MS2 RNA, just upstream of the maturation start codon. A hairpin at the 5′ end is followed by a "cloverleaf" (three "leaves," one stem) structure in which the Shine-Dalgarno sequence (SD) of the maturation gene is base-paired to an upstream complementary sequence (UCS). This long-distance interaction effectively prevents ribosome binding. How then is the maturation protein made? The answer is that ribosomes only bind to the maturation start site on RNA chains that are in the process of being synthesized, that is, "nascent" chains. Furthermore, these nascent RNAs must not yet have reached their equilibrium folding configuration.

Consider a growing plus-strand RNA being made by polymerase from a minus-strand RNA template. By the time nt 123 is added, all nucleotides needed to build the cloverleaf shown in Figure 5.8a have been incorporated. However, translation cannot yet start because ribosome binding needs the start codon and sequences downstream of the initiator codon, up to about nt 140–145. It is known that small RNAs such as tRNA (~75 nts) fold up correctly within milliseconds after denaturation. On the other hand, phage RNA replicases incorporate approximately 35 nucleotides per second into a growing chain. Thus one would expect the cloverleaf structure to be formed long before the ribosome binding site is made, as addition of a further 20 nucleotides beyond nt 123 could take as long as half a second.

Synthesis of maturation protein is controlled by delayed RNA folding

However, the folding of the 5′ untranslated region of MS2 RNA to its final equilibrium configuration is unusually slow; *in vitro* experiments show that it takes minutes to refold the heat-denatured cloverleaf structure. Generally speaking, such a delay can be caused only by a "kinetic trap." This is an alternative folding configuration that is stable enough to temporarily prevent the RNA from reaching the equilibrium structure.

Mutational analysis has indicated that the kinetic trap involves the small hairpin between nts 25 and 43 shown in Figure 5.8b. Formation of this hairpin excludes the inhibitory long-distance base-pairing between the upstream complementary sequence and the Shine-Dalgarno sequence, thus permitting ribosome binding. However, this secondary structure is not highly stable; the RNA will eventually fold up into the structure shown in Figure 5.8a, and translation will be shut

Detection of minute amounts of RNAs by using the power of Q_β replicase

Among all the single-stranded RNA phages, Q_β is famous for its replicase. This is the best-characterized RNA-dependent RNA polymerase and can be readily purified. In addition to Q_β RNA itself, a number of small (100–200 nt) RNAs can be efficiently replicated *in vitro* by this RNA polymerase. Many of these evolved spontaneously in cells infected with the phage or in amplification reactions carried out *in vitro*.

One of these templates is called midi variant-1 (MDV-1). This RNA can be amplified *in vitro* some 10^{11} times in about 30 min by Q_β replicase. Several groups are trying to exploit this amplification capacity to detect minute amounts of DNA or RNA molecules in clinical samples, for instance HIV-RNA. The concentration of these molecules is usually too small to be detected by conventional techniques. The principle is outlined below (Figure 5.9).

MDV-1 RNA is first modified to contain a short sequence complementary to the target RNA; this sequence does not interfere with the replication of the RNA by Q_β polymerase. RNAs can be produced by *in vitro* transcription of plasmids containing cloned copies of the modified MDV-1 RNA, using bacteriophage T7 RNA polymerase. The target RNA in the clinical sample is first hybridized to a complementary DNA oligonucleotide that carries a magnetic bead at one end, which allows binding to a solid support (the magnet). In this way the target can be separated from the bulk of the other RNAs in the sample.

The immobilized target RNA is then hybridized to the modified MDV-1 RNA (Figure 5.9a). After thorough washing to remove all unbound RNAs, the hybridized MDV-1 RNA is released by denaturation and amplified by Q_β polymerase. The amplified RNA can be detected by gel electrophoresis and ethidium bromide staining. The presence of the target RNA is therefore signaled by the appearance of a band of MDV-1 RNA in the gel.

False positive reactions can arise if there is partial homology between the MDV-1 RNA and some other cellular or viral RNA in the sample. To reduce such reactions, MDV-1 RNA can be added as two half molecules (Figure 5.9b), which are joined when properly aligned on the target RNA by adding RNA ligase to the reaction. Since Q_β replicase needs specific sequences at both ends of the RNA to replicate RNAs, half-molecules by themselves cannot be replicated. Any partial hybridization of the insert to an RNA molecule other than the target would be very unlikely to align both ends of the MDV-1 RNA half-molecules in such a way that ligation could occur. The ligated MDV-1 RNA is then released and amplified by Q_β polymerase.

down. It is not difficult to see what the biological purpose of this control is; it ensures that the maturation gene will be translated only once or a few times per newly synthesized plus-strand phage RNA. Since one molecule of maturation protein is packaged per plus-strand RNA molecule, this mechanism assures the appropriate level of its synthesis. Furthermore, since maturation protein cannot be translated from the full-length genome, replicase needs only to compete with ribosomes that bind at the coat start site to create a ribosome-free template, which is needed for unimpeded synthesis of minus-strand RNAs.

Assembly and release of virions

The assembly of virions has not been studied in great detail. It is known that the first step in this process must be the binding of the maturation protein to the RNA. This complex can be made *in vitro* and is infectious for F+ cells; thus it can bind to pili in the absence of the capsid, and direct the phage RNA into the cytoplasm.

The second step in virion construction is the assembly of the coat protein around the RNA to form an icosahedral shell. The nucleation point for this coating reaction is the replicase operator hairpin (Figure 5.5). The coat protein will form capsids even in the absence of phage RNA. However, capsid formation occurs at a much lower concentration of coat protein in the presence of phage RNA, assuring that empty capsids will not form and all phage RNA becomes packaged.

The infectious cycle ends when the accumulation of the lysis protein in the cytoplasmic membrane has caused the collapse of the cell wall. This 75–amino acid hydrophobic protein short-circuits the membrane potential, eventually leading to degradation of peptidoglycan. Electron micrographs show that only a very small section of the sacculus (the cell wall network) has dissolved, usually at the equatorial growth zone.

SUGGESTED READING

1. Furuse, K. (1987). Distribution of coliphages in the environment: general considerations. In Goyal, S.M., Ed., Phage ecology, p. 89. John Wiley & Sons, Inc., New York.

2. Havelaar, H.A. (1991). Bacteriophages as model viruses in water quality control. IAWPRC Study Group on Health Related Water Microbiology. *Water Research* **25,** 529–545.

3. Olsthoorn, R.C.L., Licis, N., and van Duin, J. (1994). Leeway and constraints in the forced evolution of a regulatory RNA helix. *EMBO Journal* **13,** 2660–2668.

4. Schuppli, D., Miranda, G., Tsui, H.-C.T., Winkler, M.E., Sogo, J.M., and Weber, H. (1997). Altered 3′ terminal RNA structure in phage Q_β adapted to Host Factor-less *Escherichia coli*. *Proceedings of the National Academy of Sciences of the United States of America* **94,** 10239–10242.

5. Valegård, K., Murray, J.B., Stockley, P.G., Stonehouse, N.J., and Liljas, L. (1994). Crystal structure of an RNA bacteriophage coat protein-operator complex. *Nature* **371,** 623–626.

6. Van Duin, J. (1988). The bacteriophages. In Fraenkel Conrat, H. and Wagner, R.R., Eds., The viruses, vol. 1, pp. 117–167. Plenum Press, New York.

KEY TERMS

F-pili

Shine-Dalgarno sequence

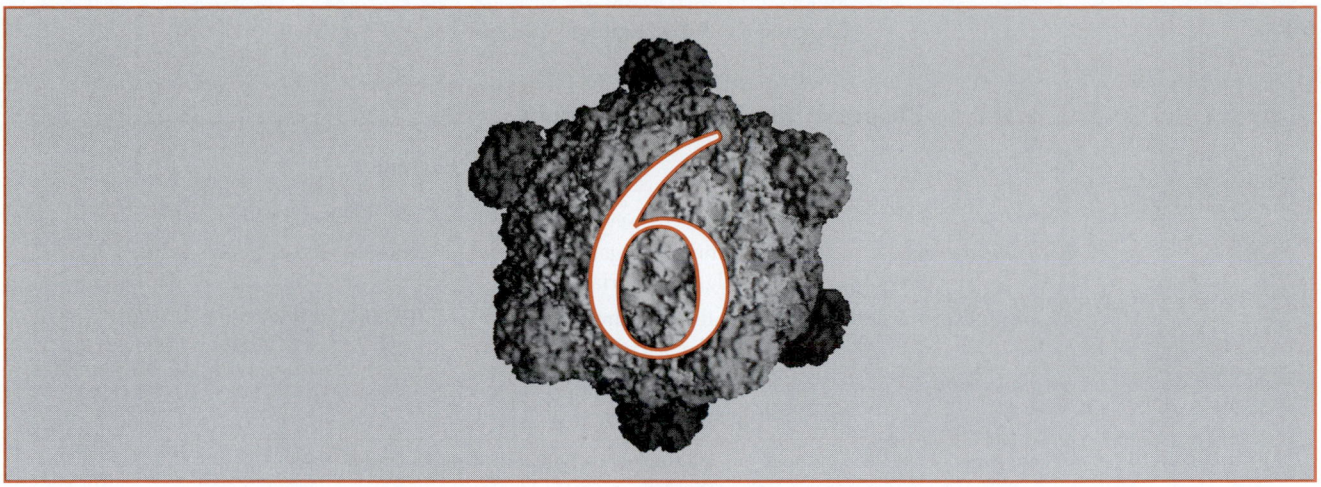

Bacteriophage φX 174

Microviridae

From Greek *micro* (small), referring to size of phage

VIRION

Naked icosahedral capsid (T = 1).
Diameter 33 nm.
Twelve spikes project from surface.

GENOME

Circular ss DNA, 4.4–6.1 Kb

GENES AND PROTEINS

Microviruses: 3 transcriptional promoters and 4 terminators control gene expression.
Code for 11 proteins including:

- *2 DNA replication/packaging proteins*
- *2 scaffolding proteins*
- *4 virion proteins*
- *1 lysis protein*

VIRUSES AND HOSTS

Microviridae: two subfamilies

- *Microviruses infect free-living bacteria.*
- *Gokushoviruses, from Japanese* gokusho *(very small), infect obligate intracellular parasitic bacteria.*

Inoviridae: a distinct and unrelated family of ss DNA phages with filamentous capsids. Examples: fd, M13.

DISTINCTIVE CHARACTERISTICS

Genomes contain extensive overlapping reading frames.
DNA delivered into cell through spikes on virion surface.
DNA becomes double-stranded before being transcribed.
Rolling circle DNA replication.
Scaffolding proteins used for virion assembly.
DNA inserted into procapsids during synthesis.

φX174: a tiny virus with a big impact

Although viruses were first discovered and distinguished from other microorganisms by Martinus Beijerinck and Dimitrii Ivanowski in the late 1890s (tobacco mosaic virus), the first bacteriophage was not identified until 1917 by Felix d'Herelle. Throughout the 1920s and 1930s, many different bacterial viruses were identified and classified, including the *Microviridae* (Table 6.1). The name of the best known family member, φX174, reflects its history. Without electron microscopes, Sertic and Bulgakov were able to distinguish different

TABLE 6.1 Phages in the *Microviridae* family

Genus	Hosts	Examples
Microvirus	*E. coli*, Shigella, Salmonella	φX174, G4, α3, φR
Bdellomicrovirus[a]	Bdellovibrio (obligate intracellular bacteria)	φMK2K
Chlamydiamicrovirus[a]	Chlamydia (obligate intracellular bacteria)	CHP1 and CHP2
Spiromicrovirus[a]	The Mollicute bacteria, Spiroplasma	SpV4

[a]These genera have been grouped into a distinct subfamily, called the Gokushoviruses.

groups, or "races," of bacteriophage, primarily by plaque morphology and ultrafiltration. The φ (Greek letter *phi*) in φX174 refers to "phage"; the X is the Roman numeral 10; "race X" bacteriophages were unusually small, as determined by ultrafiltration, and the first isolate was placed in a test tube labeled 174.

This small virus would have a big impact on molecular biology. Recognizing its unique properties, Robert Sinsheimer began detailed investigations in the late 1950s. The first surprise was the discovery that its genome consisted of single-stranded DNA; other known phage genomes were made of double-stranded DNA, as are all genomes of cellular organisms. Masaki Hayashi and colleagues defined the φX174 assembly pathway, and demonstrated that viral DNA could be packaged into capsids in vitro. Using φX174, Arthur Kornberg and co-workers elucidated many of the fundamental mechanisms of DNA replication.

In 1978, Fred Sanger and colleagues were the first to determine the nucleotide sequence of the entire genome of an organism, and that organism was φX174. The results were astounding: many of the genes overlapped! One region of DNA could even encode part of four different genes (Figure 6.1). This finding so challenged the concept of the gene, reported the *New York Times*, that some scientists were searching for hidden messages in the DNA sequence, hypothesizing that φX174 was a genetically engineered extraterrestrial organism! More recently, the atomic structure of the φX174 **procapsid**, an assembly intermediate containing a full complement of **scaffolding proteins**, was solved in 1997 by a collaborative effort of the laboratories of Michael Rossmann, Bentley Fane, and Nino Incardona.

Overlapping reading frames allow efficient use of a small genome

Seven of the 11 φX174 genes (genes A through E and gene K) lie in a region of extensive overlapping reading frames that covers nearly one-half the genome (Figure 6.1). Most of these genes code for nonstructural proteins (Table 6.2). The remaining four genes (F through J) code for the structural proteins of φX174, and do not overlap. The use of overlapping reading frames can be understood as a way of increasing the amount of genetic information encoded in a small viral genome. The acquisition of new genes is achieved by nesting them within

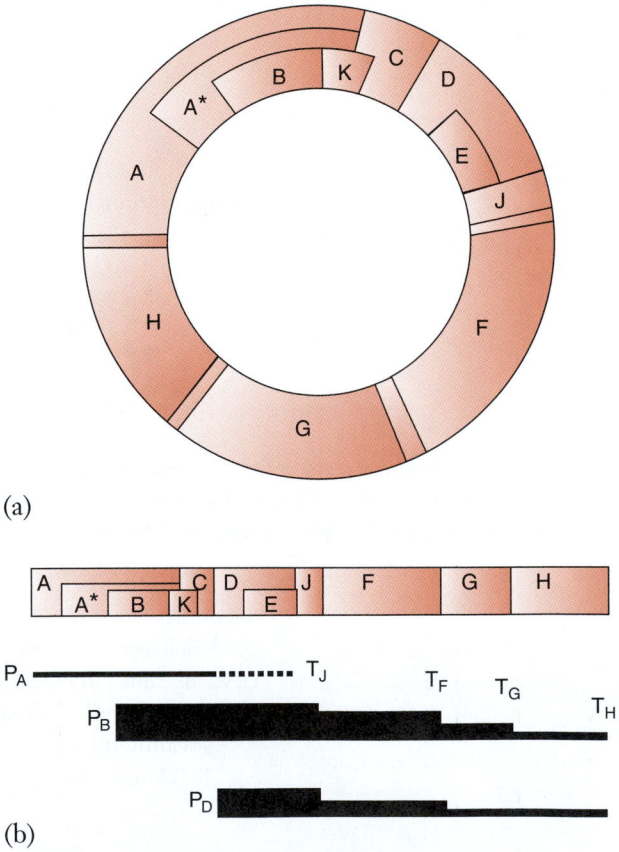

FIGURE 6.1 The φX174 genome. (a) Map of genes on the circular DNA genome (see Table 6.2 for functions of gene products). (b) Transcriptional map. Promoters (P) and transcription terminators (T) are indicated on a linearized map of φX174 DNA. The major transcripts are illustrated below. Line thickness indicates the relative abundance of the transcripts.

TABLE 6.2 Gene products of φX174

Protein	Function
A	DNA replication. Cleaves plus-strand viral DNA at a specific site, initiating rolling circle replication. Ligates linear plus-strand DNA to form full-length single-stranded circles.
A*	Not essential for virus growth. May play a role in the inhibition of host cell DNA replication and in exclusion of superinfecting phage.
B	Internal scaffolding protein, required for procapsid morphogenesis; 60 copies in procapsid.
C	DNA packaging. Binds to Stage II DNA replication complex, directing concomitant stage III replication and entry of plus-strand viral DNA into the procapsid.
D	External scaffolding protein, required for procapsid morphogenesis; 240 copies in procapsid.
E	Host cell lysis.
F	Major capsid protein; 60 copies in virion.
G	Major spike protein; 60 copies in virion.
H	DNA pilot protein, required for DNA delivery; also called the minor spike protein; 12 copies in virion.
J	Small DNA-binding protein, needed for DNA packaging; 60 copies in virion.
K	Not essential for virus growth. May play a role in the optimization of burst size in various hosts.

the sequences of preexisting genes rather than inserting additional DNA sequences in the genome.

The A gene overlaps with genes A*, B, and K. The A protein plays a critical role in genome replication and interacts with many host cell proteins. Therefore, it must be able to co-evolve with its host. Adaptive mutations in this gene can simultaneously create mutations in the overlapping genes and change their functions. However, the A* and K proteins are not essential for virus replication, and the B protein has a flexible structure that is highly tolerant of amino acid substitutions (up to 70% of its sequences can be altered with minimal effects on its functions). Similarly, the E gene, which overlaps with the essential D gene, leads to host cell lysis; it can be considered a "luxury" gene, whose protein product is not required for replication, but increases the efficiency of release of newly replicated phage.

φX174 binds to glucose residues in lipopolysaccharide on the cell surface

The icosahedral capsid (Figure 6.2) contains 60 copies of the viral capsid protein F. Each large protrusion found at the 12 icosahedral vertices is composed of five copies of the major spike protein G (total 60 copies) and one copy of the DNA pilot protein H. In addition, there are 60 copies of the DNA binding protein J within the virion.

φX174 attaches to host cells via sugar residues, most likely glucose, in the **lipopolysaccharide** that is a part of the outer membrane of gram-negative bacteria. A site on the surface of the capsid, near the threefold axes of symmetry, binds reversibly to glucose. The glucose-binding site was identified by accident: Large quantities of virions were purified by sucrose gradient sedimentation for crystallization and subsequent x-ray structure determination. The structure showed that sucrose molecules (a dimer of glucose and fructose) were bound via the glucose moiety to a distinct site on the major capsid protein F. Members of the Gokushovirus subfamily appear to recognize proteins rather than sugars on host cell surfaces.

FIGURE 6.2 The atomic structure of φX174. Spikes at fivefold symmetry axes are shown in orange.

φX174 delivers its genome into the cell through spikes on the capsid surface

When φX174 virions are placed in solutions of high ionic strength, the DNA genome is extruded from the capsid via the spikes, located at the axes of fivefold symmetry. This implies that during infection of *E. coli*, genome DNA may be delivered into the host cell through the spike. Host range mutations change amino acids primarily in the spike proteins G and H, indicating that the spikes are also involved in attachment to host cells, and arguing that a second host cell receptor may be required for DNA penetration. The identity of this second factor is unknown. However, penetration may follow a pathway similar to that of the large, tailed bacteriophage T4.

T4 interacts reversibly with lipopolysaccharide via its long tail fibers. The phage then "walks" along the surface of the cell, binding to lipopolysaccharide residues one after the other, until it finds a second receptor, which leads to delivery of the viral DNA. Instead of walking, the tailless, spherical φX174 may roll along the cell surface in a similar fashion. Electron micrographs show that the vast majority of adsorbed φX174 particles are imbedded at points of adhesion between the cell wall and the inner membrane. This suggests the location of the second receptor and indicates that the genome is delivered directly into the cytoplasm, where replication occurs, and not into the periplasmic space. Penetration requires the viral H protein, also called the DNA pilot protein, which is delivered into the cell along with the genome.

Stage I DNA replication generates double-stranded replicative form DNA

Because microviruses depend on the host RNA polymerase to synthesize their mRNA, they must first make a double-stranded DNA before any viral genes can be transcribed. This double-stranded DNA, called replicative form (RF) DNA, is synthesized during Stage I DNA replication (Figure 6.3). A host cell protein, single-stranded DNA binding protein (SSB), coats the phage DNA except for a specific short, highly base-paired hairpin region. That region serves as a signal for the assembly of a **primosome** by seven different host cell proteins. The primosome enables primase, one of its components, to generate short complementary RNA primers at different locations around the single-stranded phage DNA. These are then extended by cellular DNA polymerase III to make a complete complementary copy of the phage DNA. DNA polymerase I subsequently removes the RNA primers and fills in the gaps, and DNA ligase joins the DNA chains to form a covalently closed, circular, double-stranded RF DNA. The stage is now set for expression of phage genes that allow further genome replication and capsid production.

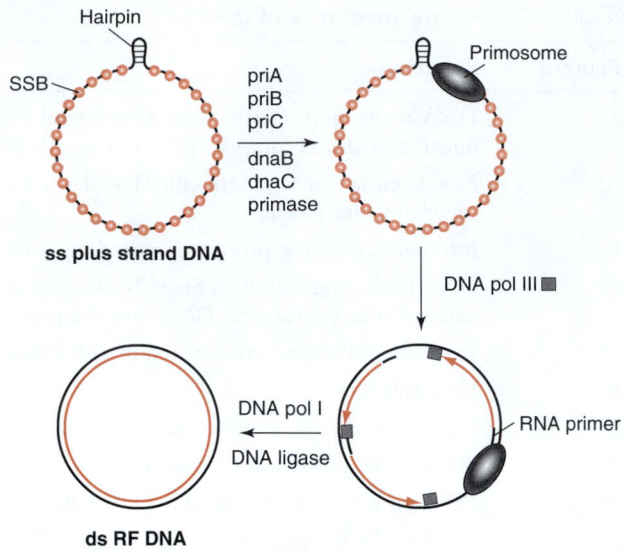

FIGURE 6.3 **Stage I DNA replication.** Formation of double-stranded replicative form DNA from single-stranded genome DNA, carried out entirely by host cell proteins. SSB: single-strand DNA binding protein; priA-C and dnaB-C: primosome proteins. Newly synthesized DNA strand shown in orange.

Gene expression is controlled by the strength of promoters and transcriptional terminators

Unlike most DNA viruses, microviruses do not produce regulatory proteins to ensure temporal control of gene expression. The production of different amounts of viral proteins is entirely dependent on transcriptional promoters, transcription terminators, and ribosome binding sites (Figure 6.1b). Three promoters (P_A, P_B, and P_D) and four terminators (T_J, T_F, T_G, and T_H) have been detected. None of the terminators is highly efficient, leading to the production of a variety of transcripts with the same 5′ ends but extending for different lengths.

In general, higher levels of messenger RNAs are made for proteins that are required in greater abundance. Transcripts initiated at P_A are unstable, keeping the levels of the A and A* proteins low; the terminator for P_A transcripts is unknown. Gene D mRNAs are more abundant than any other transcript, and more of this protein is needed than any other; 240 copies of the D scaffolding protein are required to build one virion. Similarly, there are more mRNAs that encode the J, F, and G proteins than mRNAs that encode the H protein, because transcription is successively reduced at a series of terminators at the ends of each of these four genes. The numbers of these proteins per virion are 60:60:60:12, respectively.

Regulation is also achieved on the translational level. Early and abundant synthesis of the E protein would lead to premature cell lysis, before sufficient infectious progeny could be made. Gene E resides in an overlapping reading frame with gene D, and therefore many E transcripts are made. However, few E proteins are synthesized because the E gene has a very weak ribosome binding site, and therefore initiation of protein synthesis is strongly reduced at this site. Furthermore, the presence of many nonoptimal codons in both the E and H genes most likely reduces the level of synthesis of these proteins.

Replicative form DNAs are amplified via a rolling circle mechanism

A single molecule of replicative form DNA cannot generate sufficient mRNAs and progeny single-stranded DNAs to construct the 100 phage particles made in a productive infection. Stage II DNA replication (Figure 6.4) amplifies the number of replicative form DNA molecules to about 10 per cell. This happens in two distinct steps. First, progeny plus-strand DNA is made by a **rolling circle** mechanism using replicative form DNA as a template. The viral A protein cleaves the plus strand on replicative form DNA at a specific site and is covalently linked to the resulting 5′ end via a tyrosine residue in the protein. The host cell rep protein acts together with the A protein as a **helicase** to unwind the plus strand from its association with its complementary minus strand. Simultaneously, host cell DNA polymerase III extends the new 3′ end created by cleavage of the plus strand, and thereby synthesizes more plus-strand DNA. The polymerase and helicase thus make their way around the circular double-stranded DNA, eventually passing through the site where the A protein originally cleaved the plus strand.

At this point, the A protein cleaves off a full-length plus strand and ligates it to form a single-stranded circle. The A protein is simultaneously transferred to the newly formed 5′ end resulting from this cleavage, recreating the original replication complex. The result is the synthesis of a progeny viral plus-strand DNA circle and regeneration of the double-stranded replicative form DNA each time the replicative machinery makes a full turn on the template DNA. The second phase of Stage II DNA replication is the formation of more replicative form DNA from the progeny plus-strand viral DNA. This is carried out by host cell proteins exactly as in Stage I DNA replication (Figure 6.3).

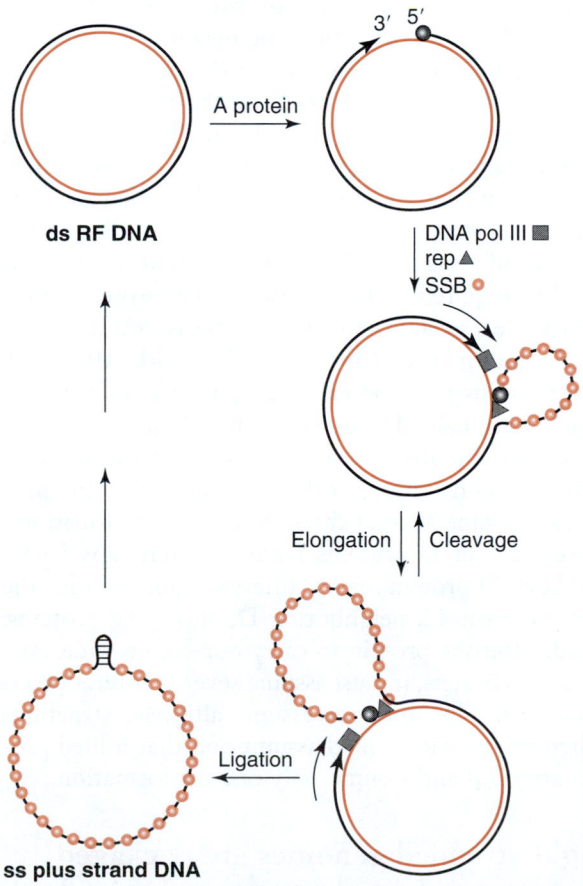

FIGURE 6.4 **Stage II DNA replication.** Rolling circle replication forms single-stranded DNA from double-stranded replicative form DNA, using viral protein A (black circle) and three host cell proteins.

Summary of viral DNA replication mechanisms

Figures 6.3 and 6.4 illustrate the contrasting mechanisms used for plus and minus strand synthesis. To summarize: minus-strand DNA is made on a plus-strand template by a complex RNA-primed mechanism utilizing a host cell primosome, DNA polymerase III, DNA polymerase I, and DNA ligase (Figure 6.3). The progeny minus-strand DNA remains associated with its template plus strand in a double-stranded replicative form. In contrast, plus-strand DNA is made from the double-stranded replicative form template by a continuous rolling circle mechanism utilizing the viral A protein, DNA polymerase III, the single-stranded DNA binding protein SSB, and rep helicase (Figure 6.4). The primer for elongation by DNA polymerase III is the 3′ end of plus-strand DNA cleaved by the viral A protein. Progeny plus-strand DNA molecules are cleaved from the growing strands on the rolling circle by the A protein, and are ligated to form single-stranded circles. The A protein remains on the replicative form DNA to help synthesize further progeny plus-strand DNA.

Procapsids are assembled by the use of scaffolding proteins

The assembly pathway of φX174 (Figure 6.5) was elucidated by analyzing the intermediate capsid structures that accumulate in cells infected with virus mutants. Small particles denoted "9S" and "6S" (named for their relative sedimentation rate in sucrose gradients) accumulate in cells infected with phage mutated in the internal scaf-

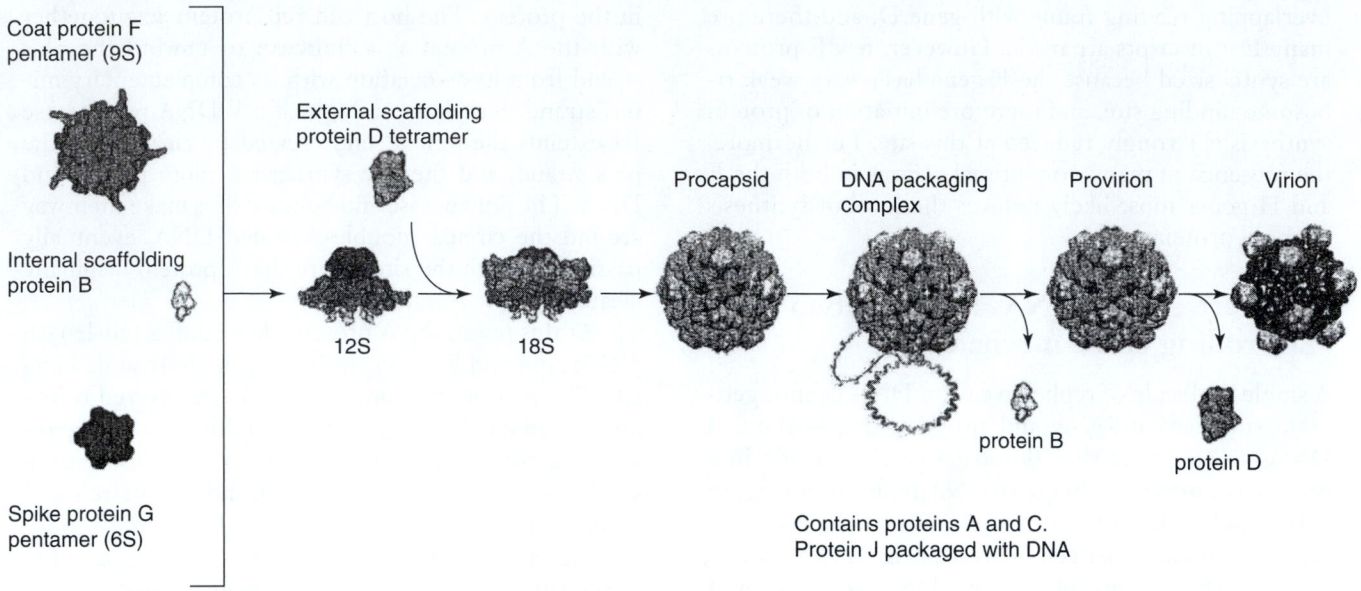

FIGURE 6.5 The φX174 morphogenetic pathway. Procapsids are assembled with the help of scaffolding proteins, and single-stranded genome DNA is inserted into procapsids during Stage III DNA replication.

folding protein B. The 9S particles are pentamers of the major capsid protein F, and the 6S particles are pentamers of the major spike protein G. This finding demonstrates that the 9S and 6S particles are the first intermediates in the assembly pathway and that the internal scaffolding protein B is required for capsid assembly to proceed. Much like the scaffolds used in construction, these scaffolding proteins facilitate capsid assembly but are removed when construction is complete.

When protein B is present, five copies will bind to the underside of a 9S particle, triggering conformational changes on the upper surface of the particle. These conformational changes allow the 9S capsid protein pentamers to interact with spike and external scaffolding proteins and also prevent the premature association of 9S particles into aggregates, a function reminiscent of molecular chaperones. The two pentamers are brought together, forming the 12S particle. The 18S particle results from the addition of 20 copies of the external scaffolding protein D (Figure 6.5). The scaffolding proteins within 18S particles subsequently mediate inter-pentamer contacts, organizing 12 pentameric particles into a "procapsid." Inside the procapsid, B proteins interact with each other across the twofold symmetry axes. On the outer surface of the procapsid, the D proteins self-associate across both the twofold and threefold axes of symmetry.

Scaffolding proteins have a flexible structure

Several lines of evidence suggest that scaffolding proteins have structures that allow a high degree of adaptability. For example, the internal scaffolding proteins of the related phages φX174 and α3 have diverged 70% in amino acid sequence, yet the proteins of each phage can efficiently function in assembly of the other. A similar phenomenon has been observed for the scaffolding proteins of various herpesviruses. These findings suggest that internal scaffolding proteins may be inherently flexible and may undergo folding directed by their contacts with the substrate structural proteins. Considering the dynamics of virus assembly, some inherent flexibility is probably required. Internal scaffolding proteins must first assume a structure that directs the assembly of pentameric intermediates into a rigid capsid. Afterward, these proteins must be extruded through holes in the capsid, presumably by assuming an alternate structure.

The atomic structure of the φX174 procapsid (Figure 6.6) demonstrates that the external scaffolding protein can assume at least three different conformations. There are four D proteins for every viral capsid protein. Each D protein makes different contacts with the underlying capsid, neighboring D, and spike proteins. In order for one protein to carry out such a wide variety of interactions, it must assume several different conformations. The ability to assume alternate structures challenges the widely held assumption that folded proteins are rigid and assume only one conformation.

Single-stranded genomes are packaged into procapsids as they are synthesized

Genome biosynthesis and packaging occur concurrently. In what is called Stage III replication, the viral

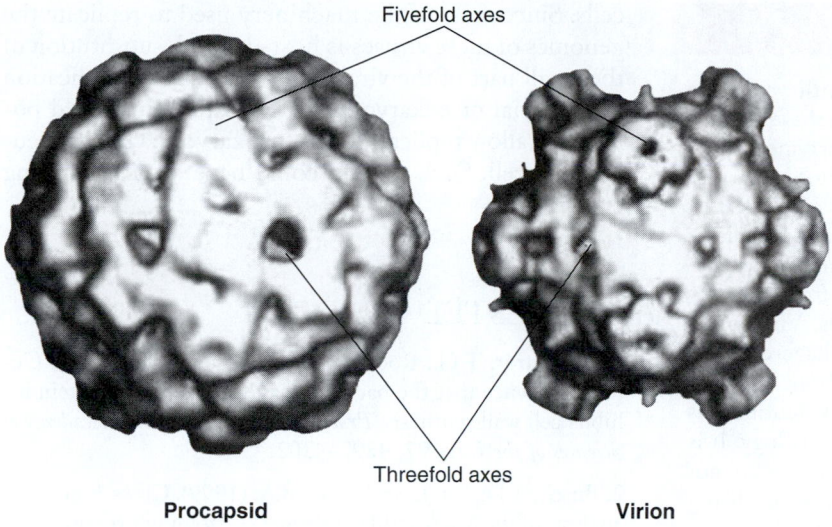

FIGURE 6.6 The φX174 procapsid and virion. Image reconstruction from cryo-electron microscopy.

C protein redirects plus strand synthesis by binding to the replicative complex (shown in Figure 6.4) consisting of double-stranded DNA, the A protein, the rep protein, and DNA polymerase III. Instead of releasing free progeny plus strands, this complex now associates with preformed procapsids (Figure 6.5) and inserts the plus-strand DNA into the procapsid as this DNA strand is displaced from the double-stranded template. The J protein (not SSB, as in normal replication) binds to the phage DNA being packaged and accompanies it into the procapsid. Packaging most likely occurs through one of the pores at the threefold axes of symmetry found in the procapsid but not in the virion (Figure 6.6). During packaging, B proteins are extruded from the procapsid, probably through the same pores.

After one round of replication, the viral A protein cleaves the plus-strand DNA and ligates the 3′ end of the cleaved genome to its 5′ end, creating a single-stranded circular molecule as in Stage II replication, but this time wholly contained within a procapsid. The replication complex, which retains proteins A and C, rep, and DNA polymerase III, can presumably bind to another procapsid to synthesize and package another phage genome DNA. The packaged particle is called the provirion because it still contains the external scaffolding protein D. The provirion may represent the end of the intracellular assembly pathway. Upon cell lysis, the influx of divalent cations probably dissociates protein D, yielding the virion.

Role of the J protein in DNA packaging

The J protein is not only required for DNA packaging but may also play a role in organizing the DNA within the symmetrical capsid. Microvirus J proteins are small, very basic polypeptides ranging from 20 to 40 amino acids whose amino termini are rich in lysine and arginine residues. These positively charged amino acids bind to the negatively charged phosphate backbone of the viral DNA. There is also a cluster of basic amino acid acids in the viral capsid (F) protein that forms a DNA-binding pocket. The carboxy termini of the J proteins are aromatic and hydrophobic. In the virion these hydrophobic carboxy termini lodge into a cleft in the capsid protein, tethering the genome to the inner surface of the capsid. This tether prevents the single-stranded genome from forming secondary structure by base-pairing with itself. Biophysical characterization of virions packaged with mutant J proteins indicates that particles have altered dimensions, which leads to a decrease in infectivity.

Cell lysis caused by E protein leads to release of phage

The E protein is responsible for cell lysis. The mechanism of E-protein mediated lysis is similar to that of antibiotics like penicillin, which inhibits cell wall biosynthesis. Recent studies demonstrate that the E protein most likely inhibits the host cell Mra Y protein. This protein is a translocase that catalyzes the formation of the first lipid-linked intermediate in cell wall biosynthesis. Penicillin, on the other hand, inhibits the formation of cross-links within peptidoglycan. In either case, the weakening of the cell wall causes an increased sensitivity to osmotic pressure, which leads to cell lysis.

Did all icosahedral ssDNA virus families evolve from a common ancestor?

There are four icosahedral single-stranded DNA virus families: *Microviridae* (bacteriophages), *Geminiviridae* (plant viruses), and *Circoviridae* and *Parvoviridae* (animal viruses). These families have remarkably similar fea-

φX174 as a mutation detector in mammals

The φX174 genome is being developed as a shuttle vector for the study of *in vivo* mammalian mutagenesis. In these studies, a double-stranded copy of a φX174 genome carrying known mutations is inserted into mammalian germline cells and transgenic animals are produced. The inserted φX174 DNA is subject to the same DNA replication and repair mechanisms as the animal genome. After exposure of the animals to possible carcinogens, the φX174 DNA is excised from the animal genome and transfected into bacterial cells. Viruses are then assayed for the reversion of the various phage-encoded mutations. This work, being conducted in the laboratories of Heinrich Malling, James Burkhart, and Carrie Valentine, has very powerful applications within the field of toxicology. It is easy and rapid, because labor-intensive animal crosses are not needed to obtain results. Since many of the mutations being used can revert only by certain base substitutions, the mutagenic specificity of carcinogens or drugs in various tissues within the animal can be determined.

tures. They all have very small T = 1 capsids. Similarities between the atomic structure of parvovirus and microvirus capsid proteins are striking, as are the mechanisms employed for DNA packaging (however, parvovirus genomes are linear, not circular). Are parvoviruses and microviruses, one family replicating in animal cells, the other in bacteria, distantly related? Are they two extremes of the same superfamily?

Answers to these questions may likely be found by studying the recently discovered and poorly understood gokushoviruses (Table 6.1). The genomes of these viruses are smaller than the microvirus genomes. Their genomes encode primarily an A protein homologue, needed for DNA replication, and a capsid protein, although other genes are present. Parvoviruses (Chapter 9) similarly contain one DNA replication gene and one capsid protein gene, although alternative initiation sites and splicing produce several proteins from each gene. The prominent G protein spikes seen in the coliphages are not present in the gokushoviruses, nor is the external scaffolding protein. An added domain of the capsid protein performs some of the functions of the external scaffolding protein. This domain, located at the threefold axes of symmetry, forms protrusions called mushrooms on the surface of the virion, similar to those seen in some parvoviruses.

The hosts of gokushoviruses are obligate intracellular parasitic bacteria, providing a potential vehicle for bacteriophage genomes to gain access to eukaryotic cells. Since most of the machinery used to replicate the genomes of these viruses is host-derived, substitution of the small part of the viral DNA that directs replication by bacterial or eukaryotic host cell proteins could potentially allow replication of a prokaryotic virus in a eukaryotic cell. Such a jump could have occurred during evolution.

SUGGESTED READING

1. Bernhardt, T.G., Roof, W.D., and Young, R. (2000). Genetic evidence that the bacteriophage φX174 lysis protein inhibits cell wall synthesis. *Proceedings of the National Academy of Sciences of the USA* **97,** 4297–4302.

2. Burch, A.D., Ta, J., and Fane, B.A. (1999). Cross-functional analysis of the Microviridae internal scaffolding protein. *Journal of Molecular Biology* **286,** 95–104.

3. Clarke, I.N., Cutcliffe, L.T., Everson, J.S., Garner, S.A., Lambden, P.R., Pead, P.J., Pickett, M.A. Brentlinger, K.L., and Fane, B.A. (2004). Chlamydiaphage Chp2, a skeleton in the φX174 closet: scaffolding protein and procapsid identification. *Journal of Bacteriology* **186,** 7571–7574.

4. Crill W.D., Wichman, H.A., and Bull, J.J. (2000). Evolutionary reversals during viral adaptation to alternating hosts. *Genetics* **154,** 27–37.

5. Dokland, T., Bernal, R.A., Burch, A.D., Pletnev, S., Fane, B. A., and Rossmann, M.G. (1999). The role of scaffolding proteins in the assembly of the small, single-stranded DNA virus φX174. *Journal of Molecular Biology* **288,** 595–608.

6. Fane, B.A., Brentlinger, K.L., Burch, A.D., Chen, M., Hafenstein, S., Moore, E., Novak, C.R., and Uchiyama, A. (2005). Phi X 174 et al., the Microviridae. In Calendar, R., ed., The Bacteriophages, 2nd Ed., pp. 128–144. Oxford University Press, UK.

7. Kornberg, A. and Baker, T.A. (1992). Chapters 8 and 17 in DNA Replication, 2nd Ed. W.H. Freeman & Co., New York.

8. Valentine C.R., Montgomery, B.A., Miller, S.G., Delongchamp, R.R., Fane, B.A., and Malling, H.V. (2002). Characterization of mutant spectra generated by a forward mutational assay for gene A of PhiX174 from ENU-treated transgenic mouse embryonic cell line PX-2. *Environment and Molecular Mutagenesis* **39,** 55–68.

KEY TERMS

Helicase
Lipopolysaccharide
Primosome
Procapsid
Rolling circle
Scaffolding proteins

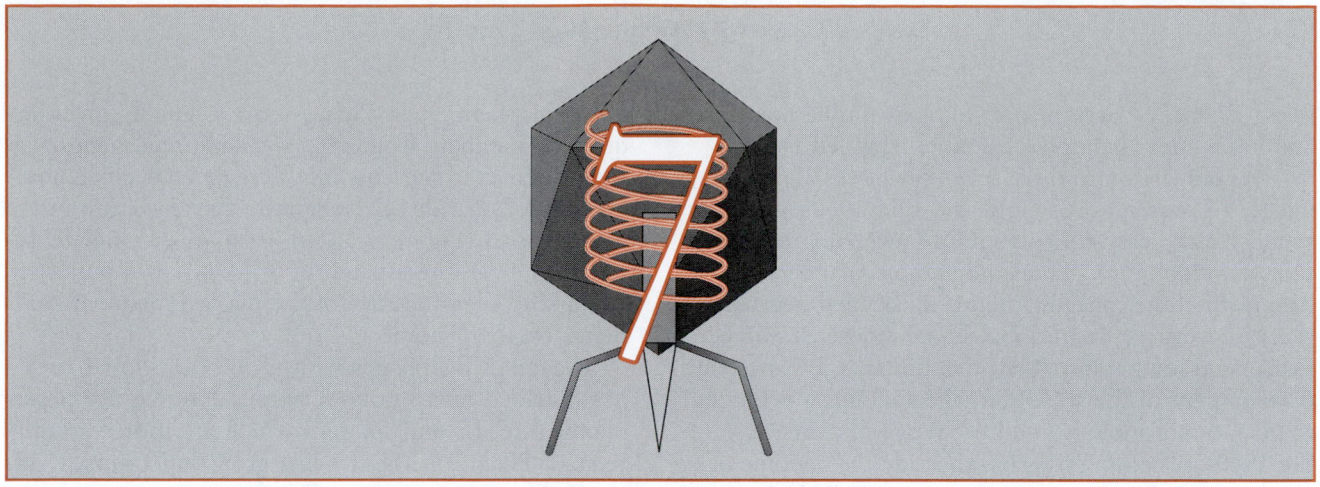

Bacteriophage T7

T7-like bacteriophages

T7 ("type 7"), from the collection of phages chosen for study by the phage group

VIRION

Naked icosahedral capsid (T = 7).
Diameter 60 nm.
Short tail with short tail fibers.

GENOME

Linear ds DNA, 40 Kbp.
Terminal direct repeat sequence of 150 bp.

GENES AND PROTEINS

56 genes encode for 59 proteins.
Three sequentially expressed classes of genes organized from left to right end of genome:

- *Class I ("early") genes: control of phage and host gene expression*
- *Class II ("middle") genes: phage DNA replication*
- *Class III ("late") genes: structural and assembly proteins, DNA packaging, cell lysis*

VIRUSES AND HOSTS

A genus within family *Podoviridae*. From Greek *podos* (foot), refering to short tail.
Similar phages infect a variety of bacterial hosts, including marine bacteria.
Examples: bacteriophages T3, φII, H, SP6, and gh-1.

DISTINCTIVE CHARACTERISTICS

Entry of DNA into cell powered by transcription.
Novel 100 KDa phage-coded RNA polymerase transcribes class II and III genes.
Phage transcripts are cleaved by host RNAse III to small, stable mRNAs.
DNA synthesis initiated by phage RNA polymerase.
Phage-coded DNA polymerase.

T7: a model phage for DNA replication, transcription, and RNA processing

Bacteriophage T7 is one of seven phages selected by a group of pioneer molecular biologists, headed by Max Delbrück, in the early 1940s to serve as prototypes on which the phage research effort should be focused. Some of these phages, including T7, were isolated from a commercially available mixture of phages being sold for treatment of intestinal infections. Each phage was designated by T (for "type") and arbitrarily assigned a number. T7 and its relative, T3, are the smallest of the T phages. They have a 60-nm icosahedral capsid and a short, stubby tail attached to one of the icosahedral vertices. T7 makes large plaques on susceptible host cells and has a linear genome of double-stranded DNA.

In comparison to the T-even phages (T2, T4, T6), phage T7 was ignored until the mid-1960s when its small size and easy-to-isolate DNA attracted the attention of both physical chemists and molecular geneticists. About this time, phage genetics blossomed because of the discovery of conditional-lethal nonsense (amber, ochre) and temperature-sensitive mutants. Prior to the development of these tools, phage genetics was limited to plaque-morphology and host-range mutants. Also in the 1960s, it became possible to study the genome of a phage by physicochemical techniques. T7 DNA became one of the first well-studied phage genomes (along with that of phage φX174) and its molecular weight was established as 25×10^6 daltons both by classical physical determinations as well as by measurement of its contour length (12 μm) in electron microscope images.

Chemical analysis indicated that the ends of T7 DNA have unique sequences; this was a major achievement prior to the development of DNA sequencing methods. Thus T7 DNA does not have a circularly permuted sequence, as does the DNA of the unrelated phage T4. The separation of the complementary strands of T7 DNA was particularly easy by equilibrium sedimentation in CsCl in the presence of excess poly riboG. This technique facilitated the use of T7 DNA in many experiments requiring intact, complementary DNA strands. F. William Studier and Rudolf Hausmann independently obtained large collections of conditional-lethal T7 mutants, which they mapped and classified in terms of phenotypes. Genes were originally given sequential numbers from 1 to 19; additional genes that were discovered subsequently were given decimal numbers such as 0.3 or 1.2 to indicate their position in the genetic map (Figure 7.1). In 1983, the complete sequence (39,937 base pairs) of T7 DNA was published along with identification of the major functional landmarks on the genome.

Searches for phages with properties similar to T7 have yielded many isolates, some of which are nearly identical to T7 and some of which are more distantly related. Phages in the T7-like group infect many different host bacteria, including marine organisms, the plague bacillus (*Yersina pestis*), *Serratia*, *Klebsiella*, and *Pseudomonas*.

T7 genes are organized into three groups based on transcription and gene function

The sequence of T7 DNA predicts that it can encode 56 proteins, defined by open reading frames with recognizable ribosome binding sites and protein synthesis initiation sequences. Fifty-nine phage proteins have been identified by genetic or biochemical evidence; some genes give rise to more than one protein by initiation of translation at internal AUG codons. T7 has a logical arrangement of genes that could have been designed by an engineer: three groups of genes are organized in continuous blocks from left to right on the genome (Figure 7.1). These genes are clustered both by function and by the timing of their expression during

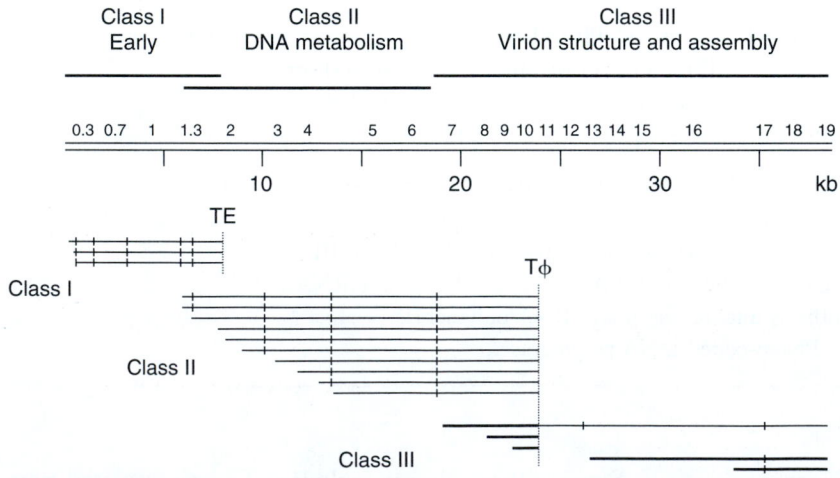

FIGURE 7.1 Genetic and transcriptional map of the T7 genome. The 40 Kb T7 genome is shown with the positions of selected genes immediately above the map. Positions of major gene classes I, II, and III are indicated by heavy lines. Major transcripts from the 3 class I promoters, the 10 class II promoters, and the 5 class III promoters are shown below, with vertical hash marks indicating ribonuclease III cleavage sites. TE and Tϕ show positions of transcriptional terminators. Thickness of lines indicates approximate transcript abundance. Low level of readthrough of Tφ is indicated on one class III transcript but can occur on all transcripts.

the life cycle of the phage. Class I ("early") genes, located in the left-hand one-fifth of the genome, are expressed during the first 6 to 8 minutes after infection (at 30 C). Class I gene products are needed for expression of the remainder of phage genes and to shut down host cell functions. Class II ("middle") genes occupy the next one-third of the genome. Most of the class II gene products are concerned with phage DNA replication. Class III ("late") genes, encoded in the right half of the genome, code for phage structural proteins and proteins needed for DNA maturation and packaging. All T7 genes are transcribed from left to right on the map as shown.

Entry of T7 DNA into the cytoplasm is powered by transcription

The icosahedral T7 capsid (Figure 7.2) is constructed from a single capsid protein, gp10. The tail is attached at one vertex to a portal consisting of 12 molecules of gp8. The tail consists of two phage proteins, gp 11 and 12, as well as six fibers made from trimers of gp17. Three additional proteins (gp14, gp15, and gp16) form an internal core around which the DNA is wrapped. The complex structures of the tail and the internal core provide the machinery for packaging the viral genome and for inserting it into cells during infection.

T7 phage binds to lipopolysaccharide molecules in the outer cell membrane via its tail fibers (Figure 7.2). The three internal core proteins are then inserted via the tail into the cell, forming a channel through the outer membrane, the peptidoglycan cell wall, and the cytoplasmic membrane. There is evidence that one of these proteins, gp16, is able to enzymatically degrade the peptidoglycan layer as it bores through, creating a hole for the passage of the DNA. Gp16 is also responsible for transporting the first 850 base pairs of the left end of T7 DNA through this channel into the cytoplasm, exposing three closely spaced promoters recognized by the host *E. coli* RNA polymerase. At this point no more DNA enters the cell unless transcription is allowed to proceed. The host cell RNA polymerase transcribes the class I genes, simultaneously spooling the DNA out of the capsid into the cytoplasm at a rate of 40 bp per second. Thus RNA polymerase serves as a winch to transfer the DNA from the capsid into the cell while making messenger RNAs (Figure 7.2). *E. coli* RNA polymerase encounters a strong termination signal (TE) about 8 Kb into the genome from the left end (Figure 7.1), where it releases the mRNA and detaches from the DNA.

Transcription of class II and III genes requires a novel T7-coded RNA polymerase

The initial steps of infection described above can take place in the presence of an inhibitor of protein synthesis, but further transcription and entry of the remainder of the genome is blocked unless phage class I proteins are synthesized. Systematic examination of infections by phage with mutations in class I genes revealed that the protein product of gene 1.0 is essential for class II and class III gene transcription. This protein was isolated and found to be a monomeric, 100 KDa RNA polymerase that is highly specific for class II and class III T7 phage promoters. Up to that time, all known RNA polymerases were multi-subunit enzymes, so the discovery of this single-subunit RNA polymerase, which does not depend on any host proteins for function, galvanized the transcription field. It has since become an important part of the toolbox of molecular biologists (see below).

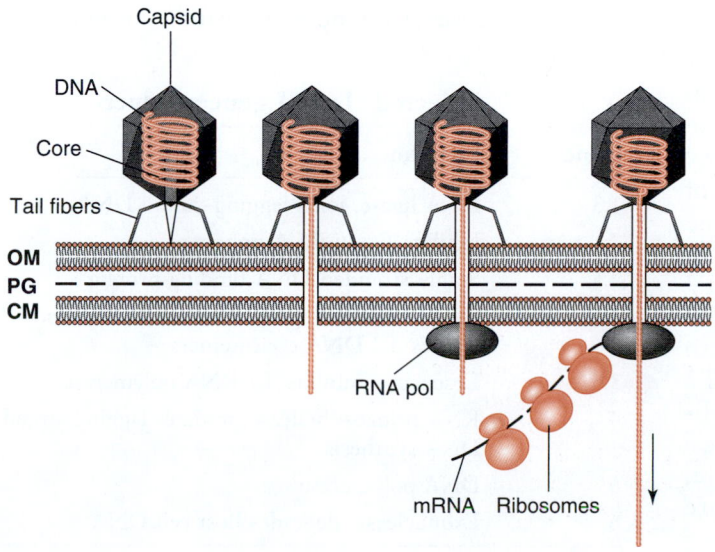

FIGURE 7.2 **Entry of T7 genome powered by transcription.** T7 virion is shown attached to outer cell membrane via tail fibers (left image). Subsequent images show channel formation and entry of the end of the DNA (orange), followed by binding of RNA polymerase and transcription-mediated entry of the remainder of the DNA. OM: outer membrane; PG: peptidoglycan layer; CM: cytoplasmic membrane.

TABLE 7.1 Selected class I gene products

Gene	Function
0.3	Inhibits host type I restriction endonucleases
0.7	Inhibits *E. coli* RNA polymerase
1.0	T7-specific RNA polymerase

T7 RNA polymerase not only takes over transcription of the phage genome, but it also continues the work begun by the host RNA polymerase of spooling the phage DNA into the cell. DNA entry at this stage progresses at about 250 bp/second, reflecting the high rate at which the phage RNA polymerase transcribes DNA.

Two other class I gene products play important roles in the infection (Table 7.1). The product of gene 0.3 inactivates host type I restriction endonucleases EcoB and EcoK, protecting the infecting DNA from cleavage and degradation. Notably, this is the first T7 gene transcribed as the DNA enters the cell, so the protection is established rapidly. Gp 0.7 inactivates the host cell RNA polymerase, limiting the amount of class I mRNAs synthesized and interrupting further class I gene expression. This also shuts off cellular mRNA synthesis, freeing up energy and nucleotides for phage transcription.

Class II genes code for enzymes involved in T7 DNA replication

Promoters recognized by the phage RNA polymerase are highly conserved, extending from 17 nt before to 6 nt after the transcription start site. Ten different class II promoters are used to initiate transcription over a 10-Kb region of the T7 genome (Figure 7.1). Most of the RNAs initiated at these promoters end at a single, stong terminator (Tφ) located 24 Kb from the left end of the genome. A minority of polymerases reads through this terminator to form longer RNAs. The use of multiple promoters leads to the production of a **nested set** of overlapping primary transcripts of the class II region with a common 3′ end. This allows the phage to make lower levels of gene products coded at the left end, and higher levels of gene products coded at the right end of this part of the genome. Twenty-nine class II gene products are made from the resulting mRNAs.

Most of the class II proteins participate in T7 DNA replication and recombination (Table 7.2). The products of genes 3 and 6 degrade the host DNA to provide nucleotide precursors for the synthesis of T7 DNA. Gene products 1.3, 2.5, 4, and 5 act at the replication fork to synthesize progeny DNA molecules. Gene 2, like gene 0.7, inhibits host RNA polymerase, ensuring that class I genes and host transcription are shut down.

T7 RNAs are cleaved by host cell ribonuclease III to smaller, stable mRNAs

The primary RNA transcripts of T7 phage undergo post-transcriptional processing by the host enzyme, ribonuclease III. This ribonuclease recognizes double-stranded RNA regions and cleaves them at specific sites. The long class I transcripts synthesized by the host RNA polymerase are cleaved into 2 monocistronic mRNAs and 3 polycistronic mRNAs (Figure 7.1). The class II and class III primary transcripts are also cleaved at specific sites by ribonuclease III to form smaller functional mRNAs. The stem structures that remain at the 3′ ends of the mature T7 mRNAs after cleavage render them relatively resistant to degradation by other host nucleases, making the phage mRNAs substantially more stable and long-lived than most rapidly turning-over host cell mRNAs.

Regulation of class III gene expression

As the phage DNA enters the cell via transcriptional winching, the five class III promoters become available for transcription by T7 RNA polymerase. At an entry rate of 250 bp/second, it takes about 1 minute of transcription starting at the first class II promoter (5.9 Kb from the left end) until the first class III promoter enters the cell, and about 2 minutes until the most distal class III promoter enters (Figure 7.1). Thus transcription-dependent DNA entry is one factor responsible for the delayed transcription of class III genes compared with class II genes. However, class III promoters have a higher affinity for T7 RNA polymerase than class II promoters due to slight sequence differences, and therefore higher levels of class III transcripts are made. Furthermore, the class II gp3.5, which has lysozyme activity, binds specifically to T7 RNA polymerase and inhibits its activity. This can reduce class II promoter usage by lowering the overall availability of active T7 RNA polymerases, with the result that the stronger class

TABLE 7.2 Selected class II gene products

Gene	Function
1.3	DNA ligase, joins lagging-strand DNAs
2	Inhibits *E. coli* RNA polymerase
2.5	Single-stranded DNA binding protein
3	Endonuclease, degrades host cell DNA and cleaves T7 DNA concatemers
3.5	Lysozyme, inhibits T7 RNA polymerase
4	RNA primase/helicase, initiates lagging-strand DNA synthesis
5	DNA polymerase
6	Exonuclease, degrades host cell DNA

TABLE 7.3 Selected class III gene products

Gene	Function
8	Portal protein, connector for tail
9	Scaffolding protein for formation of capsids
10	Capsid protein
11	Tail protein
12	Tail protein
14, 15, 16	Core proteins, form channel for DNA entry
17	Tail fiber, attachment to cell
17.5	Holin, access of lysozyme to cell wall for lysis
18, 19	Maturation proteins, DNA packaging

III promoters are favored over the weaker class II promoters. This allows the phage to concentrate its efforts on production of large amounts of virion structural and maturation proteins (Table 7.3) during the final stages of infection.

DNA replication starts at a unique internal origin and is primed by T7 RNA polymerase

T7 DNA is replicated bidirectionally starting at an origin about 5.9 Kb from the left end of the DNA. Alternate origins are known to exist because T7 mutants that have lost the primary origin are able to replicate well. Short DNA fragments joined to RNA primers are found in replicating T7 DNA, so it is clear that T7 DNA replication resembles that of the *E. coli* host cell, using continuous synthesis on the leading strand and discontinuous synthesis with Okazaki fragments on the lagging strand (Figure 7.3). However, T7 uses its own replication proteins.

Surprisingly, transcription by T7 RNA polymerase was found to be necessary for the initiation of T7 DNA replication. It turns out that the RNA polymerase provides the first primer for leading-strand synthesis. The two class II promoters closest to the left end of the DNA, promoters 1.1A and 1.1B, precede an A/T-rich region on the DNA. When sufficient levels of class II proteins have been made, the T7 DNA polymerase can displace the RNA polymerase in the A/T-rich region shortly after it has begun transcription at one of these promoters. The DNA polymerase will then use this nascent RNA chain as a primer, and begin synthesizing DNA in a rightward direction. The gene 4 primase/helicase protein subsequently binds to this replication fork, laying down a short RNA primer on the opposite ("lagging") DNA strand and simultaneously unwinding the DNA to allow the T7 DNA polymerase to continue forward on the leading strand. Thus a replication fork is born. T7 DNA polymerase can then extend the new primer toward the left, setting up a second replication fork and establishing bidirectional replication. Note that initiation of DNA replication at the leftmost T7 RNA polymerase promoter avoids head-on collisions between DNA polymerases and RNA polymerases, because they all travel in the same direction; transcription from class I genes, which could affect replication moving leftward, has ceased.

The T7 DNA polymerase (gp5) has structural similarity to mammalian DNA polymerase β and HIV reverse transcriptase. It forms a stable complex with a normal host cell protein, *E. coli* thioredoxin. While the gene 5 DNA polymerase alone can catalyze the addition of dNTPs to a growing DNA chain, the binding of thioredoxin greatly increases its rate and **processivity** (the tendency to continue on the same DNA template-primer complex). Thioredoxin interacts electrostatically with the DNA polymerase to tether the polymerase to its template DNA strand. The T7 DNA polymerase can also act as a 3′ to 5′ exonuclease to effect proofreading removal of misincorporated nucleotides.

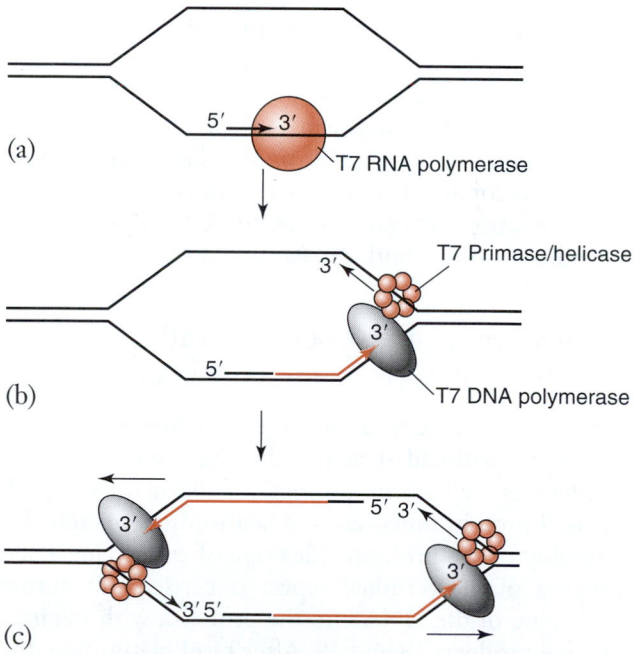

FIGURE 7.3 Initiation of DNA replication at the leftward T7 promoter. (a) T7 RNA polymerase initiates an RNA chain (black arrow) at promoter 1.1. RNA polymerase is displaced by T7 DNA polymerase in an A/T-rich region shortly downstream of the promoter. DNA polymerase uses the RNA as a primer for synthesis of DNA (orange arrow). T7 primase/helicase loads onto the replication fork and generates short RNA primers (black arrows) for DNA synthesis on the opposite strand. (c) Primers are extended by DNA polymerase and replication forks move toward opposite ends of DNA (horizontal arrows).

Large DNA concatemers are formed during replication

A classical problem with RNA primer-directed replication of linear DNAs is that the last primer laid down at the 3′ end of each template DNA strand cannot be removed and replaced with DNA, because there is no sequence beyond the 3′ end to which the next primer could bind (see also Chapter 9). Thus during each round of replication, a short sequence would be lost at each end of the DNA. Some viruses form circular intermediates that overcome this problem, but no circular forms of T7 DNA have been found.

T7 solves this problem by forming large linear DNA **concatemers** via a terminal repeat sequence of about 150 bp at each end of its DNA. This repeated sequence may be a site of recombination between genomes to generate the concatemers, which contain only one copy of the repeat between each copy of genome-length DNA. Concatemeric DNAs accumulate in cells infected with T7 phage mutated in gene 3 (endonuclease) or genes 18 or 19 (DNA maturation). These gene functions are therefore implicated in processing concatemers to unit-length molecules in preparation for packaging into phage heads. Endonucleolytic cleavage near the repeat, followed by copying of the repeated sequence, allows formation of a second repeat and release of a full-length DNA molecule containing repeats at both ends. For an idea of how this is carried out, see Figure 9.4.

Recombination between both input and progeny T7 genomes occurs and allows standard genetic mapping by recombination frequency measurements. This recombination depends on the products of genes 3, 4, 5, and 6 as shown both *in vivo* and *in vitro*.

Concatemers are processed and packaged into preformed proheads

Assembly of progeny virions begins by the formation of an empty **prohead** structure that includes the gene 9 product as well as the proteins of the mature capsid. Gene 9 protein functions as a **scaffolding protein** for assembly of the prohead. Cleavage of concatemers and copying of the terminal repeat is carried out during packaging of the DNA into the prohead, with the help of gene products 18 and 19. After head maturation, the scaffolding protein dissociates, leaving T7 virions.

The lysis of the infected cell to liberate approximately 100 progeny phage particles is dependent on two T7 functions. The gene 17.5 protein is a small membrane-specific protein called a **holin**. Under the appropriate conditions it assembles in the cytoplasmic membrane of the infected cell to create a hole that allows access to the bacterial cell wall of the phage endolysin, the product of gene 3.5 (T7 lysozyme). The endolysin, once in the periplasmic space, degrades the cell wall, leading to cell lysis.

Special features of the T7 family of phages

T7 and related phages have three properties that make them objects of special interest to biologists as well as to biotechnologists. The first feature is the phage-encoded RNA polymerase, which is highly specific for phage promoters. The second feature is a mechanism for modulating the classic host restriction-modification system to overcome restriction. The third feature is that T7 and many of its relatives are able to grow efficiently only in *E. coli* lacking the fertility plasmid F; that is, they are female-specific or "F-restricted" phages.

Genetic engineering with T7. The unique properties of the RNA polymerases of the T7 family of phages make them very useful tools for biotechnological and genetic engineering applications. T7 RNA polymerase is easily purified, is a very efficient and rapid enzyme, and is highly specific for its own late promoter sequences. Therefore, insertion of a T7 late promoter sequence upstream of any DNA sequence allows *in vitro* synthesis of substantial quantities of homogeneous RNA from the desired DNA sequence. This strategy has been used to prepare specific RNA substrates for study of RNA conformation, ribozymes, and RNA processing.

Because the RNA polymerases from the T7 group (T7, T3, and SP6 phages) differ in promoter recognition, it has been possible to make plasmids with several different promoters, so that transcription patterns can be controlled by the choice of which RNA polymerase is used. For example, cloning vectors with both T7 and T3 promoters arranged to promote convergent transcription from opposite DNA strands have been useful in applications of sequence analysis to both DNA strands of the same fragment and in preparation of antisense RNAs.

The T7 RNA polymerase/late phage gene promoter system has been exploited to generate a family of highly efficient gene expression vectors, called pET-vectors (for *p*lasmid, *e*xpression, *T7* (Figure 7.4). This expression system has the dual advantages of strict regulation of expression and high yields upon induction. The gene of interest, say gene X, is transcriptionally silent until expression is activated, at which time a massive takeover of the cellular transcription and translation capacities occurs. Gene X is placed under the control of a late T7 phage promoter and thus cannot be expressed in a normal bacterial cell. This is an important advantage if the product of gene X is toxic or even mildly detrimental to the growth of the host bacterium (in which case there is selection against plasmid retention and/or gene X functionality).

A plasmid containing the gene X sequence under the control of the T7 promoter is introduced into a bac-

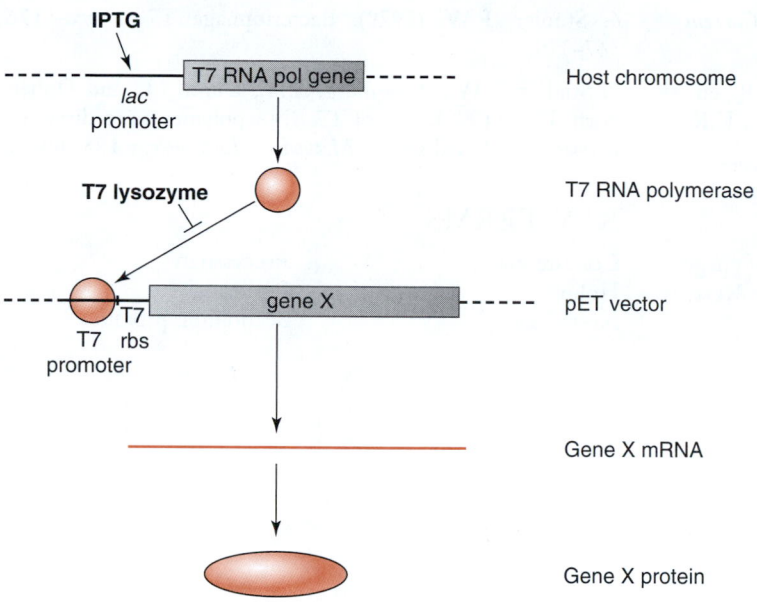

FIGURE 7.4 Expression of foreign genes in T7-derived expression vectors. The gene to be expressed (X) is inserted in a pET vector downstream of a T7 late promoter and a T7 gene 10 ribosome binding site (rbs). This vector is introduced into a host cell that contains the T7 RNA polymerase gene under the control of the *lac* operon regulatory region (*lac* promoter). An inducer of the *lac* operon (IPTG) turns on T7 RNA polymerase expression and leads to expression of gene X in the pET vector. A second plasmid (not shown) expresses low levels of T7 lysozyme, which inhibits the activity of the small amount of T7 RNA polymerase made in absence of the inducer, allowing genes coding for toxic proteins to be cloned.

terium in which the T7 RNA polymerase is made from an inducible *lac* operon promoter. When an inducer of the *lac* promoter is added, T7 RNA polymerase is made and gene X is transcribed. Often, a very efficient class III T7 ribosome binding site is included in the plasmid construction. Under these conditions, not only is the mRNA for gene X the major RNA in the cell, but it also is able to compete effectively for most of the protein synthetic apparatus, so that gene X product becomes the major protein in the cell. It is common to achieve expression levels so high that the desired protein product represents 25 to 50 percent of the total cell protein. Thus, only a fewfold purification is needed to obtain homogeneous protein X for further study.

Overcoming host restriction/modification. Because of the widespread possibility of horizontal gene transfer by exchange of DNA between bacteria, many bacteria have developed mechanisms to recognize and degrade foreign DNA, which might otherwise carry genes of lesser fitness from unrelated organisms. One such mechanism employs endonucleases that recognize specific DNA sequences as sites for degradation. These so-called restriction enyzmes are prevented from acting on the cell's own DNA by modification of the host cell DNA, frequently by methylation of one of the nucleotides in the restriction enzyme recognition sequence. Since phage infections involve introduction of foreign DNA into the host cell, phages are often the target of this sort of restriction. Unless the immediately prior host cell has "marked" the phage DNA with the host-specific modification, allowing it to escape the restriction endonucleolytic attack, the phage may be restricted and the infection aborted. Indeed, it was the discovery of this host-controlled modification in the early 1950s by Salvador Luria and Mary Human and by Giuseppe Bertani and Jean Weigle that led to the identification and characterization of site-specific restriction endonucleases.

Bacteriophage T7 and related phages have evolved a strategy to avoid these methylation-dependent restriction systems by inactivating the restriction endonuclease. Some members of the T7 family of phages supplement this strategy by destroying the methyl group donor substrate (*S*-adenosyl methionine) present in the host cell. The product of the class I phage gene 0.3 possesses both restriction nuclease inhibitory activity and (in some phages, but not T7) *S*-adenosylmethionine hydrolase activity (SAMase). By this strategy T7 and its relatives have extended their fitness in environments with diverse host strains containing mutually incompatible restriction-modification specificities.

F-restriction. T7 (but not T3) is unable to replicate in cells containing the F-factor. The abortive phage infection progresses more or less normally in the early phase, but the class II and class III mRNAs fail to express proteins. This restriction is under the control of several genes carried by the F-factor. These genes are designated *pif*A and *pif* B (for *p*hage *i*nhibition by *F*). The molecular mechanism of this apparently translational inhibition is not yet clear.

SUGGESTED READING

1. Dunn, J.J., and Studier, F.W. (1983). Complete nucleotide sequence of bacteriophage T7 DNA and locations of T7 genetic elements. *Journal of Molecular Biology* **166,** 477–535.

2. Hausmann, R. (1976). Bacteriophage T7 genetics. *Current Topics in Microbiology and Immunology* **75,** 77–110.

3. Molineux, I.J. (2005). The T7 group. In Calendar, R., ed., The Bacteriophages, 2nd Ed. Oxford University Press, U.K.

4. Krüger, D.H. and Schroeder, C. (1981). Bacteriophage T3 and bacteriophage T7 virus-host cell interactions. *Microbiological Reviews* **45,** 9–51.

5. Molineux, I.J. (2001). No syringes please, ejection of phage T7 DNA from the virion is enzyme driven. *Molecular Microbiology* **40,** 1–8.

6. Studier, F.W. (1972). Bacteriophage T7. *Science* **176,** 367–376.

7. Studier, F.W., Rosenberg, A.H., Dunn, J.J., and Dubendorff, J.W. (1990). Use of T7 RNA polymerase to direct expression of cloned genes. *Methods in Enzymology* **185,** 60–89.

KEY TERMS

Concatemer
Holin
Nested set
Processivity
Prohead
Scaffolding protein

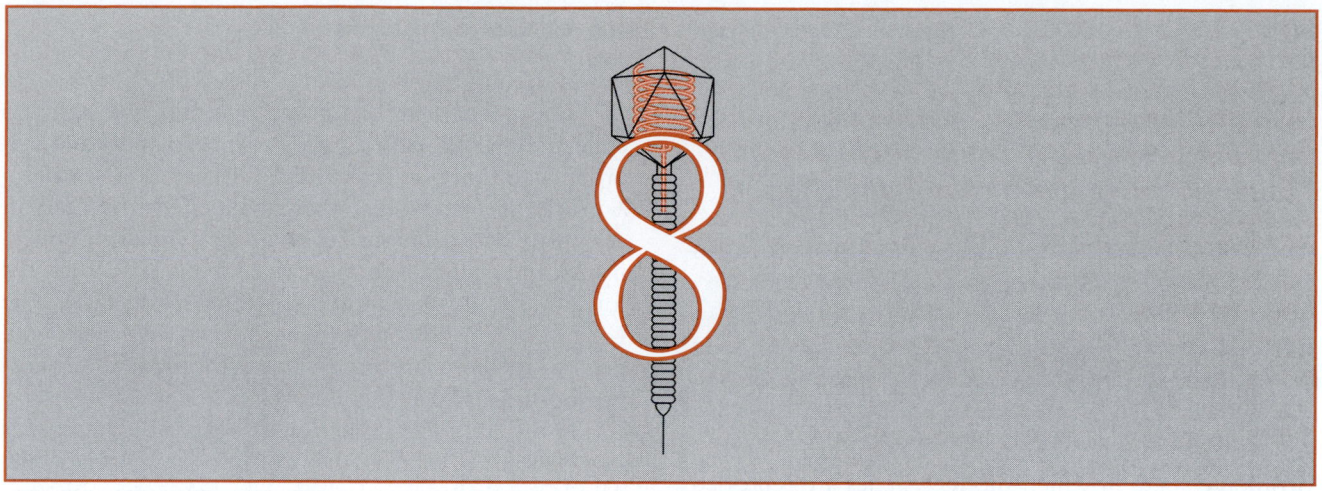

Bacteriophage Lambda: A Piñata of Paradigms

λ-like bacteriophages

λ is the Greek letter "lambda"

VIRION

Head is a naked icosahedral capsid (T = 7).
Diameter 63 nm.
Flexible tube-shaped tail, 135 nm long.
Tail fiber.

GENOME

Linear ds DNA, 40-60 Kb.
Ends have complementary single-stranded extensions (cohesive ends).

GENES AND PROTEINS

Codes for approximately 60 proteins.
Transcribed by cellular RNA polymerase.
Genes grouped by function.
Sequential expression of "immediate-early," "early," and "late" genes.
Transcription of DNA controlled by

- *CI, cro: transcription repressors*
- *N, Q: transcription antiterminators*
- *CII: transcription activator*

VIRUSES AND HOSTS

A genus within family *Siphoviridae*. From Greek *siphon* (tube), referring to long, noncontractile tail.
Related phages: 434, φ80, 21, HK022, N15.
Some *Podoviridae* share limited genome homology to λ phage. Examples: P22, L, LP-7.

DISTINCTIVE CHARACTERISTICS

DNA circularizes upon entry via cohesive ends.
A temperate phage: DNA can integrate into host chromosome, creating lysogen.
Complex lysogenic and lytic regulatory circuits.
DNA packaged by cleavage from linear concatemers.
Engineering of lambda DNA has generated a variety of powerful cloning vectors for genetic and genomic studies.

"For strange effects and extraordinary combinations we must go to life itself, which is always far more daring than any effort of the imagination." S. Holmes

In the beginning . . .

Infection of *E. coli* by phage λ can have two possible outcomes. The lytic growth cycle produces up to 200 progeny viruses per infected cell and causes cell lysis. Alternatively, the infecting λ DNA can enter a **prophage** state, in which the lytic genes are repressed and the viral DNA integrates into and is passively replicated with the host chromosome (Figure 8.1). Phages that can either grow lytically or exist as a repressed prophage are called **temperate phages**. The prophage makes a repressor that turns off expression of the λ lytic genes. In addition to inhibiting prophage expression, the repressor turns off the lytic genes of any incoming λ chromosome. Thus an *E. coli* cell carrying a λ prophage is immune to infection by λ. Such an *E. coli* cell is called a **lysogen**, because the expression of the prophage can be induced, causing prophage excision, genome replication, and cell lysis (*lysogen* means "able to generate lysis").

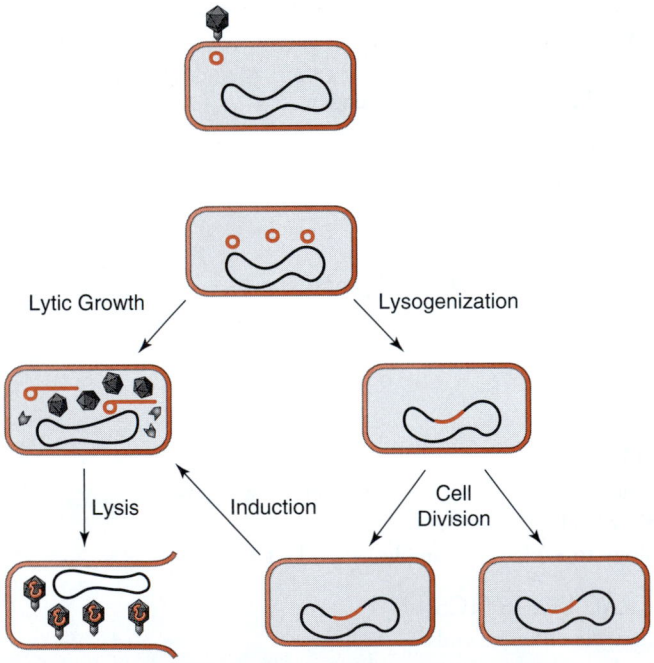

FIGURE 8.1 **Outline of λ biology.** A bacterial cell is infected by a λ phage particle. Phage DNA is in orange; phage heads are shown as hexagons, tails as bullets, and the bacterial chromosome is shown as a large black circle. A decision is made during infection to progress to lytic growth (virus production and lysis) or lysogeny (integration of viral DNA and continued growth of the cells). Induction is the switch from lysogeny to lytic growth.

The standard K-12 laboratory strain of *E. coli* was collected in 1922 from a patient with an intestinal disorder. It was not known until 1951 that K-12 carries a λ prophage. Esther Lederberg happened to spread an agar plate with a mixture of a K-12 (λ^+) culture supernatant and cells of a strain derived from K-12 that had fortuitously lost its λ prophage, that is, was (λ^-). The K-12 (λ^-) cells grew to form a cell lawn, but there were plaques in the lawn caused by λ virions in the supernatant of the K-12 (λ^+) culture.

In the late 1950s, Dale Kaiser began studying genes involved in the establishment of lysogeny. Meanwhile, Allan Campbell isolated a large set of mutants with conditional lethal mutations. Campbell proposed in 1961 that the prophage inserts into the bacterial chromosome by site-specific recombination between a circular form of the λ chromosome and the bacterial chromosome. In 1963, Alfred Hershey and co-workers, and Hans Ris, showed that the ends of λ chromosomes could bind to each other, cyclizing the molecule. In 1971, Wu and Kaiser showed that the cohesive ends were complementary, single-stranded, 12 base-long extensions at the 5′ ends of each λ DNA strand. In 1982, Fred Sanger and co-workers published the sequence of the 48,502 bp λ DNA molecule. The genetic map (Figure 8.2) shows that λ genes are arranged in functional blocks. For example, the 10 head assembly genes are followed by 11 tail assembly genes, and so on.

"The game's afoot!" S. Holmes

Uptake of λ DNA depends on cellular proteins involved in sugar transport

Infection begins with binding to the cell surface and injection of the DNA into the cell. At the tip of the λ tail (Figure 8.3) is a single fiber, the J protein. J binds to an *E. coli* outer membrane protein, LamB, named because it confers susceptibility to λ infection. LamB is a maltose-inducible porin for transport of maltose and maltose oligosaccharides. Injection of λ DNA through the tail and into the cell requires the cytoplasmic membrane components of another sugar transport system, mannose permease. Injection is fascinating and more than a little mysterious. How λ DNA traverses the outer and inner lipid membranes and the periplasm of the Gram-negative *E. coli* cell is by no means clear. Following injection, λ DNA circularizes via the *cos* ("cohesive") sites at its ends, and the circle is sealed by host DNA ligase. Transcription by host RNA polymerase then begins at two promoters, P_L and P_R (initiating transcription toward the left and right, respectively), located on either side of the repressor gene, *cI* (Figure 8.4).

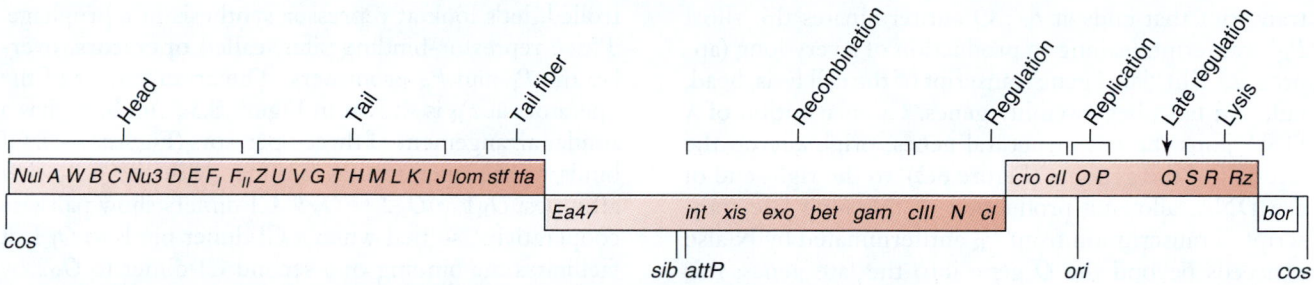

FIGURE 8.2 **The genetic map of λ DNA.** Genes are functionally grouped. Genes boxed above the central line are transcribed from left to right, those below are transcribed from right to left. *cos*: cohesive end site; *sib*: site encoding a transcription termination sequence, or if antiterminated, a site for RNAseIII attack; *attP*: phage attachment site for integration; *ori*: origin of DNA replication.

The λ lytic transcription program is controlled by termination and antitermination of RNA synthesis at specific sites on the genome

Transcripts from P_L and P_R express the *N* and *cro* genes; these "immediate-early" transcripts end at transcription terminators t_L1 and t_R1 (Figure 8.4a). N protein is a transcriptional antiterminator that, with a recruited ensemble of four host proteins, modifies RNA polymerase so that transcripts initiated at P_L and P_R are no longer terminated. For N to function, it binds to specific mRNA sequences, called NUT (*N-ut*ilization) sequences (Figure 8.4a). Because NUT sequences are found only on transcripts made from P_L and P_R, these are the only transcripts that are antiterminated by N. N acts by a mechanism analogous to that of the Tat protein of human immunodeficiency virus (see Chapter 26).

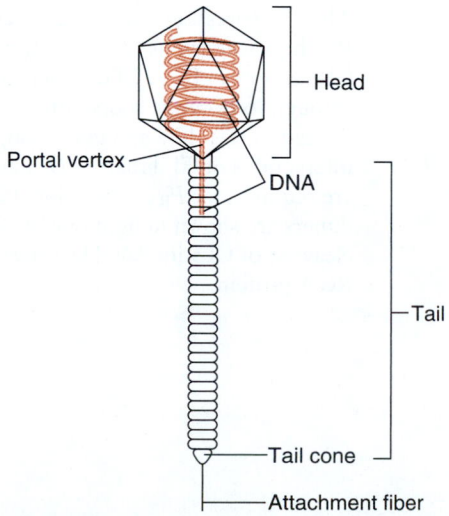

FIGURE 8.3 **Schematic drawing of λ virion.** DNA is shown as a orange coil within the capsid.

Once N is made in sufficient quantity, "early" transcription can take place. N-modified RNA polymerase simply transcribes through terminators t_L1 and t_R1 without taking any notice of them. The N-antiterminated transcript from P_L extends leftward through the general recombination genes *gam*, *bet*, and *exo*, the site-specific recombination genes *xis* and *int*, across the locus for site-specific recombination, *attP*, all the way to *Ea47* (Figures 8.2 and 8.4a). Similarly, the N-antiterminated transcript from P_R extends through *cII* and the replication genes *O* and *P* all the way to gene *Q*, which encodes the late gene activator.

Q is also a transcriptional antiterminator. The P_R' promoter (Figure 8.4b) constitutively expresses a short

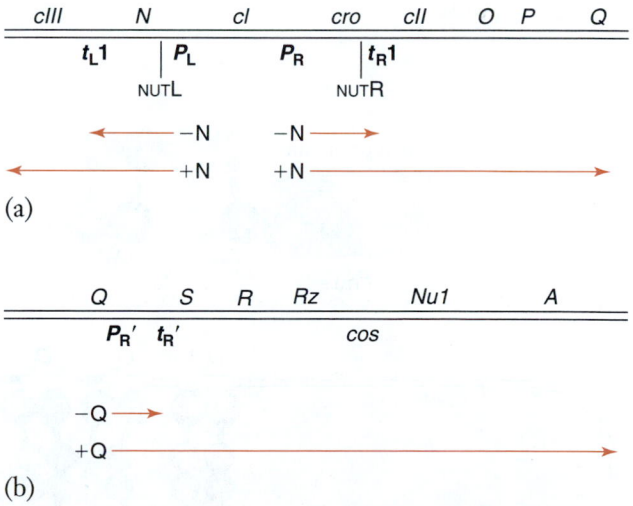

FIGURE 8.4 **The lytic transcription program.** Genes encoding proteins are drawn above the double line, and promoters and terminators are below the double line. Orange arrows represent transcripts. (a) Immediate early transcription (−N) and early, N-antiterminated transcription (+ N) initiated at P_L and P_R. (b) Late transcription, sponsored by Q-dependent antitermination of the P_R' transcript (which continues beyond gene *A*, through all late genes).

transcript that ends at t_R'. Q antiterminates this short P_R' transcript, leading to production of a very long (approx. 25 Kb) "late" gene transcript of the cell lysis, head, tail, and tail fiber assembly genes. Circularization of λ DNA joins the late structural genes, originally on the left half of the genome (Figure 8.2), to the right end of the DNA, allowing production of this long late transcript. Transcription from P_R antiterminated by N also proceeds beyond the *Q* gene into the late genes; this transcription gives a level of late gene expression equal to about 20% of the P_R', Q-sponsored level.

The CI repressor blocks expression of the lytic program by regulating three nearby promoters: P_L, P_R, and P_{RM}

When λ infects active cells in logarithmic growth, the phage grows lytically in most of the cells. However, when λ infects metabolically sluggish, stationary phase cells, most of the infected cells are lysogenized. The decision to go lytic or lysogenic rests on whether sufficient repressor is produced to shut down transcription of the lytic genes. The repressor is the CI protein (pronounced "see-one"), product of gene *cI*. As a first step in understanding how the level of repressor is controlled, let's look at repressor synthesis in a prophage. Three repressor-binding sites, called **operators**, overlap the P_L and P_R promoters. The arrangement of the operators at P_R is shown in Figure 8.5a and b; P_L has a similar arrangement of three operators (Figure 8.5c). CI binds to the operators as a dimer with the following affinities: $O_R1 > O_R2 \approx O_R3$. CI dimers show pairwise cooperativity, so that when a CI dimer binds to O_R1, it facilitates the binding of a second CI dimer to O_R2 by protein–protein interactions between the two repressor dimers. O_R1 and O_R2 overlap the −10 and −35 segments of P_R, so that CI bound to these operators blocks transcription from P_R (Figure 8.5b). Because the CI dimers cooperate only in pairs, binding to O_R3 is not facilitated by repressors cooperatively bound at O_R1 and O_R2. It has been postulated that CI cooperativity is pairwise because the cooperating repressors must lean toward each other, so CI at O_R2 leans toward CI at O_R1, and cannot interact with a third repressor at O_R3.

A second promoter, called P_{RM} for repressor maintenance, is located near P_R but initiates transcription in the opposite direction (Figure 8.5a); this promoter directs transcription of *cI* mRNA. P_{RM} is a weak promoter, but in a prophage, P_{RM} is activated to produce sufficient CI repressor to shut down P_R and P_L, as follows.

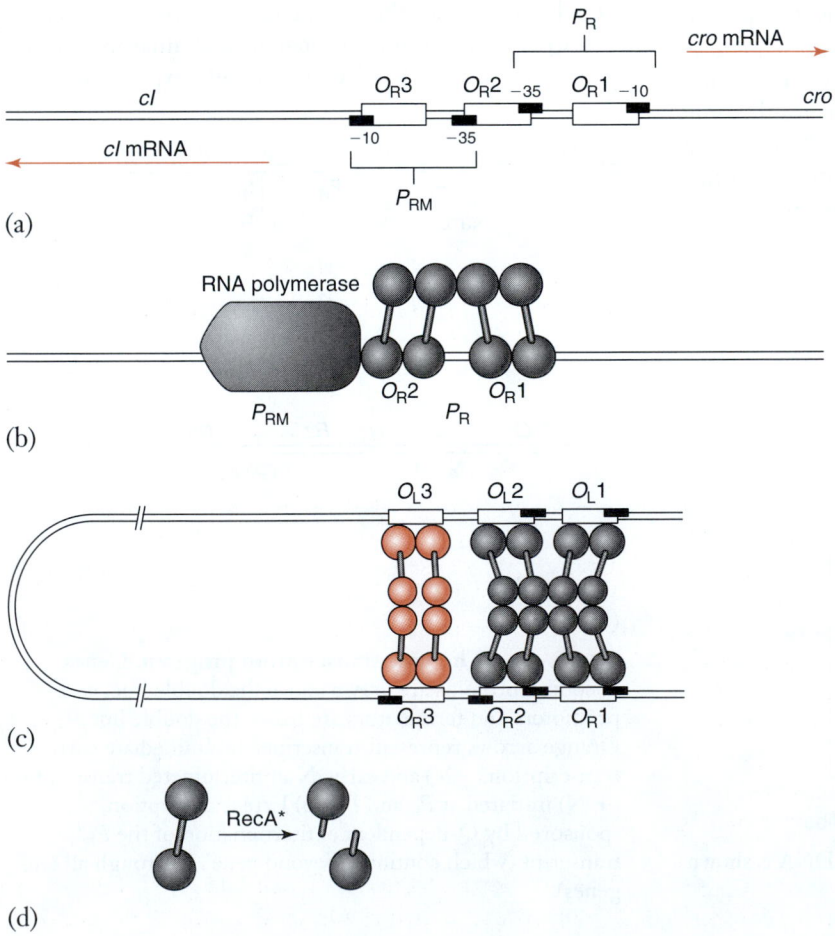

FIGURE 8.5 Transcription and its repression at P_{RM} and P_R. (a) Region coding for CI and Cro. Boxes indicate operator sites for binding by CI repressor. Brackets indicate promoters P_{RM} and P_R, with their −10 and −35 boxes shown. Orange arrows denote the P_{RM} (CI) and P_R (Cro) transcripts. (b) Activation of P_{RM} by CI prophage repressor. CI monomers are shown as dumbbells, and RNA polymerase is shown as a large asymmetric arrow. (c) The CI repression loop. The upper arm is P_L; the lower arm is the P_{RM}–P_R region (also shown in B). A CI octamer formed through long-range cooperative interactions is shown in gray. Cooperative interactions of CI dimers at O_L3 and O_R3 are required for P_{RM} repression; these dimers are shown in light orange. (d) The cleavage of CI stimulated by activated RecA protein.

TABLE 8.1 Control of P_{RM} and P_R by CI and Cro repressors

		Occupancy at operator			*Activity of promoter*	
Protein	**Concentration**	O_R3	O_R2	O_R1	P_{RM}	P_R
CI	None	−	−	−	Weak	On
	Low	−	+	+	Activated	Off
	High	+	+	+	Off	Off
Cro	None	−	−	−	Weak	On
	Low	+	−	−	Off	On
	High	+	+	+	Off	Partially off

CI bound to O_R2 is physically near P_{RM} and forms favorable contacts with RNA polymerase (Figure 8.5b); these contacts make P_{RM} a much better promoter. Stated another way, CI activates its own gene to ensure that sufficient repressor is produced to maintain prophage repression. However, should the CI concentration become very high, O_R3 is also occupied and transcription from P_{RM} is turned off until the CI concentration returns to a normal level.

It turns out that effective repression of P_{RM} requires long-range interactions of the CI tetramers at O_R1 and O_R2 with CI tetramers at O_L1 and O_L2. These interactions result in formation of a CI octamer (Figure 8.5c). Formation of the octamer results in the formation of a large (2.4 Kb) DNA loop. The looping that accompanies octamerization also brings O_R3 and O_L3 near to each other, enabling cooperative interactions between CI dimers at O_R3 and O_L3 (Figure 8.5c), which leads to repression of P_{RM}. Thus, effective repression of P_{RM} depends on O_R1, O_R2, O_L1, and O_L2 for CI octamerization, and on O_L3 and O_R3 for CI dimerization. Occupancy of the O_R operators by CI is summarized in Table 8.1.

Cleavage of CI repressor in cells with damaged DNA leads to prophage induction

Prophage λ DNA within a lysogen can be induced, or derepressed, so that lytic growth ensues. Induction is a way for λ to bail out of a damaged cell and move on to more fertile hunting grounds. Induction occurs in cells that have suffered DNA damage caused by ultraviolet light or some other agent. When damaged nucleotides such as pyrimidine dimers are removed by the cell's excision repair system, a set of bacterial genes is induced whose products are involved in helping the cell to survive irradiation; these genes are called the **SOS genes**. The SOS genes of undamaged cells are repressed by the cellular LexA repressor. Induction of the SOS genes occurs as follows. Excision repair excises single-stranded DNA oligonucleotides, which in turn bind to and activate the cell's RecA recombination protein such that it acquires a "co-protease" activity. The LexA repressor has a weak autoprotease activity that is enhanced by binding to the activated RecA protein, with the result that the LexA repressor cleaves itself into two domains and loses repressor activity, allowing induction of SOS gene transcription.

The λ CI repressor is a mimic of LexA; both consist of two globular domains connected by a linker segment. CI's weak autoprotease activity is also stimulated by activated RecA, leading to linker cleavage that inactivates CI (Figure 8.5d). Inactivation of CI leads to transcription from P_R (see Table 8.1) and P_L, so that the early genes *cro* and *N* are transcribed. N-antiterminated transcription from P_L leads to expression of both Xis (for "excision") and Int (for "integration"), which in turn sponsor excision of the prophage DNA from the bacterial chromosome, and the lytic transcription pattern ensues.

The Cro repressor suppresses CI synthesis and regulates early gene transcription

As CI is the repressor used to establish and maintain a lysogenic state, the dimeric Cro protein is the repressor that establishes and controls a lytic infection by turning off the synthesis of CI. Cro binds to the same operators as CI repressor, but with different affinities, and without cooperativity. The operator affinities for Cro are $O_R3 > O_R2 \approx O_R1$. Early during lytic growth, either during an infection or induction of a lysogen, Cro binds to O_R3 and blocks P_{RM}. This prevents CI expression and channels the infected cell to lytic growth. Later during lytic growth, sufficient Cro accumulates to partially fill O_R2 and O_R1, along with O_L2 and O_L1. Partial repression of P_R and P_L by Cro reduces synthesis of early proteins that are already present in sufficient amounts to allow the lytic phase to proceed. Occupancy of the O_R operators by Cro is summarized in Table 8.1.

Making the decision: go lytic or lysogenize?

You might say, "This is all well and good, but when a λ chromosome is injected into a cell, there is no re-

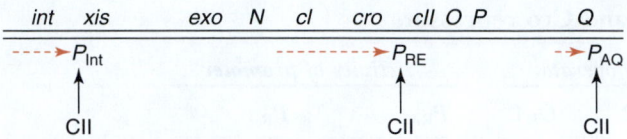

FIGURE 8.6 CII-activated promoters in lysogenization. See text for details.

pressor in the cell! If CI activates its own gene, whence comes the repressor that initially activates P_{RM}?" The answer is that λ has a system, involving a second *cI* promoter, to provide the CI repressor needed to activate P_{RM}. P_{RE}, the promoter for repression establishment, is used to make the initial dose of CI repressor that in turn activates P_{RM}. P_{RE} is located downstream of the *cro* gene, so that RNA polymerase initiating at P_{RE} transcribes *cro* backward, prior to transcribing *cI* (Figure 8.6). The antisense *cro* transcript may pair with the sense *cro* transcript to lower Cro expression. Whether a sufficient amount of CI is produced from P_{RE} transcripts is the crux of the choice to go lytic or lysogenic. Sufficient CI ensures that P_R and P_L are turned off and P_{RM} is activated, producing more CI and leading to stable repression of the lytic cycle. Insufficient CI results in the lytic program.

How much CI is produced from P_{RE} depends in turn on the level of the P_{RE} activator protein, CII. The *cII* gene, located just downstream from *cro* (Figure 8.6), is transcribed from the P_R promoter and sponsored by N antitermination. Like P_{RM} (activated by low levels of CI, see Table 8.1), P_{RE} is a weak promoter that, in this case, is activated by CII. Since the level of CI produced from P_{RE} depends on the level of CII, the lysis/lysogeny decision depends finally on the intracellular concentration of CII protein. In addition to activating P_{RE}, CII activates two additional promoters (Figure 8.6). P_{Int} is a CII-dependent promoter for a gene that encodes Int, the site-specific recombinase that inserts the λ prophage into the bacterial chromosome. P_{Int} lies within the coding region for the *xis* gene, so its product, the protein that directs excision of λ prophage from the cell chromosome, is not expressed during lysogenization. The

TABLE 8.2 Activities of λ promoters

Promoter	Phase of growth cycle	Genes transcribed	Control mechanism
P_R	Prophage	None	Repressed by CI
	Immediate early	*cro*	Terminated at t_R1
	Early	*cro*, *cII*, *O*, *P*, *Q*	Antiterminated by N
	Late	" (reduced levels)	Repressed by Cro
P_L	Prophage	None	Repressed by CI
	Immediate early	*N*	Terminated at t_L1
	Early	*N*, *cIII*, recombination genes	Antiterminated by N
	Late	" (reduced levels)	Repressed by Cro
P_R'	Prophage	None	Terminated at t_R'
	Immediate early	None	Terminated at t_R'
	Early	Lysis and structural protein genes (low)	Antiterminated by N
	Late	" (high)	Antiterminated by Q
P_{RE}	Prophage	*cI*, anti-*cro* (very low)	Weak constitutive activity
	Immediate early	" (very low)	Weak constitutive activity
	Early	" (high)	Activated by CII
	Late	" (very low)	Weak constitutive activity
P_{RM}	Prophage	*cI* (moderate)	Activated by CI
	Immediate early	" (very low)	Weak constitutive activity
	Early	None	Repressed by Cro
	Late	None	Repressed by Cro
P_{int}	Prophage	*int* (low)	Weak constitutive activity
	Immediate early	" (low)	Weak constitutive activity
	Early	" (high)	Activated by CII
	Late	" (low)	Weak constitutive activity
P_{AQ}	Prophage	Antisense *Q* (low)	Weak constitutive activity
	Immediate early	" (low)	Weak constitutive activity
	Early	" (high)	Activated by CII
	Late	" (low)	Weak constitutive activity

TABLE 8.3 Roles of λ regulatory gene products

Gene product	Action	Result
CI	Represses P_R and P_L	Shuts off early transcription, excludes exogenous λ phages ("immunity")
	Activates P_{RM}	Maintains CI expression during lysogeny
	At high concentration, represses P_{RM}	Controls CI levels during lysogeny
Cro	Represses P_{RM}	Shuts off CI expression during lytic growth
	Turns down P_R and P_L	Reduces early transcription during late phase
CII	Activates P_{RE}	Turns on CI expression, directs cell to lysogeny
	Activates P_{int}	Makes Int protein, leads to prophage integration
	Activates P_{AQ}	Reduces Q expression, lowers late transcription
CIII	Blocks proteolysis of CII	Directs cell to lysogeny
N	Antiterminates at t_R1, t_L1, and downstream terminators	Activates early transcription from P_R and P_L
Q	Antiterminates at t_R'	Activates late transcription from P_R'

third CII-activated promoter, P_{AQ}, is located within gene *Q*. Transcription from P_{AQ} produces an anti-sense transcript of *Q* that lowers Q expression, delaying late gene expression.

Since the level of CII determines the lysis/lysogeny decision, an important question becomes: What controls CII? CII is inactivated by host cell proteases, and has a half-life of only a few minutes. It is even less stable in metabolically active cells than in inactive cells. The stability of CII is enhanced by CIII, which inhibits the host proteases. The web of interactions between CII, CIII, and the proteases is not completely defined. However, it is clear that the level of CII in the cell, and hence the decision to lysogenize, depends on the metabolic state of the cell.

A quick review

The events of the lytic/lysogenic decision may be summarized as follows (see Tables 8.2 and 8.3). During an infection, immediate early transcription from P_L and P_R produces N and Cro. Antitermination sponsored by N leads to early gene expression that includes CII and CIII, along with recombination and replication proteins. At this point, two outcomes are possible, depending on the level of CII in the cell.

On the one hand, if the host cell is metabolically active, instability keeps the CII concentration low, resulting in insufficient CI production from P_{RE} to repress P_L and P_R. Continued transcription from P_L and P_R produces high levels of early proteins, including the late gene activator Q, and the lytic program ensues in >90% of infected cells. On the other hand, if the host cell is metabolically inactive, resulting in fairly stable CII, sufficient CI is produced from P_{RE} to repress P_L and P_R. Repression arrests production of early proteins (including Cro and CII) and activates P_{RM} for prophage expression of CI. Int produced from CII-activated P_{Int} inserts the λ chromosome into the bacterial chromosome. As CII decays, P_{RE}, P_{Int}, and P_{AQ} return to their unactivated states. CI produced from P_{RM} maintains prophage repression. More than 90% of infected stationary phase cells become lysogenized.

"The plot thickens." S. Holmes

Breaking and entering: the insertion of λ DNA into the bacterial chromosome

To be stably inherited as a prophage, the circularized λ chromosome must integrate into the bacterial chromosome; otherwise, the nonreplicating circles are simply diluted out as the cells divide. So λ carries a site-specific recombination system for integration into and excision from the bacterial chromosome (Figure 8.7). Int is a phage protein that catalyzes crossing-over between the phage *attP* site, near the *int* gene, and the bacterial *attB* site, located between the galactose (*gal*) and biotin (*bio*) operons. *attP* is ~250 bp long, with multiple binding sites for Int; in contrast, *attB* is about 25 bp long. *attP* is also designated POP′, to symbolize its three parts. O is a 15-bp-long "core" sequence found in both *attP* and *attB*, where the DNA crossing-over reaction takes place. Flanking the *attP* core are P and P′ arms.

Int binds to two different DNA sequence motifs within the arms and the core of *attP*. Int binds to the arm sequences strongly; the arms also have binding sites for the *E. coli* DNA bending protein, IHF, which bends the arms to allow Int to make bridging contacts with

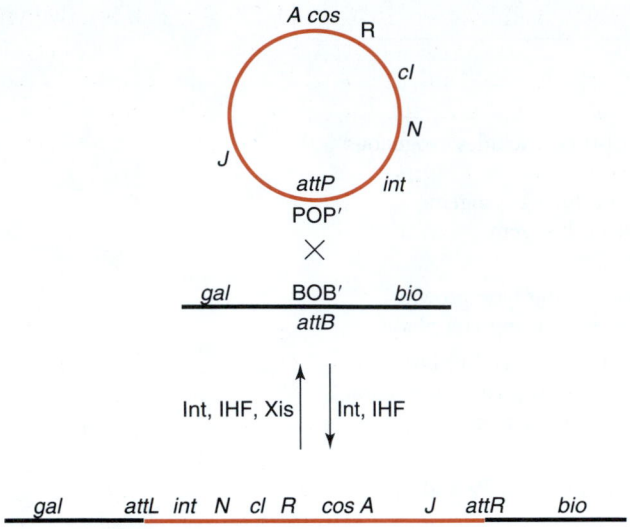

FIGURE 8.7 Integration and excision of λ DNA. Circular λ DNA is shown above the *E. coli* chromosome with the *attP* and *attB* sites aligned. Integration (downward arrow) and excision (upward arrow) are catalyzed by the proteins shown next to the arrows. The integrated λ DNA, flanked by *attL* and *attR*, is shown in orange at bottom.

the core sequences. The result is a complex nucleoprotein structure called an **intasome**. In the intasome, some of the Int proteins have free binding domains; these free motifs enable the intasome to capture an *attB* site on a bacterial chromosome by binding to its core sequence. Then the crossover occurs, inserting λ into the bacterial chromosome.

The great escape: The liberation of λ DNA from the bacterial chromosome

The same two proteins, Int and IHF, plus an excision protein named Xis, are involved in the reverse reaction that generates a circular DNA molecule from the integrated λ prophage. The *attP–attB* exchange during integration shuffles the *att* sequences, creating *attL* (BOP′) and *attR* (POB′) sites at the left and right junctions between the prophage DNA and the bacterial chromosome (Figure 8.7). When an integrated prophage is induced, N-antiterminated mRNA from P_L produces Int and Xis, the latter a protein that binds to sites only in the P arm of *attR*. Nucleoprotein structures assembled at *attL* (IHF, Int) and *attR* (Int, IHF, Xis) combine, and excision occurs.

In an intriguing variation, the prophage DNA of the λ-like phage N15 is linear, and is not integrated into the cell chromosome; such linear plasmids are rare in *E. coli*. Like λ, N15 virion DNA has cohesive ends that cyclize the molecule upon injection. However, instead of an *attP* site and Int protein, N15 has a *telRL* site and a **telomerase** enzyme. This enzyme opens the dsDNA at *telRL* and seals the resulting ends, creating covalently closed hairpins at both ends of the duplex. Upon induction, these ends must be opened and joined to recreate the *telRL* site.

Int synthesis is controlled by retroregulation

After excision of a prophage induced to lytic growth, and also during a lytic infection, high levels of Int are no longer necessary. An interesting mechanism called *retroregulation* limits Int expression during lytic growth only. A site named *sib*, located just beyond the *attP* site downstream of the *int* gene (Figure 8.2), contains a transcription termination sequence. When RNA polymerase reads through the *sib* sequence during N-antiterminated transcription from P_L, the mRNA forms a hairpin structure, SIB, that is a site for attack by double-stranded RNA-specific ribonuclease III (Figure 8.8; segments 1 and 2 pair with segments 3 and 4).

RNAse III cleavage at the SIB hairpin leads to further degradation of the mRNA into the adjacent *int* coding sequence, limiting expression of the Int protein. On the other hand, transcripts initiated at the P_{Int} promoter, which are not subject to N antitermination, terminate upstream of segment 1 in *sib*. This RNA cannot form the complete SIB hairpin; therefore these transcripts are not cleaved by RNAse III, and are more stable. Thus, during lysogenization, CII-activated expression of Int from P_{Int} is not limited by retroregulation. Note also that in the prophage state, *sib* is separated from *int* by site-specific recombination at *attP; sib* is only downstream of *int* after the prophage has been excised. Therefore N-antiterminated transcription from P_L on prophage DNA is not subject to retroregulation, and sufficient amounts of Int and Xis are produced for prophage excision.

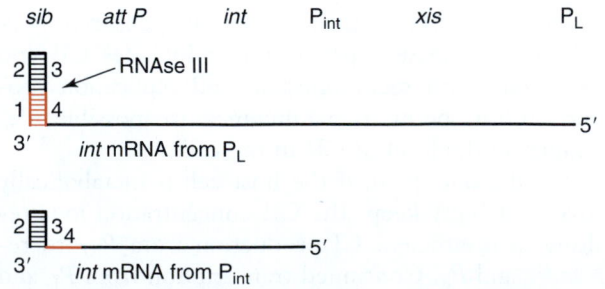

FIGURE 8.8 Retroregulation of int synthesis. Hairpin 1,2/3,4 at the 3′ end of int mRNA made from P_L is long enough to be cleaved by RNAse III, resulting in instability, but hairpin 2/3 at the 3′ end of int mRNA made from P_{int} escapes RNAse III cleavage and this mRNA is more stable.

λ DNA replication is directed by O and P, but carried out by host cell proteins

Expression of the early genes *O* and *P* leads to λ DNA replication. The O protein binds to multiple sites in the replication origin, *ori*λ (actually located within the *O* gene; Figure 8.2), forming a nucleoprotein structure, the "O-some." As a result, the two strands of an adjacent AT-rich segment of the DNA become unpaired. The P protein recruits the *E. coli* replication helicase DnaB to the O-some. In hijacking DnaB, P binds so tightly that *E. coli* chaperone proteins DnaJ and DnaK are needed to release it from the O-some. Then growing points for DNA replication are set up using *E. coli* replication proteins. RNA primers are laid down on the separated DNA strands, and DNA polymerase extends these primers to begin copying the DNA. Bidirectional replication produces progeny circles at early times, prior to the lysis/lysogeny decision (Figure 8.1). During lytic growth, there is a shift to **rolling circle** replication (see Chapters 6 and 13). This mode produces multimeric double-stranded DNAs, called **concatemers**, that are the DNA packaging substrate.

Assembly of λ heads involves chaperone and scaffolding proteins

A large number of λ proteins are involved in the construction of virions; some are incorporated into the head or tail structures intact, some are cleaved during incorporation, and some are used as **scaffolding proteins** but are not incorporated into the mature virus particle. Assembly of the icosahedral capsid is initiated on a donut-shaped dodecamer of B, the "portal protein." B is modified by addition of small numbers of proteins X_1 and X_2, which are formed by a protein fusion reaction between C and E. It is thought that the X_1 and X_2 proteins form a "bushing" on the B donut, to which E can bind and form the head shell. The host GroES and GroEL chaperone complex assists in the formation of the portal vertex. In fact, these host chaperone proteins were first discovered by isolating *E. coli* mutants that did not support growth of λ phage. The GroE gene was so named because certain mutations in the λ E gene allowed growth on these *E. coli* mutants. Chaperones have since been shown to play a major role in the proper folding of many proteins and in the assembly of multisubunit protein complexes.

The λ scaffolding protein Nu3 helps E, the major head protein, to form a closed icosahedral shell containing many (~405) E subunits. B molecules at the portal vertex are simultaneously cleaved to remove 20 amino acids from their amino termini. The single B-containing portal is located at one of the 12 vertices of the icosahedral shell, called a prohead.

Recently the first crystal structure of the head shell of a tailed dsDNA bacteriophage was determined for the λ-like phage HK97. The HK97 shell is remarkable in that neighboring subunits of gp5, its major head protein, become covalently joined in pentameric and hexameric rings. The rings, in addition, loop through rings formed by neighboring pentamers and hexamers, generating protein chainmail!

DNA is inserted into preformed proheads by an ATP-dependent mechanism

Concatemeric λ DNA generated by rolling circle replication is cut to generate unit-length virion DNA molecules during packaging of the DNA into the prohead shell (Figure 8.9). Cutting requires the introduction of nicks, staggered 12 bp apart, so that the single-stranded cohesive ends are made. Terminase, the viral cutting enzyme, contains multiple Nu1 and A subunits. Nu1 binds to an anchoring site *cosB*, and protein A cleaves the DNA at an adjacent nicking site, *cosN*. Following nicking of a concatemer, terminase separates the cohesive ends, remaining bound to the end of the DNA at *cosB*.

This DNA-protein complex docks at the portal vertex of a prohead, and DNA packaging proceeds. ATP hydrolysis by λ terminase powers translocation of the DNA into the prohead. The mechanics of the DNA packaging motor are not fully understood. Models include myosin-

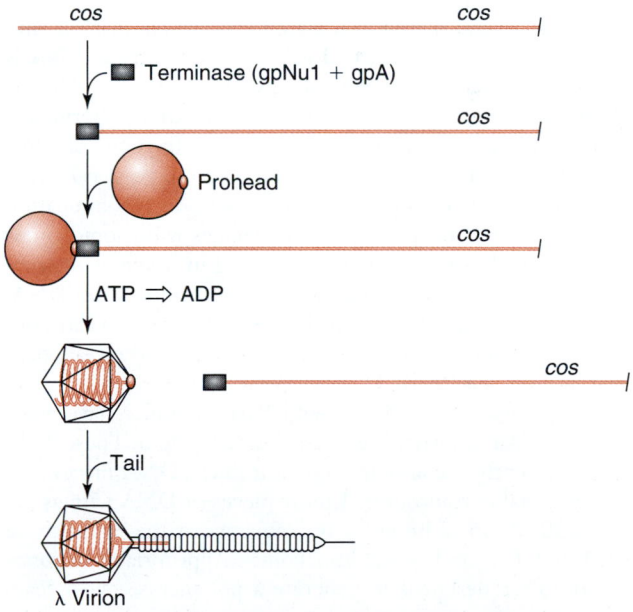

FIGURE 8.9 DNA packaging and virion assembly. A segment of concatemeric λ DNA is shown cleaved at a *cos* site by λ terminase, which then binds to a prohead. ATP then powers translocation of the DNA into the prohead, changing the prohead's shape. As the DNA is packaged, the downstream *cos* site is brought to, and cut by, terminase, which can then package another genome into another prohead. The filled head binds a tail to generate an infectious λ virion.

like contractions of terminase subunits, or rotation of the portal protein to winch or screw the DNA into the head shell. During translocation of the DNA into the prohead, the E shell subunits undergo rearrangement, causing the shell to expand and become more angular. This rearrangement creates openings in the shell lattice for D trimers, that fill the sites and greatly stabilize the shell. As the shell is filled, the next *cos* site of the concatemer approaches the terminase located at the portal. Cutting the second *cos* sequence requires recognition of *cosQ*, a site located just upstream of the nicking site *cosN*, and this completes packaging of a chromosome. Terminase undocks from the filled head, remaining bound to the end of the concatemer to form a new complex that is ready to bind to another prohead. Proteins W and FII bind to the portal vertex, stabilizing the filled head and creating a binding site for tail attachment.

Tail formation starts with J, the attachment fiber. Proteins J, K, I, H, M, and L assemble into the cone-shaped initiator, and then hexameric disks of V, the major tail subunit, stack on the initiator. The number of the hexameric V disks, 32, is dictated by the length of H, which Isao Katsura and Roger Hendrix have shown acts as a tape measure by extending up the central hole of V hexamers. When the stack of V disks reaches to the end of H, U binds, terminating tail polymerization. Finally Z is added and the tail is complete. About 200 bp of the *cosQ*-containing end of the λ DNA protrude down into the tail of a virion from within the head (Figures 8.3 and 8.9). The DNA may interact with the protein Z of the tail, because the tails of Z-less virions do not have inserted DNA.

Recombineering: a powerful use for Red, λ's general recombination system

Although several methods allow one to put circular plasmid DNA into *Escherichia coli* and other bacteria, linear DNA is rapidly degraded by the cell's RecBCD exonuclease. The inability to use linear DNA in transformation experiments makes it difficult to create gene knockouts. In yeast cells, linear DNA transformation works well, because yeast cells very efficiently carry out double-strand break repair recombination to join the ends of linear DNA molecules with homologous chromosomal DNA sequences, resulting in frequent replacement of chromosomal sequences by the incoming linear DNA. The Exo and Beta proteins of the λ Red system also carry out double-strand break repair recombination, and the Gam protein inhibits the RecBCD exonuclease.

The labs of Don Court and Barry Wanner have engineered cells to transiently express the Red system. These Red^+ cells efficiently recombine incoming linear DNA pieces into the bacterial chromosome. Linear pieces of DNA with as little as 30–40 bp of homologous sequence at the ends can be used for bacterial gene knockouts. Appropriate oligonucleotides are designed to generate a polymerase chain reaction product containing a chloramphenicol resistance gene flanked by short segments of homology to the ends of the target gene. The DNA piece is used to transform Red^+ *E. coli* cells to chloramphenicol resistance, resulting in cells in which a segment of the target gene is replaced by the chloramphenicol resistance gene. Since replacement occurs in about 0.1% of the cells, knockouts and other manipulations can also be done without a selectable marker. Tests of the system's applicability to other bacterial species should be forthcoming.

Host cell lysis

At the end of lytic growth, the host cell is lysed to release progeny phages. Three λ gene products govern lysis. Two of the proteins, R, a transglycosidase, and Rz, an endopeptidase, attack the cell's peptidoglycan layer. The third, S, provides a route through the cytoplasmic membrane for R and Rz. S has been dubbed a *holin*, as it multimerizes in the membrane, forming a pore. S assembly is regulated so that the pore forms at the proper time.

SUGGESTED READING

1. de Beer, T., Meyer, J., Ortega, M., Yang, Q., Maes, L., Duffy, C., Berton, N., Sippy, J., Overduin, M., Feiss, M., and Catalano, C.E. (2002). Insights into specific DNA recognition during the assembly of a viral genome packaging machine. *Molecular Cell* **9,** 981–991.

2. Hendrix, R.W., Roberts, J.W., Stahl, F.W., and Weisberg, R.A. Eds. (1983). Lambda II. Cold Spring Harbor Laboratory Press, Cold Spring Harbor, NY.

3. Ptashne, M. (2004). A genetic switch: Phage lambda revisited, 3rd Ed. Cold Spring Harbor Laboratory Press, Cold Spring Harbor, NY.

4. Dodd, I.B., Shearwin, K.E., Perkins, A.J., Burr, T., Hochschild, A., and Egan, J.B. (2004). Cooperativity in long-range gene regulation by the lambda CI repressor. *Genes and Development* **18,** 344–354.

5. Gan, L., Conway, J.F., Firek, B.A., Cheng, N., Hendrix, R.W., Steven, A.C., Johnson, J.E., and Duda, R.L. (2004). Control of crosslinking by quaternary structure changes during bacteriophage HK97 maturation. *Molecular Cell* **4,** 559–569.

6. Wang, I.N., Deaton, J., and Young, R. (2003). Sizing the holin lesion with an endolysin-beta-galactosidase fusion. *Journal of Bacteriology* **185,** 779–787.

KEY TERMS

Concatemer
Intasome
Lysogen
Operator
Prophage
Rolling circle
Scaffolding proteins
SOS genes
Telomerase
Temperate phage

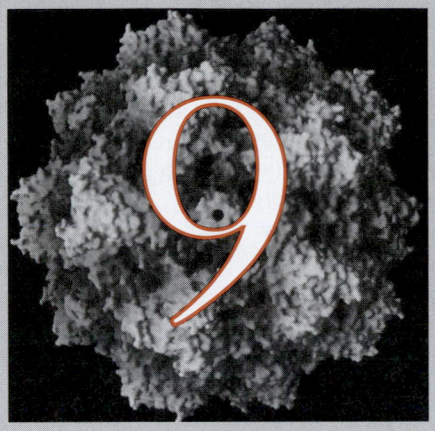

9

Parvoviruses

Parvoviridae

From Latin *parvus* (small), referring to size of virion

VIRION

Naked icosahedral capsid (T = 1).
Diameter 28 nm.

GENOME

Linear ss DNA, 5 Kb.
Self-complementary hairpin structures at genome termini.

GENES AND PROTEINS

1–3 transcriptional promoters on same DNA strand; transcribed by cell RNA polymerase II.
Splicing generates up to six different mRNAs.
2–4 replication (Rep) or nonstructural (NS) proteins: DNA replication and regulation.
1–3 capsid (Cap) proteins: formation of virion.

VIRUSES AND HOSTS

Autonomous parvoviruses: minute virus of mice, canine and feline parvoviruses.
Dependoviruses: adeno-associated viruses (human, bovine, canine).
Erythroviruses: B19 (humans).
Densoviruses: infect insects.

DISEASES

Humans: B19 causes erythema infectiosum (rash/fever), aplastic crisis in anemia.
Leukopenia and enteritis in dogs, cats, mink.
Dependovirus infection has not been associated with disease.

DISTINCTIVE CHARACTERISTICS

Strong dependency on cell cycle.
Autonomous parvoviruses replicate only in S-phase cells.
Dependoviruses require a helper virus to replicate.
DNA must become double-stranded before being transcribed.
Site-specific integration into host genome (adeno-associated virus).
Selectively replicate in and kill tumor cells.
Used as vectors for gene transfer into human cells.

Parvoviruses have very small virions and a linear, single-stranded DNA genome

Parvoviruses, as their name implies, are among the smallest of all viruses. Although their genome (5 Kb) is actually larger than that of hepatitis B virus (3 Kb) and is approximately the same size as the genomes of the polyomaviruses, parvoviruses have small virions because their DNA is single-stranded, and can therefore be packaged very compactly. Their 28 nm diameter, nonenveloped T = 1 icosahedral capsid (Figure 9.1) resembles the capsid of the single-stranded DNA bacteriophage PhiX 174 (Chapter 6).

In spite of their small size, parvoviruses have a number of unique properties that set them apart from other viruses. Parvovirus genomes are linear, not circular like those of the single-stranded DNA bacteriophages. The ends of their linear DNA molecules have self-complementary nucleotide sequences that can anneal to form hairpin structures. These unusual stuctures enable DNA replication by serving as primers for DNA polymerase. Certain parvoviruses can integrate their DNA into a specific region of a human chromosome, in a manner reminiscent of site-specific integration of bacteriophage lambda (Chapter 8). This ability to integrate has stimulated interest in using these viruses as vectors for the transfer of genes into human cells.

Parvoviruses replicate in cells that are going through the cell cycle

The small size and limited genetic coding capacity of parvoviruses make them exceptionally dependent on host cell functions. In contrast to the polyomaviruses and adenoviruses (Chapters 10 and 12), parvoviruses are unable to induce cells to enter the DNA synthesis (S) phase of the cell cycle. Since parvoviruses nevertheless need the cellular DNA replication machinery for their own replication, they are able to productively infect only cells that are going through the cell cycle. This and their cell-type dependence largely account for the biology and pathology of parvovirus infections.

Parvoviruses (Table 9.1) have been isolated from numerous mammals and birds, and members of a distinct subfamily, the densoviruses, infect a variety of insects. Parvoviruses that infect animals can be divided into two major groups: the autonomous parvoviruses and the dependoviruses. The autonomous parvoviruses can replicate in cells that spontaneously enter the DNA replication phase of the cell cycle. However, the dependoviruses require coinfection with an adenovirus or a herpesvirus to undergo productive replication; these viruses are therefore named adeno-associated viruses (AAV). Several autonomous parvoviruses can cause disease in animals; canine parvovirus, feline panleukopenia virus, and Aleutian mink disease virus are well-known examples. No diseases have been associated with infection by dependoviruses. The erythrovirus B19, named for its specificity for red blood cell precusors, is noteworthy as a parvovirus that is pathogenic for humans (see below).

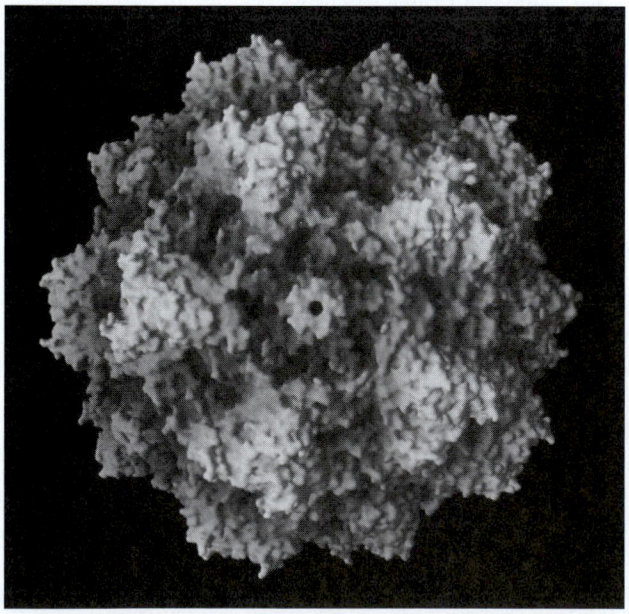

FIGURE 9.1 **Virion of minute virus of mice.** Structure determined by x-ray crystallography. The virion is oriented so that one of the 12 fivefold axes of symmetry is visible.

Discovery of mammalian parvoviruses

Several autonomous parvoviruses were discovered during the hunt for tumor viruses. Material from tumors was used to infect cell cultures or laboratory rodents and the viruses that emerged were considered as possible candidate oncogenic viruses. However, it soon became clear that these parvoviruses do not cause cancer. Rather, they usually replicate more efficiently in tumor cells. This so-called **oncotropism** of parvoviruses may well turn out to be useful because it could be a basis for targeting a lytic virus to cancer cells.

Other parvoviruses were discovered by chance. For example, the B19 virus was found during screening for an unrelated virus, hepatitis B virus, and only later was it shown to cause human disease. Adeno-associated viruses were originally found contaminating preparations of adenovirus; they are fairly widespread since the majority of people have antibodies against them. Minute virus of mice (MVM), an autonomous parvovirus, was first discovered contaminating a preparation of a mouse adenovirus, suggesting that these viruses, though competent for replication, are not above using some help from a larger virus if it is available. In much of what follows, we will use minute virus of mice and adeno-associated virus type 2 (AAV2) as examples, these being the two parvoviruses whose replication is best understood at the molecular level.

TABLE 9.1 Some representative parvoviruses

Subfamily	Example	Host
Autonomous parvoviruses	Minute virus of mice (MVM)	Mouse
	Canine parvovirus	Dog
Dependoviruses	Adeno-associated virus (AAV)	Human
Erythroviruses	B19	Human
Densoviruses	Culex pipiens densovirus	Mosquito

Parvoviruses have one of the simplest known virion structures

The structures of the spherical capsids of several parvoviruses have been solved at atomic resolution by x-ray crystallography (Figure 9.1). The capsid is constructed from 60 protein subunits arranged with T = 1 icosahedral symmetry. Each subunit contains the same 8-stranded antiparallel β barrel shape that is found in capsid proteins of many other icosahedral viruses (see Chapter 2). The single-stranded DNA molecule is tightly compacted within the capsid. The x-ray diffraction structure shows that some of the viral DNA is distributed with the same icosahedral symmetry as the capsid proteins. This implies that there are DNA recognition sites on the inside of the parvovirus capsid, and these are probably involved in packaging the viral DNA.

Parvoviruses have very few genes

The coding region of parvovirus DNA (Figure 9.2) can be divided into two approximately equal sections. The left half, from 5 to 50 map units, codes for the nonstructural (NS) or replication (Rep) proteins (a map unit is 1% of the length of the viral genome, or approximately 50 nucleotides). The right half, from about 50 to 90 map units, codes for the viral capsid (Cap) proteins. Despite the overall similarity of coding regions among genomes of different parvoviruses, they use different strategies for controlling the expression of viral genes. MVM uses one promoter (p4, located 4 map units from the left end) to synthesize the mRNA encoding the nonstructural proteins and a second promoter (p38) for the mRNA encoding the capsid proteins. AAV makes use of an additional promoter (p19) within the left open reading frame to generate mRNAs encoding shorter versions of its replication proteins. Finally, B19 seems to use predominantly a single promoter for transcription of the entire genome (not shown). In each case, differential splicing then generates the specific mRNAs that the virus needs.

The three capsid proteins, VP1, 2, and 3, have overlapping amino acid sequences, because they are translated in the same reading frame (Figure 9.2). For MVM, VP2 is the major capsid protein; only a few copies of VP1 are incorporated into the capsid. VP1 has additional N-terminal sequences and is translated from an alternatively spliced mRNA. VP3 is formed from VP2 by proteolytic cleavage of about 20 N-terminal amino acids, after its incorporation into the capsid. By contrast, the major AAV capsid protein is VP3; the larger VP2 is made by occasional initiation of protein synthesis at an ACG initiation codon upstream of the VP3 AUG start codon but in the same reading frame. The distinct roles of the minor capsid proteins are not well understood, although their absence can reduce infectivity of virions.

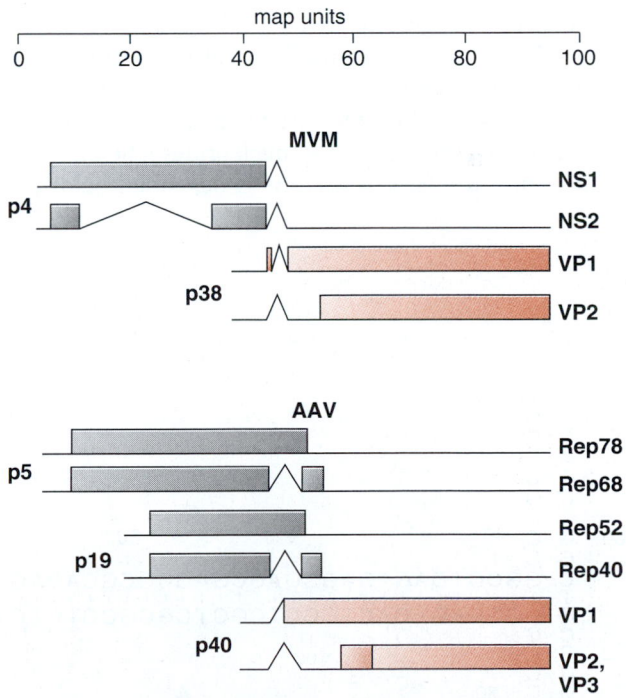

FIGURE 9.2 Genes and transcripts: minute virus of mice and adeno-associated virus. Messenger RNAs are shown with open reading frames as rectangular boxes (gray for early proteins, orange for late proteins) and introns as inverted V's. Transcriptional promoters are noted at the left; proteins coded by each mRNA are shown at the right. Map units divide the viral genome into 100 equal divisions of approximately 50 nt.

Single-stranded parvovirus DNAs have unusual terminal structures

One of the most distinctive features of parvoviruses is the structure of the linear, single-stranded viral DNA. Most autonomous parvoviruses (e.g., MVM) encapsidate only the DNA strand complementary to the viral mRNA. On the other hand, dependoviruses (e.g., AAV) encapsidate equally well strands of both polarities, but each virion contains only one of the two DNA strands. As a result, DNA extracted from AAV virions can self-anneal *in vitro*, producing a fully double-stranded DNA molecule. Researchers who first extracted AAV DNA were puzzled by this, because the AAV capsid is the same size and densisty as the MVM capsid, but their experiments suggested that AAV DNA was double-stranded and therefore AAV capsids should contain twice as much DNA as MVM capsids. AAV DNA can be extracted from virions under conditions that do not allow self-annealing, and it then remains single-stranded.

The nucleotide sequence at each end of parvovirus DNAs contains self-complementary regions. This allows one part of the terminal sequence to base-pair with the other part, enabling the molecule to fold back on itself to form hairpin-shaped or T-shaped structures (Figure 9.3). The length of the self-complementary sequence element ranges from 115 nucleotides at the 3′ end of MVM to 300 nt in B19. The sequences at the 3′ and 5′ ends of MVM are different from each other. However, the 125-nt self-complementary sequences at the two ends of AAV DNA are identical, except for the orientation of the internal 42 nucleotides labeled B and C in Figure 9.3 (see below). As is described below, 3′ terminal hairpins function as primers for the initiation of viral DNA replication.

Uncoating of parvovirus virions takes place in the nucleus and is cell-specific

Parvoviruses are particularly dependent on host cell functions, a result of their small genomes and limited coding capacity. Successful infection also depends on the differentiated state of the host cell. Parvoviruses bind to cells via receptors that have in some cases been identified: globoside (blood group P antigen) for B19 virus; transferrin receptor for canine parvovirus; and heparin sulfate proteoglycan, fibroblast growth factor receptor 1, and integrin $\alpha_v\beta_5$ for AAV2. The virions subsequently enter the cell by endocytosis. They migrate to the nucleus, where they enter via the nuclear pores and are uncoated, releasing the single-stranded viral DNA. At this point, autonomous parvoviruses like MVM must wait until the cell enters the S phase to begin their replication cycle. Dependoviruses like AAV rely on coinfection by a helper virus to push the cell into the S phase and for other functions.

Unexpectedly, the ability of parvoviruses to infect certain cell types or species seems be determined at the stage of virus uncoating. The specificity of canine parvovirus for dog cells rather than cat cells maps to the capsid protein gene. Likewise, the **tropism** of different MVM strains for fibroblasts or lymphocytes is determined by the capsid protein. In both cases, the amino acids important for this function are located at the surface of the virion. It was therefore expected that virus-receptor interactions would be the factor that deter-

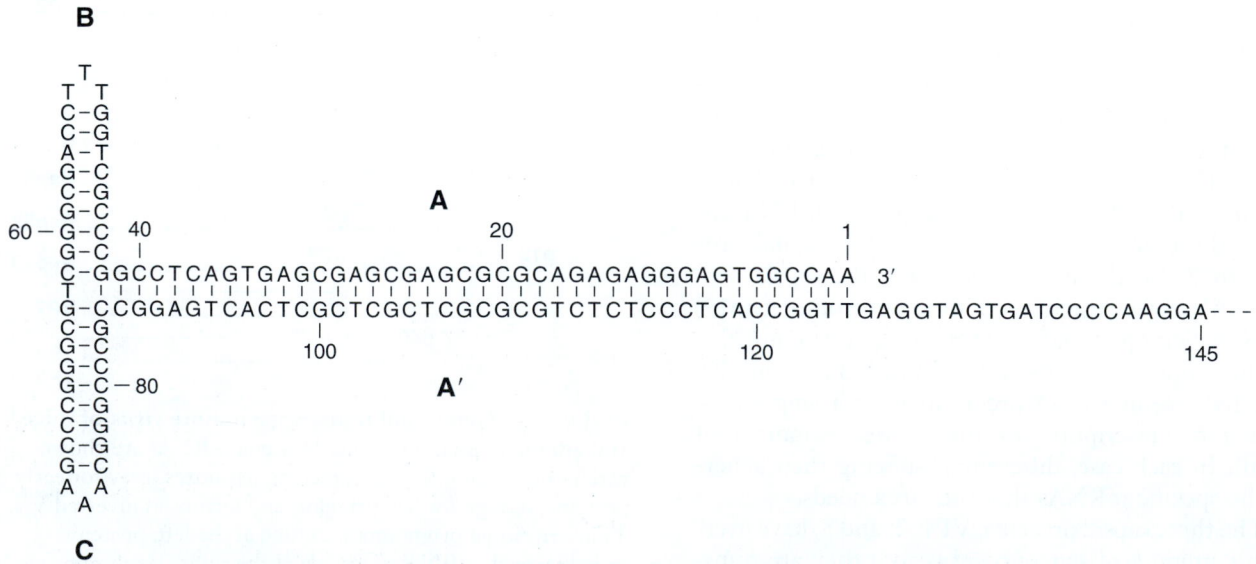

FIGURE 9.3 **DNA hairpin at the 3′ end of the AAV genome.** Nucleotides 1-41 (A) and 85-125 (A′) are base-paired, as are nucleotides 42-50/54-62 (B) and nucleotides 64-72/76-84 (C).

mines the specificity of infection. It was a surprise to discover that these viruses bind to and enter **permissive** and **nonpermissive cells** equally well. But when cells are transfected with naked viral DNAs instead of being infected with virions, the cell-specific block is overcome and normally nonpermissive cells produce virus. These findings point to uncoating, rather than entry, as the level of the cell-specific block to replication of certain MVM strains. This may involve an intracellular receptor involved in uncoating.

DNA replication begins by extension of the 3′ end of the terminal hairpin

Incoming parvovirus single-stranded DNA must first be converted to a double-stranded form to be able to function as a template for transcription. This is achieved by using the terminal hydroxyl group of the 3′ hairpin as a primer for host cell DNA polymerase. Because of the hairpin, this 3′ end is hydrogen-bonded to a complementary sequence further along on the same DNA strand (Figure 9.3), allowing the same DNA molecule to serve both as primer and template for DNA replication. Beginning at the 3′ end, the polymerase copies the single-stranded genome DNA up to its 5′ end, generating a double-stranded DNA molecule with an intact hairpin at one end (step 2 in Figure 9.4). This reaction is similar to that used to repair gaps in double-stranded cellular DNA.

Cellular RNA polymerase II then transcribes this template starting from the leftmost promoter (p4 in MVM, or p5 in AAV, see Figure 9.2). The strong p4 promoter contains binding sites for cellular transcription factors E2F, Ets, Sp1, and TATA-binding protein TBP (Figure 9.5). Transcription from this promoter generates mRNAs encoding the NS1 and NS2 proteins. NS1 is the main regulatory protein of MVM and has several functions to ensure efficient viral replication. NS1 binds to viral DNA and stimulates transcription from the p38 promoter, which controls synthesis of the mRNAs for the coat proteins. NS1 also further stimulates transcription from the p4 promoter, leading to increased production of its own mRNA. However, it is not clear exactly how NS1 activates transcription. The p38 promoter includes a binding site for the NS1 protein (ACCAACCA), in addition to Sp1 and TATA sites, but apparently activation can also occur without this sequence.

The DNA "end replication" problem

Besides acting as transcriptional activators, the NS proteins also directly control the replication of viral DNA. Parvoviruses are faced with an important general problem in the replication of DNA: the "end replication" problem. All known DNA polymerases need a primer, usually RNA, to initiate DNA synthesis, which proceeds in the 5′ to 3′ direction. RNA primers are synthesized by copying a short stretch of the template DNA strand, and are then extended by DNA polymerase to make a full complementary DNA strand (Figure 9.6). The RNA primer is later removed and replaced with DNA, whose synthesis is initated upstream of where the primer originally lay. However, an RNA primer at the very 5′ end of a linear DNA molecule cannot be replaced, because there is no complementary DNA strand to copy beyond the end of the template. Therefore, without a special trick, the very ends of a linear DNA cannot be duplicated, and the DNA would lose a short sequence each time it replicates ("X" in Figure 9.6). This problem also applies to linear cellular chromosomes. Many organisms, including bacteria such as *E. coli*, mitochondria, plasmids, and a number of DNA viruses, solve this problem by circularizing their DNA—thereby eliminating the ends. Linear DNAs need another way to solve this problem.

A British cell biologist, Thomas Cavalier-Smith, suggested how cells could solve the end replication problem for their linear chromosomes. If linear DNAs are provided with hairpin ends, they effectively contain two tandem inverted copies of the sequences at the ends of their DNA. Because the hairpin end can fold back on itself, it can serve as a primer for DNA polymerase, eliminating the need for an RNA primer. Some DNA gymnastics will then reproduce the original ends. It turns out that Cavalier-Smith's proposal is not actually used by eukaryotic cells; instead cells use the enzyme **telomerase** to extend and maintain the ends of their linear chromosomes. But his proposal does explain exactly the principle by which parvoviruses duplicate the ends of their linear DNA. Besides hairpin sequences at the ends of the DNA, the mechanism requires a site-specific endonuclease that cuts DNA within the hairpin—and both are present in parvoviruses. The steps are shown in Figure 9.4, using AAV as the example. Although all parvoviruses employ the same principle, the scheme is simpler with AAV since the two ends of the viral DNA are identical.

Steps in DNA replication

1. The palindromic sequences fold to form hairpins at the ends of the DNA.

2. The 3′-hydroxyl group (arrow) of the left hairpin is used as primer for DNA chain elongation, generating a fully double-standed DNA with one hairpin end. This can be used for transcription by RNA polymerase II.

3. The fully double-stranded DNA is nicked at a specific site, just opposite the original 3′ end, by Rep78/68 of AAV (or its equivalent NS1 in the case of

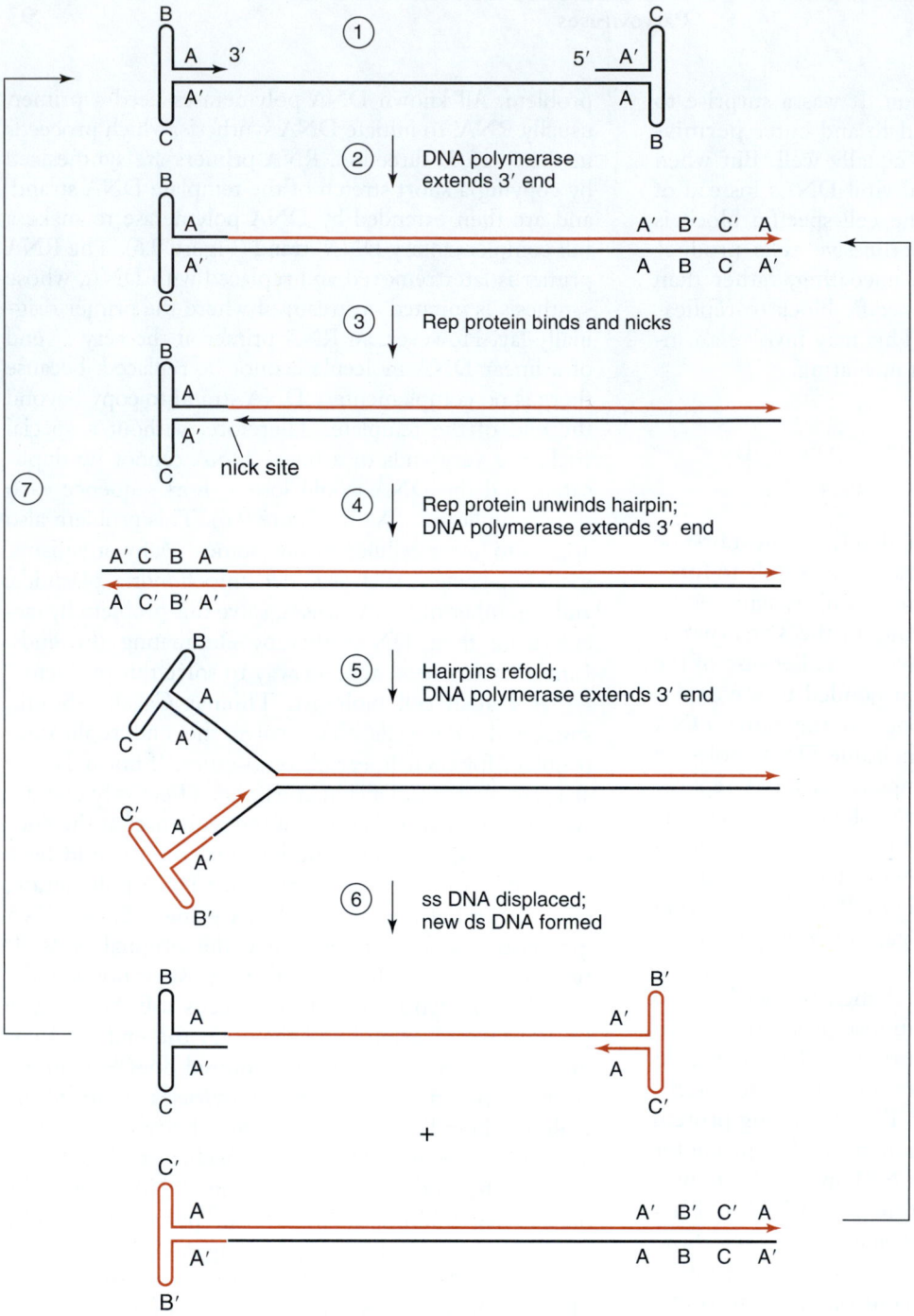

FIGURE 9.4 Replication strategy of AAV DNA. See text for details. Orange lines: newly synthesized DNA strands. Arrowheads show 3′ ends. Not drawn to scale: the hairpin ends are 125 nt in length, and the entire genome is 5 Kb long.

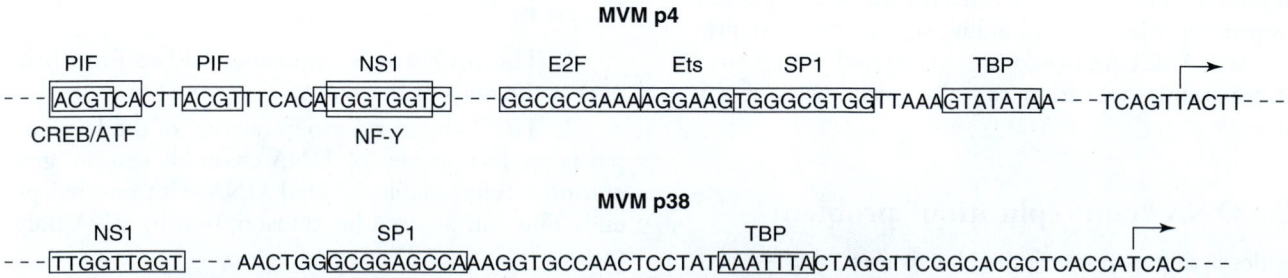

FIGURE 9.5 Binding sites for transcription factors in MVM promoters p4 and p38. Binding sites are boxed and the factor is noted above or below the box. Start sites for transcription are shown by arrows.

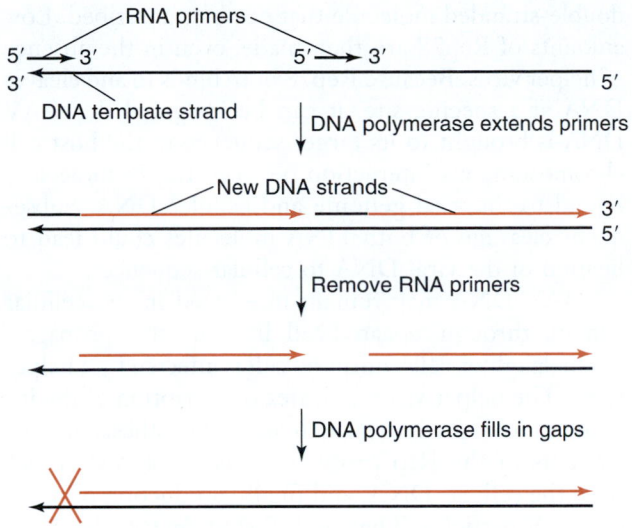

FIGURE 9.6 RNA primers and the DNA "end replication" problem. RNA primers (black) initiate DNA replication by copying a short sequence on the DNA template strand, and primers are then extended (orange) by DNA polymerase. The RNA primers are subsequently removed and DNA polymerase extends upstream DNAs, filling in the gaps. Adjacent DNA chains are then sealed with DNA ligase (not shown). However, the RNA primer at the very 5′ end of a linear DNA (orange "X") cannot be replaced because there is no DNA upstream.

MVM). This protein remains covalently attached to the 5′ end of the DNA it has cut and is incorporated into new virions along with the progeny DNA (and therefore it is not strictly a nonstructural protein).

4. Starting from the 3′ end at this nick, the cellular DNA polymerase then elongates the DNA chain in the other direction, to fully copy the left end hairpin. The NS1 or Rep78/68 protein helps by unwinding the hairpin.

5. The palindromic sequences at one end then reform hairpins to prime further DNA synthesis.

6. DNA synthesis using the newly formed hairpin generates a new double-stranded DNA molecule with a hairpin end, and also results in displacement of progeny single-stranded DNA. Thus the DNA replication cycle is complete. Note that refolding and priming can happen at either end, or at both ends simultaneously, in which case two double-stranded molecules are formed.

7. The displaced single-stranded DNA can begin a new replication cycle, or can be packaged if capsid proteins are present. The newly synthesized double-stranded DNA can be transcribed to provide more messenger RNA, or can again be nicked and extended, to generate further progeny DNAs.

The attentive reader will notice that the sequence at the 3′ end of the original AAV DNA molecule shown in Figure 9.4 reads ABCA′ in the 3′ to 5′ direction, and the sequence at the 3′ end of the progeny single-stranded DNA shown at the bottom of Figure 9.4 reads AC′B′A′, so that the order and orientation of B and C are inversed. This "flip-flop" sequence inversion, resulting from the nicking, unfolding, and copying of the hairpin, can be detected in AAV DNA, and was instrumental in testing the validity of the replication model.

The 3′ and 5′ ends of some parvovirus DNAs, such as MVM, have different sequences, which leads to a modified and more complex replication scheme, notably involving a dimer-length replicative intermediate. For details the reader is referred to the review by Cotmore and Tattersall (1995).

Nonstructural proteins are multifunctional

NS1 of the autonomous parvoviruses and its equivalent in AAV, Rep78/68, are the principal viral regulatory proteins and play several roles in parvovirus replication. Each protein has a number of functions and biochemical activities (Figure 9.7). They: (1) bind in a sequence-specific manner to viral DNA origins and promoters; (2) have ATP-binding, ATPase, and helicase activities; (3) are sequence-specific endonucleases; (4) are transcriptional activators; and (5) prevent infected cells from progressing through the cell cycle. The biological importance of the last activity may be to keep infected cells in S phase and so increase the time available for viral replication.

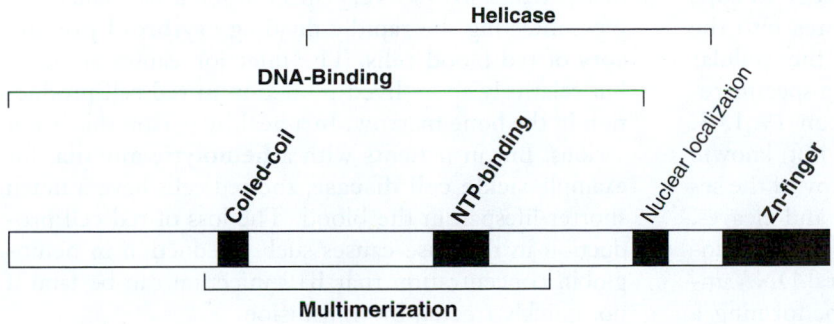

FIGURE 9.7 Functional domains in AAV Rep78 protein. The protein is shown with the N terminus at the left and the C terminus at the right. NTP = nucleoside triphosphate.

Much less is known about the functions of the minor nonstructural proteins. MVM NS2 protein is needed for efficient viral replication only in certain cell types, but the reasons and mechanism are not clear. Likewise, AAV Rep52 may be involved in encapsidation and interactions with cellular regulatory pathways, but again little is known.

Adenovirus functions that help AAV replication

For efficient replication, AAV requires functions provided by a helper virus. If the helper is adenovirus (see Chapter 12), the adenovirus genes involved are E1A, E1B, E2, E4, and the VA RNA genes. The combined action of these gene products ensures that the conditions in the cell are right for AAV replication. Herpesviruses can also serve as helper viruses, and exposure of the cell to certain agents that affect cell gene expression (ultraviolet light, chemical carcinogens, etc.) can allow virus replication to proceed in the absence of a helper virus. Examples of how some adenovirus gene products help AAV replication are:

1. Adenovirus E1A protein acts as a transcriptional activator both of AAV genes and of cellular genes involved in DNA replication, by binding to the retinoblastoma (Rb) protein and releasing it from the cellular transcription factor E2F (see Figure 12.6). As we saw above, E2F binds to the strong p4 promoter and helps activate gene expression.

2. Adenovirus E1B 55K protein and E4 orf6 protein combine to form a complex that transports AAV mRNAs from the nucleus, where they are made, to the cytoplasm.

3. VA RNA is a short, double-stranded RNA that binds to and inhibits RNA-dependent protein kinase (PKR). VA RNA therefore suppresses inhibition of viral protein synthesis directed by PKR when it is activated by longer, viral double-stranded RNA (see Chapter 31).

In the absence of helper virus, AAV DNA can integrate into the cell genome

What happens if AAV infects unstressed cells in the absence of a helper? AAV has evolved a strategy to cope with this situation: the viral DNA integrates into the cellular DNA and replicates as part of the cellular genome. The genome usually integrates in a specific region of human chromosome 19, at position 19q13.3. The exact mechanism of integration is not known. However, the integration site contains a copy of the sequence in AAV DNA where Rep78 binds and cleaves. It is also known that Rep78 is required to promote integration. It would appear that infecting viral DNA undergoes the first step in DNA replication, forming a double-stranded molecule that can be transcribed. Low amounts of Rep78 are then made, even in the absence of helper virus. Because Rep78 both binds to and cleaves DNA at a specific site, it can be imagined that AAV DNA is brought to its target sequence in the host cell chromosome via interaction between Rep78 molecules bound to the viral genome and cellular DNA. Subsequent cleavage of both DNA molecules could lead to ligation of the viral DNA to cellular sequences.

AAV DNA may remain integrated in the cellular genome through repeated cell divisions, thus propagating in daughter cells, until the cell is infected by a helper virus. The helper virus activates transcription of the integrated AAV genome, which leads to synthesis of large amounts of the Rep proteins, excision of AAV DNA from the cellular DNA, and finally production of progeny AAV particles. The survival strategy used by AAV, and its ability to integrate at a specific site into the human genome, are unique among animal viruses and resemble the strategy used by bacteriophage lambda (Chapter 8). It may be possible in the future to exploit site-specific integration for gene transfer applications using AAV-based vectors (see Chapter 33).

Parvovirus pathogenesis: the example of B19 virus

In a remarkable way the biological properties of parvoviruses help to determine the diseases they cause. Parvoviruses replicate in rapidly growing cells that are in specific states of differentiation, but not in resting cells that make up most of the body tissues. Typically parvoviruses will infect cells of the intestine, the hematopoietic system, and the fetus. Several parvoviruses of animals are responsible for serious diseases. Canine parvovirus infects puppies and causes disease and death. Feline **panleukopenia** virus, as the name suggests, affects blood formation in cats. Other parvoviruses that are not covered here, the densoviruses, infect invertebrate species. But there is just one parvovirus known to cause disease in humans, and that is the B19 virus.

B19 was discovered by accident during screening of blood donations. It was then realized that this virus was associated with a serious illness—**aplastic crisis** in anemia patients. B19 is very specific for a particular cell type, infecting the rapidly dividing **erythroid precursors** of red blood cells. The infection causes an acute but relatively short-lived reduction in red cell production in the bone marrow. In a healthy person this is not serious. But in patients with a **hemolytic anemia**, for example **sickle cell disease**, the red cells have a much shorter lifespan in the blood. The loss of red cell production in this case causes such a reduction in hemoglobin concentration that B19 infection can be fatal if not quickly treated by transfusion.

Antitumor effects of parvoviruses

Parvoviruses have a long-standing relation with cancer. Although some parvoviruses were originally isolated from tumor-bearing animals or tumor cells, they are not oncogenic. Rather, they grow productively in tumor cells, killing these cells. Parvoviruses have repeatedly been shown to reduce the formation of tumors that arise spontaneously in animals or after treatment with chemicals or other viruses. But how do they do this?

The antitumor effects of parvoviruses could be exerted in several ways. Autonomous parvoviruses grow better in transformed cells than in normal cells, partly because transformed cells are continually passing through the cell cycle. More NS1, which is toxic for cells, is produced in transformed cells. Transformed cells are, in addition, more sensitive to the toxic effects of NS1. The Rep78/68 proteins of AAV also have antiproliferative effects on cells.

However, a new mechanism recently came to light. It appears that the unusual structure of the AAV DNA, which is mostly single-stranded with terminal hairpins, looks like damaged DNA to the infected cell. This elicits a DNA damage response, even though the cellular DNA is not actually damaged. In normal cells, DNA damage leads to arrest of the cell cycle by a mechanism dependent on the cellular p53 protein. However, most tumor cells lack the cellular p53 protein or proteins that it activates, and in these cells, the DNA damage response caused by AAV infection leads to apoptosis and cell death. If this mechanism can be exploited in the future for treatment of cancer, then infection with certain parvoviruses could turn out to be of benefit for the host.

In addition to its involvement in aplastic crisis, B19 is also known to cause a common childhood disease called **erythema infectiosum** ("infectious rash"). Although usually not severe, this disease may lead to serious birth defects if a fetus is infected during pregnancy. The disease spectrum of B19 therefore well illustrates the basic principle of parvovirus pathogenesis resulting from infection of and damage to dividing cells.

SUGGESTED READING

1. Faisst, S. and Rommelaere, J. Eds. (2000). Parvoviruses: From molecular biology to pathology and therapeutic uses. Karger, Basel.

2. Cavalier-Smith, T. (1974). Palindromic base sequences and replication of eukaryote chromosome ends. *Nature* **250,** 467–470.

3. Cotmore, S.F. and Tattersall, P. (1995). DNA replication in autonomous parvoviruses. *Seminars in Virology* **6,** 271–281.

4. Heegaard, E.D. and Brown, K.E. (2002). Human parvovirus B19. *Clinical Microbiology Reviews* **15,** 485–505.

5. Raj, K., Ogston, P., and Beard, P. (2001). Virus-mediated killing of cells that lack p53 activity. *Nature* **412,** 914–917.

6. Rommelaere, J. and Cornelis, J. (1991). Antineoplastic activity of parvoviruses. *Journal of Virological Methods* **33,** 233–251.

7. Vihinen-Ranta, M., Suikkanen, S., and Parrish, C.R. (2004). Pathways of cell infection by parvoviruses and adeno-associated viruses. *Journal of Virology* **78,** 6709–6714.

8. Young, S.M. Jr., McCarty, D.M., Degtyareva, N., and Samulski, R.J. (2000). Roles of adeno-associated virus Rep protein and human chromosome 19 in site-specific recombination. *Journal of Virology* **74,** 3953–3966.

KEY TERMS

Aplastic crisis
Erythema infectiosum
Erythroid precursors
Hemolytic anemia
Nonpermissive cells
Oncotropism
Panleukopenia
Permissive cells
PKR
Sickle cell disease
Telomerase
Tropism

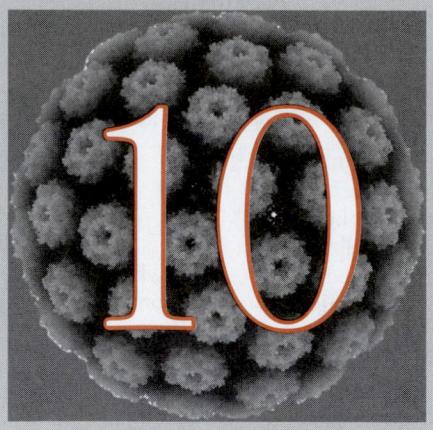

Polyomaviruses

Polyomaviridae

From Greek *poly* (many) and *-oma* (tumor)

VIRION

Naked icosahedral capsid (T = 7). Diameter 45 nm.
Formed from 72 capsomers, pentamers of VP1.

GENOME

Circular double-stranded DNA, 5.3 Kb.
DNA is packaged as "minichromosome" with nucleosomes formed from cellular histones.

GENES AND PROTEINS

Two transcription units: early and late.
Divergent transcription from a central control region containing origin of DNA replication.
Differential splicing produces 3–4 mRNAs from each transcription unit.
Early proteins (T antigens) regulate cell cycle and direct DNA replication.
Late proteins (VP1, 2, 3) make virus capsid.

VIRUSES AND HOSTS

Mouse polyomavirus; simian virus 40; polyomaviruses infecting birds, rodents, cattle;
and two human viruses: BK and JC virus.

DISEASES

Usually persistent, nonsymptomatic infections.
Progressive multifocal leukoencephalopathy, a rare demyelinating disease caused by human polyomavirus JC.

DISTINCTIVE CHARACTERISTICS

Polyomaviruses are dependent on cellular RNA synthesis and RNA processing enzymes, and DNA replication systems.
T antigens interact with multiple signaling pathways and activate the cell cycle to facilitate viral DNA replication.
T antigens are oncogenes that can transform nonpermissive cells *in vitro*.
Polyomaviruses are tumorigenic in animals when administered at high concentrations, but rarely cause tumors in nature.

Mouse polyomavirus was discovered as a tumor-producing infectious agent

While studying the transmission of leukemia in mice in the early 1950s, Ludwig Gross unexpectedly discovered that extracts of mouse tissues could cause salivary gland tumors when injected into baby mice. His group and that of S. Stewart and Bernice Eddy soon showed that these tumors were caused by a filterable agent. This virus was distinguished from retroviruses that cause mouse leukemia by its smaller size and greater stability. It was subsequently named polyomavirus (Greek *poly* for many and *-oma* for tumor), because it can cause tumors at numerous sites in experimental animals. Mouse polyomavirus is commonly found in wild mice and their inbred laboratory cousins but causes little or no overt disease. However, when injected into susceptible rodents and other mammals at high concentrations, mouse polyomavirus can cause multiple tumors in parotid glands, skin, kidneys, bone, lungs, liver, and many other tissues. These tumors grow rapidly and can eventually kill the host animal.

Simian virus 40 was found as a contaminant of Salk poliovirus vaccine

Simian virus 40 (SV40) was discovered some 10 years later as a contaminant of rhesus monkey kidney cell cultures that were being used for production of human poliovirus vaccines. Injection of simian virus 40 into newborn hamsters also produced tumors. It was soon recognized that mouse polyomavirus and simian virus 40 are closely related, and they were for some time grouped together with the papillomaviruses (Chapter 11), in a family called the *Papovaviridae*. However, differences in genome size and organization and in virus replication strategies led scientists to consider the polyomaviruses as members of a family (*Polyomaviridae*) distinct from the papillomaviruses (*Papillomaviridae*). A number of related polyomaviruses have been discovered in rodents, birds, monkeys, and cattle.

Before simian virus 40 was identified, a number of batches of the Salk inactivated poliovirus vaccine unknowingly contaminated with live simian virus 40 were administered by injection to millions of people. It has been estimated that virtually the entire population of the United States aged 20 and below by 1961 received such contaminated vaccine at some point. There has been controversy among scientists about the possibility that some human tumors may have been caused by simian virus 40 injected during poliovirus vaccination. Causality is hard to prove; some human tumors were shown to contain simian virus 40 DNA and express viral early proteins. However, this could be fortuitous, as simian virus 40 appears to circulate among the human population, causing no obvious symptoms. No clear excess levels of cancer or of specific tumors have been detected in people during the 45 years since they received contaminated poliovirus vaccine.

This vast, unplanned experiment on the effect of simian virus 40 infection on humans, as well as the general absence of tumors in naturally infected mice or monkeys, show that these viruses usually cause tumors only under very specific conditions: injection of high titers of virus into particularly susceptible animals shortly after birth. Polyomaviruses are therefore not "professional" tumor viruses. However, their complex interactions with cellular signaling pathways, developed to enable high levels of viral DNA replication and progeny virus production, can occasionally transform a normal cell into a tumor cell.

Two human polyomaviruses, named BK and JC after the initials of patients from whom they were isolated, are widespread in the human population but usually cause no apparent disease. A rare but fatal demyelinating disease, **progressive multifocal leukoencephalopathy (PML)**, is caused by JC virus, particularly in immunosuppressed individuals, including patients suffering from acquired immune deficiency syndrome (AIDS).

Polyomaviruses are models for studying DNA virus replication and tumorigenesis

Both mouse polyomavirus and simian virus 40 have been studied intensively starting shortly after their discovery because they can transform cells *in vitro* and cause tumors, and therefore they were correctly perceived as holding answers to basic questions about tumorigenesis in mammals. These viruses were also studied because their small number of genes and their strong dependence on cellular functions make them useful model systems for studying mammalian RNA synthesis and processing, DNA replication, and signal transduction. In this chapter, emphasis in different sections is placed on results found with either mouse polyomavirus or simian virus 40, but these two viruses have a common overall replication strategy and they share many details of structure, genome organization, and gene function.

Polyomavirus capsids are constructed from pentamers of the major capsid protein

The virions of polyomaviruses are naked capsids approximately 45 nm in diameter. Their surface is constructed from 72 capsomers visible by electron microscopy, each made of five molecules of the major capsid protein VP1 (Figures 10.1 and 10.2). Twelve of these capsomers are surrounded by five nearest neighbors, and define the fivefold symmetry axes typical of capsids with icosahedral symmetry. The rules of symmetry would suggest that the remaining 60 capsomers should be made from three or six molecules of VP1, because each such capsomer has six nearest neighbors. However, VP1 forms stable pentamers, not hexamers, and these pentameric capsomers fit into "sixfold holes" on the surface of the capsid. Capsomers are linked by the flexible carboxy-terminal arms of VP1, which extend across the gap and embrace VP1 molecules in neighboring capsomers, stabilizing the capsid structure (see Figures 2.8a and b). Every capsomer also contains a single molecule of one of the minor capsid proteins

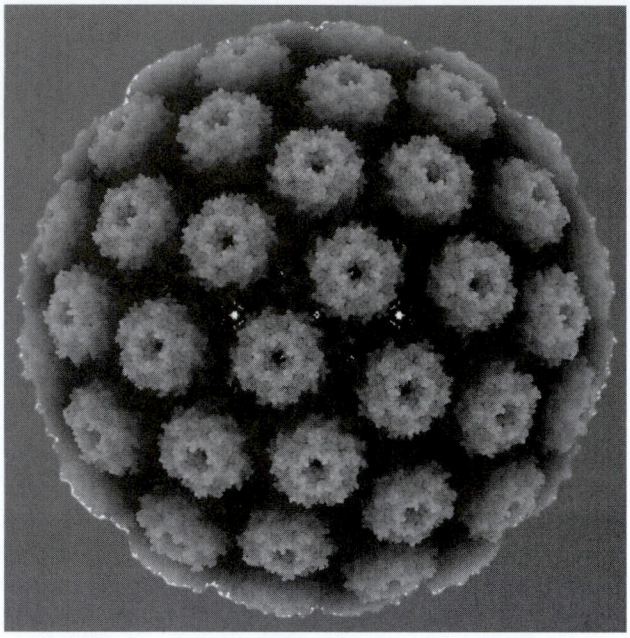

FIGURE 10.1 Computer-generated image of mouse polyomavirus capsid. The 72 capsomeres, each composed of pentamers of VP1 and a single molecule of VP2 or VP3, are organized with T = 7 icosahedral symmetry.

VP2 or VP3, located below the capsid surface. The N-terminal amino acid of VP2 is linked to **myristate**, a C_{14} saturated fatty acid. This common protein modification enables interaction of proteins with lipid membranes (see below).

The circular DNA genome is packaged with cellular histones

Packaged within the capsid is the viral genome, a 5.3-Kb circular, double-stranded DNA molecule (Figure 10.2a). The viral genome is associated with **nucleosomes** formed from two molecules each of the four cellular histones H2A, H2B, H3, and H4. These are the same histones that package and condense cellular chromosomal DNA. A length of approximately 200 bp of DNA is wrapped as a **toroidal coil** around the outside of each cylindrical nucleosome (Figure 10.2b). This condenses the viral DNA into a "minichromosome" containing about 25 such nucleosomes (Figure 10.3), giving a total of 200 histone molecules per viral DNA molecule. The use of cellular histones by polyomaviruses (as well as papillomaviruses) for DNA condensation and packaging is a striking example of genetic economy. The small genomes of these viruses have limited coding capacity, and instead of coding for their internal structural proteins, they simply pick them up in the cell nucleus, where viral DNA replicates.

Circular DNA becomes supercoiled upon removal of histones

When viral genomes are gently released from the capsid without denaturing the bound nucleosomes, they can be visualized as circular molecules by electron microscopy, and the minichromosome resembles a necklace of "beads on a string" (Figure 10.3). However, when the histones are removed by treating with the strong protein denaturant **sodium dodecyl sulfate** and extraction of the proteins with the organic solvent phenol, the circular DNA now becomes coiled about itself like a rubber band held between the thumbs and forefingers of both hands and twisted. This form of DNA is called **supercoiled**, and results from the toroidal coiling of the double-helical DNA around the nucleosomes in the intact minichromosome. These toroidal coils distribute themselves around the now-naked DNA molecule (Figure 10.3).

If the phosphodiester bond between two nucleotides of one of the DNA strands is cleaved ("nicked"), this

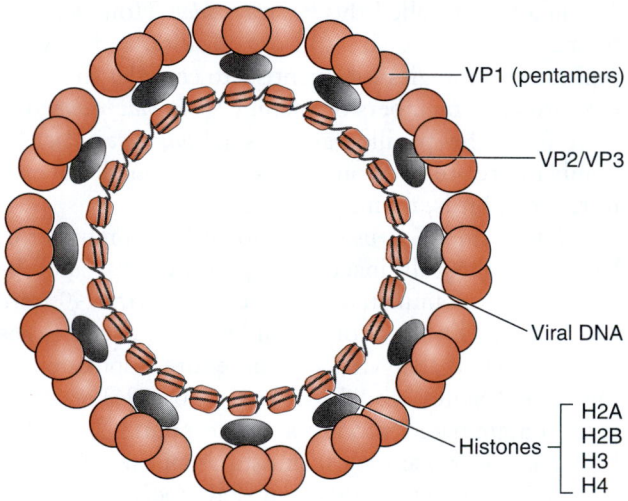

(a)

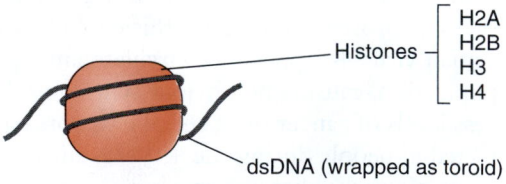

(b)

FIGURE 10.2 Components of polyomavirus virion. (a) A schematic diagram of a virion in cross-section, showing pentamers of VP1 (orange spheres), with single molecules of VP2 or VP3 (gray ovals) underneath the capsid surface. Circular viral DNA is bound to nucleosomes formed from cellular histones. (b) A larger view of a nucleosome, around which DNA is wrapped as a toroidal coil.

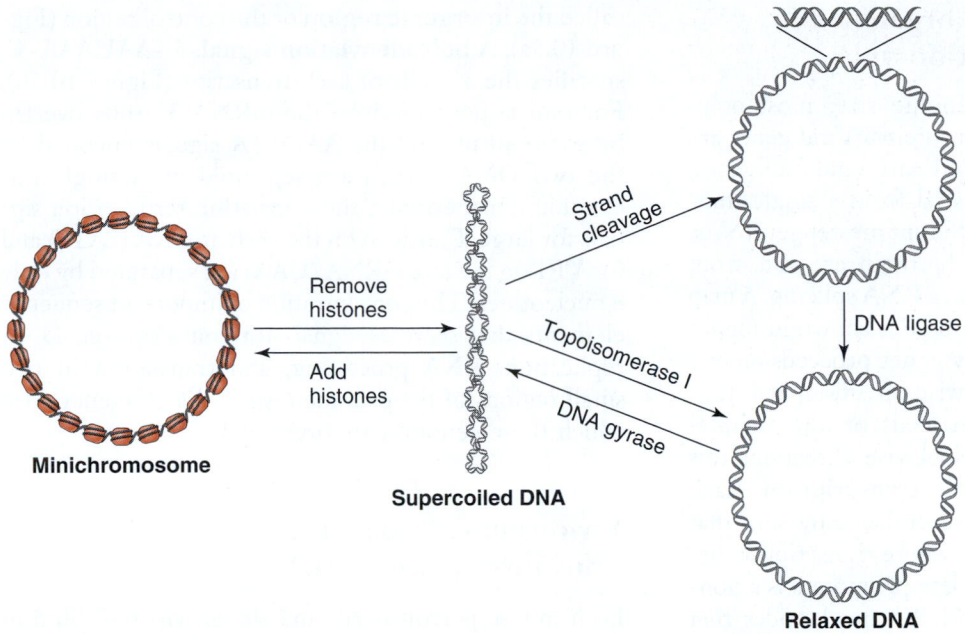

FIGURE 10.3 **Gymnastics of circular DNA molecules.** A circular, double-stranded DNA molecule is shown as a minichromosome, bound to nucleosomes. Removal of histones generates a supercoiled molecule due to the toroidal coils around each nucleosome. Nicking of one strand allows the supercoiled form to unwind to its lowest-energy relaxed circular state; a portion of the double-helix containing a nick is shown enlarged above. If the nick is sealed with DNA ligase, this becomes a relaxed, closed circular DNA molecule. Topoisomerase I can produce the same relaxed molecule by cleaving a phosphodiester bond on one strand of supercoiled DNA allowing the strands to rotate around each other, and resealing the cleaved bond. DNA gyrase uses energy to catalyze the reverse reaction, resulting in a supercoiled DNA molecule. Addition of nucleosomal histones to a supercoiled DNA molecule can regenerate a circular minichromosome.

supercoiled molecule will unwind and become a relaxed circular DNA, as one of the DNA strands can now rotate freely about the other strand, allowing the DNA to come to the lowest-energy topological state (Figure 10.3). Alternatively, the enzyme **topoisomerase I** can bind to the DNA and unwind the supercoils by breaking a phosphodiester bond, allowing one of the broken ends to rotate around the opposite DNA strand, and then resealing the phosphodiester bond. However, both strands of this relaxed molecule are covalently closed circles, as in the original supercoiled molecule but unlike nicked relaxed circular DNA. Supercoiling of relaxed DNA molecules can be catalyzed by specialized enzymes found in bacteria, called **DNA gyrases**, which carry out the reverse reaction of topoisomerase I, using ATP as an energy source.

Supercoiled DNA can be separated from relaxed or linear DNA molecules

The properties of circular DNA molecules were first understood by studying the genomes of polyomaviruses. Supercoiled DNA has different physical properties than relaxed circular DNA and linear DNA of the same nucleotide sequence and molecular weight. For example, supercoiled DNA migrates more rapidly upon electrophoresis in agarose gels than does the same DNA when it is a relaxed circle or a linear molecule, because the supercoiled form is more compact. Also, nicked circular DNA or linear DNA will bind more of the fluorescent dye **ethidium bromide** than covalently closed circular DNA. Binding and insertion of ethidium bromide molecules between adjacent nucleotide base pairs unwinds the double helix by a small amount. Nicked circular or linear DNA can be unwound to a greater extent than covalently closed circular DNA, because the strands can freely rotate about each other and therefore unwinding is less constrained. These DNA-ethidium bromide complexes are less dense than complexes with supercoiled DNA, and the two forms can be separated when centrifuged in a cesium chloride density gradient. Both of these methods allow scientists to purify supercoiled DNA away from other forms.

Polyomavirus genes are organized in two divergent transcription units

As a result of their small genome size, most polyomaviruses code for only 6–7 proteins. Viral genes are organized as two families, called early and late genes, and each gene family is expressed from a single transcription unit. Three or four different messenger RNAs, each of which codes for one viral protein, are made from each gene transcript by alternative RNA splicing. A map of the mouse polyomavirus genome is shown in Figure 10.4. Transcription of the early genes proceeds on one DNA strand in a counterclockwise direction, and transcription of the late genes proceeds on the complementary DNA strand in a clockwise direction; this pattern is said to be **divergent transcription**. Each transcription unit has one or several nearby sites that serve as promoters (P_E and P_L) where transcription begins. In between the early and late promoters is a noncoding region of approximately 400 nucleotides that contains several important sequence elements involved in both transcription and genome replication; this is called the **intergenic region** or the control region (Figure 10.5a). A **polyadenylation signal**, 5′-AAUAAA-3′, specifies the 3′ ends of each transcript (Figure 10.5b). For mouse polyomavirus, the mRNA 3′ ends overlap by some 40 nt, and the AAUAAA signals encoded by the two DNA strands are separated by a single nucleotide. Furthermore, the translation termination signals for large T antigen on the early mRNA (UGA) and for VP1 on the late mRNA (UAA) are separated by only 8 nucleotides. The concentration of important sequence elements that serve as signals for transcription, DNA replication, RNA processing, and translation in two small regions of the genome reveals the efficiency with which these genomes are organized.

Virions enter cells in caveolae and are transported to the nucleus

Both mouse polyomavirus and simian virus 40 bind to **sialic acid**–containing glycoproteins and **gangliosides** on the cell surface and are subsequently internalized in

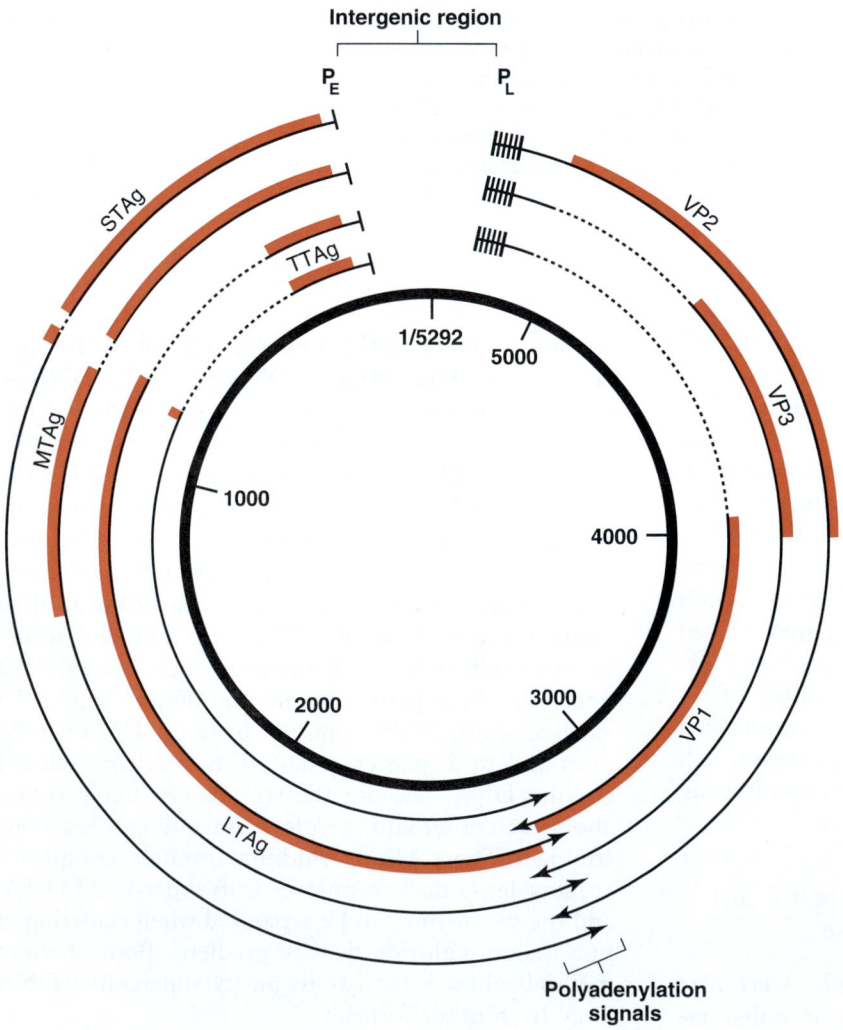

FIGURE 10.4 Map of mouse polyomavirus genome. Nucleotide numbers for the A2 strain of mouse polyomavirus are shown inside the circular map. Early mRNAs are shown on left side, late mRNAs on right side. The 5′ ends (transcription initiation sites) are shown as vertical lines and 3′ ends (polyadenylation sites) are shown as arrowheads on mRNA maps. P_E and P_L: early and late promoters. Introns removed during alternative splicing are shown as dotted lines. Coding regions are shown in orange. Sequence elements in the intergenic region and the polyadenylation region are shown in Figure 10.5.

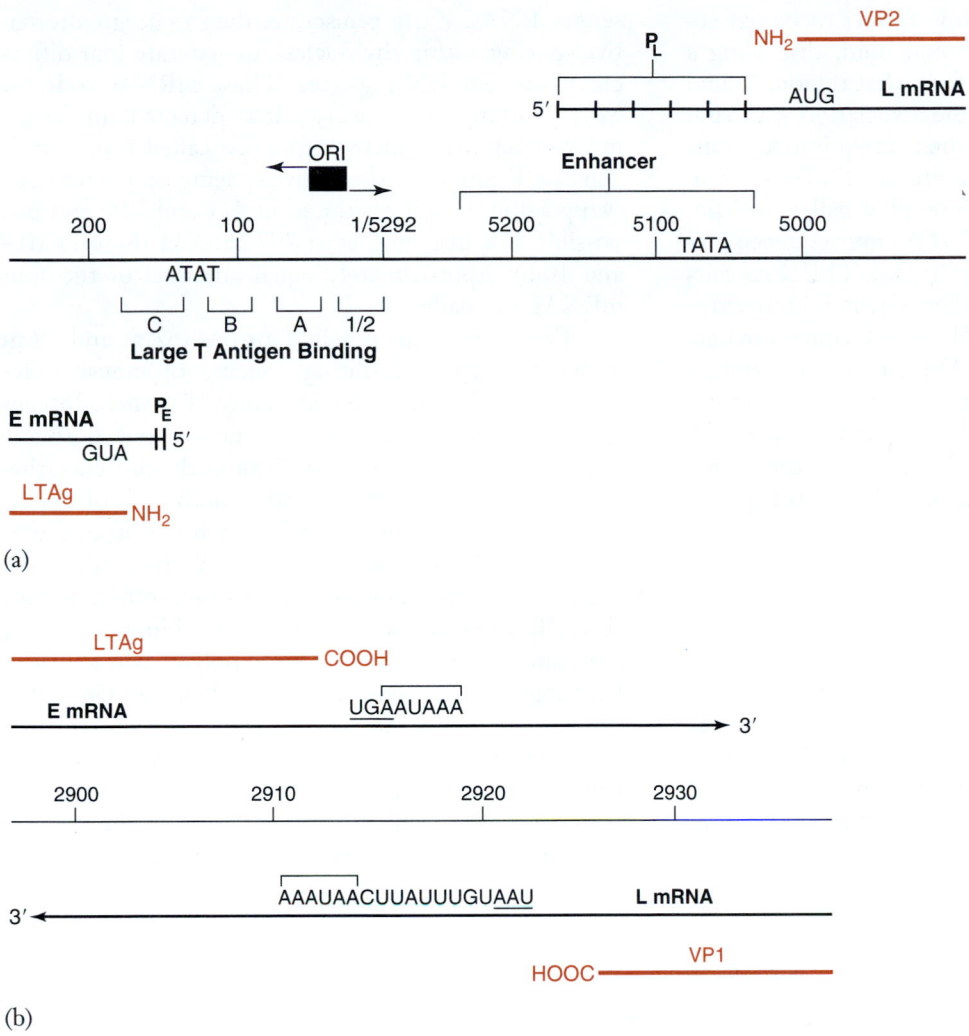

FIGURE 10.5 Control regions on mouse polyomavirus genome. Expanded views of control regions shown in Figure 10.4. (a) The intergenic region, a noncoding region of ~400 nt, bounded by the divergent promoters for early and late transcription (P_E and P_L), contains sequences that control transcription and DNA replication. Transcriptional enhancer, origin of DNA replication (ORI), and large T antigen binding sites are shown above and below map. TATA boxes that direct initiation of transcription at selected promoters are shown. Vertical lines on mRNA maps show transcription start sites. N-terminal parts of the coding regions for T antigens and VP2 are shown above and below in orange. (b) Region where overlapping 3′ ends of mRNAs are located. Polyadenylation signals (AAUAAA) are bracketed translation termination signals for large T antigen (UGA) and VP1 (UAA) are underlined.

caveolae, which are specialized vesicles formed at the plasma membrane. Virions are then transported inside these vesicles through the cytosol to the lumen of the endoplasmic reticulum, near the nuclear membrane (see Figure 4.5). From there, viral DNA must be transported to the nucleus. Passage of virions across the endoplasmic reticulum membrane appears to be mediated by VP2 molecules, which lie below the surface of each capsomere. Conditions within the endoplasmic reticulum lead to extrusion of the N-terminus of VP2 through the capsid surface, allowing the N-terminal myristate to interact with lipid membranes and leading to release of the virions into the cytosol.

Transport and release of viral DNA into the nucleus appears to be a two-step process. Capsids can be loosened or dissociated *in vitro* by treatment with reducing agents, which break disulfide bonds between cysteines, and chelating agents, which remove divalent cations bound to capsid proteins. Similar processes probably lead to uncoating and release of the viral genome. Upon entry into the cytosol, the capsid dissociates but VP1 remains associated with the viral minichromosome. The nuclear localization signal on VP1 directs this complex to the nucleus via the nuclear pores. Once in the nucleus, VP1 dissociates from the viral minichromosome, which remains intact during transcription and DNA replication.

The viral minichromosome is transcribed by cellular RNA polymerase II

Initiation of transcription of viral minichromosomes is a complex process that involves numerous cellular proteins. A **transcriptional enhancer** within the intergenic region (Figure 10.5a) contains a number of consensus DNA sequences that are recognized by a variety of cellular **transcription factors**. Multicellular organisms encode hundreds of different transcription factors, which control gene expression in many circumstances, including embryogenesis and cellular differentiation. Each cell type in an organism may harbor a different combination of transcription factors, each of which recognizes specific binding sites on DNA.

Certain of these transcription factors recognize sequences in the viral enhancer and bind, displacing a small number of nucleosomes from that region. Bound transcription factors facilitate the association with viral DNA of the multiple cellular proteins required for transcription. Among these proteins are the TATA-binding protein (TBP), present in a complex called TFIID, which binds to a consensus **TATA box** sequence element on the viral DNA (Figure 10.5a). TFIID recruits RNA polymerase II and a number of general transcription factors, named TFII A to H, to a site approximately 30 nt beyond the TATA box. Alternative transcription start sequences called initiators are present at the beginning of many viral and cellular genes that lack TATA boxes, but play a similar role. When all of these proteins are in place, transcription of early genes begins.

Four early mRNAs are made by differential splicing of a common transcript

The 2.7-Kb early transcripts of mouse polyomavirus extend about half way around the circular viral DNA (Figure 10.4). They are capped at their 5′ ends, and cleaved and polyadenylated at their 3′ ends, by protein machinery used by the cell for generating its own messenger RNAs. Early transcripts then undergo alternative splicing within the nucleus to generate four different messenger RNA species. These mRNAs code for early proteins that were first detected in virus-induced tumors, and therefore are called tumor antigens, or T antigens. Alternative splicing occurs between two possible 5′ splice sites, at nt 411 and 748, and two possible 3′ splice sites, at nt 797 and 811 (Figures 10.4 and 10.6). Approximately equal amounts of the four mRNAs are made.

Two of the four possible introns (62 nt and 48 nt in length) removed during splicing of mouse polyomavirus early transcripts are unusually small. Introns smaller than approximately 70 nt are not normally found in mammalian cells. Ribonucleoproteins that make up the **spliceosome** bind to both ends of the intron as well as to an internal branch site near the 3′ splice site. These factors cannot bind efficiently if the ends of the intron are too close to each other, as they mutually interfere with each other's binding ("steric hindrance"). The unusually small polyomavirus introns have apparently evolved to reduce splicing efficiency between the 5′ splice site at nt 748 and either of the two 3′ splice sites. If these small introns are artificially lengthened by insertion of as little as 40 nt into the viral genome, almost all splicing events target the 5′ splice site at nt 748, and no large T antigen mRNA is made.

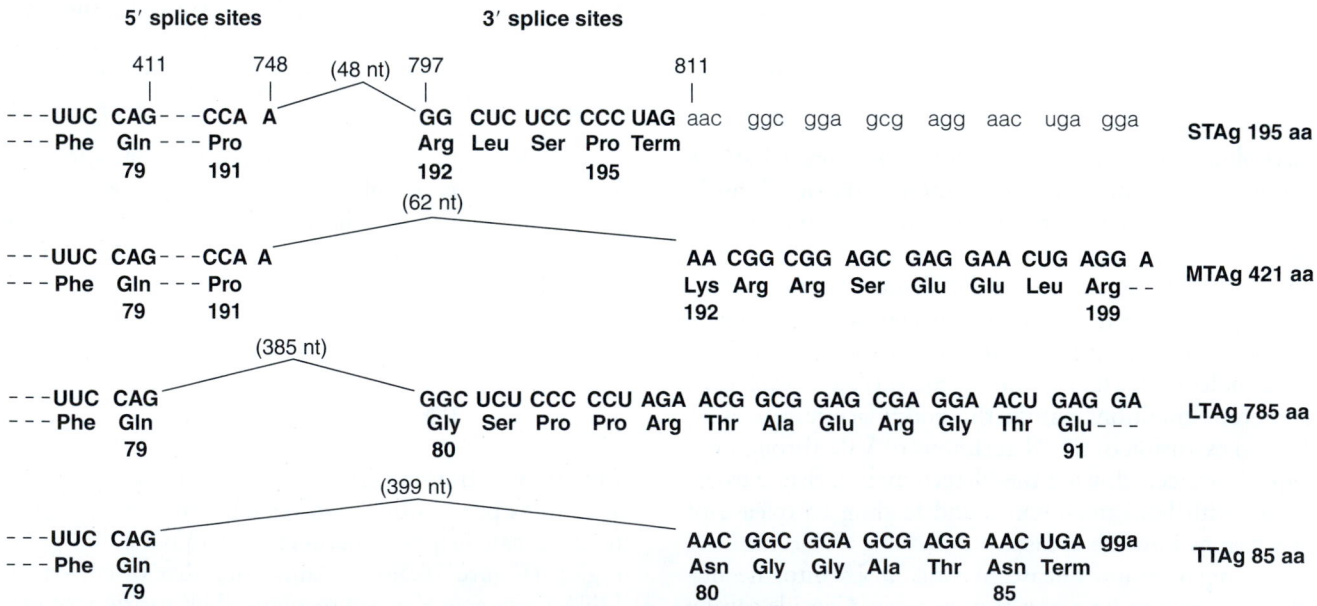

FIGURE 10.6 Alternative splicing of early mouse polyomavirus transcripts generates four different T antigens. Shown are the nucleotide sequences of parts of the four mouse polyomavirus early mRNAs and their translation products. Translated codon triplets are shown in capitals in bold type; nucleotides that lie beyond a translation termination signal are shown in small, unbolded type. The amino acid coded by each triplet is shown directly below. Excised introns are shown as tent-shaped lines, with the number of nucleotides excised shown at the apex. Positions of 5′ and 3′ splice sites are noted. Translation of all four early mRNAs starts at an AUG at nt 175. Note that all three reading frames are used between nt 811 and 831.

Besides being economical in the use of space in a small genome, these small introns therefore ensure a balanced production of the different early mRNAs.

T antigens share common N-terminal sequences but have different C-terminal sequences

The four mouse polyomavirus early mRNAs code for four different proteins. Ribosomes initiate translation on all four early mRNAs at the first AUG codon they encounter, at nt 175 on the polyomavirus genome. This defines reading frame 0. As shown in Figure 10.6, excision of the smallest intron (48 nt) removes exactly 16 triplet codons. Ribosomes translate the first 191 codons and continue translation of this mRNA in the same reading frame downstream of the splice junction. They encounter a UAG termination codon only 12 nt after the splice junction, leading to production of a 195-amino acid protein called small T antigen (STAg).

In contrast, excision of the 62-nt intron removes 20 triplet codons plus 2 nt. Ribosomes translating this mRNA make a protein that contains the same first 191 amino acids as small T antigen, but they continue in a different reading frame (−1) downstream of the splice junction. In reading frame −1, the protein is extended to a size of 421 amino acids before a termination codon is encountered. This protein is called middle T antigen (MTAg).

Excision of the 385-nt intron removes 128 codons plus 1 nt. Ribosomes translating this mRNA make a protein that shares the same first 79 amino acids with small and middle T antigens, because the 5′ splice site lies exactly 79 triplet codons beyond the initiator AUG. However, the reading frame downstream of the splice junction is now +1; although this sequence overlaps with parts of the small and middle T antigen reading frames, it codes for a totally different amino acid sequence because different triplet codons are recognized by the ribosome. This reading frame does not terminate until very near the 3′ end of the mRNA (Figure 10.5b). The result is large T antigen (LTAg), a protein 785 amino acids in length (708 amino acids for simian virus 40).

The fourth mouse polyomavirus early mRNA makes a protein with the same 79 N-terminal amino acids as the other T antigens. A 399-nt intron (133 triplet codons) keeps the ribosome in reading frame 0, as with small T antigen mRNA. However, the 3′ splice site at nt 811 lies just downstream of the UAG that terminates small T antigen translation. Another termination codon, UGA, is present just 7 codons beyond the splice junction, making an 85-amino acid protein called tiny T antigen (TTAg). This protein has not been carefully studied. Note that between nt 797 and 811, two different reading frames are used to make parts of STAg and LTAg, and between nt 811 and 831, all three reading frames are used to make parts of MTAg, LTAg, and TTAg.

Simian virus 40 has two 5′ splice sites analogous to those of mouse polyomavirus, but only one 3′ splice site. As a result, it makes a small and large T antigen, but no middle T antigen. However, a rare doubly spliced simian virus 40 mRNA codes for a protein similar to mouse polyomavirus tiny T antigen.

T antigens bring resting cells into the DNA synthesis (S) phase of the cell cycle

The principal function of the early proteins is to enable viral DNA replication. Most cells in an organism are not dividing, but are differentiated and in a resting (G_0) state. Under these conditions, the enzymes and precursors needed to support cellular or viral DNA synthesis are not normally present in the cell nucleus. The early proteins of DNA viruses induce entry of the cell into the DNA synthesis (S) phase of the cell cycle. The T antigens of both mouse polyomavirus and simian virus 40 have been intensively studied and have been shown to interact with numerous cellular signaling pathways, leading to stimulation of the cell cycle. Only a selection of these interactions will be discussed here.

Small T antigen inhibits protein phosphatase 2A and induces cell cycling

Small T antigen forms a complex with a subunit of the trimeric cellular regulatory protein, protein phosphatase 2A (PP2A), inhibiting the phosphatase activity of this protein. PP2A plays an important regulatory role in several cellular signaling pathways by removing phosphate groups that stimulate the activities of proteins in those pathways. One set of cellular targets of PP2A are mitogen-activated protein kinases (MAP kinases). When a MAP kinase is phosphorylated at specific serine and threonine residues (by MAP kinase kinase), it can in turn activate the nuclear transcription factor AP-1, which activates transcription of the cyclin D1 gene among others (Figure 10.7). **Cyclin** D1 is an important activator of the cell cycle. Inhibition of PP2A activity by small T antigen allows MAP kinases to remain phosphorylated and therefore enzymatically active. Thus small T antigen indirectly stimulates the MAP kinase pathway and helps bring the cell into the S phase.

Middle T antigen stimulates protein tyrosine kinases that signal cell proliferation and division

Middle T antigen is localized at the plasma membrane of infected cells, where it is inserted via a hydrophobic

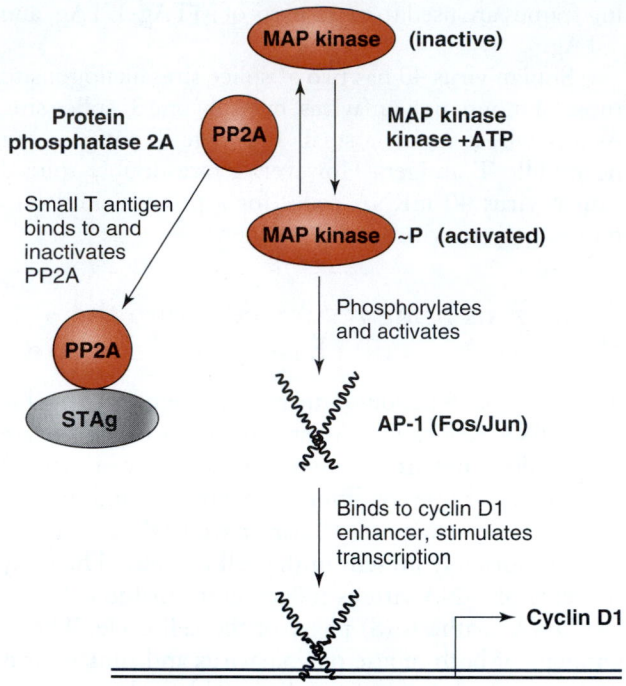

FIGURE 10.7 Small T antigen binding to protein phosphorylase 2A stimulates MAP kinase pathway. MAP kinase pathway, leading to activation of transcription of the cyclin D1 gene, is shown. Protein phosphatase 2A inactivates MAP kinase by removing specific phosphate molecules. Binding of small T antigen to protein phosphatase 2A inhibits its activity, allowing phosphorylated MAP kinase to accumulate and to activate transcription factor AP-1, which enhances cyclin D1 transcription.

region near its C-terminus (Figure 10.8). Because it contains most of the amino acid sequences of small T antigen, middle T antigen can also bind to and inactivate PP2A. However, its major functions are carried out by association with several cellular **protein tyrosine kinases**, of which the best known is c-Src. C-Src can modify a number of other cellular proteins by phosphorylating specific tyrosine residues on those proteins. A mutated and highly active form of c-Src is encoded by the acutely transforming retrovirus Rous sarcoma virus, which produces cancer in chickens (see Chapter 25).

When associated with middle T antigen, c-Src loses a phosphate group at tyrosine 527 and phosporylates itself at tyrosine 416. These changes greatly stimulate its protein kinase activity. C-Src then phosphorylates several tyrosine residues on middle T antigen itself. These phosphotyrosines act as part of binding sites for a set of cellular proteins including Src-homology 2 domain protein (Shc), phosphatidylinositol 3′-kinase (PI3K), and phospholipase C (PLC). When bound to middle T antigen in the presence of activated c-Src, these proteins are in turn phosphorylated by c-Src and therefore activated. This leads to a cascade of signaling events too complex to detail here. Phosphorylated Shc leads, through an activated Ras protein, to the activation of the MAP kinase pathway referred to above. PI3 kinase synthesizes phosphatidylinositol 3-phosphate, a phospholipid that controls such cellular functions as proliferation, vesicle trafficking, and organization of the cytoskeleton. Phospholipase C cleaves phosphatidyl in-

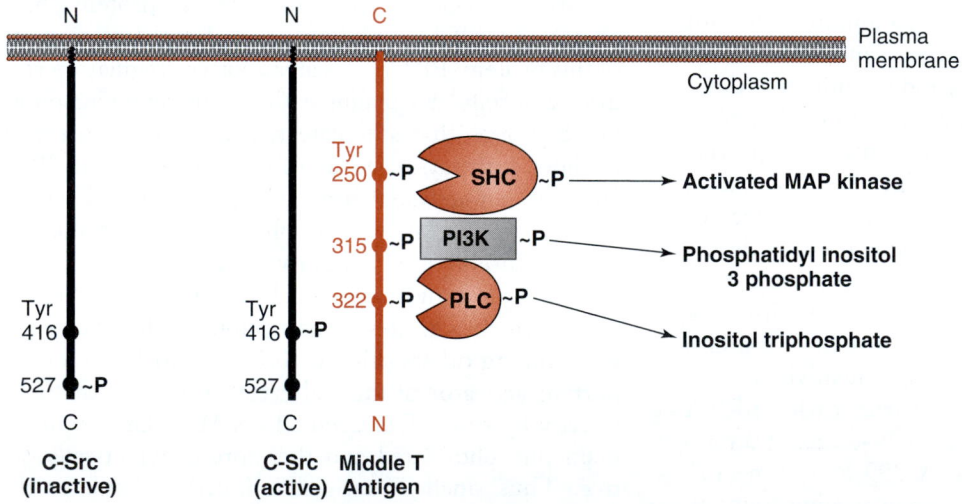

FIGURE 10.8 Middle T antigen binding to c-Src activates pathways that stimulate cell metabolism and the cell cycle. Shown at left is c-Src, a protein tyrosine kinase inserted into the plasma membrane via a N-terminal myristate residue. When phosphorylated at tyrosine 527, c-Src is inactive. Middle T antigen, bound to plasma membrane via hydrophobic residues near its C-terminus, associates with c-Src. This leads to removal of the phosphate residue at tyrosine 527, and autophosphorylation of c-Src at tyrosine 416, activating its protein tyrosine kinase activity. Middle T antigen is phosphorylated at three specific tyrosine residues, which then serve as binding sites for cellular proteins involved in cell cycle signaling: Shc, phosphatidyl inositol 3 kinase (PLC), and phospholipase C (PI3K). These are in turn phosphorylated by c-Src and activated, leading to increase in cell metabolism and cell cycling.

ositol, generating inositol triphosphate, which releases calcium ions from the endoplasmic reticulum. Many of these pathways activate cell metabolism and the cell cycle.

Large T antigen activates or suppresses transcription of cellular genes by binding to a number of important cellular regulatory proteins

Large T antigen is a complex protein with a surprisingly large number of different functions. Most large T antigen is localized in the nucleus of infected cells, although some is found in the cytoplasm and at the plasma membrane. The nuclear localization signal of simian virus 40 large T antigen was one of the first such signals identified. A schematic diagram showing functional domains of simian virus 40 large T antigen is shown in Figure 10.9. Like small and middle T antigens, large T antigen interacts with a number of cellular proteins that control cell metabolism and the cell cycle.

1. Retinoblastoma protein (pRb) controls cell-cycle and S-phase gene expression. The retinoblastoma protein (pRb), so-named because mutations in the Rb gene can lead to retinal tumors in humans, is one of the major cell cycle control proteins. pRb and related proteins p107 and p130 repress transcription from a large number of cellular genes when bound to members of the E2F family of transcriptional activator proteins. When phosphorylated, pRb dissociates from its E2F partner, allowing transcription of these cellular genes, entry into the cell cycle, and expression of proteins required for both cellular and viral DNA replication.

Large T antigen has a conserved binding site for the Rb protein (Figure 10.9). Binding to pRb begins a process that mimics pRb phosphorylation, releasing E2F and allowing transcription of cellular genes targeted by E2F. These genes include cyclins, DNA polymerase alpha, thymidine kinase, DNA ligase, histones, and others. A number of DNA viruses make early proteins that interact with Rb and stimulate the cell cycle. For a more complete description of pRb interactions, see chapters on papillomaviruses and adenoviruses (Figures 11.4, 11.5, and 12.6).

2. The DNA J domain acts as a co-chaperone to dissociate pRb from E2F. The N-terminal region of large T antigen, called the DNA J domain, is also required for gene activation via the Rb pathway. DNA J is a **co-chaperone** protein, first described in *E. coli* as a bacterial protein required for replication of bacteriophage lambda DNA (Chapter 8). DNA J and its eukaryotic analogues interact with a cellular chaperone called Hsp70 (known as DNA K in *E. coli*) and take part in ATP-dependent protein folding and unfolding reactions that are vital in assembling or disassembling multiprotein complexes. It was surprising to discover that the N-terminal region of large T antigen can, by itself, complement DNA J-minus mutants of *E. coli* and enable bacteriophage lambda DNA replication! This shows the extraordinary conservation of protein functions in widely separated organisms, and reveals a connection across perhaps billions of years of evolution between a bacteriophage and a mammalian DNA tumor virus. The DNA J domain of large T antigen is believed

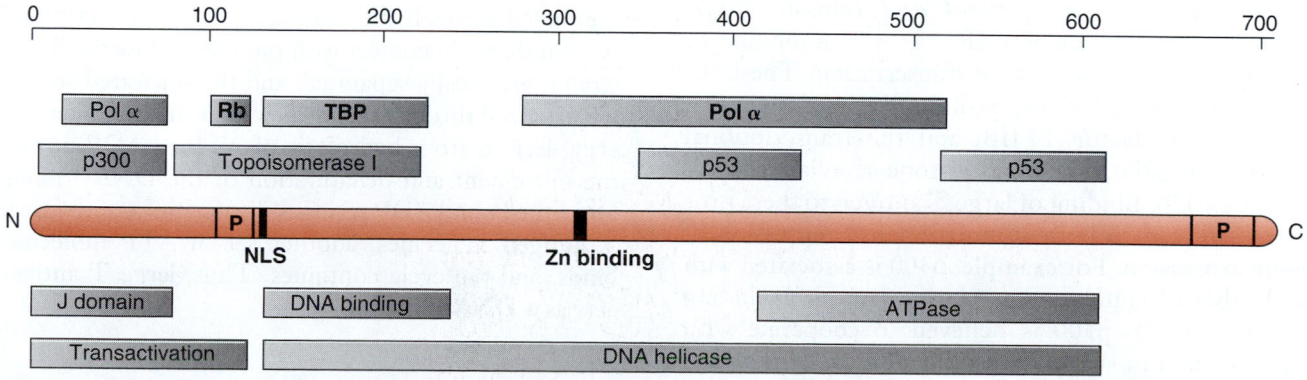

FIGURE 10.9 Functional and protein-binding domains of simian virus 40 large T antigen. Shown in orange is the 708-amino acid large T polypeptide. **P** designates regions that are phosphorylated, **NLS** designates the nuclear localization signal, and **Zn** designates a Zinc-binding site. Above are shown regions involved in binding to a number of cellular proteins involved in cell signaling or DNA replication: **Pol α** = DNA polymerase α/RNA primase; **Rb** = retinoblastoma protein; **TBP** = TATA-binding protein. Below are shown functional domains; see text for details.

to act by helping to separate the Rb protein from a complex containing E2F and other proteins, thus stimulating transcription of cell cycle and S-phase genes.

3. p53 blocks the cell cycle and induces apoptosis in response to virus infection. The cellular control protein p53 was first discovered as a 53-KDa protein present in immunoprecipitates of simian virus 40 large T antigen. p53 responds to a variety of signals by binding to and activating transcription of genes that can lead either to blockage of the cell cycle or **apoptosis** (programmed cell death). Its central role in controlling the cell cycle is emphasized by the fact that a majority of human tumors have undergone mutations in the p53 gene. Therefore, for a tumor cell to thrive, it must rid itself of the control exercised by the p53 protein.

Normally p53 is rapidly degraded, is present in very low concentrations in the cell, and is inactive as a transcription factor. Various insults to the cell, including infection by polyomaviruses, can lead to increases in the concentration and the activity of p53. This is at least partly mediated by large T antigen interacting with pRb, which stimulates transcription of genes that raise levels of p53. However, binding of large T antigen to p53 suppresses its transcriptional activation function, allowing the cell to enter S phase and avoiding p53-mediated apoptosis. This allows the infected cell to survive long enough to produce progeny virus particles. Both papillomaviruses and adenoviruses make analogous proteins that interact with p53. For a more complete description of p53 control and function, see chapters on papillomaviruses and adenoviruses (Figures 11.6 and 12.7).

4. p300 and general transcription factors control levels of transcription of viral and cellular genes. Large T antigen interacts directly with a number of cellular proteins involved in transcription. These include the TATA binding protein (TBP), the general transcription factor TFIIB, and the transcriptional coactivator p300, which has histone acetylase activity (Figure 10.9). Binding of large T antigen to these proteins can lead to both stimulation and repression of gene expression. For example, p300 is associated with a number of cellular transcription factors, including p53 and AP-1. p300 is believed to cooperate with these factors in stimulating transcription by acetylating nucleosomal histones, therefore reducing their affinity for DNA and making DNA more accessible to components of the transcription machinery. Binding of large T antigen to p300 can inhibit its transcriptional coactivation function and therefore reduce transcription of genes targeted by these factors. However, large T antigen is known to increase transcription of many other genes, probably via its interactions with the machinery that assembles RNA polymerase II at promoters.

Large T antigen hexamers bind to the origin of DNA replication and locally unwind the two DNA strands

Once the cell has entered the S phase, viral DNA replication can begin. Large T antigen now plays an entirely different role: instead of interacting with cellular regulatory proteins, it binds to viral DNA and directs the assembly of a number of cellular proteins that carry out DNA replication. The DNA-binding domain of large T antigen (Figure 10.9) recognizes G(A/G)GGC pentanucleotide sequences on double-stranded DNA. Both simian virus 40 DNA and mouse polyomavirus DNA have 12 such sequences that lie within a 160-nt span in the intergenic region. In mouse polyomavirus, these 12 binding sequences are grouped in four sites denoted 1/2, A, B, and C (Figure 10.5a). Large T antigen molecules bind cooperatively to these sequences and assemble at a specific site denoted the DNA replication origin (Ori) to form two circular hexamers that enclose the DNA like a wheel on an axle.

Electron microscopy and x-ray crystallographic studies have revealed the detailed structure of the hexamer form of large T antigen (Figure 10.10). Each large T antigen monomer in a hexamer binds a molecule of ATP. Hydrolysis to ADP and release of the ADP molecule lead to a conformational change that rotates adjacent protein domains with respect to each other and changes the size of the central opening of the hexamer that is in contact with the DNA. This has been likened to an iris that can change its diameter by contracting or expanding. As a result of this change, 5–6 bp of ds DNA are pulled through the hexamer by the motion of a β hairpin that is in contact with the DNA. The two DNA strands are locally separated, and the separated strands are extruded through channels within the hexamer. Energy derived from hydrolysis of ATP to ADP powers the movement and denaturation of the DNA strands. Once the bound ADP residues dissociate from the large T antigen molecules, another set of ATP molecules binds, and the cycle continues. Thus, large T antigen acts as a **DNA helicase**.

Large T antigen hexamers assemble cellular DNA synthesis machinery to initiate viral DNA replication

This process generates a localized region of single-stranded DNA that can now serve as a template for DNA

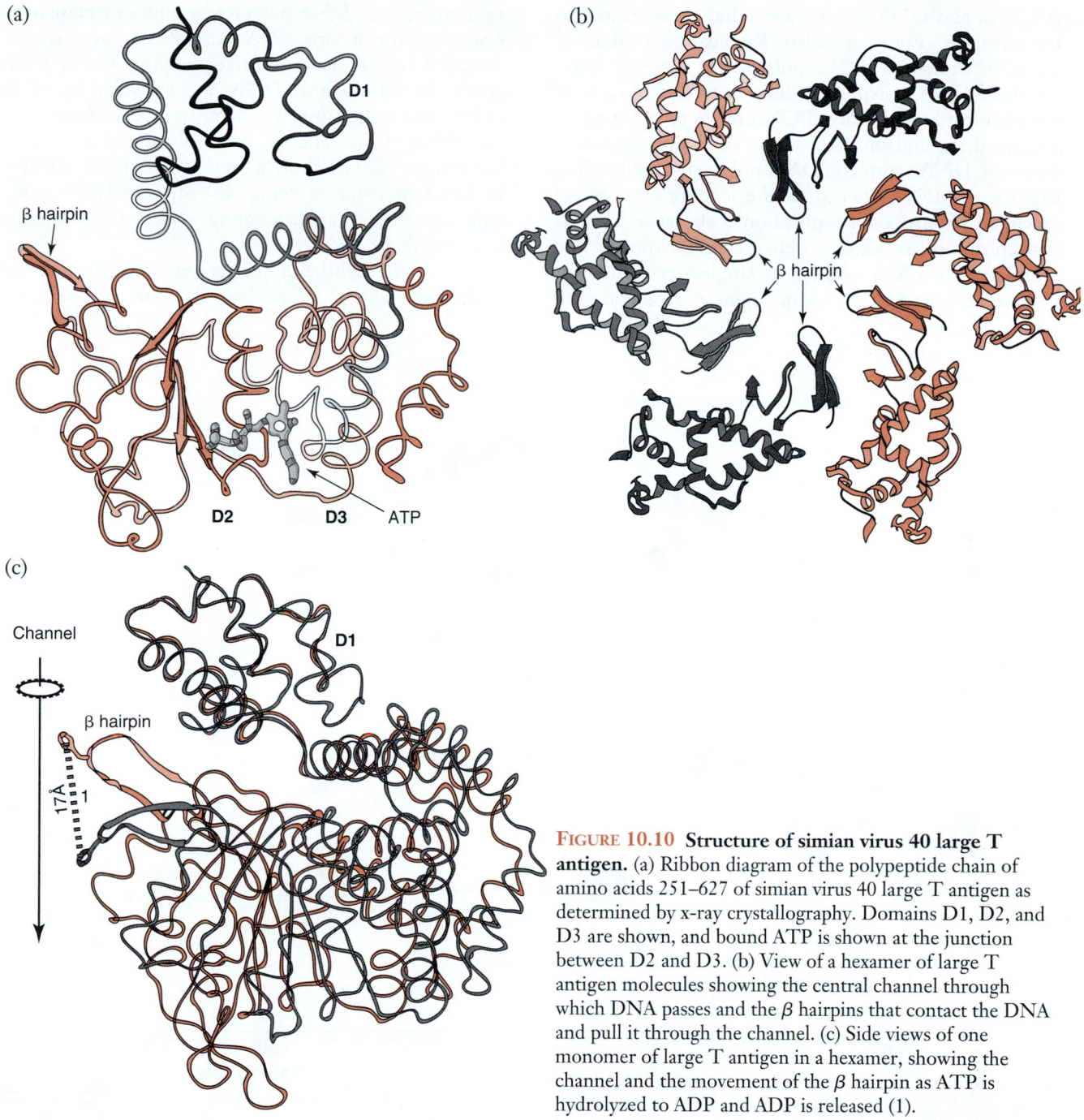

FIGURE 10.10 Structure of simian virus 40 large T antigen. (a) Ribbon diagram of the polypeptide chain of amino acids 251–627 of simian virus 40 large T antigen as determined by x-ray crystallography. Domains D1, D2, and D3 are shown, and bound ATP is shown at the junction between D2 and D3. (b) View of a hexamer of large T antigen molecules showing the central channel through which DNA passes and the β hairpins that contact the DNA and pull it through the channel. (c) Side views of one monomer of large T antigen in a hexamer, showing the channel and the movement of the β hairpin as ATP is hydrolyzed to ADP and ADP is released (1).

replication. Here again large T antigen has a vital function. DNA-bound large T antigen hexamers bind to three cellular proteins required for initiation of DNA replication: replication protein A, DNA polymerase alpha/primase, and topoisomerase I (Figure 10.11a). Replication protein A binds to the unwound DNA strands and keeps them from renaturing into a double-helical form. DNA polymerase alpha/primase binds to the single-stranded DNA and lays down a short (6–9 nt) RNA primer that remains hydrogen-bonded to the DNA template. The same enzyme then extends the primer by adding a short (20–30 nt) DNA molecule. Primers are formed and elongated on both DNA strands, and as a result, bidirectional DNA synthesis is initiated. Topoisomerase I is required to help release the tension created in the circular, double-stranded DNA molecule as the DNA polymerase and helicase proceed along the DNA molecule, twisting the double helix in front as it is unwound behind.

Finally, three other cellular proteins, replication protein C (RPC), proliferating cell nuclear antigen

(PCNA), and DNA polymerase delta, are recruited to the complex (Figure 10.11b). Replication protein C helps to dissociate DNA polymerase alpha/primase and load DNA polymerase delta onto the 3′ ends of the growing DNA chains. PCNA binds to DNA polymerase delta and ensures that it does not dissociate from the DNA template. DNA polymerase delta then extends the DNA chains. As the leading strands are extended toward each replication fork, large T antigen continues to act as a helicase by unwinding the template DNA. As a result, more single-stranded DNA becomes exposed on the complementary strand at each replication fork. DNA polymerase alpha/primase molecules lay down new RNA primers on these single-stranded regions and reinitiate lagging strand DNA synthesis. These short DNAs are extended by DNA polymerase delta up to the RNA primers of the adjacent DNA chains, and the primers are removed by ribonuclease H. The short gaps that remain are filled in by DNA polymerase delta, and adjacent DNA molecules are then ligated, forming a continuous lagging strand (Figure 10.11c).

As DNA synthesis progresses, cellular histones present in the S-phase nucleus assemble on the grow-

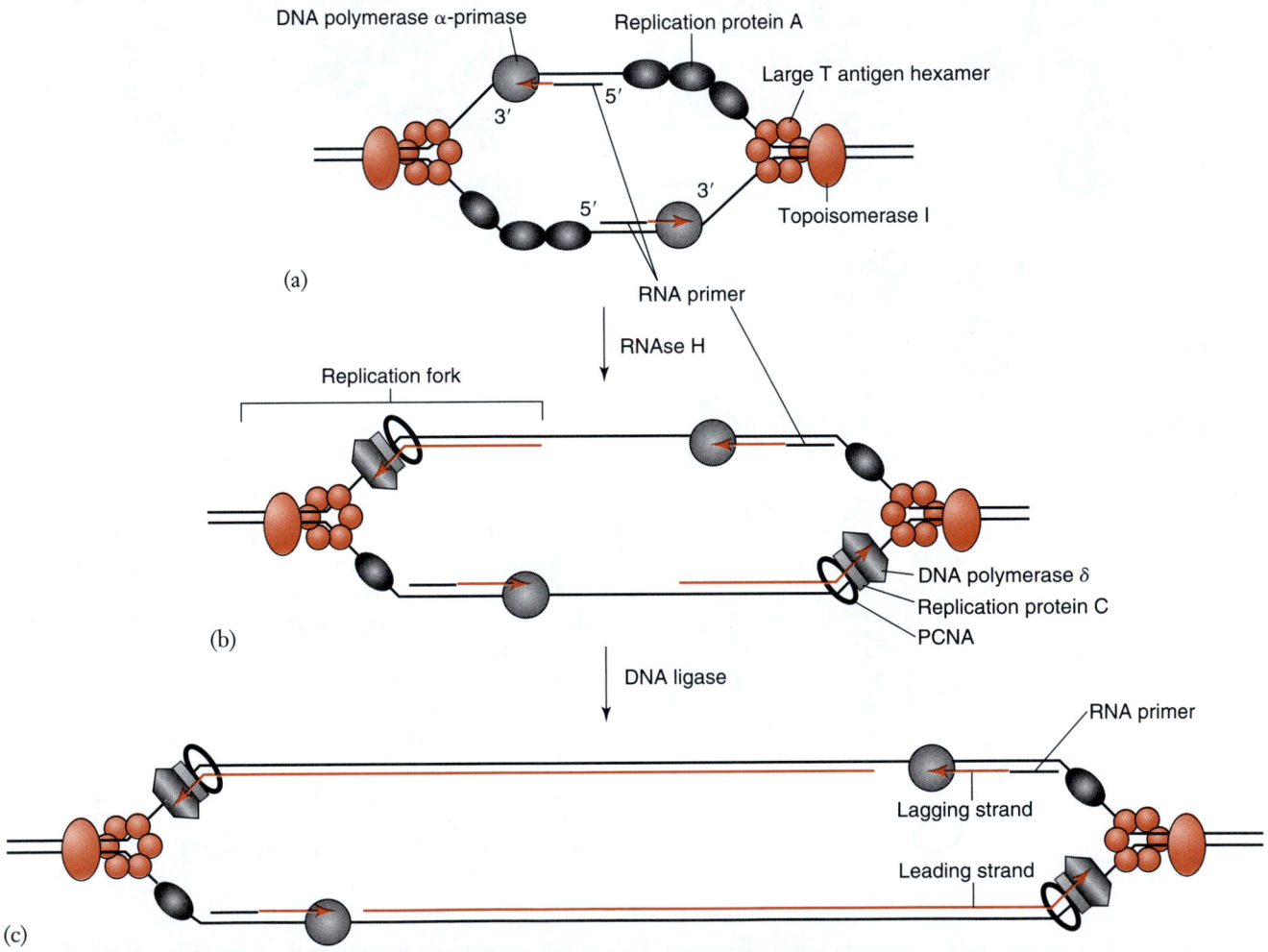

FIGURE 10.11 Mechanism of bidirectional replication of polyomavirus DNA. (a) Initiation of DNA synthesis at viral replication origin. DNA polymerase α/RNA primase binds to a region of single-stranded DNA unwound by the DNA helicase activity of large T antigen hexamers and bound by replication protein A. (b) Extension of DNA chains beyond the initiation region by DNA polymerase δ associated with replication protein C and proliferating cell nuclear antigen (PCNA). Ribonuclease H (RNAse H) has removed initial RNA primers. DNA polymerase α/RNA primase reinitiates synthesis on single-stranded DNA generated by continued helicase action of large T antigen as the two replication forks travel in opposite directions on the DNA template. (c) Joining of growing DNA strands. Lagging strand DNAs are extended up to the 5′ end of the leading strand DNA and the strands are linked by DNA ligase to produce a single, covalently joined daughter strand on each template strand. Needy synthesized DNA is shown in orange.

ing DNA chains to form nucleosomes. The final result is the synthesis of two circular double-stranded viral minichromosomes. The progeny DNA can be used as a template for further transcription, or can be replicated.

High levels of late transcripts are made after DNA replication begins

Initiation of late transcription on both mouse polyomavirus and simian virus 40 DNA occurs at multiple promoter sites within a 100-nt region that overlaps with the transcriptional enhancer. Most of these late promoter sequences are not associated with a consensus TATA box. Before DNA replication begins, low but detectable levels of late transcripts are made. After DNA replication begins, much higher levels of late transcripts are made, and synthesis of early transcripts is reduced. Five different reasons for the increase in the ratio of late to early transcription have been suggested.

1. Repression of early transcription by large T antigen. The early promoters of both simian virus 40 and mouse polyomavirus overlap with one or more of the G(A/G)GGC binding sites for large T antigen. Binding of large T antigen to these sites can interfere with the binding of RNA polymerase and general transcription factors such as TFIID, reducing the amount of early RNA synthesized. Therefore, when large T antigen concentration increases, early transcription decreases. In this way, large T antigen controls its own levels of synthesis.

2. Activation or induction of new transcription factors. The combined effects of small, middle, and large T antigen on cellular signaling pathways (see above), and the ability of large T antigen to bind to a variety of transcriptional repressors and activators, changes the mix of transcription factors available to bind to the transcriptional enhancer. These changed factors specifically activate the late promoters. Because numerous transcription factors are involved, and because the same enhancer region powers both early and late transcription, the exact mix of transcription factors that activate late transcription is not fully understood.

3. Derepression of late transcription by dilution of a repressor. Simian virus 40 late transcription is initially blocked by binding of a cellular protein to viral DNA near the late promoter. The concentration of this repressor in the cell is limited. As the number of viral DNA molecules increases due to DNA replication, a smaller proportion of viral DNA molecules contain bound repressor. This allows activation of the late promoter and increased levels of late transcription.

4. Increased efficiency of processing and export of late RNAs. Before the onset of DNA replication, most mouse polyomavirus late transcripts initiate at a single promoter sequence. RNA transcripts with this 5′ end are not efficiently exported to the cytoplasm, perhaps because they are missing a critical RNA export sequence. Activation of late transcription leads to initiation at multiple sites upstream of this promoter sequence, and RNAs with these 5′ ends are more efficiently processed into mRNAs and exported to the cytoplasm.

5. Readthrough RNAs hybridize with early transcripts and lead to their degradation. Most RNA polymerases that transcribe the late genes of mouse polyomavirus do not terminate transcription after passing through the late polyadenylation signal, but continue transcription along the DNA template. Because the DNA is circular, this leads to the production of very long late RNAs that can contain multiple, tandem repeats of the DNA sequence, similar to the production of long DNAs by rolling circle DNA replication. Some RNA polymerases can pass 10 or more times around the circular DNA before they terminate! These giant viral RNAs contain sequences that are complementary to early transcripts, and they are made in large amounts for reasons explained above. The presence of complementary early and late RNAs in the nucleus leads to the formation of double-stranded RNAs, which are degraded by cellular enzymes. Because late transcripts are in excess, most of the early transcripts but only a fraction of late transcripts are degraded.

Three late mRNAs are made by alternative splicing

Three different late messenger RNA species are formed from mouse polyomavirus late transcripts by alternative splicing (Figure 10.4). An unspliced late mRNA is translated to produce viral capsid protein VP2. Splicing removes either a small or a large intron, starting at the same 5′ splice site but cutting at two possible 3′ splice sites. Removal of the small intron generates VP3 mRNA. Translation of this protein starts in the same reading frame as VP2 but at an AUG that codes for an internal methionine in VP2; thus VP3 is identical to the C-terminal two-thirds of VP2. Removal of the large intron generates VP1 mRNA. This is coded in a different reading frame that overlaps with the end of the VP2/VP3 reading frame.

Note that alternative splicing of early transcripts removes introns downstream of the AUG start codon, within open reading frames, and generates distinct proteins by changing reading frames. Alternative splicing of late transcripts, in contrast, generates distinct proteins by removing AUG start codons, leading to initiation of translation at different start sites, in the same or different reading frame.

Capsid proteins are imported into the nucleus, where they form capsids that package viral minichromosomes. Empty capsids can also form. Under certain conditions, fragmented cellular DNA, resulting from cleavage during cell death, can also be packaged into capsids to form "pseudovirions." Because these particles can enter into other cells or be transmitted to other animals, pseudovirions may enable passage of certain cellular DNA sequences from one host to another.

How do polyomaviruses transform cells *in vitro* and cause tumors *in vivo*?

As we have seen, productive infection of permissive cells by polyomaviruses leads to cell death and release of progeny virions. However, polyomaviruses were first discovered as agents able to induce formation of tumors in mice and other rodents. Furthermore, these viruses can transform normal rat or hamster fibroblasts grown *in vitro* into cell lines that are immortal, form dense and rapidly growing colonies, have reduced requirements for growth factors, and in some cases can cause tumors when injected into animals. Paradoxically, polyomaviruses cause few if any tumors in their natural hosts.

The ability to study cell transformation *in vitro* by these viruses allowed scientists to study the biochemical and signaling pathways that are altered upon infection by polyomaviruses. It was quickly discovered that polyomavirus early genes act as **oncogenes** both in cells cultured *in vitro* and in intact animals. Study of the structure and activity of the early gene products (T antigens) led to the discovery of a number of important cellular regulatory pathways that are disrupted or modified during transformation and tumor formation. Some of these pathways have been outlined above. Their discovery has helped provide a detailed understanding of the process of tumorigenesis.

Transformation of cells by polyomaviruses can take place only if the cells are **nonpermissive** for viral replication. This means that the virus can enter the cell and express its early genes, but can neither replicate its DNA nor express late proteins. Therefore, the infectious cycle is **abortive** and does not result in the production of progeny virus or kill the infected cell. The inability of rat and hamster cells to replicate mouse polyomavirus or simian virus 40 DNA has been traced to a lack of interaction between some elements of the cellular DNA replication machinery and the viral large T antigen.

Only 1 to 10 out of every 10^5 infected nonpermissive cells end up being transformed; these can be detected as dense, rapidly growing colonies on a monolayer of normal fibroblasts growing on the surface of a Petri dish. Such transformed cells invariably contain most or all of the early region of the viral genome, integrated into the cellular genome at apparently random positions. The low frequency of cell transformation *in vitro* (and tumor formation *in vivo*) is probably due to the low probability of integration of the early region of viral DNA into the cellular genome. Unlike retroviruses (Chapter 25), polyomaviruses do not normally integrate their DNA into the host cell chromosome during replication; this rare event is carried out by cellular enzymes that pick up degraded fragments of viral DNA and insert them into the host genome.

Cells that have integrated a functional early region can continue to express T antigens, and all daughter cells inherit this ability. Therefore the various signaling pathways that turn on cell cycling, reduce dependence on growth factors, and increase cell growth rate are permanently altered. When this occurs *in vivo*, such cells can eventually evolve to form tumors. Transgenic animals that express one or more of the early genes of polyomaviruses in specific tissues are being studied as model systems for tumor formation.

SUGGESTED READING

1. Dang-Tan, T., Mahmud, S.M., Puntoni, R., and Franco, E.L. (2004). Polio vaccines, simian virus 40, and human cancer: The epidemiologic evidence for a causal association. *Oncogene* **23,** 6535–6540.

2. Dilworth, S.M. (2002). Polyoma virus middle T antigen and its role in identifying cancer-related molecules. *Nature Reviews Cancer* **2,** 951–956.

3. Gai, D., Zhao, R., Li, D., Finkielstein, C.V., and Chen, X.S. (2004). Mechanisms of conformational change for a replicative hexameric helicase of SV40 large tumor antigen. *Cell* **119,** 47–60.

4. Kumar, M. and Carmichael, G.G. (1997). Nuclear antisense RNA induces extensive adenosine modifications and nuclear retention of target transcripts. *Proceedings of the National Academy of Sciences USA* **94,** 3542–3547.

5. Lanoix, J., Tseng, R.W., and Acheson, N.H. (1991). Production of polyomavirus late mRNAs requires sequences near the 5′ end of the leader but does not require leader-to-leader splicing. *Journal of Virology* **65,** 4728–4734.

6. Peng, Y.C. and Acheson, N.H. (1998). Polyomavirus large T antigen binds cooperatively to its multiple binding sites in the viral origin of DNA replication. *Journal of Virology* **72,** 7330–7340.

7. Simmons, D.T. (2000). SV40 large T antigen functions in DNA replication and transformation. *Advances in Virus Research* **55,** 75–134.

8. Sullivan, C.S. and Pipas, J.M. (2002). T antigens of simian virus 40: Molecular chaperones for viral replication and tu-

morigenesis. *Microbiology and Molecular Biology Reviews* **66**, 179–202.

KEY TERMS

Abortive
Apoptosis
Caveolae
Co-chaperone
Cyclins
Divergent transcription
DNA gyrase
DNA helicase
Ethidium bromide
Gangliosides
Intergenic region
Myristate
Nonpermissive
Nucleosomes
Oncogene
Polyadenylation signal
Progressive multifocal leukoencephalopathy (PML)
Protein tyrosine kinase
Sialic acid
Sodium dodecyl sulfate
Spliceosome
Supercoiled
TATA box
Topoisomerase I
Toroidal coil
Transcription factors
Transcriptional enhancer

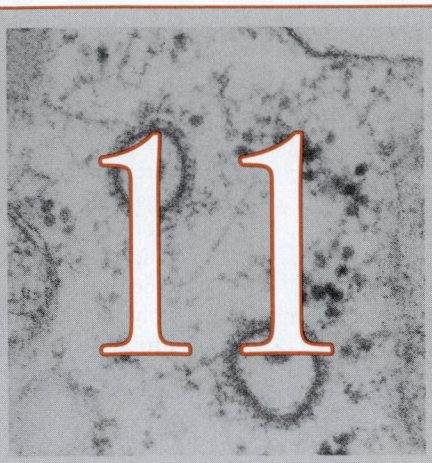

11

Papillomaviruses

Papillomaviridae

From Latin *papilla* (nipple) and Greek *–oma* (tumor) referring to warts (papillomas)

VIRION

Naked icosahedral capsid (T = 7).
Diameter 55 nm.
Formed from 72 capsomers, pentamers of L1.

GENOME

Circular ds DNA, 8 Kb.
DNA is packaged as "minichromosome" with cellular nucleosomal histones.

GENES AND PROTEINS

Two transcriptional promoters and two polyadenylation signals (early and late) on same DNA strand.
Splicing generates 10 or more different mRNAs.
Early proteins E1 to E7 stimulate cell proliferation and enable viral DNA replication.
Late proteins L1 and L2 form capsid.

VIRUSES AND HOSTS

Over 100 known human papillomaviruses.
Other hosts: cattle, dogs, deer, rabbits, etc.

DISEASES

Benign warts at specific sites (skin, mucosa, larynx) depending on virus strain.
Transmitted by direct contact.
Some types cause cervical carcinoma, a sexually transmitted disease and major cause of cancer in women.

DISTINCTIVE CHARACTERISTICS

Difficult to grow *in vitro*: require specialized raft cultures of epithelial cells.
Replication pattern follows differentiation of epithelial cells in skin or mucosa.
Early proteins E6 and E7 can transform cells *in vitro*, are expressed in cervical carcinomas.

Papillomaviruses cause warts and other skin and mucosal lesions

The first papillomavirus was discovered in 1933 by Richard Shope. He showed that warts on wild cottontail rabbits could be transmitted to domestic rabbits by inoculation with extracts of the warts. It is now recognized that papillomaviruses are widespread in animals and humans. Bovine papillomavirus, which causes warts in cattle, has served as a prototype for the study of the molecular biology and transforming genes of papillomaviruses because bovine warts are easily available, and because this virus can transform cultured rodent cells *in vitro*.

Over 100 types of human papillomaviruses have been identified (Table 11.1). Because they are difficult to grow *in vitro*, most of these viruses were identified by direct cloning of their DNA in bacterial plasmids, followed by nucleotide sequence analysis. Papillomavirus genotypes are considered distinct if the nucleotide sequences of specific regions of the viral genome (L1, E6, and E7 genes) differ by more than 10%. Different human papillomavirus types tend to cause lesions in different anatomical sites. For example, plantar warts of the feet are associated with human papillomavirus type 1, and common hand warts are associated with human papillomavirus types 2, 4, and 7.

Human papillomaviruses tend to infect either mucosa (genital tract, oral and nasal cavity) or skin. Types that infect skin are likely transmitted through skin contact with an infected individual or through contact with a virus-containing surface. Transmission of viruses that infect genital tract mucosa is largely through sexual contact with an infected individual. The epithelium of the male genital tract (penis and urethra) likely serves as the virus reservoir. Papillomaviruses have been shown to exist in apparently normal mucosal sites of the penis and urethra and possibly the prostate. The existence of such a reservoir provides a means of sexual transmission of the virus.

Oncogenic human papillomaviruses are a major cause of genital tract cancers

Since 1976, it has been recognized that human papillomavirus infections of the genital tract frequently lead to benign genital warts, neoplasias, and invasive **squamous cell carcinomas**. Cervical cancer in women is associated with infection with the human papillomavirus types 16 and 18 and some other types. By contrast, infection of the genital tract with nononcogenic human papillomavirus types 6 and 11 can produce cervical warts (also known as **condylomas**), which are benign and only rarely progress to cancer. Because of the link with cancer, the focus of research on human papillomaviruses is now largely centered on the oncogenic human papillomavirus types 16, 18, and 31.

Many human papillomavirus genital infections are transient and result in no disease. This can result in widespread virus dissemination, because infected individuals are not aware of their infections. Only a minority of infections with oncogenic human papillomaviruses result in cervical warts or cancer, and many of the lesions regress spontaneously without treatment. Nevertheless, infection with these viruses is the most significant and important risk factor for the development of cervical cancer. Although these viruses infect both males and females, papillomavirus-associated genital cancer is much less prevalent in men than in women; the reason for this is not completely understood. In women, the cells that become infected are nondifferentiated basal cells of the epidermis within a region known as the "transformation zone." These cells are particularly susceptible to oncogenic transformation by the high-risk human papillomavirus types such as 16 and 18.

Papillomaviruses are not easily grown in cell culture

A major problem in studying papillomaviruses is the difficulty in propagating them in cultured cells *in vitro*. Much of what is known about papillomavirus replication is derived from studies on warts, or on cells transformed by papillomaviruses and propagated *in vitro*. However, transformed cells do not generally produce virus, and express a limited range of viral proteins. There has been recent success in propagating small amounts of several human papillomaviruses in specialized **raft culture** systems, which are capable of maintaining fully differentiated epithelial cells over long periods of time, and this is providing valuable insight into the life cycle of these viruses.

TABLE 11.1 Sites of infection of selected human papillomaviruses

Virus type	Site of infection	Clinical manifestation
1	Soles of feet	Deep plantar warts
2, 4, 7	Hands	Common warts
13, 32	Oral cavity	Focal epithelial hyperplasia
3, 8, 9, 10, 12, 14, 15, 17, 19–25, 36, 38	Arms, forehead, trunk	Flat warts, skin cancer
6, 11	Genital tract, larynx, nasal cavity	Cervical warts, condyloma
16, 18, 31, 33, 35, 39, 41–45	Genital tract	Cervical cancer, atypical condyloma

Papillomavirus genomes are circular, double-stranded DNA

Papillomavirus virions are small, nonenveloped particles with T = 7 icosahedral symmetry (Figure 11.1). Virions are constructed from 72 capsomeres, each containing 5 molecules of the major capsid protein L1. The viral genome, an 8 Kbp molecule of double-stranded circular DNA, is associated with cellular nucleosomal histones. Papillomaviruses share many of these properties with the polyomavirus family (Chapter 10), with which they were previously grouped. However, papillomaviruses have a significantly larger genome, their gene map is organized in a different fashion, and they share a distinct lifestyle, all justifying their classification in their own family.

Most human and animal papillomavirus genomes contain 8 to 10 identifiable open reading frames. The human papillomavirus type 16 genome contains six early genes (E1, E2, E4, E5, E6, E7) and two late genes (L1, L2). These open reading frames are shown on a linearized genome map in Figure 11.2. Unlike the polyomaviruses, all papillomavirus mRNAs are transcribed from only one of the DNA strands. Note that several protein coding regions overlap, allowing the efficient use of a small genome. A brief description of the function of each viral gene is presented in Table 11.2.

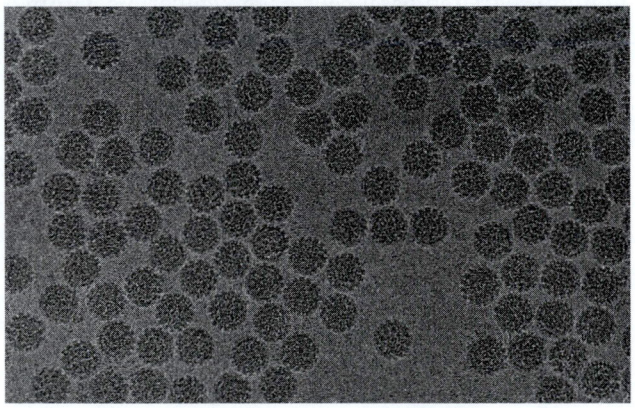

(a)

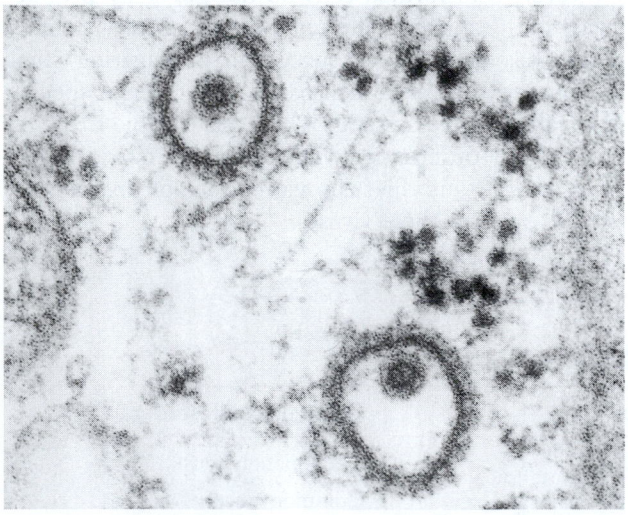

(b)

FIGURE 11.1 Papillomavirus virions. (a) Transmission electron micrograph of negatively stained bovine papillomavirus type 1. (b) Three-dimensional image reconstruction of virions derived from cryoelectron microscopy analysis.

The infectious cycle follows differentiation of epithelial cells

During infection of the **mucosal epithelium** (Figure 11.3), virions must first reach the nondifferentiated **basal cells** of the epithelium where they bind to heparin and to the cellular surface protein, α_6-integrin. Virions are then taken up by endocytosis and viral DNA enters the nucleus, where it is maintained as a free, circular minichromosome in the basal cells. Early genes are transcribed, permitting an initial replication phase that results in 50 to 100 copies of DNA per cell. Genomes then replicate on average about once per cell cycle as the basal cells divide, and the viral genomes are distributed equally among the daughter cells. This type of viral DNA replication is referred to as "plasmid replication."

When the basal cells become committed to the pathway toward terminally differentiated epithelial cells (known as **keratinocytes**), there is a burst of viral DNA synthesis, known as "vegetative replication." At the same time, late genes L1 and L2 are expressed, producing the viral capsid proteins. Progeny virus particles are assembled in the nucleus, and are released upon cell death and shedding at the surface of lesions. Little is presently known about the intracellular signals that regulate the switch from plasmid to vegetative viral DNA replication during differentiation in keratinocytes.

Viral mRNAs are made from two promoters and two polyadenylation signals

Numerous viral messenger RNAs have been described. Two major promoters are used by human papillomavirus types 16 and 31: an early promoter located upstream of the E6 gene, and a late promoter located upstream of the E1 gene (Figure 11.2). In undifferentiated basal epithelial cells, the early promoter directs transcription of RNAs that code for all six early gene products. The early primary transcript is polyadenylated at a site downstream of the E5 gene, and these transcripts are spliced in a variety of ways to generate numerous early mRNAs. Many of these RNAs contain multiple open reading frames, but it is not clear how these multiple coding regions are translated by ribosomes.

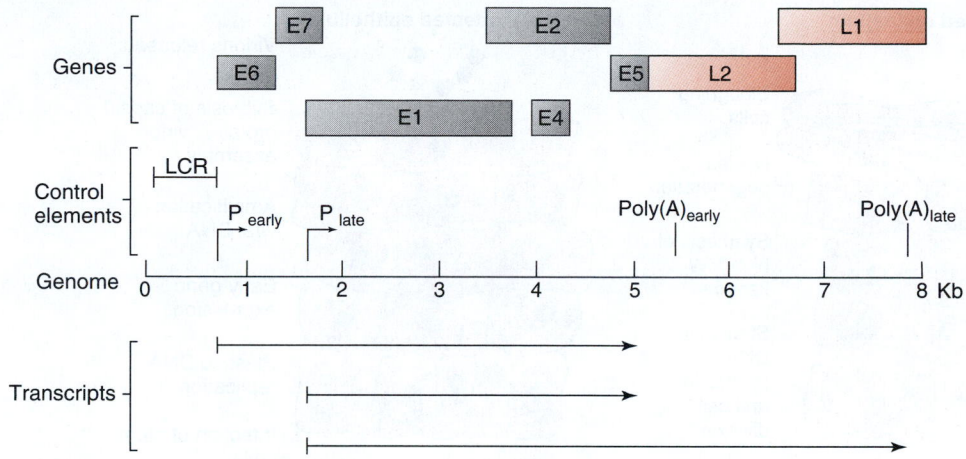

FIGURE 11.2 Genetic and transcriptional map of human papillomaviruses. A linear representation of the circular genome of papillomaviruses, based on information from human papillomavirus types 16 and 31. Open reading frames of the six early (E) genes and two late (L) genes are shown in boxes. LCR: long control region. Two promoters (P_{early} and P_{late}) and two polyadenylation signals (Poly A_{early} and Poly A_{late}) control production of transcripts, shown at bottom. These transcripts are spliced in multiple alternative ways to produce a large number of mRNAs (not shown), many of which contain multiple open reading frames.

The late promoter is active in differentiated keratinocytes (Figure 11.3). Some of these transcripts are polyadenylated at the early polyadenylation site, giving rise to abundant mRNAs that express the E1 and E2 proteins. The increased level of these proteins contributes to the shift to vegetative DNA replication, during which many thousands of viral DNA molecules per cell are made. Other transcripts from the late promoter extend through most of the viral genome to a second major polyadenylation site. Again, the alternative use of several splice sites leads to the production of a number of late mRNAs. Only mRNAs that start at the late promoter and end at the late polyadenylation site code for L1 and L2 capsid proteins. These mRNAs undergo splicing that removes most of the 3–5 Kb region that lies between the late promoter and the beginning of the L1 and L2 open reading frames.

Viral E1 and E2 proteins bind to the replication origin and direct initiation of DNA replication

Regulatory DNA sequences known as the long control region (LCR) of human papillomaviruses extend from the end of the L1 gene to the beginning of the E6 gene (Figure 11.2). The long control region contains the viral origin of DNA replication as well as enhancer se-

TABLE 11.2 Functions of human papillomavirus genes

Gene	Function
E1	DNA helicase, binds to replication origin, initiates DNA replication
E2	Binds to specific sites on viral DNA, enhances binding of E1 to replication origin, activates or represses transcription of viral genes
E4	Function largely unknown, may help virus maturation and release
E5	Small, membrane-associated protein, enhances growth factor responses in infected cell, stimulates cell proliferation
E6	Directs ubiquitin-mediated degradation of cellular p53, inhibits cell cycle block and apoptosis (oncogenic papillomaviruses only)
E7	Binds to Rb protein, releases E2F, activates cell cycling and DNA replication genes, stimulates cell and viral DNA replication
L1	Major capsid protein
L2	Minor capsid protein, helps encapsidate viral DNA

Note: Other human and animal papillomaviruses have additional genes designated E3 and E8.

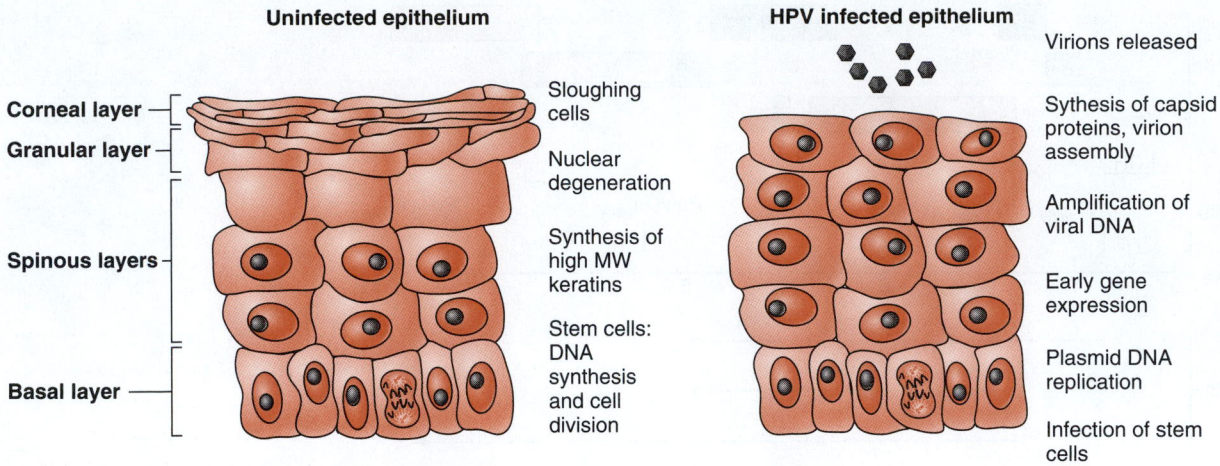

FIGURE 11.3 Papillomavirus infection of differentiating epithelial cells. Cells differentiate as they pass from the basal layer (bottom) to the surface (top). Activities in different cell layers are shown in the middle. HPV infection leads to increased cell division and retention of cell nuclei in outer layers. Viral functions that take place in different layers are shown at the right.

quences that control viral gene expression. The viral E1 protein is a DNA helicase that binds to the origin of replication and locally unwinds the viral DNA, allowing cellular DNA polymerases and other cellular proteins to carry out viral DNA replication. E1 is therefore the papillomavirus homologue of the polyomavirus large T antigens (Chapter 10). The viral E2 protein is also involved in viral DNA replication. It binds to distinct sites near the replication origin, and forms a complex with the E1 protein. Binding of E2 to E1 increases the affinity of E1 binding to the replication origin.

Numerous cellular transcription factors including NF1, SP1, AP1, Oct1, YY1, and the DNA-binding glucocorticoid receptor protein have been shown to interact with the regulatory regions of papillomavirus DNA and to play important roles in activating viral transcription. The E2 protein also acts as a transcriptional activator of viral promoters through its ability to bind to specific sequences within the long control region. However, in some cases, E2 may also inhibit viral gene transcription.

Viral E7 protein interacts with cell cycle regulatory proteins, particularly Rb

Papillomaviruses do not code for their own DNA replication enzymes. Therefore, in order to replicate their double-stranded DNA genomes, they must use the machinery of the host cell. Viral DNA replication occurs in fully differentiated keratinocytes, which do not normally divide, and therefore DNA synthesis does not normally occur in these cells. It is the role of the viral E7 protein to induce cellular DNA synthesis.

The E7 proteins of all papillomaviruses are small multifunctional proteins approximately 100 amino acids in length. E7 proteins share some structural homology with proteins from other viruses, including the adenovirus E1A protein and polyomavirus large T antigens. E7, E1A, and large T antigens have been shown to interact with many of the same cellular targets, and these interactions result in deregulation of the normal cell cycle, resulting in the induction of cellular DNA synthesis (Figure 11.4). The most significant cellular targets of E7 are the **retinoblastoma (Rb) tumor suppressor protein** and its functionally related proteins, p107 and p130. Rb is considered a tumor suppressor protein since loss of the corresponding Rb genes is associated with the development of different forms of cancers, most notably retinoblastoma, a childhood tumor of the eye.

The Rb protein is a major regulator of the cell cycle, and functions through its ability to bind to and repress the activity of a family of cellular transcription factors known as the E2F family (Figure 11.5). When E2F is freed from bound Rb, it can induce the expression of a number of genes whose products initiate and carry out cellular DNA synthesis. Consequently, activation of the E2F transcription factors results in progression of the cell cycle from the G1 phase to the S phase, during which DNA synthesis takes place.

It is noteworthy that the transition from G1 to S phase is a crucial point in the cell cycle; once cellular DNA synthesis begins, the cell is committed to undergo mitosis. The expression of E7 alone in resting cells can stimulate cellular DNA replication, and this activity correlates with the ability of E7 to bind to Rb, resulting in the release of E2F. The E7 proteins derived from the highly oncogenic human papillomavirus types 16 and 18 bind more strongly to Rb than E7 proteins derived from nononcogenic virus

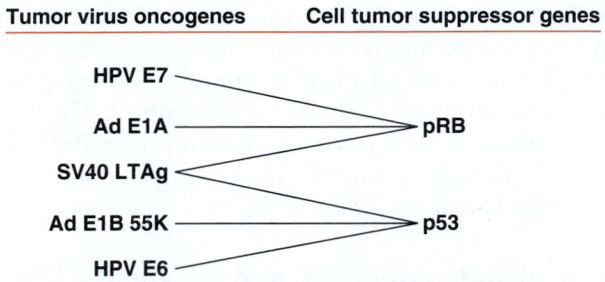

FIGURE 11.4 Common pathways used by viral oncogenes. Human papillomaviruses (HPV), adenoviruses (Ad), and polyomaviruses (SV40) express proteins that bind to and inhibit the activity of cellular tumor suppressor proteins involved in controlling cell cycle progression (pRb) and apoptosis (p53).

types 6 and 11. This may account in part for the differences in their respective transforming activities.

Therefore, the major activity of E7 is its ability to induce unscheduled DNA synthesis in fully differentiated infected keratinocytes. This allows the viral genome to be replicated to a high copy number by the cellular DNA replication machinery. However, unscheduled DNA synthesis in fully differentiated cells is clearly not a normal cellular event; mechanisms have developed to eliminate such runaway cells from the body. Unscheduled DNA synthesis results in activation of **apoptosis,** leading to cell death, and this process is dependent on the activity of the cellular p53 protein. In fact, not only does E7 induce cell cycle progression, but this also results in an increase in p53 levels. Therefore, in order for the virus to keep the cell alive long enough to complete its replication, it has developed a means to counteract p53-mediated apoptosis. The virus expresses another protein, E6, which mediates the degradation of cellular p53 protein in the infected cell (Figures 11.4 and 11.6).

Viral E6 protein controls the level of cellular p53 protein

Human papillomavirus E6 proteins are approximately 150 amino acids in length. The most notable feature of E6 proteins made by oncogenic papillomaviruses such as types 16 and 18 is their ability to mediate the destruction of p53 through the **ubiquitin** proteolysis pathway (Figure 11.6). E6 first associates with another cellular protein termed E6AP. The E6-E6AP complex then binds to p53, stimulating the **ubiquitination** of p53 and its subsequent degradation in **proteasomes.** In comparison, E6 proteins made by nononcogenic human papillomaviruses such as types 6 and 11 are unable to mediate the degradation of p53.

Cellular p53 is a prototype tumor suppressor protein. This is best illustrated by experiments showing that p53 knockout mice, genetically engineered to contain no p53 genes, develop normally in every respect except that they succumb to a variety of tumors at a young age (a few months old). Clinical studies have shown that mutations in the human p53 gene are present in over 50% of all human tumors. The majority of cancers retaining wildtype p53 genes suffer mutations in genes whose prod-

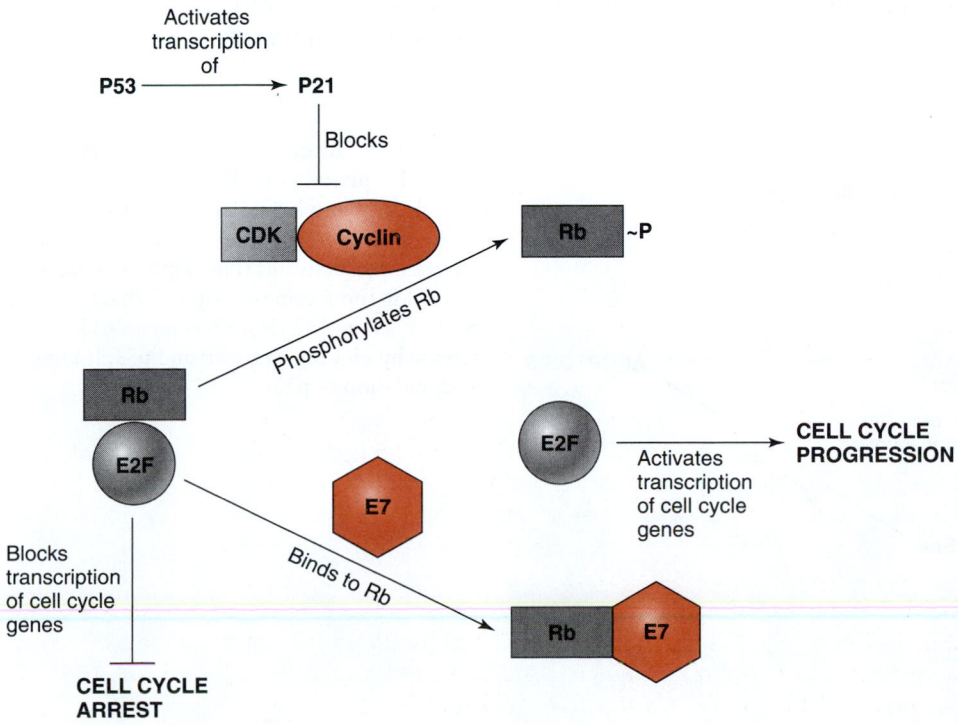

FIGURE 11.5 Papillomavirus E7 protein activates cell cycle progression. Rb protein binds to E2F, blocking activation of cell cycle genes. In normal cell cycle (top half), CDK/cyclin complex phosphorylates Rb, releasing it from E2F and leading to expression of cell cycle genes. This can be blocked by activated p53 via p21. In HPV-infected cells (bottom half), E7 binds to Rb, activating E2F and leading to cell DNA replication and cell cycle progression.

ucts are required to activate p53 function. Consequently, the majority of all cancers have undergone either a mutation in the p53 gene or alterations in genes encoding p53 regulatory molecules. Oncogenic papillomaviruses, simian virus 40, and adenoviruses have all evolved to express proteins that target and inhibit the activity of p53, albeit through different mechanisms (Figure 11.4).

The p53 protein is normally present at low levels in the cell in a latent, inactive form. When the cell is exposed to stressful conditions such as viral infection or DNA damage, p53 becomes stabilized by phosphorylation, and its level increases. The accumulated p53 either induces cell cycle arrest in G1, allowing sufficient time to repair damaged DNA, or induces apoptosis, which eliminates these abnormal cells. In a multicellular organism, the loss of abnormal cells through apoptosis can be considered as a defense mechanism against disease, including virus infection and cancer.

The biological effects of p53 are largely due to sequence-specific transcriptional activation of target genes whose products mediate cell cycle arrest or apoptosis (Figure 11.6). Bax and p21 are two examples of such genes. However, p53-mediated apoptosis of virus-infected cells could limit the ability of viruses to complete their life cycle and thus restrict the spread of the virus. The major role of papillomavirus E6 protein is to direct the degradation of p53, therefore inhibiting apoptosis in the infected cell and allowing cell survival until sufficient new progeny virus is made. Although the p53 gene is commonly mutated in the majority of human cancers, it is not mutated in cervical cancers caused by papillomaviruses. This makes sense: There should be no selective pressure to mutate the p53 gene in these cancers, since the p53 protein is eliminated by the activity of the viral E6 protein.

E6 proteins from high-risk human papillomaviruses have also been shown to interact with several other cellular proteins involved in signal transduction, apoptosis, DNA replication, and calcium metabolism. The full significance of these interactions and how they affect viral replication are unknown. However, they may also contribute to tumorigenesis.

Synergism between E6 and E7 and the predisposition to cancer

Although p53 and Rb are multifunctional tumor suppressor proteins, they intersect in a common regulatory pathway (Figures 11.5 and 11.6). Stabilization of p53 in response to DNA damage or viral oncogenes such as E7 results in transcriptional activation of the p21 gene. The resulting p21 protein is a potent inhibitor of Rb phosphorylation, and this prevents the expression of E2F-responsive genes. Therefore, p53-activated p21 expression inhibits progression of the cell cycle from G1 to S phase, inhibiting cell proliferation. When this occurs in response to the presence of damaged DNA, cells will not replicate until the damaged DNA has been repaired. However, in cells infected with oncogenic papillomaviruses, this pathway is not under normal control since p53 levels are suppressed by E6, and the pRb protein is bound by E7, resulting in activation of E2F-mediated genes (Figures 11.5 and 11.6). The papillomavirus oncoproteins are therefore able to act at two distinct points in the same pathway by abrogating both p53 and Rb activity. This effectively results in the uncontrolled proliferation of the virus-infected cell and, together with the anti-apoptotic activity of p53, is a potent oncogenic combination. As a result of these host–virus pro-

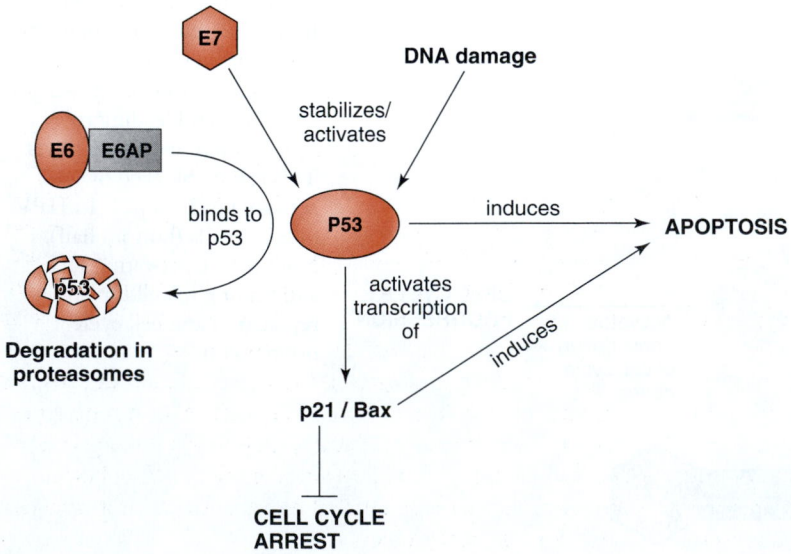

FIGURE 11.6 Interaction of papillomavirus E7 and E6 proteins with p53. Following DNA damage or virus infection, p53 becomes stabilized and activated, and its concentration increases. p53 activates transcription of genes involved in induction of apoptosis (bax) and arrest of cell cycle (p21). E6 reverses p53 increase by binding to E6AP and p53, leading to degradation of p53.

tein interactions, cervical cancer is among the most common forms of malignancy in the world.

However, the ultimate objective of viruses is not to induce malignant transformation, but to replicate by completing a productive growth cycle. To replicate in resting cells, the virus must target the growth regulatory pathways described above. Cells transformed by papillomaviruses are in fact unable to produce new virus particles, because only a part of their genome is present in the transformed cell, and furthermore it is integrated into the cellular genome and can no longer replicate independently. Cancers mediated by oncogenic papillomaviruses occur in only a small percentage of people who become infected with these viruses, because integration of viral DNA is a rare event. Nevertheless, these viruses are widespread, and therefore oncogenic papillomaviruses are important human carcinogens.

Cells transformed by papillomaviruses express E6 and E7 gene products from integrated viral DNA

The most important and interesting disease aspect of human papillomaviruses such as types 16 and 18 is their involvement in cervical cancer. Transformed cells derived from cervical carcinomas can be propagated *in vitro*. In addition, some cell types cultivated *in vitro* can be transformed by exposing them to isolated viral DNA, allowing their study. Observations made with both kinds of transformed cells have led to a detailed understanding of the molecular mechanisms by which viral gene products induce cell transformation.

Cell lines derived from cervical cancers contain copies of a part of the human papillomavirus genome integrated into the cellular genome at sites that differ in each cell line. Integration of viral DNA into the cellular genome only occurs during the development of the cancer; viral DNA in productively infected cells is present as a free plasmid. Integration in cancer cells often results in the loss of all viral genes except the E6 and E7 genes. These two viral genes are expressed in most cervical cancers. Indeed, one of the most widely used human cell lines, HeLa, was derived from a cervical cancer many years ago and has been shown to contain papillomavirus DNA and express the E6 and E7 proteins.

No virus replication takes place in transformed cells. The process of integration of viral DNA into the host cell therefore has no obvious benefit for the virus, and can be considered as an "unplanned" byproduct of the infection process that has unfortunate consequences for the host. Much research on these viruses has focused on defining the biological activity of the E6 and E7 viral proteins. These proteins function to induce cellular DNA replication while preventing the cell from undergoing apoptosis (programmed cell death). This is indeed a fine balance, since induction of unscheduled DNA synthesis in fully differentiated cells usually induces apoptosis.

Future prospects for diagnosis and treatment of diseases caused by papillomaviruses

The cervical smear test (**Pap test**, named for its inventor, George Papanicolaou) has proven to be a very effective means of reducing the incidence of death caused by cervical cancer in developed countries. This test is based on the recognition of morphologically abnormal cervical cells indicating the presence of premalignant or malignant lesions. Cervical warts and precancerous tumors are routinely removed by surgery or physically destroyed by freezing with liquid nitrogen or laser vaporization. Therefore, with appropriate screening and effective treatment options, the development of malignant cervical cancer is relatively rare in developed countries. However, human papillomavirus-associated cervical cancer is one of the most common forms of cancer in developing countries (several hundred thousand new cases per year). The incidence of this cancer could be dramatically reduced with appropriate health care intervention such as a routine Pap smear and follow-up treatment. Universal screening has the potential for decreasing the worldwide incidence of cervical cancer.

Treatment of papillomavirus-induced precancerous lesions involves the removal of the infected lesions. To a large extent this is made possible because these are localized surface lesions, and papillomaviruses do not give rise to systemic infections. Several topical creams are also available; these have poorly understood antiviral activity or work by stimulating a local cellular immune response against the virus.

Diagnosis of papillomavirus infections by traditional methods such as detection of specific antibodies has limited value because virus types cannot be effectively distinguished by this means. The Pap smear is still the most widely used and effective test to detect cellular changes indicative of disease such as cervical neoplasia. However, it is a screening method, not a viral diagnostic test. Diagnostic tests are now available and are being further developed to detect oncogenic papillomaviruses such as types 16 and 18 in Pap smear samples. Detection of an oncogenic papillomavirus type in patients with an abnormal Pap smear may become valuable in patient management to direct the patient toward further tests and treatments, and more stringent subsequent monitoring. In future, diagnostic testing in concert with the routine Pap smear may result in better detection and management of those individuals at greatest risk for cervical cancer.

Human papillomavirus vaccines

Recent progress in vaccine development and testing is very encouraging. Two different human papillomavirus vaccines have been produced by using viral capsid protein L1 assembled into virus-like particles that do not contain the viral genome. These vaccines carry no risk of infection but can induce antibody responses to the virus. Both of these vaccines contain L1 of papillomavirus types 16 and 18, responsible for most cervical carcinomas; one of the vaccines also contains L1 of types 6 and 11, responsible for most genital warts or condylomas. Large-scale trials are underway, and have already shown strong antibody production in vaccinated individuals and nearly 100% protection against infection by targeted viruses. Furthermore, vaccinated individuals were protected against development of cervical carcinomas. Based on these results, both vaccines are expected to be licensed for use in the general population soon. The consensus is that girls 11–12 years old should be the principal target for vaccination; young people develop the highest antibody responses to the vaccine, and vaccinating people before most sexual activity begins will obviously be most effective.

Another approach involves the use of the human papillomavirus L2 protein, which may provide protection against a wider range of virus types. A vaccine based on the L2 protein will soon be tested. However, vaccines based on virus-like particles are expensive to manufacture and require injection of several doses, and poorer countries may have difficulties in adopting their widespread use. Therefore one group is engineering harmless bacteria to express papillomavirus capsid proteins and generate virus-like particles. A vaccine based on live baacteria would be inexpensive to produce and could be administered orally. If effective, such a vaccine could eventually replace the more expensive injected vaccines and contribute to the prevention of a major cancer worldwide.

SUGGESTED READING

1. Franco, E.L. (1995). Cancer causes revisited: Human papillomavirus and cervical neoplasia. *Journal of the National Cancer Institute* **87,** 779–780.

2. Longworth, M.S. and Laimins, L.A. (2004). Pathogenesis of human papillomaviruses in differentiating epithelia. *Microbiology and Molecular Biology Reviews* **68,** 362–372.

3. Matlashewski, G., Schneider, J., Banks, L., Jones, N., Murray, A. and Crawford, L. (1987). Human papillomavirus type 16 DNA cooperates with activated ras in transforming primary cells. *EMBO Journal* **6,** 1741–1746.

4. Munger, K., Phelps, W., Bubb, V., Howley, P., and Schlegel, R. (1989). The E6 and E7 genes of human papillomavirus type 16 together are necessary and sufficient for transformation of primary human keratinocytes. *Journal of Virology* **63,** 4417–4421.

5. Schaller, J.T. and Davies, P. (2004). Delivering on the promise: HPV vaccines and cervical cancer. *Nature Reviews Microbiology* **2,** 343–347.

6. Scheffner, M., Werness, B., Huibregtse, J., Levine, A., and Howley, P. (1990). The E6 oncoprotein encoded by human papillomavirus types 16 and 18 promotes the degradation of p53. *Cell* **63,** 1129–1136.

7. Zur Hausen, H. (1996). Papillomavirus infection: A major cause of human cancers. *Biochimica et Biophysica Acta* **1288,** 55–78.

KEY TERMS

Apoptosis
Basal cells
Condyloma
Keratinocytes
Mucosal epithelium
Pap test
Proteasome
Raft culture
Retinoblastoma (Rb)
Squamous cell carcinoma
Tumor suppressor protein
Ubiquitin
Ubiquitination

Adenoviruses

Adenoviridae

From Greek *adenos* (gland), for site of first isolation, adenoids

VIRION

Naked icosahedral capsid (T = 25) with 20 triangular faces.
Diameter 70–90 nm.
Knobbed fibers protruding from each of 12 vertices.
Eleven proteins in virion.

GENOME

Linear ds DNA, 30–36 Kb.
Short inverted terminal repeat sequence.
Both 5′ ends covalently bound to a virus-coded terminal protein.

GENES AND PROTEINS

Codes for up to 50 proteins.
Most regions of both DNA strands are transcribed by cellular RNA polymerase II.
Six early and two delayed early transcription units: E1A, E1B, E2A, E2B, E3, E4; IVa2, IX.
One major late transcription unit: five subclasses of late mRNAs, L1 through L5.
Each transcription unit gives rise to multiple mRNAs by RNA splicing.
Two small VA RNAs transcribed by RNA polymerase III.

VIRUSES AND HOSTS

Fifty human serotypes in six subgroups (A through F).
Other adenoviruses in cattle, mice, birds, etc.

DISEASES

Respiratory syndromes including pneumonia, but not common colds.
Eye and gastrointestinal infections.

DISTINCTIVE CHARACTERISTICS

Multiple mRNAs arise by extensive alternative splicing.
Late transcripts made from single strong promoter and five alternative polyadenylation signals.
DNA synthesis primed by terminal protein, uses viral DNA polymerase.
Can cause tumors in experimental animals but not known to cause tumors in humans.
Adenoviruses are widely used as gene therapy and anticancer virus vectors.

Adenoviruses cause respiratory and enteric infections in humans

Adenoviruses were first isolated by Wallace Rowe in 1953 from tissue culture cell lines derived from tonsils and **adenoid** tissues of sick children. Shortly afterward, Maurice Hilleman and colleagues identified the same agent as the cause of a respiratory epidemic in army recruits in the United States. Originally (and incorrectly) postulated to be a cause of the common cold, adenoviruses were subsequently shown to be responsible for a number of other respiratory syndromes including pneumonia; they can also cause eye and gastrointestinal infections. The current adenovirus name and nomenclature were adopted in 1956, acknowledging the original source of the virus from adenoid tissue.

Adenoviruses can be oncogenic, but not in humans

Human adenovirus type 12 can induce malignant tumors in hamsters. This was the first human virus shown to be oncogenic. Although no evidence has linked adenoviruses to human cancer, these viruses have been used as models for understanding the molecular biology of cancer. Furthermore, their study has helped uncover basic mechanisms of gene expression in mammalian cells. In 1977, the laboratories of Philip Sharp and Richard Roberts discovered RNA splicing by studying adenovirus late mRNAs, and were jointly awarded the Nobel prize for this discovery. In 1979, Robert Roeder used the adenovirus major late promoter to demonstrate specific initiation of transcription by RNA polymerase II *in vitro*, and he and others subsequently discovered many of the protein factors involved in transcription in mammalian cells by using this model system. Human adenoviruses have recently become of great importance as gene therapy vectors and are promising anticancer agents.

The 100 known adenoviruses include 50 human virus serotypes, distinguished by their sensitivity to neutralizing antibodies. The human viruses are further classified into six subgroups (A to F) based on their ability to agglutinate red blood cells. This chapter will deal with the best-studied human adenoviruses: types 2 and 5, nononcogenic members of group C, and type 12, a highly oncogenic member of group A.

Virions have icosahedral symmetry and are studded with knobbed fibers

Adenovirus virions have a remarkable resemblance to a regular icosahedron, with 20 equilateral triangular faces and 12 vertices (Figures 12.1 and 12.2). This handsome capsid contains a total of 11 viral proteins, but is constructed from two basic building blocks, the **hexon** and the **penton base** (Figure 12.2). Hexons consist of trimers of the hexon protein (polypeptide II); each hexon has six nearest neighbors, hence the name. A total of 240 hexons cover the 20 faces of the T = 25 icosahedral capsid. Penton base subunits, each made of five molecules of the penton protein (polypeptide III), lie at the 12 vertices of the icoashedron. Projecting from the penton bases like antennae on a space satellite are fibers made of trimers of polypeptide IV. The fiber forms a triple helical structure anchored to the penton base at one end and has a terminal knob at the other end. A number of "glue" proteins (polypeptides IIIa, VI, VIII, and IX) mediate interactions between hexons and pentons, and with the viral DNA.

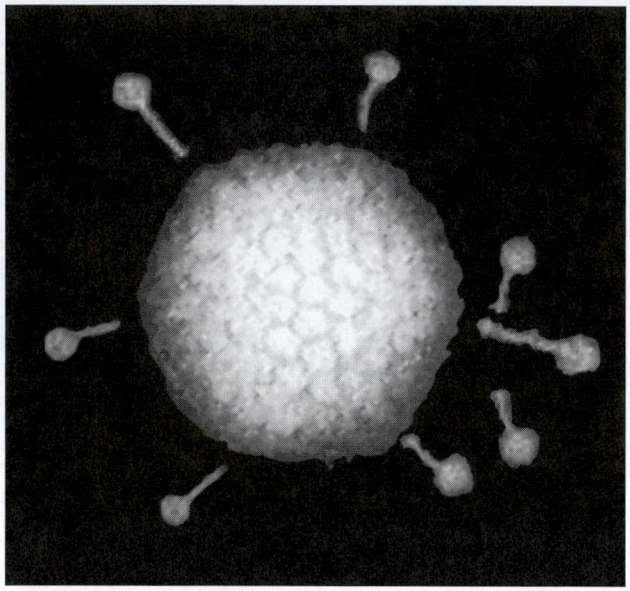

FIGURE 12.1 Adenovirus virion. Electron micrograph of negatively stained virion, showing fibers radiating out from the capsid.

The 30–36 Kb linear, double-stranded viral DNA genome is organized into coiled domains by the histone-like polypeptide VII. Basic polypeptides V and X are also bound to the viral DNA. A 55 kDa virus-coded **terminal protein** is covalently linked via a serine residue to each of the 5′ ends of the viral DNA (Figure 12.3). The viral DNA contains **inverted terminal repeat** sequences of approximately 100 bp (Figure 12.3). This means that the first 100 nt at the 5′ and 3′ ends of each DNA strand are complementary and can base-pair, allowing single DNA strands to circularize, forming a short panhandle end. Both the terminal protein and the inverted terminal repeats are involved in viral DNA replication (see below).

Fibers make contact with cellular receptor proteins to initiate infection

Adenoviruses attach to specific cellular receptors via the knob of the fiber protein. Most adenovirus serotypes,

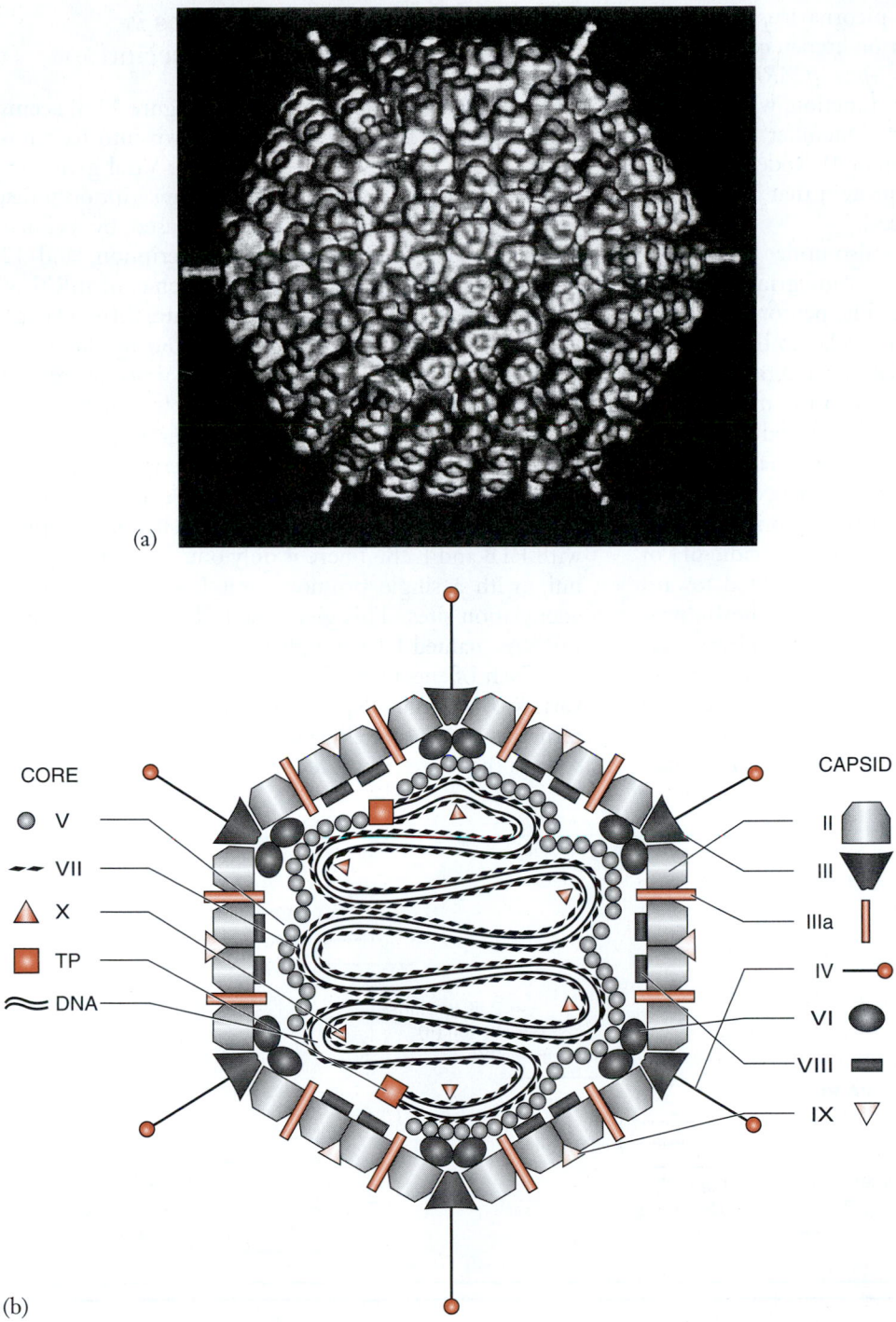

FIGURE 12.2 Structure of adenovirus virion. (a) A computer-generated image of adenovirus type 2 virion, viewed along a threefold symmetry axis. Reconstructed from cryoelectron microscopy images. (b) A schematic cross-sectional view of a human adenovirus type 5 virion, with components shown as indicated.

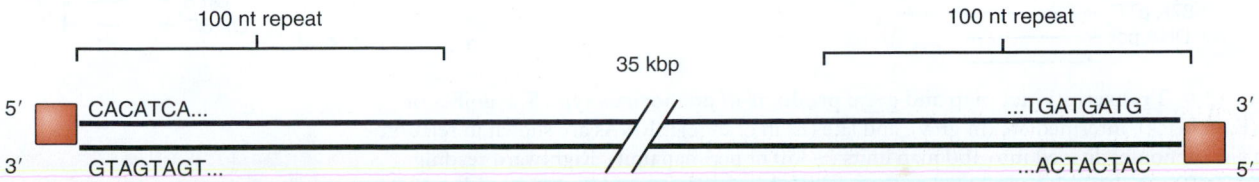

FIGURE 12.3 Terminal inverted repeats on adenovirus DNA. Linear double-stranded adenovirus type 5 DNA showing the first 8 nt of the inverted terminal repeats. Terminal protein shown as a orange square.

as well as Coxsackievirus B3 (a picornavirus; see Chapter 16) utilize the same receptor on human cells, termed *Coxsackievirus and adenovirus receptor (CAR)*. CAR is a cell surface protein of unknown function, whose amino acid sequence shows that it is a member of the **immunoglobulin family** (see Chapter 4). Rodent cells also possess a closely related CAR protein that allows their infection by human adenoviruses.

Adenoviruses types 2 and 5 also utilize certain cell surface **integrin** proteins, including integrins $\alpha_V\beta3$ and $\alpha_V\beta5$, as **secondary receptors**. The penton base binds to these integrins via a domain similar to that used by extracellular adhesion molecules that interact with integrins. Binding to integrins leads to the migration of virus-receptor complexes to clathrin-coated pits, where they are taken up into endosomes (see Chapter 4). Internalization within an endosomic vesicle occurs quickly (within 10 minutes), after which the virion escapes to the cytosol in a process dependent on the acidic pH of the endosome. Virus particles are transported toward the nucleus using interactions between the hexon protein and microtubules. They localize at nuclear pores where they are disassembled in a sequential manner, and the DNA is released into the nucleus. There the viral DNA associates with the nuclear matrix via the terminal protein, and transcription of viral genes commences.

Expression of adenovirus genes is controlled at the level of transcription

Although the adenovirus gene map (Figure 12.4) seems complicated at first, when broken down into its component parts it is easier to understand. Viral genes are arranged in a well-conserved set of transcription units; eight transcriptional promoters are used by cellular RNA polymerase II to initiate transcription, and 12 polyadenylation signals define the 3′ ends of mRNAs. Transcription units can be divided into three classes, depending on the timing of their use during the infectious cycle. The six *early* transcription units are named E1A, E1B, E2A, E2B, E3, and E4. Note that E2A and E2B use a common promoter but different polyadenylation sites. The two *intermediate* transcription units are named IX and IVa2. Note that these genes use distinct promoters but share polyadenylation sites in common with E1B and E2B. There is only one *late* transcription unit, with a single promoter but five different polyadenylation sites. This gives rise to five families of late mRNAs, named L1 through L5.

Each of the 13 viral transcription units contains a variety of often overlapping protein coding regions. Alternative RNA splicing generates different mRNA species that together code for a grand total of about 50

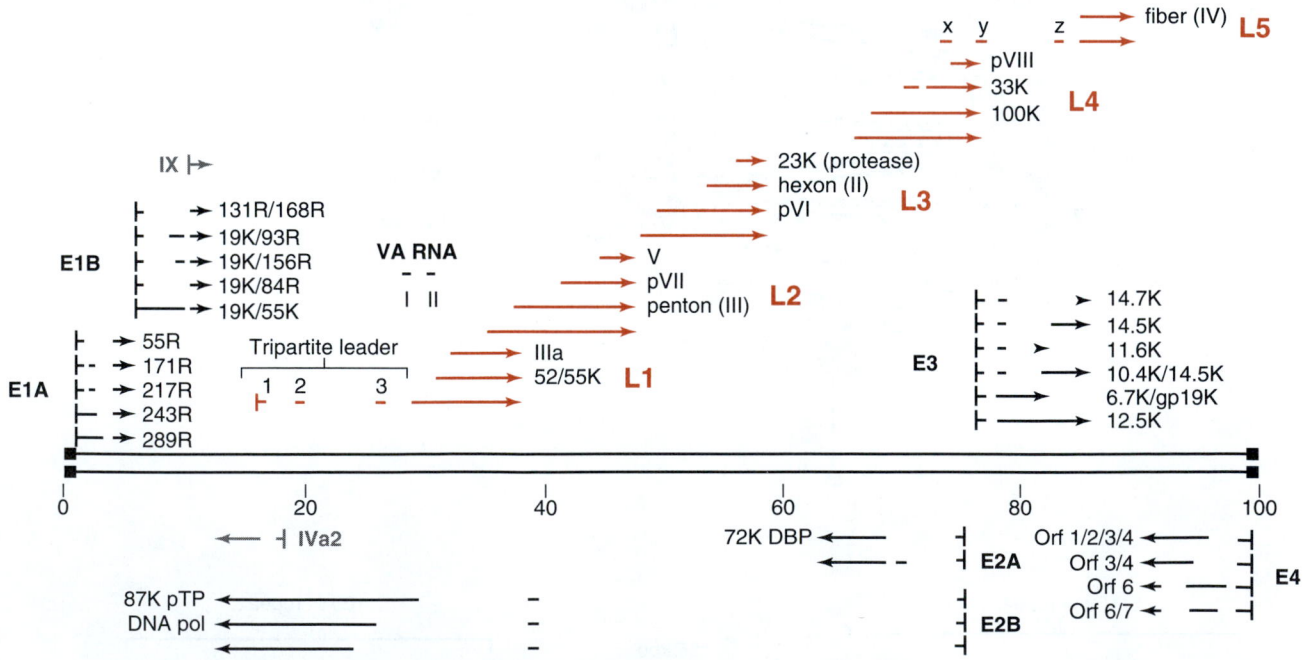

FIGURE 12.4 Transcriptional map and gene products of adenovirus type 5. Families of early (E, in black), intermediate (in gray), and late (L, in orange) mRNAs are shown in relation to the viral genome, divided into 100 map units (~350 nt per map unit). Rightward reading transcripts shown at top, leftward reading transcripts shown at bottom. Vertical bars show 5′ ends of RNAs, corresponding to transcriptional promoters; arrowheads represent 3′ ends of mRNAs, corresponding to polyadenylation sites. Introns are shown as interruptions in mRNAs. Names of proteins coded by mRNAs are shown to the right of the corresponding mRNAs.

distinct adenovirus proteins. Adenovirus genomes can be seen as constructed from modular units similar to the early and late transcription units of the smaller and simpler polyomaviruses (Chapter 10). Adenovirus transcription units are interspersed in such a way that virtually all of the viral genome is utilized for coding proteins on one or the other DNA strand.

Adenovirus genomes also contain one or two copies of VA genes, transcribed at high levels by cellular RNA polymerase III, the enzyme used by the cell to transcribe tRNA and other small RNA genes. The small, highly base-paired VA (for "virus-associated") RNAs are involved in translational control by inhibiting activity of the cellular double-stranded RNA-dependent protein kinase **(PKR)**, one of the pathways by which **interferon** acts (see Chapter 31).

E1A proteins are the kingpins of the adenovirus growth cycle

E1A is the first viral gene to be transcribed. The E1A promoter, located near the left end of the genome (Figure 12.4), is driven by a transcriptional enhancer that requires only cellular transcription factors. Viruses that contain mutations in E1A genes make strongly reduced amounts of other early gene products, and therefore little DNA or progeny virions. Two major E1A RNAs, denoted 13S and 12S (for their slightly different sedimentation rates in sucrose gradients), encode proteins of 289 and 243 amino acid residues, respectively (289R and 243R, Figure 12.5). These two proteins are identical except for a 46–amino acid region in the 289R protein that is missing in the 243R protein because of removal of a longer intron by alternative splicing. Alignment of E1A proteins from different adenovirus serotypes reveals three regions with highly conserved amino acid sequences, termed CR1 (for "conserved region 1"), CR2, and CR3 (Figure 12.5); the 243R protein is missing CR3.

The two major E1A proteins are powerful transcriptional activators of a number of viral and cellular genes. The natural target cells of adenoviruses are terminally differentiated, nondividing epithelial cells. Adenoviruses rely on a variety of host proteins for optimal DNA replication and production of adequate levels of precursor deoxynucleotides. The E1A proteins help provide these required cellular proteins by inducing nondividing cells to enter the S, or DNA synthesis, phase of the cell cycle. An unintended result is the ability of adenoviruses to transform nonpermissive cells and to induce tumor formation in experimental animals, thus making E1A an **oncogene**.

E1A proteins bind to the retinoblastoma protein and activate E2F, a cellular transcription factor

How do E1A proteins induce cells to enter the S phase of the cell cycle? The best-understood mechanism involves an interaction between E1A proteins and the

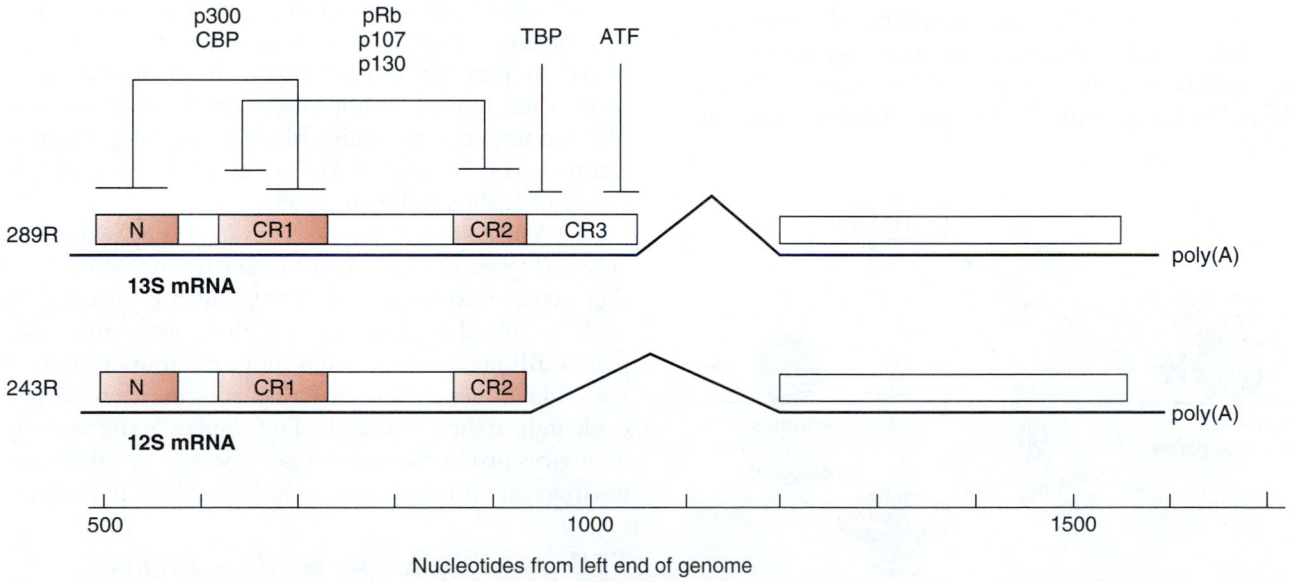

FIGURE 12.5 Major E1A proteins and interactions with cellular proteins. The two major E1A mRNAs are shown; introns removed by splicing are shown as inverted V's. Reading frames for the 289-amino acid and 243-amino acid protein products are shown (boxes) with the critical amino terminal region (N) and conserved regions (CR1, 2, and 3). The domains involved in binding by histone acetyl transferases (p300/CBP), Rb family members, and cellular transcription factors TATA-binding protein (TBP) and ATF are indicated.

retinoblastoma (Rb) family of cellular tumor suppressor proteins (Figures 12.5 and 12.6). Rb was first identified as a gene that is deleted or mutated in human retinoblastomas and other cancers. Proteins of the Rb family bind to and inactivate a family of cellular transcription factors, termed E2F. E2F was first described as a transcription factor regulating the adenovirus E2 promoter (hence its name), but it also regulates promoters of many genes coding for cellular proteins involved in DNA synthesis and cell cycle progression factors.

Rb, and its family members p107 and p130, contain a *pocket domain* that binds to both the E2F and E1A proteins. In the G1 phase of normally cycling cells, Rb is phosphorylated by **cyclin-dependent protein kinases** (Cdks), and the phosphorylated Rb no longer binds to E2F. Free E2F then activates expression of S-phase genes, allowing progression into S phase. The E1A proteins activate E2F by a similar mechanism: they bind to the Rb pocket, releasing E2F from the embrace of Rb. E2F is therefore free to activate transcription of its target genes (Figure 12.6). At the same time, the freed E2F activates both viral E2 genes by binding to a site near their common promoter. This leads to production of three early viral proteins that direct adenovirus DNA replication.

E1A proteins also activate other cellular transcription factors

The E1A 289R protein independently activates transcription of adenovirus E3 and E4 genes by interactions between its unique CR3 domain and cellular transcription factors. CR3 contains two regions involved in transactivation: a zinc finger that interacts with the TATA-binding protein (TBP) of the basal transcription machinery, and a promoter-targeting region that activates cellular transcription factors such as ATF (Figure 12.5). These interactions stabilize the transcription complex and stimulate viral gene expression.

Finally, the E1A amino terminus and part of CR1 engage in a complex with a distinct set of cellular proteins, the **histone acetyl transferases**, which also help to induce cellular DNA synthesis. These enzymes, including p300 and the cyclic AMP-response element-binding protein, CBP (Figure 12.5), play an important role in the transcriptional activation of large blocks of cellular genes. This is carried out via the acetylation of nucleosomal histones and other proteins, resulting in a more relaxed chromatin structure ("chromatin remodeling") that is more accessible to specific transcription factors and is more easily transcribed. Binding of E1A proteins to histone acetyl transferases increases histone acetylation and thus promotes entry into the cell cycle by enhancing expression of cell cycle genes.

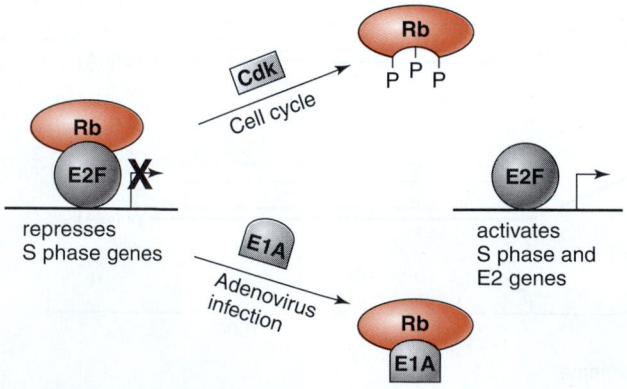

FIGURE 12.6 **S-phase gene activation by E1A proteins.** A complex between Rb and E2F is disrupted either by cyclin-dependent kinase (Cdk) phosphorylation of Rb during the cell cycle (above) or by binding of adenovirus E1A proteins during infection (below), allowing E2F to transcriptionally activate cellular and viral genes.

E1A proteins indirectly induce apoptosis by activation of cellular p53 protein

Expression of E1A proteins leads indirectly to stabilization and activation of the cellular **tumor suppressor protein**, p53 (Figure 12.7; see also Figure 11.6). p53 is a transcription factor that both activates and represses transcription, and functions to police the natural progression of the cell cycle. Activation of p53 serves as a warning signal to the cell that something is wrong, and activated p53 induces growth arrest and cell death by **apoptosis**. Apoptosis is a regulated form of cell suicide invoked when cells undergo extensive DNA damage, virus infection, or other insults. p53 activity is regulated in part by at least two cellular proteins that control its stability in opposite ways. Mdm-2 binds to p53 and targets it for **ubiquitin**-mediated degradation, whereas $p14^{Arf}$ binds to Mdm-2 and inhibits its binding to p53, thus stabilizing p53.

E1A proteins stabilize and activate p53 by two mechanisms (Figure 12.7). First, by binding to Rb (Figure 12.6) they activate expression of $p14^{Arf}$, which is encoded by an E2F-inducible gene. Second, their association with p300/CBP prevents p53 from activating transcription of the Mdm-2 gene. The resulting accumulation of p53 could induce the early death of infected cells and severely limit virus production were it not reversed by other adenovirus gene products, particularly two E1B proteins.

E1B proteins suppress E1A-induced apoptosis, allowing viral replication to proceed

The E1B region also produces multiple mRNAs by alternative splicing. The two best characterized proteins, denoted 19K and 55K (for their molecular weights), are

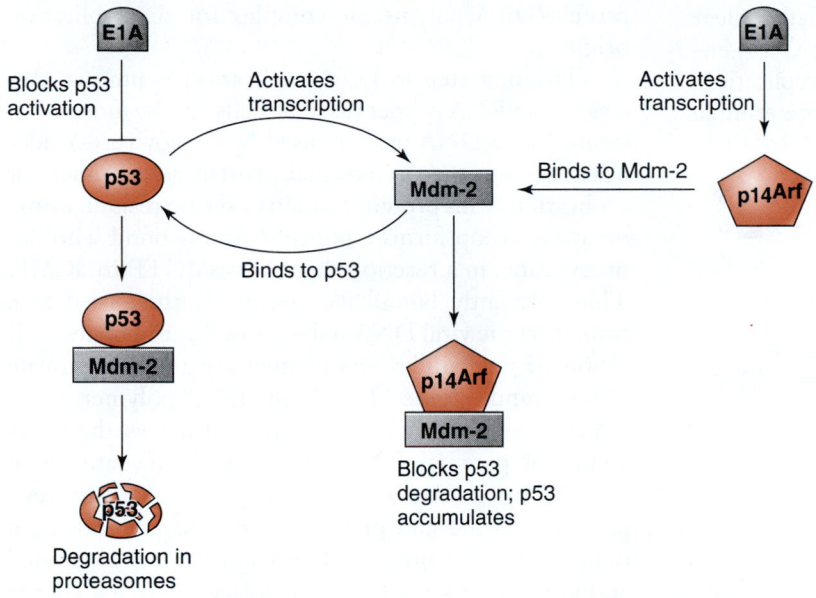

FIGURE 12.7 Degradation pathway of p53 is blocked by E1A proteins. See text for details.

produced from a single 2.2 Kb mRNA by initiation of translation in two different reading frames. 19K is a functional homologue of the cellular protein Bcl-2, which is a suppressor of apoptosis. Homodimers of another Bcl-2 family member, called Bax, stimulate the **caspase** cascade and lead to cell death. Bcl-2 reverses this effect by binding to Bax, preventing dimer formation. Both Bcl-2 and Bax are associated with mitochondrial membranes in a complex that regulates apoptosis. The adenovirus 19K protein also binds to Bax, thereby blocking apoptosis (Table 12.1).

The 55K protein binds directly to p53 and blocks transcription from p53-dependent promoters; therefore 55K inhibits p53-dependent apoptosis. It also forms a complex with a protein coded by the E4 transcription unit, called E4 orf6 (for "open reading frame 6"). This complex binds to p53 and induces its rapid degradation, lowering the concentration of p53 in the cell (Table 12.1). Thus the two major E1B proteins protect infected cells from both p53-dependent and p53-independent apoptosis, reversing activities stimulated by E1A proteins. This allows the cell to survive long enough to produce high levels of progeny adenovirus virions before eventually succumbing to the effects of virus infection.

The preterminal protein primes DNA synthesis carried out by viral DNA polymerase

Once E1A has induced the cell to enter S phase and the early viral gene products have accumulated, viral DNA replication begins. Three essential proteins made by E2 genes carry out viral DNA replication. A single-stranded DNA binding protein (ssDBP) is produced from the E2A region, while both a viral DNA polymerase and an 80 KDa preterminal protein (pTP) are encoded by E2B (Figure 12.4).

DNA replication can begin at either end of the linear DNA molecule (Figure 12.8). The replication origins, including the first 50 nt of each inverted terminal repeat, contain binding sites for a complex between the adenovirus DNA polymerase and preterminal protein.

TABLE 12.1 Functions of adenovirus E1B proteins

Protein	Function	Cellular proteins involved
19K	• Suppresses apoptosis	• Binds to Bax (homologue of Bcl-2)
55K	• Suppresses apoptosis	• Binds to p53 (transcriptional repression)
	• Induces p53 turnover (with E4 orf6)	• Binds to p53
	• Promotes transport, stabilization and translation of viral mRNAs (with E4 orf6)	• Unknown
	• Host cell shut-off (with E4 orf6 and L4-100K)	• eIF4E, eIF2α-, eIF4G

Binding by cellular transcription factors called nuclear factor I and nuclear factor III (also known as Oct I) enhances the efficiency of initiation of DNA replication, perhaps by increasing the affinity of the preterminal protein/DNA polymerase complex for the replication origin.

The first step in DNA replication is unusual. Instead of an RNA primer (used by cells and by most DNA viruses) or a DNA primer (used by parvoviruses), adenoviruses use the preterminal protein as a primer for replication. This protein contains a serine residue whose hydroxyl group forms a phosphodiester bond with deoxycytidine, in a reaction that cleaves dCTP to dCMP. This covalently bound dC residue is then used as a primer by the viral DNA polymerase for the further addition of nucleotides complementary to the template DNA strand (Figure 12.8). As the DNA polymerase extends the growing DNA chain, it displaces the other strand of parental DNA, which is then coated with the single-stranded DNA binding protein. Synthesis proceeds to the end of the linear DNA, producing a double-stranded progeny DNA containing a parental and a daughter strand, and a displaced parental single-stranded DNA molecule.

Single-stranded DNA is circularized via the inverted terminal repeat

At this point the double-stranded progeny DNA can serve as a template for further replication, but in each round only one double-stranded DNA molecule is formed, and there would therefore be no net accumulation of viral DNA. To solve this problem, we come back to the inverted terminal repeat sequences on adenovirus DNA. These sequences allow the two ends of a full-length single-stranded DNA molecule to anneal over the first 100 nt, forming a panhandle structure with a short double-stranded stem. This stem recreates the replication origin found at both ends of the parental double-stranded DNA molecule. Therefore, the DNA polymerase-preterminal protein complex can bind and initiate DNA replication. Once the DNA polymerase has copied the first 100 nt of the double-stranded DNA stem, it proceeds along the single-stranded DNA, displacing the single-stranded DNA-binding protein as it goes. The final result is a second double-stranded DNA molecule (bottom of Figure 12.8). Therefore, beginning with one adenovirus DNA molecule, these two pathways together produce two double-stranded progeny molecules.

In an alternative pathway (not shown), replication can begin at both ends of a double-stranded DNA mol-

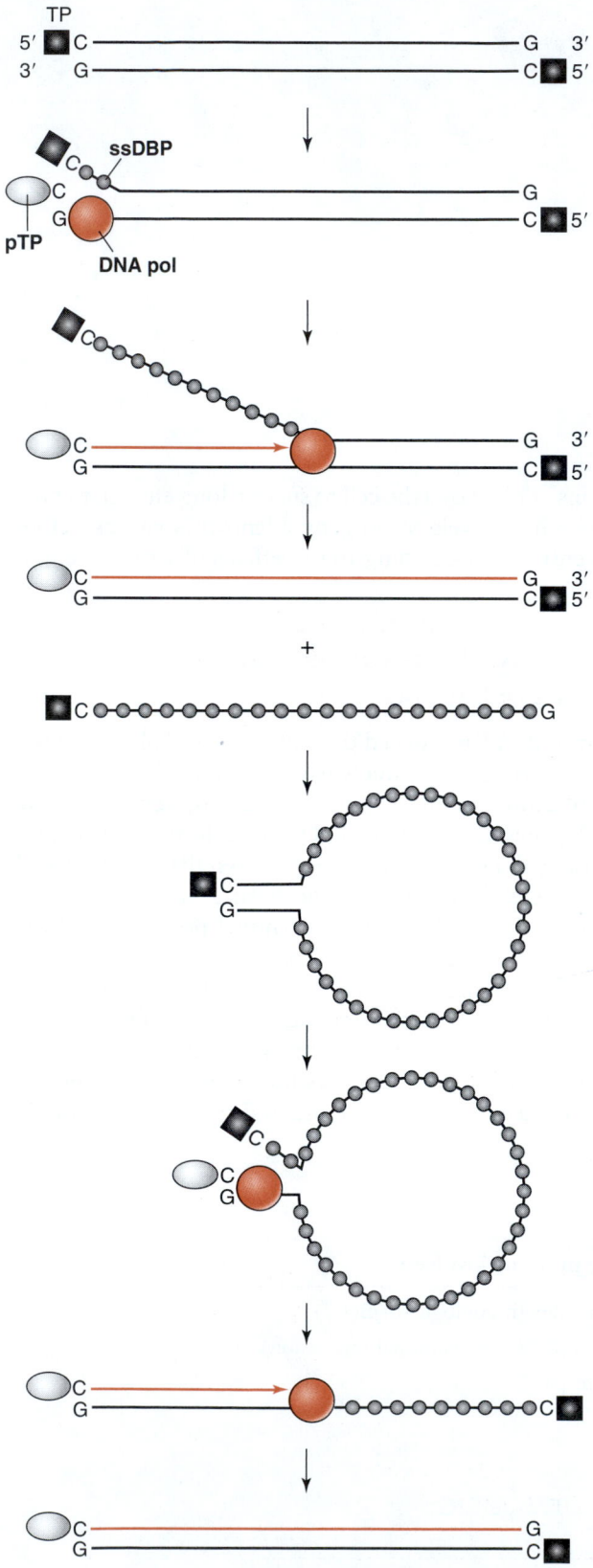

FIGURE 12.8 Adenovirus DNA replication. Viral DNA is replicated by three viral proteins: preterminal protein (pTP, white oval), DNA polymerase (DNA pol, orange circle), and single-stranded DNA binding protein (ssDBP, small gray circles). The cleaved, 55 KD terminal protein (TP) is represented by a black square. Newly synthesized DNA is shown in orange. See text for details.

ecule, as both ends are identical. This also gives rise to two progeny genomes, as the DNA polymerase molecules pass each other on the complementary DNA strands like trains on separate tracks. There is evidence that both pathways are used. During assembly of virions the 80 KDa preterminal protein is cleaved to generate the 55 KDa terminal protein, which remains covalently attached to the genome after packaging.

The major late promoter is activated after DNA replication begins

More than half of the adenovirus genome is dedicated to coding for at least 18 different late mRNAs. These mRNAs encode structural proteins, scaffolding proteins, a protease, and other products required for packaging of progeny virus. All late mRNAs are produced from a single primary transcript that extends 29 Kb in the rightward direction from the highly active "major late promoter," at 16.5 map units (Figures 12.4 and 12.9). During the early phase this promoter has a low activity, and the rare transcripts terminate prematurely, so that only low levels of L1 mRNAs are made. Transcription of late genes is activated after the onset of viral DNA replication, and late transcripts become the major RNAs made in the infected cell.

At least two factors control the increase in late transcription. First, newly replicated viral DNA interacts with a different set of DNA binding proteins, thus altering DNA structure to make the major late promoter more accessible to cellular transcription factors. Second, polypeptide IVa2, which is encoded by an intermdiate gene, binds to viral DNA at a site shortly downstream of the position where late transcription begins. IVa2 cooperates with factors bound to an upstream enhancer to activate transcription. Together, these steps activate the major late promoter by a factor of 100 or more.

Five different poly(A) sites and alternative splicing generate multiple late mRNAs

There are five families of late mRNAs, denoted L1 to L5, each defined by a unique polyadenylation site (Fig-

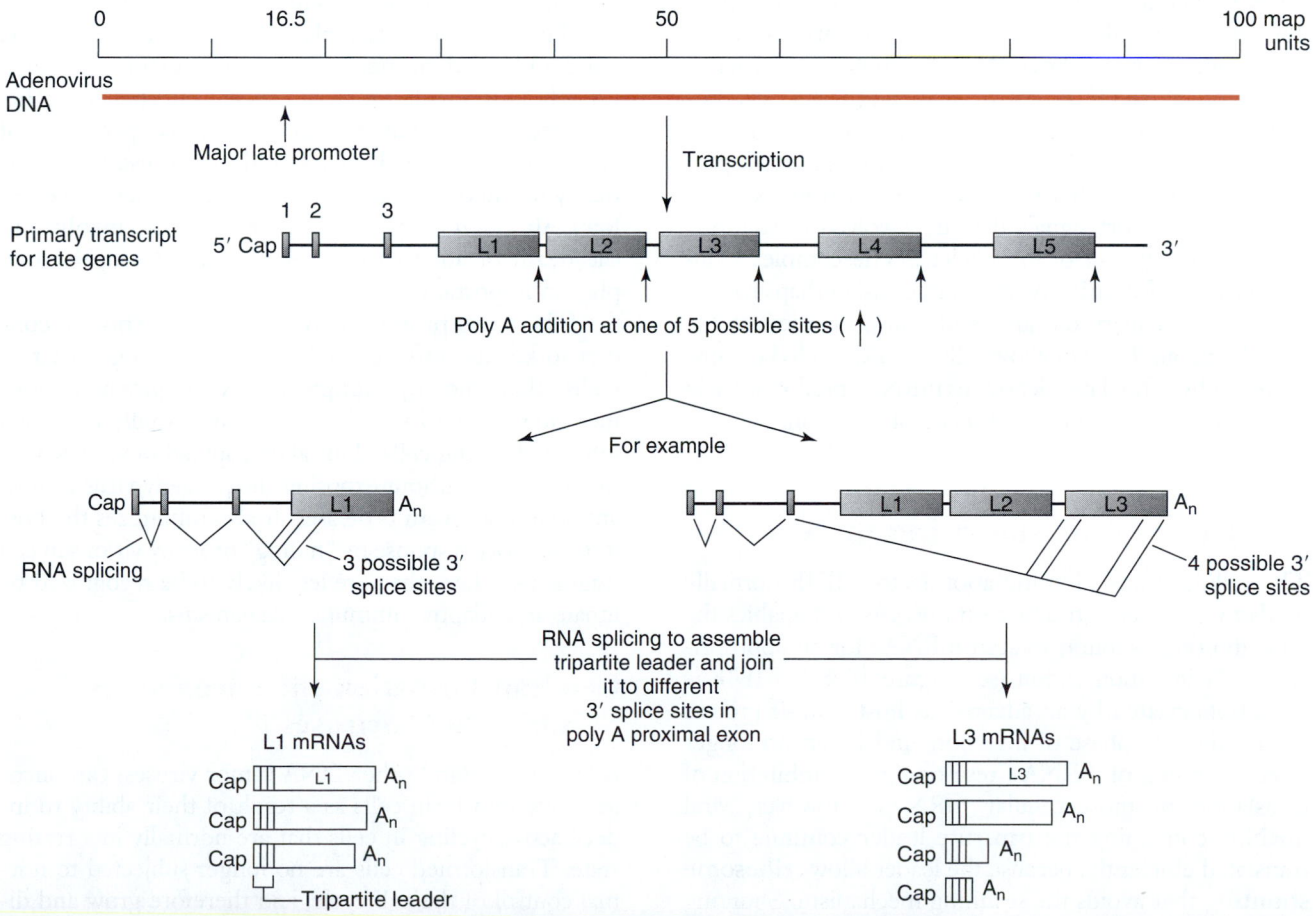

FIGURE 12.9 Production of late adenovirus mRNAs. Transcription initiates at the major late promoter, and the RNA is cleaved and polyadenylated at one of 5 possible sites. The tripartite leader is formed by splicing of 3 small exons; it is spliced to the mRNA body at one of several alternative 3′ splice sites. Exons are shown as shaded rectangles.

ures 12.4 and 12.9). Each primary transcript from the major late promoter is cleaved at only one of these sites, and the resulting polyadenylated RNA is spliced in a complex fashion to generate one of the 18 possible late mRNAs. Three short exons in the first 4 Kb of these transcripts are spliced together to form the noncoding 200-nt "tripartite leader." This leader is then spliced to an mRNA "body"; multiple alternative 3′ splice sites give rise to several different mRNAs for each of the five late gene families. As a result, all late mRNAs have the same tripartite leader at their 5′ ends. Each of these mRNAs is translated into a unique protein by use of reading frames beginning at different AUG initiation codons within the mRNA body.

The tripartite leader ensures efficient transport of late mRNAs to the cytoplasm

What are the advantages to adenovirus in expressing its late mRNAs in such a fashion? First, all late mRNAs are made by turning on a single, highly active promoter, allowing coordinate expression of late proteins when they are needed, after DNA synthesis begins. Second, all late mRNAs contain the same leader sequence at their 5′ ends. This allows the mRNA transport and translation machinery to concentrate on these RNAs. A complex between two early proteins, the E1B 55K protein and the E4 orf6 protein, is responsible for selective transport of late mRNAs to the cytoplasm (Table 12.1). E4 orf6 targets the 55K protein to the nucleus, but both 55K and E4 orf6 also contain nuclear export signals that are involved in shuttling these complexes out of the nucleus. The complex binds specifically to late mRNAs in the nucleus, perhaps by recognizing the tripartite leader, and transports them to the cytoplasm, but does not allow cellular mRNAs to be transported. Therefore late adenovirus mRNAs predominate in the cytoplasm during the late phase of infection.

The tripartite leader directs efficient translation of late adenovirus proteins

The cellular translation initiation factor eIF4F normally binds to the 5′ cap structure on mRNAs and enables the 40S ribosomal subunit to scan mRNAs for an appropriate AUG initiation codon (see Figure 16.4A). EIF4F is dephosphorylated by an adenovirus host shutoff protein during the late phase of infection, and it can no longer direct scanning of mRNAs, resulting in the inhibition of translation of most cellular mRNAs. However, viral mRNAs containing the tripartite leader continue to be translated efficiently, because the leader allows **ribosome shunting** that avoids the scanning mechanism. Shunting is directed by complementary regions between the tripartite leader and the 18S RNA, which is part of the 40S ribosomal subunit. Ribosome shunting has parallels to internal initiation of protein synthesis at "internal ribosome entry sites" (IRES elements) of picornaviruses and flaviviruses (see Chapters 16 and 17).

Adenoviruses make different amounts of the various late mRNAs (and therefore of their protein products) by controlling which polyadenylation sites and 3′ splice sites are recognized most efficiently. This system, using a single transcription unit and producing multiple mRNAs by different RNA processing reactions, may be a simpler and more economical means than having distinct transcriptional enhancers and promoters for each of the many late mRNAs needed.

Adenoviruses kill cells by apoptosis, aiding virus release

Two viral proteins have been implicated in cell killing and virus release. One is the E3 11.6 KDa "death protein." Although it is nominally an early protein, the 11.6 KDa protein is expressed at high levels during the late phase when its mRNA is made using the major late promoter and an E3 polyadenylation site. The death protein is a membrane glycoprotein found in the Golgi membranes, endoplasmic reticulum, and most predominantly in the nuclear membrane; how it kills cells is not known.

The E4 orf4 protein also can induce cell death, via a p53-independent mechanism. When expressed alone, E4 orf4 kills a wide range of cancer cells, but it has little effect on normal human cells. This specificity of killing suggests that E4 orf4 could be of use therapeutically in the treatment of human malignancies (see below). Although it has not been proven to be involved in the death of adenovirus-infected cells, E4 orf4 could play an important role.

These two proteins may therefore function in concert to kill the infected cell and release progeny virus. Cells that die by apoptosis are fragmented into membrane-bound vesicles, which are rapidly absorbed into neighboring cells. This allows spread of virus within the organism without exposing the progeny virus to host antibodies or serum proteases. It also minimizes the host inflammatory response by "hiding" progeny virions in cell fragments, where they are less likely to be recognized by innate and adaptive immune mechanisms.

Cell transformation and oncogenesis by human adenoviruses

Adenoviruses (and other DNA tumor viruses) can oncogenically transform cells as a result of their ability to induce active cycling in cells that are normally in a resting state. Transformed cells are no longer subjected to normal control of the cell cycle, and therefore grow and divide in an unregulated fashion. Adenovirus infection of human cells usually results in cell death, and therefore no transformed cells can survive. However, the human fetal embryonic kidney cell line 293 was generated *in vitro* by

Therapeutic applications of human adenoviruses

In the near future we will almost certainly see adenoviruses used in a variety of clinical applications. Their ability to infect a wide range of cells and to express genes at high levels make them ideal tools to fight disease in humans and other mammals. Adenoviruses can be readily manipulated to allow expression of almost any gene by insertion of the gene under control of a suitable promoter into the viral genome.

Vaccines and gene therapy. As discussed in Chapter 33, human adenoviruses are already being used experimentally in gene therapy. In this application they function primarily as vectors to deliver genes that correct genetic deficiencies associated with human diseases. Most of the commonly used adenovirus vectors are missing E1A and E1B genes and therefore cannot replicate in normal cells. Replication-competent adenovirus vectors could also be used as vaccines to enhance the immune response against any desired target. Numerous genes have been inserted into adenovirus vectors for experimental vaccination, including genes encoding hepatitis B surface and core antigens, herpes simplex glycoprotein, rabies virus glycoprotein, respiratory syncytial virus F protein, and several HIV-1 proteins.

New cancer therapies. In a novel approach currently being tested in clinical trials, a replication-competent adenovirus mutant is injected into human tumors. The idea is that the virus will replicate in and spread through the tumor tissue, killing the cancerous cells. The virus strain, ONYX-015, is a mutant that does not express the E1B 55K protein, which plays a vital role in inactivating and degrading p53. In normal cells infected with this virus p53 will remain active, leading to poor replication of the virus, little spread, and only limited pathological consequences. However, more than 50% of human tumors lack p53; in such cells, inactivation of p53 by the 55K protein is not required for virus replication, and therefore the ONYX-015 virus replicates actively in tumor cells and kills them selectively.

Clinical use of other adenovirus functions. Viral proteins have evolved to modulate many aspects of cell growth, apoptosis, and the immune system. It should be possible to take advantage of these viral proteins to develop new therapies. Such proteins could be expressed in target cells using viral vectors. In addition, understanding their mechanism of action could identify new targets for the development of drugs that mimic the action of the viral protein. Both E1B 19K and 55K proteins are potent inhibitors of apoptosis, and could be of value in diseases characterized by excess or unwanted apoptotic cell death, such as certain neurological disorders (e.g., Alzheimer's disease) or stroke. The E4 orf4 protein kills cancer cells selectively and could be applied to the treatment of human malignancies. Finally, several E3 proteins reduce the immune response, and could be used in immune suppression therapy. It seems almost certain that disease-causing viruses such as human adenoviruses will soon be developed into valuable new therapeutic agents.

transfection with a fragment of adenovirus DNA containing only the E1A and E1B genes, leading to cell transformation. Rodent cells can be readily infected by human adenoviruses, but they are nonpermissive for viral replication. Virtually all human adenoviruses are capable of transforming rodent cells in culture, and such cells usually form tumors in newborn or immunologically compromised rodents. However, only group A adenoviruses yield tumors by direct injection into animals.

In most cases, E1A is the oncogene that stimulates uncontrolled cell cycle progression. Expression of E1A alone usually produces abortive transformation. In this process, transformants begin to proliferate, but die soon afterward due to E1A-induced apoptosis. Formation of stable transformants requires co-expression of E1A and either E1B 19K or E1B 55K proteins, preferably both. As discussed previously, these E1B proteins blunt the pro-apoptotic effects of E1A, and inhibit p53-induced growth arrest and apoptosis. This system has provided valuable tools to understand the more general roles of the Rb and p300/CBP families, cell cycle progression genes, and p53.

SUGGESTED READING

1. Barbeau, D., Charbonneau, R., Whalen, S.G., Bayley, S.T., and Branton, P.E. (1994). Functional interactions within adenovirus E1A protein complexes. *Oncogene* **9,** 359–373.

2. Bayley, S.T. and Mymryk, J.S. (1994). Adenovirus E1A proteins and transformation. *International Journal of Oncology* **5,** 425–444.

3. Bischoff, J.R., Kirn, D.H., Williams, A., Heise, C., Horn, S., Muna, M., Ng, L., Nye, J.A., Sampson-Johannes, A., Fattaey, A., and McCormick, F. (1996). An adenovirus mutant that replicates selectively in p53-deficient human tumor cells. *Science* **274,** 373–376.

4. Branton, P.E. (1998). Early gene expression. In Prem Seth, ed. (NCI, NIH, Bethesda, MD), Adenoviral vectors for gene therapy. R.G. Landes, Austin, TX.

5. Harris, S.L. and Levine, A.J. (2005). The p53 pathway: Positive and negative feedback loops. *Oncogene* **24,** 2899–2908.

6. Ko, L.J. and Prives, C. (1996). p53: Puzzle and paradigm. *Genes and Development* **10,** 1054–1072.

7. McConnell, M.J. and Imperiale, M.J. (2004). Biology of adenovirus and its use as a vector for gene therapy. *Human Gene Therapy* **15,** 1022–1033.

8. Roulston, A., Marcellus, R.C., and Branton, P.E. (1999). Viruses and apoptosis. *Annual Reviews of Microbiology* **53,** 577–628.

KEY TERMS

Adenoid
Apoptosis
Caspase
Cyclin-dependent protein kinase
Hexon
Histone acetyl transferase
Immunoglobulin gene superfamily
Integrins
Interferon
Inverted terminal repeat
Oncogene
Penton base
PKR
Retinoblastoma (Rb)
Ribosome shunting
Secondary receptor
Terminal protein
Tumor suppressor protein
Ubiquitin

Herpes Simplex Virus

Herpesviridae

From Greek *herpein* (creep), referring to spreading of lesions

VIRION

Enveloped icosahedral capsid (T = 16).
Diameter of capsid 125 nm.
Diameter of enveloped virion 200 nm.
Capsid contains six proteins, envelope contains at least ten glycoproteins.
Material between capsid and envelope is called "tegument," contains at least fourteen viral proteins.

GENOME

Linear ds DNA, 125–250 Kb.
Contains unique regions, inverted repeat elements, and terminal direct repeat sequences.

GENES AND PROTEINS

Herpes simplex virus type I: 90 different transcriptional units, transcribed by cellular RNA polymerase II.
Most mRNAs are unspliced.
Most transcriptional units encode only one protein.
Genes expressed in three temporal classes:

- *α (immediate early): early gene activation (six genes)*
- *β (early): DNA replication, late gene activation*
- *γ (late): virion proteins, regulatory proteins*

VIRUSES AND HOSTS

Three subfamilies: alpha, beta, and gammaherpesvirus.
Nine human herpesviruses, including herpes simplex virus, varicella virus, Epstein-Barr virus, cytomegalovirus.
Over 100 known herpesviruses from many animal species.

DISEASES

Chickenpox, mononucleosis, pneumonia, hepatitis, encephalitis.
Recurrent eye, mouth, and genital lesions.
Kaposi's sarcoma and Burkitt's lymphoma.
Life-threatening infections in immune-suppressed individuals.
Neonatal infections, birth defects.

DISTINCTIVE CHARACTERISTICS

Tegument proteins function in early steps of virus replication: activation of transcription and host shutoff.
Virions formed by budding through inner nuclear membrane.
Many viral genes counter host defenses against virus infection.
Most herpesviruses become latent in the body and can be reactivated months or years after primary infection.

Herpesviruses are important human pathogens

Herpesviruses are best known as the cause of discomfort from recurring "cold sores," lesions of the mucosa of the mouth and lips. Genital herpes lesions are a common sexually transmitted disease, and chickenpox is a common childhood disease that can be serious particularly when people are first exposed as adults. These three relatively mild but often annoying human diseases are caused, respectively, by herpes simplex virus types 1 and 2, and by varicella-zoster virus.

However, there are many other herpesviruses; the *Herpesviridae* family includes more than 100 known members, infecting a large variety of animal species including mammals, birds, reptiles, fish, and oysters. Among these are nine distinct herpesviruses that infect humans (Table 13.1). Herpesviruses are grouped in three subfamilies denoted α, β, and γ, based on biological and physical properties including cell **tropism** and genome organization. Under conditions of reduced immune response found in newborns, transplant recipients, and AIDS victims, herpesviruses can cause serious and often fatal disease, including **encephalitis**, pneumonia, and **hepatitis**. Also, some cancers are associated with infection by Epstein-Barr virus, a widespread herpesvirus that causes mononucleosis in young adults, and by Kaposi's sarcoma virus (human herpesvirus 8). These cancers are rare or localized to specific populations, and other factors probably contribute to their incidence.

Most herpesviruses can establish latent infections

One of the most intriguing characteristics of herpesviruses is their ability to establish **latent infections**. After an initial infection, viral DNA is harbored in a latent state in neurons, **B** or **T lymphocytes**, or other cell types. Latently infected individuals can remain without symptoms for months or years, or even their entire life. Reactivation of latent virus can lead to recurrent disease, such as repeated outbreaks of labial or genital herpes, or herpes zoster ("shingles"), a localized rash that can occur years after the initial chickenpox infection. People whose immune systems are impaired can suffer reactivation of cytomegalovirus or Kaposi's sarcoma virus with dire results. The mechanisms by which herpesviruses establish latent infections and reactivate are presently under intense study. A number of effective chemotherapeutic agents are commonly used against several herpesviruses (see Chapter 32), but mutants that are resistant to chemotherapy often develop.

The large number of herpesviruses and their complexity makes it unwieldy to consider more than one virus in any detail in this chapter. Although a great deal of fascinating information has recently become available on cytomegalovirus, varicella-zoster virus, Epstein-Barr virus, and human herpesvirus 8, the present chapter will concentrate on the best-studied herpesviruses, the closely related types 1 and 2.

Herpes simplex virus genomes contain both unique and repeated sequence elements

All herpesviruses contain a double-stranded linear DNA genome, which is between 120 and 230 Kb in length depending on the virus species. The 150 Kb herpes simplex virus genome has an unusual structure; it consists of two covalently linked components, long (L) and short (S) (Figures 13.1 and 13.2). Each of these components contains a stretch of unique sequences, U_L, and U_S, flanked by inverted repeats. The sequence repeat at the left end of U_L is designated *ab*, and the internal inverted repeat is designated *b′a′*. The inverted repeats flanking U_S are designated *a′c′* and *ca*. In shorthand, the linear genomic sequence can be written $a_n b\text{-}U_L\text{-}b'a'_n c'\text{-}U_S\text{-}ca$, where the subscript n designates one or more copies of the *a* sequence and the prime designates complementary sequences.

TABLE 13.1 Human herpesviruses

Subfamily and characteristics	Members	Diseases
α Short growth cycle, rapid spread. Latent in sensory neurons.	Herpes simplex virus 1 & 2	Labial and genital lesions
	Varicella-zoster virus	Chickenpox, zoster (shingles)
β Long growth cycle, slow spread, restricted host range. Latent in secretory glands and lymphoreticular cells.	Cytomegalovirus	Mononucleosis, hepatitis, fetal abnormalities
	Human herpesvirus 6A, 6B	Roseola
	Human herpesvirus 7	Roseola
γ Growth primarily in epithelial cells and B or T lymphocytes. Latent in lymphocytes.	Epstein-Barr virus	Mononucleosis, Burkitt's lymphoma
	Human herpesvirus 8	Kaposi's sarcoma

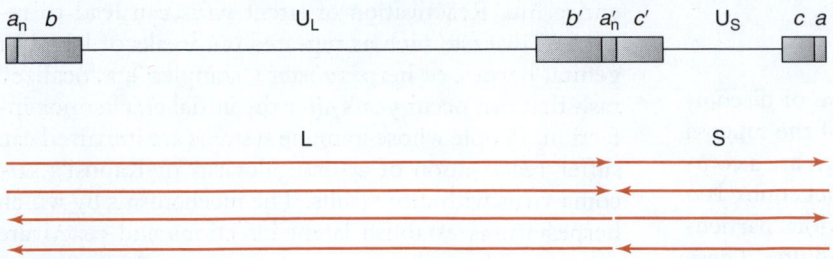

FIGURE 13.1 The genome of herpes simplex virus. Top: organization of unique and repeated sequence elements. Repeated sequence elements are shown in small letters (a, b, c) and unique sequence elements are shown in capital letters (U_L, U_S). Elements shown with a prime are inverted. Below: four isomeric forms of viral DNA with different orientations of the long (L) and short (S) components due to frequent recombination at the repeated sequence elements.

FIGURE 13.2 Map of herpes simplex virus genes. Genes are shown on the circularized 152 Kb map of herpes simplex virus type 1 DNA. Arrows indicate direction of transcription and location of polyadenlyation sites. Introns are shown as interruptions connected by a V. The positions of the three DNA replication origins (two OriS and one OriL) are shown on the map.

The *a* sequence varies from 200 to 500 bp in length in virus isolates, and contains no open reading frames. Different virus strains contain different numbers (n) of tandemly repeated copies of the *a* sequence. The *a* sequence contains signals for packaging viral DNA in capsids, and enhances recombination between genomes. The *b* sequence contains four open reading frames. Since the genome contains two identical copies of the *b* sequence, these four viral genes are therefore present in two copies per genome, albeit in inverted orientation. The *c* sequence contains a single open reading frame, also present in two copies per genome. Most viral genes are present in only one copy per genome; the U_L region contains 65 protein coding sequences, and the U_S region contains 14. Thus herpes simplex viruses contain at least 84 different genes.

Recombination takes place during viral DNA replication to generate four isomers of the genomic DNA, each with a different relative orientation of the U_L and U_S segments (Figure 13.1). Each isomer accounts for one-quarter of the DNA packaged into virions. DNA in any one of these orientations is infectious, and genomes frozen into only one orientation by removing the inverted repeats are also able to generate infectious virus. The genomes of other herperviruses show variations on the theme of interspersed unique and repeated sequences.

Nomenclature of herpes simplex virus genes and proteins

Partly because of the size and complexity of the viral genome, a number of different naming conventions have been used to identify herpesvirus genes and proteins. Presently, most herpes simplex viral genes are numbered sequentially from one end of the unique sequence in which they are found. Thus genes U_L1 to U_L56, and $U_S1.5$ to U_S12, are located respectively in the L and S unique segments. Genes lying within the *b* and *c* segments have names related to their time of expression ($\alpha 0$, $\alpha 4$, and $\gamma_1 34.5$) or other properties (LAT = latency associated transcript; ORF P/O = open reading frames P and O). Some proteins were first identified as *v*irion *p*roteins (e.g., VP5) or *g*lycoproteins (e.g., gp), or were designated as *i*nfected *c*ell *p*roteins found in cell extracts (e.g., ICP25). Other proteins, first identified by function (e.g., Vhs for viral host shutoff protein), have kept their functional names. A number of herpes simplex virus proteins are denoted by several different names because of these historical conventions. A circular map of the herpes simplex virus genome is shown in Figure 13.2.

The icosahedral capsid is enclosed in an envelope along with tegument proteins

The herpesvirus genome is wrapped in a bilayered capsid constructed from six proteins, of which the major one, VP5 (150 KDa), makes up the 162 capsomers (Figure 13.3). These are arranged as 150 hexamers on the faces of a T = 16 icosahedron, and 12 pentamers at the vertices. Another capsid protein, VP26, is located at the tips of the capsomers. The capsid is enclosed in an envelope that contains at least 10 different glycoproteins (gB through gM) and several other nonglycosylated membrane proteins. Unlike most enveloped viruses, there is a large space between the capsid and the envelope, called the **tegument**, which contains about 14 virus-coded proteins. Tegument proteins are introduced into the cell upon infection by herpesviruses, and some of them carry out crucial functions in the virus growth cycle. Even some viral messenger RNAs are packaged in the tegument and can be translated in newly infected cells before most viral genes have been transcribed.

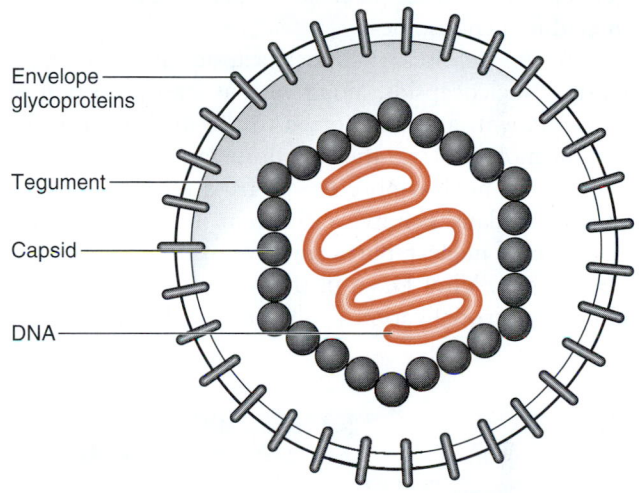

(a)

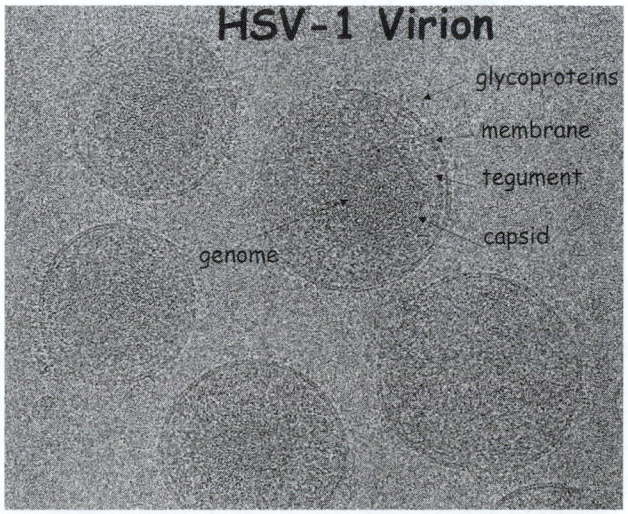

(b)

FIGURE 13.3 Structure of the herpesvirus virion. (a) Diagram of herpesvirus particle with major elements denoted. (b) Cryoelectron microscopy image of virion of herpes simplex virus type 1.

Entry by fusion is mediated by envelope glycoproteins and may occur at the plasma membrane or in endosomes

Herpes simplex virus initially binds to **heparan sulfate proteoglycans** at the host cell surface via viral envelope glycoproteins gB and gC. Subsequently, envelope glycoprotein gD binds to at least two alternative receptors: **nectin** 1, a member of the nectin family of intercellular adhesion molecules; and herpesvirus entry mediator (HVEM), a member of the **tumor necrosis factor (TNF)** receptor family. The two receptors are differently distributed in human cells and tissues. Receptor binding triggers fusion of the viral envelope with a cellular membrane. This requires three conserved glycoproteins: gB, gH, and gL. Of these, gH appears to serve as the actual fusion glycoprotein. Depending on the target cell, fusion takes place at the plasma membrane or in acidified endosomes.

Fusion releases the nucleocapsid into the cytoplasm. Nucleocapsids, along with at least a portion of the tegument, are transported to the nuclear pores by dynein motors along the microtubular network of the cell. At the nuclear pore the DNA is released and enters the nucleus. Empty capsids remain at the cytoplasmic side of nuclear pores for several hours until they disintegrate. Viral DNA in the nucleus is circularized, either by direct ligation of the ends or by recombination between the *a* sequences at the ends. The circularized DNA is localized near nuclear structures known as nuclear domain 10 (ND10), where early transcription takes place. Steps along the entry pathway are shown in Figure 13.4.

Viral genes are sequentially expressed during the replication cycle

Like the smaller DNA viruses of eukaryotes discussed in this book, transcription of herpesvirus genes is carried out in the nucleus by cellular RNA polymerase II. Viral genes can be classified in at least four groups based on the time of their expression during the replication cycle. Transcription of the *immediate early* or α genes takes place during the first several hours after infection. Some α gene products are required for the subsequent transcription of *early* or β genes, which are expressed between about 4 and 8 h after infection. The β gene products are mostly involved in viral DNA replication. The vast majority of viral genes are *late* genes, designated γ_1 and γ_2. Genes in the γ_1 group begin to be transcribed early but their transcription is stimulated severalfold after the onset of viral DNA replication; expression of γ_2 genes begins only after DNA replication starts. Although β and γ genes are spread throughout the viral genome, it is notable that all six α genes are localized within or near the boundaries of the in-

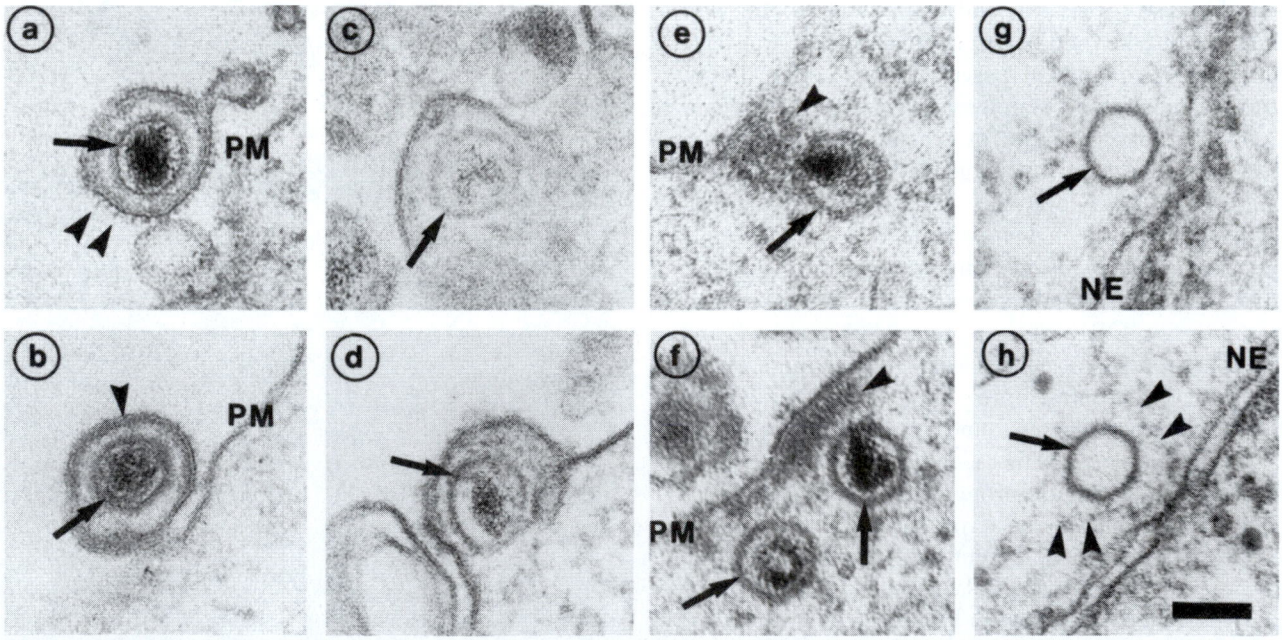

FIGURE 13.4 Entry and uncoating of herpes simplex virus. (a) and (b) show virus bound to cell surface; (c) and (d) show fusion of the viral envelope with the plasma membrane; (e) and (f) show nucleocapsids (arrows) detaching from the tegument (arrowhead) at the plasma membrane; (g) and (h) show empty capsids (arrows) at the nuclear pores (fibers shown by arrowheads) after release of DNA into the nucleus. PM = plasma membrane; NE = nuclear envelope.

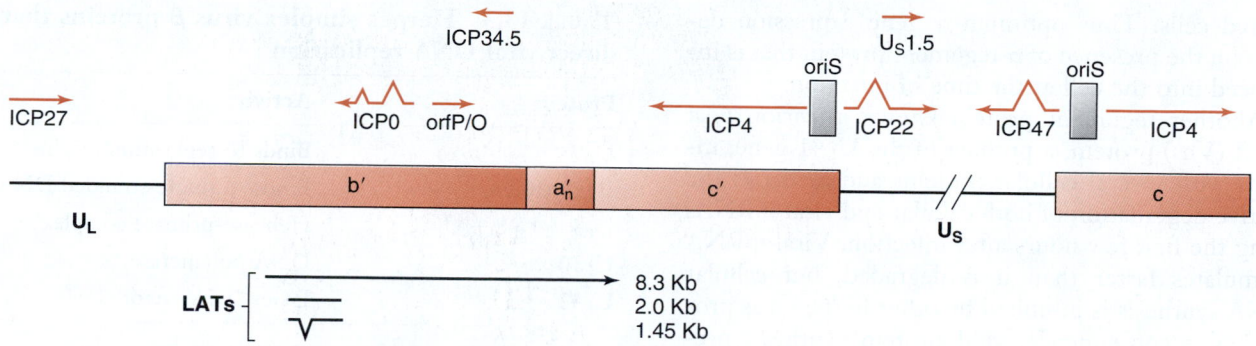

FIGURE 13.5 Transcriptional map of the *b* and *c* regions. Above: transcripts of the six α genes (a0, 4, 22, 27, 47, and U_S1.5) and three other genes (ICP34.5, ORF P/O) are shown as orange arrows. Introns are shown as interruptions connected by a V. The two copies of oriS are shown flanked by α genes. The remainder of the U_S region, containing γ genes, is not shown. Below: latency-associated transcripts (LATs).

verted repeat sequences *b* and *c* (Figure 13.5). Furthermore, three out of the six α genes contain introns, but only one of the 79 β and γ genes contains an intron.

Tegument proteins interact with cellular machinery to activate viral gene expression and to degrade cellular messenger RNAs

The precise functions of only a few of the tegument proteins are known. Although the α genes can be transcribed at a low level by basal cell transcription factors, expression is not optimal. A viral tegument protein known as α-trans-inducing factor (α-TIF) or VP16 forms a complex with at least two key cellular proteins, Oct-1 and HCF-1. This complex binds to specific response elements in viral DNA that have a central consensus sequence TAATGARATT. These elements, along with binding sequences for other cellular transcription factors, are found upstream of the promoters of all α genes (Figure 13.6). The bound protein complex acts as transactivator to enhance the transcription of these genes—the first set of viral genes expressed in

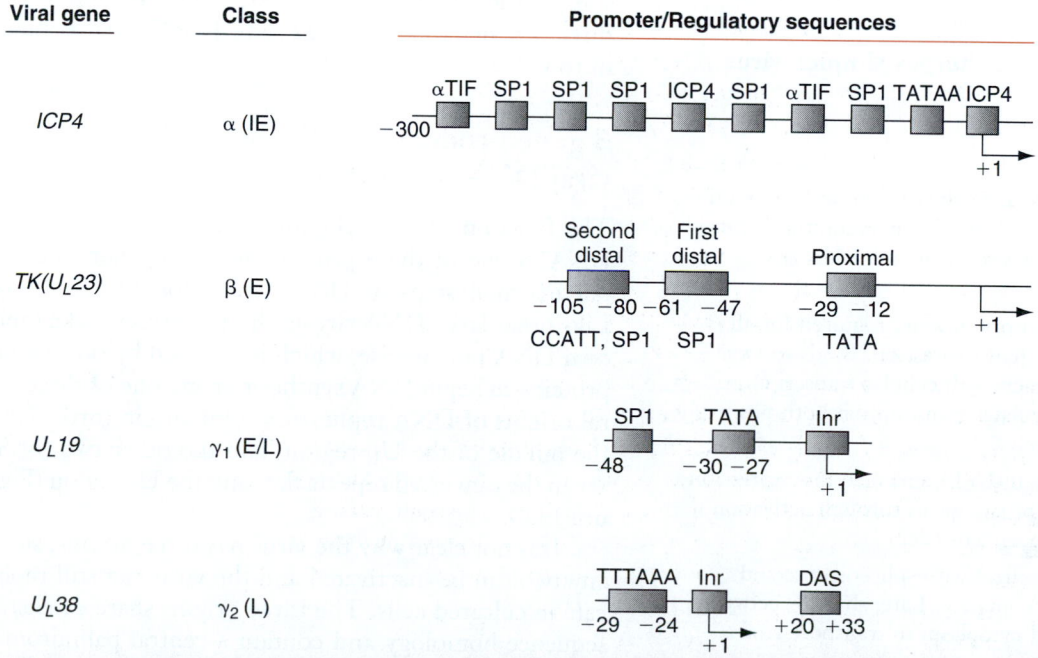

FIGURE 13.6 Structure of promoter regions of different classes of herpes simplex virus genes. A gene from each class (α, β, γ_1, and γ_2) is shown. +1 = transcription start site. Boxes indicate TATA boxes or sequence elements that interact with cellular (SP1, CCAAT) or viral (αTIF, ICP4) proteins. Inr = initiator element; DAS = downstream sequence activator element.

infected cells. Thus optimum α gene expression depends on the presence of α tegument protein that is introduced into the cell at the time of infection.

Another tegument protein known as *v*irion *h*ost *s*hutoff (Vhs) protein, a product of the U_L41 gene, interacts with several cellular proteins and mediates the selective degradation of both cellular and viral mRNAs during the first few hours after infection. Viral mRNA accumulates faster than it is degraded, but cellular mRNA synthesis is inhibited by other herpesvirus proteins; as a consequence, viral protein synthesis predominates. Vhs is also responsible for the rapid disappearance of immediate early viral RNAs once their synthesis stops. Interestingly, the RNA degradation activity of Vhs made during the late stages of infection is suppressed by binding to newly synthesized VP16, and this inactive complex is incorporated into progeny virions. This mechanism allows large amounts of late viral mRNAs to be made and translated even though Vhs is being made at the same time.

Immediate early (α) genes regulate expression of other herpesvirus genes

A major function of the α genes is to set up a regulatory cascade that leads to properly timed expression of other viral genes. The products of five of the six α genes (Table 13.2) are involved in the regulation of viral and cellular gene expression. Mutations in the genes that code for ICP0 and ICP4 strongly reduce expression of β and γ genes. ICP0 is considered a promiscuous activator protein, because its expression will activate transcription not only of other viral genes but also of a number of cellular genes. Both ICP0 and ICP4 activate transcription of β genes by relieving the silencing of these genes and by recruiting cellular transcription factors to their promoters (Figure 13.6). ICP0 also activates transcription of β genes by a number of other pathways, not fully understood, involving interactions with a variety of cellular proteins.

The decrease in the rate of synthesis of α gene products at 4 hours postinfection involves several proteins in addition to the tegument protein Vhs. ICP4 can repress transcription of its own gene by direct binding to the ICP4 promoter region (Figure 13.6). Another α gene product, ICP27, colocalizes with spliceosomes in the nucleus and blocks RNA splicing. This not only inhibits expression of cellular genes, most of which make spliced mRNAs, but also inhibits the expression of the three α genes (ICP0, ICP22, and ICP47) that contain introns.

TABLE 13.2 Functions of herpes simplex virus α proteins

Protein	Activities
ICP0	Promiscuous activator of gene transcription (viral and cellular). Acts as ubiquitin ligase (E3). Localizes to nuclear domain 10 structures.
ICP4	Major regulatory protein, required for all β and γ virus gene expression. Binds to viral DNA; interacts with cellular transcription factors. Regulates transcription both positively and negatively.
ICP22 $U_S1.5$	Together with U_L13, enhance the expression of a subset of late genes through activation of cyclin-dependent cdc2.
ICP27	Early: colocalizes with spliceosomes and blocks RNA splicing. Late: shuttles between nucleus and cytoplasm to transport viral mRNAs.
ICP47	Binds to TAP1/TAP2 transporter, blocking presentation of antigenic peptides to immune system via MHC class I protein.

TABLE 13.3 Herpes simplex virus β proteins that direct viral DNA replication

Protein	Activity
U_L9	Binds to replication origins
U_L29 (ICP8)	Binds to single-stranded DNA
$U_L5 + U_L8 + U_L52$	Helicase-primase complex
U_L30	DNA polymerase
U_L42	Processivity factor

β gene products set the stage for viral DNA replication

The function of most β genes is to replicate the viral DNA. Seven of these proteins work together to carry out essential steps in DNA replication (Table 13.3). Like other large DNA viruses, herpesviruses make their own DNA polymerase, which is directed by other viral proteins to begin DNA synthesis at any one of three viral origins of DNA replication. One origin (oriL) is in the middle of the U_L region, and the other two (oriS) are in the *c* inverted repeats flanking the U_S region (Figure 13.2).

It is not clear why the virus has three origins, since any two can be inactivated and the virus can still replicate in cultured cells. The three origins share extensive sequence homology and contain a central palindromic sequence of 144 bp (oriL) or 45 bp (oriS) that can form a stem structure on single-stranded DNA. All three origins are flanked by genes whose promoters direct divergent transcription away from the origin, reminiscent of the replication origins of polyomaviruses. Interestingly, oriL is flanked by genes coding for the viral single-stranded DNA binding protein (U_L29) and the viral DNA polymerase (U_L30), both intimately connected to DNA replication. The two copies of oriS are flanked by the important regulatory α genes coding for ICP4, ICP22, and ICP47 (Figure 13.5).

Herpesvirus begins with bidirectional DNA replication

The mechanism of herpesvirus DNA replication can be reconstructed only from knowledge of the proteins involved and in comparison with better-studied systems, because it has so far been impossible to reconstitute origin-dependent herpesvirus DNA replication *in vitro*. Replicating DNA in infected cells is found as high molecular weight, branched head-to-tail multimers called **concatemers**, suggesting that DNA is made by a **rolling circle mechanism**, which can produce linear multimers from a circular template. The branching could result from initiation of replication at several origins on the concatemers, as well as intermolecular recombination between replicating DNAs. However, herpesvirus initially sets up bidirectional replication on the circular viral DNA, and only later shifts to rolling circle replication.

An origin-binding protein (U_L9) binds to specific 10 bp sequences in oriS or oriL and unwinds the two DNA strands with the help of the single-stranded DNA binding protein ICP8 (Figure 13.7). A DNA helicase-primase complex made from three other β proteins (U_L5, U_L8, and U_L52) then binds to the single-stranded DNA and synthesizes RNA primers, the first step in the initiation of DNA replication. This is similar to the mechanism by which polyomaviruses and eukaryotic cells initiate their DNA replication, but polyomaviruses use the cellular DNA polymerase α/RNA primase.

At this point the herpesvirus DNA polymerase, product of the U_L30 gene, binds to the RNA primers and extends them as DNA chains. U_L42, a viral protein that is bound to the DNA polymerase, increases the **processivity** of the polymerase, allowing it to extend growing DNA strands over considerable distances without dissociating. These two proteins have functions similar to those of the cellular proteins DNA polymerase δ and PCNA (see Chapter 10). A structure is set up in which two replication forks progress in opposite directions from the origin (Figure 13.7).

Rolling circle replication subsequently produces multimeric concatemers of viral DNA

At some point one of the strands in the parental DNA is nicked, resulting in rolling circle replication (Figure, 13.8). In this mode, only one of the leading DNA strands is extended, as the DNA polymerase copies the remaining intact circular template strand. As it progresses, the DNA polymerase encounters the other end of the same DNA strand and displaces it from the template as it continues around the circular DNA. The DNA helicase-primase complex, along with DNA polymerase, fills in the complementary lagging strand on the extruded DNA to produce a double-stranded linear concatemer.

In addition to the proteins listed above, herpes simplex virus encodes other proteins concerned with DNA metabolism. Thymidine kinase (U_L23), ribonucleodide reductase (U_L39/U_L40), and deoxyuridine triphosphatase (U_L50) are involved in producing nucleoside triphosphates used as building blocks for DNA replication, while uracil N-glycosylase (U_L2) is involved in DNA repair. These genes are essential for virus replication in nondividing cells (e.g., neurons) but not in dividing cells, which can make cellular analogues of these enzymes.

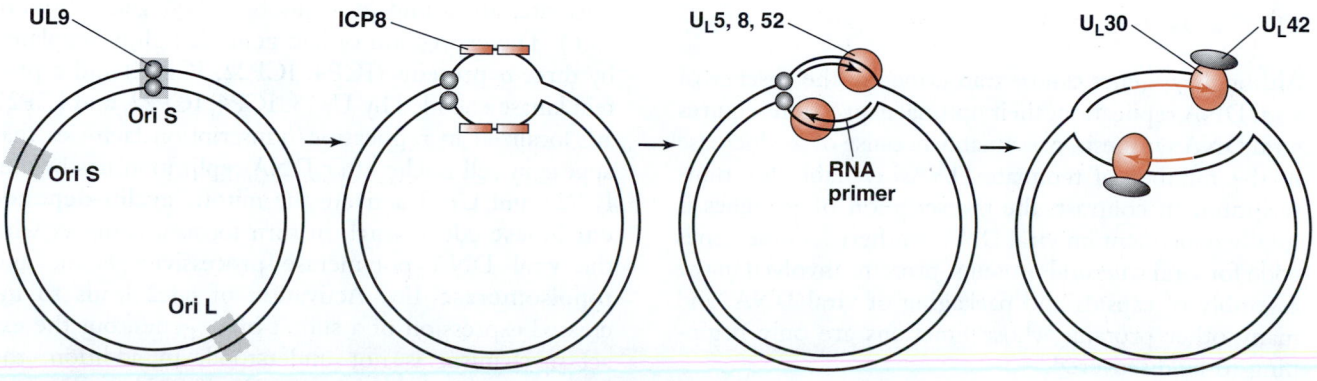

FIGURE 13.7 Establishment of bidirectional DNA replication. Replication is shown beginning at one of the three possible origins. RNA primers are black; newly-synthesized DNAs are orange.

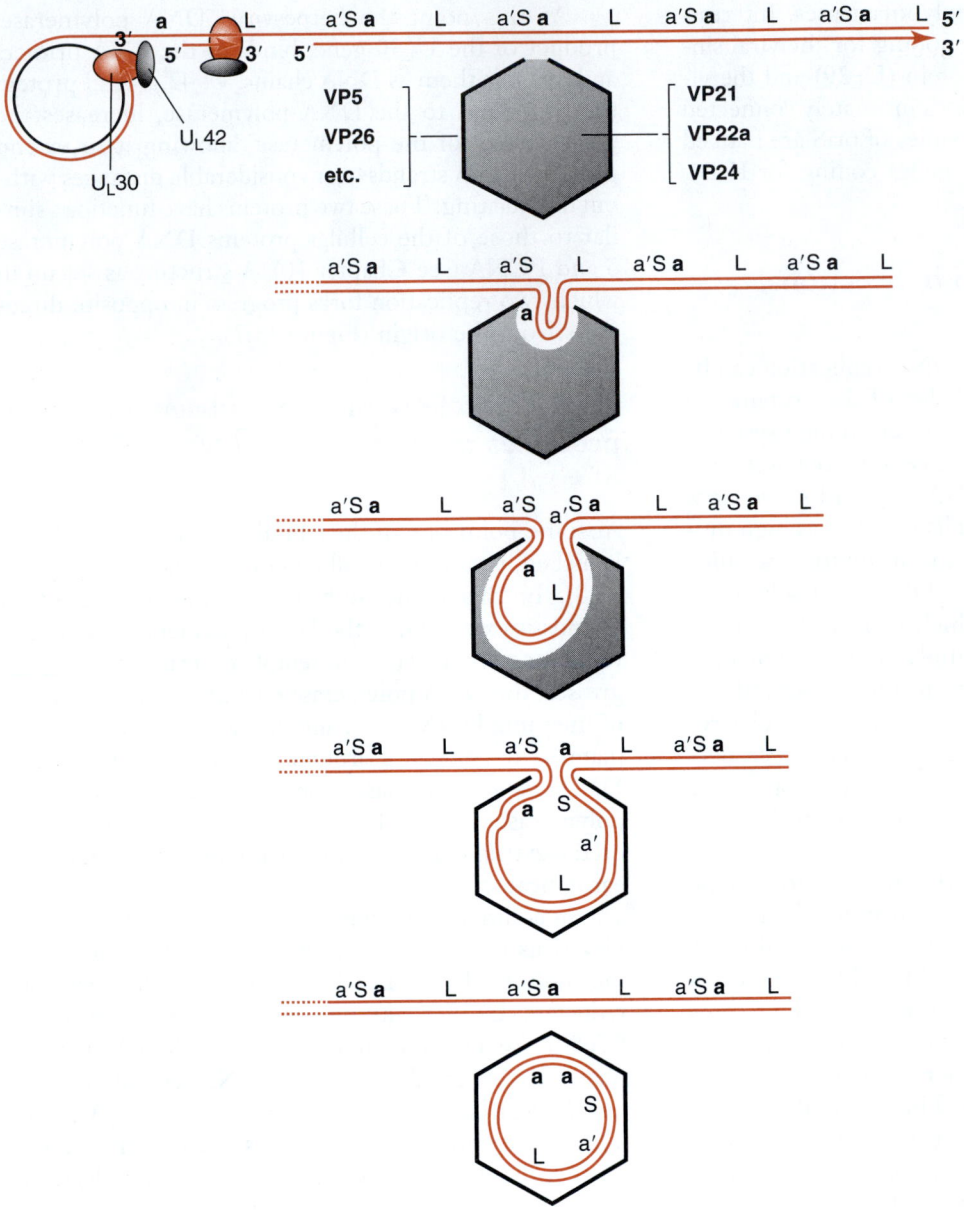

FIGURE 13.8 Packaging of herpesvirus DNA. Top: rolling circle replication mediated by herpesvirus DNA polymerase (U_L30) and processivity factor (U_L42) generates head-to-tail concatemers of herpesvirus DNA. **a** = terminal *a* repeats; L and S = long and short components; *a'* = internal *a* repeats (*b* and *c* regions are not shown). Below: precapsids made from VP5 and other capsid proteins are formed around scaffolding proteins (VP21, VP22a, VP24, shown in gray). Precapsids package viral DNA under the direction of packaging proteins that recognize the terminal *a* repeats and cleave off genome-length linear DNA fragments. Scaffolding proteins are extruded from the capsids in the process.

DNA replication leads to activation of γ_1 and γ_2 genes

Although γ_1 genes can be transcribed in the absence of viral DNA replication, their optimal expression requires viral DNA synthesis, presumably because of the increase in the number of template DNAs available for transcription. In contrast, the transcription of γ_2 genes is totally dependent on viral DNA synthesis. These genes code for viral stuctural proteins, proteins involved in the assembly of capsids and packaging of viral DNA, and many other proteins whose functions are only beginning to be discovered.

Most late gene promoters contain few or no binding sequences for cellular transcription factors upstream of their transcription start sites. However, downstream activation sequences, located beyond the transcription start site, are found in a number of γ_2 genes (Figure 13.6). The expression of late genes is tightly regulated by three α proteins (ICP4, ICP22, ICP27) and a protein kinase encoded by U_L13. ICP4, ICP22, and ICP27 are localized in replication/transcription factories that appear in cell nuclei after DNA replication has begun. ICP22 and U_L13 activate the mitotic **cyclin-dependent kinase** cdc2, which in turn forms a complex with the viral DNA polymerase processivity factor and **topoisomerase** IIα. Activation of cdc2 leads to increased expression of a subset of γ_2 genes, but the exact mechanism is not understood. In addition, the carboxy-terminal domain (CTD) of RNA polymerase II is phosphorylated by a reaction mediated by cdc9, ICP22, and both U_S2 and U_L13 protein kinases. Coincident with the onset of synthesis of late proteins, ICP27 begins to shuttle between the nucleus and cytoplasm, exporting viral mRNAs.

Viral nucleocapsids are assembled on a scaffold in the nucleus

An immature capsid lacking DNA is assembled by accumulation of the major capsid protein VP5 and several other proteins around a scaffold (Figure 13.8). This is reminiscent of the way in which bacteriophage lambda constructs and fills its capsid (Chapter 8). Three late proteins—VP21, pre-VP22a, and VP24—form the scaffold. VP24 is actually a viral protease that cleaves itself from a precursor protein, producing VP21 as another cleavage product.

Viral DNA in the form of head-to-tail concatemers is cleaved and stuffed into the preformed capsids by the interaction of several viral late proteins with highly conserved packaging sites (pac1 and pac2) located in the *a* sequence near the termini of the DNA. Entry of viral DNA into the capsids is carried out by a multimer of U_L6 protein located at one vertex of the capsid, and involves several other nonstructural proteins. DNA packaging is complete when a sufficient length of the DNA is inserted and when the packaging machinery encounters the next *a* sequence along the DNA in the same orientation (Figure 13.8). The packaged DNA is cleaved from the concatemer at a specific site within a 20 nt direct repeat flanking the terminal *a* sequence. In the process the scaffolding is dismantled and the scaffolding proteins are ejected from the capsid. Following packaging of the DNA the capsid changes conformation, sealing the DNA inside the capsid.

Envelopment and egress: three possible routes

Most enveloped viruses bud from cytoplasmic membranes or the plasma membrane to form infectious virions. Herpesvirus nucleocapsids are assembled in the nucleus, and need to exit this compartment and acquire an envelope. It is generally thought that herpesviruses use the nuclear membrane to form an envelope as a means of escape from the nucleus. They bud through the inner nuclear membrane into the **lumen** between the inner and outer nuclear membranes, acquiring an envelope and a layer of tegument proteins (Figure 13.9).

Virions are then transported from the lumen to the outside of the cell. Three distinct theories account for the mechanism by which virus particles exit the cell. According to one theory (Model 1 in Figure 13.9), virions retain their envelopes and are transported to the Golgi membranes within vesicles that bud from the outer nuclear membrane. According to another theory (Model 2 in Figure 13.9), virions lose their envelope by fusion with the outer nuclear membrane, releasing nucleocapsids into the cytoplasm. These nucleocapsids subsequently reacquire an envelope by budding into Golgi membranes. In either case, the envelope proteins would be glycosylated within the Golgi, and further transport of mature virions to the extracellular space is via secretory vesicles.

A third theory is based on the observation that nuclear pores become enlarged and the Golgi membranes are fragmented during viral replication. A hypothesis that is gaining support is that nucleocapsids exit through the enlarged nuclear pores without an envelope (Model 3 in Figure 13.9), and become enveloped at cytoplasmic vesicles derived from fragmented Golgi membranes.

Many viral genes are involved in blocking host responses to infection

Virus replication in cells cultured *in vitro* requires a relatively small number of genes—at most 37 out of the 84 identified to date. Among the genes dispensable for replication *in vitro* are a number that block host responses to infection. This is an active subject of research and is important for understanding how herpesviruses can survive and thrive in the host animal.

Vhs and ICP27. By degrading existing and induced cellular mRNAs, Vhs impedes infected cells from expressing or responding to **interferons**, a major line of host defense against viruses. ICP27 inhibits splicing of mRNAs during the early stages of infection, also contributing to blocking host responses. Transcription of cellular genes is also grossly reduced after infection, possibly as a result of changes in the chromatin structure of cellular DNA and in RNA polymerase II.

ICP47. Herpes simplex virus blocks the presentation of antigenic peptides to the immune system (see Chapter 31). A fraction of newly synthesized viral proteins are normally degraded to peptides in **proteasomes**, especially as a consequence of misfolding. These peptides are translocated to the endoplasmic reticulum and are ultimately presented to the immune system by **major histocompatibility (MHC)** class I proteins at the cell surface. An α protein, ICP47, binds to the transporter of antigenic peptides (TAP1/TAP2) and blocks the translocation of peptides into the endoplasmic reticulum, therefore inhibiting antigen presentation.

ICP34.5. Herpesvirus genes are tightly packed and are transcribed from both DNA strands. During the late phase of infection, polyadenylation signals are frequently misread, resulting in extended downstream transcription and the accumulation of large amounts of complementary RNAs that can anneal to each other. The resulting double-stranded RNA activates a cellular

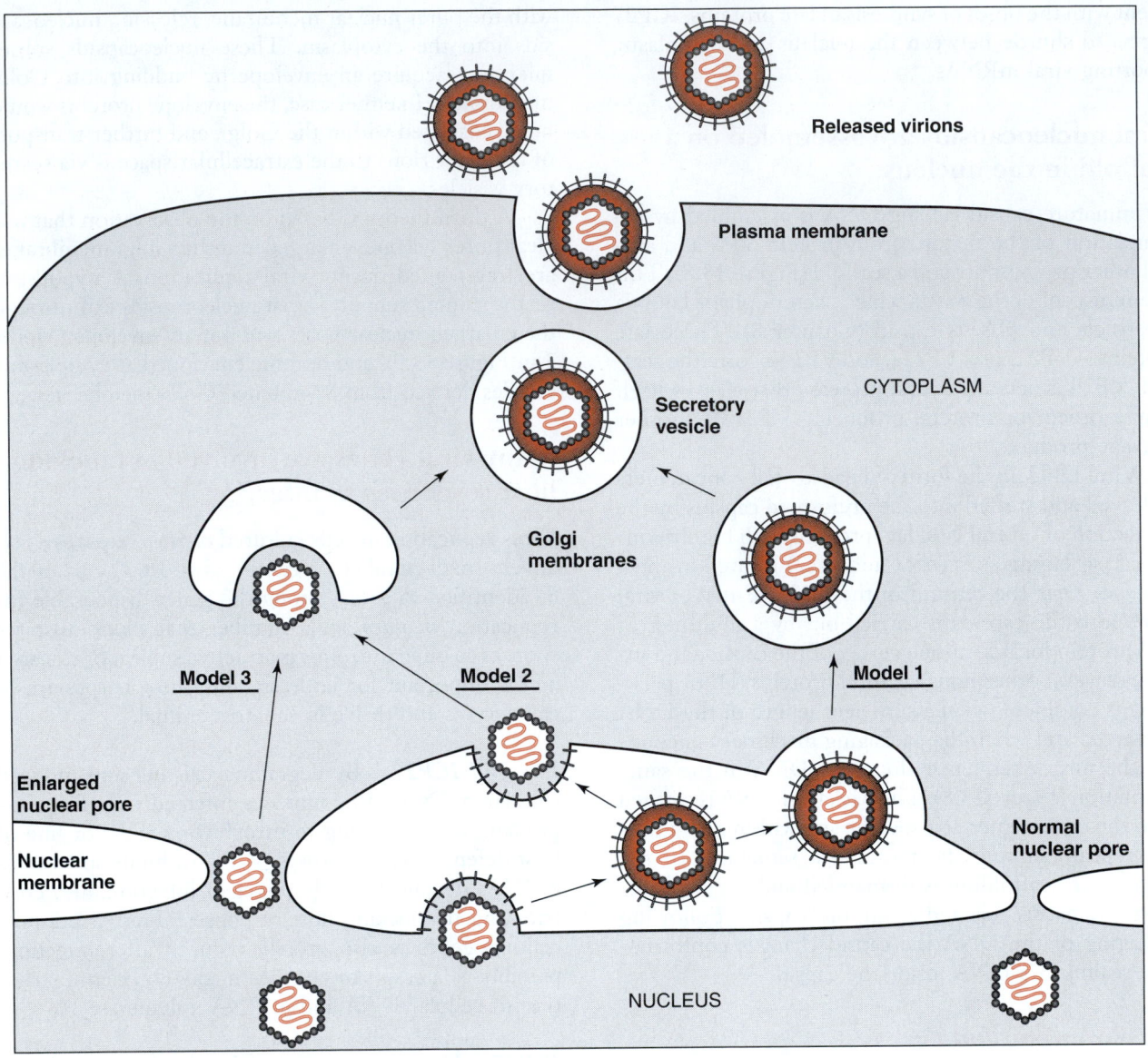

FIGURE 13.9 **Envelopment and egress of herpesvirus virions.** Nucleocapsids are assembled in the nucleus. Three possible routes of acquiring an envelope and escaping from the cell are shown. The "single nuclear envelopment" model (1) predicts that after budding through the inner nuclear membrane, virions are enclosed in vesicles derived from the outer nuclear membrane, and are transported to the plasma membrane without further envelopment. The "envelopment/de-envelopment" model (2) predicts that capsids acquire an envelope by budding through the inner nuclear membrane, become de-enveloped at the outer nuclear membrane, and are subsequently re-enveloped at Golgi membranes. The "single cytoplasmic envelopment" model (3) predicts that nuclear pores become enlarged and allow translocation of capsids to the cytoplasm, where they become enveloped at Golgi membranes. In all three models the transport vesicles fuse with the plasma membrane to release virions into the extracellular space.

protein, double-stranded **RNA-dependent protein kinase (PKR)**. Activated PKR phosphorylates the α subunit of eukaryotic initiation factor 2 (eIF-2α), which in turn shuts off protein synthesis (see Chapter 31). Activated PKR can also lead to activation of the important cellular transcription factor, **NFκB**.

ICP34.5 binds to protein phosphatase 1α and redirects it to efficiently dephosphorylate eIF-2α, thereby reversing the effect of the activated PKR. The carboxy-terminal domain of ICP34.5 is homologous to a corresponding domain of a cellular protein, GADD34. This suggests that herpes simplex virus acquired part of a cellular gene whose function is to control protein synthesis in infected cells. PKR is a key factor in determining host susceptibility to infection. In fact, herpesvirus mutants lacking the $\gamma_1$34.5 gene are highly attenuated. These mutants, however, become fully virulent in mice lacking components of the interferon pathway that control PKR.

ICP0. In addition to its function of inducing viral and cellular gene expression (see above), this protein mediates the degradation of several cellular proteins involved in response to interferon, cell cycle events, and chromosome partitioning. It acts as a **ubiquitin ligase** (E3), capable of directing **ubiquitination** of proteins, leading to their destruction in proteasomes. At present two ubiquitin ligase sites have been identified in ICP0. ICP0 can also lead to the degradation of surface **receptor tyrosine kinases**, blocking cellular signaling pathways.

Anti-apoptosis genes. Programmed cell death, or **apoptosis**, results from a complex cascade of events in which a cell can self-destruct. Viral replication is one stimulus that can induce apoptosis. Recent observations suggest that apopotosis is a primary line of host defense and that herpes simplex virus, like many other viruses, has evolved a number of mechanisms designed to block this host response to infection, allowing infected cells to survive long enough to produce progeny virus.

Wildtype herpes simplex virus blocks apoptosis caused by viral gene expression or by exogenous agents such as osmotic shock, thermal shock, and Fas ligand interaction. However, at least four different viral mutants have been shown to allow apoptosis to proceed. These include mutants that (1) lack glycoprotein D, an envelope protein; (2) have a temperature-sensitive U_L36 tegument protein that does not properly assemble virions; (3) lack ICP4 or ICP27, important viral regulatory proteins, and (4) lack the major subunit of ribonucleotide reductase, encoded by U_L39. Analysis of these mutants led to the discovery that the virus encodes at least three genes whose function, in part, is to block apoptosis. Cell death induced by mutants lacking the gD gene is blocked by expressing either glycoprotein D or glycoprotein J in infected cells. Apoptosis induced by a mutant lacking the gene encoding ICP4 is blocked by a protein kinase encoded by the U_S3 gene. This kinase prevents apoptosis at a premitochondrial stage by impeding release of cytochrome c. Its target is a protein called Bad, a pro-apoptotic member of the Bcl-2 family.

The establishment and maintenance of virus latency

Latency is a hallmark of all known herpesviruses in their natural hosts. In humans and experimental animal systems, herpes simplex virus replicates in skin or mucosal cells at the portal of entry, producing infectious virus. The progeny virions, and more rarely the virus present in the initial inoculum, infect nerve endings and undergo **retrograde transport** along axons to the cell bodies of **dorsal root neurons** innervating the site of viral entry. Viral DNA enters the nucleus of the neuron, is circularized, and remains dormant; transcription is repressed and there is no DNA replication.

Several factors contribute to the lack of expression of viral genes in neurons. As described above, expression of herpesvirus α genes is activated by a complex between viral tegument protein VP16 and two cellular transcription factors, Oct-1 and HCF-1. This complex binds to α gene promoters and leads to their transcription. Because VP16 is a tegument protein that enters neurons with the infecting virus, little VP16 actually makes it to the nucleus of the neuron, which is a long distance along the nerve axon from the site of infection. Furthermore, neurons contain two transcription factors named Luman and Zhangfei, which bind to HCF-1 and prevent formation of the active transcription factor complexes. In the absence of α gene transcription, the rest of the genome remains silent, with the exception of a class of RNAs called latency-associated transcripts (LATs).

Latency-associated transcripts include stable introns

Latency-associated transcripts are a family of RNAs (8.3, 2, and 1.45 Kb) that map to the *b* repeat sequences and are antisense to ICP0 mRNA (Figure 13.5). A surprising finding was that the 2 Kb and 1.45 Kb species are stable, unresolved lariat structures that accumulate in the nucleus. Lariat structures are generated from introns during RNA splicing; these two LATs are therefore stable introns. It is very unusual that introns accumulate in cells, because they are usually rapidly degraded after splicing occurs. The 2 Kb and 1.45 Kb RNAs are probably spliced from the 8.3 Kb LAT transcript, but very little full-length transcript accumulates. Although they contain substantial open reading frames, there is no evidence that LAT RNAs express any proteins, and their accumulation as introns in the nucleus argues against a role as mRNAs.

The role of LATs remains controversial. It has been suggested that they inhibit ICP0 expression by an antisense mechanism, therefore inhibiting viral gene expression. Extension of transcription beyond the polyadenylation site could also produce RNA antisense to ICP4 mRNA. LATs are not essential for establishment of latency, although the presence of the LAT transcriptional unit can increase the efficiency of latent infection in some model systems, perhaps by reducing productive infection. LATs have been shown to protect cells against apoptosis, perhaps contributing to the survival of neurons that express low levels of viral proteins.

In humans, local or systemic stimuli—such as physical or emotional stress, hyperthermia, exposure to ultraviolet light, menstruation, and so on—induce virus reactivation from latency. Low levels of viral transcripts are made in sensory neurons, leading to formation of small amounts of progeny virions. Virus is transported along the axon to mucosal sites where the virus can replicate. This causes the appearance of local lesions, whose spread is usually limited by host inflammatory and immune responses.

SUGGESTED READING

1. Campadelli-Fiume, G., Cocchi, F., Menotti, L., and Lopez, M. (2000). The novel receptors that mediate the entry of herpes simplex viruses and animal alphaherpesviruses into cells. *Reviews of Medical Virology* **10,** 305–319.

2. Johnson, D.C. and Hill, A.B. (1998). Herpesvirus evasion of the immune system. *Current Topics in Microbiology and Immunology* **232,** 149–177.

3. Jones, C. (2003). Herpes simplex virus type 1 and bovine herpesvirus 1 latency. *Clinical Microbiology Reviews* **16,** 79–95.

4. Kay, M.A., Glorioso, J.C., and Naldini, L. (2001). Viral vectors for gene therapy: The art of turning infectious agents into vehicles of therapeutics. *Nature Medicine* **7,** 33–40.

5. Mettenleiter, T.C. (2002). Herpesvirus assembly and egress. *Journal of Virology* **76,** 1537–1547.

6. Roizman, B. and Whitley, R.J. (2001). The nine ages of herpes simplex virus. *Herpes* **8,** 23–27.

7. Advani, J.S. and Roizman, B.(2005). The strategy of conquest: The interaction of herpes simplex virus with its host. In Pales, P., ed., Modulation of host gene expression and innate immunity by viruses, pp. 141–161. Springer.

KEY TERMS

Apoptosis
B lymphocyte
Concatemer
Cyclin-dependent kinase
Dorsal root neurons
Encephalitis
Heparan sulfate
Hepatitis
Interferons
Latency
Latent infection
Lumen
Major histocompatibility complex (MHC) proteins
Nectin
NFκB
PKR
Processivity
Proteasomes
Proteoglycans
Receptor tyrosine kinases
Retrograde transport
Rolling circle mechanism
Tegument
T lymphocyte
Topoisomerase
Tropism
Tumor necrosis factor
Ubiquitination
Ubiquitin ligase

Baculoviruses

Baculoviridae

From Latin *baculum* (stick), referring to shape of virion

VIRION

Enveloped rod-shaped nucleocapsid.
Diameter of nucleocapsid 30-60 nm, length 300 nm.
Nucleocapsid is capped asymmetrically.
Material between capsid and envelope is called "tegument."
Two types of virion: budded and occlusion-derived.

GENOME

Circular ds DNA, 100-180 Kb.
Arranged in virions as a cylindrical core.
Condensed with a small basic virus-coded protein.

GENES AND PROTEINS

Autographa californica MNPV codes for over 150 proteins.
Most mRNAs are unspliced.
Genes expressed in four temporal classes:

- *Immediate early: early gene activation*
- *Early: DNA replication, late gene activation (viral RNA polymerase)*
- *Late: assembly of budded virions (capsid and envelope proteins)*
- *Very late: assembly of occluded virions (polyhedrin)*

VIRUSES AND HOSTS

Genus *Nucleopolyhedrovirus* (from polyhedron, shape of occlusion body). Example: *Autographa californica* multiple nucleopolyhedrovirus (AcMNPV).
Genus *Granulovirus* (from granule, shape of occlusion body). Example: *Cydia pomonella* granulovirus (CpGV).
Hosts: arthropods, mostly insects of the order Lepidoptera (butterflies and moths).

DISEASES

Important pathogens of commercial silkworms, larvae of butterflies and moths.

DISTINCTIVE CHARACTERISTICS

Occlusion bodies contain virions in a form that allows insect-to-insect transmission.
Widely used for high-level expression of foreign proteins.
Also used as biological insecticides.

Insect viruses were first discovered as pathogens of silkworms

Throughout the ages, economic pressures have helped to establish the direction of basic research. Before viruses were recognized as infectious agents, scientific efforts were focused on establishing a cause for a particular disease. Following the arrival of silk production in Europe in the thirteenth century, most likely as a result of smuggling silkworms out of China, the rearing of silkworms became a huge industry, especially in Italy, and outbreaks of disease in silkworms were financially devastating to the silk industry. Early accounts of silkworm diseases are found in both ancient Chinese and Western literature. Marco Vida of Cremona, an Italian bishop of the sixteenth century, described in poetic form an infection of silkworm that is reminiscent of baculovirus pathogenesis. In 1856, A. Maestra and Emilio Cornalia independently observed polyhedral bodies in silkworm larvae and made the connection between the presence of polyhedra and the symptoms in the diseased silkworms.

Stanislaus Von Prowzek showed in the early 1900s that jaundice of the silkworm is caused by a filterable agent, but he believed the agent to be a bacterium and the occlusion bodies to be by-products of the disease. Shortly thereafter, Rudolf Glaser and J.W. Chapman (in gypsy moths) and C. Acqua (in silkworms) independently established the viral etiology of polyhedrosis. In 1943, Glaser and Wendell Stanley concluded, based on biochemical analysis, that polyhedra contain nucleoproteins. In 1947, Gernot von Bergold identified virus-like rod-shaped particles from polyhedral bodies by electron microscopy, and subsequently found that they contained nucleic acids and proteins. These viruses were named baculoviruses because of their shape (*baculo* means "stick" or "rod" in Latin).

Baculoviruses are used for pest control and to express eukaryotic proteins

Baculoviruses replicate only in invertebrates, primarily insects of the order Lepidoptera (butterflies and moths). Infections of other insects, including mosquitoes (Diptera), sawflies (Hymenoptera), and caddis flies (Trichoptera), have been reported. Baculovirus-like particles have been found in shrimp (Decapoda), but there are no equivalent baculoviruses that infect vertebrates, plants, bacteria, or archae. The important role played by baculoviruses in insect diseases suggested that they might be harnessed to control insect pests. Rather than a disease agent that needed to be eradicated, baculoviruses were recognized as beneficial.

The first practical application of an insect virus is credited to Edward Steinhaus, for the control of the alfalfa caterpillar in California in the 1940s. In 1950, the Insect Pathology Research Institute in Sault Ste-Marie, Ontario, successfully controlled sawfly populations with applications of baculovirus. Animal tests demonstrated that mice and guinea pigs were unaffected by exposure to baculoviruses while experimental or natural exposure of humans to baculoviruses demonstrated no adverse effect on human health or signs of infection. For example, a survey of naturally occurring baculoviruses on market cabbage demonstrated that over 110 million virus-containing polyhedra might be consumed in one serving of coleslaw. Therefore, in contrast to research on animal viruses, where the goals are to cure or prevent infection, research on baculoviruses eventually focused on enhancing viral pathogenesis and virulence.

Much of the earlier work on baculoviruses centered on their potential use as species-specific insecticides. Each virus was named (using the Latin binomial) according to the insect from which it was isolated, leading to the description of hundreds of different baculoviruses isolated from as many species of insects (see Table 14.1). However, few of these viruses have been studied in detail. Exceptions are viruses that infect insects of some economic importance and that were readily adapted to cell culture. One virus in the latter group was originally isolated from an alfalfa looper (*Autographa californica*) and was named *Autographa californica* multicapsid nucleopolyhedrovirus (AcMNPV). Try to say that before brushing your teeth in the morning!

AcMNPV replicates well in *in vitro* cell cultures from the cabbage looper (*Trichoplusia ni*) and the fall armyworm (*Spodoptera frugiperda*, Sf). Since the late 1970s, the AcMNPV-Sf virus-cell system has become

TABLE 14.1 Some representative baculoviruses

Virus	Host
Autographa californica multinucleocapsid nucleopolyhedrovirus (AcMNPV)	alfalfa looper
Orgyia pseudosugata multinucleocapsid nucleopolyhedrovirus (OpMNPV)	tussock moth
Bombyx mori nucleopolyhedrovirus (BmNPV)	silkworm
Lymantria dispar multinucleocapsid nucleopolyhedrovirus (LdMNPV)	gypsy moth
Spodoptera exigua multinucleocapsid nucleopolyhedrovirus (SeMNPV)	beet army worm

the model for the study of baculoviruses. The results of basic research with AcMNPV led to a renewed interest in using genetically modified baculoviruses for insect pest control and to the exciting application of baculovirus and insect cells as an efficient eukaryotic gene expression system. As more detailed information has accumulated about baculovirus gene structure and function in the last decade, baculoviruses have also emerged as important potential vectors for human gene therapy. Two baculovirus genera are recognized: the nucleopolyhedroviruses (type species AcMNPV), and the granuloviruses (type species *Cydia pomonella* granulovirus, or CpGV).

Baculoviruses produce two kinds of particles: "budded" and "occlusion-derived" virions

Baculovirus virions consist of rod-shaped nucleocapsids 30–60 nm in diameter and 250–300 nm in length, surrounded by a lipid-containing envelope (Figure 14.1). The space between the envelope and the nucleocapsid is denoted the "tegument." The nucleocapsid contains a single molecule of circular, double-stranded DNA arranged as a cylindrical core. The two ends of the nucleocapsid are capped asymmetrically, giving polarity to the virion. The core contains a small, highly basic protein, p6.9, bound to the DNA. This protein likely helps to condense the genome DNA during packaging into capsids. The nucleocapsid also contains a major capsid protein, vp39; orf 1629 protein, which may be a capsid end structure; two capsid-associated proteins, vp80 and p24, of unknown function; a protein/RNA phosphatase; and a protein kinase. One of the consequences of having a tubular nucleocapsid is the potential for packaging DNA longer than the viral genome. Virions of various lengths have been observed by electron microscopic examination of AcMNPV mutants that produce genomes of different sizes. This characteristic has been used to incorporate extra DNA into virions when using baculovirus as an expression vector. However, there is no information on how the length of the nucleocapsid is regulated during production in the nucleus.

The baculovirus infectious cycle is unique because two types of virions are produced: **budded virions** and **occlusion-derived virions**. Budded virions (Figure 14.1) consist of single nucleocapsids surrounded by an envelope. They are formed by budding of nucleocapsids through the plasma membrane, which is modified by incorporation of a virus-encoded protein. In contrast, occlusion-derived virions (Figure 14.2) are formed in the nucleus where they are occluded in a crystalline protein matrix. Their envelope is acquired either from a modified nuclear membrane or from newly-synthesized membranes within the nucleus. Nucleopolyhedrovirus **occlusion bodies** are polyhedral, 0.15 to 15 μm in diameter, and can contain many virions; granulovirus occlusion bodies are ovicylindrical, 0.3 by 0.5 μm, and contain only one virion.

The occlusion bodies consist largely of a single viral-encoded polypeptide called **polyhedrin** (nucleopolyhedroviruses) or **granulin** (granuloviruses). Poly-

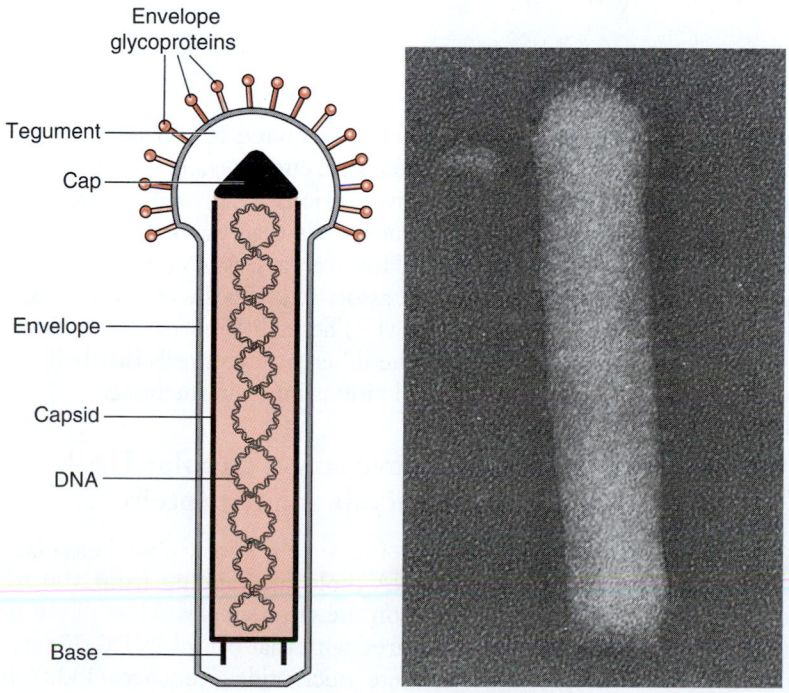

FIGURE 14.1 **Budded baculovirus virion.** At right, an electron micrograph of a negatively stained *Autographa californica* MNPV virion. At left, a diagram showing structural elements.

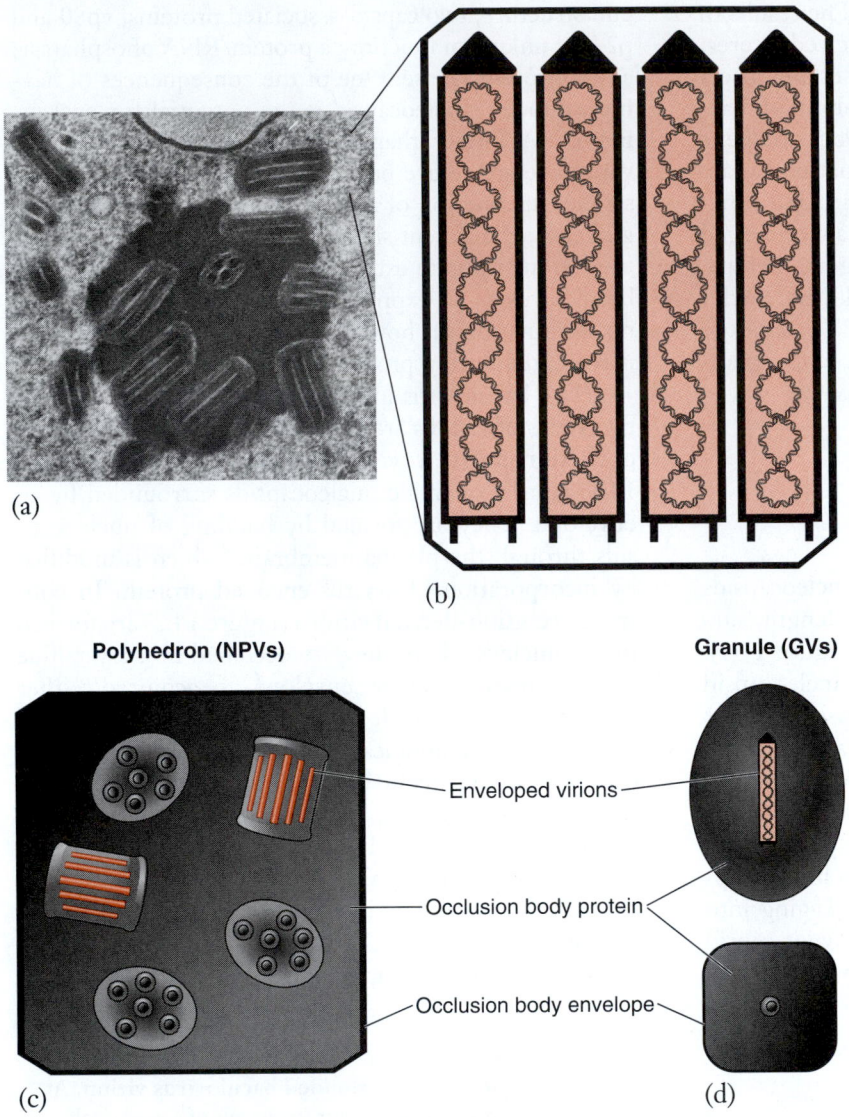

FIGURE 14.2 Occluded baculoviruses. (a) Electron micrograph of a thin section through a polyhedron in which baculovirus particles are occluded. (b) A diagram showing occluded virions. (c) and (d) Diagrams showing a polyhedron (nucleopolyhedrovirus) and a granule (granulovirus) with occluded virions.

hedrins and granulins from different baculoviruses have highly conserved amino acid sequences. The occlusion bodies are very characteristic of baculovirus infections and represent an obvious cytopathic effect easily seen by light microscopic examination of infected insect cells. Because occluded and budded virions are enveloped at distinct sites in the cell, there are differences in the protein composition of the two kinds of virions. For example, the tegument of occluded virions contains a 41 KDa glycoprotein, gp41, not present in budded virions; however, gp41 may be involved in the egress of nucleocapsids from the nucleus, prior to the formation of budded virions. Occlusion-derived virions initiate infection in the insect gut epithelium (see below).

The virion envelope has a trilaminar structure typical of lipoprotein unit membranes. The loose-fitting envelope of budded virions of AcMNPV (Figure 14.1) contains terminal spikes composed of a single virus-coded glycoprotein, gp64. This protein is required for infection of cell cultures by fusion of the viral envelope and the cell membrane. Occlusion-derived virions are several hundredfold less infectious in cultured cells than budded virions, and do not contain gp64. However, at least six other viral proteins are known to be associated with the envelope of occlusion-derived virions. These differences in envelope composition reflect the different target cells that budded and occlusion-derived virions infect in the insect.

Baculoviruses have large, circular DNA genomes and encode many proteins

The baculovirus genome is a covalently closed circular, double-stranded DNA molecule ranging from 100 to 180 Kbp depending on the virus species. Most physical maps are now compared with that of AcMNPV (Figure 14.3), whose complete nucleotide sequence (133,894

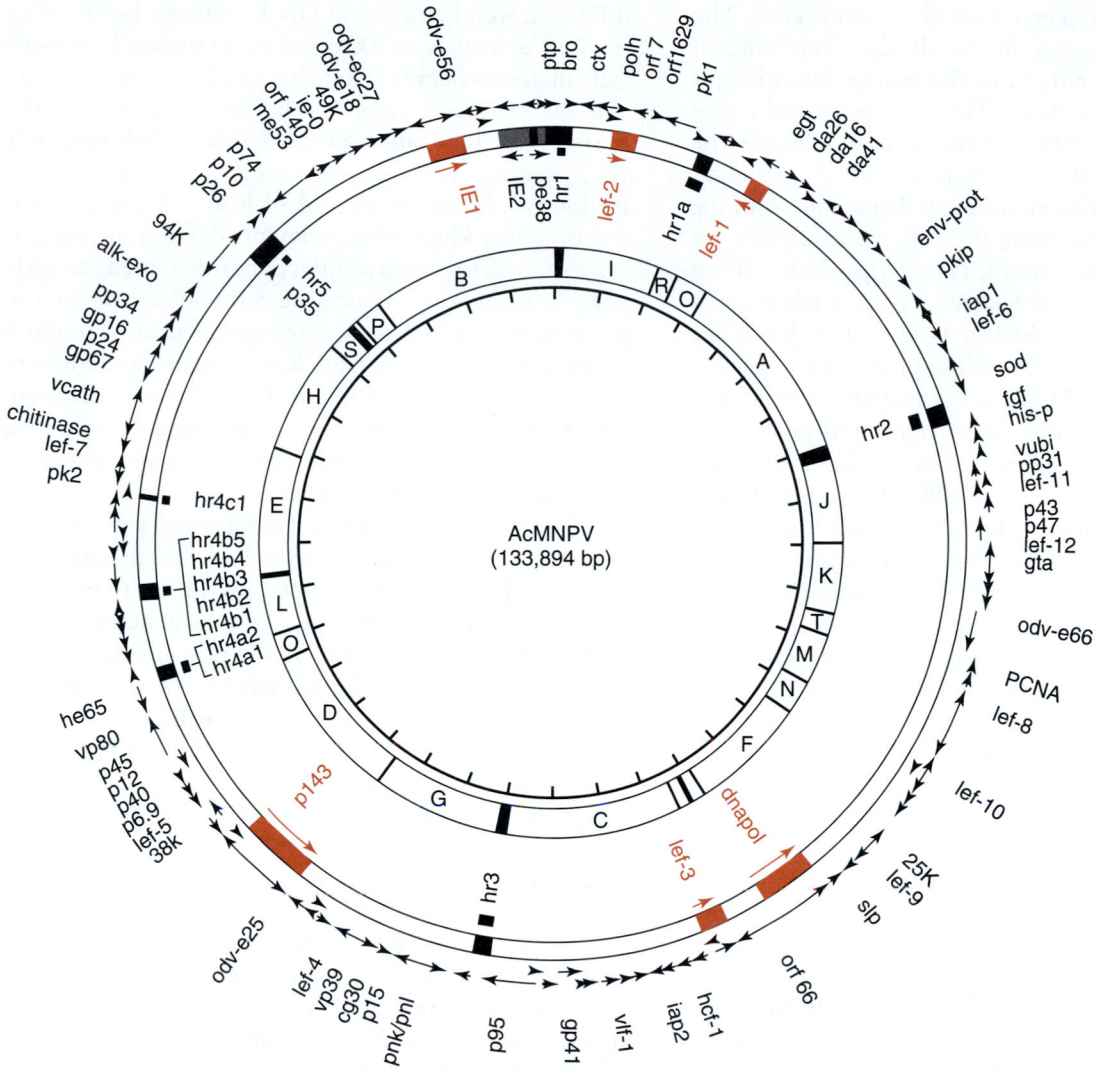

FIGURE 14.3 Genetic map of *Autographa californica* MNPV. The inner circle shows the locations of EcoRI restriction sites; the 20 largest EcoRI fragments are labeled alphabetically, according to size. The locations of homologous repeat sequences (hrs) usually associated with multiple EcoRI sites are labeled on the outer circle and shown in black. The locations of genes essential for viral DNA replication are shown in orange, and immediate early genes are shown in gray, in the same circle. Arrows in the periphery show the location and orientation of identified coding regions. Known genes are labeled.

bp) has been determined. Based on analysis of open reading frames and transcriptional promoters, it has been estimated that AcMNPV encodes more than 150 proteins. Some of the immediate early genes including ie-0, ie-1, ie-2, and pe38 are clustered within a 7 Kbp region, but there do not appear to be other significant groupings of genes by function or time of expression.

The sequences of five other baculovirus genomes, ranging from 128 to 179 Kbp, are also known. Baculovirus genomes carry repeated nucleotide sequences (hrs) that are transcriptional enhancers and may also function as initiators of DNA replication. Computer analysis of baculovirus genome sequences revealed a common set of about 60 homologous genes that are present in all species. These genes are probably the minimum set required for virus replication and form a pool that are the essence of what constitutes the baculovirus family. However, some genes are unique to particular species within the family; these genes likely determine individual species adaptation including host range or tissue tropism, virulence factors, and differences in virus morphology.

Insects are infected by ingesting occlusion bodies; infection spreads within the insect via budded virions

In natural populations, insects become infected with baculoviruses by feeding on cadavers of infected insects or on leaves contaminated with occlusion bodies. The occlusion bodies dissolve in the alkaline conditions of the insect gut (the acid pH of the human gut does not dissolve occlusion bodies). The released virions pass through the **peritrophic membrane** and invade the columnar epithelial and regenerative cells of the midgut by attachment and fusion of the viral envelope with the cell **microvilli**. Once inside the cell, nucleocapsids are transported to nuclear pores. Either the viral DNA is released directly into the nucleus, or the nucleocapsids first enter through the nuclear pore and then are uncoated in the nucleus. The eclipse phase of the replication cycle follows, resulting in the appearance of a ring-like **virogenic stroma** in the nucleus, and production of viral nucleocapsids. These nucleocapsids exit the nucleus and travel to the basal cell plasma membrane, where they form virions by budding. These virions subsequently infect the midgut connective tissue, tracheal epidermal cells and tracheoblasts, initiating a second round of replication.

The budded virions produced by these infections are released into the **hemocoel** and spread to infect other tissues including hemocytes, fat body, and epidermis. In these tissues, infection initially results in release of budded virus but there is subsequently a switch in virus production, carried out by an unknown mechanism. Assembled nucleocapsids are retained in the nucleus of infected cells where they become associated with membranes, eventually becoming enveloped either singly or as groups of multiple nucleocapsids. In nucleopolyhedroviruses, these enveloped virions then become associated in the nucleus with crystallizing polyhedrin protein, forming the distinctive occlusion bodies or polyhedra associated with infection by these viruses. Maturation of the polyhedra involves the deposition of an outer envelope (**calyx**) over the surface of the occlusion body, which may be facilitated by fibrillar structures composed of a very late gene product called p10.

The end result of virus infection is death and liquefaction of the insect from two to nine days after infection, depending on virus dosage and insect host. The final stages of granulovirus infection are different and not as well understood as for nucleopolyhedroviruses. During granulovirus infection, the cytoplasm merges with the nucleus and a distinct nuclear membrane structure is no longer present. Normally, single granulovirus virions are occluded within small occlusion bodies called granules. It is not clear if an outer envelope (calyx) is present on granules.

Viral proteins are expressed in a timed cascade regulated at the transcription level

In Sf21 cells infected with AcMNPV, several virus-induced proteins can be distinguished by 3 hours after infection, well before viral DNA synthesis begins. The number of virus-specific proteins synthesized dramatically increases beginning at about 6 hours postinfection, and the synthesis of host cell proteins gradually decreases from that time onward. By about 24 hours, only viral proteins are made, indicating that baculovirus infection results in the shutoff of host cell protein synthesis. These kinds of experiments defined the temporal phases of baculovirus infection: the immediate early phase is from 0 to 3 hours; the early phase is from 3 to 6 hours; viral DNA synthesis begins between 6 and 8 hours; the late phase is from 6 to 24 hours; and the very late phase begins at 24 hours after infection. Different viral proteins are synthesized during these specific time frames, although some viral proteins expressed early continue to be synthesized late.

The timing and levels of baculovirus gene expression are transcriptionally controlled. Immediate early and early genes are transcribed by the host cell RNA polymerase II, but late and very late genes are transcribed by a virus-coded RNA polymerase. These RNA polymerases can be distinguished based on their differential susceptibility to **α-amanitin**, a specific inhibitor of cellular RNA polymerase II; α-amanitin inhibits early transcription but not late or very late transcription. The RNA polymerases recognize DNA promoter elements in the viral genome. Many of the early genes carry promoters and transcription start sites that mimic normal host insect gene promoters. These include TATA-box-like elements and initiator motifs related to the sequence ATCA(G/T)T(C/T). However, the late and very late AcMNPV gene promoters usually contain the motif (A/G/T)TAAG, which serves as the transcription initiator for the virus-coded RNA polymerase.

Immediate early gene products control expression of early genes

The viral genes responsible for temporal control of baculovirus gene expression were discovered by using reporter genes whose transcription is driven by specific baculovirus promoters. Insect cells were cotransfected with plasmids containing a variety of specific viral DNA fragments and a reporter plasmid driven by an early promoter. One of the viral DNA fragments enabled transactivation of the reporter plasmid, and was found to contain an immediate early viral gene called ie-1. The IE-1 protein is responsible for transactivating other early genes and is considered the major regulator of early baculovirus gene expression. Other immediate early baculovirus genes, including ie-2, pe38, and ie-0, were identified based either on their ability to transactivate other viral genes or on the time of their expression in infected cells. IE-0 is the only baculovirus protein known to be made as a result of RNA splicing; a small exon located about 4 Kb upstream of the ie-1 start codon is spliced to the ie-1 open reading frame, producing a protein identical to IE-1 except for an additional 54 amino acids at its N-terminus.

Early gene products regulate DNA replication, late transcription, and apoptosis

A large number of early baculovirus proteins are made. These are responsible for three major activities: replication of viral DNA, expression of late and very late viral genes, and modification of cell and host physiology via cellular signaling pathways. Much remains to be learned about the detailed functions of these proteins. Table 14.2 lists some early viral proteins known to be involved in viral DNA replication. It is notable that baculoviruses make their own DNA polymerase, as do most large DNA viruses (including adenoviruses, herpesviruses, and poxviruses), even though baculoviruses replicate their DNA in the cell nucleus where cellular DNA polymerases are located. The exact mechanism of baculovirus DNA replication is not well understood, although recent data suggest that homologous recombination may be involved.

DNA replication. Baculoviruses contain multiple copies of highly homologous DNA sequences located in several different specific regions of the genome. In AcMNPV, these homologous regions (hrs, see Figure 14.3) consist of multiple copies of a 60–70 bp repeat carrying an imperfect inverted repeat sequence of 30 bp with an EcoRI restriction site at its center. Several groups have shown that hrs act as transcriptional enhancers and as initiators of DNA replication.

Replication assays rely on the ability of the restriction endonuclease DpnI to cleave methylated DNA but not unmethylated DNA. Plasmid DNA produced in bacteria is methylated, but when it is introduced into eukaryotic cells and undergoes replication, the DNA is no longer methylated. Therefore, DpnI can be used to distinguish between input plasmid DNA (methylated) and progeny unmethylated DNA. Plasmids carrying either AcMNPV hrs or early promoter regions have been shown to replicate by this assay when transfected into insect cells that are subsequently infected with AcMNPV. Therefore, there is a strong link between induction of early viral transcription and initiation of viral DNA replication.

Apoptosis. Baculoviruses also code for proteins that inhibit **apoptosis** (programmed cell death). These inhibitors are divided into two major groups called p35 and inhibitors of apoptosis (IAP). P35 has only been found in AcMNPV and its close relatives, whereas iap genes are apparently present in most baculoviruses and they are often present as multiple copies. The expression of these genes inhibits the normal process of apoptosis, allowing for greater production of budded virus and development of the very late stage of virus infection.

Acquired cellular genes. A number of baculovirus-coded proteins such as the viral proliferating cell nuclear antigen (PCNA), **ubiquitin** (vUBI), and superoxide dismutase (SOD) share a high degree of amino acid sequence similarity with homologous cellular genes. This suggests that these genes have been acquired by the virus from the host cell genome. Although these genes may not be essential for virus replication, they likely enhance some stage of virus multiplication in their natural hosts. In addition, some baculoviruses have amplified copy numbers of certain genes, or carry additional unique genes that are not similar to any other known genes. This partly accounts for the variety of baculovirus genome sizes.

Late genes are transcribed by a novel virus-coded RNA polymerase

Once DNA replication begins, another set of baculovirus genes is expressed. The late genes code for viral proteins involved in assembly of budded virions, while the very late genes are involved in the formation of occluded virions. The virus-coded RNA polymerase

TABLE 14.2 Early *Autographa californica* MNPV proteins involved in DNA replication

Gene	Mol. wt. (KDa)	Function
lef-1	31	Viral DNA replication; possible RNA primase
lef-2	24	Viral DNA replication; associates with lef-1
lef-3	45	Single-stranded DNA binding protein/helicase
p143	143	Helicase, host range factor, transcription factor
ie-1	67	May bind to replication origin
dnapol	114	DNA polymerase

TABLE 14.3 ***Autographa californica*** **MNPV genes that regulate late transcription**

Gene	Mol. wt. (KDa)	Function
p47	47	Viral late RNA polymerase subunit
lef-4	54	Viral late RNA polymerase subunit
lef-8	102	Viral late RNA polymerase subunit
lef-9	59	Viral late RNA polymerase subunit
lef-7	27	Viral late RNA polymerase subunit?
lef-5	31	Transcription elongation factor?
lef-6	20	?
lef-10	9	?
lef-11	13	?
lef-12	21	Absent in some baculoviruses
39K/pp31	31	?

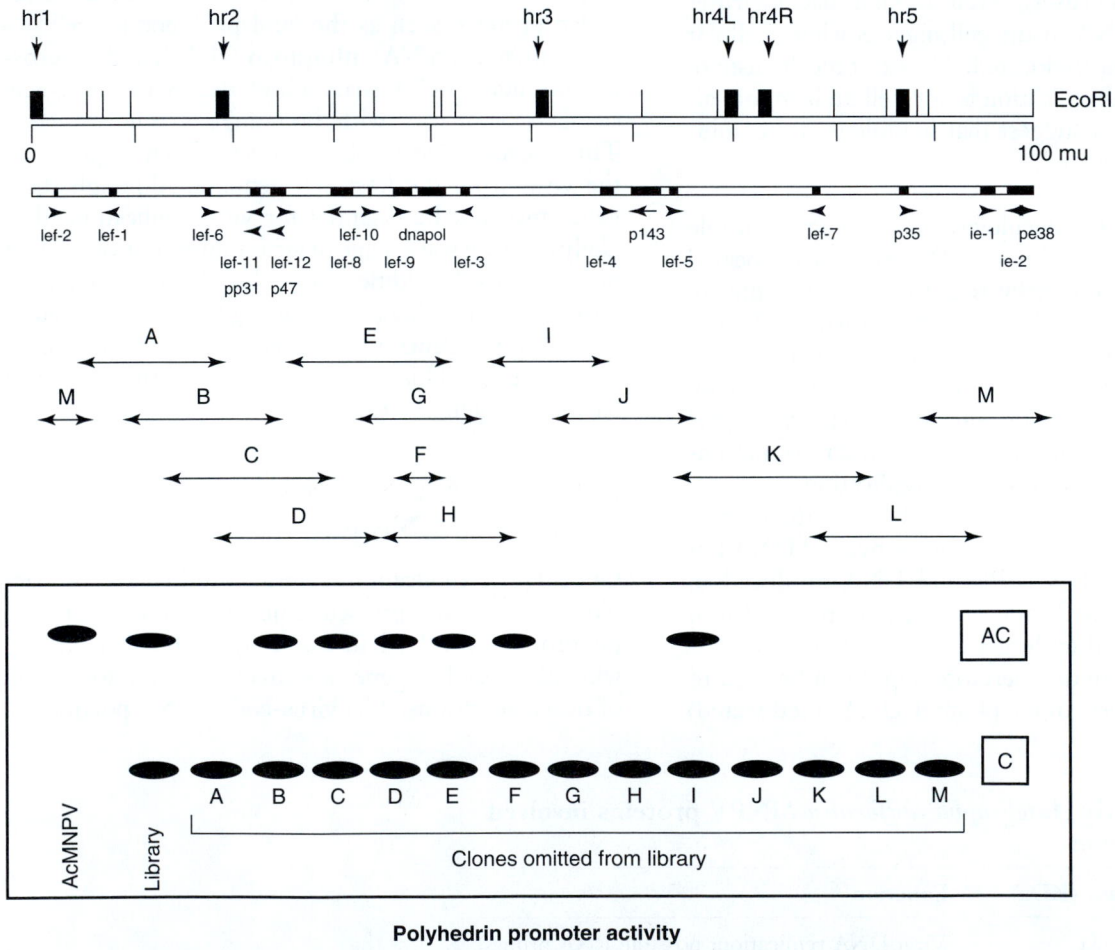

FIGURE 14.4 Identification of baculovirus late expression factors (LEFs) by a transient expression assay. Top: locations of hrs superimposed on the EcoRI fragment map. Middle: locations of the *Autographa californica* MNPV late expression factors and a set of overlapping cloned DNA fragments (A-M) used to transfect cells. Bottom: radioactive chloramphenicol is incubated with extracts of transfected cells; acetylated chloramphenicol (AC) is then separated from unacetylated chloramphenicol (C) by thin-layer chromatography. Data adapted from Passarelli and Miller, *J. Virol.* 67, 3481–3488 (1993).

is responsible for transcription of late and very late genes. Additional sequences downstream of the late initiator motif appear to be responsible for the high transcriptional activity of very late genes such as polyhedrin and p10. The product of the vlf-1 gene, very late transcription factor, is involved in the high-level expression of very late genes.

The development of *in vitro* transcription assays led to the identification of viral proteins required for transcription from late and very late promoters (Table 14.3). The viral RNA polymerase contains four or five virus-coded subunits. These genes were also implicated in the regulation of late and very late transcription by transient *in vivo* expression assays (Figure 14.4). A chloramphenicol acetyl transferase reporter gene under the control of a baculovirus late promoter was transactivated by a combination of viral genes expressed on different transfected plasmids. This assay is based on a set of overlapping clones (A through M) that together represent the entire AcMNPV genome. Omission of certain clones (B, C, D, E, F, and I) had no effect on reporter gene activity (shown by presence of a band of acetylated chloramphenicol, labeled AC). However, omission of other clones (A, G, H, J–M) resulted in the absence of chloramphenicol acetyl transferase activity (band representing unmodified chloramphenical, labeled C). Therefore, these clones must contain factors essential for transcription of late genes.

By substitution of individual clones with smaller fragments, a total of 19 specific genes were identified that were capable of transactivating the expression of late but not early promoters. These genes include many of those shown in Tables 14.2 and 14.3. A library of plasmids, each carrying one of these genes, is sufficient

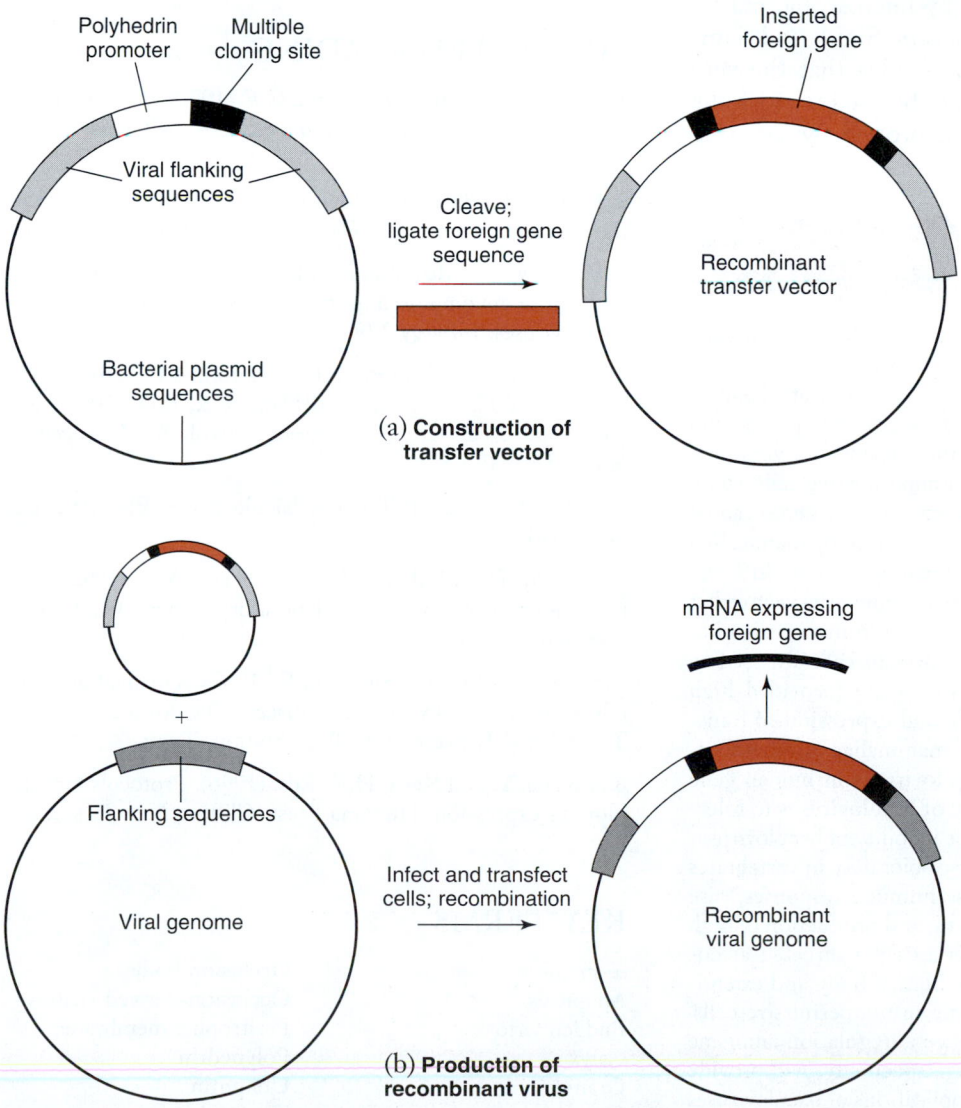

FIGURE 14.5 Preparation of a recombinant baculovirus designed to express a foreign gene product. See text for details.

to support expression of a transfected reporter gene under the control of an AcMNPV late or very late promoter. As discussed above, several late expression factors are also required for replication of transfected reporter plasmids, supporting the hypothesis that DNA replication is required for efficient late gene expression. It is not known whether genes required for viral DNA replication also play some role in regulating late or very late gene transcription. Once viral structural proteins are synthesized, assembly of nucleocapsids occurs in the nucleus.

Baculoviruses are widely used to express foreign proteins

In the early 1980s, several lines of evidence indicated that the very late gene product, polyhedrin, is not essential for virus replication in cell culture. Viruses missing the polyhedrin gene could be easily distinguished from polyhedrin-positive virus by microscopic examination of plaques in cell monolayers. Smith and Summers came up with the ingenious idea that the very strong polyhedrin promoter could be used to drive the high-level synthesis of a foreign protein by infecting cells with recombinant viruses. The production of the recombinant virus occurs in two stages (Figure 14.5). First, the foreign gene is cloned adjacent to the polyhedrin promoter in a plasmid transfer vector, replacing the polyhedrin-coding region (step a). Second, insect cells are cotransfected with infectious, wildtype viral DNA and the transfer vector carrying the foreign gene (step b). During replication of the viral DNA, homologous recombination can occur between sequences flanking the polyhedrin gene on the transfer vector and the identical sequences on viral DNA, generating recombinant genomes with the foreign gene in place of the polyhedrin gene. These recombinant genes are incorporated into progeny virions. Recombinant viruses can be easily identified by screening for their polyhedra-negative phenotype. Since their invention, recombinant baculoviruses have become one of the most widely used vehicles for the high-level expression of foreign proteins.

Recombinant baculoviruses as gene therapy vectors

In 1985, Lois Miller's group demonstrated that foreign gene expression from recombinant baculoviruses is cell type dependent. Baculoviruses can attach to and penetrate nonpermissive cells from a wide variety of tissues and species, and the viral genome is transported to the nucleus, but viral promoters are not recognized in these nonpermissive cells, so no viral gene expression occurs. However, reporter genes can be expressed in mammalian cells if they are driven by mammalian promoters engineered into the baculovirus genome. In 1995, Strauss reported the development of a baculovirus vector that could be used for efficient gene transfer into mammalian hepatocytes *in vitro*. Nearly 100% of human cells exposed to baculoviruses can express the desired gene (provided high multiplicities of infection are used), and expression of transduced genes under the control of mammalian promoters is high. The advantages of using baculoviruses in human gene therapy include the known capacity of baculoviruses to tolerate large inserts of foreign DNA, the inability of baculoviruses to replicate in vertebrates, a history of biosafety in vertebrates including a lack of toxic or adverse immune responses, ease of manipulation in insect cell cultures, and production of high virus titers. Current efforts in this area are directed at enhancing baculovirus stability in the human body and extending the life span of the viral genome in nonpermissive cells. Continued research on baculovirus gene regulation and gene function, and the basis of host range specificity, will provide important information for future applications of baculoviruses in biotechnology.

SUGGESTED READING

1. Blissard, G.W. and Rohrmann, G.F. (1990). Baculovirus diversity and molecular biology. *Annual Review of Entomology* **35,** 127–155.

2. Clem, R.J. (2001). Baculoviruses and apoptosis: The good, the bad, and the ugly. *Cell Death & Differentiation* **8,** 137–143.

3. Hayakawa, T., Rohrmann, G.F., and Hashimoto, Y. (2000). Patterns of genome organization and content in lepidopteran baculoviruses. *Virology* **278,** 1–12.

4. Inceoglu, A.B., Kamita, S.G., Hinton, A.C., Huang, Q., Severson, T.F., Kang, K., and Hammock, B.D. (2001). Recombinant baculoviruses for insect control. *Pest Management Science* **57,** 981–987.

5. Miller, L.K. Ed. (1997). The baculoviruses. Plenum Press, New York.

6. O'Reilly, D.R., Miller, L.K., and Luckow, V.A. (1992). Baculovirus expression vectors: A laboratory manual. W.H. Freeman and Company, New York.

7. Summers, M.D. and Smith, G.E. (1987). A manual of methods for baculovirus vectors and insect cell culture procedures. Texas A&M University, College Station, Texas.

8. Tanada, Y. and Kaya, H.K. Eds. (1996). Protocols for baculovirus expression. Humana Press, Clifton, New Jersey.

KEY TERMS

α-amanitin
Apoptosis
Budded virions
Calyx
Granulin
Hemocoel
Microvilli
Occlusion body
Occlusion-derived virions
Peritrophic membrane
Polyhedrin
Ubiquitin
Virogenic stroma

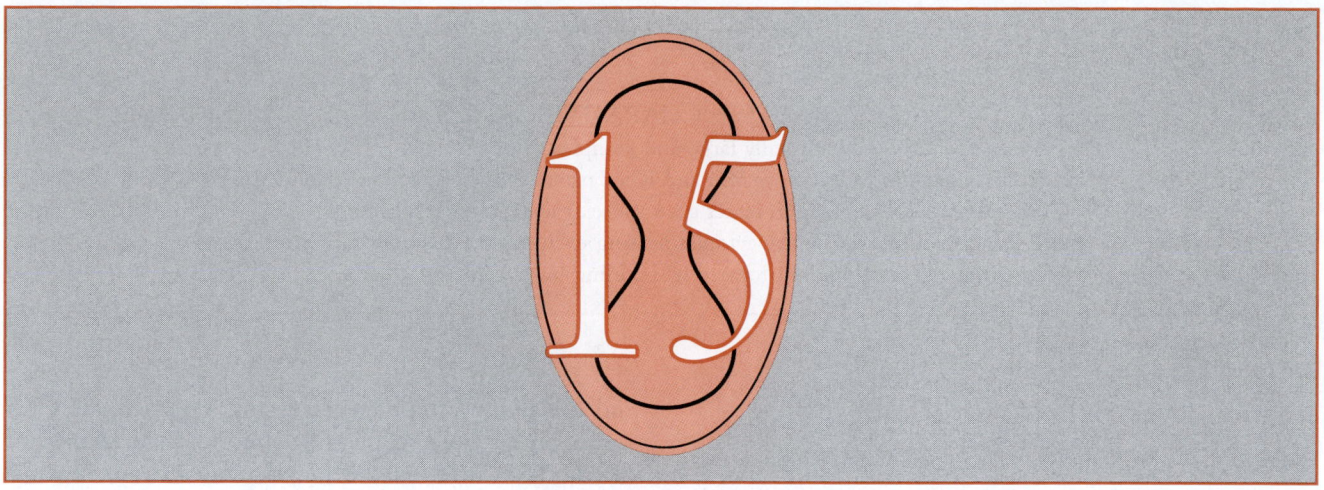

Poxviruses

Poxviridae

From English *pocks* (pox), referring to blistering skin lesions

VIRION

Complex, ovoid or brick shaped, 300 × 250 × 150 nm
Surface ridges or "tubules."
No typical symmetry elements.
Internal core and lateral bodies.
Virions exist in two forms:

- *Intracellular mature virus (IMV): one lipid membrane*
- *Extracellular enveloped virus (EEV): two lipid membranes*

GENOME

Linear ds DNA, 150–250 Kb (vaccinia virus: 200 Kb).
Covalently closed hairpin ends: no free 3′ and 5′ ends.
10 Kb inverted terminal repeats.

GENES AND PROTEINS

150–250 genes (vaccinia virus: 200).
Each gene has own promoter.
No spliced mRNAs (no introns).
Genes distributed on both DNA strands.
Genes expressed in three temporal classes:

- *Early: dissolve core, direct DNA replication and intermediate gene expression*
- *Intermediate: direct late gene expression*
- *Late: virion structural proteins, RNA polymerase, initiation factors, RNA processing enzymes*

VIRUSES AND HOSTS

Two subfamilies:

Chordopoxvirinae: infect vertebrates:

- *Humans: variola, vaccinia (vaccine strain), molluscum contagiosum*
- *Viruses that infect a variety of birds and mammals: monkeys, cattle, etc.*

Entomopoxvirinae: infect insects (beetles, butterflies, flies, etc.).

DISEASES

Smallpox: high fatality rate, now eradicated.
Molluscum contagiosum: the only existing natural infection of humans; relatively rare but more common in immunocompromised patients.
Monkeypox, cowpox, tanapox are animal poxviruses that occasionally infect humans.

DISTINCTIVE CHARACTERISTICS

Unusually large and complex virus.
Virus-coded enzymes for transcription and RNA processing packaged in virion core.
Early RNAs made in intact core and extruded into cytoplasm.
DNA genome replicates in the cytoplasm using exclusively virus-coded enzymes.
Internal envelope not formed by budding but assembled *de novo*.

Smallpox was a debilitating and fatal worldwide disease

Smallpox, caused by the poxvirus variola, embodies simultaneously one of humankind's greatest trials and one of its greatest triumphs. The human disease was called "*small*pox" in fifteenth century France and England to distinguish it from "large pox," or syphilis, which caused larger lesions and infected older (larger) people. Once a gruesome and lethal disease worldwide, smallpox has now been eradicated from the planet in a magnificent and unprecedented international collaboration of science and society.

Smallpox plagued humans for at least 3000 years. Scarred Egyptian mummies provide evidence for the existence of smallpox as early as the second millennium B.C. Written records suggest that smallpox was **endemic** in regions of Egypt and India by the first century A.D., from where it spread throughout China, southwestern Asia, Europe, and northern Africa by the tenth century A.D. European exploration and colonization in the sixteenth, seventeenth, and eighteenth centuries brought the disease to North and South America, southern Africa, and Australia, so that by the eighteenth century smallpox was common in virtually all densely populated areas worldwide. Smallpox influenced the course of history, killing rulers, disabling attacking armies, and decimating native populations infected by advancing conquerors.

The human toll of smallpox is almost unimaginable today. Although in rural areas the incidence of smallpox may have been relatively low, in large cities in seventeenth century Europe, smallpox probably behaved like a typical childhood disease, so that virtually every child could be expected to contract it. Smallpox strains varied in mortality between 1% and 40%, but the average case fatality rate in endemic countries was approximately 25%. The infection was horrific in its presentation. The disease usually began as a respiratory infection, transmitted by virus contained in nasopharyngeal secretions, pus, or scabs from an actively infected person, or by virus-contaminated bedding or clothing. The infection progressed though an asymptomatic phase lasting approximately 12 days, followed by sudden onset of high fever, splitting headache, and vomiting lasting four days, a slight improvement, and then the development of the characteristic smallpox skin rash and the resumption of fever, lasting approximately 12 days. Painful, isolated, pus-filled blisters covered the entire body as well as the mouth and airway, making breathing difficult and eating and drinking virtually impossible. In severe cases, the rash became confluent and/or hemorrhagic. Most survivors were scarred for life, many were blinded, and some suffered debilitating limb deformities. Comte de la Condamine, an eighteenth-century French mathematician and scientist, wrote: "No man dared to count his children as his own until they had had the disease."

Variolation led to vaccination, which has eradicated smallpox worldwide

Early attempts at smallpox prevention, called **variolation**, took advantage of the fact that infection with variola virus by an unnatural route resulted in a disease of reduced severity and decreased mortality. In China in the tenth century A.D., practitioners pulverized **pock** scabs from individuals undergoing mild smallpox infections, and blew the powder through a tube into the nostrils of the recipient, resulting in a generalized but nevertheless relatively mild infection. In India at about the same time, variolation was accomplished by seeding freshly scratched skin of a recipient with pus from an active smallpox infection. This cutaneous route of variolation resulted in primary and satellite lesions at the site of inoculation, but a much milder generalized infection, and an overall mortality rate of approximately 1%, significantly lower than the natural infection. The practice of cutaneous variolation spread through southwest Asia and northern Africa, and by the mid- to late eighteenth century became well established in Europe and North America, reducing significantly the impact of smallpox where practiced, and setting the stage for the discovery of vaccination at the turn of the nineteenth century.

Immunization against smallpox by inoculation with cowpox was introduced by Edward Jenner, an English physician, in 1796. Jenner's bold experiment represented a synthesis of three concepts, all of which were fairly common knowledge at the time: (1) cutaneous variolation produced resistance to smallpox, (2) milkmaids were rarely pockmarked and often resistant to variolation, and (3) milkmaids often acquired pock-like lesions on their hands, called cowpox disease, presumably from milking cows bearing similar lesions on their teats. Reasoning that the apparently protective effects of cowpox could be artificially induced using the procedures of variolation, Jenner inoculated eight-year-old James Phipps with cowpox from the hand of a milkmaid, Sarah Nelmes, on May 14, 1796. The procedure excited a limited cowpox disease on Phipps's arm, but otherwise "the indisposition attending it was barely perceptible." Most importantly, six weeks later Jenner challenged Phipps with smallpox by variolation, and found that he was resistant. To distinguish his procedure from variolation, Jenner named it **vaccination** (*vacca* means "cow" in Latin), and within 10 years vaccination against smallpox had been adopted worldwide.

Despite vigorous individual efforts over the next 150 years to control smallpox with vaccination, in 1958 the disease was still endemic in numerous countries in South America, Africa, southwestern Asia, the Indian subcontinent, and China. In that year, the World Health Organization launched the Global Smallpox Eradication Program, a massive, internationally coordinated program of surveillance and vaccination, which ultimately lead to the eradication of smallpox worldwide, certified in 1979. Vaccination for smallpox has now been discontinued, since the small chance of serious complications from vaccination exceeds the negligible risk of contracting smallpox.

Poxviruses remain a subject of intense research interest

Although smallpox has been eradicated, interest in poxviruses persists, and with good reason. Poxviruses encode and package in virions a full complement of enzymes required for viral mRNA synthesis and modification, and thus provide an ideal model system for studying the basics of mRNA metabolism. Poxviruses also encode several proteins that have evolved to confound the host inflammatory and immune responses. This allows researchers to study the immune response to viral infections, and may lead to development of useful immunotherapeutic and anti-inflammatory reagents. Foreign genes can be spliced into poxviruses and expressed during viral infection, providing research vectors for eukaryotic gene expression as well as promising recombinant vaccines. Finally, recent fears that smallpox might be used as a biological weapon have spawned renewed interest in development of both improved vaccines and antipoxvirus drugs.

Two major subfamilies of poxviruses can be distinguished based on host range: *Chordopoxvirinae* infect vertebrates and *Entomopoxvirinae* infect insects. Most of the vertebrate poxviruses cause blistering skin diseases, while the insect poxviruses cause a generalized necrosis of infected larvae. Some poxviruses are transmitted among different animal species, resulting in sometimes misleading nomenclature. For example, cowpox virus actually uses rodents as a reservoir host, cattle being only an occasional and accidental host for the virus. A list of some poxviruses is shown in Table 15.1.

Linear vaccinia virus genomes have covalently sealed hairpin ends and lack introns

Vaccinia virus, used for vaccination against smallpox, is used widely in research laboratories. Because vaccinia virus is the vaccine strain, it is relatively safe to work with and has been studied in clinical and laboratory situations for over 200 years. Although it was originally isolated as a poxvirus that infected cows and/or dairy workers, vaccinia virus is distinct from cowpox and cannot presently be found in nature. Vaccinia virus may be a variant of cowpox virus that has drifted significantly during two centuries of artificial propagation, or it may represent a species of poxvirus that has become extinct since its initial isolation.

The genome is a 200 Kb double-stranded DNA molecule that contains 10 Kb inverted repeats at each

TABLE 15.1 Some poxviruses and their hosts

Virus	Reservoir host	Other hosts
Variola (smallpox)	Humans	None
Molluscum contagiosum	Humans	None
Vaccinia	Unknown	Humans, cows, buffaloes
Monkeypox	Squirrels	Humans, monkeys
Cowpox	Rodents	Humans, cows, cats, zoo animals
Orf	Sheep	Humans, other ruminants
Fowlpox	Birds	Humans (as vaccine vector)
Myxoma	Rabbits	None
Entomopox	Insects	None

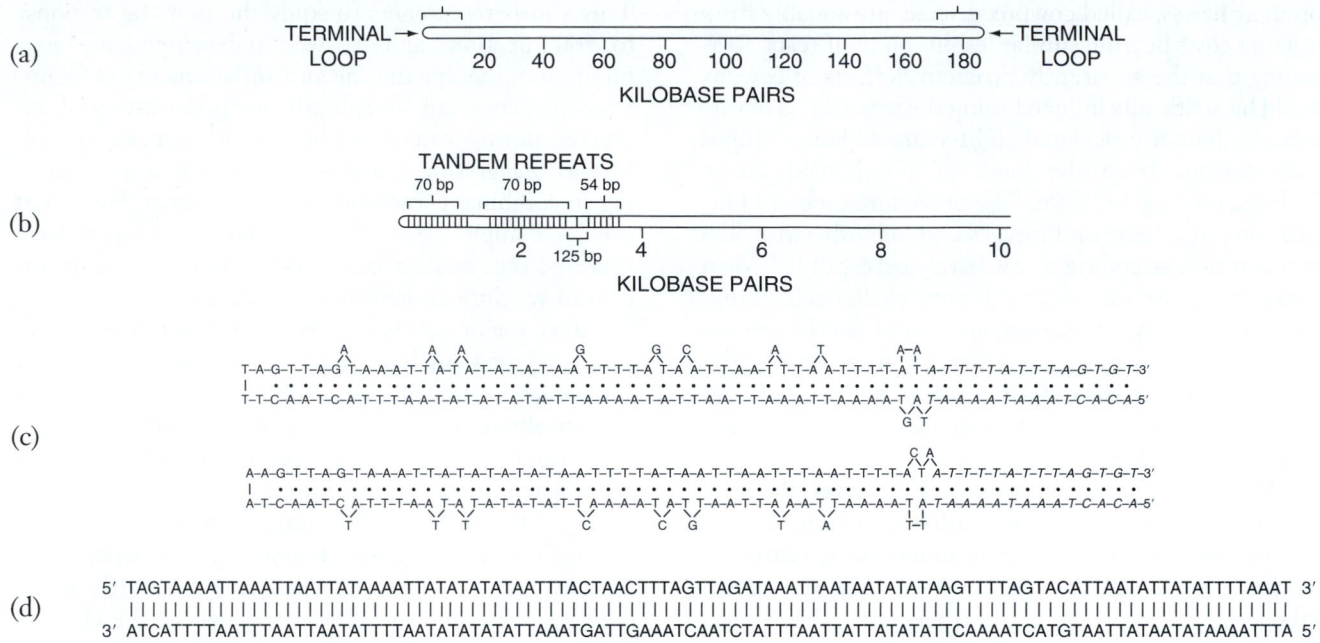

FIGURE 15.1 Vaccinia virus genome structure. From the top down are shown: (a) The double-stranded DNA genome terminal loops; (b) an enlargement of the inverted terminal repeat, emphasizing the tandem repeats within the terminal repeat; (c) the two hairpin loops (nonitalic font) as they are joined to the terminal repeats (italic font); (d) the duplex structure formed at a concatemer junction, consisting of the unfolded hairpins base-paired to each other.

end (Figure 15.1a). Several genes lie within these repeats and are therefore present in two copies per genome. The last 4 Kb of the repeats at each end lack coding sequences, but consist of about 200 short tandem sequence repeats of unknown function (Figure 15.1b). The genome, although linear, contains no free 5′ and 3′ ends; each terminus is sealed in a hairpin structure. The extreme terminal 100 bp exist in two related forms (Figure 15.1c). Each of these regions is imperfectly base-paired, and in their unfolded, single-stranded form, they represent inverse complements of each other (Figure 15.1d). The hairpins and immediate flanking sequences have important functions in DNA replication.

Poxvirus genes do not contain introns and therefore viral mRNAs are not spliced. Each gene contains its own transcriptional promoter, and genes are closely packed; intergenic regions larger than 100 bp are rare, and coding sequences sometimes overlap. By convention, genes are numbered sequentially within individual Hind III restriction endonuclease fragments, and designated as being transcribed rightward or leftward (Figure 15.2). Thus gene D8L is the eighth gene from the left end of the Hind III D DNA fragment, and is transcribed in a leftward direction. Genes near the ends of the genome are usually transcribed toward the nearest end, while genes in the center of the genome can be transcribed in either direction. "Housekeeping" genes, involved in processes common to all poxviruses such as transcription, DNA replication, and morphogenesis, are conserved among the poxviruses and are generally clustered in the center of the genome (Figure 15.2). Genes located near the termini vary among poxviruses, and they encode functions that are distinct for each virus such as host range and pathogenesis.

Two forms of vaccinia virions have different roles in spreading infection

Mature vaccinia virions are large (300 × 250 × 150 nm) ovoid or brick-shaped particles (Figure 15.3a) containing as many as 100 proteins. Thin sections of virions (Figure 15.3b, c) reveal a bioconcave core flanked by lateral bodies. The lateral bodies are trypsin-sensitive and therefore contain protein, but otherwise are of unknown composition and function. The core, which contains the virus genome, is surrounded by a bilayer structure of unknown composition. Virion cores also contain a spectacular array of virus-coded enzymes required for early mRNA synthesis and modification (Table 15.2).

Mature virions exist in two forms, called **intracellular mature virus (IMV)** and **extracellular enveloped virus (EEV)**, which differ in membrane composition as a result of differences in morphogenesis. The IMV particle contains an outer membrane composed of lipid and viral proteins. EEV is an IMV particle wrapped

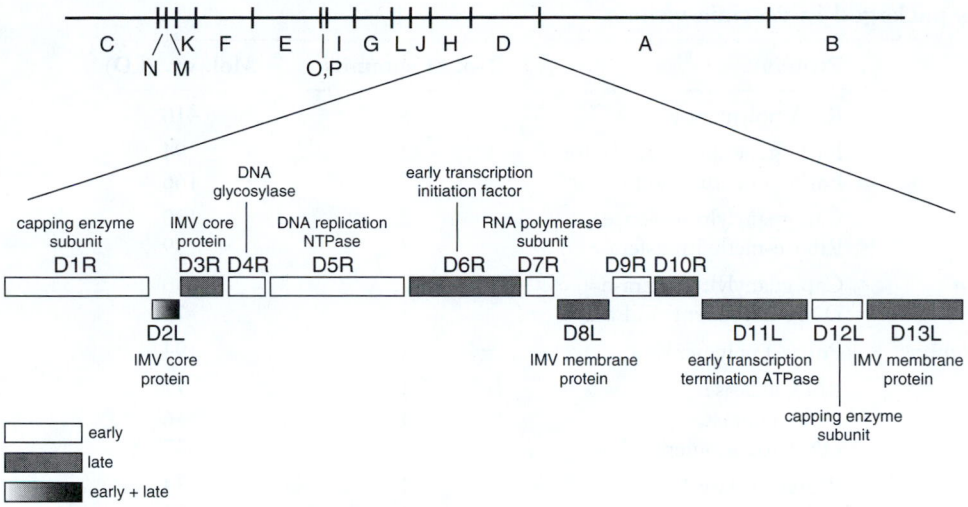

FIGURE 15.2 Functional organization of the vaccinia virus genome. Top: Hind III map of the entire genome. Below: exploded map of the 16 Kb Hind III D fragment. Individual genes are represented as rectangles (direction of transcription: R: rightward, L: leftward). Early and late genes are indicated by shading, and a function for each gene, where known, is indicated.

in an additional lipid bilayer containing a distinct set of embedded virus-coded proteins. Both forms of the vaccinia virion are infectious.

IMV and EEV use distinct cell attachment proteins in the virus particle and different receptors on the cell surface. At least three different IMV membrane proteins that may facilitate attachment have been described, and each of these binds to ubiquitous cell surface **glycosaminoglycans**, consistent with the broad host range of the virus. EEV particles are taken up by endocytosis and the EEV membrane ruptures due to a decrease in pH in the endosome, releasing an IMV within the vesicle. IMV then fuses with the vesicle membrane to release the virion core into the cytoplasm. Infection of cells with IMV may occur by direct fusion of the IMV membrane with the plasma membrane, or alternatively by endocytosis followed by fusion with the membrane of the endosome. In either case, the end result is the release of the virus core into the cytoplasm.

The different forms of the poxvirus virion probably serve different roles in nature. The EEV particle is required for spread of virus from cell-to-cell and tissue-to-tissue in an infected animal. However, the EEV membrane is fragile, and probably does not survive well in the environment. In contrast, the IMV particle is stable, and is likely the form responsible for transmission of infections between invididuals. Poxviruses are notoriously stable in the environment; the smallpox vaccine can be freeze-dried, and dried pox scabs or virus-contaminated surfaces can retain infectivity for prolonged periods of time. Thus the IMV particle may provide for stability in the environment while EEV allows efficient spread in the body.

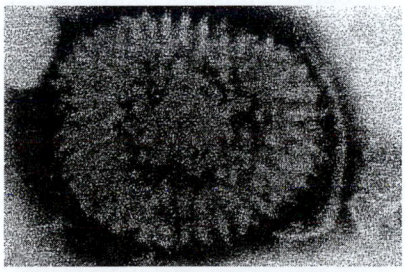

(a)

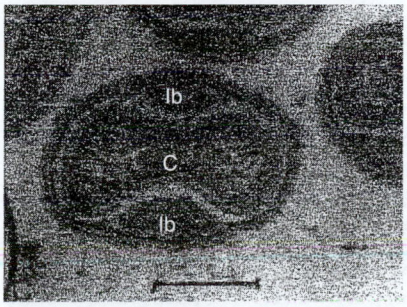

(b)

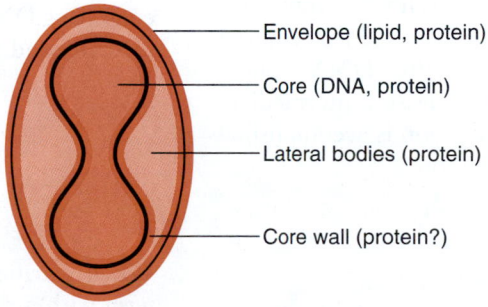

(c)

FIGURE 15.3 Vaccinia virus virions. a. Electron micrograph of negatively stained vaccinia IMV, showing surface ridges, a drying artifact. b. Thin section of purified vaccinia IMV, showing the lateral bodies (lb) and core (c). c. A diagram of the sectioned IMV particle, showing internal structures.

TABLE 15.2 Enzymes packaged in vaccinia virions

Function	Protein	No. of subunits	Mol. wt. (kD)
RNA synthesis	RNA polymerase	8	410
Transcription initiation	Early gene specificity factor	1	94
	Early gene initiation factor	2	166
mRNA 5′ end capping	Cap guanylyltransferase	2	130
	Ribose-methyltransferase	1	39
Transcription termination	Cap guanylyltransferase	2	130
	DNA-dependent ATPase	1	72
mRNA 3′ end polyadenylation	Poly(A) polymerase	2	94
Nucleic acid topology	RNA helicase	1	77
	DNA helicase	1	56
	DNA topoisomerase I	1	37
Protein modification	Protein kinase I	1	34
	Protein kinase II	1	52
	Protein phosphatase	1	20

Poxviruses replicate in the cytoplasm

Unlike other eukaryotic DNA-containing viruses covered in this book, poxviruses replicate exclusively in the cytoplasm of infected cells. Because most of the cellular enzymes required for RNA synthesis and processing and DNA replication are located in the nucleus, this implies that poxviruses must provide virus-coded enzymes to carry out these processes. Their large genome size enables poxviruses to code for a variety of enzymes involved in RNA and DNA metabolism.

Vaccinia virus genes required for DNA replication are expressed immediately following infection, and are therefore called early genes. Early genes are expressed within virus cores, which are released into the cytoplasm after fusion of the virion envelope with cellular membranes during virus entry (Figure 15.4). Early messenger RNAs are extruded from cores into the cytoplasm, where they are translated into early virus proteins. Some of these early proteins dissolve the core and release viral DNA into the cytoplasm, where further transcription and DNA replication take place. Genes required for virion formation are expressed after DNA replication has begun, and comprise two classes, intermediate and late. Regulation of gene expression is accomplished primarily at the level of transcription.

Poxvirus genes are expressed in a regulated transcriptional cascade controlled by viral transcription factors

The timing of gene expression is dictated by the sequential synthesis of virus-coded transcription initiation factors (Figure 15.4), which bind to specific promoter elements on viral DNA. Initiation factors required for transcription of early genes (virus early transcription factors, VETF) are packaged in infecting virions and act within the core upon its release into the cytoplasm. Certain early genes encode distinct intermediate transcription factors (VITF) that trigger intermediate gene expression. Late transcription factors (VLTF) are in turn encoded by some of the intermediate genes, and finally, late genes encode early transcription factors, which are packaged into virions in preparation for a subsequent round of infection. This organization of gene expression assures that sufficient newly-replicated virus genomes will accumulate before synthesis of structural proteins begins. Inhibition of DNA replication prevents expression of both intermediate and late genes.

Roughly one half of vaccinia genes are controlled by early promoters, the other half being controlled by intermediate or late promoters. A certain number of genes have compound promoters, allowing for their expression throughout infection. Vaccinia promoters are approximately 35 nucleotides in length regardless of gene class. Mutational analysis has defined optimum consensus DNA sequences that distinguish early, intermediate, and late promoters.

Virus-coded enzymes packaged in the core carry out early RNA synthesis and processing

When purified vaccinia virions are permeabilized *in vitro* with a detergent and incubated with ribonucleoside triphosphates and S-adenosylmethionine, early genes are transcribed, and the resulting mRNAs are capped, methylated, polyadenylated, and extruded from the virion core. Thus the vaccinia virion is a "mini-nucleus," a simple and easy-to-purify mRNA synthesis machine that has proved to be of extraordinary value for dis-

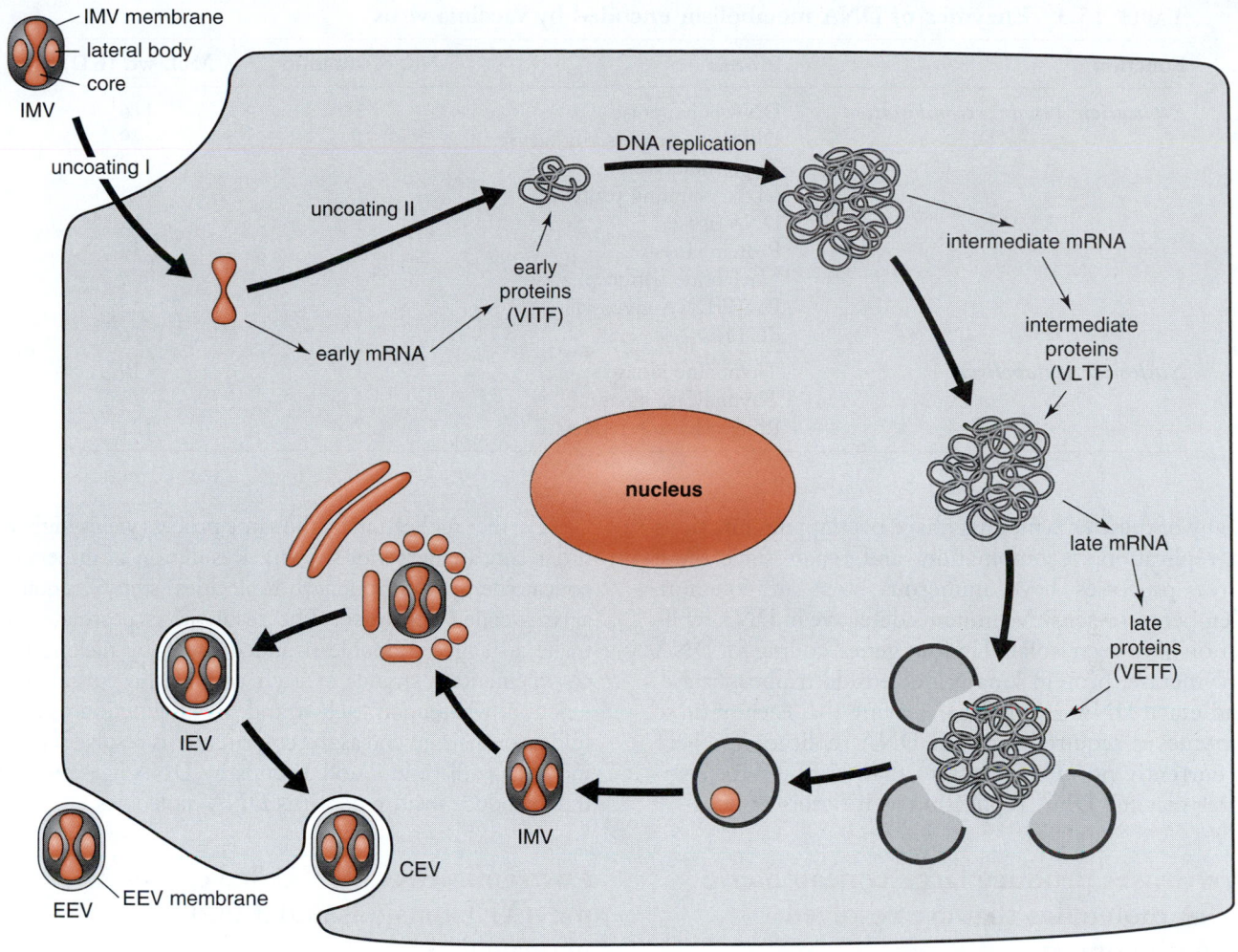

FIGURE 15.4 **The vaccinia virus life cycle.** IMV: intracellular mature virus; IEV: intracellular enveloped virus; CEV: cell-associated enveloped virus; EEV: extracellular enveloped virus; VETF: viral early transcription factors; VITF: viral intermediate transcription factors; VLTF: viral late transcription factors.

secting the biochemical details of transcription, mRNA 5′ end capping, and mRNA 3′ end polyadenylation.

Enzymes representing each of these steps can be purified from virions, so that the enzymology of each process can be dissected in detail (Table 15.2). The vaccinia virus RNA polymerase is a complex, eight-subunit enzyme that bears some resemblance to other eukaryotic RNA polymerases. Initiation of early transcription requires two additional proteins: a specificity factor that binds to RNA polymerase, and a 2-subunit initiation factor that binds to the promoter. Together they load the RNA polymerase at the transcription start site. Capping of mRNAs also requires two enzymes. A 2-subunit capping enzyme (the cap **guanylyltransferase**) adds 7-methylguanosine to the 5′ triphosphate end of the nascent RNA in a 5′-5′ triphosphate linkage, and a distinct (nucleoside-2′-O-)-methyltransferase (the cap **ribose-methyltransferase**) methylates the ribose of the first transcribed base in the RNA.

Transcription termination occurs in response to a sequence element, U_5NU, in the growing RNA chain, and requires two protein factors: a DNA-dependent ATPase and the same cap guanylyltransferase that modifies the mRNA 5′ end. Polyadenylation of freshly terminated mRNA 3′ ends is catalyzed by a 2-subunit poly(A) polymerase, one subunit of which is the same cap ribose-methyltransferase that participates in mRNA 5′ cap formation. Thus there is a remarkable economy of function, with two enzymes, the cap guanylyltransferase and the cap ribose-methyltransferase, performing dual functions essential for both 5′ end and 3′ end maturation of the early mRNA.

Enzymes that direct DNA replication are encoded by early mRNAs

Table 15.3 lists several viral early gene products with known or presumed functions in DNA metabolism.

TABLE 15.3 Enzymes of DNA metabolism encoded by vaccinia virus

Function	Protein	No. of subunits	Mol. wt. (KD)
Replication, repair, recombination	DNA polymerase	1	116
	DNA pol. processivity factor	1	49
	Topoisomerase I	1	36
	ssDNA binding protein	1	34
	DNA ligase	1	63
	Protein kinase	1	34
	Nucleoside triphosphatase	1	90
	Uracil DNA glycosylase	1	25
	dUTPase	1	16
Nucleotide metabolism	Thymidine kinase	1	19
	Thymidylate kinase	1	23
	Ribonucleotide reductase	2	121

Many of these enzymes may have overlapping functions in replication, recombination, and repair since these three processes have numerous steps in common. Temperature-sensitive mutants defective in DNA replication have been isolated in four genes, coding for DNA polymerase, protein kinase, nucleoside triphosphatase, and uracil DNA glycosylase, proving that each of these enzymes is required for viral DNA replication. There is currently no *in vitro* system that faithfully reconstitutes vaccinia DNA replication as it occurs *in vivo*.

Poxviruses produce large concatemeric DNA molecules that are resolved into monomers

The current model describing poxvirus DNA replication (Figure 15.5) is fashioned after the "rolling hairpin" model of DNA replication in parvoviruses (see Chapter 9). This model is based on the structure of the virus genome and on two additional observations: (1) replication appears to initiate within or near the hairpin termini; and (2) head-to-head and tail-to-tail **concatemers** (oligomers of genome-length units) accumulate during infection. Replication is initiated by a site-specific nick in one of the terminal hairpins, revealing a 3′ end that primes DNA synthesis (step 1). Extension from this 3′ end (step 2) copies the sequence of the melted hairpin to its 5′ end, approximately 100 nt. This nascent DNA chain then folds into a new hairpin (step 3) that can prime further DNA synthesis (step 4). Synthesis continues through the entire length of the genome (200,000 nt) using one DNA strand as template, then through the distal hairpin, and back through the entire length of the genome using the complementary strand as template (step 5), to generate a head-to-head dimer. This process forms a "concatemer junction" that contains the two complementary hairpin ends in a perfectly base-paired duplex (see Figure 15.1d).

Further replication by the same process yields higher-order concatemers (not shown). Resolution of dimers or concatemers into unit-length molecules (step 6) requires a virus-coded **resolvase**. The resolvase is postulated to make a staggered double-stranded break by nicking the complementary strands at each end of the concatemer junction, whereupon each strand of the junction can refold into a hairpin end as the concatemer is resolved. Joining of the adjacent 5′ and 3′ ends by DNA ligase would then produce mature poxvirus DNA molecules.

Postreplicative mRNAs have 5′ end poly(A) extensions and 3′end heterogeneity

Intermediate and late genes are called postreplicative genes because their transcription begins only after DNA replication has started. They use distinct transcriptional promoters and initiation factors, but are otherwise similar and can be considered together. Postreplicative RNA synthesis, capping, and polyadenylation are carried out by the same enzymes used for early mRNA synthesis (see Table 15.2), freshly synthesized from early mRNAs. Initiation of transcription of intermediate genes requires two virus-coded factors (VITF) synthesized from early viral mRNA, and one cellular factor. The virus factors are the cap guanylyltransferase referred to above, and one of the smaller RNA polymerase subunits. The identity of the cellular factor is unknown. Initiation of transcription of late genes requires at least three viral protein factors (VLTF) produced from intermediate mRNAs, and may require in addition an early gene product and a cellular factor.

Intermediate and late mRNAs share two unique structural features that distinguish them from early mRNAs. First, 5′-terminal "poly(A) heads" are added to these mRNAs via a slippage mechanism during the initiation of transcription. Both intermediate and late promoters contain the sequence TAAA at the RNA start

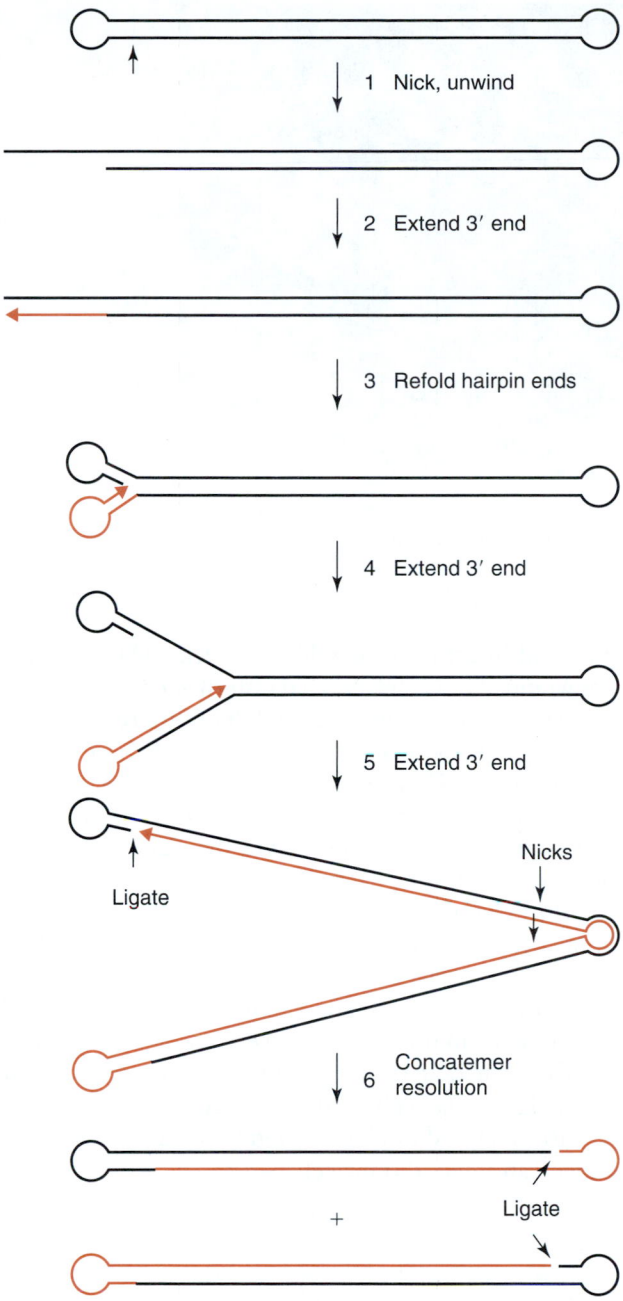

FIGURE 15.5 A model for vaccinia virus DNA replication. See text for details. Newly synthesized DNA is shown in orange.

site, usually located just upstream of the translation initiation ATG for a given gene. The viral RNA polymerase presumably synthesizes the initial AAA sequence, then backs up while still gripping the nascent RNA, adds another AAA sequence, then backs up and repeats the process several more times before clearing the promoter and continuing downstream. The resulting RNA 5′ end contains 30–50 A residues upstream of the first AUG codon. This extra 5′ noncoding region may aid in recognition of the AUG by ribosomes.

Second, these mRNAs have heterogeneous 3′ ends, resulting from a transcription termination mechanism that is fundamentally different from the process used during early transcription. The RNA polymerase ignores the early termination sequence U_5NU, and instead terminates inefficiently at multiple sites far downstream, yielding a family of transcripts 2–4 Kb in length with heterogeneous 3′ ends. At least three viral genes that influence the length at the 3′ end of postreplicative mRNAs have been identified. One of these genes codes for the same ribose-methyltransferase that participates both in mRNA 5′ cap formation and in 3′ end polyadenylation.

Just in case you lost track, the cap guanylyltransferase participates in mRNA capping, early mRNA transcription termination, and intermediate gene transcription initiation; and the cap ribose-methyltransferase participates in mRNA capping and polyadenylation, and in postreplicative gene transcription elongation or termination. Although the details of this remarkable multifunctionality remain to be resolved, it appears that each protein has multiple independent activities rather than multiple uses of a single enzymatic active site.

Mature virions are formed within virus "factories"

Morphogenesis of vaccinia virions is a unique and complex process that provides fascinating insights into intracellular protein trafficking and organelle biogenesis. Morphogenesis is studied by electron microscopic examination of cells infected with wildtype or mutant viruses (Figure 15.6); viral structural and assembly proteins are localized using labeled antibodies. Viral DNA replication results in the formation of cytoplasmic structures called **viroplasm** or DNA factories, recognizable as large areas of uniform density, cleared of normal cellular organelles. The first evidence of morphogenesis is the appearance within viroplasm of rigid, crescent-shaped membrane structures (Figure 15.6 and lower-right corner of Figure 15.4). These presumably contain both lipid and viral protein and are destined to become the IMV membrane. Crescent formation is directed by viral proteins that are synthesized on ribosomes bound to the endoplasmic reticulum and are subsequently transported specifically to viroplasm.

The exact mechanism of crescent formation is controversial. Some data indicate that the crescents may be modified continuations of endoplasmic reticulum or Golgi intermediate compartment cisternae, in which case the IMV membrane would actually be composed of two tightly apposed lipid bilayers. Other data indicate that crescents are at no time directly connected to a preexisting intracellular membrane, and that the IMV membrane is a single bilayer. Lipids would then have to be delivered to viroplasm from the

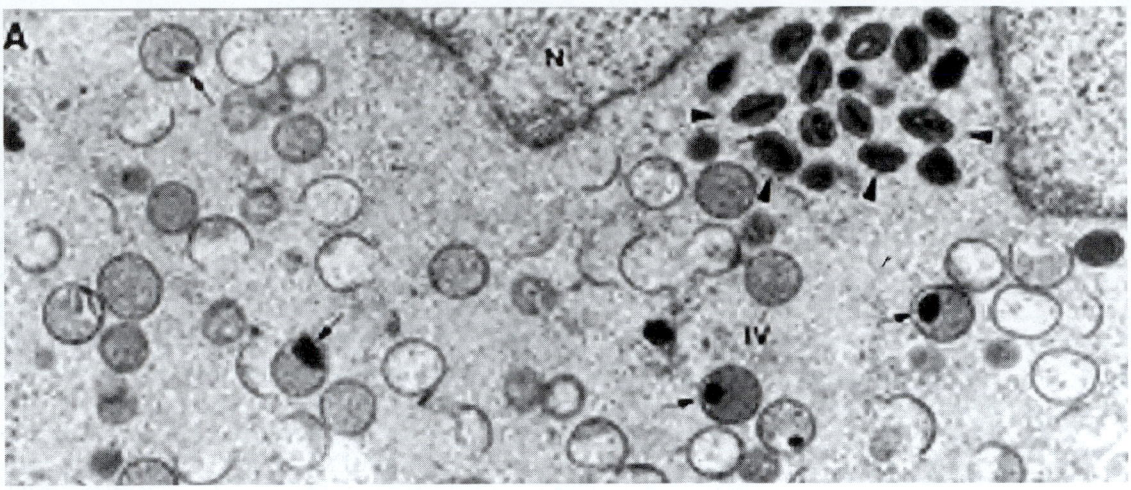

FIGURE 15.6 Electron micrograph of a vaccinia virus–infected cell. N: nucleus; IV: immature virions; arrows: condensing nucleoids within immature virions; arrowheads: mature virions.

endoplasmic reticulum in a novel form, perhaps as small vesicles or **micelles**, for assembly into crescents. The crescents ultimately mature into spheres called immature virions, and viral chromatin condenses within the spheres. Immature virions then develop the distinguishing internal core and lateral body structures by rearrangement and assembly of the material captured inside spheres, and in the process acquire the final brick shape to become IMV. In a normal infection the vast majority of mature virions remain intracellular IMV.

Extracellular virions are extruded through the plasma membrane by actin tails

A small fraction of IMV undergoes further maturation to become EEV, which are specifically designed for export and infection of neighboring cells. IMV particles are wrapped in Golgi-derived cisternae containing specific viral EEV proteins (center left of Figure 15.4). The resulting intracellular enveloped virus (IEV) therefore consists of an IMV encased in two additional bilayer membranes. IEV are transported to the cytoplasmic side of the plasma membrane, where the outermost membrane of IEV fuses with the plasma membrane. This results in an EEV particle that is attached to the cell membrane, called cell-associated enveloped virus (CEV). Actin tails are assembled within the cell beneath the CEV particles, resulting in the formation of **microvilli** extending from the cell into the extracellular space, with CEVs perched at their tips. These microvilli can contact neighboring cells, infecting them with CEV. Alternatively, CEV can be released from the cells as EEV, which can infect distant cells.

TABLE 15.4 Poxvirus immune-modulating genes

System	Virus gene	Activity	Protection against host defense
Interferon (IFN)	K3L	Homologue of eukaryotic initiation factor 2α (eIF-2α)	Pseudosubstrate for interferon-induced protein kinase (PKR)
	E3L	Double-stranded RNA binding protein	Sequesters dsRNA, an activator of interferon-induced PKR and 2′-5′A
	B8R	Secreted, soluble IFN-γ receptor	Binds to and inhibits IFN-γ
Interleukin-1 (IL-1)	B14R	Serine protease inhibitor (serpin)	Inhibits proteolytic activation of IL-1, decreasing inflammatory response
	B16R	Soluble IL-1 receptor	Binds to and inhibits IL-1β
Tumor necrosis factor (TNF)	B28R, A53R	Soluble TNF receptor	Binds to and inhibits TNF-α and β
Complement	C3L	C4B and C3B binding protein	Inhibits classical and alternative complement pathways

Poxviruses make several proteins that target host immune defenses

As obligate intracellular parasites, viruses have no choice but to confront the host immune response, which has two fundamental branches: innate nonspecific immunity includes the **interferon** response (Chapter 33), and induced adaptive immunity leads to production of **antibodies** and **cytotoxic lymphocytes** (see also Chapter 33). Importantly, both the innate and adaptive immune responses are regulated by **cytokines**, peptide hormones produced by numerous cells of the immune system. Poxviruses have evolved a number of genes that cleverly combat the host immune response (Table 15.4).

Interferon response. The first line of defense against virus infections is the interferon response, triggered by the synthesis of double-stranded RNA (dsRNA) in infected cells. A significant quantity of poxvirus dsRNA is produced during the late phase of infection as a result of synthesis of transcripts with heterogeneous 3′ ends, made on opposite DNA strands from convergent viral promoters. Poxviruses synthesize two proteins that inhibit the interferon-induced antiviral state. The K3L gene product, a homologue of eukaryotic protein synthesis initiation factor eIF-2α, competitively inhibits the dsRNA-activated protein kinase **PKR**, relieving the inhibition of protein synthesis induced by interferon. The E3L gene product binds to and sequesters dsRNA, interrupting both the PKR and the parallel **2′, 5′-oligoadenylate synthetase** antiviral pathways.

Cytokine signalling. Virus infection triggers synthesis of a large number of cytokines, several of which are important in coordinating the innate immune response, including **interleukin-1**, **tumor necrosis factor**, and interferon γ. Poxviruses have at least two different strategies for interfering with cytokine function. In one method, the virus produces a serine protease inhibitor (**serpin**), which specifically blocks a proteolytic cleavage required for activation of interleukin-1β. In another strategy, the virus produces soluble receptors for tumor necrosis factor, interleukin-1 and interferon γ, which bind to these cytokines and prevent them from interacting with their intended receptors. Importantly, while all three cytokines influence primarily the innate response, interferon γ also activates critical components of the adaptive immune response.

Complement. **Complement** plays a role in both the innate and adaptive immune responses. In both cases, interaction of virus surface components with elements of the complement system triggers a cascade of complement reactions that results ultimately in **phagocytosis** of the virus particle. The innate ("alternative") pathway involves a nonspecific interaction of the complement system with the virion surface, while the adaptive ("classical") pathway involves a reaction of the complement system with antibody-coated virus particles.

The vaccinia-T7 system

In an intriguing technological perversion of nature, vaccinia virus has been adapted for use as an efficient and versatile tool for expression of foreign proteins in eukaryotic cells. The system is composed of two recombinant vaccinia viruses. The first contains a copy of the bacteriophage T7 RNA polymerase gene (see Chapter 7) under transcriptional control of a vaccinia virus promoter, and the second contains the coding sequence for the gene of choice to be expressed cloned downstream of a T7 RNA polymerase promoter. Coinfection of cells with the two viruses results in synthesis of T7 RNA polymerase from the first virus, which then can transcribe the gene of choice from the T7 promoter in the second virus. The same outcome can be achieved without even constructing the second virus, simply by infecting cells with the virus that expresses T7 RNA polymerase, followed by transfection with a DNA molecule containing any coding sequence cloned downstream from a T7 promoter.

The scheme has several advantages as an expression system: (1) recombinant vaccinia viruses are relatively easy to prepare, essentially by homologous recombination in virus-infected cells between an infecting vaccinia genome and a transfected DNA that contains the foreign gene embedded within a nonessential virus gene; (2) the entire process takes place in the cytoplasm, obviating the need to be concerned with RNA splicing or nuclear-cytoplasmic transport of any of the component RNAs or proteins; (3) the system is very specific and efficient, owing to the remarkable activity and promoter specificity of the T7 RNA polymerase and to the amplification of expression by synthesis of numerous T7 RNA polymerase molecules, which efficiently and repeatedly transcribe the gene of choice.

The system has found utility in two areas: overexpression of foreign proteins for purification and biochemical analysis, and expression of foreign proteins for assay of their intracellular function. An interesting use of the system is the construction of recombinant vesicular stomatitis virus (see Chapter 20). In this example, cells are infected with vaccinia virus expressing T7 RNA polymerase, then are transfected simultaneously with four plasmid DNAs: one contains the vesicular stomatitis virus genome sequence downstream from a T7 promoter; and the other three contain the coding sequences of the three vesicular stomatitis virus proteins (P, L, and N) required for viral genome replication, each cloned downstream from a T7 RNA polymerase promoter. The T7 RNA polymerase supplied by the vaccinia infection transcribes all four genes simultaneously, providing the RNA genome of vesicular stomatitis virus and messenger RNAs for all three replication proteins. This allows replication, expression and packaging of the genome of vesicular stomatitis virus.

The vaccinia virus C3L gene product, synthesized and secreted during infection, binds to and inactivates two components of the complement system, C3B and C4B, which are critical to both the innate and adaptive complement pathways.

In summary, even a partial list of the abundant poxviral immune defense functions reveals an ingenious multifarious attack on the host defense system. Some have likened the poxvirus immune defense to "star wars." Extending this metaphor, the virus carries a MIRV (multiple independently targeted reentry vehicle) warhead, and launches this missile in a preemptive strike on the host's first line of defense. Importantly, the virus probably has evolved to target the pressure points of the host immune defenses and thus has much to tell us about the relative priorities of immune function.

SUGGESTED READING

1. Moss, B. (1996). Genetically engineered poxviruses for recombinant gene expression, vaccination, and safety. *Proceedings of the National Academy of Sciences of the USA* **93,** 11341–11348.

2. Rodriguez, J.R., Risco, C., Carrascosa, J.L., Esteban, M., and Rodriguez, D. (1998). Vaccinia virus 15-kilodalton (A14L) protein is essential for assembly and attachment of viral crescents to virosomes. *Journal of Virology* **72,** 1287–1296.

3. Sanz, P. and Moss, B. (1999). Identification of a transcription factor, encoded by two vaccinia virus early genes, that regulates the intermediate stage of viral gene expression. *Proceedings of the National Academy of Sciences of the USA* **96,** 2692–2697.

4. Seet, B.T., Johnston, J.B., Brunetti, C.R., Barrett, J.W., Everett, H., Cameron, C., Sypula, J., Nazarian, S.H., Lucas, A., and McFadden, G. (2003). Poxviruses and immune evasion. *Annual Review of Immunology* **21,** 377–423.

5. Smith, G.L., Vanderplasschen, A., and Law, M. (2002). The formation and function of extracellular enveloped vaccinia virus. *Journal of General Virology* **83,** 2915–2931.

6. Traktman, P. (1996). Poxvirus DNA replication. In DePamphilis, M., Ed., DNA replication in eukaryotic cells, pp. 775–798. Cold Spring Harbor Laboratory Press, Cold Spring Harbor, New York.

7. Upton, C., Slack, S., Hunter, A.L., Ehlers, A., and Roper, R.L. (2003). Poxvirus orthologous clusters: Toward defining the minimum essential poxvirus genome. *Journal of Virology* **77,** 7590–7600.

KEY TERMS

Antibody
Complement
Concatemer
Cytokine
Cytotoxic lymphocyte
Endemic
Extracellular enveloped virus (EEV)
Glycosaminoglycan
Guanylyltransferase
Interferons
Interleukin-1
Intracellular mature virus (IMV)
Micelle
Microvilli (singular: microvillus)
2′, 5′oligoadenylate synthetase
Phagocytosis
PKR (double-stranded RNA-dependent protein kinase)
Pock
Resolvase
Ribose methyltransferase
Serpin
Tumor necrosis factor (TNF)
Vaccination
Variolation
Viroplasm

Picornaviruses

Picornaviridae

From Spanish *pico* (point; 10^{-12}), referring to small size of virions, and *RNA*

VIRION

Naked icosahedral capsid (T = 1).
Diameter 30 nm.

GENOME

Linear ss RNA, positive sense.
7–8.5 Kb.
Viral protein VPg covalently bound to 5′ end.
Short, genome-encoded poly(A) tail at 3′ end.

GENES AND PROTEINS

All genes translated as single polyprotein followed by proteolytic cleavage.
Four capsid proteins (VP1–4).
one–three proteinases.
six–eight replication proteins.

VIRUSES AND HOSTS

Six genera, including *Enterovirus* (poliovirus, coxsackieviruses), *Rhinovirus* (common cold viruses), *Cardiovirus*, *Hepatovirus* (hepatitis A virus).
Hundreds of known virus species.
Infect humans, mammals, birds, fish, etc.

DISEASES

Humans: poliomyelitis (paralysis), myocarditis, hepatitis, meningitis, common cold.
Cattle: foot-and-mouth disease.

DISTINCTIVE CHARACTERISTICS

Translation is initiated at internal ribosome entry sites (IRES).
Self-cleavage of polyprotein by viral proteinases.
RNA replication is associated with vesicles in cytoplasm.

Picornaviruses cause a variety of human and animal diseases including poliomyelitis and the common cold

Picornaviruses (pronounced "pic-or-na-viruses") were so named because they are small (*pico* comes from Spanish and means "point") RNA-containing viruses. They cause a variety of diseases, including **hepatitis** (hepatitis A virus), **myocarditis** (coxsackievirus), the common cold (rhinoviruses), and paralytic **poliomyelitis** (poliovirus). Foot-and-mouth disease virus, caused by a picornavirus, was the first animal virus discovered in 1898. This virus causes economically important outbreaks of disease in cattle and sheep; recent outbreaks in Britain required extensive quarantines and slaughter of infected animals.

Hepatitis A virus is a picornavirus that causes acute liver disease, which is usually self-limiting, but occasional fatalities do occur. It is spread by direct contact between individuals and by contamination of food or water. Contaminated food in restaurants can lead to a local outbreak of many infections in a short time. This virus is more commonly encountered in countries with poorly-developed sewage sytems and low hygiene. Unlike disease caused by hepatitis B and C viruses (see Chapters 28 and 17), hepatitis A infection does not lead to chronic disease or liver cancer. An inactivated vaccine protective against hepatitis A virus infection is available.

Rhinoviruses replicate preferentially at 33°C, the temperature of the nose; they are believed responsible for approximately 50% of common colds. Over 100 different types of human rhinoviruses have been described, making it unlikely that an effective vaccine against these cold viruses would ever be developed. Recently, antiviral agents active against rhinoviruses have been developed (see Chapter 32), and there is a large potential market for an effective and inexpensive agent against the common cold.

Poliomyelitis is a particularly feared disease because it causes disabling paralysis, particularly in children and adolescents. Figure 16.1 shows an ancient Egyptian stele that is believed to show an early case of poliomyelitis. Franklin D. Roosevelt, President of the United States of America from 1933 to 1945, used a wheelchair because of paralysis caused by poliomyelitis, although many people were unaware of the extent of his disability. His administration helped support the "March of Dimes," an organization that funded successful efforts to develop vaccines against poliomyelitis and other childhood diseases. Table 16.1 shows the current genera in the *Picornaviridae* family and some representative members in each genus.

FIGURE 16.1 Egyptian stele (Astarte Temple, Memphis, 1580–1350 B.C.). Man in center with paralysis of right leg, perhaps from poliomyelitis.

Poliovirus: A model picornavirus for vaccine development and studies of replication

John Enders and co-workers showed in the late 1940s that poliovirus and other human viruses could be grown in cells cultured *in vitro*. This led the way to biochemical and genetic analyses allowing detailed understanding of picornavirus gene expression. It also provided the basis for the generation of poliovirus vaccines during the 1950s. The formalin-inactivated vaccine developed by Jonas Salk and the live, attenuated vaccine developed by Albert Sabin are two of the most successful vaccines ever produced (see Chapter 34). Poliomyelitis is now virtually absent from the developed world thanks to the Salk and Sabin vaccines, and the World Health Organization is using these vaccines in a far-reaching plan to eradicate poliovirus worldwide.

Because poliovirus had been so intensively studied for vaccine development, it and several related picornaviruses became model systems during the 1960s for understanding how RNA viruses replicate in eukaryotic cells. David Baltimore and Richard Franklin were the first to demonstrate the presence of an RNA-dependent RNA polymerase in virus-infected cells in the early 1960s. This showed that viral RNAs could replicate directly rather than being transcribed from a DNA tem-

TABLE 16.1 Selected members of the picornavirus family

Genus	Members	Characteristics
Enterovirus	Poliovirus Coxsackie virus (A, B) Human echovirus	Replicate in gastroenteric tract; can cause meningitis, encephalitis, rashes, cardiac and muscular disease
Rhinovirus	Human rhinovirus	Over 100 serotyptes; cause common cold; sensitive to low pH
Cardiovirus	Encephalomyocarditis virus Theiler's murine encephalomyelitis virus	Infect mice and other mammals; cause cardiac or neurologic disease
Aphthovirus	Foot-and-mouth-disease virus	Epidemics in domestic ruminants; sensitive to low pH
Hepatovirus	Hepatitis A virus	Causes acute hepatitis in humans

Note: There are presently four additional picornavirus genera that are not listed here.

plate. Donald Summers and Jacob Maizel found that picornavirus proteins are synthesized as large polyproteins, which are subsequently cleaved into the various structural and enzymatic viral proteins. This foreshadowed discoveries that numerous other animal viruses also make polyproteins that are cleaved by viral proteinases, now prime targets for antiviral drugs. More recently, independent work from the laboratories of Nahum Sonenberg and Eckard Wimmer showed that initiation of protein synthesis is cap-independent and occurs at **internal ribosome entry sites** on picornavirus RNAs. This is in contrast to most cellular mRNAs, on which ribosomes recognize the capped 5′ end of the RNA and begin protein synthesis at the first AUG codon encountered.

Picornavirus virions bind to cellular receptors via depressions or loop regions on their surface

Picornaviruses have a nonenveloped, icosahedral capsid that encloses a single-stranded, positive-sense genome RNA. The three-dimensional structures of the virions of many different picornaviruses have been determined by x-ray crystallography (Figure 16.2). Mature virions contain 60 copies of each of four proteins: VP1, VP2, VP3, and VP4. Although very different in primary amino acid sequence, VP1, VP2, and VP3 all fold into a common structure that resembles a wedge shape, or a **jelly roll**, comprised of an eight-stranded **β-barrel** structure (see Figure 2.2e). These three proteins combine to form the capsid shell of picornaviruses, and VP4 remains buried beneath the shell. Similar folds are shared by capsid proteins of other animal and plant viruses, suggesting a common functional motif required for subunit packing and assembly of icosahedral virus particles.

Table 16.2 lists some cell surface receptor molecules used by various picornaviruses to bind to target cells. Some picornaviruses have depressions (called "canyons" or "pits") on the capsid surface that are involved in binding to receptors. This is unusual; most viruses bind to cellular receptors via proteins that protrude from the surface of the virion. Other picornaviruses do not have these pronounced surface depressions but rely on surface-exposed loop regions of the capsid proteins for recognition of cellular receptors.

Genome RNA may pass through pores formed in cell membranes by capsid proteins

Rhinoviruses and foot-and-mouth disease virus have been shown to enter the cell by **receptor-mediated endocytosis** (see Chapter 4). Acidification within the en-

FIGURE 16.2 Picornavirus virion structure. Model of poliovirus virion determined by x-ray crystallography, showing a 5-fold symmetry axis.

TABLE 16.2 Cell receptors and co-receptors used by some picornaviruses

Virus	Receptor or co-receptor	Characteristics
Poliovirus	Poliovirus receptor (Pvr or CD155)	Immunoglobulin superfamily
Coxsackievirus	Decay accelerating factor (CD55)	Complement cascade
	Vitronectin receptor ($\alpha_v\beta_3$)	Integrin
	Coxsackievirus-adenovirus receptor (CAR)	Immunoglobulin superfamily
	Intercellular adhesion molecule-1 (Icam-1)	Immunoglobulin superfamily
	$\alpha_v\beta_6$	Integrin
Human rhinoviruses	Icam-1	Immunoglobulin superfamily
	Low-density lipoprotein receptor	
Foot-and-mouth disease virus	Heparan sulfate	Glycosaminoglycan
	Vitronectin receptor ($\alpha_v\beta_3$)	Integrin
Encephalomyocarditis virus	Vcam-1	Immunoglobulin superfamily
	Sialylated glycophorin A	Carbohydrate

dosomal vesicle causes a conformational change in the capsid, resulting in the dissociation of capsid subunits and release of the viral RNA. Exactly how the RNA subsequently passes from the vesicle into the cytoplasm is not known.

Other picornaviruses, including poliovirus, undergo a major conformational change directly at the cell surface upon binding to their receptor. Virions lose the small internal protein VP4, and the hydrophobic N-terminus of VP1 is extruded to the surface of the particle. This is reminiscent of the conformational changes that take place in fusion proteins of enveloped viruses like influenza and HIV-1 upon receptor binding (Chapter 4). The altered picornavirus virions may participate in the formation of a pore within the plasma membrane, allowing the genome RNA to pass directly from the virion through the membrane into the cytoplasm (Figure 16.3). Pore formation has not been directly demonstrated, but this is an attractive hypothesis to explain entry of viral RNA from altered virions at the cell surface.

Translation initiates on picornavirus RNAs by a novel internal ribosome entry mechanism

Once in the cytoplasm, the genome RNA must be translated by cellular ribosomes to produce viral proteins. Purified genome RNA can be used to infect cells, showing that none of the structural proteins present in the virion is needed to initiate the replication cycle once the RNA has entered the cell. How picornavirus genome RNAs engineer their translation and how the virus eventually takes over the cell's translation machinery is a fascinating story.

The positive-sense genome RNAs of picornaviruses range from 7 to 8.5 Kb in length (Figure 16.4). A small

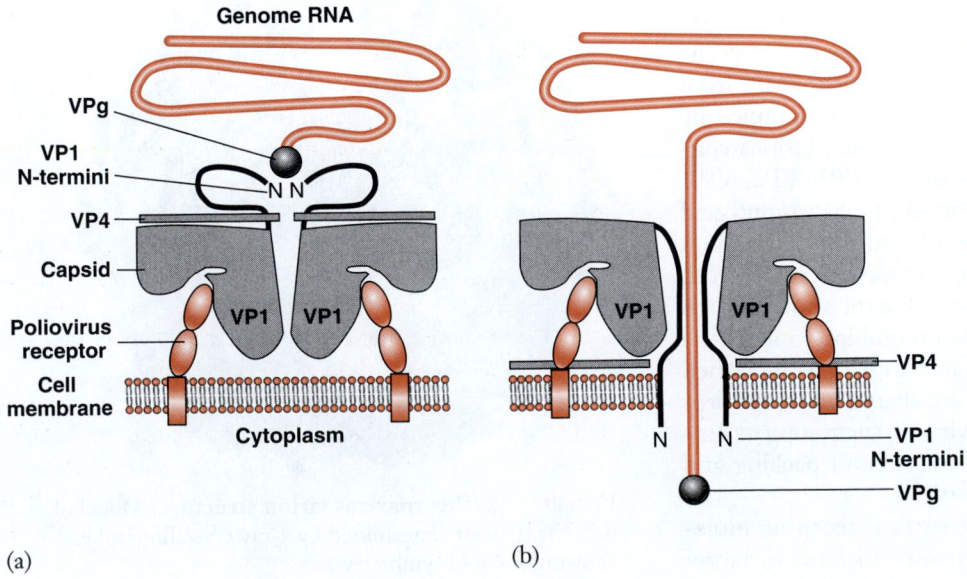

FIGURE 16.3 Model for entry of poliovirus RNA into the cytoplasm. (a) Shown is a cross-section of one center of 5-fold symmetry on a virion docked at the cell surface by binding to the poliovirus receptor. VP4, the N-termini of VP1, and the genome RNA bound at its 5′ end to VPg are all located in the interior of the capsid (above). (b) Binding leads to a major rearrangement during which VP4 is externalized and the hydrophobic N-termini of the five VP1 molecules (only two shown here) insert into the membrane to form a channel through which the viral RNA can pass into the cytoplasm.

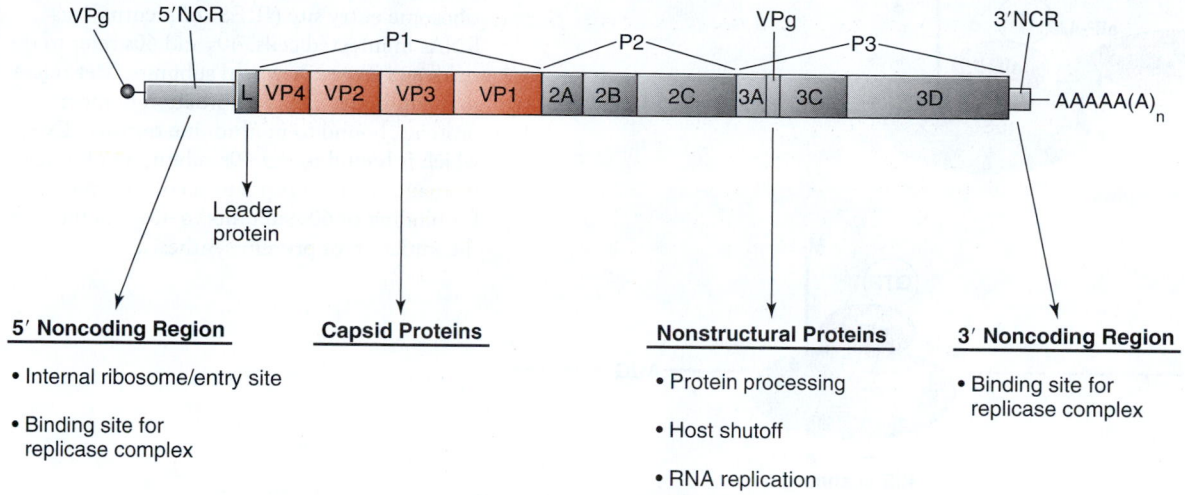

FIGURE 16.4 Picornavirus genome organization. Shown are the major landmarks on the positive-strand RNA genome of picornaviruses. The single open reading frame is translated into a large polyprotein that is cleaved by viral proteinases into intermediates P1–P3 and finally into the capsid and replication proteins shown. VPg: viral protein bound to 5′ end of RNA.

(22 to 24 amino acid) virus-encoded protein called VPg is covalently attached to the 5′ end of genome RNA, and a poly(A) tract of 30 to 100 nucleotides (depending on the virus strain) is present at the 3′ end. Unlike most cellular mRNAs, the poly(A) tract of picornaviruses is part of the genome sequence and is not added by a separate poly(A) polymerase. Picornaviruses have an unusually long (600–1200 nt) stretch of 5′ noncoding sequences upstream of the AUG codon at which protein synthesis begins. These noncoding sequences contain extensive secondary and tertiary structure (see below). Although multiple AUG codons are located between the 5′ end and the authentic initiator AUG, these are not used by ribosomes to initiate protein synthesis as would be expected if ribosomes were "scanning" this region searching for appropriate initiation sites.

Most cellular mRNAs employ cap-dependent mechanisms to initiate translation near their 5′ end. In this mode (Figure 16.5a), the eIF-4F complex, consisting of eukaryotic initiation factors eIF-4A, eIF-4E, and eIF-4G, recruits a 40S ribosomal subunit to the capped 5′ end of an mRNA. The 40S subunit scans the mRNA, searching for an AUG codon in an appropriate sequence context. Then a 60S ribosomal subunit joins to form an 80S complex, and protein synthesis begins.

Following infection by a picornavirus (Figure 16.5b), cap-dependent translation in the host cell is abolished by the proteolytic cleavage of eIF-4G or by sequestration of the cap-binding protein eIF-4E. Because picornavirus RNAs are not capped, a novel mechanism is used to direct ribosomes to the initiator AUG codon. All picornaviruses initiate protein synthesis via an internal ribosome entry site (IRES) that lies within the highly structured 5′ noncoding region. Host cell proteins bind to the IRES and help to dock the 40S ribosomal subunit onto the RNA. In this mode, the 40S subunit bypasses multiple upstream AUG sequences, moving by translocation or scanning from its site of initial contact with the IRES to reach the initiator AUG. The dynamics of this movement are not understood, but the ultimate outcome is the assembly of 80S ribosomes at the initiator AUG and the start of polypeptide chain synthesis. IRES sequences have also been found in other virus mRNAs and in a number of cellular mRNAs.

Essential features of picornavirus IRES elements

Secondary structure elements in RNA (stem-loops) can be detected both by computer prediction from the primary sequence and by treating RNAs *in vitro* with structure-specific chemical and enzymatic probes. These procedures indicate the presence of six stem-loop structures (I–VI) in enterovirus and rhinovirus 5′ noncoding regions, and 12 such structures (A–L) in cardiovirus and aphthovirus 5′ noncoding regions (Figure 16.6). These two types of structure have therefore been named type I and type II internal ribosome entry sites, respectively. The 5′ noncoding region of hepatitis A virus RNA shares little primary sequence or structure homology with that of other picornaviruses, and this virus has been suggested to have a Type III IRES element.

All picornavirus IRES elements contain a pyrimidine-rich tract 20–25 nucleotides from a conserved AUG codon. The distance between the pyrimidine-rich tract and the conserved AUG appears to be critical; if it is much smaller or larger than 20–25 nt, initiation of translation is reduced. For viruses with a type II IRES ele-

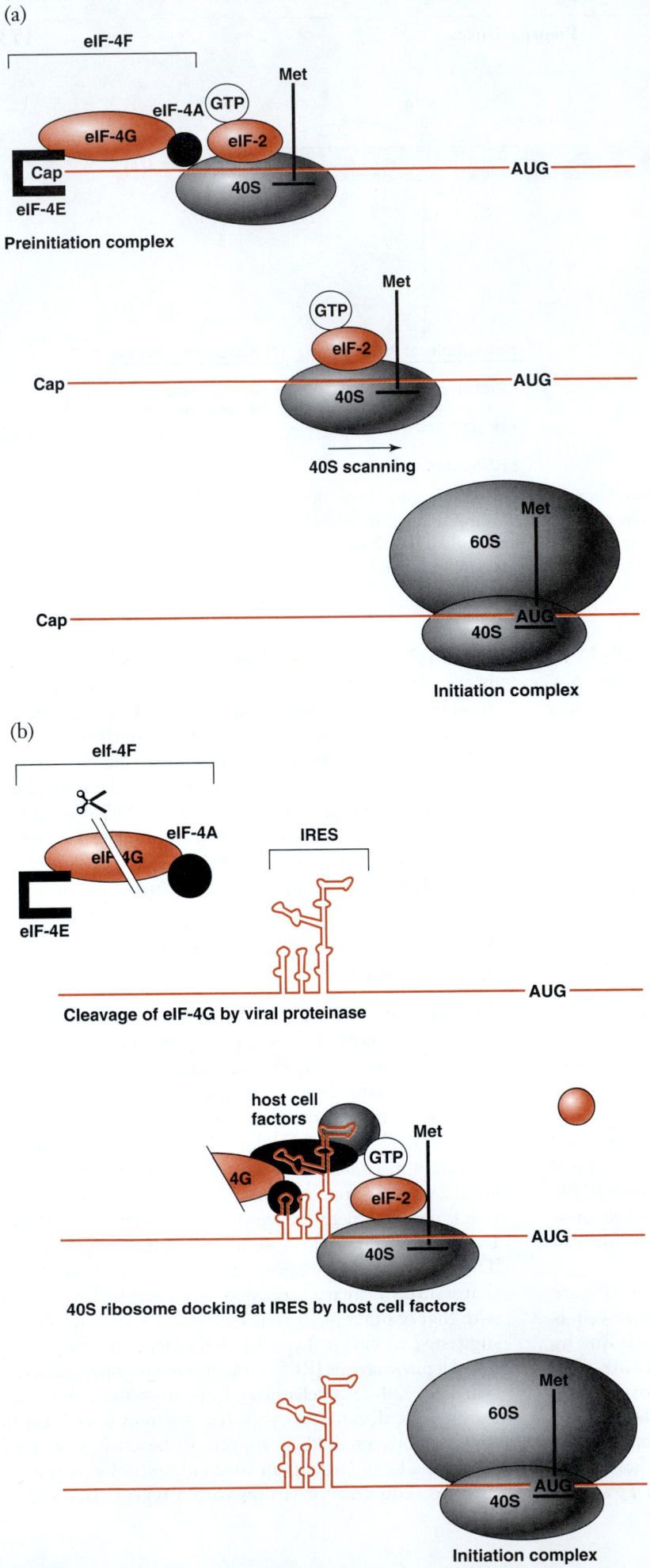

FIGURE 16.5 Initiation of protein synthesis. (a) Cap-dependent initiation directed by EIF-4F and other initiation factors (not all shown). (b) Cap-independent initiation at the internal ribosome entry site (IRES) of picornavirus RNAs in infected cells. 40s and 60s refer to the small and large ribosomal subunits. Met refers to methionine, used for initiation of most proteins, bound to methionine transfer RNA, which is bound to the 40s submit. GTP refers to guanosine triphosphate, a cofactor required for binding of 60s subunits to 40s subunits and the initiation of protein synthesis.

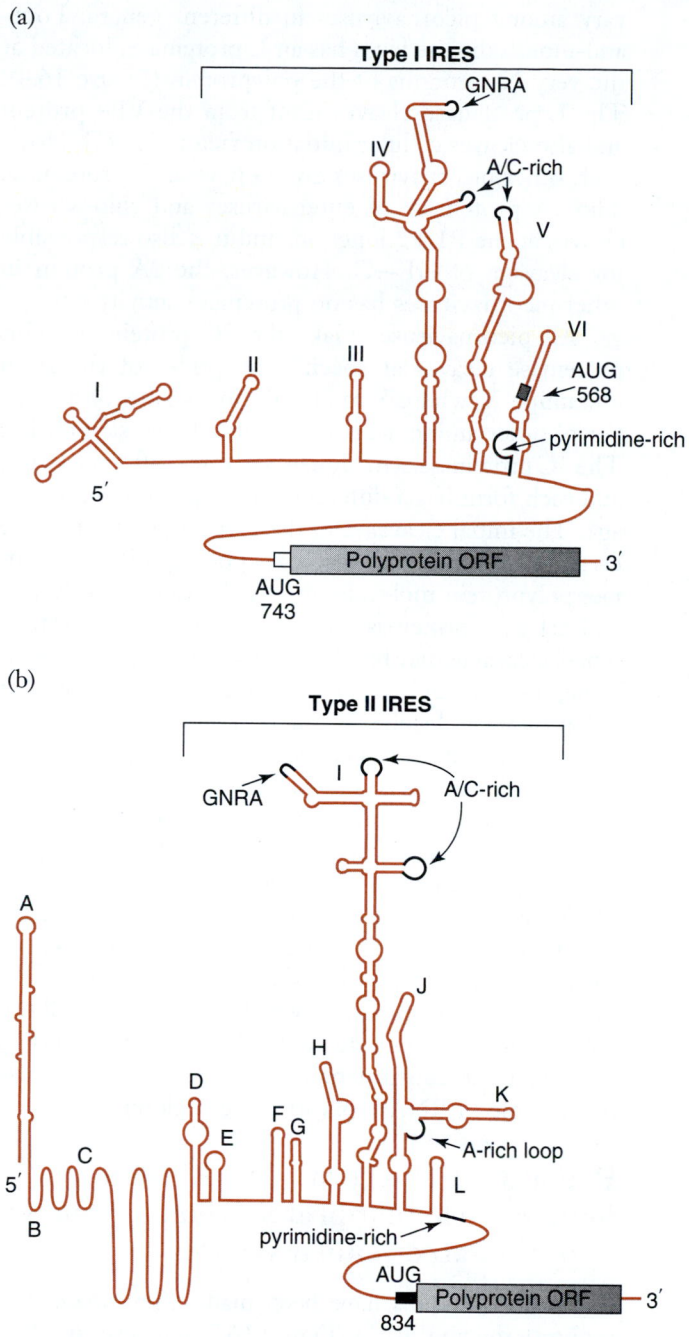

FIGURE 16.6 Secondary structures of internal ribosome entry sites. (a) Type I IRES, found in genome RNAs of enteroviruses and rhinoviruses. (b) Type II IRES, found in genome RNAs of cardioviruses, aphthoviruses, and hepatoviruses. The positions of conserved sequence elements are shown in black. The polyprotein open reading frame is shown at reduced scale.

ment, the conserved AUG is used for initiation of translation. For viruses with a type I element, translation is initiated at an AUG located 30–150 nucleotides downstream from the conserved AUG. Additional conserved features in picornavirus IRES elements include a four-nt GNRA sequence (where N = any nucleotide and R = A or G), and (A + C)-rich sequences in specific loop regions of the predicted stem-loop structures. The roles of these conserved elements have not been precisely defined, but they may provide important RNA structure and sequence information required for binding of cellular proteins and the 40S ribosomal subunit.

Interaction of picornavirus IRES elements with host cell proteins

Because no viral proteins other than VPg accompany picornavirus RNA upon entry into a cell, the factors that direct translation at an IRES must be preexisting cellular proteins. Cellular proteins known to interact with picornavirus IRES elements include the La autoantigen, polypyrimidine tract binding protein, poly (rC) binding protein 2, and the product of the unr gene. The precise roles of these proteins in IRES function are not known. Cap-independent translation also utilizes some of the same eukaryotic initiation factors that are involved in cap-dependent translation. Eukaryotic initiation factors eIF-2α and eIF-4B are known to interact with picornavirus IRES elements, and eIF-2, eIF-4E, and eIF-4F all stimulate cap-independent translation. Binding of eIF-4F to an IRES is specified by the central portion of its eIF-4G subunit, replacing the function of eIF-4E, which normally recognizes the cap and directs binding of this complex to capped mRNAs.

Picornavirus proteins are made as a single precursor polyprotein that is autocatalytically cleaved by viral proteinases

Many bacterial messenger RNAs contain several different protein coding regions and can initiate protein synthesis at various internal sites. For example, the genome RNAs of bacteriophages such as Qβ and MS2 express their four viral proteins from three or four different initiation sites, and control of RNA phage gene expression is largely carried out at the level of initiation of protein synthesis (see Chapter 5). In contrast, most eukaryotic messenger RNAs code for a single protein. This poses a problem for positive-strand RNA viruses such as the picornaviruses: How can they express their numerous genes from a single RNA molecule that serves both as a genome and as a messenger RNA, in a cellular environment that favors mRNAs that code for only one protein?

The picornaviruses have solved this problem by linking all of their protein-coding regions into a single unit, and synthesizing a single, large polyprotein (Figure 16.4). This polyprotein is subsequently cleaved into distinct functional viral proteins by proteinases that are themselves part of the polyprotein; presumably, these proteinases fold into enzymatically active structures while still part of the polyprotein. The initial cleavages are rapid, so that little if any full-length polyprotein can be found in an infected cell, unless amino acid analogues that inhibit proper protein folding are incorporated during synthesis. The polyprotein is first cleaved into three fragments, called P1, P2, and P3 (Figure 16.4). P1 is subsequently cleaved into capsid proteins VP0, VP1, and VP3; VP0 is cleaved into VP2 and VP4 during virion maturation. P2 and P3 are also cleaved into a number of smaller proteins whose functions are described below.

Picornaviruses make a variety of proteinases that cleave the polyprotein and some cellular proteins

Picornaviruses encode three distinct proteinases: the L, the 2A, and the 3C proteinases. The cleavage pathways vary among picornaviruses in different genera. Foot-and-mouth disease virus has an L proteinase, located at the very N-terminus of the polyprotein (Figure 16.4). The L proteinase cleaves itself from the VP4 protein and also cleaves cellular initiation factor eIF-4G. However, other picornaviruses do not have an L proteinase. The 2A proteinase of enteroviruses and rhinoviruses cleaves at the P1/P2 junction, and it is also responsible for cleavage of eIF-4G. However, the 2A protein in other picornaviruses has no proteinase activity.

All picornaviruses make the 3C proteinase. This proteinase cleaves at specific dipeptides of the form glutamine-X, where X can be glycine, serine, or a number of other amino acids with hydrophobic side chains. The 3C proteinase actually exists in three different forms, and each form has a different role in polyprotein cleavage. The initial cleavages of the nascent polyprotein are carried out by the uncleaved 3C polypeptide in the intact polyprotein molecule; that is, 3C can properly fold and act as a proteinase while part of the polyprotein. These cleavages may be within the same polyprotein molecule, or one 3C region can cleave within another polyprotein molecule. 3C then cleaves at its own N-terminus within P3 to generate 3AB and 3CD.

For rhinoviruses and enteroviruses, further processing of the P1 region is carried out primarily by 3CD. The molecular chaperone Hsp70, a cellular protein, has been implicated in P1 processing; it is possible that Hsp70 assists in the interaction between 3CD and P1. Alternatively, Hsp70 could help fold P1 so that its cleavage sites are recognized by 3CD. Some of 3CD is eventually cleaved to form 3C, which, along with 3CD, is able to carry out the remaining cleavages of P2 and P3 to form their constituent mature proteins. 3CD also plays a role in RNA replication (see below).

Replication of picornavirus RNAs is initiated in a multiprotein complex bound to proliferated cellular vesicles

Once viral proteins have been made, they proceed to replicate the viral RNA (Figure 16.7). Genetic and biochemical studies show that most of the proteins made from the P2 and P3 regions are involved in picornavirus RNA replication. Replication starts with the synthesis of a full-length negative-strand RNA, using the initial positive-strand RNA as a template. This negative-strand RNA is used as a template for the synthesis of additional positive-strand RNAs via multistranded replicative intermediate structures (diagrammed in Figure 16.7). These multistranded structures result from highly efficient reinitiation of synthesis of positive strands on the template RNA, before the synthesis of previously initiated positive strands is completed. The number of positive strands synthesized in infected cells exceeds the number of negative strands by a ratio of 40 to 1.

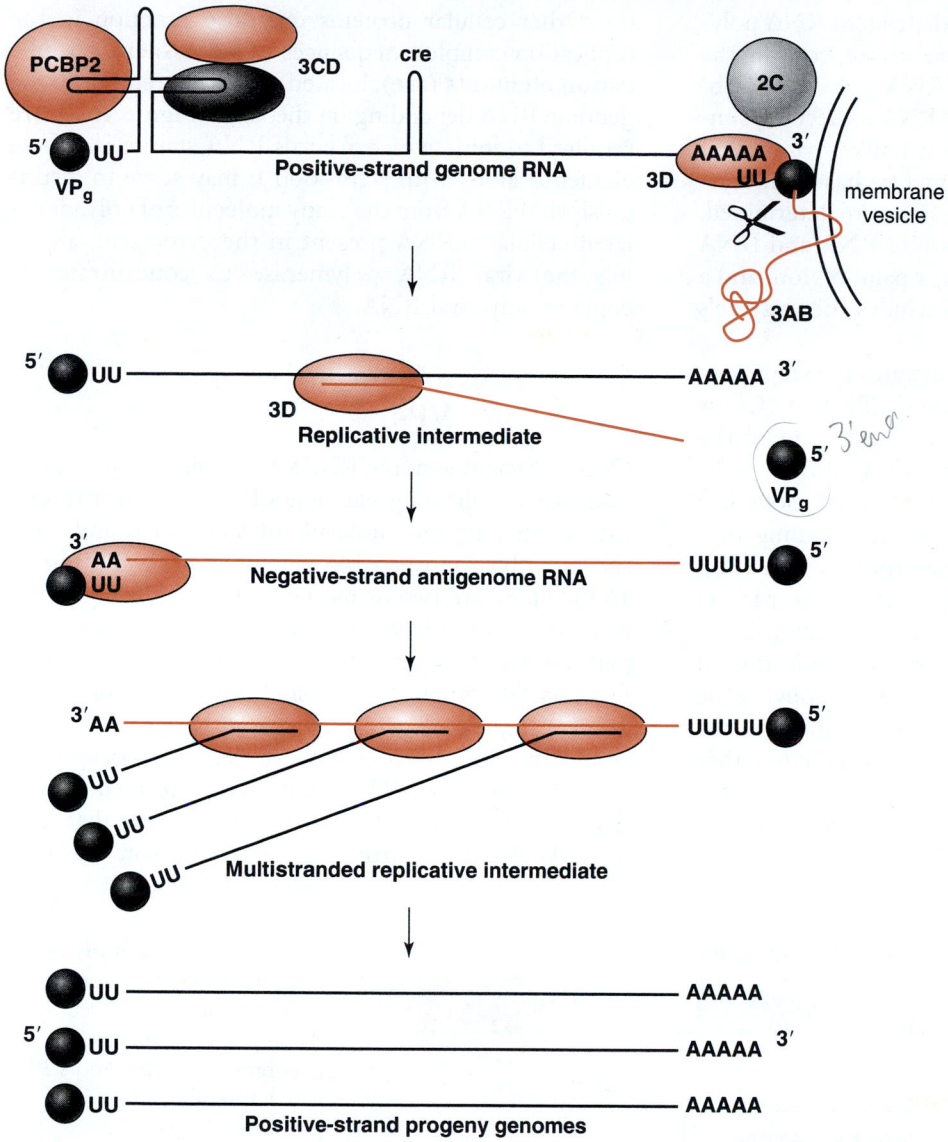

FIGURE 16.7 Proposed mechanism for replication of picornavirus RNA. Top: genome RNA is shown with sequence elements believed to participate in formation of replication complex (not drawn to scale). Cre = *cis*-acting replication element. Viral proteins VPg (black circle), 3AB (orange line plus black circle), 3CD, 3D, and 2C are shown bound to RNA or cytoplasmic vesicles. Poly C binding protein 2 (PCBP2) is one of several cellular proteins believed to be involved in viral RNA replication. VPg (protein 3B) linked to U residues acts as a primer for the 3D RNA polymerase, and is rapidly cleaved from protein 3A (scissors). Positive-strand RNA is shown in black, negative-strand RNA in orange. Although shown in the figure as free, single-stranded RNA, the negative-strand antigenome RNA may also exist in a double-stranded form, hybridized to positive-strand viral RNA.

Viral RNA is synthesized in association with membrane vesicles that are induced during virus infection. These vesicles are required for picornavirus RNA synthesis and may serve as a nucleation site for the formation of replication complexes. Association of viral RNA synthesis with membranous vesicles in the cytoplasm is a common theme encountered with a variety of positive-strand RNA viruses (see also Chapters 17–19); these vesicles help to localize all the elements required for efficient viral RNA replication.

Picornavirus proteins 2B, 2C, and 3A participate in the induction of these vesicles. Proteins 2B and 3A both mediate the inhibition of cellular protein secretion, perhaps by blocking membrane trafficking from the endoplasmic reticulum through the Golgi apparatus to the plasma membrane. This could contribute to the buildup of vesicles in infected cells. In addition to inducing vesicle formation, 2C is an RNA-binding protein implicated specifically in the initiation of negative-strand RNA synthesis. It may be involved in bringing genome RNAs in contact with the vesicles.

RNA synthesis is primed by VPg covalently bound to uridine residues

The proteins that actually direct viral RNA synthesis are derived from P3. Protein 3AB is anchored to the proliferated cytoplasmic vesicles via a hydrophobic region in the 3A portion (Figure 16.7, first step). The C-terminal part of 3AB, protein 3B, is identical to VPg, a protein that is covalently attached to the 5′ end of genome and antigenome viral RNAs. VPg becomes covalently attached to uridine residues via a tyrosine side chain near its N-terminus. These protein-bound uridine residues hybridize to the poly(A) tail at the 3′ end of viral RNA, and serve as a primer to initiate RNA synthesis.

The 3D protein is the RNA-dependent RNA polymerase that adds nucleotides to the primer, copying the template RNA strand. The 3D RNA polymerase absolutely requires a primer to copy RNA molecules, similar to most DNA polymerases but unlike many other RNA polymerases. The 3D polymerase has been crystallized and its three-dimensional structure determined. It has the familiar "hand" structure of RNA and DNA polymerases, with a finger region, a palm region, and a thumb region; the palm region includes the enzyme's active site.

The model in Figure 16.7 illustrates the steps postulated to lead to viral RNA synthesis. Protein 3CD is believed to bind to stem-loop I at the 5′ end of the positive-strand RNA, bringing the RNA into the replication complex. The binding of 3CD to this cloverleaf stem-loop structure is facilitated by the binding of a cellular protein, poly r(C) binding protein 2 (PCBP2). PCBP2 was previously mentioned as taking part in recognition of the IRES by 40S ribosomes, and therefore it has a role in both translation and replication of viral RNA. Since the 3D polymerase begins copying the positive-strand RNA at its 3′ end, it is possible that the 5′ and 3′ ends of the genome RNA are brought together at this point. Protein 2C is also believed to bind to the RNA and bring it to the vesicles where RNA synthesis takes place.

Other cellular proteins may also function in the replication complex. Sequences called ***cis*-acting replication elements (cre)**, located in various regions of the genome RNA depending on the virus (Figure 16.7), are required to initiate picornavirus RNA synthesis. These elements, along with stem-loop I, may serve to distinguish viral RNA from the many molecules of polyadenylated cellular mRNA present in the cytoplasm, allowing the viral RNA polymerase to concentrate on copying only viral RNA.

Virion assembly involves cleavage of VP0 to VP2 plus VP4

Once sufficient genome RNAs are made, virions can be assembled. Following cleavage of P1, five **protomers**, each containing one molecule of VP0, VP1, and VP3, self-assemble to form a 14S pentamer structure (Figure 16.8). There are two proposed pathways leading to the production of infectious picornavirus particles. One pathway has 12 of the pentamers coming together to form an 80S empty "procapsid" (lower part of Figure 16.8). A single molecule of newly synthesized viral RNA would then be threaded into the empty procapsid to form a "provirion." This short-lived intermediate undergoes one final proteolytic cleavage event that produces VP4 and VP2 from the VP0 precursor, thus pro-

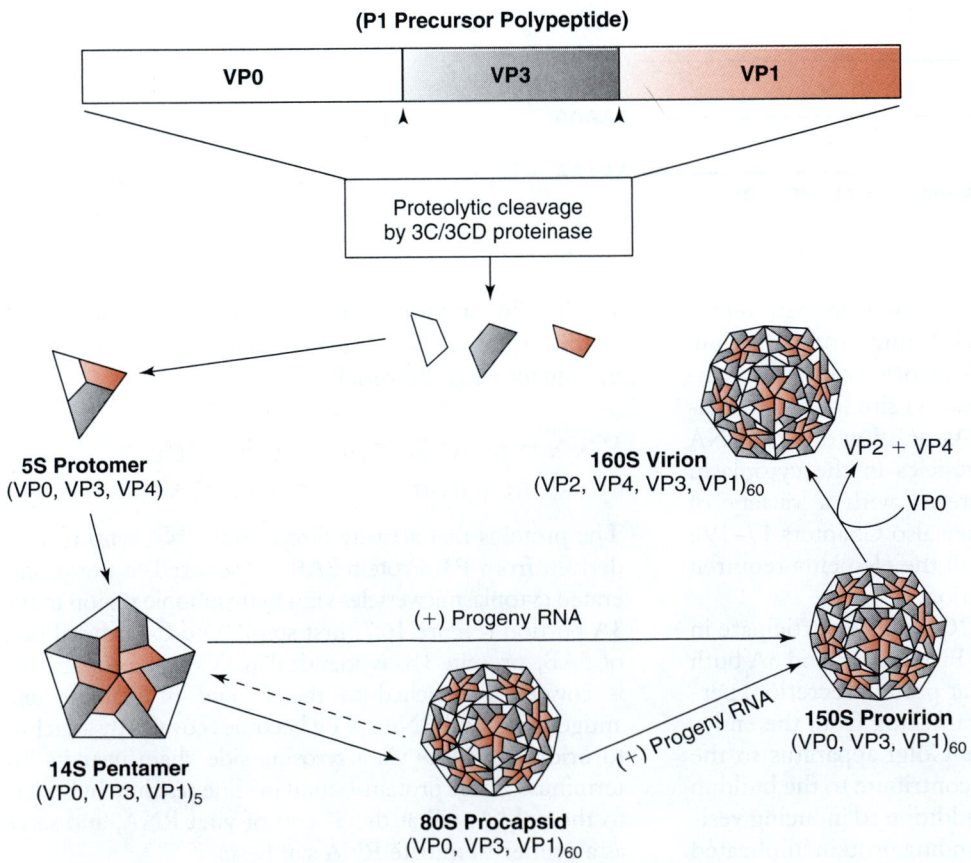

FIGURE 16.8 Assembly of the picornavirus virion. The P1 polypeptide is cleaved by the 3C/3CD proteinase to generate VP0, VP3, and VP1, which assemble into a 5S protomer. Five of these protomers then assemble into a 14S pentamer. Capsids are formed by the assembly of 12 such pentamers into a closed shell by one of two pathways; they may assemble directly around viral RNA (upper pathway) or they may form an empty 80S procapsid, which then incorporates viral RNA (lower pathway). VP0 subsequently is cleaved into VP2 and VP4, giving rise to the mature virion.

Picornavirus pathogenesis and disease

Picornaviruses cause a number of important and well-studied diseases in humans and animals. Most people infected with poliovirus have few or no symptoms and the infection is limited to the enteric tract, but a small percentage suffer from muscle paralysis and brain stem involvement. A major determinant of neurovirulence for all three poliovirus serotypes is found in the stem-loop V region of the IRES (Figure 16.6). Poliovirus RNA containing a single point mutation in stem-loop V, present in attenuated Sabin vaccine strains, shows pronounced translation defects in neuroblastoma cells compared to the wildtype RNA. These differences in protein synthesis are not observed in nonneuronal cells, indicating that a translation factor in neurons may have a defective interaction with the mutated RNA. This translation defect may therefore be a major cause of the inability of the Sabin vaccine strains to cause paralytic diesase, while replicating well in intestinal cells and therefore inducing a strong immune response.

Cell-free translation of hepatitis A virus RNA is stimulated by the addition of mouse liver extract but not by other tissue extracts, indicating that specific liver cell translation factors may contribute to the preference of this virus for liver cells. Coxsackie B virus can infect heart muscle cells, causing severe and sometimes fatal damage in humans. In animal models of myocarditis, RNA sequences in the 5′ noncoding regions of coxsackie B virus have been shown to be involved in virulence, again implicating specific cellular factors in translation of viral RNA. Theiler's murine encephalomyelitis virus causes a neurologic disease in mice that ranges from paralytic polio-like symptoms to a persistent infection accompanied by chronic demyelination, depending on the strain of virus. Unlike the other picornaviruses discussed here, important disease determinants for this virus appear to reside in the viral capsid proteins.

Knowledge of the cellular factors involved in initiation of translation of viral RNA and of RNA synthesis, as well as the roles of viral proteins such as the RNA-dependent RNA polymerase and the 3C and 2A proteinases, may help in the development of novel antiviral agents that can target these critical steps in virus replication and reduce or eliminate the most serious disease symptoms.

ducing the fully infectious virion. One problem with this pathway is that the threading of genome RNA into the empty capsid would require a complex mechanism that is not known to exist for other RNA viruses.

The alternative assembly pathway proposes that the 14S pentamers first associate with viral genome RNA and subsequently combine with other 14S pentamers to form the provirion directly. Whether the viral RNA is threaded into an empty capsid or is bound by 14S pentamers prior to particle assembly, only positive-strand viral RNAs with a VPg protein linked to the 5′ end are packaged.

The nature of the proteinase that carries out the final (maturation) cleavage of VP0 is unknown. This cleavage is not carried out by the known viral proteinases (L, 2A, 3C, or 3CD) and, because the cleavage site is found on the inside of procapsids and provirions and would not therefore be accessible to proteolytic enzymes, there has been speculation that viral RNA or the capsid proteins themselves participate in the cleavage reaction.

Inhibition of host cell macromolecular functions

Infection by nearly all picornaviruses leads to a number of dramatic effects on host cell macromolecular functions. These include shutoff of cap-dependent translation, shutoff of host cell RNA synthesis, induction of cytoplasmic vesicles, and alteration of intracellular transport pathways between the endoplasmic reticulum and Golgi apparatus.

As noted above, some picornaviruses encode a proteinase (protein 2A for the enteroviruses and rhinoviruses or the L protein for aphthoviruses) that cleaves eIF-4G, one of the essential protein components required for recognition of the 5′ end of capped cellular mRNAs. Infected cells are therefore unable to translate most cellular mRNAs, allowing the cap-independent translation of viral RNAs to take over the cellular protein synthesis machinery. Other picornaviruses inhibit cap-dependent translation by binding and sequestration of eIF-4E, the protein subunit of the eIF-4F complex that binds directly to the 5′ cap structure of eukaryotic mRNAs.

Infection of cultured mammalian cells by most picornaviruses causes a shutoff of host cell RNA synthesis. Experiments performed with extracts from poliovirus-infected HeLa cells showed that RNA synthesis by all three classes of mammalian RNA polymerases (pol I, II, and III) is inhibited. These defects have been traced to the proteolytic cleavage of specific subunits of the different RNA polymerases by the 3C viral proteinase. The foot-and-mouth disease virus 3C proteinase also cleaves histone H3, a cellular protein associated with transcriptionally active chromatin in the cell nucleus.

Collectively, these effects on host cell functions ensure that the infected cell devotes its metabolic activities primarily to the expression and replication of picornavirus RNAs. It is notable that hepatitis A virus, a picornavirus with a very slow replication cycle, does not inhibit host cell protein or RNA synthesis and, for the most part, does not produce observable cytopathic effects during infection of cells in culture.

SUGGESTED READING

1. Bedard, K.M. and Semler, B.L. (2004). Regulation of picornavirus gene expression. *Microbes and Infection* **6**, 702–713.

2. Hansen, J.L., Long, A.M., and Schultz, S.C. (1997). Structure of the RNA-dependent RNA polymerase of poliovirus. *Structure* **5,** 1109–1122.

3. Jang, S.K., Krausslich, H.G., Nicklin, M.J., Duke, G.M., Palmenberg, A.C., and Wimmer, E. (1988). A segment of the 5′ nontranslated region of encephalomyocarditis virus RNA directs internal entry of ribosomes during *in vitro* translation. *Journal of Virology.* **62,** 2636–2643.

4. Molla, A., Paul, A.V., and Wimmer, E. (1991). Cell-free, de novo synthesis of poliovirus. *Science* **254,** 1647–1651.

5. Palmenberg, A.C. (1990). Proteolytic processing of picornaviral polyprotein. *Annual Reviews of Microbiology* **44,** 603–623.

6. Pelletier, J. and Sonenberg, N. (1988). Internal initiation of translation of eukaryotic mRNA directed by a sequence derived from poliovirus RNA. *Nature* **334,** 320–325.

7. Stewart, S.R. and Semler, B.L. (1997). RNA determinants of picornavirus cap-independent translation initiation. *Seminars in Virology* **8,** 242–255.

KEY TERMS

cis-acting replication elements (cre)
Hepatitis
Internal ribosome entry site (IRES)
Jelly roll β-barrel
Myocarditis
Poliomyelitis
Protomers
Receptor-mediated endocytosis

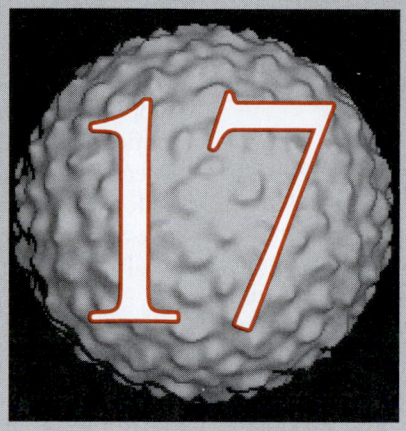

Flaviviruses

Flaviviridae

From Latin *flavus* (yellow), referring to jaundice caused by yellow fever

VIRION

Spherical enveloped particle.
No projections, "golf ball" appearance.
Diameter 50 nm.
Spherical nucleocapsid (25–30 nm diameter) with icosahedral symmetry (T = 3).
Envelope glycoproteins also arranged with icosahedral symmetry.

GENOME

Linear ss RNA, positive sense.
10–12.3 Kb.
Capped at 5′ end but no poly(A) tail at 3′ end.

GENES AND PROTEINS

All genes translated as single polyprotein followed by proteolytic cleavage.
One capsid protein (C).
Two envelope proteins (M and E).
Seven nonstructural proteins.

VIRUSES AND HOSTS

Three genera:

- *Flavivirus*: infect humans, monkeys, bats, birds, etc.
- *Pestivirus*, from Latin *pestis* (plague): infect pigs, cattle, other ruminants.
- *Hepacivirus*, from Greek *hepatos* (liver): hepatitis C virus, in humans only.

DISEASES

Many are associated with human disease: yellow fever, dengue hemorrhagic fever, tick-borne encephalitis, West Nile virus, hepatitis C virus.
Also affect domestic or wild animals and cause economically important diseases: swine fever, bovine viral diarrhea.

DISTINCTIVE CHARACTERISTICS

Many flaviviruses (but not hepatitis C virus) are transmitted from host to host by mosquitoes or ticks.
The major envelope protein of flaviviruses lies parallel to lipid membrane.
Viral proteins are synthesized as a single polyprotein that is cleaved by viral and cell proteases.
Pestiviruses and hepaciviruses have an internal ribosome entry site (IRES) for initiation of protein synthesis.
Virions are assembled by budding at endoplasmic reticulum and are released by exocytosis.

Flaviviruses cause several important human diseases

Flaviviruses have had an important impact on humans over the last several hundred years. This has been especially true in the Americas, which suffered dramatic outbreaks of yellow fever, transmitted by mosquitoes, in the 1700s and 1800s. Dengue virus and Japanese encephalitis virus, also transmitted by mosquitoes, cause serious human disease throughout tropical regions of the world. Hepatitis C virus, unknown prior to 1989, has infected 4 million people in the United States alone and is responsible not only for acute and chronic **hepatitis**, but also for **hepatocellular carcinoma**. West Nile virus, originally discovered in Uganda and prevalent in Africa, southern Europe, and the Middle East, suddenly appeared in North America in 1999. Although not a major public health threat, infection of humans with this mosquito-borne virus can lead to neurological effects and is occasionally fatal. West Nile virus caused a mini-epidemic as it spread from the East coast to the West coast of North America over a five-year period via infected birds. Although certain flaviviruses have been brought under control through vaccination and vector management, a large number remain significant threats to human health.

The *Flaviviridae* family presently contains three genera. The *Flavivirus* genus includes at least 73 distinct viruses, including many human pathogens, that are transmitted by bloodsucking arthropods (Table 17.1). Flaviviruses were previously grouped with the togaviruses (Chapter 18), as they are both small enveloped viruses, and many are transmitted by arthopod vectors, within which they also replicate. However, the gene organization and replication strategies of flaviviruses and togaviruses are quite distinct. Members of the *Pestivirus* genus cause economically important disease in cattle, pigs, and sheep. The *Hepacivirus* genus contains a single member, hepatitis C virus. Pestiviruses and hepatitis C virus are not transmitted by arthopod vectors; a major transmission pathway for hepatitis C virus was via blood transfusions due to contamination of the blood supply. Screening of the blood supply has led to dramatic drop in spread of transfusion-related hepatitis C infections. The type member of the flavivirus genus, yellow fever virus, will be used as an example throughout this chapter due to its historical importance and the wealth of information available on its biology.

Yellow fever is a devastating human disease transmitted by mosquitoes

Yellow fever was first recognized in 1648 in the Yucatan. The disease progresses in three stages: (1) an initial stage marked by high fever and accompanied by muscle pain, headaches, and severe vomiting; (2) a quiescent stage in which the fever dissipates and the patient appears to be in remission; and (3) a final stage marked by a repeat of the original symptoms in a more severe form that results in multi-organ failure with a mortality rate of between 20 and 50%. Many victims suffer from acute hepatitis, which gives the skin a yellow tinge, hence the name.

The origin of the virus was most certainly West Africa, with multiple reintroductions of the virus in the New World due to the extensive slave trade. Despite quarantine efforts, which required ships of African and Caribbean origin to anchor for several weeks prior to entry into North American ports, outbreaks of yellow fever were frequent and severe for ports as far north as Boston. The epidemic that struck Philadelphia in 1793 is of particular note due to its impact on the fledgling government of the United States located in that city. The disease was imported from the Caribbean in July of that year and quickly established itself in the city due to the presence of an unusually high number of mosquitoes. All told, greater than 10% of the city's population of 40,000 died during the summer epidemic and a mass exodus reduced Philadelphia's population even further. President Washington and many other prominent government leaders and staff left the city for fear of infection.

TABLE 17.1 Members of the *Flaviviridae* family

Genus	Virus name	Hosts	Diseases
Flavivirus	Yellow fever	Monkeys, humans	Hemorrhagic fever
	Dengue (types 1–4)	Monkeys, humans	Fever, arthralgia, rash, hemorrhagic fever
	Japanese encephalitis	Birds, pigs, humans	Encephalitis
	West Nile	Birds, humans	Fever, arthralgia, rash
	Tick-borne encephalitis	Rodents, birds, humans	Encephalitis
Pestivirus	Bovine viral diarrhea	Ruminants	Persistent infection, mucosal disease
	Border disease	Sheep, goats	Acute and chronic hemorrhagic symptoms
	Classical swine fever	Pigs	
Hepacivirus	Hepatitis C	Humans	Acute and chronic hepatitis

The scourge of yellow fever outbreaks in the United States continued through the nineteenth century and was not brought under control until the Yellow Fever Commission headed by Walter Reed produced two significant results. First, they showed that the agent responsible for yellow fever could be passed to uninfected individuals using infected patient serum, and established that this agent was a virus by filtration methods. Second, they demonstrated that mosquitoes are capable of transmitting the virus after obtaining a blood meal from an infected host. Given the nature of the infecting agent and its transmission route, preventive steps could be taken to reduce the spread of the virus. Thus mosquito control in Panama overcame the most significant impediment to construction of the Panama Canal.

A live, attenuated yellow fever virus vaccine is available and widely used

In the 1930s, a safe and effective vaccine (strain 17D) against yellow fever virus was developed by Max Theiler. Although this turned out to be one of the best viral vaccines known, yellow fever continues to inflict misery and death in unvaccinated populations in South America and Africa. Comparison of the nucleotide sequences of the virulent Asibi strain and the 17D vaccine strain derived from it provided insight into the determinants of attenuation. Of the 10,862 nucleotides in the yellow fever genome, 68 nucleotide differences resulting in 32 amino acid substitutions were noted between the two strains. Although these mutations are spread across the genome, the envelope protein has a high number with 12 amino acid substitutions. In 1989, a full-length cDNA clone of yellow fever virus was constructed that could be used to generate infectious viral RNA *in vitro*. This allowed scientists to introduce mutations into the viral genome and therefore to study the functions of each gene. Since then, several other flavirus cDNA clones that are capable of producing infectious virus have been constructed.

Hepatitis C virus: A recently discovered member of the *Flaviviridae*

Hepatitis C virus was first identified, not by the traditional method of isolation and growth of virus in culture, but rather by direct cDNA cloning of the viral genome. The use of molecular clones of the virus genome greatly facilitated the study of hepatitis C virus, because up until 2005, virus replication could not be carried out in cells cultured *in vitro* or in nonprimate animals. Although hepatitis C virus will not be discussed in detail in this chapter, it is perhaps the most interesting member of the family from a clinical perspective. It has been estimated that there are over 170 million hepatitis C seropositive individuals in the world. Acute infection with hepatitis C virus is similar to other forms of acute viral hepatitis. However, the development of chronic hepatitis, which occurs in up to 85% of infected individuals, results in an increased risk of liver cirrhosis and hepatocellular carcinoma. Therefore, it is not surprising that hepatitis C infection is one of the primary indicators for a liver transplant. Hepatitis C virus is transmitted by blood transfusion, by contaminated needles of intravenous drug users, and by sexual contact; unlike most flaviviruses, it is not known to be transmitted by an arthropod host.

The flavivirus virion contains an icosahedral nucleocapsid wrapped in a tightly fitted envelope

Flaviviruses have a small (25–30 nm) nucleocapsid with T = 3 icosahedral symmetry, enclosed within a lipid-containing envelope. In this respect they are similar to togaviruses (Chapter 18), but flavirus particles are 50 nm in diameter compared with 70–80 nm for togavirus particles, and togaviruses have a T = 4 icosahedral structure. Nucleocapsids are constructed from 180 copies of the capsid (C) protein, and each virion also contains 180 copies of two envelope proteins, M and E. Cryoelectron microscopy reveals a spherical particle with a smooth, golf-ball like appearance (Figures 17.1, 17.2a), in contrast to the fringed envelope of togaviruses and many other enveloped viruses. A high-resolution map showing a cross-section of the virion (Figure 17.2b) reveals concentric layers of density, from the outer layer of envelope proteins, through two layers of the lipid membrane, to the surface of the capsid, within which lies the viral RNA.

The 101–amino acid C protein encapsidates the genome RNA and serves as a scaffold around which the

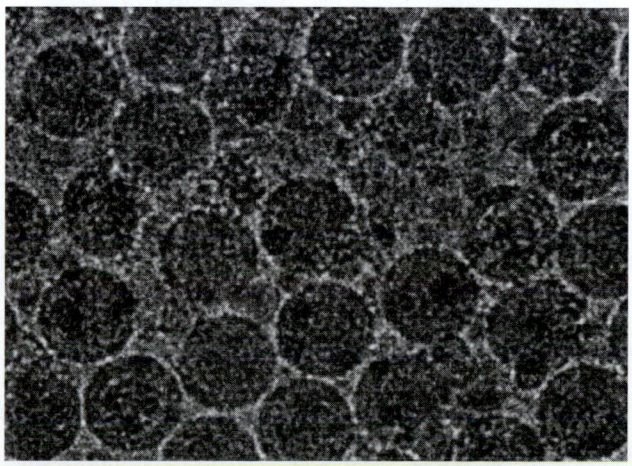

FIGURE 17.1 Dengue virus type 2 particles visualized by cryoelectron microscopy. The sample is slightly out of focus, as is required to obtain useful data for image reconstruction of the particles.

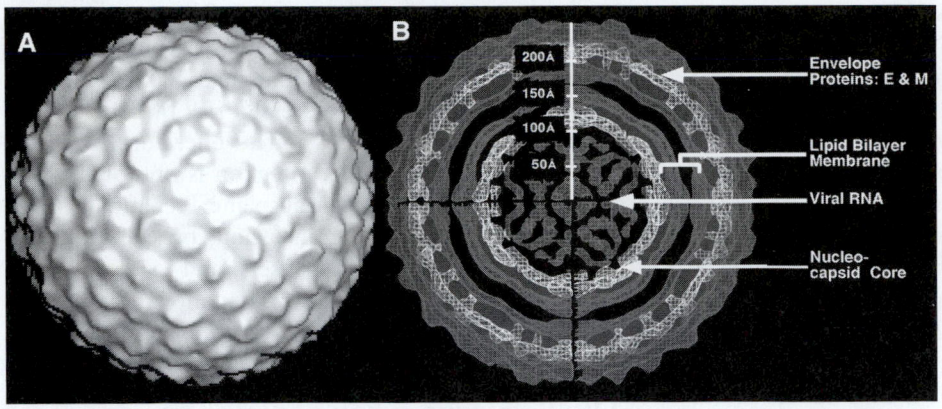

FIGURE 17.2 Morphology of dengue virus type 2 virions. (a) A surface shaded view of dengue virus type 2 as determined by cryoelectron microscopy and image reconstruction. The surface of the particle is completely covered by proteins, and the underlying lipid bilayer is not visible. The virion is 50 nm in diameter. (b) Cross-section of the virus from cryoelectron microscopy density maps showing layers of electron density. The outermost (gray) shell of density is comprised of the E protein and the underlying (white) layer contains a mixture of E and M proteins. The polar head groups of the lipid bilayer give rise to two clear layers of density (gray). Inside the nucleocapsid core layer (white), ordered RNA (gray) can be seen.

envelope proteins and the lipid bilayer are organized. It is the least conserved protein of the flaviviruses, but has a high number of basic amino acid residues and a central hydrophobic region. The basic amino acids probably interact with the negatively charged phosphate groups of the genome RNA inside the capsid. The hydrophobic region may be involved in binding the capsid to the viral envelope.

Flavivirus E protein directs both binding to receptors and membrane fusion

The E protein is a **type I transmembrane protein** of approximately 500 amino acids, anchored in the membrane via two C-terminal hydrophobic regions. Its amino terminus and most of its mass reside on the outside of the envelope. This protein plays a double role in entry: it binds to cell surface receptors and directs fusion of the viral envelope with host cell membranes. The M protein is a smaller type I transmembrane protein of approximately 80 amino acids. Its precursor interacts with E during virus assembly (see below).

A fragment of the E protein of tick-borne encephalitis virus (a close relative of yellow fever virus) has been crystallized and its structure was determined by x-ray diffraction analysis. This fragment is missing the C-terminal membrane anchor and an additional 50 amino acids, called the stem region. The remaining 400 amino acids are arranged as an elongated three-domain structure (Figure 17.3). Examination of this structure revealed several unusual features for a viral envelope protein. Unlike other viral glycoproteins and many cellular membrane proteins, which protrude outward from membranes, the long axis of the E molecule lies parallel to the lipid bilayer and has a convex shape that follows the curved surface of the viral envelope. This explains the smooth, golf ball–like surface of the virion. More recently, the structures of the E proteins from

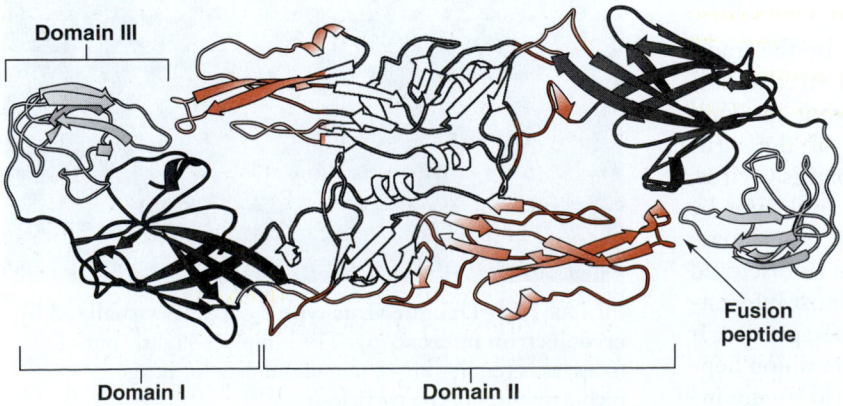

FIGURE 17.3 The atomic structure of the tick-borne encephaltitis virus E protein dimer as determined by x-ray crystallography. Each monomer consists of three domains. Note the position of the fusion peptide, as it resides in a protected niche covered by the neighboring E monomer.

dengue viruses type 2 and 3 have been solved by x-ray crystallography; these proteins have a very similar architecture to the tick-borne encephalitis E protein.

The E protein is found as a dimer, with domain II forming the dimer interface. At the outer ends of the dimer lies domain III, which has an immunoglobulin-like fold, and has been implicated in receptor binding. Many cell surface proteins that are used as virus receptors have immunoglobulin-like folds, but few viral structural proteins are known to have such folds. Joining these two domains is the central domain I, which contains the amino terminus of the protein. A short hydrophobic sequence, known as the fusion peptide, is found at one end of domain II, protected from the aqueous environment by the position of the neighboring E molecule within the dimer. The fusion peptide is predicted to play a major role in the fusion of the viral and cellular membranes that leads to entry of the nucleocapsid into cells. Its position within a region of the dimer interface suggests that upon rearrangement of the dimer at low pH, the fusion peptide is exposed, enabling its interaction with cellular membranes.

Flaviviruses enter the cell by pH-dependent fusion

Although possible cellular receptor molecules have been suggested for several flaviviruses, none has been clearly identified. In some flaviviruses, binding of antibody to the virus particle can result in enhanced uptake into cells expressing immunoglobulin Fc receptors. This process has been observed *in vitro* and is known as **antibody-dependent enhancement**. In the case of dengue virus, such a mechanism has been proposed to explain the disease known as dengue **hemorrhagic fever** or dengue shock syndrome. This disease can occur when a patient previously infected with one dengue virus serotype is subsequently infected with a second serotype within a period of one to two years. The first infection is believed to expand the population of cells that can be infected by a second serotype via antibody-dependent enhancement of virus entry.

Following attachment of virions to cell surface receptors, entry into the cell is achieved by endocytosis within clathrin-coated vesicles. These vesicles fuse with endosomes, which subsequently undergo acidification. The reduction in pH results in a rearrangement of the E dimer into the fusion-active trimeric state. Recent studies have demonstrated that the E protein of flaviviruses and the E1 glycoprotein of togaviruses have a similar structural fold and share similar functions in both fusion and icosahedral particle formation. These proteins have been termed **class II fusion proteins**, as they differ in several physical and functional aspects from the **class I fusion proteins** represented by influenza virus HA protein (see Chapters 4 and 23).

The structures of the low pH post-fusion form of the E protein have been determined for dengue and tick-borne encephalitis viruses using x-ray crystallography. This form of the protein is arranged as a trimer with the distal end of domain II, containing the fusion peptide, exposed and oriented toward the target membrane. Domain III has packed against domain I, suggesting that the stem region can pack against domain II, promoting the close approach of the viral and endosomal membranes and leading to their fusion.

Flavivirus genome organization resembles that of picornaviruses

The modern era of flavivirus research began with the determination of the nucleotide sequence of the 17D strain yellow fever virus RNA genome in 1985. The organization of the genome clearly established the flaviviruses as a group of enveloped, positive-strand RNA viruses different from the togaviruses, with which they had been previously grouped because of similar virion morphology, genome size, and transmission by arthropod vectors. Unlike the togaviruses, which encode their structural proteins on a subgenomic RNA, the flaviviruses have only one messenger RNA, identical to genome RNA. This is translated into a single, long **polyprotein** that undergoes proteolytic processing to generate the individual proteins. Coding regions for virion structural proteins are in the 5′-proximal region of the viral RNA and are therefore present in the N-terminal part of this polyprotein; the viral RNA polymerase is coded at the most distal 3′ region of the viral RNA. This gene arrangement closely resembles that used by picornaviruses, a family of nonenveloped positive-strand RNA viruses (Chapter 16).

The genome of yellow fever virus (Figure 17.4) is an RNA molecule 10,862 nucleotides in length. Like cellular messenger RNAs, the viral RNA has a methylated cap at its 5′ end, but it is not polyadenylated at its 3′ end. The 118-nt 5′ noncoding region is followed by a single, long open reading frame of 3411 codons. Three tandem stop codons are used to ensure termination of translation. The 511-nt 3′ noncoding region contains an 87-nt sequence that is predicted to form a stem-loop structure. Similar stem-loop structures are present in other flavivirus genomes, and biochemical and genetic experiments suggest a role for this region in viral RNA replication. A complementary sequence in the 5′ end of the genome is involved in the cyclization of the RNA, and may also be involved in RNA replication.

Other flaviviruses, including pestiviruses and hepatitis C virus, share with picornaviruses the use of an internal ribosome entry site (IRES) to direct initiation of viral protein synthesis. This further underlines similarities between the *Flaviviridae* and the *Picornaviridae*, and suggests that these two virus families may share a

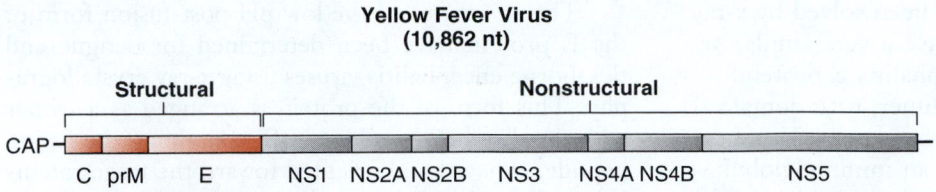

FIGURE 17.4 Genome organization of yellow fever 17D vaccine strain. Noncoding regions at each end of the genome are indicated by a solid line. Proteins are shown in shaded boxes. The sizes of the individual proteins are listed in Table 17.1. The genomes of the pestiviruses and hepatitis C virus are similar in layout but differ in the number and type of proteins. In addition, they have IRES elements in the 5′ noncoding region that direct initiation of translation.

common evolutionary origin: flaviviruses can be seen as picornaviruses that have acquired an envelope.

The polyprotein is processed by both viral and cellular proteinases

After virus uptake and release of the genome RNA in the cytosol, the RNA is bound by ribosomes and translated, giving rise to the polyprotein. Proteolytic processing yields 10 mature virus proteins (Figure 17.4 and Table 17.2). The three structural proteins C, prM, and E occupy the first quarter of the polyprotein, followed by the seven nonstructural proteins (NS1-5). Both viral and cellular proteases are required to process the viral polyprotein, which is inserted into the membrane of the endoplasmic reticulum. The viral protease resides within the protein NS3; however, an accessory protein, NS2B, is also required to carry out proteolytic processing with high efficiency. The cellular proteases responsible for processing include **signal peptidase** and **furin**.

The capsid protein is inserted into the endoplasmic reticulum via a 20–amino acid signal sequence at its carboxy terminus (Figure 17.5). The signal sequence is cleaved in the **lumen** of the endoplasmic reticulum by a cellular signal peptidase, releasing the remainder of the polyprotein from the capsid protein precursor (anchC or "anchored C protein"). The amino-terminal end of the signal sequence, on the cytoplasmic side of the endoplasmic reticulum membrane, is eventually cleaved by the viral protease NS2B/NS3A, releasing the mature capsid protein.

Following the capsid protein is the precursor to the membrane protein, prM (Figure 17.5). This small protein contains a transmembrane domain at its C-terminus followed by a signal sequence that is required to insert

TABLE 17.2 Proteins of Yellow Fever Virus

Protein	Size (aa)	Function
anchC	121	"Anchored" capsid protein. Precursor to capsid protein; signal sequence inserted in membrane is cleaved by viral protease.
C	101	Encapsidation of genome RNA.
prM	164	Binds to E glycoprotein, inhibits membrane fusion during transit through Golgi. Cleaved to produce mature M protein, releasing pr.
M	75	Small envelope glycoprotein. Function in the mature virion is unknown.
E	493	Major envelope glycoprotein. Receptor binding, fusion of virus envelope with cell membrane.
NS1	352	RNA replicase component, possibly involved in replication complex formation.
NS2A	224	RNA replicase component, either direct role in replication or targeting replicase complex to membranes.
NS2B	130	Part of viral protease that cleaves viral polyprotein.
NS3	623	Viral serine protease involved in polyprotein cleavage; RNA replicase component; nucleoside triphosphatase and helicase activities.
NS4A	149	RNA replicase component, may recruit NS1 and other proteins into replicase complex.
NS4B	250	Unknown role. Associated with membranes in cytoplasm; may translocate to nucleus.
NS5	905	RNA-dependent RNA polymerase; methyltransferase (methylates cap structure).

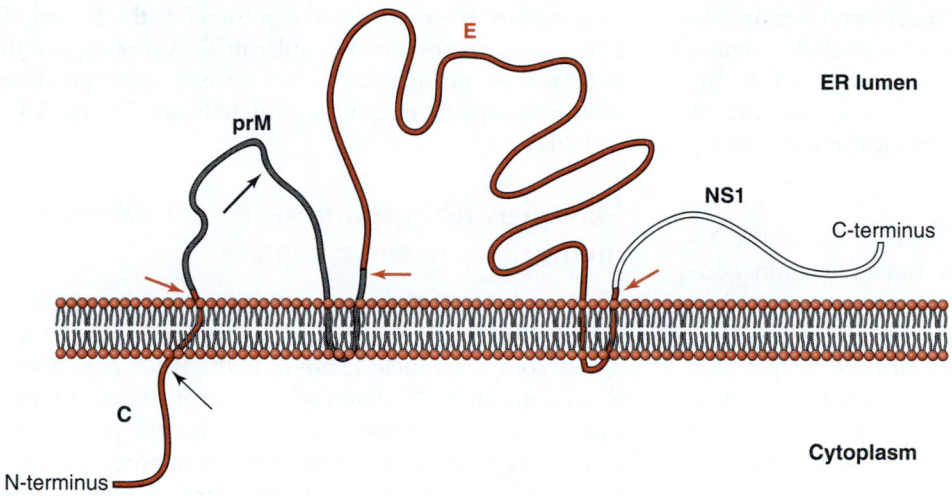

FIGURE 17.5 **Membrane topology of the flavivirus structural proteins.** Cleavages carried out by cellular signal peptidase are shown as thick orange arrows. The cleavage of the C-terminal signal sequence of the capsid protein by viral protease NS2B-NS3 is shown by the thin black arrow. Both prM and E have two C-terminal domains that span the lipid bilayer. The maturation cleavage of prM, carried out by furin (thick black arrow), releases the C-terminal fragment as mature M protein. The NS1 protein and other nonstructural proteins that follow E are not shown in their entirety.

the E protein in the correct orientation in the membrane. The prM protein associates with the E protein in the endoplasmic reticulum to form a heterodimer. This association apparently protects the E protein from a premature conformational change that would lead to exposure of the fusion peptide in the low pH environment in Golgi vesicles. At a late stage in the maturation process of the virion, a cellular furin protease cleaves prM, releasing the "pr" peptide into the extracellular medium, and leaving the small M protein associated with the virus particle. This cleavage causes a conformational change in the virion, resulting in a conversion of the prM/E heterodimer into an E homodimer as shown in Figure 17.3.

Analogous to prM, the E protein contains both a transmembrane domain and a signal sequence at its C-terminus. The signal sequence functions to insert the NS1 protein into the endoplasmic reticulum, where it functions in the early steps of RNA synthesis. Except for the NS1 protein, all other nonstructural flavivirus proteins are believed to reside within the membrane or on the cytoplasmic side of the membrane.

Nonstructural proteins organize protein processing, viral RNA replication, and capping

The nonstructural proteins (NS1 through NS5) are involved in the replication of the genome RNA. Seven different polypeptides are derived from this region. Of these seven, three are known to have specific enzymatic activities: NS5 is an RNA-dependent RNA polymerase; NS3 is a multifunctional, multidomain protein that contains both a protease activity and a predicted RNA helicase acitivity; NS2B has been shown to complex with NS3 and is a required cofactor for protease activity.

NS1. NS1 is translocated into the lumen of the endoplasmic reticulum, where it is cleaved from the polyprotein at its N-terminus by signal peptidase (Figure 17.5) and at its C-terminus by an unknown protease. The protein is glycosylated, forms dimers, and is found both secreted and in association with membranes in infected cells. The extracellular form of NS1 is capable of eliciting a protective humoral immune response. The function of the NS1 protein in replication remains obscure, but several lines of experimental evidence suggest that it functions in an early step in RNA synthesis.

NS2A. The NS2A protein has also been implicated in RNA synthesis, as it has been shown to be localized in RNA replication complexes. NS2A binds to the NS3 and NS5 proteins, as well as to sequences within the 3′ noncoding region of the genome RNA. Cleavage at its N-terminus occurs in the lumen of the endoplasmic reticulum, but cleavage at its C-terminus is carried out in the cytoplasm by the NS2B/NS3 protease. Therefore, this protein must cross the membrane of the endoplasmic reticulum at least once, and it may have additional membrane spanning sequences.

NS2B. The short NS2B protein has several membrane-spanning domains and a conserved hydrophilic domain of about 40 amino acids. NS2B is a cofactor for the protease activity found within NS3. The association of NS2B with NS3 provides enhanced cleavage site specificity to the protease.

NS3. The protease activity of NS3 maps to the N-terminal one-third of the protein. In its C-terminal domain there are amino acid sequence motifs found in RNA helicases as well as in nucleoside triphosphatases. The x-ray structure for the dengue virus NS3A protease domain confirmed sequence analyses and genetic studies that suggested that NS3 is a serine protease having a catalytic triad consisting of His, Asp, and Ser residues.

NS4A and NS4B. The functions of the small, hydrophobic NS4A and NS4B proteins are not known. Based on their subcellular localization, it is likely that they play a role in RNA replication. Like the NS2A and NS2B proteins, they contain several transmembrane domains, and NS4A also contains a signal sequence. Thus, it is likely that these proteins remain associated with the endoplasmic reticulum.

NS5. By far the largest of the flavivirus nonstructural proteins is the RNA-dependent RNA polymerase, NS5. It has the canonical Gly-Arg-Arg motif found in other RNA virus RNA polymerases, and it also contains sequences that have homology with **methyltransferases**. This activity, along with the **nucleoside triphosphatase** in the C-terminus of NS3, is involved in the formation of the methylated cap found at the 5′ end of flavivirus genomes. Since cellular mRNAs are capped in the nucleus, cytoplasmic RNA viruses must produce their own capping enzymes (for details, see Chapters 15 and 24).

Flavivirus RNA synthesis is carried out on membranes in the cytoplasm

The synthesis of the nonstructural proteins results in the establishment of active RNA replicase complexes. These complexes include not only the RNA-dependent RNA polymerase NS5 and other viral proteins, but presumably also host proteins recruited for the purpose of genome replication. RNA replication requires the synthesis of a complementary (minus-strand) copy of the plus strand RNA. This is followed by the synthesis of plus strands that are then either translated, making more viral proteins; replicated, making more RNA; or packaged, making virions. The synthesis of RNA is asymmetric, generating more than 10 times as many plus strands as minus strands.

Significant alterations of subcellular structure have been observed in flavivirus-infected cells. New accumulations of endoplasmic reticulum membranes appear early in the infectious cycle, and numerous smooth vesicles can be found later in infection. RNA synthesis is associated with these structures, and most of the nonstructural proteins are localized there. A number of these proteins contain predicted membrane-spanning sequences. RNA synthesis directed by most positive-strand RNA viruses, including picornaviruses, to-

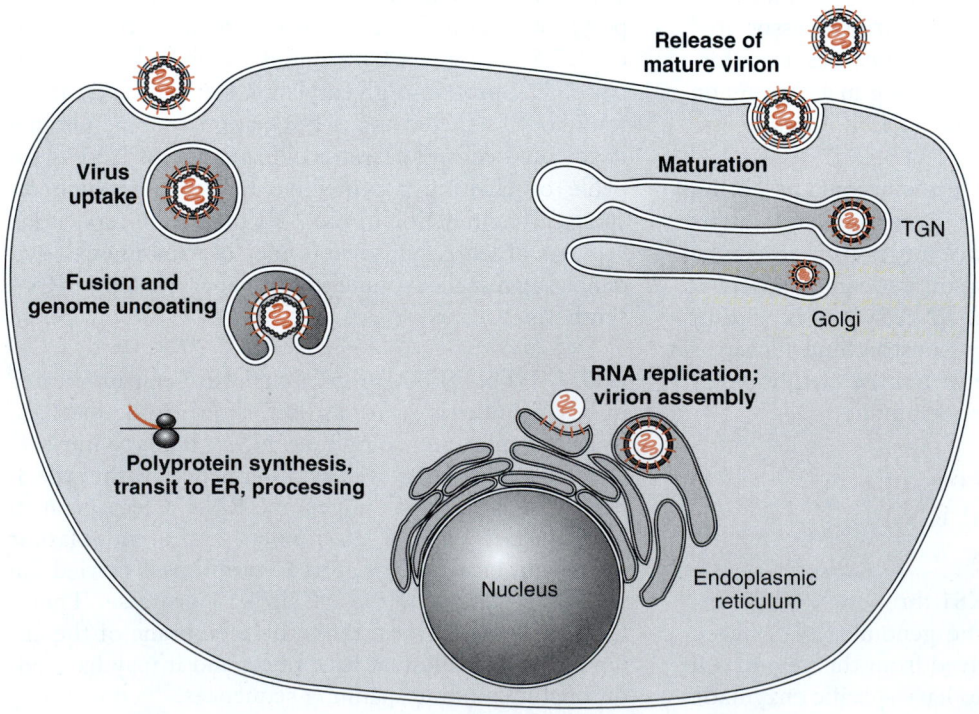

FIGURE 17.6 **Flavivirus replication pathway.** Viral RNA is shown as a gray line; viral envelope proteins are in orange. TGN = *trans*-Golgi network.

gaviruses, and coronaviruses (Chapters 16, 18, and 19), is also associated with membranes in the cytoplasm of infected cells, and in some of these cases is understood in more detail.

Virus assembly also takes place at intracellular membranes

Unlike many other enveloped viruses, which derive their envelope from the plasma membrane of the cell, flaviviruses assemble their virions inside the cell by using intracellular membranes for envelope formation (Figure 17.6). Electron microscopy has shown that immature virions can be found in the lumen (interior) of the endoplasmic reticulum.

Assembly of the nucleocapsid takes place at the cytoplasmic face of membranes with which prM and E proteins are associated. The carboxy-terminal signal sequence of the precursor to the C protein anchors that protein to the membrane (see Figure 17.5); this sequence penetrates into the lumen of the endoplasmic reticulum or smooth vesicles. This enables interactions between the C protein and the envelope proteins, which are also anchored to the membrane but reside within the lumen. Multiple copies of the C protein assemble into nucleocapsids. Subsequent cleavage of the membrane anchor of the C protein by the viral protease (see above) releases the capsid from direct contact with the membrane.

The coupling of protein synthesis, RNA synthesis, and virion assembly on membranous structures assures that newly synthesized genome RNA can associate with C protein and initiate the assembly process. Encapsidation of the RNA initiates the budding of particles into the endoplasmic reticulum. Particles that have budded into the endoplasmic reticulum are then processed by carbohydrate addition and modification as they proceed through the Golgi membrane system. It is likely that transport to the Golgi and into the *trans*-Golgi network requires the presence of the glycosylated prM protein. Virions follow the **exocytosis** pathway and are released to the extracellular space by fusion of vesicles containing virions with the plasma membrane (Figure 17.6). The cleavage of the prM protein by host-encoded furin occurs just prior to virion release and converts the particle to its mature form.

Introduction of West Nile virus into the United States of America

In the summer of 1999, residents of Queens, New York, experienced an outbreak of an encephalitic disease reminiscent of a flavivirus infection. Concurrently, the local population of crows in the greater New York City area suffered from disease, and many were found dead. First analyses of the infectious agent suggested that these diseases might be caused by a flavivirus known as St. Louis encephalitis virus. This virus appears sporadically in the eastern half of the United States, with a number of notable outbreaks in Florida near Orlando. However, it was subsequently shown that the infectious agent affecting both the human and crow populations was West Nile virus, a virus not previously seen in North America. This prompted substantial discussion about how the virus was introduced and whether it would establish itself in the United States and become an **endemic** disease. West Nile virus was originally isolated in Uganda in 1937 and is found in Israel and throughout the Middle East, as well as in parts of Europe and Asia. The virus was likely accidentally brought into the United States via an infected bird or mosquito, perhaps from an airplane that landed at a New York airport.

The next year (2000) proved pivotal for the presence of West Nile virus in the United States. The virus reappeared with the advent of the mosquito season and began its expansion outside of the New York City region, most likely carried by birds. Over the next few years, the virus expanded its presence into all of the contiguous 48 states and the adjacent Canadian provinces. The initial human cases were widely reported in the press, leading the public to fear a widespread **epidemic**. However, a year or two after the virus became established in a given area, the number of human cases of encephalitis declined. In 2002, it was shown that West Nile virus could also be transmitted by blood transfusions and tissue transplantation. This quickly led to the establishment of a screening assay for the blood supply, causing a dramatic reduction in transfusion-associated West Nile virus disease. However, its presence in wild populations of mosquitoes and birds remains a threat for transmission to humans.

The ecology and epidemiology of West Nile virus is not yet understood well enough to predict its future effects on either birds or humans in North America. The level of **seroconversion** (an indicator of exposure to the virus) in the human population has remained relatively constant and low (~3%), but an increase in infections of the host (bird) or vector (mosquito) population could have a significant impact on the number and perhaps severity of human infections. We do know for certain that West Nile virus is now endemic in most of the United States and much of Canada, and is capable of causing occasional severe human disease and death.

SUGGESTED READING

1. Khromykh, A. and Westaway, E.G. (1997). Subgenomic replicons of the flavivirus Kunjin: Construction and applications. *Journal of Virology* **71**, 1497–505.

2. Mukhopadhyay, S., Kuhn, R.J., and Rossmann, M.G. (2005). A structural perspective of the flavivirus lifecycle. *Nature Reviews Microbiology* **3**, 13–22.

3. Monath, T.P. (1996). Milestones in the conquest of yellow fever. In Koprowski, H. and Oldstone, M.B.A., eds., Microbe hunters, now and then, pp. 95–111. Medi-Ed Press, Bloomington, Illinois.

4. Strauss, J.H. and Strauss, E.G. (1996). Current advances in yellow fever research. In Koprowski, H. and Oldstone, M.B.A., eds., Microbe hunters, now and then, pp. 113–137. Medi-Ed Press, Bloomington, Illinois.

KEY TERMS

Antibody-dependent enhancement
Class I fusion proteins
Class II fusion proteins
Endemic
Epidemic
Exocytosis
Furin
Hemorrhagic fever
Hepatitis
Hepatocellular carcinoma
Lumen
Methyltransferase
Nucleoside triphosphatase
Polyprotein
Seroconversion
Signal peptidase
Type I transmembrane protein

Togaviruses

Togaviridae

From Latin *toga* (gown or cloak), referring to virus envelope

VIRION

Spherical enveloped particle with a fringe of projections.
Diameter 70 nm.
Nucleocapsid has icosahedral symmetry (T = 4).
Envelope glycoproteins also arranged with icosahedral symmetry, T = 4.

GENOME

Linear ss RNA, positive sense.
Alphaviruses: 11 Kb; rubella virus: 9 Kb.
5′ end is capped and methylated.
3′ end has a 70-nt poly(A) tail.

GENES AND PROTEINS

Four nonstructural proteins (for viral RNA synthesis) translated directly from genome RNA.
Five structural proteins (one capsid protein, three envelope proteins, a small hydrophobic protein) translated rom a subgenomic mRNA.

VIRUSES AND HOSTS

Two genera: *Alphavirus* (27 members) and *Rubivirus* (rubella virus).
Alphaviruses (Sindbis, Semliki Forest, Ross River) infect birds and mammals.
Most alphaviruses are transmitted by mosquitoes.

DISEASES

Symptoms include fever, rash, arthritis, and encephalitis; can be fatal.
Rubella infection (German measles) can lead to fetal abnormalities.

DISTINCTIVE CHARACTERISTICS

A subgenomic mRNA codes for virion structural proteins.
Structural proteins are synthesized as a polyprotein that undergoes cleavage and membrane association.
RNA replication is associated with cytoplasmic membranes.
Alphaviruses have the unusual ability to replicate in both invertebrate and vertebrate hosts.
Lipid envelope derived from the plasma membrane (vertebrate cells) or internal membranes (insect cells).

Most togaviruses are arthropod borne, transmitted between vertebrate hosts by mosquitoes

Walter Reed, working in Cuba in 1900, demonstrated that yellow fever virus (see Chapter 17) was transmitted to humans by mosquitoes. This startling finding led to efforts by many laboratories, particularly in tropical regions of the world, to discover other mosquito-borne viruses. In the next half-century scientists isolated a large number of viruses that infect mammals and birds and are transmitted by a variety of arthropod vectors, including mosquitoes, ticks, and sandflies. Such viruses were originally classified as **arthropod-borne viruses** ("arboviruses"). Most of these viruses can replicate in both cold-blooded arthropods and warm-blooded vertebrate hosts, and thus have an unusual ability to adapt to different growth conditions.

When studies on the molecular biology of arthropod-borne viruses were undertaken, it became apparent that not all such viruses are closely related. Group A arboviruses were reclassified into the *Alphavirus* genus of the *Togaviridae* family (Table 18.1), so named in reference to their envelope ("toga" means "cloak" in Latin). Most other arthropod-borne viruses are members of the *Flaviviridae* family (Chapter 17; named after yellow fever virus) or the *Bunyaviridae* family (Chapter 22); these viruses have distinct genome organization and replication strategies. Some members of the *Reoviridae* and *Rhabdoviridae* are also arthropod-borne.

Several alphaviruses, including eastern and western equine encephalitis viruses, Venezuelan equine encephalitis virus and Ross River virus, cause disease in animals and humans. Symptoms range from rashes and high fevers to joint pains and **encephalitis**, and the outcome can be fatal in both humans and domestic animals such as horses. The incidence of human disease in North America is low except for occasional local outbreaks. In most cases, the natural vertebrate hosts of these viruses are wild mammals and birds; humans (and horses) are usually dead-end hosts as far as the virus is concerned, because they do not transmit disease to other individuals. Rubella virus, the cause of German measles, is the only member of the togavirus family that is not transmitted via arthropod vectors. It causes fetal abnormalities and birth defects when women are infected in the first trimester of pregnancy.

Many togaviruses were isolated from mosquitoes or infected animals in tropical regions of the world, and their names often recall the localities where they were first found. Sindbis virus (named for an Egyptian village) and Semliki Forest virus (from Eastern Uganda) are two togaviruses that have been studied in great detail in the laboratory. They have been valuable models for investigating togavirus uptake, structure, and replication, and have also been used as important tools for studying cellular pathways such as cleavage and modification of glycoproteins and their transport to the plasma membrane. These viruses are able to infect cultured cells from a wide variety of animal species and replicate rapidly, producing high titers of progeny (about 10^4 virions per infected cell in 12 to 16 hours). Infection of vertebrate cells quickly inhibits host cell protein synthesis and leads to cell death. Infection of mosquito cells is often not **cytopathic** but instead causes a **persistent infection**. Why infections with the same virus can lead to these two very different outcomes is not known. Sindbis and Semliki Forest viruses are not highly pathogenic in humans, allowing their safe and easy use in the laboratory.

Togavirus virions contain a nucleocapsid with icosahedral symmetry wrapped in an envelope of the same symmetry

Togavirus virions are enveloped spherical particles with a diameter of 70–80 nm. The outer surface of the virion consists of 80 flower-like structures each made up of three subunits, projecting outward in the form of petals (Figure 18.1a). The 240 petals are heterodimers composed of one molecule each of the glycoproteins E1 and E2. They are organized with icosahedral symmetry in a T =

TABLE 18.1 Representative togaviruses

Genus and virus name	Vertebrate host	Arthropod host
Alphavirus		
Sindbis virus	Birds	Mosquitoes
Semliki Forest virus	Birds, rodents, primates	Mosquitoes
Ross River virus	Marsupials, rodents	Mosquitoes
Venezuelan equine encephalitis virus	Rodents, horses, birds	Mosquitoes
Western and Eastern equine encephalitis viruses	Birds	Mosquitoes
Rubivirus		
Rubella virus	Humans	None

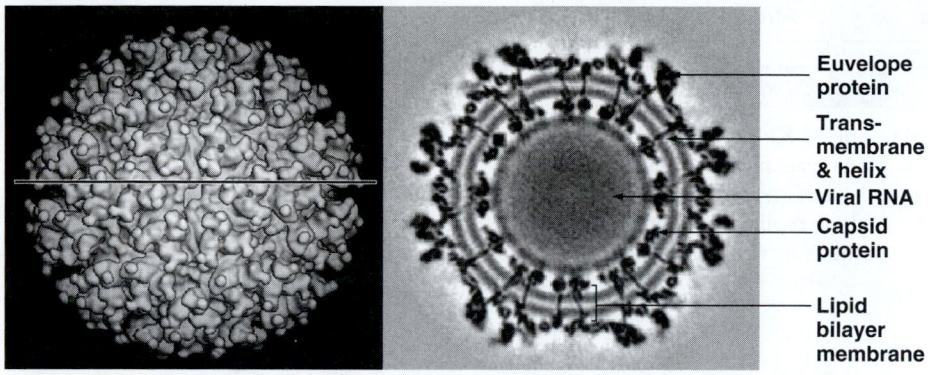

FIGURE 18.1 **Cryoelectron microscopic image reconstruction of a togavirus virion.** Left: a view of a Sindbis virion at 20 Å resolution. The outer surface is composed of flower-like projections of the E1/E2 heterodimers arranged as trimers. The interconnective portions form a protein shell that covers most of the lipid bilayer except for small holes. Right: a cross-section along the black line shown in the picture on the left. The lipid bilayer and transmembrane alpha helices crossing the membrane are clearly seen. The black images above and underneath the bilayer are envelope and capsid protein molecules.

4 lattice. The E1 polypeptide comprises most of the protein shell or "skirt" that covers the underlying lipid membrane (Figure 18.1b); its structure is similar to that of the flavivirus envelope protein (Chapter 17). **Cryoelectron microscopy** data indicate that E2 has a shape similar to that of E1, but E2 forms the external part of the petals that project outward. Both E1 and E2 are anchored in the lipid membrane with carboxy-terminal portions of their polypeptide chains protruding into the interior of the particle, as can be seen in Figure 18.1b.

The envelope surrounds the virus nucleocapsid, which contains a single molecule of virus genome RNA. The nucleocapsid has a diameter of 40 nm and is composed of 240 copies of the capsid protein also arranged in a T = 4 icosahedral lattice. The 240 capsid protein molecules are each in direct contact with one of the overlying 240 envelope protein heterodimers, leading to identical symmetry arrangements in both nucleocapsid and envelope, unusual for enveloped viruses (but present also in flaviviruses). The envelope is tightly apposed to the surface of the nucleocapsid, resulting in a very uniform diameter of virions, unlike many enveloped RNA viruses with helical nucleocapsids, which can have variable size and shape. Rubella virus has not been studied in as much detail, but is likely to have similar morphology and symmetry.

Togaviruses enter cells by low pH-induced fusion inside endosome vesicles

The E2 glycoprotein binds to cellular receptors. No single cellular receptor has been unambiguously identified for alphaviruses; different proteins have been identified as receptors for Sindbis virus in different host cells. A high-affinity **laminin receptor** was found on rodent and primate cells, and a laminin-like receptor is also believed to function as the receptor for Venezuelan equine encephalitis virus on mosquito cells. Mutations leading to amino acid changes in the viral E2 glycoprotein rapidly accumulate during passage in cultured cells; these mutations allow the virus to bind to **heparan sulfate**, which is found on many cell surface proteins. The finding that mutations in E2 can alter receptor binding confuses attempts to identify specific cellular receptors.

Uptake of virions occurs in vertebrate cells via **receptor-mediated endocytosis**. Togaviruses, particularly Semliki Forest virus, served as models to establish this means of entry, which is used by many enveloped RNA-containing viruses (see Chapter 4). After uptake of virus into endosomal vesicles, protein pumps in the vesicule membrane generate a drop in intravesicular pH, leading to conformational changes in the viral envelope glycoproteins. The flower-like envelope projections, which consist of trimers of E1-E2 heterodimers, reorganize to form trimers of E1, exposing a hydrophobic fusion domain in that protein. Fusion of the endosome membrane with the viral envelope leads to the release of the nucleocapsid into the cytoplasm. The pH-dependence of envelope-membrane fusion was shown directly by lowering the pH of medium bathing cells to which large numbers of virions were bound; virus envelopes fused directly with the plasma membrane when the pH was reduced.

Nonstructural proteins are made as a polyprotein that is cleaved by a viral protease

Once in the cytoplasm, nucleocapsids release the viral genome RNA so that it can be translated by cellular ri-

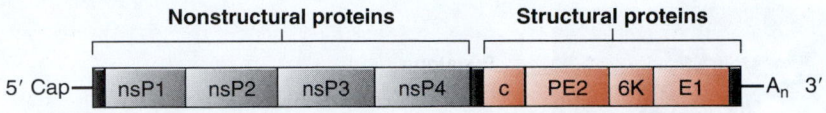

FIGURE 18.2 Alphavirus genome RNA. The 11 Kb positive-strand RNA contains a methylated cap at its 5′ end and is polyadenylated at its 3′ end. Regions coding for nonstructural (ns) proteins and structural proteins are shown in orange boxes. Noncoding regions that include replication promoters and the internal promoter for subgenomic RNA synthesis from antigenome RNA are shown in black.

bosomes. Togavirus nucleocapsids are permeable to ribonuclease *in vitro*, suggesting that the RNA in nucleocapsids may be accessible to other proteins within the cytoplasm. It has been shown that large ribosomal subunits can bind to capsid proteins and disrupt nucleocapsids, liberating genome RNA. Interaction with ribosomes within the cytoplasm may therefore initiate nucleocapsid breakdown and release the genome RNA, allowing its translation.

The 11 Kb single-stranded alphavirus RNA genome (Figure 18.2) contains two separately translated modules. The 5′-proximal two-thirds codes for the nonstructural proteins required for replication and transcription of the RNA (Table 18.2), and the 3′-proximal one-third codes for the capsid and envelope proteins (Table 18.3). These structural proteins are not translated directly from genome RNA, but from a subgenomic mRNA made during viral infection. Both the genome RNA and the subgenomic mRNA have methylated caps at their 5′ ends and are polyadenylated at their 3′ ends, resembling cellular messenger RNAs.

Translation of genome RNA begins at an initiator AUG near its 5′ end. In Sindbis virus, translation terminates most of the time at a UGA codon positioned close to the end of the nsP3 coding region, generating a P123 **polyprotein** (Figure 18.3). The complete P1234 polyprotein is formed only when there is readthrough of this codon, which occurs 10 to 20% of the time, depending on the host cell. The P1234 polyprotein is cleaved in stages to yield initially a complex of P123 and nsP4, followed by a complex consisting of nsP1, nsP23, and nsP4 and finally by a complex consisting of the individual polypeptides nsP1, nsP2, nsP3, and nsP4. The protease responsible for these cleavages is encoded in a domain at the C-terminal end of nsP2.

Partly cleaved nonstructural proteins catalyze synthesis of full-length antigenome RNA

The nonstructural proteins carry out synthesis of the different viral RNAs, a highly regulated event in infected cells. During the early phase of the infectious cycle, full-length negative-strand **antigenome** RNAs are synthesized by a replicase complex formed between P123 and nsP4 (Figure 18.3). Interestingly, sequences at both the 3′ and 5′ termini of genome RNA are required for synthesis of negative-strand antigenome RNA, implying that the replication complex may be bound to both ends of the RNA.

At this stage, little or no positive-strand RNA is produced. In contrast, during the late phase mostly positive-strand genome and **subgenomic RNAs** are made. The switch from synthesis of negative to positive strands is correlated with proteolytic processing of P123 (Figure 18.3). P123 does not undergo autoproteolysis and it is not cleaved during the early phase, suggesting that the protease domain in nsP2 is only able to cleave other polyprotein molecules than the one in which it lies. During the late phase, when the concentration of

TABLE 18.2 Sindbis virus nonstructural proteins

Protein	Length (aa)	Properties and functions
nsP1	540	RNA capping enzyme; has methyltransferase activity. Required for initiation of transcripion of minus-strand RNA and modulation of nsP2 protease activity. Palmitoylated and membrane-associated.
nsP2	807	Cysteine protease, RNA helicase. Required for synthesis of subgenomic RNA. Mutations in nsP2 show reduced cytopathic effects and lower shutoff of host cell protein synthesis. A fraction of nsP2 is found in the nucleus of infected cells.
nsP3	556	Unknown function. Phosphorylated at Thr and Ser residues.
nsP4	610	RNA polymerase. Unstable, degraded by the ubiquitin pathway.

TABLE 18.3 Sindbis virus structural proteins

Protein	Length (aa)	Properties and functions
Capsid	264	Forms nucleocapsid; contains serine-like protease that autocatalytically cleaves itself from the polyprotein. Binds to genomic RNA packaging signal via positively charged amino terminus.
PE2	487	Precursor to E2; palmitoylated and glycosylated. Also referred to as p63, based on its molecular weight. Cleaved in *trans*-Golgi vesicles by a furin-like cellular protease; rarely incorporated into virions.
E3	64	Amino-terminal part of PE2 containing the signal sequence. Glycosylated; retained in virions of Semliki Forest but not Sindbis virus.
E2	423	Carboxy-terminal part of PE2, a component of virus envelope projections. C-terminal cytoplasmic domain binds to capsid during assembly. Binds to cellular receptors; contains epitopes for neutralizing antibodies. Palmitoylated.
6K	55	Membrane associated; palmitoylated; C-terminus is signal sequence for E1. Small amounts in virions; enhances virus assembly and budding.
E1	439	Component of virion envelope projections; type 1 transmembrane glycoprotein with two positively charged C-terminal amino acids in the cytoplasm; palmitoylated and glycosylated; functions in low-pH activated membrane fusion.

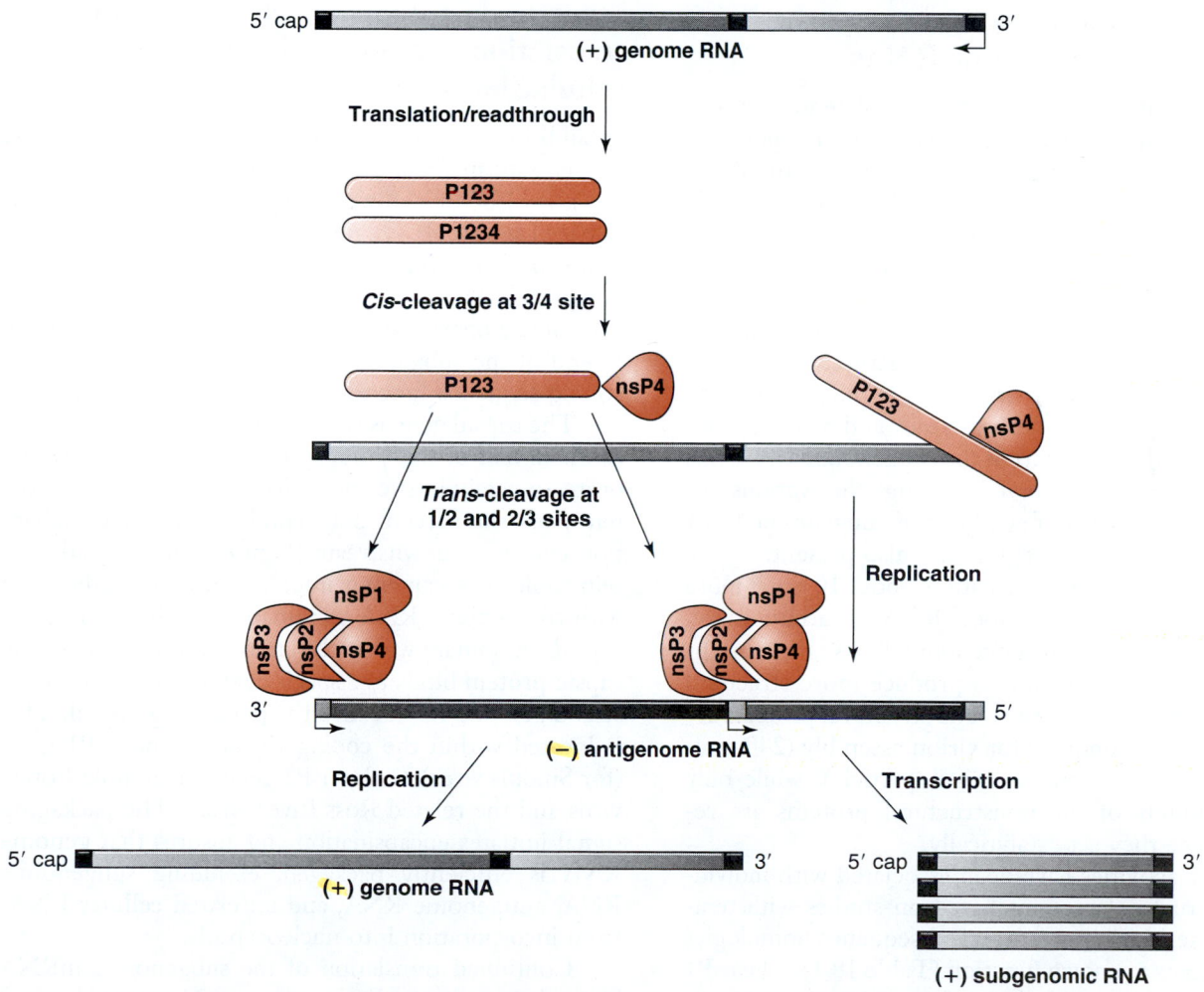

FIGURE 18.3 A model for the temporal regulation of minus and plus-strand RNA synthesis. Nonstructural proteins are translated from genome RNA. A complex of proteins P123 and nsP4 initiates minus-strand RNA synthesis (replication) from a promoter (black with directional arrow) at the 3′ end of genome RNA; 5′ terminal sequences are also required. Once sufficient P123 has been made, trans-cleavage generates nsP1, P23, and finally nsP2 and nsP3. These proteins, together with nsP4, form complexes that efficiently initiate synthesis of positive-strand genome RNA (replication), and larger amounts of subgenomic mRNA (transcription), from promoters (white squares with directional arrows) on the antigenome RNA template. Host cell proteins are probably also involved in these reactions.

P123 becomes high enough, P123 molecules can encounter each other and undergo mutual trans-cleavage.

One of the most convincing arguments for the importance of intact P123 in the synthesis of negative strands is based on a mutation in the nsP2 protease that leads to unusually rapid proteolytic processing of P123. RNAs containing this mutation, when transfected into cultured cells, are unable to produce new virus progeny, suggesting that intact P123 is essential for negative-strand RNA synthesis. Replicase complexes containing this nsP2 mutation regain the ability to carry out negative-strand RNA synthesis when combined with mutations in P123 that destroy the sites of cleavage between the individual nonstructural proteins nsP1, nsP2, and nsP3.

Replication and transcription: synthesis of genome and subgenomic RNAs

The antigenome RNA serves as a template for the synthesis of both full-length and subgenomic positive-strand RNAs. Genome RNA synthesis (replication) initiates at the 3′ end of the antigenome RNA and proceeds to its 5′ end, to make a complete complementary copy of the negative-strand RNA (Figure 18.3). Synthesis of the subgenomic RNA (transcription) initiates at a promoter that lies at the junction between the genes for nonstructural proteins and those for structural proteins. The minimal promoter sequence required for transcription is 19 nucleotides upstream and 5 nucleotides downstream from the start of the subgenomic RNA; this region is highly conserved among the various alphaviruses. The level of synthesis of subgenomic RNA is increased if flanking regions are also present.

At the peak of virus formation, about 10 times more subgenomic RNA than genome RNA has accumulated. Making higher levels of subgenomic RNA is one means by which the virus is able to produce more structural than nonstructural proteins. The structural proteins are needed in large amounts for virion assembly (240 molecules of each protein per virus particle), while only small amounts of the nonstructural proteins are required since they act catalytically.

Other activities have been associated with individual nonstructural proteins based on studies with temperature-sensitive mutants and on sequence homologies with proteins of known function (Table 18.1). The nsP1 protein contains a **methyltransferase** domain and functions as a capping enzyme, adding a methylated GTP cap to the 5′ end of positive-strand viral RNAs (for details on enzymes involved in RNA capping, see Chapters 15 and 24). In addition to its protease activity, nsP2 contains nucleoside triphosphate-binding motifs in its amino terminal region that are homologous to those found in bacterial helicases. When nsP2 was expressed in bacteria in the absence of other viral proteins, it was shown to have an **RNA helicase** activity that may be needed for optimal RNA replication/transcription. The nsP4 protein has a Gly-Asp-Asp motif common to the active site of a number of viral RNA polymerases. It has been identified as the polymerase responsible for both genome replication and subgenomic mRNA synthesis. A specific role for nsP3 has not yet been identified.

Proteolytic processing of the rubella virus nonstructural polyprotein (p200) also appears to regulate replication and transcription of the viral RNA. The intact polyprotein can synthesize negative-strand antigenome RNA, but cleavage of p200 by a virus-encoded protease to produce two proteins named p150 and p90 is required for efficient formation of new positive-strand rubella virus genome RNA and subgenomic RNA.

Structural proteins are cleaved during translation and directed to different cellular locations

In addition to being produced in large amounts, the subgenomic mRNA, which encodes viral structural proteins, is also very efficiently translated. Translation is initiated at a single site near its 5′ end and proceeds uninterrupted for some 3000 nucleotides to produce a polyprotein that is co- and post-translationally cleaved. For some alphaviruses, a hairpin-like structure near the 5′ end of the subgenomic RNA and within the capsid coding sequence functions to enhance its translation.

The capsid gene is translated first, and the carboxy-terminal half of this polypeptide folds to form a serine protease catalytic site that self-cleaves the capsid protein from the nascent polypeptide chain while translation continues downstream (Figure 18.4). Capsid protein molecules can then interact with each other and with the genome RNA to initiate assembly of nucleocapsids. A domain within the amino-terminal half of the capsid protein binds to a specific region of the positive-strand genome RNA called the **packaging signal**. This is located within the coding region of the nsP1 gene (for Sindbis virus) or the nsP2 gene (for Semliki Forest virus and the related Ross River virus). The packaging signal initiates encapsidation and ensures that genome RNA is efficiently packaged, excluding subgenomic RNA, antigenome RNA, and unrelated cellular RNAs from incorporation into nucleocapsids.

Continued translation of the subgenomic mRNA beyond the capsid gene produces a N-terminal signal sequence that translocates the PE2 polypeptide to the rough endoplasmic reticulum (Figure 18.4). Additional stop-transfer and signal sequences in the PE2, 6K, and E1 regions of the polyprotein lead to insertion of the 6K protein and translocation of the E1 portion into the endoplasmic reticulum membrane. A host cell **signal peptidase** cleaves the polyprotein at two sites that de-

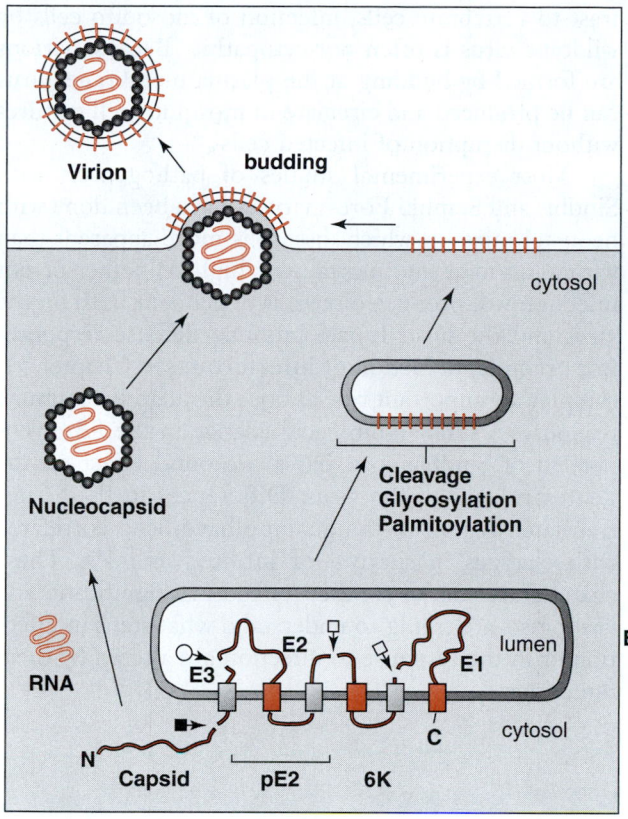

FIGURE 18.4 Processing and modification of Sindbis virus structural proteins. During translation of subgenomic mRNA, the capsid protein is released from the growing polypeptide chain by self-cleavage (arrow with black square). Capsid proteins oligomerize and bind to genome RNA in the cytoplasm to form nucleocapsids, which migrate to the plasma membrane, as shown here, or alternatively form during association with viral glycoproteins at the plasma membrane. The remainder of the structural polyprotein, including PE2, 6K, and E1, is translocated as a unit to the endoplasmic reticulum by means of a signal sequence at its N-terminus. Proteolytic processing by cellular signal peptidase (arrows with white squares) produces PE2, 6K, and E1. Additional protein modifications carried out during vesicle transport to the cell surface include cleavage by furin protease (arrow with white circle) in a *trans*-Golgi compartment to form E2 and E3, as well as glycosylation. Signal sequences: white rectangles; transmembrane segments: orange rectangles; processed viral envelope proteins: orange spikes viral RNA: orange curved line.

fine the carboxy termini of the PE2 and 6K proteins. In the endoplasmic reticulum, PE2 spans the membrane twice at its carboxy terminus; the 6K protein spans the membrane at both its amino and carboxy termini; and E1 spans the membrane once at its carboxy terminus (Figure 18.4). As these proteins are transported to the Golgi stacks, they are modified by **palmitoylation**, the addition and trimming of oligosaccharides, and a reorientation of the carboxy-terminus of PE2 from a membrane-embedded to a cytosolic location. Prior to their appearance in the plasma membrane, a host-cell **furin protease** converts PE2 to E2 and E3.

Assembly of virions and egress at the plasma membrane

Two models have been proposed to explain virus assembly and release from the plasma membrane. The first model proposes that preformed nucleocapsids migrate to the plasma membrane, where they bind to sections of the membrane that are densely populated with virus envelope glycoproteins, and the enveloped particle is formed by budding from the surface of the infected cell (Figure 18.4; see also Figure 2.9a). This model is suggested by the presence in the cytoplasm of free, fully formed nucleocapsids, and by electron microscopic images showing budding virus particles that contain such nucleocapsids.

Initiation of the budding event is thought to occur when a nucleocapsid in the cytoplasm recognizes the cytoplasmic tails of arrays of glycoprotein trimers localized in the plasma membrane. There is a conserved motif (Tyr-X-Leu) in the carboxy-terminal tail of the E2 glycoprotein that interacts with a hydrophobic pocket on the surface of the capsid protein. As budding begins, each of the 240 capsid protein subunits of the core binds to an E2 protein molecule in the membrane. This forces bending of the membrane around the icosahedral nucleocapsid, culminating in fusion of the highly curved membrane and release of the virion from the cell. The precise pairing of envelope glycoproteins to capsid proteins in the core precludes incorporation of host cell surface proteins into the released virus particles.

A different assembly model proposes that budding progresses through "shell interactions" of the glycoproteins. In this model, capsid proteins bind to genome RNA in a complex that is not a structured nucleocapsid. The complex encounters the cytoplasmic tails of viral glycoproteins anchored in the plasma membrane, and the capsid proteins in the complex bind on a 1:1 basis to these tails. It is this binding that leads to the formation of the inner shell (made of capsid proteins) and outer shell (made of envelope glycoproteins), and membrane bending and budding produces virions. Evidence for this model comes from studies with capsid protein mutants in which preformed nucleocapsids are not detected in the cytoplasm of infected cells, but intact enveloped virions are still able to bud from the plasma membrane.

Effects of mutations in viral proteins on cytopathic effects and on pathogenesis

Infection of vertebrate cells with alphaviruses is usually cytopathic. Host cell protein synthesis is inhibited and in many cell types there appears to be induction of programmed cell death, or **apoptosis**. Alphavirus gene expression vectors, described in the next section, are also cytopathic. However, some mutants of these vectors are able to persist in mammalian cells without causing cytopathic effects. The mutants isolated so far all contain a single amino acid change in the C-terminal half of nsP2 and have a decreased level of viral RNA synthesis. There appears to be a correlation between the decrease in cytopathogenicity and the decrease in viral RNA synthesis, but it is not known which factors are most critical in producing cytopathic effects nor why mutations in the nsP2 gene allow cell survival. In contrast to vertebrate cells, infection of mosquito cells by wildtype virus is often not cytopathic. Because virions are formed by budding at the plasma membrane, virus can be produced and circulate in mosquito cell cultures without disruption of infected cells.

Most experimental studies of pathogenesis with Sindbis and Semliki Forest viruses have been done with neonatal mice, in which disease is more apparent than with adult mice and mortality is high. Whether or not infection will produce disease is dependent both on the virus and the host. Innate immune defense responses (for example, induction of **interferon**; see Chapter 31) can play an important role as does the adaptive immune response. A single amino acid change in the E2 glycoprotein of Sindbis virus has a profound effect on the neurovirulence of the virus. Differences in the 5′ untranslated region of the genome have been correlated with changes in survival of infected animals. These changes are subtle, yet their effects are significant, and virologists are trying to understand what steps in infection or in the response to infection are altered by these mutations.

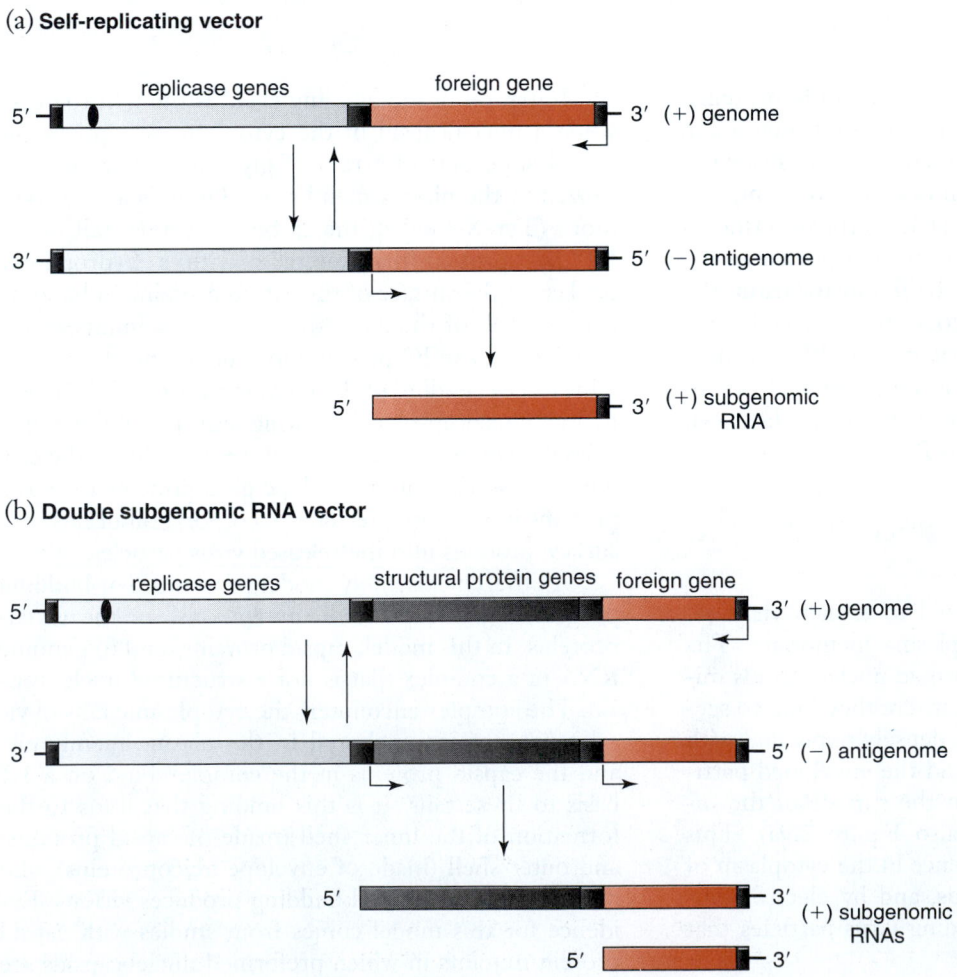

FIGURE 18.5 Alphavirus expression vectors. Black oval near left end of genomic RNA is packaging signal. The arrow on the antigenome pointing toward the 5′ terminus indicates the subgenomic RNA promoter and the start of the subgenomic RNA.

Alphaviruses have been modified to serve as vectors for the expression of heterologous proteins

The use of viruses as gene expression vectors is now a well-established technology (see Chapter 33). The first step in the generation of a viral vector from a positive-strand RNA virus is to convert the RNA genome to a DNA copy (cDNA) that can be cloned into a bacterial plasmid, modified in various ways, and then transcribed back into an infectious RNA. Alphavirus cDNAs were initially placed under the control of a promoter for bacteriophage T7 or SP6 DNA-dependent RNA polymerase. The cDNA was then transcribed *in vitro* and the RNA was introduced into cultured cells by transfection, leading to virus replication. Subsequently, viral cDNAs were put under the control of a eukaryotic promoter, enabling production of viral RNA by transcription of DNA introduced into the nucleus of a eukaryotic cell by transfection.

In the most widely used "self-replicating" alphavirus vectors (Figure 18.5a), the structural protein genes are replaced by a heterologous gene—initially, **reporter genes** such as the gene producing **β-galatcosidase** or the gene encoding **chloramphenicol acetyltransferase**, and subsequently any foreign gene of interest. The modified genome RNAs (referred to as **replicons**) serve as mRNAs for the translation of the nonstructural proteins, and their replication is carried out by these proteins. They produce subgenomic RNAs, which express the foreign protein.

Since replicons lack structural protein genes, production of virus particles requires cotransfection with defective helper genomes. These are alphavirus RNAs that have undergone large deletions in the nonstructural protein genes, but retain the promoter for the subgenomic RNA and the structural protein genes. They can be replicated by the nonstructural proteins made by the replicon RNA. If the defective helper genomes are mutated so that they lack signals for RNA packaging, only replicon genomes will be packaged into virions. These viruses have been termed "suicidal," as they are able to infect cells and produce a particular protein of interest within those cells, but do not produce new particles in the absence of helper genomes, and therefore cannot spread to neighboring cells.

In the "double subgenomic" alphavirus gene expression vector (Figure 18.5b), coding regions for the structural proteins are retained and a second subgenomic RNA promoter is inserted into the genome, allowing the synthesis of both the viral structural proteins and a foreign protein, each on different subgenomic RNAs (Figure 18.5). This type of vector produces a viable virus that can spread from cell to cell either *in vitro* or in an animal or human. These vectors tend to be less stable than replicons, particularly when the inserted foreign genes are large (greater than 2 Kb). The size limitation may reflect the packaging preference of alphavirus capsids for RNAs the size of genome RNA.

Alphavirus vectors may be useful for treatment of neurological diseases

Virologists who study alphaviruses have an interest in understanding the ability of these viruses to infect neurons. Many alphaviruses cause encephalitis; a better understanding of how these viruses enter the brain and why only some strains are able to cause symptoms may help in prevention of disease. The ability of alphaviruses to infect neurons may turn out to have some benefits. Studies with rat **hippocampus** slice cultures show that neurons are selectively infected by Sindbis and Semliki Forest virus replicons. Live neurons expressing **green fluorescent protein** produced by engineered viral replicons can be observed for up to five days postinfection. Because the infected neurons eventually die, the current alphavirus vectors would not be appropriate for any type of treatments in the human nervous system. Will it be possible to isolate replicon expression vectors that are not cytopathic to neurons? If so, then such vectors could be considered as a means to treat neurological diseases by replacement gene therapy. These vectors, however, can already provide valuable tools for expressing proteins and for identifying and studying neuronal functions.

Alphavirus vectors have multiple potential uses

Alphavirus vectors can be used to express a variety of important cellular proteins in different cell types. They can also be used for production and purification of selected proteins, because the levels of foreign proteins synthesized in infected cells can be quite high. These vectors are particularly useful for production of proteins that have post-translational modifications (such as phosphorylation or the addition of carbohydrates) that occur in vertebrate or insect cells but not in bacterial cells. Alphavirus vectors have also been engineered to express proteins of unrelated viruses, with the prospect of developing them as vaccines. In initial studies, proteins derived from influenza virus, herpes virus, and human immunodeficiency virus were expressed. Injection of these vectors into animals showed them to be effective in inducing an immune response and, in the case of influenza and herpes viruses, protecting the animal from subsequent infection.

Since alphaviruses also infect insects, particularly mosquitoes, vectors are being developed for use in insects. One approach is to use a Sindbis virus strain that is not pathogenic for insects to express an interfering RNA (RNAi) directed against human pathogens such as yellow fever virus or dengue virus (both flaviviruses). Experiments have shown that infection of mosquitoes with such a recombinant vector will effectively inhibit replication of yellow fever virus or dengue virus in the mosquito. This opens up the possibility of disseminating DNAs that code for such RNAi in mosquito populations, perhaps by introducing them into transposons that will spread within natural populations of mosquitoes. Any such strategy would be effective only if a significant proportion of mosquitoes received the gene that expresses the antisense RNA; infection by the Sindbis vector itself would not likely achieve this result.

Another approach is to have Sindbis virus express a protein toxic to the insect, serving essentially as an insecticide. Such a vector would generate an infectious virus that contains two subgenomic RNA promoters, one for the expression of structural proteins and another for expression of the toxic protein. However, release of such a virus into nature may be difficult to justify unless extensive testing shows its safety for humans and other animals. Because these viruses circulate between vertebrate and arthropod hosts, introduction of engineered viruses into mosquito populations could involve substantial risks to the vertebrate host.

SUGGESTED READING

1. Ehrengruber, M.U. (2002). Alphaviral vectors for gene transfer into neurons. *Molecular Neurobiology* **26,** 183–201.

2. Garoff, H., Sjöberg, M., and Cheng, R.H. (2004). Budding of alphaviruses. *Virus Research* **106,** 103–116.

3. Gibbons, D.L., Vaney, M.C., Roussel, A., Vigouroux, A., Reilly, B., Lepault, J., Kielian, M., and Rey, F.A. (2004). Conformational change and protein-protein interactions of the fusion protein of Semliki Forest virus. *Nature* **427,** 320–325.

4. Gorchakov, R., Hardy, R., Rice, C.M., and Frolov, I. (2004). Selection of functional 5′ *cis*-acting elements promoting efficient Sindbis virus genome replication. *Journal of Virology* **78,** 61–75.

5. Lemm, J.A., Rumenapf, T., Strauss, E.G., Strauss, J.H. and Rice, C.M. (1994). Polypeptide requirements for assembly of functional Sindbis virus replication complexes: A model for the temporal regulation of minus-strand and plus-strand RNA synthesis. *EMBO Journal* **13,** 2925–2934.

6. Schlesinger, S. and Dubensky, T.W., Jr. (1999). Alphavirus vectors for gene expression and vaccines. *Current Opinion in Biotechnology* **10,** 434–439.

7. Strauss, J.H. and Strauss, E.G. (1994). The alphaviruses: Gene expression, replication and evolution. *Microbial Reviews* **58,** 491–562.

8. Weaver, S.C. and Barrett, A.D. (2004). Transmission cycles, host range, evolution and emergence of arboviral disease. *Nature Reviews Microbiology* **2,** 789–801.

9. Zhang, W., Mukhopadhyay, S., Pletnev, S.V., Xiao, C., Baker, T.S., Kuhn, R.J., and Rossmann, M.G. (2002). Placement of the structural proteins in Sindbis virus. *Journal of Virology* **76,** 11645–11658.

KEY TERMS

Antigenome
Apoptosis
Arthropod-borne viruses
β-galactosidase
Chloramphenicol acetyltransferase
Cryoelectron microscopy
Cytopathic
Encephalitis
Furin protease
Green fluorescent protein
Heparan sulfate
Hippocampus
Interferon
Laminin receptor
Methyltransferase
Packaging signal
Palmitoylation
Persistent infection
Polyprotein
Receptor-mediated endocytosis
Replicon
Reporter gene
RNA helicase
Signal peptidase
Subgenomic RNAs

Coronaviruses

Coronaviridae

From Latin *corona* (crown), referring to spikes projecting from envelope

VIRION

Spherical enveloped particles studded with clubbed spikes.
Diameter 120–160 nm.
Coiled helical nucleocapsid.
Core shell may be icosahedral.

GENOME

Linear ss RNA, positive sense.
27–32 Kb.
5′ terminal cap.
3′ poly(A) tail.

GENES AND PROTEINS

Six–nine genes code for more than twenty proteins.
Replicase genes translated directly from genome RNA:

- *Two large open reading frames joined by frameshift*
- *Translated as polyprotein, cleaved to fourteen–sixteen proteins*

Other genes translated from multiple 3′-coterminal, subgenomic mRNAs:

- *Three–four envelope proteins (HE, S, E, M)*
- *One nucleocapsid protein (N)*
- *Four–six nonstructural or accessory proteins*

VIRUSES AND HOSTS

Three subgroups based on antigenic and genome homologies.
Humans: human coronaviruses 229E, OC43, NL63, HKU-1; SARS coronavirus.
Animals including mice, dogs, cats, pigs, cattle, turkeys, bats: murine hepatitis virus (MHV), transmissible gastroenteritis virus (TGEV), avian infectious bronchitis virus (IBV).

DISEASES

Responsible for up to 30% of common colds in humans.
Severe acute respiratory syndrome (SARS): pneumonia in humans with 10% fatality rate.
Important veterinary diseases: severe endemic and epidemic respiratory, hepatic, gastrointestinal, and neurological diseases in mammalian and avian species.
Virus spreads between hosts by direct contact, aerosol and fecal-oral transmission, and contact with contaminated surfaces.

DISTINCTIVE CHARACTERISTICS

Largest known RNA virus genome.
Large replicase polyprotein cleaved by viral proteinases to numerous functional proteins.

Subgenomic mRNAs made by unusual discontinuous transcription mechanism.
RNA replication occurs on virus-induced double-membrane vesicles in cytoplasm.
Error-prone RNA polymerase and high rate of RNA recombination result in rapid adaptation and changes in host range and disease.

Coronaviruses cause common colds in humans and important veterinary diseases

The search for viruses that cause upper-respiratory infections in humans (the "common cold") was intense in the mid 1960s. The laboratories of David Tyrrell, Dorothy Hamre, and Robert Chanock isolated viruses from humans that could replicate on fragments of tracheal tissue or in human cells in culture. When examined by electron microscopy, the enveloped virion was seen to be surrounded by a crown or *corona* of studded projections, and was therefore named a coronavirus. Two distinct human coronaviruses, called 229E and OC43, were isolated and studied. These viruses typically cause mild to moderate respiratory illness in healthy individuals, being responsible for up to 30% of colds in some outbreaks. They only rarely cause more severe respiratory diseases, **gastroenteritis**, and **encephalitis**. These human coronaviruses can cause repeated respiratory infections on an alternating 2–3-year basis in otherwise healthy hosts, and do not appear to induce long-lasting immunity.

A number of veterinary diseases of economic importance were shown to be caused by similar viruses. One of the first coronavirus diseases to be recognized (1912) was feline infectious **peritonitis**, although the virus was isolated much later. Coronaviruses cause devastating outbreaks of gastroenteritis and respiratory infections in cattle (bovine coronavirus), pigs (transmissible gastroenteritis virus), and chickens (infectious **bronchitis** virus), with mortality rates as high as 100% in newborn animals. These viruses have been targets for vaccine development and use, and have served as models for study of coronavirus biology. Murine (mouse) **hepatitis** virus (MHV), a coronavirus, was isolated from laboratory mouse strains in the early 1960s. This virus has been very useful for experimental studies, as it grows to high titers in cultured cells and its molecular biology and pathogenesis can be readily studied.

A newly emerged coronavirus caused a worldwide epidemic of severe acute respiratory syndrome (SARS)

The field of coronavirus research took an abrupt turn when, in early 2003, an epidemic of severe and often fatal pneumonia broke out in southeast China, Hong Kong, and Vietnam, and subsequently spread to Toronto, Canada. This new disease was named severe actute respiratory syndrome (SARS), and the causative agent was quickly identified as a previously uncharacterized coronavirus. The SARS coronavirus was isolated from people of all age groups who presented with fever and progressive pneumonia, leading to **hypoxemia** and death. Over a six-month period this epidemic led to more than 8000 illnesses and nearly 800 deaths worldwide. Mortality was highest (up to 44%) in individuals over 65 years of age.

The spread of SARS infections usually took place among family members of victims or in hospital care settings, and occured principally during symptomatic illness. Virus was transmitted by direct contact with patients, but also via aerosol droplets and contaminated stool. Air travel by infected individuals quickly spread the disease to 32 countries, resulting in the first pandemic of the twenty-first century. Public health isolation measures appear to have interrupted the chain of person-to-person infection, but not without serious economic loss for the involved cities, a significant drop in travel to southeast Asia and China, and the illness and death of hospital personnel caring for victims. Although human transmission and epidemic disease appears to have been eradicated, additional cases in 2003 and 2004 arose in laboratory settings and from animal-to-human transmission.

The SARS coronavirus may have passed from animals to humans via direct contact

The SARS coronavirus was likely introduced into humans from contact with infected animals that serve as a virus **reservoir**. As suggested for avian influenza, Ebola virus, and several other viruses, this emerging coronavirus probably crossed from animals to humans as a result of repeated direct contact. Shortly after the appearance of the SARS epidemic, a SARS coronavirus was isolated from palm civets, carnivores that are captured or farmed and sold in markets for food in Guangdong province, in southern China. Subsequent field studies of wild and farmed palm civets suggest that the SARS virus is not commonly present in these animals, and that they may have served as intermediate or "amplifier" hosts rather than the natural reservoir. A SARS-like coronavirus was more recently discovered in a number of species of Chinese horseshoe bats. Because a significant proportion of these bats show indication of present or previous SARS virus infection, and because virus isolates from these bats are more genetically diverse than those from humans or palm civets, Chinese horseshoe bats are now considered the natural host of this virus. Because coronaviruses can undergo high rates of recombination, it is feared that other coronaviruses might cross the species barrier as a result of the generation of recombinants containing both animal and human coronavirus genes.

Improved technologies and diagnostic tools in combination with the renewed interest in coronaviruses have resulted in the identification of several novel human viruses. Human coronaviruses isolated in the Netherlands (NL and NL63) and in Hong Kong (HKU1) were both associated with patients with pneumonia. As better diagnostic tools are used to screen humans and animals for infectious diseases, additional members will likely be added to the *Coronaviridae*.

Coronaviruses have large, single-stranded, positive-sense RNA genomes

Coronaviruses have the largest RNA genomes (27–32 Kb) of any known virus family. The single-stranded, positive-sense RNA genome contains a 5′ methylguanosine cap and a 3′ poly(A) tail. The RNA genome contains 6 to 10 genes, depending on the virus strain. Figure 19.1 shows a map of the genome of murine hepatitis virus. Like other positive-sense RNA viruses, coronavirus genomes are infectious; introduction of purified genome RNA into a cell will initiate infection and result in production of new virus particles.

The order of genes is highly conserved among different coronaviruses. Remarkably, the first gene from the 5′ end occupies two-thirds of the viral genome (22 Kb in the case of murine hepatitis virus), and is translated into **polyproteins** that are cleaved by viral proteinases into 16 or more different proteins required for genome replication. The distal one-third of the genome contains genes coding for virion structural proteins. These are found in the following order: **hemagglutinin**-esterase (HE), spike (S), small envelope protein (E), membrane protein (M), and nucleocapsid protein (N). Hemagglutinin-esterase is not present in all coronaviruses. The conservation of gene order is notable given the high frequency of recombination observed in coronaviruses and the experimental demonstration that genes can be artificially reordered and still allow virus replication. A variable number of accessory or group-specific genes are interspersed between these structural protein genes.

Coronaviruses fall into three groups based on genome sequences

Sequence analysis of coronavirus genomes and studies of their serotypic (antigenic) relationships have led to division of most coronaviruses into three groups of closely related viruses (Table 19.1). The best-studied viruses are: porcine transmissible gastroeneteritis virus and human coronavirus 229E (group I); murine hepatitis virus, bovine coronavirus, and human coronavirus OC43 (group II); and avian infectious bronchitis virus (group III). Study of these viruses has provided the basis for the understanding of replication and pathogenesis of *Coronaviridae* and has provided most of what is known about this family of viruses as a whole.

The SARS coronavirus does not fit neatly within one of three established phylogenetic groups of coronaviruses. Instead, its genome sequence suggests that the SARS coronavirus is a product of a recombination event between an ancestral group II mammalian and a group III avian coronavirus, resulting in a **chimeric** virus. The recombination event is believed to have occurred in an ancestral form, because viruses isolated from palm civets are nearly identical to the SARS coronaviruses isolated

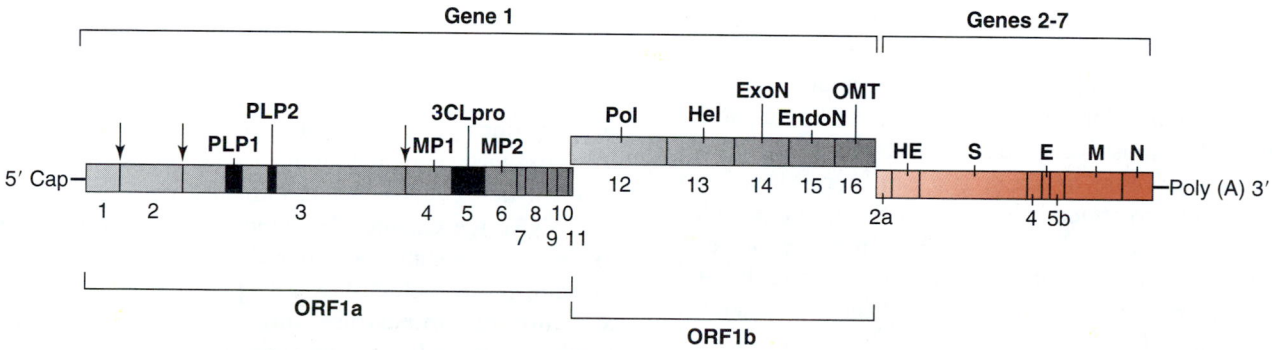

FIGURE 19.1 **Organization of the coronavirus genome.** The 31.5 Kb genome of murine hepatitis virus genome is shown. A short 5′ nontranslated leader region (gray) is followed by gene 1, which consists of two overlapping reading frames (ORF1a and 1b) that are translated as polyproteins from genome RNA. Genes 2–7 (orange) are expressed from subgenomic messenger RNAs. Where known, the names or function of gene products are shown above the genome map. The nonstructural proteins nsp1–16 (shown below the map) are generated by cleavage from the 1a/1b fusion protein by viral proteinases. Cleavage by proteinases PLP1 and PLP2 (black, cleavage sites shown as vertical arrows), yields nsp1–3, and cleavage by 3CLpro (black) yields nsp4–16.

TABLE 19.1 Human and animal coronaviruses

Group	Virus	Disease
I	Human coronavirus-229E	Colds, pneumonia
	Human coronavirus-NL63	Bronchiolitis, colds, pneumonia
	Human coronavirus-NH	Systemic vasculitis (Kawasaki disease)
	Porcine transmissible gastroenteritis virus	Pneumonia, gastroenteritis
	Canine coronavirus	Gastroenteritis
	Feline infectious peritonitis virus	Peritonitis, enteritis
	Bat coronavirus	Asymptomatic (respiratory and enteric)
II	Human coronavirus-OC43E	Colds, pneumonia
	Human coronavirus-HKU1	Pneumonia
	Murine (mouse) hepatitis virus	Hepatitis, encephalitis
	Bovine coronavirus	Pneumonia, gastroenteritis
IIb	SARS coronavirus	Pneumonia, gastroenteritis
III	Avian infectious bronchitis virus	Tracheitis, kidney infection

from humans, and these isolates are more closely related to each other than to any other coronaviruses.

Coronaviruses have enveloped virions containing helical nucleocapsids

The coronavirus virion is a spherical enveloped particle, ranging in size from 75 to 160 nm, that carries a characteristic crown of spikes that gives the virus family its name (Figure 19.2). Electron micrographs of thin sections of infected cells reveal virions with a donut-like core structure that represents the helical nucleocapsid underlying the viral envelope; the helical nucleocapsid can also be seen in negatively stained disrupted virions. The nucleocapsid is formed from multiple copies of the nucleocapsid protein (N) bound to the viral RNA genome. Most other positive-strand RNA viruses have nucleocapsids with icosahedral symmetry; coronaviruses thus resemble negative-strand RNA viruses in having helical nucleocapsids. However, there is evidence that mature virions of at least some coronaviruses, including porcine transmissible gastroenteritis virus, have a stable, spherical core structure that can be released from the virion by treatments that remove the envelope. This core is associated with the virus M (membrane) protein. These coronaviruses may be unique among known viruses in having both a spherical (perhaps icosahedral) core shell and an internal helical nucleocapsid.

Coronavirus virions contain multiple envelope proteins

Spike protein. The spike (S) glycoprotein is the major surface transmembrane protein protruding from the viral envelope (Figure 19.2b). S forms trimers whose stalk-and-club morphology gives the virion a crown-like fringe of proteins. The spike protein is responsible for virus entry and spread, and is the primary determinant of host range. It is a target for neutralizing antibody and cytotoxic T lymphocyte responses to virus infection.

The spike protein is synthesized as a single polypeptide chain that in some coronaviruses is cleaved by a cellular proteinase to yield an amino-terminal S1 domain and a carboxy-terminal S2 domain. The S1 subunit, which forms the clubbed end of the spike, recognizes specific cellular receptors and initiates attachment. The S2 subunit, which forms the stalk, has a short C-terminal tail, a hydrophobic transmembrane domain, and an exterior domain with two long alpha-helical regions. Adjacent alpha helices within the trimer interact with each other to form a structure called a **coiled-coil**, similar to the transmembrane portion of the envelope glycoproteins of a number of other RNA viruses.

Hemagglutinin-esterase protein. Some coronaviruses, notably group II viruses, possess a hemagglutinin-esterase (HE) protein on the surface of the virion. HE has functional similarities both to the receptor-binding influenza virus hemagglutinin protein (HA), and to the receptor-destroying enzyme, influenza virus neuraminidase (NA) (see Chapter 23). Since not all coronaviruses express HE, it is not considered an essential structural protein. Nevertheless, HE may play a role in binding and entry of coronaviruses into target cells, and could facilitate virus release from the cell surface.

Membrane protein. The membrane (M) glycoprotein is the most abundant structural protein in the coronavirus virion. M is a highly conserved, triple membrane-spanning protein containing three hydrophobic domains separated by short hydrophilic regions. Rigid constraints on structure are likely essential for the functional requirements of M in virus assembly. The amino

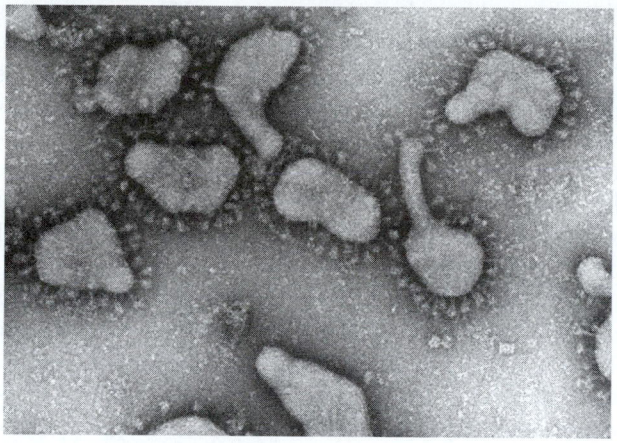

(a)

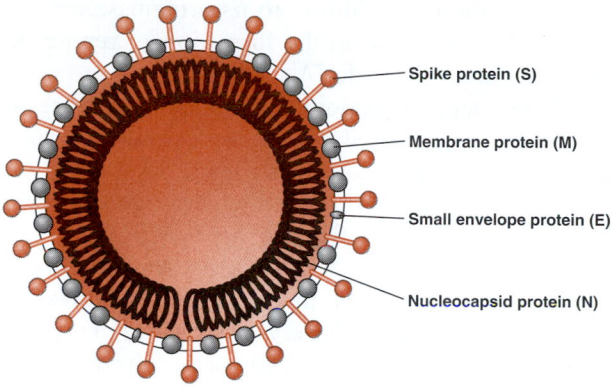

(b)

FIGURE 19.2 Coronavirus virion. (a) Electron micrograph of negatively stained murine hepatitis virions shows spikes protruding from the envelope, providing its crown-like appearance. (b) Schematic diagram. The three envelope proteins spike (S), membrane (M), and envelope (E) are shown. The hemagglutinin-esterase (HE) is only present in some coronaviruses and is not shown here. The helical nucleocapsid, made from the RNA genome complexed with N protein, is shown just underneath the envelope. Some coronaviruses have a spherical core.

terminus of M is exposed on the outer surface of the envelope and is in contact with the spike protein; the carboxy terminus extends within the virion underneath the envelope, and is in contact with the viral nucleocapsid.

E: Many form virion
But no E still OK, ∴ not essential

Small envelope protein. The small envelope (E) protein is a 9–12 KDa protein that is present only in very limited numbers in the virion. Virus-like particles can be formed by expressing only the E and M proteins together in cells, suggesting a critical role for E in virion formation and budding. The E protein may serve to facilitate budding of the virion into the lumen of the endoplasmic reticulum and Golgi membranes during particle maturation (see below). Interestingly, mutant coronaviruses lacking the E protein are viable, indicating that this protein is not absolutely required for virion formation.

Coronavirus spike proteins bind to a variety of cellular receptors

The spike proteins of group I coronaviruses bind to aminopeptidase-N, a family of zinc-binding **metalloproteinases** found in cellular plasma membranes. These are broadly distributed in epithelial and fibroblast cells in the small intestine, renal epithelium, and central nervous system, and on monocyte and granulocyte lineages. Despite the high degree of similarity among members of the aminopeptidase-N family, the utilization of these receptors by coronaviruses is for the most part species-specific.

Group II coronaviruses utilize a variety of cell receptors. Murine hepatitis virus binds to carcinoembryonic antigen-related cell adhesion molecule 1 (CEACAM1 or CD66a). CEACAM1 is a member of the immunoglobulin family and is widely distributed in all mammalian species. The SARS coronavirus uses angiotensin-converting enzyme 2 (ACE2), another metalloproteinase, as a receptor. Because the SARS coronavirus can also infect certain cell types that lack ACE2, additional or alternate receptors are being sought. Recent studies indicate that L-SIGN, a cell surface **lectin**, may also serve as a receptor for the SARS coronavirus.

Viruses that make hemagglutinin-esterase (HE) can also bind to cells by interaction between HE and **sialic acid**, found on a variety of cell membrane glycoproteins and glycolipids. Sialic acid is a common target bound by surface proteins of many animal viruses. This esterase can cleave an acetyl group from sialic acid, resulting in the release of bound virus from the cell surface. This may help in the release and dispersal of virions from some cell types.

The virus envelope fuses with the plasma membrane or an endosomal membrane

Following virion attachment, the coronavirus S protein mediates fusion of the virus envelope with cellular membranes. While the external S1 subunit is responsible for attachment, the S2 subunit that forms the stalk facilitates fusion through a series of conformational changes that result in the insertion of S2 into the target cell membrane, bringing the cell membrane and the viral envelope into close contact. This mechanism appears to parallel the better-studied fusion mechanisms of influenza virus (Chapters 4 and 23) and human immunodeficiency virus (Chapter 26). These proteins are called **class I fusion proteins**, and are structurally distinct from the **class II fusion proteins** of togaviruses and flaviviruses.

Some coronavirus S proteins can induce cell-cell fusion at neutral or alkaline pH values, suggesting that these virions fuse directly with the plasma membrane. When these S proteins are incorporated into the plasma membrane of an infected cell, cell fusion can occur, resulting in the formation of large multinucleated giant cells, or **syncytia**, which are observed in some coronavirus infections. Other coronavirus S proteins require low pH to induce fusion, suggesting that they enter via endosomes and fuse with the endosomal membrane once the internal pH has dropped.

The replicase gene is translated from genome RNA into a polyprotein that is processed by viral proteinases

Because coronaviruses have positive-sense RNA genomes, the first step in their replication cycle after entry is the synthesis of viral proteins that organize and catalyze genome replication. All known coronaviruses have a similar genome organization, in which the 20–22 Kb replicase gene (gene 1) comprises the 5′ two-thirds of the RNA genome. Strikingly, this gene is organized as two partially overlapping open reading frames, called ORF1a and ORF1b (Figure 19.1). Translation by cellular ribosomes begins at an AUG initiation codon shortly beyond the 60–100 nt untranslated leader and progresses through ORF1a, producing a polyprotein that is subsequently cleaved into 11 smaller proteins in the case of murine hepatitis virus. Some of the ribosomes translating ORF1a pause on a complex RNA structure known as a **pseudoknot** in the overlap between ORF1a and 1b, just upstream of the translation termination codon at the end of ORF1a. This leads to a shift in the reading frame, allowing continued translation through ORF1b and production of a larger ORF1a/1b polyprotein, which is cleaved into as many as 16 mature proteins. It is not clear why coronaviruses have these two overlapping reading frames; it may be a mechanism for ensuring production of higher levels of the proteins coded by ORF1a compared with proteins coded by ORF1b.

The polyproteins are processed by virus-coded proteinases at cleavage sites flanking the protein domains. These proteinases act autocatalytically, as they are part of the polyprotein, and therefore they can be enzymatically active before they are released by self-cleavage. Polyprotein synthesis and cleavage is a familiar pattern for other positive-strand RNA animal viruses (picornaviruses, flaviviruses, and togaviruses: Chapters 16–18). The viral proteinases include one or two **papain-like cysteine proteinases** (PLP1 and PLP2) and a chymotrypsin-like cysteine proteinase that resembles the **picornavirus 3C proteinases** (3CLpro, also known as Mpro or nsp5) (Figure 19.1). The most consistent nomenclature for the replicase protein domains is "nonstructural protein" (nsp), although roles in virus structure have not been exhaustively studied. The SARS coronavirus and murine hepatitis virus have analogous replicase proteins that will be referred to as nsp1 through nsp16 (Figure 19.1 and Table 19.2).

The papain-like proteinases (PLPs) are subdomains of nsp3, a 195 to 213 KDa replicase protein (Figure 19.1). The PLPs catalyze cleavage between nsp1-nsp2, nsp2-nsp3, and nsp3-nsp4. Interestingly, cleavage between nsp1-nsp2 and nsp2-nsp3 does not appear to be required for viral replication in cultured cells *in vitro*, as these cleavage sites can be mutated without effect on virus replication. In addition to its proteinase activity, nsp3 is likely involved in the formation of replication complexes and in viral RNA synthesis.

The remaining portions of the polyproteins are autocatalytically processed by the nsp5 proteinase, producing proteins nsp4 through nsp16. The activity of nsp5 is thought to be regulated by cleavage site specificity as well as by the flanking nsp4 (MP1) and nsp6 (MP2) proteins, both of which are hydrophobic membrane-spanning proteins that likely are involved in membrane association. The crystal structure of nsp5 for several coronaviruses has recently been solved, and is a target for development of antiviral agents against the SARS coronavirus.

RNA polymerase, RNA helicase, and RNA modifying enzymes are coded by the replicase gene

The coronavirus replicase gene encodes multiple proteins that are known or predicted to direct viral RNA synthesis or to interact with or modify RNAs (Table 19.2). The predicted RNA-dependent RNA-polymerase (nsp12) is a 100 KDa protein encoded in ORF1b. ATPase and RNA helicase activities of nsp13 may be critical in replication and packaging of genome RNA. Proteins with RNA modifying activity have been predicted for nsp14 (exonuclease) and nsp16 (methyltransferase, which methylates 5′ cap structures), and demonstrated *in vitro* for nsp15 (endoribonuclease). The activities of nsp12 through nsp16 during infection remain to be experimentally confirmed.

The functions of nsp1, 2, 4, and 6–11 are not known, but there are data suggesting roles for these replicase proteins in virulence, cell cycle control, cell membrane modification, and induction of **apoptosis** and **autophagy** (see below). Thus during evolution of these viruses, a large range of protein functions important for replication have been assembled and incorpo-

TABLE 19.2 Group II coronavirus proteins and functions

Protein	Functions
nsp1, 2, 7-10, 11	Unknown functions in replication; nsp1 may mediate cell-cycle arrest during overexpression in culture.
nsp3	One or two papain-like proteinase domains (PLP1, PLP2) responsible for cleavage of nsp1, nsp2, and nsp3; zinc ribbon motifs with predicted transcription factors; transmembrane sequences with membrane integration.
nsp4, nsp 6	Membrane-spanning proteins, may localize replication complexes to membranes.
nsp5	Picornavirus 3C-like proteinase (3CLpro or Mpro) responsible for cleavage of nsp4 through nsp16.
nsp12	Predicted RNA-dependent RNA polymerase responsible for genome replication and transcription.
nsp13	RNA helicase, nucleoside triphosphatase activity *in vitro*. Likely involved in genome unwinding, separation, and packaging; may be virulence factor.
nsp14, 15, 16	Predicted RNA modifying enzymes: 3′ to 5′ exonuclease (14), endoribonuclease (15), and O-methyl transferase (16).
Hemagglutinin-esterase (HE)	Envelope protein in some but not all coronaviruses. Virus entry, pathogenesis, virus release.
Spike (S)	Envelope protein. Receptor binding, fusion, tropism.
Envelope (E)	Minor envelope protein. Required for envelope formation.
Membrane (M)	Major envelope protein that contacts both spike and nucleocapsid protein. Required for envelope formation.
Nucleocapsid (N)	Forms helical nucleocapsid with genome RNA. Interacts with M protein; may form icosahedral shell.
Accessory or group-specific proteins	Three to five proteins, dispensible for replication *in vitro*. Functions unknown.

rated as the result of the plasticity and adaptability of the coronavirus genome.

Replication complexes are associated with cytoplasmic membranes

All known coronavirus replicase proteins can be found in membrane-associated complexes in the cytoplasm. These complexes are the sites of viral RNA synthesis, and are thus referred to as **replication complexes**. Nucleocapid protein is also found in abundance in replication complexes, suggesting that encapsidation of newly synthesized genome RNA for packaging into virions also occurs at these sites.

Replication complexes are observed as double-membrane vesicles by electron microscopy. The formation of these unusual structures likely involves both direct membrane modifications by viral replicase proteins and the induction of a cellular organelle-recycling pathway known as autophagy (*auto* = self, *phagy* = eating). Autophagy can serve as a cellular defense response against invading pathogens, but the coronaviruses may usurp components of this pathway for generation of membrane structures required for replication. Membrane association of viral RNA synthesis is a common pattern seen among positive-strand RNA viruses of eukaryotes (see Chapters 16–18).

Genome replication proceeds via a full-length negative-strand intermediate

The proteins in replication complexes direct both genome replication and messenger RNA synthesis, or transcription. Genome replication requires the synthesis of a full-length negative-strand copy of the viral genome, which in turn is used to direct synthesis of full-length positive-strand genome RNAs. Negative-strand templates account for only 1–2% of total viral RNA produced in an infected cell. Most negative-strand RNAs are associated with one or more growing positive-strand molecules in a **replicative intermediate**.

Transcription produces a nested set of subgenomic mRNAs

The expression of all coronavirus genes beyond the replicase gene occurs from a series of **subgenomic mRNAs** (Figure 19.3). These mRNAs have a unique structure. The sequence of the first 60–100 nucleotides starting from their capped 5′ ends is identical to the same region at the 5′ end of genome RNA. This untranslated "leader" sequence is joined to an mRNA "body" sequence that includes one or more of the open reading frames in genes 2 through 7 (for murine hepatitis virus). All six of these subgenomic RNAs extend to the 3′ end of the genome RNA and, like genome RNA, have poly(A) tails. Because they have different lengths but share common sequences at their 3′ ends, these are referred to as a **nested set** of mRNAs. Each one of the subgenomic mRNAs is translated to produce one or more proteins coded by the open reading frames closest to the 5′ end of the mRNA. These include the virion envelope proteins HE, S, E, and M, and the nucleocapsid protein N, as well as several nonstructural or accessory proteins that are distinct to each virus species.

Subgenomic mRNAs are most likely transcribed from subgenomic negative-sense RNA templates

How are these subgenomic mRNAs synthesized by coronaviruses? Most experimental evidence suggests that each subgenomic mRNA is transcribed from a corresponding subgenomic negative-strand RNA. The negative-strand subgenomic RNA templates are themselves generated by **discontinuous transcription** of full-length positive-strand genome RNA. This model is outlined in Figure 19.4. Briefly, the viral RNA polymerase initiates transcription at the 3′ end of (positive-strand) genome RNA and makes a (negative-strand) RNA copy up through one of the sequences known as transcription-regulating sequences (TRS). These repeated, short TRS (5′-UCUAAAC-3′ for murine hepatitis virus) are located at the 5′ ends of each of genes 2 through 7; there is an additional copy at the 3′ end of the leader sequence in genome RNA.

At this point the RNA polymerase pauses, dissociates the nascent RNA chain from the TRS, and brings the nascent chain to the TRS located at the end of the leader sequence in the template RNA. The copy of the TRS at the 3′ end of the nascent RNA chain forms an RNA-RNA hybrid with the complementary TRS at the end of the leader sequence on the genome RNA. This allows the RNA polymerase to extend the nascent negative-strand RNA through the leader sequence up to the 5′ end of the genome. This completes the synthesis of the subgenomic RNA.

As the RNA polymerase traverses the genome from right to left (Figure 19.4), it can pause and dissociate at any one of the TRS located at the beginning of genes 2 through 7, or it can skip them all and continue to the end of the genome, making a full-length negative-strand antigenome RNA. As a result, a nested set of negative-strand RNAs with common 5′ ends but different lengths is produced. Full-length negative-strand RNA (not shown in Figure 19-4) can serve as a template for production of genome RNA by the viral RNA polymerase. Each of the negative-strand subgenomic RNAs is copied by viral RNA polymerase to generate a nested set of positive-strand subgenomic mRNAs (bottom right of Figure 19.4).

Discontinuous transcription can also occur by the viral RNA polymerase switching between two differen positive-strand RNA genome templates within the same cell. If these two genomes are not identical, because of mutation or because two virus strains infected the same cell, the result can be homologous recombination. This template switching may be a normal feature of coronavirus replication and may be important for genome repair and generation of new strains of coronaviruses. It is reminiscent of similar recombination events between retrovirus genomes that can occur during reverse transcription (Chapter 25).

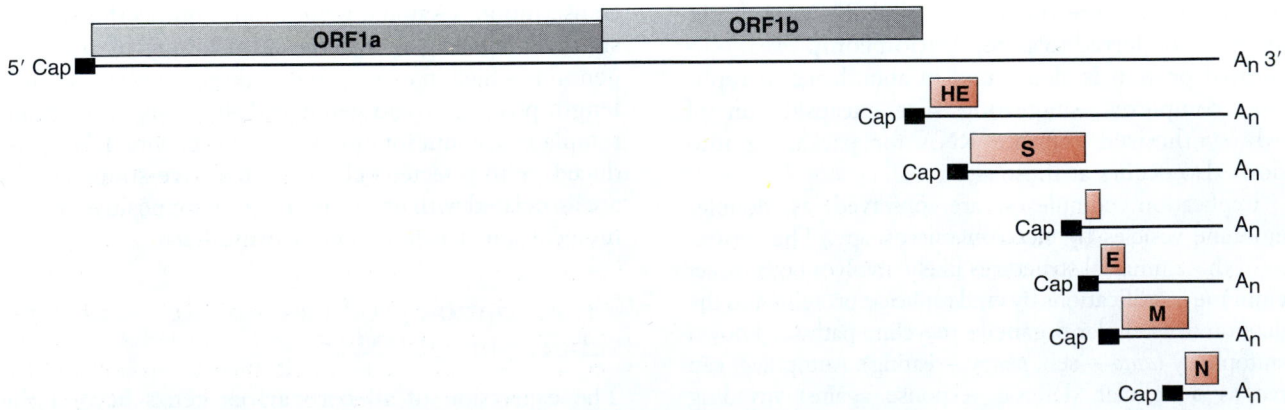

FIGURE 19.3 Coronavirus messenger RNAs: a nested set. The seven messenger RNAs of murine hepatitis virus are shown, along with the open reading frames they use to synthesize viral proteins. All share the same 5′ leader sequence and all contain overlapping sequences at their 3′ ends.

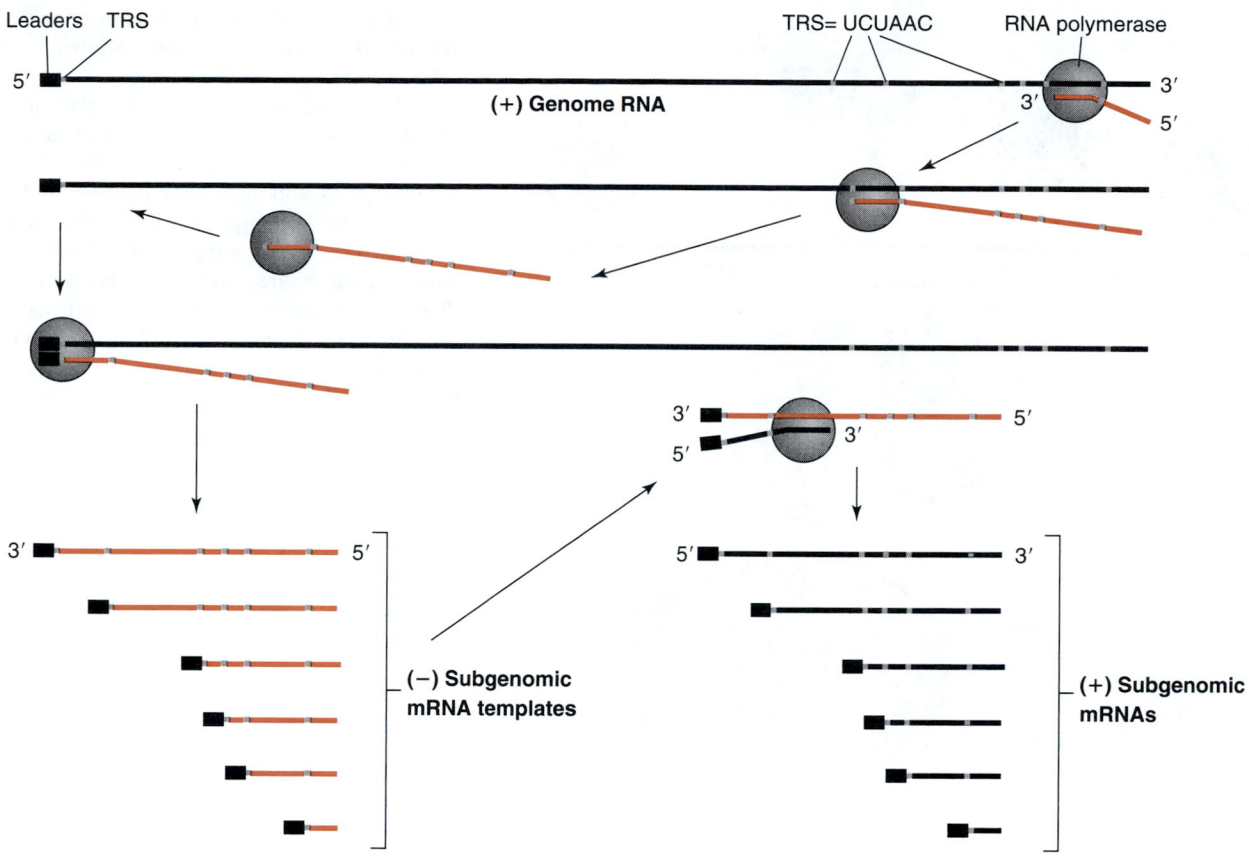

FIGURE 19.4 **Discontinuous transcription of genome RNA: a model for coronavirus mRNA production.** Top: viral RNA polymerase initiates transcription at the 3′ end of genome RNA and progresses (below) to one of six transcription regulatory sequences (TRS: 5′-UCUAAC-3′ for murine hepatitis virus), where it pauses and then interacts with the TRS at the 3′ end of the leader sequence. The two RNAs base-pair at the TRS and the RNA polymerase extends the RNA chain to the 5′ end of genome RNA, copying the leader sequence. Depending on where the RNA polymerase pauses, the product is either full-length antigenome RNA (no pause; not shown) or one of the six subgenomic negative-strand RNAs shown at bottom left. These are in turn copied by RNA polymerase to make more full-length genome RNA (not shown) or the six subgenomic mRNAs.

The alternative model of discontinuous transcription of antigenome RNA is unlikely to be correct

An alternative model proposes that subgenomic mRNAs are made directly by discontinuous transcription of full-length, negative-sense antigenome RNA. In this model, the 5′ leader sequence and its associated TRS would be synthesized first; the RNA polymerase would then switch to one of the six downstream TRS sequences and continue transcription to the end of the antigenome template, giving rise to the nested set of mRNAs. However, several lines of evidence suggest that the model shown in Figure 19.4 is correct:

1. Replicative intermediate forms corresponding to subgenomic negative-sense RNAs being transcribed into their corresponding mRNAs can be detected in the cell.

2. Equine infectious anemia virus, a member of the related *Arterivirus* family, produces a similar set of subgenomic mRNAs. When a single internal TRS was mutated in this arterivirus genome RNA, the mutation was retained in the corresponding subgenomic messenger RNA. However, when the TRS next to the leader sequence at the 5′ end of the genome RNA was mutated, the mutation was not retained in any of the subgenomic mRNAs. This implies that discontinuous transcription occurs on genome RNA (as shown in Figure 19.4); the TRS associated with the mRNA body sequence is made first as the RNA polymerase passes from the 3′ end to the 5′ end of genome RNA. This TRS would then be base-paired to its complementary sequence at the end of the leader region, allowing subsequent copying of the leader sequence. The TRS at the leader region would not be retained in the negative-strand subgenomic RNA made during discontinuous transcription of genome RNA.

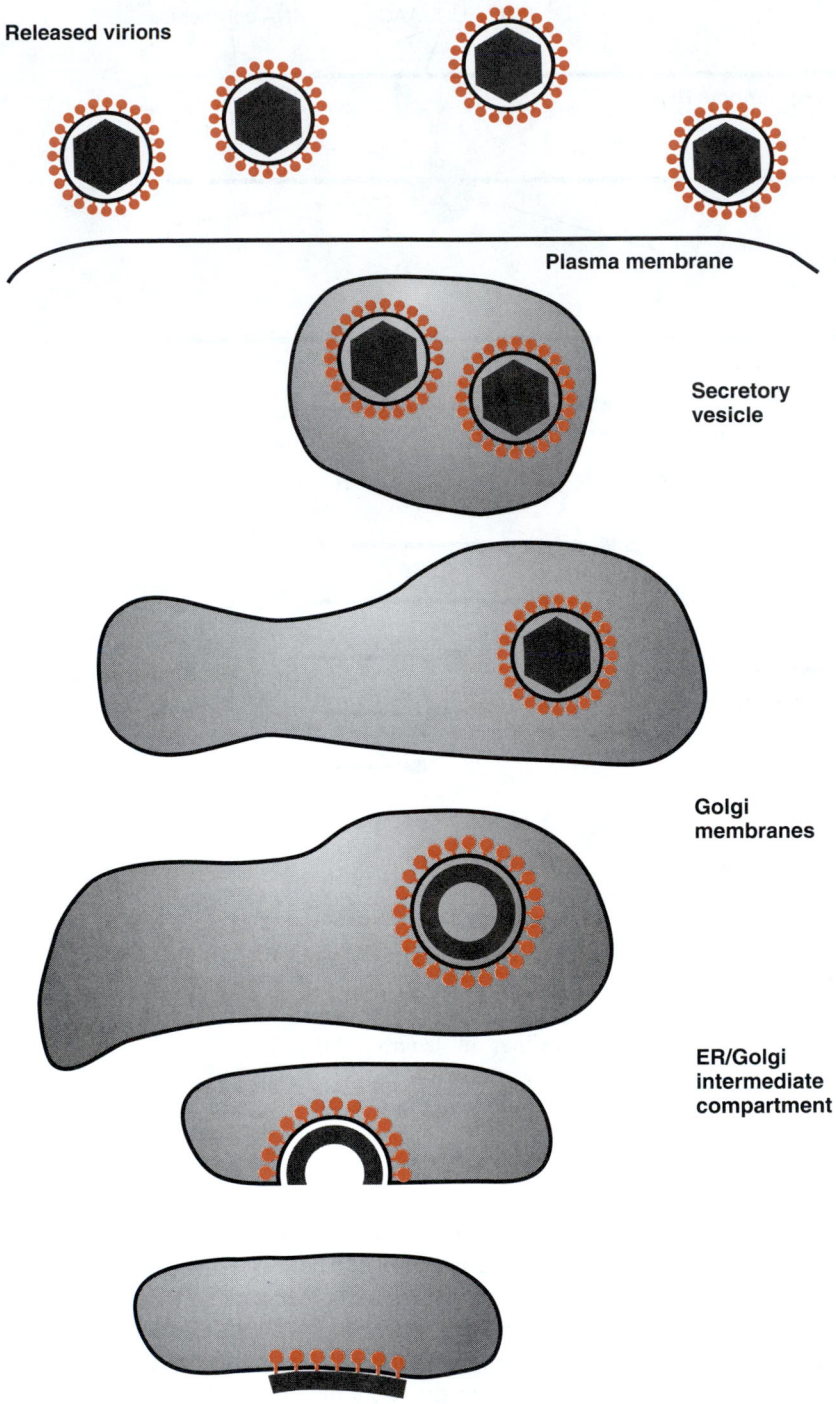

FIGURE 19.5 Assembly, maturation, and release of coronavirus virions. Shown schematically is the pathway of virion assembly at membranes of the endoplasmic reticulum–Golgi intermediate compartment (below): the thick gray lines represent helical nucleocapsids, and the orange knobbed stalks represent envelope proteins. As they bud into the lumen, virions acquire donut-shaped cores: These progress to a more uniformly dense and smaller core as virions go through the Golgi membranes, where glycosylation of envelope proteins occurs. Secretory vesicles transport virions to the cell surface, where the vesicles fuse with the plasma membrane and release virions.

3. The most abundant subgenomic mRNAs are the smallest ones, and the least abundant mRNAs are the largest ones. This can be most easily explained by the model in Figure 19.4. RNA polymerases all begin transcription at the 3′ end of genome RNA; they pause and dissociate with a certain probability as they encounter each TRS site, so the number of transcribing polymerases should decrease as they progress from the 3′ end toward the 5′ end of genome RNA. Note that this allows the virus to control the abundance of its mRNAs, and therefore its proteins. The genes for the N and M proteins, which are needed in highest numbers for virion formation, are nearest to the 3′ end of the genome, and these are the most abundant mRNAs made. A similar gradient in the abundance of mRNAs, also dependent on RNA polymerase pausing between genes, is found with the negative-strand paramyxoviruses and rhabdoviruses, but a very different transcription mechanism is involved (Chapter 20).

Assembly of virions takes place at intracellular membrane structures

Once coronavirus genome replication and transcription have taken place, the action shifts to a different set of intracellular membranes, the **endoplasmic reticulum–Golgi intermediate compartment (ERGIC)**. These membranes, located in the perinuclear region of the cell, lie in between the endoplasmic reticulum and the Golgi apparatus, and are part of an interconnecting system of membranes that is involved in modification and secretion of cellular proteins. Virions are assembled at this membrane system (Figure 19.5). Helical nucleocapsids containing genome RNA are delivered from their site of synthesis to these membranes for packaging. While the viral and cellular determinants of this delivery system remain to be confirmed, N protein and replicase proteins nsp1 and nsp13 of murine hepatitis virus have been shown to relocalize from replication complexes to the intermediate compartment, suggesting possible roles in the movement of nucleocapsids.

Electron micrographs of infected cells show densely staining material, probably helical nucleocapsids, lying on the cytosolic face of membranes in the intermediate compartment. Virus particles can be seen forming by budding into the lumen of these membranes. These are spherical particles that usually have a donut-like core morphology and are larger than mature virions (Figure 19.5, bottom). The M and E proteins play important roles in the formation of virus envelopes by budding. Empty, spikeless virus envelopes (virus-like particles) can be formed in the intermediate compartment if M and E proteins are expressed together in cells, showing that these proteins are sufficient for assembly and budding. The C-terminal cytoplasmic tail of M is thought to interact with packaging signals in the nucleocapsid to ensure specific incorporation of only full-length viral genome RNA into virions. HE and S proteins are incorporated into the membrane via interactions with the M protein. These immature virions are transported within vesicles through the Golgi apparatus, where the envelope proteins are glycosylated. As they mature, virions become smaller and the donut-shaped core becomes more compact and dense. From the *trans*-Golgi network, mature virions are packaged into vesicles that are targeted to the plasma membrane for release of progeny virus.

Adaptability of coronaviruses

The spectrum of diseases observed in coronavirus infections reflects the capacity of these viruses to adapt to changing environments. The ability of coronaviruses to undergo RNA-RNA recombination, coupled with the high theoretical error rate of the viral RNA polymerase (1 mutation in every 10,000 nucleotides), generates coronavirus genomes with changes that may be selectively advantageous in new environments or hosts. Most coronaviruses show strong species, organ, and tissue specificity, but subtle genetic changes can greatly alter their tissue

Making mutants—coronavirus reverse genetics

The ability to introduce specific mutations in a viral genome is a powerful tool to define the genetics and functions of viral proteins in replication, pathogenesis, and virulence. For positive-strand RNA viruses, mutagenesis requires a **reverse genetic** approach, in which the RNA genome is cloned as a complementary DNA (cDNA) in a plasmid vector. Mutations are then introduced in the cDNA, the mutant RNA genome is transcribed from the mutated cDNA, and the infectious RNA is introduced into cells to generate infectious mutant virus particles. The extremely large and complex coronavirus RNA genome, and the presence of "toxic regions" resistant to stable cloning, have been formidable obstacles to cDNA cloning and mutagenesis. In the past 10 years, however, several approaches have been successfully used to introduce mutations in the viral genome. These include cloning a full-length cDNA in a bacterial artificial chromosome or in a recombinant vaccinia virus. Another strategy, called "targeted recombination," uses the natural recombination ability of coronaviruses to introduce changes along with a selectable gene.

A novel reverse genetic approach that has overcome the complexity, size, and toxicity of the coronavirus RNA genome is referred to as "*in vitro* assembly" (Figure 19.6). In this system, the genome is cloned in bacterial plasmids as a set of distinct cDNA fragments. Genes within these fragments can be mutated independently, creating libraries of mutated fragments. The cDNA fragments are excised from plasmids by cleavage with restriction endonucleaes, and ligated *in vitro* into a full-length cDNA. This assembled cDNA is transcribed *in vitro* by T7 RNA polymerase to produce a full-length genome RNA, which is introduced into permissive cells by electroporation. Because the genome RNA is infectious, viable mutant viruses are produced. This approach allows the testing of "mix and match" libraries of mutations in different genes to determine phenotypic effects. The power of reverse genetics will also allow testing of changes that occur in coronaviruses over time, or in the case of SARS, changes associated with the transition between species. Finally, this approach may allow rational design of stably attenuated viruses as possible vaccine candidates.

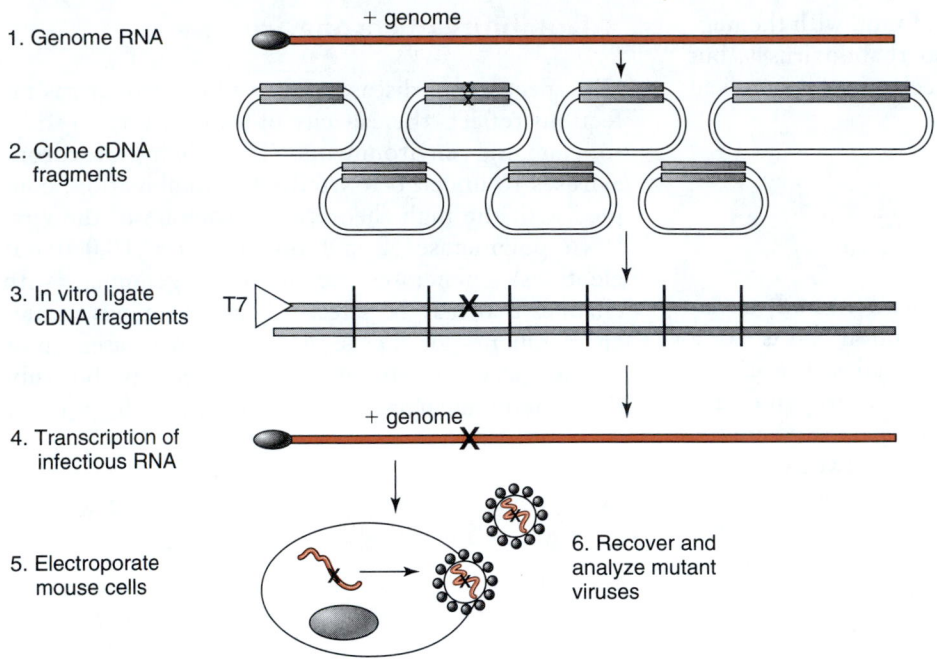

FIGURE 19.6 *In vitro* assembly of genome cDNA and production of infectious genome RNA. See text for details.

tropism, host range, and pathogenicity. This has been experimentally demonstrated for murine hepatitis virus. Co-culture of different strains of murine hepatitis virus in mixed populations of cell types generated variant viruses capable of replicating in previously nonpermissive cell types including human cell lines, thus changing host range *in vitro*. Genetic alterations in a virus can also result in changes in tissue tropism. Typically, transmissible gastroenteritis virus causes digestive tract disease in young swine. A variant form of this virus replicates preferentially in the respiratory tract. Genetic analysis has correlated this change in tropism to partial deletion of a structural protein and alterations in an accessory gene.

While some mutations may result in attenuated disease, others may increase virulence and disease. This is best illustrated by the conversion of the benign feline enteric coronavirus into lethal feline infectious peritonitis virus. These two viruses are serologically and morphologically identical. Spontaneous mutations in feline enteric coronavirus accessory proteins alter the virulence, resulting in severe immunopathology that is almost always lethal. The emergence of the SARS coronavirus provided a dramatic demonstration of the adaptability of coronaviruses, and further showed that they can cross the species barrier with sometimes devastating results.

SUGGESTED READING

1. Enjuanes, L. Ed. (2005). Coronavirus replication and reverse genetics. *Current Topics in Microbiology and Immunology* **287,** 161–197.

2. Lai, M.M. and Cavanagh, D. (1997). The molecular biology of coronaviruses. *Advances in Virus Research* **48,** 1–100.

3. Li, W., et al. (2003). Angiotensin-converting enzyme 2 is a functional receptor for the SARS coronavirus. *Nature* **426,** 450–454.

4. Navas, S. and Weiss, S.R. (2003). Murine coronavirus-induced hepatitis: JHM genetic background eliminates A59 spike-determined hepatotropism. *Journal of Virology* **77,** 4972–4978.

5. Prentice, E., Jerome, W.G., Yoshimori, T., Mizushima, N., and Denison, M.R. (2004). Coronavirus replication complex formation utilizes components of cellular autophagy. *Journal of Biological Chemistry* **279,** 10136–10141.

6. Snijder, E.J., Bredenbeek, P.J., Dobbe, J.C., Thiel, V., Ziebuhr, J., Poon L.L., Guan, Y., Rozanov, M., Spaan, W.J., and Gorbalenya, A.E. (2003). Unique and conserved features of genome and proteome of SARS-coronavirus, an early split-off from the coronavirus group 2 lineage. *Journal of Molecular Biology* **331,** 991–1004.

7. Yount, B., Curtis, K.M., Fritz, E.A., Hensley, L.E., Jahrling, P.B., Prentice E., Denison, M.R., Geisbert, T.W., and Baric, R.S. (2003). Reverse genetics with a full-length infectious cDNA of severe acute respiratory syndrome coronavirus. *Proceedings of the National Academy of Sciences of the United States of America* **100,** 12995–13000.

KEY TERMS

Apoptosis
Autophagy
Bronchitis
Chimeric
Class I fusion proteins
Class II fusion proteins
Coiled-coil
Discontinuous transcription

Encephalitis
Endoplasmic reticulum-Golgi intermediate compartment (ERGIC)
Gastroenteritis
Hemagglutinin
Hepatitis
Hypoxemia
Lectin
Metalloproteinases
Nested set
Papain-like cysteine proteinases
Peritonitis
Picornavirus 3C proteinases
Polyprotein
Pseudoknot
Replication complexes
Replicative intermediate
Reservoir
Reverse genetics
Sialic acid
Subgenomic mRNA
Syncytium
Tropism

Paramyxoviruses and Rhabdoviruses

Paramyxoviridae

From Greek *para* (alternate) and *myxa* (mucus): virions bind to mucoproteins

Rhabdoviridae

From Greek *rhabdo* (rod), referring to shape of virion

VIRION

Paramyxoviruses: spherical, diameter 150–300 nm, also found as filaments.
Rhabdoviruses: bullet-shaped rods, 180 × 75 nm.
Envelope derived from plasma membrane.
Coiled helical nucleocapsid.

GENOME

Linear ssRNA, negative sense.
Single segment.
Paramyxoviruses: 15–18 Kb.
Rhabdoviruses: 11–12 Kb.

GENES AND PROTEINS

Five to nine genes, transcribed in series from 3′ end of genome by viral RNA polymerase.
Most genes produce a single mRNA and a single protein.
Most proteins packaged in virion:

- *Nucleocapsid protein (N)*
- *RNA polymerase cofactor and accessory proteins (P/C/V)*
- *Matrix protein (M)*
- *Fusion protein (F)*
- *Hemagglutinin/neuraminidase, or envelope glycoprotein (H, HN, or G)*
- *RNA polymerase (L)*

VIRUSES AND HOSTS

Paramyxoviruses:

- *Humans: measles, mumps, respiratory syncytial, parainfluenza viruses*
- *Animals: viruses infecting birds, pigs, dogs, cats, seals, whales, etc., Sendai (mice), Newcastle disease (chickens), canine distemper, rinderpest (cattle) viruses*

Rhabdoviruses:

- *Humans: rabies virus*
- *Cattle: vesicular stomatitis virus*
- *Numerous rhabdoviruses infecting bats, fish, insects, plants*

DISEASES

Paramyxoviruses: several important childhood diseases, including measles, mumps, respiratory diseases in humans; canine distemper, rinderpest.

Emerging paramyxoviruses Hendra and Nipah cause respiratory and neurological disease in pigs, humans, horses, and bats. Rhabdoviruses: rabies (fatal encephalitis in humans), transmitted by bites of infected animals; vesicular stomatitis virus in cattle. Numerous insect and plant diseases.

DISTINCTIVE CHARACTERISTICS

RNA-dependent RNA polymerase in virion transcribes genome into mRNAs.
Single transcriptional promoter at 3′ end of RNA; mRNAs made by a start-stop mechanism.
Full-length genome or antigenome RNA is always present as assembled nucleocapsids.
Some paramyxoviruses "edit" P mRNA at a specific site, generating mRNAs with distinct protein-coding capacities.
Some paramyxoviruses induce cell-cell fusion, producing multinucleated cells (syncytia).

The mononegaviruses: a group of related negative-strand RNA viruses

Among the negative-strand RNA viruses, four families (*Paramyxoviridae*, *Rhabdoviridae*, *Filoviridae*, and *Bornaviridae*) share a number of fundamental characteristics: (1) their genome is a single RNA molecule, packaged in a helical nucleocapsid; (2) nucleocapsids are enclosed within an envelope derived from the plasma membrane of the cell; (3) a virus-coded RNA polymerase packaged in the virion synthesizes viral mRNAs by transcribing the RNA as part of the intact nucleocapsid after it enters the cell; (4) the RNA polymerase begins transcribing at the 3′ end of the genome RNA and sequentially transcribes 5–10 genes, terminating and releasing each mRNA before starting the next one. Because these viruses share the same basic genome structure and replication mechanism, they have been grouped in the order *Mononegavirales* (for "single [RNA fragment], negative strand virus"). The present chapter covers the paramyxoviruses and rhabdoviruses; filoviruses are discussed in Chapter 21.

Rabies is a fatal human encephalitis caused by a rhabdovirus

Rhabdoviruses were named because of the unusual rod- or bullet-shaped morphology of virions. Numerous different rhabdoviruses infect insects, plants, and vertebrates; this is one of the few virus families whose members infect such a wide variety of hosts. Vesicular stomatitis, an important disease of cattle, is caused by a rhabdovirus that is transmitted by insect bites. Infections of humans by this virus are mild or without symptoms. Vesicular stomatitis virus infects many different cell types, can be readily grown in cell cultures *in vitro*, and has been extensively studied.

The only rhabdovirus that is of significant concern to human health is rabies virus. Humans are infected by virus-containing saliva from bites of infected animals, often domestic dogs or bats. The virus is carried in the wild by a number of small mammals, including foxes and raccoons. Infected animals suffer from neurological symptoms that can cause tameness or unusually aggressive behavior, leading to direct contact with other animals or humans and thus helping to spread virus infection. In humans, infection results in an **encephalitis** that is invariably fatal once the first symptoms have appeared, although there may be a long latent period between exposure and disease (but one individual survived recently after drug treatment and induced coma).

Louis Pasteur developed an effective rabies vaccine in the mid-nineteenth century (see Chapter 34), and a modern, safer version is widely used. Vaccination of domestic animals and control of infections in wild animals have reduced the incidence of human rabies infections in developed countries so that only about one case per year now occurs in North America. However, there are upward of 30,000 human fatalities per year worldwide, mostly in less developed countries.

Measles is a serious childhood disease caused by a paramyxovirus

Measles virus was probably a newly emerging virus some 6000 years ago during the urbanization stage of human civilization. It may have jumped from cattle herds to humans, since measles virus is closely related to the virus that causes rinderpest, a disease of cattle. The virus is efficiently transmitted in aerosol droplets derived from coughs and sneezes. The earliest known written reference to measles is by an Arab physician, Rhazes of Bagdhad, in the fourth century A.D. Epidemics were documented during the Saracen invasions of the eighth century and continued into the Middle Ages. The virus ravaged native populations in North and South America when introduced by European explorers and colonists. Peter Panum, a Danish physician, was the first researcher to realize (in 1846) that a measles infection conveyed lifelong immunity. In 1911, Goldberger and Anderson succeeded in transmitting the virus to macaques using filtrates prepared from the blood of infected individuals.

John Enders and his team at Harvard University made a major breakthrough in 1954, when human embryonic kidney cells cultured *in vitro* were used to propagate measles virus from a young patient named David Edmonston. The virus was subsequently adapted to monkey kidney cells and chick embryo fibroblasts, and served as the basis of the Edmonston live virus vaccine that was introduced in 1961 (see Chapter 34). Introduction of the vaccine has led to a decrease from 4 million cases per year in the United States to less than 500 cases per year. However, measles remains a major problem in the developing countries of Africa and South America; there are still about 40 million cases worldwide per year resulting in nearly 1 million deaths. A rare complication of measles virus infection is a progressive, fatal neurological disease called **subacute sclerosing panencephalitis (SSPE)**. Only one in 300,000 measles victims are affected by SSPE, which seems to result from a slow, persistent infection and strikes up to 10 years after the acute illness.

Mumps virus, human parainfluenza viruses, and respiratory syncytial virus are paramyxoviruses that cause serious systemic and respiratory infections, predominantly in children. Other paramyxoviruses cause canine distemper (like rinderpest, closely related to measles virus) and diseases of mice, chickens, turkeys, seals, and dolphins. Recent epidemics in southeast Asia and Australia with a number of human fatalities were caused by two newly-discovered paramyxoviruses named Nipah and Hendra. These viruses may have emerged as a result clearing of forests for lumbering and farming; they apparently were transmitted from infected wild fruit bats to domestic livestock and then to humans. This has raised fears of wider epidemics due to these emerging agents.

This chapter will mostly use examples from the paramyxovirus family. The best-studied paramyxoviruses are Sendai virus (a mouse virus isolated in Sendai, Japan), Newcastle disease virus (a virus that infects chickens), simian virus 5 (in reality, a canine parainfluenza virus), and measles virus. The basic life cycle of rhabdoviruses is very similar to that of paramyxoviruses, justifying their inclusion together in this chapter.

Paramyxovirus and rhabdovirus virions have distinct morphologies

Virions of paramyxoviruses consist of a loose-fitting envelope surrounding a helical nucleocapsid. Electron microscopy reveals roughly spherical virions of size varying from 150 to 350 nm in diameter, with a visible fringe of glycoproteins projecting from the envelope (Figure 20.1a). However, filamentous particles are also made. Detergent treatment of virions releases the nucleocapsids, which can be purified and visualized by electron microscopy (Figure 20.1b).

In contrast to paramyxoviruses, rhabdoviruses have an unusual bullet shape (Figure 20.2), due to wrapping of the helical nucleocapsid in a supercoil or "skeleton" 50 nm in diameter, about $2\frac{1}{2}$ times the diameter of the nucleocapsid. This coiling is mediated by the rhabdovirus matrix (M) protein. The envelope forms around the coiled nucleocapsid during budding from the plasma

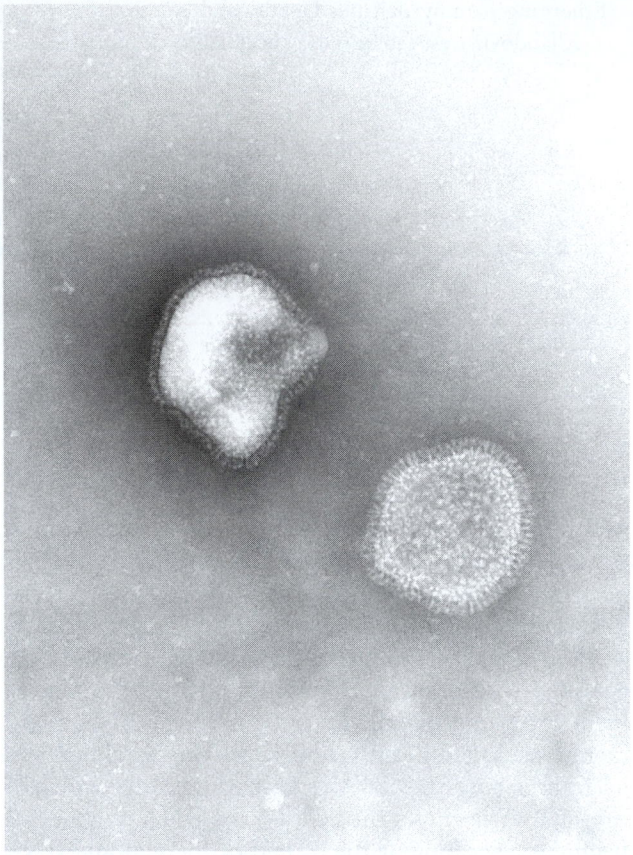

(a)

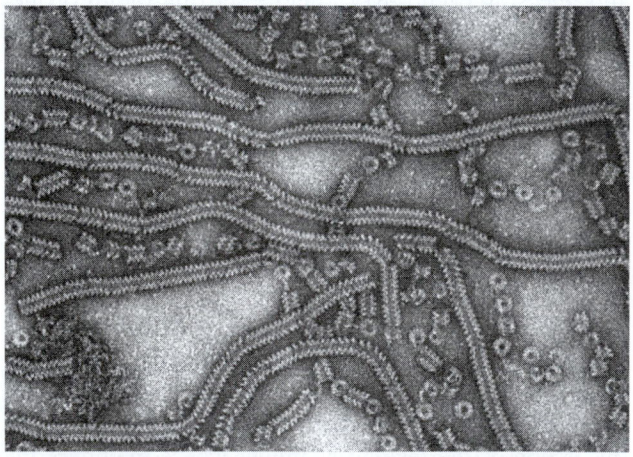

(b)

FIGURE 20.1 Structure of paramyxoviruses. (a) Electron micrograph of a negatively stained virion of simian virus 5, showing fringe of envelope proteins projecting from the surface. (b) Electron micrograph of negatively stained nucleocapsids purified from Sendai virus.

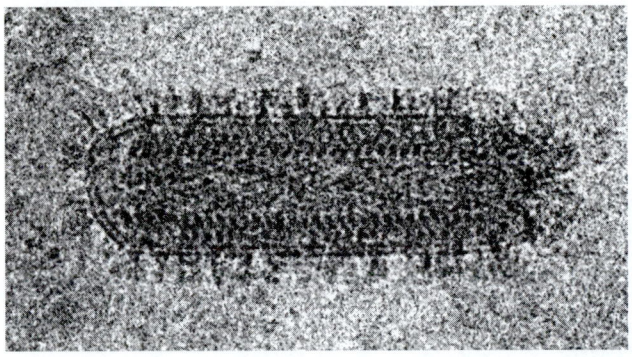

(a)

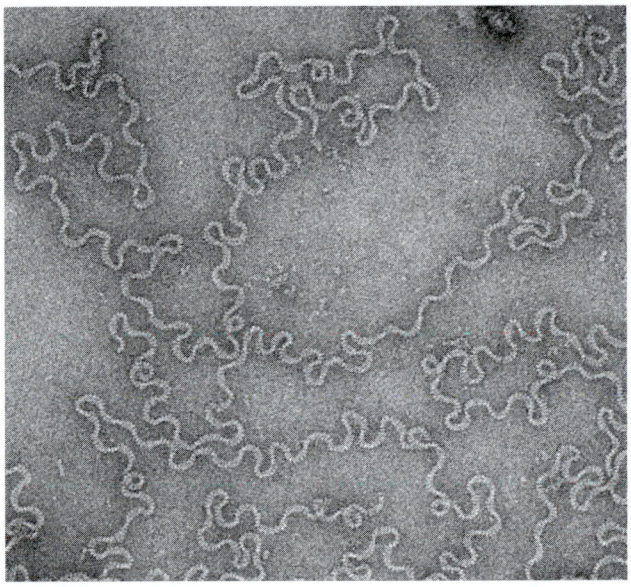

(b)

FIGURE 20.2 **Structure of rhabdoviruses.** (a) Cryoelectron micrograph of a virion of vesicular stomatitis virus. (b) Electron micrograph of negatively stained nucleocapsids purified from vesicular stomatitis virus.

membrane, giving rise to the bullet shape. The blunt end of the virion is formed by sealing of the envelope behind the coiled nuclocapsid as the virion is released from the cell membrane.

Viral envelope proteins are responsible for receptor binding and fusion with cellular membranes

Figure 20.3 shows a schematic diagram of a typical paramyxovirus virion, emphasizing localization of proteins (not drawn to scale). The virus envelope contains two or three glycoproteins. Measles virus has a **hemagglutinin** (H) and a fusion (F) protein. The H protein is responsible for binding to receptors on the cell surface, and can also bind to receptors on red blood cells and agglutinate them (**hemagglutination**). The hemagglutinin of some other paramyxoviruses also has a **neuraminidase** activity, and is therefore called HN. The F protein directs fusion with the plasma membrane. Some paramyxoviruses make an additional small (40–60 aa) hydophobic envelope protein (SH) of unknown function. SH may form pentamers and create an ion channel in membranes.

The matrix (M) protein lies on the inner surface of the envelope. M is involved in virus assembly and serves as a bridge between the nucleocapsid and the transmembrane envelope glycoproteins.

A closer view of the envelope proteins (Figure 20.4) shows that H or HN is a **type II integral membrane protein**, with a transmembrane domain located near the N-terminus, a long external C-terminal projection, and a short N-terminal sequence facing inside the virion. In contrast, the F protein is a **type I integral membrane protein**, with a long N-terminal domain facing outside and a short C-terminal internal domain. The F protein

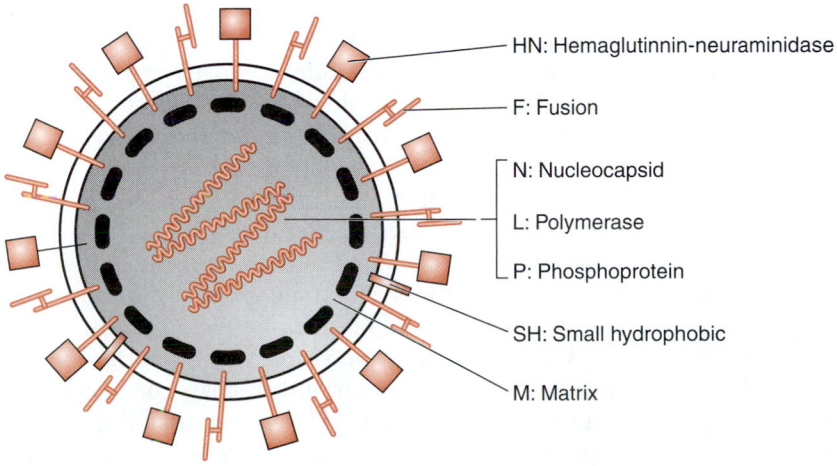

FIGURE 20.3 **Schematic diagram of paramyxovirus virion.** Lipid bilayer in which envelope proteins are inserted is shown as concentric circles. Matrix protein is shown in orange just beneath the envelope, and the helical nucleocapsid is shown in center.

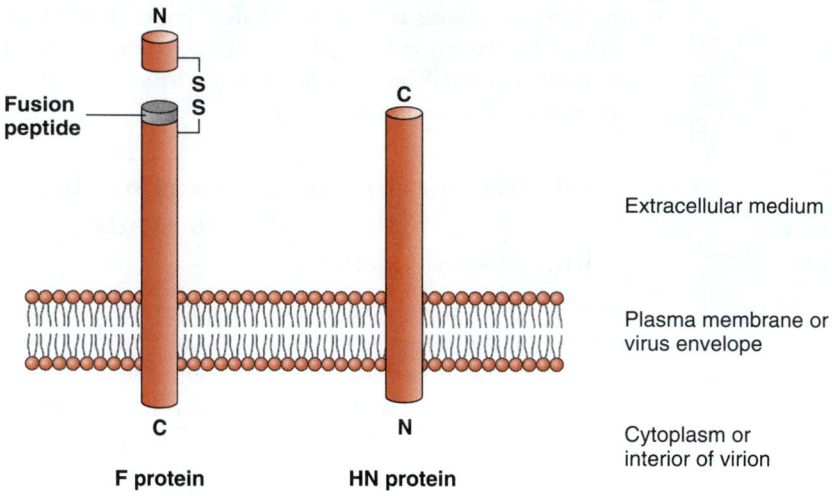

FIGURE 20.4 Envelope proteins of paramyxoviruses. The F and HN proteins are shown inserted into the plasma membrane of the cell or the virion envelope. C and N: carboxy and amino termini. S-S: cysteine bridge joining the N-terminal region of F to the remainder of the protein after cleavage.

on mature virions is cleaved into two subunits joined by a disulfide bond; this cleavage reveals the hydrophobic **fusion peptide** responsible for uptake of the virus (see below).

The best-studied rhabdovirus, vesicular stomatitis virus, has a single envelope glycoprotein (G), responsible both for receptor binding and fusion activity. Vesicular stomatitis virus can infect cells from a wide variety of species and tissues. Unlike paramyxoviruses, rhabdoviruses enter the cell by endocytosis, and fusion occurs at low pH with the endocytic membrane, releasing the nucleocapsid into the cytoplasm. The vesicular stomatitis virus G protein has been widely used to engineer recombinant enveloped viruses, particularly retroviruses, as it allows them to infect many cell types. This process of expressing a new receptor-binding protein in a recombinant virus is called **pseudotyping**.

Genome RNA is contained within helical nucleocapsids

Genomes of paramyxoviruses contain 15–18 Kb of single-stranded, negative-sense RNA. Rhabdovirus genomes are somewhat smaller (11–12 Kb). The viral N protein forms the helical nucleocapsid along with genome RNA (Figures 20.1b and 20.5). For Sendai virus and related paramyxoviruses, each N subunit binds to exactly six nucleotides in the genome; thus a 15 Kb genome contains abut 2500 copies of the N protein. The nucleocapsid of Sendai virus is wound as a 20 nm-diameter left-handed helix, with 13 subunits of N protein per turn of the helix, a pitch (distance between adjacent turns) of 5.3 nm, and a central hole 5.5 nm in diameter. In addition to the N protein, about 300 copies of P protein and 50 copies of L protein are packaged with the nucleocapsid; both are required for transcription once the virus enters the cell. Nucleocapsids protect the enclosed viral RNA from ribonuclease digestion. When serving as a template for mRNA synthesis and RNA replication (see below), the RNA genomes of paramyxoviruses and rhabdoviruses always remain associated with N protein as nucleocapsids; naked genome-length RNA is never found within the cell.

Paramyxoviruses enter the cell by fusion with the plasma membrane at neutral pH

Paramyxoviruses recognize a variety of different cellular receptors. Sendai virus and close relatives bind to terminal **sialic acid** residues on **gangliosides** (glycolipids) on the cell surface. The Sendai virus attachment protein (HN) has neuraminidase activity, which, like the influenza virus neuraminidase (Chapter 23), can release progeny virions that bind to the surface of the cell from which they are extruded. This ensures that newly made virions can leave the cell surface.

Unlike Sendai virus, measles virus H protein has no neuraminidase activity. Measles virus can bind via the H protein to at least two distinct cell surface proteins: CD46, a complement-binding protein found on most human cell types, and CD150/SLAM (signaling lymphocyte activation molecule), a member of the immunoglobulin superfamily that is expressed on activated **B lymphocytes, T lymphocytes, dendritic cells,** and some monocytes. Signal transduction mediated by CD150/SLAM may regulate T-cell proliferation and **interferon-γ** production. Because measles virus infection can lead to strong immunosuppression of the host, infection of cells in the immune system by the virus may play an important role in measles pathogenesis.

Upon binding of the virion to a cell surface receptor, the F (fusion) protein undergoes a conformational change at neutral pH that allows insertion of the fusion peptide into the plasma membrane of the cell. This

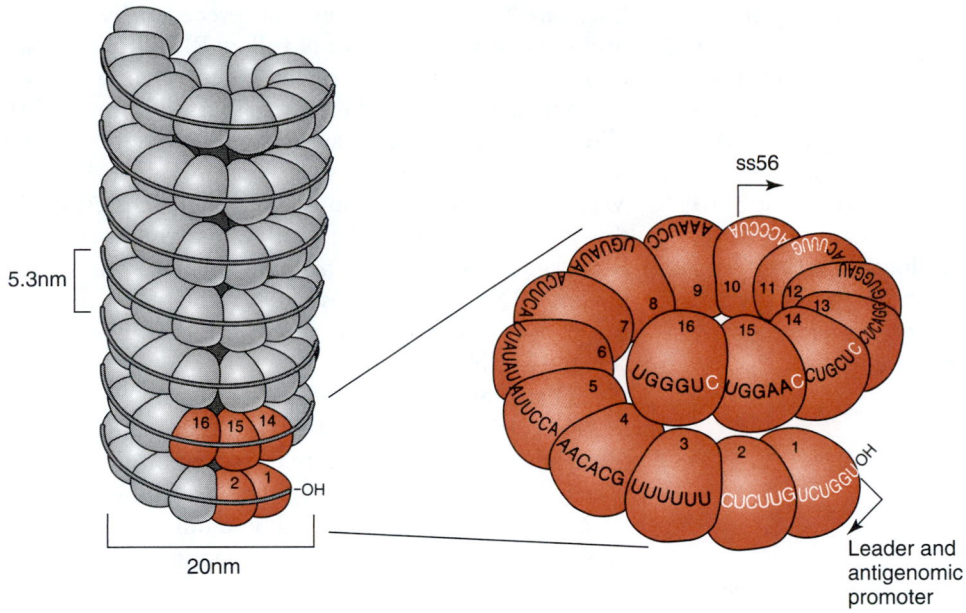

FIGURE 20.5 Structure of paramyxovirus helical nucleocapsid. Left: a model of Sendai virus nucleocapsid, showing N protein subunits (spheres) bound to genome RNA in a left-handed helix 20 nm in diameter, with 13 protein subunits per turn and a pitch of 5.3 nm. Right: expanded view of the first 16 N subunits, showing (in white) nucleotide sequences required for initiation of replication and transcription. ss56 is the start site for synthesis of N mRNA.

brings the virus envelope close to the cell membrane and leads to fusion of the two membranes, releasing the nucleocapsid in the cytoplasm. The exact nature of the conformational change that leads to fusion has not been determined for paramyxoviruses. It is known that parts of the F polypeptide chain form elongated triple-stranded **coiled-coil** complexes, similar to those formed by fusion proteins of Ebola virus, influenza virus, and human immunodeficiency virus (see Chapters 21, 23, and 26). This coiled-coil structure brings the transmembrane domain, which is anchored in the viral envelope, close to the fusion peptide, which is inserted into the cell membrane, and therefore brings the two membranes in close proximity (see Chapter 4 for a discussion of fusion by influenza HA protein). However, in most cases, the H or HN protein is also required for fusion, implying that F interacts with H or HN to produce the fusion-active state.

Gene order is conserved among different paramyxoviruses and rhabdoviruses

Paramyxovirus and rhabdovirus genomes contain between 5 and 10 genes, each coding for a specific messenger RNA. A map of the Sendai virus genome is shown in Figure 20.6. A region including the first 55 nt at the 3′ (left) end of the negative-strand genome RNA is denoted the "leader" sequence. The leader region contains signals for initiation of RNA synthesis by the viral RNA polymerase. Furthermore, its transcript contains signals that direct packaging of full-length plus-strand copies of the viral genome in nucleocapsids. The leader transcript contains no coding region, is not capped or polyadenylated, and does not serve as a viral mRNA. Immediately downstream of the leader is the N gene, followed by the P/C/V, M, F, HN, and L genes. This gene order is highly conserved, both in paramyx-

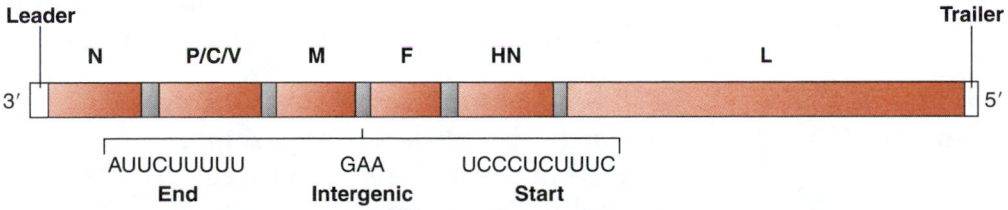

FIGURE 20.6 Paramyxovirus genome RNA. The six viral genes are shown on the map of the Sendai virus genome. Below are sequences corresponding to the end of the upstream gene, the intergenic sequences, and the start of the downstream gene.

oviruses and in rhabdoviruses (in which the G gene replaces the F and HN genes). Some paramyxoviruses have additional genes. At the junction between each gene is a highly conserved set of sequences that controls transcription termination, polyadenylation, and reinitiation, called "intergenic sequences" (see below). Beyond the end of the L gene is a short (50–150 nt) noncoding "trailer" sequence. Like the leader sequence, the (positive-strand) trailer region, located at the 3′ end of the antigenome RNA, contains the promoter for genome replication, and its negative-strand transcript contains signals for genome packaging.

Viral messenger RNAs are synthesized by an RNA polymerase packaged in the virion

The RNA of negative-strand viruses cannot be directly translated upon entry into the cell, but must first be transcribed into complementary, positive-sense messenger RNAs. However, there is no cellular RNA-dependent RNA polymerase available for production of viral mRNAs. Therefore, negative-strand RNA viruses must make their own RNA polymerase and package at least one copy of this enzyme with the nucleocapsid in the virion.

This emphasizes an important difference in replication strategy between positive-strand and negative-strand RNA viruses. Infection by positive-strand RNA viruses delivers a messenger-sense RNA directly to the cytoplasm, where it can be translated by cellular ribosomes. The translation products include a viral RNA polymerase, which can then replicate the viral genome. These viruses therefore do not package an RNA polymerase in their virions. The positive-sense genome RNA alone, in the absence of any viral proteins, is infectious when introduced artificially into susceptible cells.

In contrast, the genome RNA of negative-strand viruses is not infectious by itself because it requires the viral RNA polymerase to function. This RNA polymerase, which is packaged in the genome and enters the cell along with the genome RNA, uses ribonucleoside triphosphates provided by the cell to catalyze viral mRNA synthesis. For rhabdoviruses and paramyxoviruses, the RNA polymerase is a complex between the P and L proteins.

Viral RNA polymerase initiates transcription exclusively at the 3′ end of the viral genome

There are two general models by which viral RNA polymerases could initiate synthesis of the different viral messenger RNAs on the genomes of mononegaviruses. Each mRNA sequence could be preceded by its own promoter region, which would allow RNA polymerases to bind to each promoter region independently and begin synthesis of the different mRNAs (Figure 20.7a). These promoters could have different strengths, allowing control of the amount of each mRNA produced. Many DNA viruses organize their gene expression in this fashion. Alternatively, all RNA polymerases could begin synthesis at a single promoter at the 3′ end of the genome RNA, and progress from right to left through the genome, making the different mRNAs in sequence (Figure 20.7b).

These models can be distinguished by experimental observations. Irradiation of viral genome RNA with ultraviolet (UV) light inhibits transcription by forming dimers between adjacent thymidine or uridine residues on the template nucleic acid. These dimers block progression of the RNA polymerase along the template, and therefore inhibit all RNA synthesis downstream of the lesion. Dimers can be formed at many places along a genome RNA molecule, more or less at random.

If the multiple-promoter model were correct, inhibition of mRNA synthesis by UV irradiation of the viral genome should be proportional only to the length of the mRNA being made, because the likelihood of encountering a block to transcription should increase with distance between the promoter and the end of the mRNA. If, on the other hand, there were a single entry site for RNA polymerases at the 3′ end of the genome, synthesis of mRNAs that lie closest to the 3′ end (N and P/C/V) should be least inhibited by UV irradiation. Synthesis of mRNAs that lie more distant from the promoter, toward the 5′ end of the genome (HN and L), should be more strongly inhibited, because the polymerase would have to travel through most of the genome before it began to make those mRNAs, and therefore would be more likely to encounter a block.

Careful quantitation of the reduction of synthesis of different viral mRNAs by UV irradiation showed that the single-entry model is correct; mRNAs whose genes lie near the 3′ end are less affected by a given UV dose than mRNAs whose genes lie near the 5′ end of the genome. Independent experiments examining transcription *in vitro* with disrupted virions showed that mRNAs closest to the 3′ end of the genome appear first, and mRNAs closest to the 5′ end appear last, a result also consistent with the single-entry model.

The promoter for plus-strand RNA synthesis consists of two sequence elements separated by one turn of the ribonucleoprotein helix

All paramyxovirus and rhabdovirus genomes have conserved sequences located at the 3′ ends of their

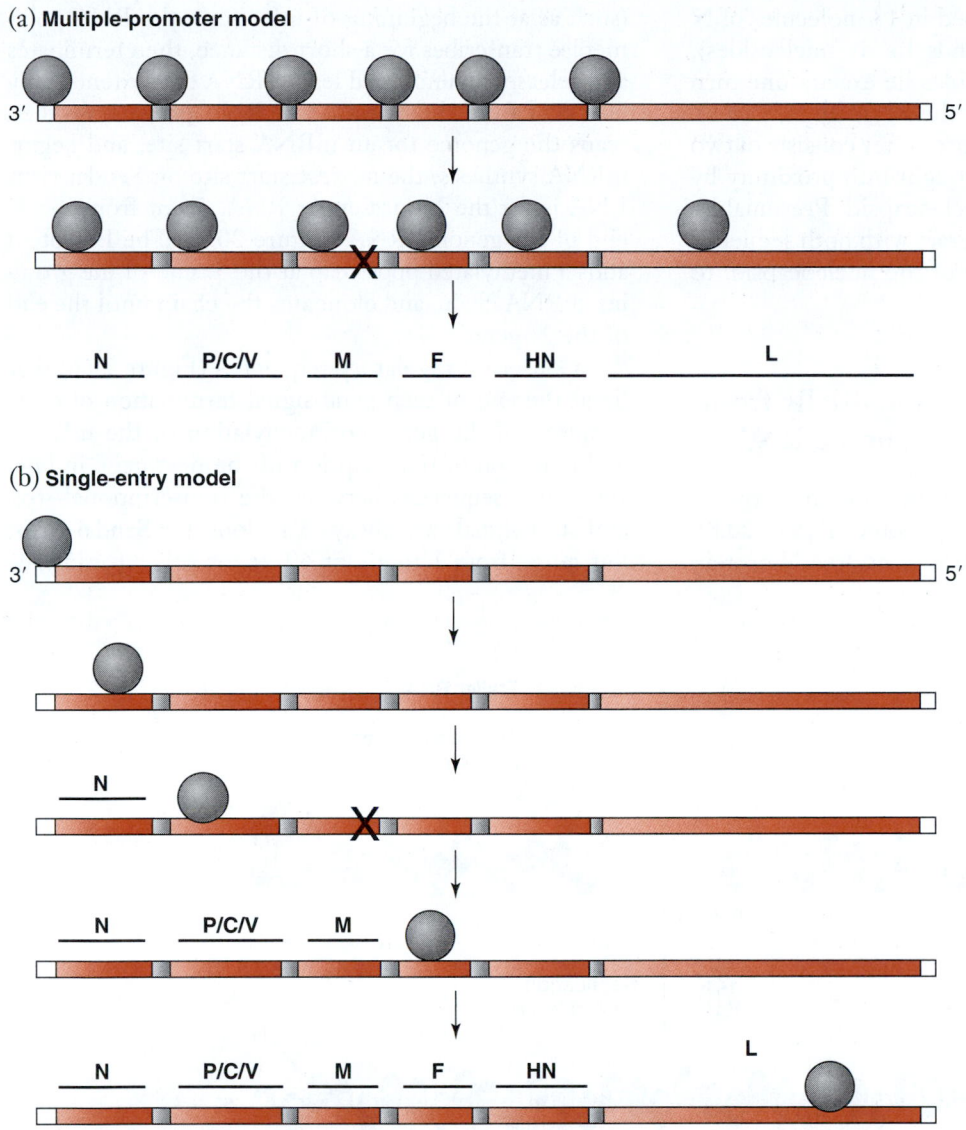

FIGURE 20.7 Alternative models for transcription of mononegavirus genome RNA. (a) Multiple-promoter model: RNA polymerases (gray spheres) initiate transcription independently at promoters just upstream of each viral gene. Growing RNA chains are shown as black lines trailing the RNA polymerases, and completed mRNAs are labeled. (b) Single-entry model: RNA polymerases initiate transcription only at the 3′ end of the genome, and transcribe each gene in sequence. X shows location of a possible UV-induced dimer that would block transcription only of M in model (a) but would block transcription of M, F, HN and L in model (b).

genomes, which are also found at the 3′ ends of the positive-strand antigenome. These conserved sequence elements are required for initiation of transcription (mRNA synthesis) and replication (genome synthesis). In the case of Sendai virus, the first 12 nucleotides at the 3′ end of the genome are identical to the first 12 nucleotides at the 3′ end of the antigenome. Mutational analysis has shown that sequences extending up to 30 nucleotides from the 3′ end are important for transcription and replication.

Surprisingly, specific sequences beginning 78 nt from the 3′ end of the genome were also discovered to be required for proper transcription and replication. In the case of Sendai virus, these nucleotides are G residues at positions 79, 85, and 91 (Figure 20.5). These three G residues are located at the first position of the 14th, 15th, and 16th hexanucleotides (groups of six nucleotides) on the genome. Their precise position and spacing is vital for optimal transcription. Since one turn of the helical nucleocapsid contains 78 nucleotides wrapped in 13 molecules of N protein (each N molecule binds to six nucleotides), these three conserved nucleotides lie exactly one turn of the helix from the 3′ end of the RNA (Figure 20.5). Therefore the transcriptional promoter consists of two distinct sequence elements, brought into proximity by the structure of the helical nucleocapsid. Presumably, the RNA polymerase must interact with both sequence elements, aligned on one side of the nucleocapsid, to initiate transcription.

mRNAs are synthesized sequentially from the 3′ to the 5′ end of the genome RNA

Viral RNA polymerase begins transcription at the 3′ terminal nucleotide of the viral genome (Figure 20.8). In conditions where there is little or no free N protein (such as at the beginning of infection), the RNA polymerase transcribes for a short distance, then terminates and releases an uncapped leader RNA that extends only 40–60 nt from the 3′ end. The RNA polymerase then scans the genome for an mRNA start site, and begins mRNA synthesis; the nearest start site on Sendai virus RNA is for the N messenger RNA, 56 nt from the 3′ end of the genome RNA (Figure 20.5). The L protein adds a methylated pppG cap at the 5′ end of the growing mRNA chain, and elongates the chain until the end of the N gene.

Conserved regulatory sequences (Figure 20.6) that lie at the end of each gene signal termination of transcription of the gene, polyadenylation of the mRNA, and initiation of transcription of the next gene in line. Intergenic sequences between the transcriptional stop and start signals are always 3 nt long for Sendai virus, but range from 1 to about 60 nt for other paramyxo-

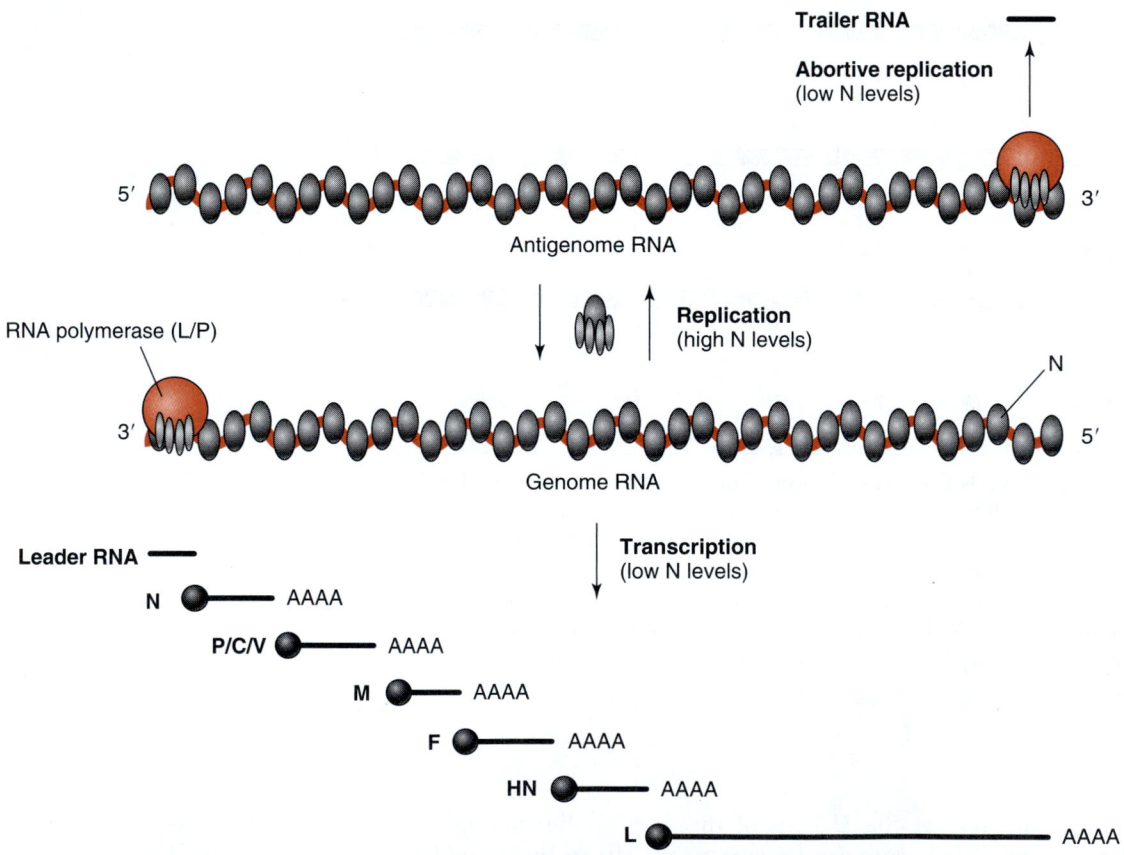

FIGURE 20.8 Paramyxovirus transcription and replication. The viral genome is shown as a helical nucleocapsid covered with N protein (gray ovals) and containing RNA polymerase components L (orange spheres) and P (thin gray ovals). When N levels are low (below), RNA polymerase terminates and reinitiates transcription at multiple sites, producing mRNAs as shown. Small gray spheres denote 5′ caps, AAAA denotes poly(A) tails. When N levels are high (above), the RNA leader is encapsidated and RNA polymerase does not terminate but copies the entire genome, to form full-length antigenome nucleocapsids. Antigenome RNA is in turn replicated to form genome RNA. When N levels drop (top), trailer RNA is not encapsidated; RNA polymerase terminates and does not complete replication of the antigenome.

viruses. At the termination site there is a run of 4 to 7 uridine residues in the template RNA, which are transcribed into adenine residues in the mRNA. The RNA polymerase repeatedly transcribes this U run by "stuttering" (see below), giving rise to a poly(A) tract some 200 nt long, and then releases the mRNA.

After releasing the first mRNA, the polymerase then has two choices: it can again scan the template for a start site and begin synthesis of the next mRNA (in the case of Sendai virus, the P/C/V gene), or it can dissociate from the template. This choice is dictated partly by the precise sequence elements in the end/start region. A certain proportion of RNA polymerases dissociates at the end of each gene, and therefore cannot reinitiate transcription at the next downstream gene. This results in diminishing quantities of mRNAs as their genes lie further from the 3′ end of the genome. This mechanism provides for the synthesis of large amounts of mRNAs for gene products needed in high concentration such as N and M, and smaller amounts of mRNAs for envelope proteins and the RNA polymerase (L).

The P/C/V gene codes for several proteins by using alternative translational starts and by mRNA "editing"

Most paramyxovirus genes produce a single mRNA species that codes for one viral protein. However, the P/C/V gene has a more complex coding strategy (Figure 20.9). First, there are multiple translational start sites between nt 81 and 203 of the Sendai virus P/C/V mRNA. Four of these are in the same reading frame, and ribosomes that initiate protein synthesis at these sites produce a **nested set** of proteins, denoted C′, C, Y1, and Y2, with different N-termini but with a common C-terminus. An alternative start site lies in a distinct and overlapping reading frame that extends through most of the mRNA to give rise to a longer protein denoted P (the polymerase cofactor). The start site for translation of the P protein is in a favorable context for recognition by the ribosome (AUG followed by a G at position 4) and is therefore most often used.

The second complexity arises at a specific *editing* site in the middle of the P reading frame. At this sequence, which contains a stretch of 6 A's followed by 3 G's, the RNA polymerase occasionally "stutters," and adds one or more extra G residues within the stretch of 3 G's (Figure 20.9). Stuttering most likely happens when the RNA polymerase pauses, and the growing RNA chain slips backward on the RNA template by one (or more) nucleotides. The RNA polymerase then resumes elongation. When this happens in the run of 3 G's, an extra G is added to the growing RNA chain. This changes the reading frame; in Sendai virus, edited mRNA with one extra G gives rise to a protein called V, and mRNA with two extra G's gives rise to a protein called W. A similar stuttering mechanism, repeated

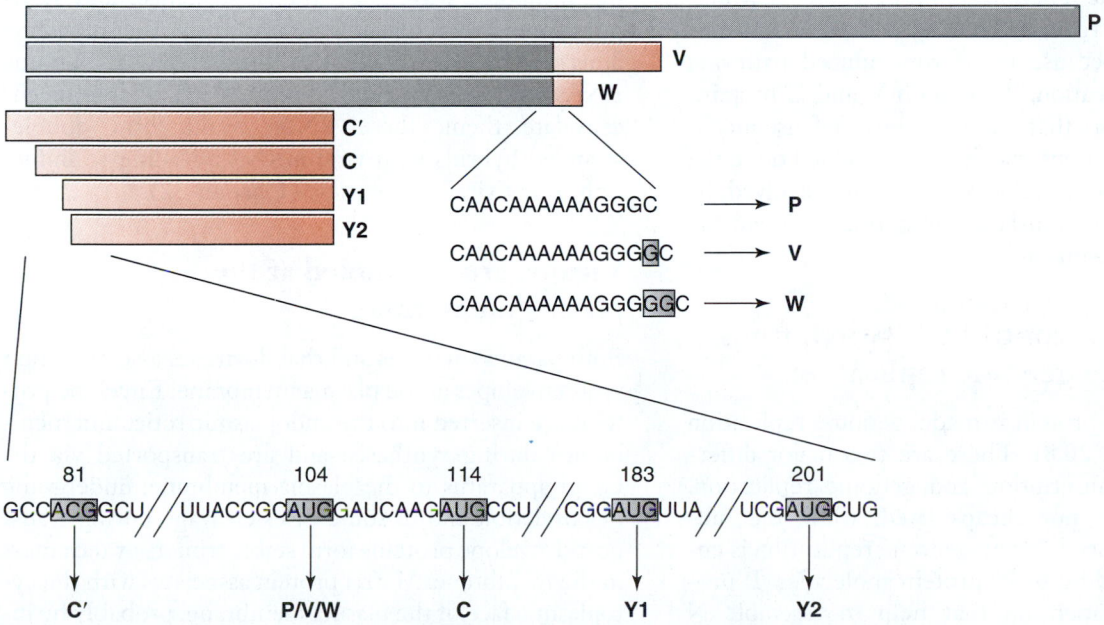

FIGURE 20.9 Sendai virus P/C/V gene: open reading frames, translational start sites, and "stuttering" site. Open reading frames are shown at top, with the protein produced shown at the right end. Five alternative start sites (ACG or AUG) in a 120 nt region are shown at bottom. Transcriptional stuttering occurs within the P reading frame at sequence shown in the middle, adding one or two G residues that change reading frame, producing proteins V or W.

many times, accounts for the production of long poly(A) tails at the end of each viral mRNA (see above).

Functions of P, C, and V proteins

The functions of the different proteins made by these mechanisms are still being worked out. Presently, it is clear that the P protein is a cofactor required for viral RNA syntheisis carried out by the L protein. P binds to L, probably as a tetramer, and directs the binding of L to the promoter at the 3′ end of the genome and antigenome RNA in nucleocapsids (see above). P remains bound to L during RNA synthesis. P also serves as a chaperone during the assembly of N protein onto RNA chains to form nucleocapsids.

The V protein, which shares common N-terminal sequences with P but has a distinct and highly conserved C-terminal region, is not normally required for replication of paramyxoviruses in cells cultured *in vitro*. However, elimination of V by mutating the editing site, or mutations in the C-terminal domain specific to V, reduce the virulence of Sendai virus in mice, indicating that V protein normally interferes with some host mechanisms that reduce virus replication or spread. V can inhibit the host interferon response and is known to stimulate degradation of STAT proteins (see Chapter 31).

C proteins are required for formation of infectious paramyxovirus virions, but their exact role in virion formation is not yet understood, and little or no C is incorporated into virions. Furthermore, like V, C proteins antagonize the antiviral effects of interferons by interfering with the function of STAT proteins. Mutations in C protein reduce virulence in experimental animals, probably because interferon-induced pathways restrict virus replication. Thus both V and C proteins are virulence factors that modulate host defense mechanisms. These functions may have been added on to the original functions of the P gene as the virus evolved, by the addition of new translation initiation sites and the RNA editing mechanism.

N protein levels control the switch from transcription to genome replication

Once sufficient N protein is made, genome replication can begin (Figure 20.8). There are two major differences between transcription and genome replication. First, mRNAs are not encapsidated, while the full-length RNA produced during genome replication is encapsidated by binding to N protein molecules. P proteins serve as chaperones that help to assemble N proteins onto the growing helical nucleocapsid. Second, the viral RNA polymerase changes its mode of synthesis, and no longer recognizes the intergenic termination-polyadenylation-reinitiation sequences. This change may be signaled by the binding of N molecules to the growing RNA chain, or it may be a result of interaction between newly synthesized molecules of P protein and the RNA polymerase. As a result, the RNA polymerase extends the growing chain to the 5′ end of the template RNA without stopping, making a full-length antigenome in a helical nucleocapsid.

This antigenome nucleocapsid can in turn be replicated by the viral RNA polymerase, producing full-length encapsidated genomes. A delicate control mechanism senses the amount of N protein present in the cell, allowing full-length genome RNA to be made only if there is sufficient N protein to encapsidate it. This mechanism works as follows (Figure 20.8, top). Like the leader sequence at the 3′ end of the genome RNA, there is a "trailer" sequence at the 3′ end of the antigenome RNA. Transcripts of both the leader and the trailer sequences contain sites that initiate nucleocapsid formation by specifically binding to N protein. Once initiated, nucleocapsid formation continues by adding N protein subunits every six nucleotides without a further requirement for specific binding sites on the RNA.

If there is not sufficient N protein to initiate encapsidation within the newly formed leader or trailer RNA chains, the RNA polymerase terminates some 40–60 nt downstream from its start site and releases a short RNA molecule. When the template is genome RNA, the RNA polymerase can then produce a set of viral mRNAs by the transcription mechanism described above; these mRNAs do not interact with N protein. When the template is antigenome RNA, there are no further transcription initiation signals downstream of the promoter within the trailer sequence, and RNA polymerase therefore dissociates from the template. This mechanism avoids the production of genome RNAs when insufficient N protein is available to encapsidate them; these RNAs could form double-stranded hybrids with viral mRNAs, leading to induction of antiviral proteins (see Chapter 32).

Virions are assembled at the plasma membrane

Both paramyxoviruses and rhabdoviruses assemble their virus envelopes at the plasma membrane. Envelope proteins are inserted into the endoplasmic reticulum membranes during synthesis, and are transported via the Golgi apparatus to the plasma membrane, undergoing glycosylation and in some cases cleavage during transport. Envelope proteins form stable trimers or tetramers in the membrane. Matrix protein associates with the cytoplasmic face of the plasma membrane, probably by interacting with the cytoplasmic tails of the viral envelope glycoproteins. Nucleocapsids, formed in the cytoplasm during genome replication, migrate to the plasma membrane where they interact with the matrix protein. When the complex between envelope proteins, matrix

The "rule of six"

Mutational analysis of paramyxovirus genomes has revealed an intriguing spatial requirement for genome replication. Many (but not all) paramyxoviruses have a strict requirement that their genomes must be an exact multiple of six nucleotides in length. This *rule of six* was discovered by making artificial minigenomes containing the two natural ends of the genome but lacking most of the internal coding sequences. When expressed in cells infected by wildtype helper virus, these minigenomes can replicate, because they contain the essential binding sequences for the viral RNA polymerase at both ends. It was found that minigenomes whose total length is an exact multiple of six nucleotides replicate much more efficiently than minigenomes of other lengths. Furthermore, most of the genomes of the *Paramyxovirinae* subgroup sequenced to date have lengths that are exact multiples of six nucleotides.

The rule of six is explained as follows. Assembly of replicating genomes and antigenomes into nucleocapsids happens during RNA synthesis, as described above. As soon as the first six nucleotides at the 5′ end of the growing viral RNA chain have been made, they bind to a molecule of N protein if available. This packaging reaction proceeds sequentially on each hexanucleotide sequence as the RNA is elongated. Therefore, RNAs whose length is an exact multiple of six nucleotides will have the sequence at their 3′ end fully bound to an N protein subunit in a precise manner. However, RNAs that have between 1 and 5 additional nucleotides will not properly bind to the last N protein subunit, leaving some RNA sequences at the 3′ end unpackaged. Presumably, when the RNA polymerase comes across a 3′ end that is not properly packaged, it cannot begin transcription or genome replication efficiently, and that RNA molecule is lost to the replication system.

protein, and nucleocapsids is formed, virions are made by budding from the plasma membrane (for a diagram showing budding, see Figure 2.9). In the case of rhabdoviruses, the matrix protein is responsible for condensing the helical nucleocapsid into a coil or "skeleton" that defines the unusual bullet shape of the virion.

Because paramyxovirus envelope glycoproteins that mediate membrane fusion at neutral pH are incorporated into the plasma membrane, infected cells often undergo fusion with neighboring cells. Cell-cell fusion can occur among numerous nearby cells, leading to the production of giant multinucleated cells, called **syncytia**. Such syncytia can form both in cells cultured *in vitro* and in tissues of infected animals. Cell fusion causes tissue destruction and may be responsible for pathogenic effects of some paramyxoviruses.

SUGGESTED READING

1. Conzelmann, K.K. (2005). Transcriptional activation of alpha/beta interferon genes: Interference by nonsegmented negative-strand RNA viruses. *Journal of Virology* **79,** 5241–5248.

2. Horvath, C.M. (2004). Silencing STATs: Lessons from paramyxovirus interferon evasion. *Cytokine Growth Factor Reviews* **15,** 117–127.

3. Kolakofsky, D., Le Mercier, P., Iseni, F., and Garcin, D. (2004). Viral DNA polymerase scanning and the gymnastics of Sendai virus RNA synthesis. *Virology* **318,** 463–473.

4. Kolakofsky, D., Pelet, T., Garcin, D., Hausmann, S., Curran, J., and Roux, L.J. (1998). Paramyxovirus RNA synthesis and the requirement for hexamer genome length: The rule of six revisited. *Journal of Virology* **72,** 891–899.

5. Lam, S.K. (2003). Nipah virus—a potential agent of bioterrorism? *Antiviral Research* **57,** 113–119.

6. Schmitt, A.P. and Lamb, R.A. (2004). Escaping from the cell: Assembly and budding of negative-strand RNA viruses. *Current Topics in Microbiology and Immunology* **283,** 145–196.

7. Schnell, M.J., Tan, G.S., and Dietzschold, B. (2005). The application of reverse genetics technology in the study of rabies virus (RV) pathogenesis and for the development of novel RV vaccines. *Journal of Neurovirology* **11,** 76–81.

8. Yin, H.S., Paterson, R.G., Wen, X., Lamb, R.A., and Jardetzky, T.S. (2005). Structure of the uncleaved ectodomain of the paramyxovirus (hPIV3) fusion protein. *Proceedings of the National Acadamy of Sciences U.S.A.* **102,** 9288–9293.

KEY TERMS

B lymphocytes
Coiled-coil
Dendritic cells
Encephalitis
Fusion peptide
Gangliosides
Hemagglutination
Hemagglutinin
Interferon
Nested set
Neuraminidase
Pseudotyping
Sialic acid
Subacute sclerosing panencephalitis (SSPE)
Synctia (singular: syncytium)
T lymphocytes
Type I integral membrane protein
Type II integral membrane protein

Filoviruses

Filoviridae

From Latin *filum* (filament or thread), referring to shape of virion

VIRION

Filamentous enveloped particles.
Diameter 80 nm, length 1000 nm, or more.
Helical nucleocapsid.

GENOME

Linear ss RNA, negative sense.
Single segment, 19 Kb.

GENES AND PROTEINS

Seven genes, transcribed in series from 3′ end of genome by viral RNA polymerase
Most genes produce a single mRNA and a single protein.
Most proteins are packaged in virion:

- *Nucleocapsid protein (NP)*
- *RNA polymerase cofactor (VP35)*
- *Matrix protein (VP40)*
- *Envelope glycoproteins (GP, cleaved into GP1 and GP2)*
- *Minor nucleocapsid protein (VP30)*
- *Membrane protein (VP24)*
- *RNA polymerase (L)*

Ebola makes an additional secreted glycoprotein (sGP).

VIRUSES AND HOSTS

Two distinct genera: *Marburgvirus* and *Ebolavirus*.
Ebolavirus species named after their original site of discovery: Ivory Coast, Reston, Sudan, and Zaire.

DISEASES

Infections with Marburg and Ebola viruses cause severe hemorrhagic fever.
Major symptoms are fever, internal bleeding, liver dysfunction, intravascular coagulation, and shock.
Lethality can be high (30 to 90%).
Local outbreaks are followed by lengthy absence of disease.

DISTINCTIVE CHARACTERISTICS

Filoviruses are prime examples of (re)emerging infectious agents.
They are probably transmitted to people from an unknown animal vector.
Gene order and expression are similar to those of paramyxoviruses and rhabdoviruses.

Marburg and Ebola viruses: sporadically emerging viruses that cause severe, often fatal disease

In August 1967, three laboratory workers of the Behringwerke in Marburg, Germany, became ill with a hemorrhagic disease after processing organs from African Green monkeys (*Cercopithecus aethiops*). In the course of the epidemic, 17 more persons engaged in this kind of work were admitted to the hospital, and two medical staff members became infected while attending the patients in the hospital. The last patient was admitted in November of the same year. Six more cases, including two people with secondary infections, occurred at the same time in Frankfurt, Germany. Furthermore, in Belgrade, Yugoslavia, a veterinarian performing an autopsy on dead monkeys became infected, together with his wife, who nursed him during the first days of the illness. Altogether, there were 32 cases, of whom seven died.

To Gustav-Adolf Martini of the Department of Internal Medicine at the University of Marburg it was immediately clear that he was faced with an unusual infectious disease. By the end of October of that year, Werner Slenczka and Rudolf Siegert of the Institute of Hygiene had identified the causative agent as a new virus that soon received the name of the town where it was discovered. The identification of Marburg virus within two months of its first appearance is even by present-day standards a remarkable accomplishment of medical virology.

Another related virus emerged in 1976, when two epidemics occurred simultaneously in Zaire (now Democratic Republic of Congo) and Sudan. The agent was isolated from patients in both countries and was named after the Ebola river in northwestern Zaire. Over 500 cases were reported, with lethality rates of 88% in Zaire and 53% in Sudan. Subsequent smaller outbreaks of Ebola disease occurred in Sudan in 1977 and 1979. In 1994, the first Ebola case was observed in West Africa

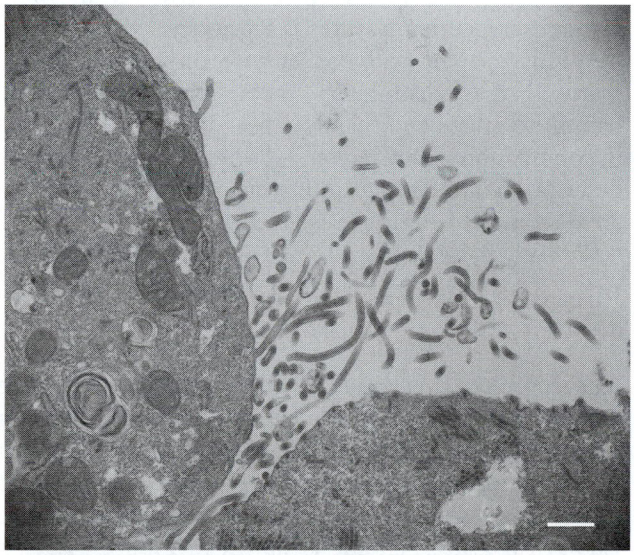

(a)

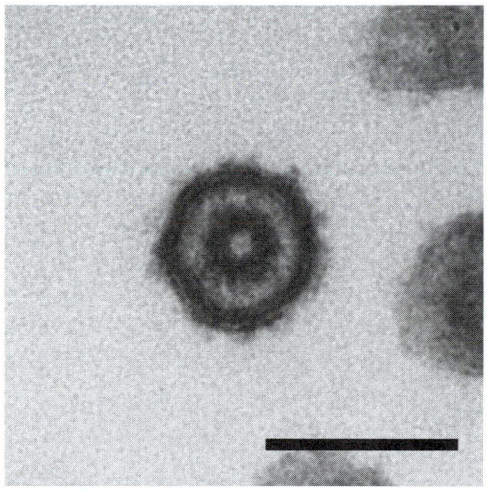

(b)

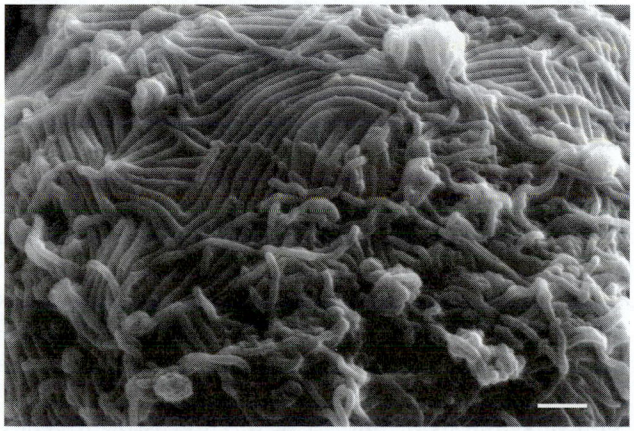

(c)

FIGURE 21.1 The structure of filoviruses. (a) Transmission electron micrograph of a thin section of Vero E6 cells from which Ebola virus particles are budding. Filamentous particles are seen in longitudinal, transverse, or cross sections. Bar: 500 nm. (b) Enlargement of a single particle in cross section. Particles consist of a nucleocapsid with a central channel, surrounded by a membrane; glycoprotein spikes are seen as a fringe outside membrane. Bar: 100 nm. (c) Scanning electron micrograph of Ebola virus particles budding from the surface of Vero E6 cells. Bar: 500 nm.

in Ivory Coast, when an ecologist was infected after having examined a dead chimpanzee.

Ebola virus reemerged in Kikwit, Democratic Republic of the Congo, in 1995 with 315 cases and 245 deaths. From 1994 to 2002, at least four outbreaks of Ebola virus disease occurred in Gabon. More recently, there have been several outbreaks in Congo-Brazzaville, mostly near the border with Gabon. A major outbreak of Ebola virus disease occurred in northern Uganda in 2000/2001.

An unexpected outbreak of Ebola virus infection occurred in cynomolgus monkeys imported from the Philippines to a facility in Reston, Virginia in 1989. This virus spread among monkeys housed in different areas within the facility, probably by aerosols spread through the ventilation system. Luckily, this species of Ebola virus, the first that originated outside of Africa, does not appear to be pathogenic to humans.

Filoviruses are related to paramyxoviruses and rhabdoviruses

Study of Marburg and Ebola viruses revealed that they produce filamentous, enveloped particles 80 nm in diameter and approximately 1000 nm (1 μm) long (Figures 21.1, 21.2). Virions contain a negative-sense, single-stranded RNA genome 19 Kb long, wrapped in a helical nucleocapsid. These viruses were first classified as rhabdoviruses, but differences in virion morphology and conserved protein sequences led to establishment of a new family, the *Filoviridae*, a name based on the elongated shape of virions. Filoviruses belong to the so-called "exotic viral pathogens" and are prime examples of (re)emerging infectious agents. *Marburgvirus* and *Ebolavirus* are two distinct genera within the filovirus family. *Ebolavirus* species are named after their original site of discovery: Ivory Coast, Reston, Sudan, and Zaire.

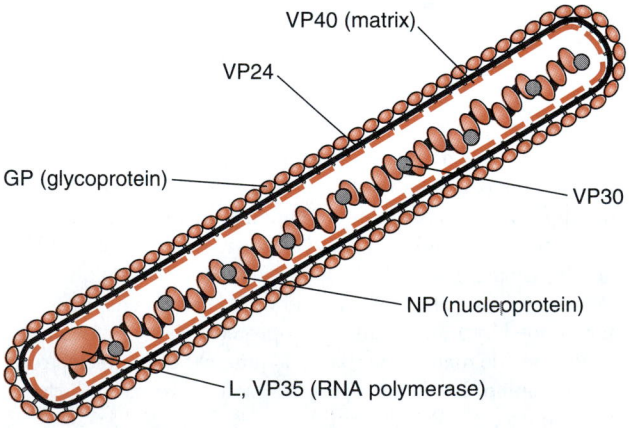

FIGURE 21.2 **Schematic view of a filovirus virion.** The RNA genome is in a helical nucleocapsid associated with 4 viral proteins: the viral polymerase (L), the nucleoprotein (NP), and proteins VP35 and VP30. VP40 is a matrix protein, the spikes are formed by trimeric GP_1/GP_2 complexes, and VP24 is associated with the envelope.

Filoviruses cause hemorrhagic fever

Infections with Marburg and Ebola viruses cause severe viral **hemorrhagic fever**. The major symptoms are fever, generalized internal bleeding, liver damage and dysfunction, disseminated intravascular coagulation, and shock. Lethality is high and varies between 30 and 90%, except for Ebolavirus Reston, which seems to have low pathogenicity for humans.

It is clear from the recorded history of filovirus outbreaks that all of them have so far been self-limited and that the total number of human infections hitherto recorded does not exceed three thousand cases. However, because of the dramatic course of the disease, the high case-fatality rates, and the lack of effective therapeutic measures, reemergence of these viruses regularly causes high public concern. Furthermore, there is concern that filoviruses could be intentionally spread as a bioterrorist measure. For these reasons, their study is important, and development of effective vaccines and chemotherapeutic measures is a priority.

This chapter will concentrate on aspects of filoviruses that distinguish them from the related paramyxoviruses and rhabdoviruses, and should be read in conjunction with Chapter 20, where a more complete description of the replication cycle of this group of negative-strand RNA viruses is given.

Filovirus genomes contain seven genes in a conserved order

The complete nucleotide sequences of both Marburg and Ebola virus strains have been determined, allowing mapping of their coding regions. These viruses contain seven genes, arranged in an order similar to those of paramyxoviruses and rhabdoviruses. A map of the Ebolavirus Zaire genome is shown in Figure 21.3. As with paramyxoviruses and rhabdoviruses, the gene coding for the nucleocapsid protein (NP) is located closest to the 3′ end of the genome, and the gene coding for the RNA polymerase (L) is located closest to the 5′ end. All genes are flanked by highly conserved sequences that signal transcription termination, polyadenylation, and transcription reinitiation. Some genes are separated by short untranscribed intergenic sequences. However, the stop and start sites (but not coding regions) of several Ebola virus genes and one Marburg virus gene actually overlap by about 20 nt, so that the 3′ end of one mRNA lies downstream from the 5′ end of the next mRNA (Figure 21.2). The effect of these overlaps on mRNA synthesis is not known, but they may control the level of initiation of synthesis of mRNA of the downstream genes. Noncoding leader and trailer sequences comparable to

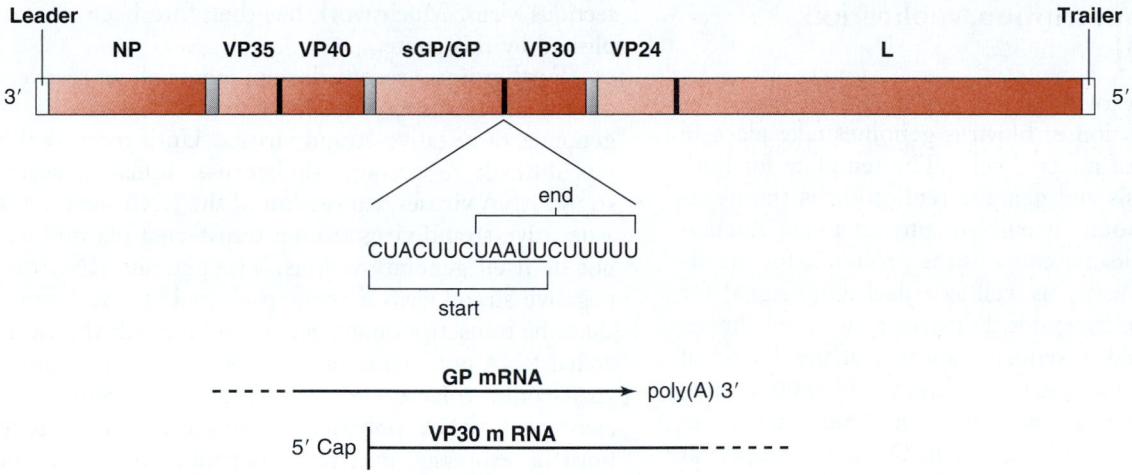

FIGURE 21.3 Map of Ebola virus genome. The seven viral genes are shown above the genome map along with the leader and trailer sequences. Nontranscribed sequences between genes are indicated by gray vertical bars; overlapping gene sequences are indicated by black vertical bars. The sequence overlap between the GP and VP30 genes, and the positions of the 3′ and 5′ ends of their mRNAs, are shown below.

those found in other mononegaviruses are found at the 3′ and 5′ ends of filovirus genomes.

With one exception, each gene codes for a single protein, and seven proteins are incorporated into virions (Figure 21.2). The proteins are named based on their function, glycosylation status, or molecular weight, and are listed in Table 21.1 in the order of their genes from the 3′ end of the genome. Four viral proteins are associated with genome RNA in nucleocapsids: NP (nucleoprotein), VP35 (RNA polymerase cofactor), VP30 (transcription activator), and L (viral RNA polymerase). The other three proteins are associated with the viral envelope: GP (envelope glycoprotein), VP40 (matrix protein), and VP24 (membrane protein).

TABLE 21.1 Ebola virus proteins

Protein	Function
NP (Nucleoprotein)	Highly phosphorylated protein that forms the helical nucleocapsid of the virus.
VP35 (Virus protein, 35 KDa)	Phosphoprotein, associated with the nucleocapsid. Required cofactor for viral RNA polymerase activity. Corresponds to the P protein of paramyxoviruses and rhabdoviruses.
VP40 (Virus protein, 40 KDa)	Associates with membranes, directs budding of virus envelopes at the plasma membrane to form virions. Corresponds to the M (matrix) proteins of paramyxoviruses and rhabdoviruses.
GP (Glycoprotein)	Heavily glycosylated type I transmembrane protein. The single envelope glyoprotein of filoviruses, responsible for both receptor binding and fusion with cellular membranes. Corresponds to the G protein of rhabdoviruses, or the H (HN) and F proteins of paramyxoviruses. GP is cleaved into two subunits: GP_1, N-terminal portion, and GP_2, C-terminal portion with membrane anchor and fusion peptide.
sGP (Secreted glycoprotein)	Glycosylated protein that shares N-terminal 295 amino acids with GP but has a distinct C-terminus because of RNA editing, and a different oligomeric structure. Lacks a membrane anchor and is secreted; not present in virions. Unknown function. Not found in Marburg virus.
VP30 (Virus protein, 30 KDa)	Associated with nucleocapsids. Has a strong activating effect on synthesis of viral messenger RNAs for Ebola but not Marburg virus.
VP24 (Virus protein, 24 KDa)	Associated with membranes, function unknown.
L (Large protein)	Viral RNA polymerase. Corresponds to L proteins of paramyxoviruses and rhabdoviruses.

Filovirus transcription, replication, and assembly

As with paramyxoviruses and rhabdoviruses, transcription and replication of filovirus genomes take place in the cytoplasm of infected cells. The template for both mRNA synthesis and genome replication is the negative-strand genome assembled into an intact nucleocapsid. The 3′ leader contains the promoter for the viral RNA polymerase as well as a packaging signal for assembly of nucleocapsids. Filovirus genes are believed to be transcribed in sequence starting at the 3′ end of the genome to yield seven different subgenomic messenger RNA species by a start–stop mechanism described in Chapter 20. These mRNAs are capped at their 5′ ends and contain a 3′ terminal poly(A) tail generated by stuttering of the viral RNA polymerase at a run of uridine residues located adjacent to each transcription termination signal sequence. Transcription efficiency may be influenced by gene order, formation of RNA secondary structures at the 3′ ends of the genes or within the intergenic sequences, overlapping genes, and presence of duplicated termination sites. These all tend to reduce the level of transcripts as the RNA polymerase passes from the 3′ end to the 5′ end of the genome. This results in higher levels of mRNAs for the nucleocapsid protein, the first gene to be transcribed, than for the L protein, the last gene.

Replication of the genome occurs via the synthesis of a full-length complementary antigenome (positive sense) RNA, which then serves as the template for the synthesis of progeny negative-stranded RNA molecules. Genome RNAs are packaged into nucleocapsids during replication. The cytoplasm of virus-infected cells contains prominent **inclusion bodies** that contain viral nucleocapsids. As infection proceeds, these bodies grow and become highly structured. Budding of completed virus particles takes place at sites on the plasma membrane that are altered by insertion of the viral glycoprotein and alignment of viral matrix protein and nucleocapsids.

Cloned cDNA copies of viral mRNAs and viral genome RNA are used for study of filoviruses

Because of the severity of human disease caused by these viruses, experimental study of infectious filoviruses is restricted to a very small number of laboratories that have high-level biological containment. This has obviously limited work on these viruses. However, complementary DNA copies of mRNA sequences for each of the viral proteins have been cloned in bacterial plasmids, and the proteins can be expressed either in bacterial or mammalian cells. This allows studies of the structure and activity of viral proteins in absence of infectious virus. Much work has therefore been accomplished by this route.

Furthermore, recent developments allow recovery of fully infectious virus from cloned DNA copies of the genomes of negative-strand viruses. Until recently this was difficult to accomplish because, unlike positive-strand RNA viruses, expression of the RNA genome of a negative-strand virus from a transfected plasmid will not by itself generate virions. The genome RNA of a negative-strand virus must be packaged in a nucleocapsid to be transcriptionally active, and it needs the virus-coded RNA polymerase and in some cases appropriate virus-coded cofactors to be expressed. Therefore, nucleocapsid, RNA polymerase, and cofactor proteins must be expressed in a cell independently of genome RNA in order to recover virus from transfected plasmids. In addition, copies of the virus genome with authentic 5′ and 3′ ends must be made, for without the authentic end sequences, transcription and replication will not occur.

Multiplasmid transfection systems allow recovery of infectious filoviruses

Both of these complications have been solved by the use of multiple-plasmid systems (Figure 21.4). A cDNA copy of the viral genome is placed under the control of a bacteriophage T7 promoter such that transcription directed by T7 RNA polymerase (see Chapter 7) initiates at the exact position of the 5′ end of the genome. A self-cleaving **ribozyme** (see Chapter 29) is engineered into the plasmid so that it is transcribed along with the genome RNA, and cleaves itself to generate the exact 3′ end of the genome. T7 RNA polymerase is expressed in a mammalian cell either from a transfected plasmid, a recombinant virus vector, or a gene in a specially engineered cell line (see Chapter 33). When the genome-expression plasmid is transfected into such a cell, authentic viral genome RNA is synthesized. Viral proteins required for packaging of the genome, transcription, and replication are expressed on a set of plasmids transfected along with the genome-expression plasmid. This allows production of nucleocapsids that can actively transcribe the genome, producing all viral messenger RNAs. In turn, newly synthesized viral proteins can replicate the genome, and viral envelope and matrix proteins incorporate nucleocapsids into infectious virus particles.

This system allows researchers to introduce directed mutations into viral genes and to study their effects on virus production, infectivity, and pathogenicity in infected animals. Some examples of such experiments are described below. However, even production of infectious filoviruses by plasmid transfection is a hazardous procedure because of the deadly consequences of infection, and must be done under biocontainment conditions. Virus-like particles that contain a mini-

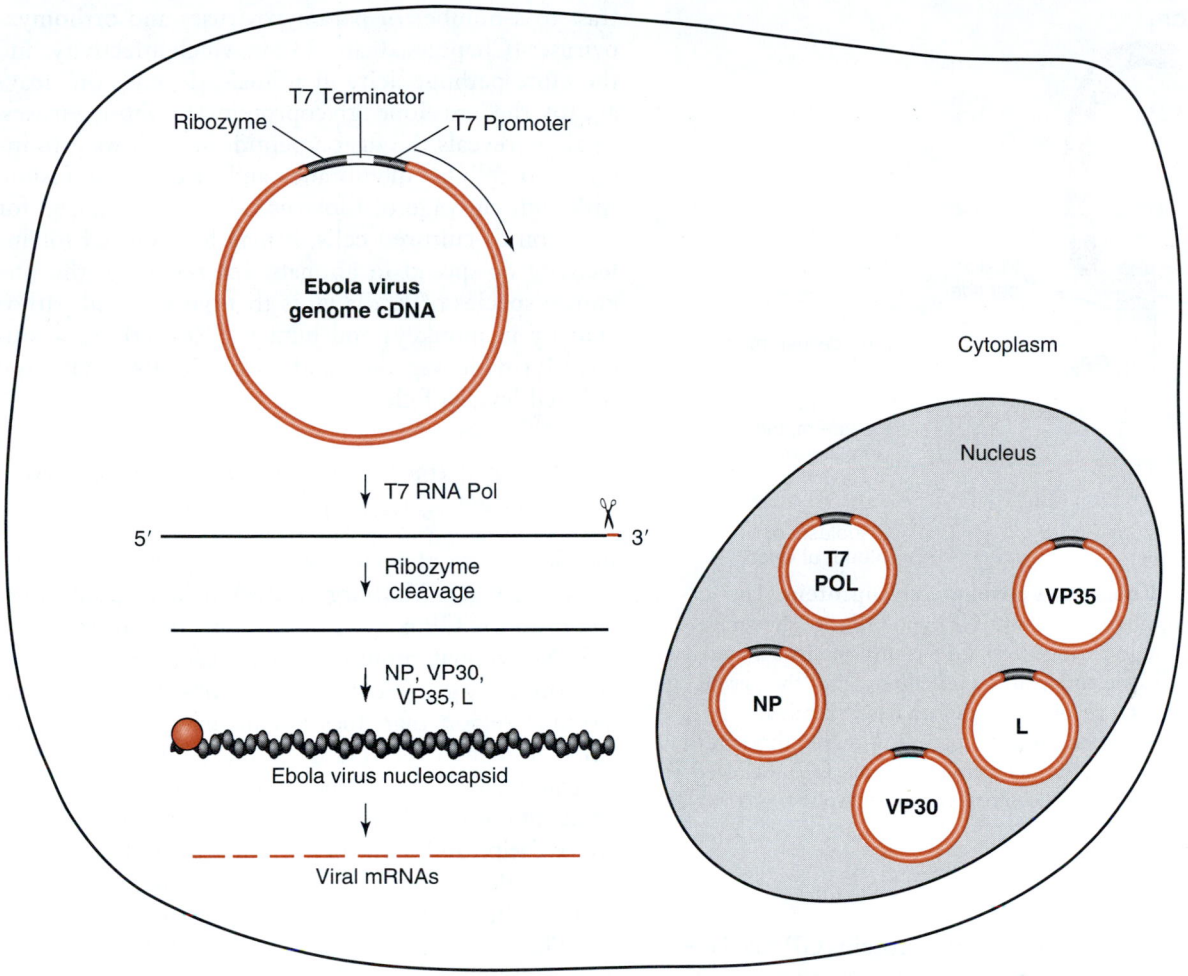

FIGURE 21.4 Production of infectious filoviruses by transfection of cells with multiple plasmids. Shown in the nucleus are plasmids used to make bacteriophage T7 RNA polymerase and four Ebola virus proteins that form nucleocapsids. These are driven by promoters (gray) that are recognized by cellular RNA polymerase II; mRNAs are exported to the cytoplasm where they are translated (a system fully driven by T7 RNA polymerase is also available; in that case, all plasmids are transcribed in the cytoplasm). Shown in the cytoplasm is a plasmid that expresses full-length Ebola virus RNA genome from a T7 promoter. Once all viral mRNAs are made from newly assembled nucleocapsids, virus replication and virion formation proceed. See text for details.

genome, coding only for **green fluorescent protein** or **chloramphenicol acetyltransferase**, convenient markers of infected cells, have been made by modifications of the above multiplasmid procedure. These particles resemble authentic Ebola virions and can enter susceptible cells, but will replicate only if filovirus nucleocapsid and envelope proteins are expressed by plasmid transfection in the same cells. Therefore these virus-like particles are not pathogenic, but can be used to study a variety of aspects of virus replication in cultured cells.

Filovirus glycoprotein mediates both receptor-binding and entry by fusion

The glycoprotein (GP, Figure 21-5) is synthesized as a 676- or 681-amino acid precursor protein that is inserted into the lumen of the endoplasmic reticulum. An N-terminal signal sequence is cleaved from the precursor during insertion into the membrane. A hydrophobic sequence near the C terminus serves to anchor the protein in the membrane; only 4–8 C-terminal amino acids project into the cytosol. GP subsequently undergoes extensive glycosylation in the endoplasmic reticulum and the Golgi compartments as it is transported to the plasma membrane; more than 50% of its mass consists of oligosaccharide residues. It is also cleaved by a cellular **furin** protease into two subunits: a large amino-terminal **ectodomain**, GP_1, and a shorter carboxy-terminal transmembrane protein, GP_2. The two subunits are held together by a cysteine bridge. GP forms trimers in the plasma membrane and in the virion envelope.

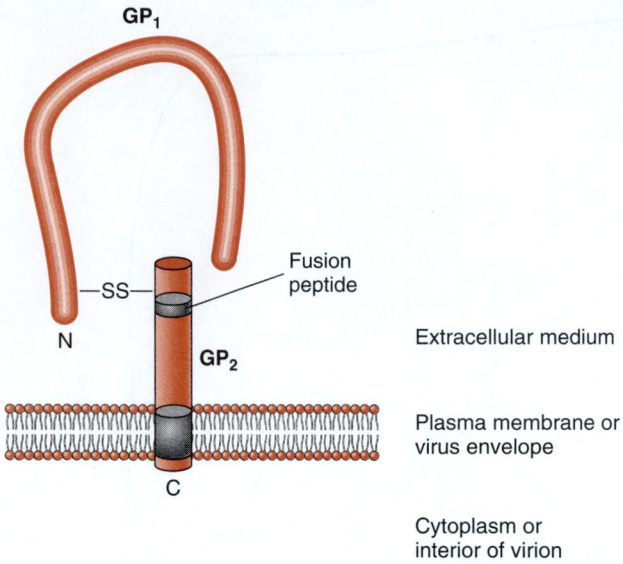

FIGURE 21.5 Ebola virus envelope glycoprotein. The cleaved glycoprotein fragments GP_1 and GP_2 are shown linked by a cysteine bridge (SS). GP_2 is inserted into a lipid bilayer (initially the endoplasmic reticulum, later the plasma membrane and the virion envelope) via its C-terminal transmembrane domain (shown in gray). A region thought to be responsible for membrane fusion is shown. GP_1, believed to interact with cellular receptors, is exposed to the extracellular medium.

Being the only virion surface protein, GP mediates binding to cellular receptors, probably via sites in GP_1. A number of different cellular proteins that mediate filovirus entry have been detected: **asialoglycoprotein receptor**, **folate receptor-α**, **integrins**, and **DC-SIGN** (*d*endritic *c*ell-*s*pecific *i*ntercellular adhesion molecule-grabbing *n*onintegrin) are among these. Most experiments on virus binding and entry have been done with replication-defective **pseudotypes**, using filovirus GP incorporated into the envelopes of unrelated recombinant viruses; these systems may not reflect the authentic entry pathways of filoviruses. However, there are indications that the virus enters **coated pits** and undergoes low pH-dependent fusion within vesicles in the cytoplasm. A region corresponding to the **fusion peptide** of other enveloped viruses lies near the N-terminus of GP_2 and is believed to mediate fusion by insertion into cellular membranes, followed by conformational changes in GP that bring the virus envelope into close contact with the membranes (see Chapter 4).

Surprisingly, cleavage of GP to produce GP_1 and GP_2 is not a requirement for production of infectious virions. The highly basic furin cleavage site of GP was mutated and an infectious virus clone was made by the multiplasmid system. Experiments confirmed that GP remained intact in this mutant virus, but the virus was able to infect monkey cells in culture. This is in contrast to a number of paramyxoviruses and orthomyxoviruses (Chapters 20 and 23), in which infectivity, and therefore pathogenicity in animals, depends on cleavage of the envelope glycoprotein. In those viruses, cleavage reveals the fusion peptide and allows it to insert into cellular membranes and direct virus fusion. Although cleavage of filovirus GP is not required for infection of cultured cells, it may be required for infectivity or spread in animals. Interestingly, the one known species of Ebola virus that has reduced pathogenicity in monkeys and humans (Reston) has a mutated furin cleavage site and its GP seems to undergo reduced levels of cleavage.

Ebola virus uses RNA editing to make two glycoproteins from the same gene

In addition to the envelope glycoprotein, all Ebola viruses (but not Marburg viruses) make a second protein from the GP gene. This protein contains the same 295 N-terminal amino acids as GP, including the N-terminal signal sequence, but has a different C-terminal region that does not include the transmembrane domain of GP (Figure 21.6). As a result, this glycoprotein (sGP) is secreted into the endoplasmic reticulum but is not anchored in the membrane. It undergoes glycosylation and transport, is released into the medium of cells infected with Ebola virus, and is found in the serum of infected patients.

The GP gene makes these two proteins by producing two messenger RNAs that differ by a single A residue, added at an editing site within the gene (Figure 21.6) in a manner similar to RNA editing of the P/C/V gene of paramyxoviruses (Chapter 20). This site contains a stretch of seven U residues, which are transcribed into A residues in positive-sense messenger RNA. The unedited mRNA has an open reading frame (frame 0) that stops at a UAA codon and produces the 364-amino acid precursor of sGP. About 20% of the time, the viral RNA polymerase pauses and "stutters" over the U stretch, adding a single additional A residue in the messenger RNA transcript. This changes the amino acid at position 295 from a threonine (ACC) to an asparagine (AAC), and changes the reading frame downstream of that site by one nucleotide (frame −1). As a result, ribosomes translating the edited mRNA make GP, a protein of 676 amino acids. The other 80% of the GP mRNA codes for sGP, which is therefore made in larger amounts than GP.

Does the secreted glycoprotein play a role in virus pathogenesis?

The role of sGP in the life cycle of Ebola virus remains obscure. All Ebola virus species tested make

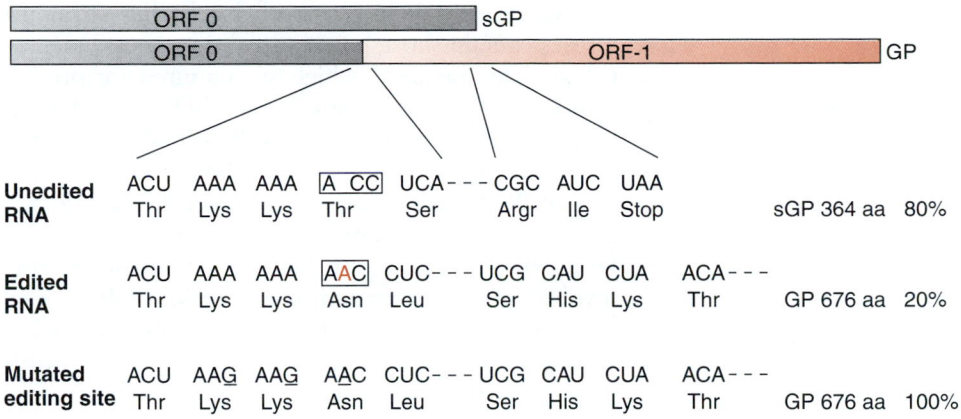

FIGURE 21.6 **Editing of Ebola virus GP mRNAs.** Shown above (in gray) is the open reading frame (ORF 0) translated to produce sGP and the N-terminal portion of GP, and below (in orange) is the open reading frame (ORF −1) for the C-terminal portion of GP. The mRNA sequences at the editing site, and their products of translation, are shown below. Editing introduces an additional A residue (shown in orange) that shifts the reading frame to that of GP. Only 20% of mRNAs are edited. Parts of the overlapping reading frames at the stop site for translation of sGP are shown at right. Below: an engineered virus mutant that does not undergo editing because the run of U residues in the editing site (A's in the complementary mRNA) is interruped by two C residues (underlined G's in the mRNA). The reading frame is shifted to −1 because of an extra U added at the end of the editing site (underlined A in mRNA). This mutant overproduces GP and makes no sGP.

sGP as well as GP, but Marburg virus does not make sGP. Because it is released into the medium of infected cells and found in the serum of patients, it has been considered that sGP may act as a soluble factor that targets elements of host defense systems. sGP could bind to **antibodies** and contribute to the immunosuppression found in patients and animals infected with Ebola virus. So far, there is no evidence that sGP is involved in activation of **macrophages** or endothelial cells, which could lead to capillary leakage and coagulation disorders.

To help understand the role of sGP, an infectious virus with a mutation in the editing site of the GP gene was made by the multiplasmid system. In this mutant genome, two U residues in the stretch of 7 were exchanged for C residues, and an additional U residue was added (Figure 21.5). As a result, editing during mRNA synthesis did not take place, and only mRNA coding for GP was made, because the extra A residue in the mRNA transcript changes the reading frame. This virus replicated and produced infectious virions, but induced cytopathic effects in cultured cells more rapidly than wild-type virus, and gave smaller plaques and lower titers of infectious virus. The increased cytopathic effects caused by the mutant virus may be due to the increased rate of synthesis of GP, which affects the protein processing and transport system of the cell. Thus RNA editing in Ebola virus limits the amount of GP made, allowing the infected cell to produce higher levels of virus. The effects of the mutant virus on pathogenesis in infected animals remain to be tested. There may be other roles that sGP plays, particularly in the (unknown) natural host of this virus.

Minor nucleocapsid proteinVP30 activates viral mRNA synthesis in Ebola virus

Filovirus nucleocapsids contain four proteins, three of which (NP, VP35, and L) correspond to their counterparts (N, P, and L) in nucleocapsids of paramyxoviruses and rhabdoviruses. The fourth nucleocapsid protein, VP30, is required for optimal levels of viral messenger RNA synthesis from Ebola virus genomes. Thus, one role of VP30 is to activate viral transcription.

Transcription of filovirus genomes begins at the NP gene start site, shortly downstream from the 3′ end of the genome (Figure 21.3). The nucleotide sequence at this start site can form a stem-loop structure in either the genome RNA or the complementary NP mRNA (Figure 21.7). When this stem-loop structure was disrupted by mutation, transcription of Ebola virus minigenomes was no longer dependent on VP30 expression. Furthermore, when the NP stem-loop was placed at the beginning of the next downstream gene, VP35, transcription of that gene also became dependent on expression of VP30. Thus, the stem-loop structure at the beginning of the NP gene inhibits viral RNA polymerase from initiating mRNA synthesis, and VP30 activates transcription by reversing this inhibition. Although VP30 is present in nu-

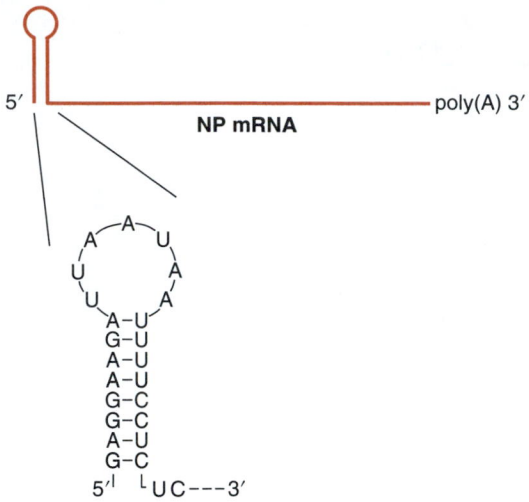

FIGURE 21.7 Stem-loop at 5′ end of NP mRNA that controls transcription of Ebola virus genome. VP30 expression activates Ebola virus transcription by counteracting the inhibitory effect of this stem-loop structure on the initiation of transcription of the NP gene by viral RNA polymerase.

cleocapsids, it is not known if VP30 directly binds to this stem-loop, either in the genome RNA or in the growing mRNA chain. VP30 is a zinc-binding protein that forms oligomers, and both zinc-binding and oligomerization are required for its transcription-stimulating activity.

Matrix protein VP40 directs budding and formation of filamentous particles

VP40, the most abundant protein in filovirus particles, is associated with the viral envelope. It does not appear to be a transmembrane protein, but is located on the cytoplasmic face of the plasma membrane, or the inside of the virus envelope. Thus it resembes the matrix proteins of other enveloped viruses. Matrix proteins often serve as a bridge between the envelope glycoproteins incorporated into the lipid membrane and the nucleocapsids enclosed within the envelope. VP40 of both Marburg and Ebola viruses binds strongly to membranes via its C-terminal region and forms oligomers when membrane-bound. These oligomers can be dimers or ring-shaped hexamers or octamers. Strikingly, when expressed by itself in mammalian cells, VP40 can direct the formation of virus-like particles that bud from the plasma membrane and are released into the medium. Co-expression of VP40 and GP increases production of such particles, which resemble mature filovirus virions but have smaller diameters. Thus VP40 appears to be the major viral protein that directs budding and virion formation.

Matrix proteins of a number of other enveloped viruses contain short, conserved amino acid sequences, called "late domains," which are required for production of virus-like particles. Both Ebola and Marburg virus have the late domain motif Pro-Pro-X-Tyr (where "X" means any amino acid) in the N-terminal region of their NP40. Mutations or deletions in this late domain motif impair or block production of virus-like particles by VP40. A number of cellular proteins known to be involved in trafficking and sorting of intracellular vesicles have been shown to bind to proteins that contain such late domains. These include Nedd4, a **ubiquitin ligase**, and Tsg101, which is part of a multiprotein complex that directs vesicles to **endosomes**, where they bud into the lumen and form **multivesicular bodies**. There is evidence that these cellular proteins are required to help VP40 and other viral matrix proteins form virions by budding at the plasma membrane. Although the exact mechanism by which these proteins interact with VP40 to form budding virions is not clear, it may involve **ubiquitination** of VP40 or associated proteins, and association with endosomes before transport to the plasma membrane.

Most filovirus outbreaks have occurred in equatorial Africa

Marburg virus and the Ebola virus species Ivory Coast, Sudan, and Zaire are indigenous to the African continent (Figure 21.8). Marburg virus was first isolated from patients in Europe, but the origin of the European cases could be traced back to foci in Uganda from where vervet monkeys compounded in Entebbe were imported into Germany and former Yugoslavia. Both index case patients of the episodes of Marburg virus hemorrhagic fever in Kenya (1980/87) traveled in the Mt. Elgon region. This region is near the shores of Lake Victoria and is close to the trapping place and holding station of the monkeys that initiated the 1967 outbreak in Marburg. The first known outbreak of Marburg virus in Africa occurred in 1998–2000 in the eastern part of the Democratic Republic of Congo close to the border with Uganda, and there is an ongoing outbreak with more than 300 deaths in northern Angola that in mid-2005 was still not under control.

Outbreaks of Ebola virus hemorrhagic fever have taken place in Gabon, Republic of Congo, Democratic Republic of Congo, Sudan, and Uganda. In 1994, a new species of Ebola virus was isolated during an outbreak of hemorrhagic fever among wild chimpanzees in a rain forest of Ivory Coast. This was the first appearance of a filovirus in West Africa.

These data strongly suggest **endemic** foci for Marburg virus and for Ebola virus species Ivory Coast, Sudan, and Zaire in an area lying between the 10th par-

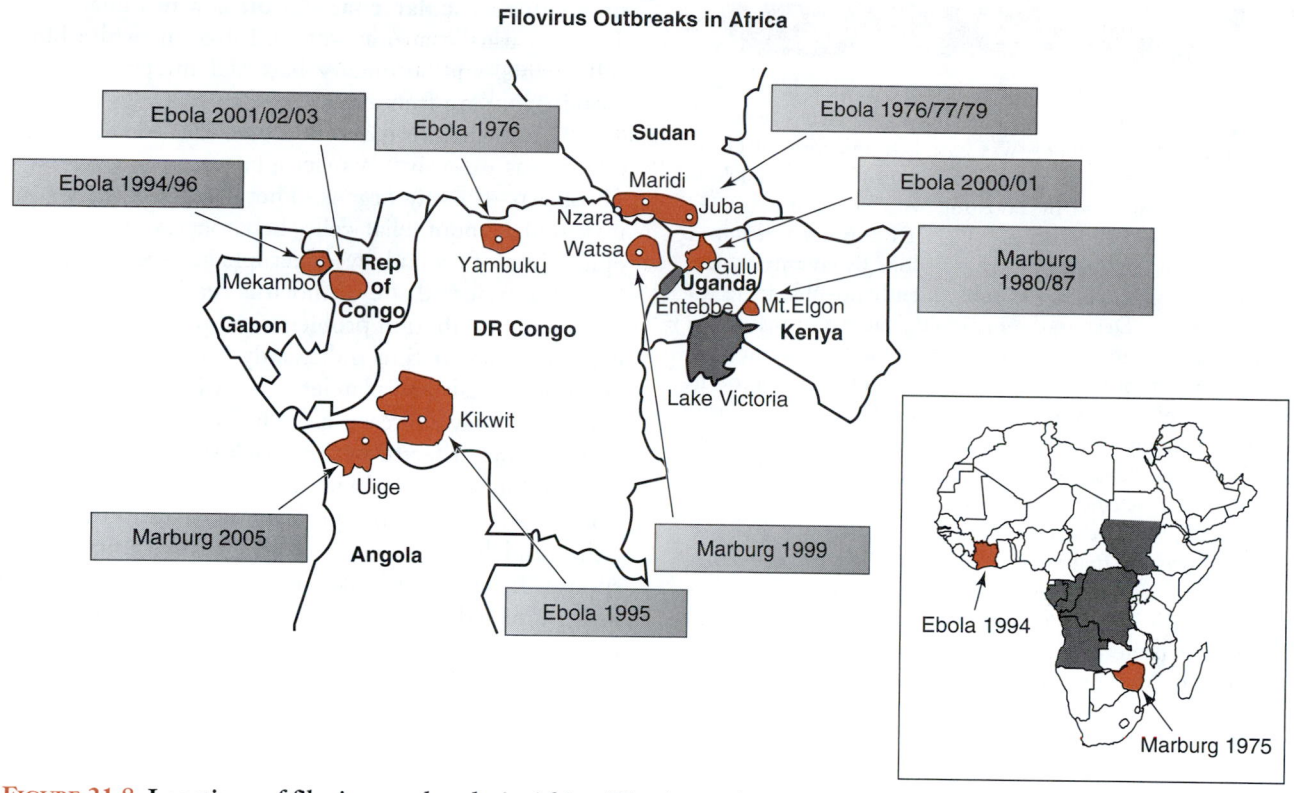

FIGURE 21.8 **Locations of filovirus outbreaks in Africa.** The sites and years of filovirus episodes and outbreaks in countries of central and west Africa are shown.

allels north and south of the equator in Central Africa. Additional endemic areas in Africa or on other continents may exist. The identification of Ebola virus species Reston in 1989 suggested for the first time the presence of a filovirus in Asia that may be associated with wild nonhuman primates.

Filovirus infections are transmitted to humans from an unknown animal origin

It is believed that filoviruses normally infect and circulate in one or more animal species in central Africa and perhaps in southern Asia. These species may or may not show overt signs of infection. They would serve as a **reservoir** for the virus, from which it occasionally emerges to infect humans, probably via direct contact. Many species have been discussed as possible natural hosts; however, no nonhuman vertebrate hosts or arthropod vectors have yet been identified. The Ebolavirus Reston infection within a colony of Asian monkeys raised the question whether nonhuman primates may serve as a reservoir. The fact that Marburg virus and Ebola virus species Sudan and Zaire are pathogenic for monkeys and apes argues against the idea that these animals may be a reservoir for the virus; they are more likely secondary hosts, as are humans.

Recent studies on experimentally infected wild animals have shown that fruit-eating and insect-eating bats support replication and circulation of high titers of Ebola virus without necessarily becoming ill. Bats have been discussed before as potential reservoirs for the Ebola virus outbreaks in Sudan and the Kenyan Marburg hemorrhagic fever cases. Since a persistently infected host is the most likely reservoir for a **zoonotic disease**, chronic infection in bats or other small mammalian species may be considered as a possible mechanism that regulates survival of filoviruses in nature.

Spread of filovirus infections among humans is limited to close contacts

The usual pattern seen in large outbreaks of filovirus hemorrhagic fever begins with a single case, who disseminates infection to several other people. Secondary and subsequent infections occur in close family members or among medical staff. The epidemic eventually ends because transmission of the infection is transient and the spread of the virus is inefficient. Person-to-person transmission by physical contact with case patients is the main route of infection in human outbreaks. Activities such as nursing and preparing bodies for burial are associated with an increased risk. During the

Filovirus vaccine candidates

Vaccine development has been an important issue in research on filoviruses in recent years. The first Ebola virus vaccine to protect nonhuman primates was based on injection of DNA plasmids coding for both the glycoprotein (GP) and nucleoprotein (NP), followed by boosting with an adenovirus recombinant vector expressing GP. This approach required several months for immunity to develop. By modifying the immunization protocol it was subsequently possible to accelerate protection. Despite the intriguing success of these vaccines, preexisting immunity to adenovirus in the human population may eventually limit the utility of this approach. Togavirus replicons expressing Marburg virus proteins protected cynomolgus monkeys from infection with Marburg virus. However, subsequent studies evaluating this strategy for an Ebola virus vaccine were less encouraging.

In 2005, live attenuated recombinant strains of vesicular stomatitis virus (a rhabdovirus of cattle) that express either Marburg or Ebola virus glycoproteins were constructed. Inoculation of cynomolgus monkeys with these vectors resulted in complete protection against disease from an otherwise lethal dose of Marburg or Ebola virus. These live virus vectors are therefore potent and promising for future vaccine development against filoviruses, and may also be useful in developing vaccines against other agents causing lethal viral hemorrhagic fevers.

Ebola virus outbreaks in 1976 and 1995, transmission within hospital settings via contaminated syringes and needles was a major problem.

Transmission is inefficient, as documented by secondary infection rates that rarely exceed 12%. Nevertheless, extreme care should be taken with blood, secretions, and excretions of infected patients. Sexual transmission has been described for Marburg virus, and neonatal transmission was reported for the 1976 outbreak in Zaire. Semen or other body fluids may be infectious for some time during convalescence. Transmission by droplets and small-particle aerosols has been observed among experimentally infected and quarantined imported monkeys. This is confirmed by identification of filovirus particles in lungs of naturally and experimentally infected monkeys and human postmortem cases. However, the contribution of aerosol transmission to the course of human outbreaks seems to be low.

Pathogenesis of filovirus infections

Marburg and Ebola viruses cause similar pathological changes in humans. The most striking lesions are found in liver, spleen, lymph nodes, and kidney. In late stages of the disease, hemorrhage occurs in the gastrointestinal tract, the spaces surrounding the lungs and heart, the peritoneal spaces, and the renal tubules. **Disseminated intravascular coagulation** is a terminal event. There is usually also a profound drop in white blood cell counts, and secondary bacterial infections in the bloodstream are frequent.

In animals infected with Ebolavirus Reston, replication was extensive in macrophages, monocytes, and fibroblasts of many organs. There was less replication in vascular endothelial cells, hepatocytes, and tubular epithelium of the kidney. Macrophages seem to be the first and preferred site of filovirus replication.

Fluid distribution problems and platelet abnormalities in disease victims indicate dysfunction of capillary endothelial cells and platelets. Virus-induced release of **cytokines** may increase endothelial permeability and may be a major factor in the **shock** syndrome regularly observed in severe and fatal cases. Fatal filovirus infections usually end with high levels of virus in the circulation and little evidence of an effective immune response. In fact, filoviruses induce immunosuppression in the infected host, which appears to be a major factor for the rapid generalization and the severity of the disease.

Clinical features of infection

The incubation period of Marburg and Ebola virus infections varies between 4 and 21 days. The onset of the disease is sudden, with fever, chills, headache, muscle pain, and loss of appetite. This may be followed by symptoms such as abdominal pain, sore throat, nausea, vomiting, cough, joint pain, and diarrhea. Prostration, lethargy, wasting, and diarrhea seem to be more severe in filovirus hemorrhagic fever than in patients suffering from hemorrhagic fevers caused by other viruses. Severe diarrhea is often misleading, since it is also found in many bacterial enteric infections.

Patients are dehydrated, apathetic, and disoriented, and they may develop a characteristic rash associated with peeling skin by day 5 or 7 of the illness. The rash is extremely useful in diagnosis. Hemorrhagic manifestations develop during the peak of the illness, and can predict the likely progress of the disease. Bleeding into the gastrointestinal tract is most prominent, and localized skin hemorrhages and hemorrhages from puncture wounds and mucous membranes are common. Nonfatal cases show fever for about five to nine days. Fatal cases develop clinical signs early during infection, and death commonly occurs between days 6 and 16 after the development of hemorrhage and shock due to low blood volume.

Patients infected with filoviruses must be isolated, and protection of medical and nursing staff is required. This can be achieved by strict barrier nursing techniques and addition of HEPA-filtered respirators for aerosol protection when feasible. Filoviruses are resistant to the antiviral effects of **interferon**. Ribavirin, a drug that is effective against other agents that cause viral hemorrhagic fever, has no effect on filoviruses. A protective vaccine (see text box) would be extremely valuable for at-risk medical personnel in Africa and researchers working with infectious filoviruses.

SUGGESTED READING

1. Feldmann, H., Jones, S., Klenk, H.D., and Schnittler, H.J. (2003). Ebola virus: From discovery to vaccine. *Nature Reviews Immunology* **3,** 677–685.

2. Feldmann, H., Volchkov, V.E., Volchkova, V.A., Stroher, U., and Klenk, H.D. (2001). Biosynthesis and role of filoviral glycoproteins. *Journal of General Virology* **82,** 2839–2848.

3. Feldmann, H., Wahl-Jensen, V., Jones, S.M., and Stroher, U. (2004). Ebola virus ecology: A continuing mystery. *Trends in Microbiology* **12,** 433–437.

4. Jasenosky, L.D. and Kawaoka, Y. (2004). Filovirus budding. *Virus Research* **106,** 181–188.

5. Klenk, H.-D., Ed. (1999). Marburg and Ebola viruses. *Current Topics in Microbiology and Immunology* **235,** 1–225.

6. Klenk, H.-D. and Feldmann, H. Eds. (2004). Ebola and Marburg viruses. *Molecular and Cellular Biology*, Horizon Bioscience, Wymondham UK.

7. Neumann, G., Feldmann, H., Watanabe, S., Lukashevich, I., and Kawaoka, Y. (2002). Reverse genetics demonstrates that proteolytic processing of the Ebola virus glycoprotein is not essential for replication in cell culture. *Journal of Virology* **76,** 406–410.

8. Peters, C.J. and Le Duc, J.W. (1999). Ebola: The virus and the disease. *Journal of Infectious Diseases* (Suppl. 1) **179,** S1–S288.

9. Preston, R.M. (1995). The Hot Zone: A terrifying true story. Anchor Books, Random House, New York.

KEY TERMS

Antibody
Asialoglycoprotein receptor
Chloramphenicol acetyltransferase
Coated pits
Cytokines
DC-SIGN
Disseminated intravascular coagulation
Ectodomain
Endemic
Endosome
Folate receptor-α
Furin
Fusion peptide
Green fluorescent protein
Hemorrhagic fever
Inclusion bodies
Integrins
Interferon
Macrophage
Multivesicular bodies
Pseudotype
Reservoir
Ribozyme
Shock
Ubiquitin ligase
Ubiquitination
Zoonotic disease

Bunyaviruses

Bunyaviridae

From *Bunyamwera*, locality in Uganda where type virus was isolated

VIRION

Spherical enveloped particles with a fringe of projections.
Diameter 90–100 nm.
Helical nucleocapsids.

GENOME

Linear ss RNA, negative or ambisense.
Three genome segments: S, 1–3 Kb; M, 3–5 Kb; L, 6–12 Kb.
Total genome length 11–19 Kb.

GENES AND PROTEINS

Each genome segment codes for 1–3 proteins.
L: RNA polymerase (L).
M: Two viral envelope glycoproteins (Gn, Gc), cleaved during translation.
S: Nucleocapsid protein (N).
Zero–three nonstructural proteins also made, depending on virus strain.

VIRUSES AND HOSTS

Viruses in four genera (*Orthobunyavirus*, *Phlebovirus*, *Nairovirus* and *Hantavirus*) infect vertebrates.
Many are transmitted between animals by arthropod vectors (La Crosse, Rift Valley Fever, Nairobi sheep disease viruses).
Hantaviruses (Hantaan, sin nombre viruses) are transmitted by rodents and are not arthropod-borne.
Viruses in *Tospovirus* genus infect plants, are transmitted by thrips.

DISEASES

Can cause severe fever, encephalitis, hemorrhagic fever, and fatal respiratory syndrome.

DISTINCTIVE CHARACTERISTICS

Can replicate in both vertebrate and invertebrate hosts.
Segmented genome.
Ambisense coding strategy in some viruses.
Viral mRNA synthesis is primed by stealing capped 5′ ends of cellular mRNAs.

Most bunyaviruses are transmitted by arthropod vectors, including mosquitoes and ticks

The *Bunyaviridae* family includes over 300 named viruses in five genera (Table 22.1). All the viruses in this family share the following common characteristics: (1) a single-stranded RNA genome composed of three unique segments, each wrapped in a helical ribonucleoprotein complex; (2) cytoplasmic site of replication; (3) viral mRNA synthesis primed by capped oligonucleotides derived from host cellular mRNAs; and (4) virion assembly at Golgi membranes.

Bunyamwera virus, isolated in 1943 in Uganda, is the prototype of both the *Orthobunyavirus* genus and the *Bunyaviridae* family. It was the first bunyavirus whose complete genome sequence was determined, and it was the first segmented negative-strand virus from which infectious virus was recovered by transfecting cells with cloned cDNAs. Rift Valley fever virus, a member of the *Phlebovirus* genus, uses an **ambisense** coding strategy on its S genome segment (see below). Ambisense coding strategies are also employed by tospovirus M and S segments, and by both the L and S segments of viruses in the *Arenaviridae* family.

Most bunyaviruses are transmitted from animal to animal by arthropod vectors, and these viruses are able to replicate in both their invertebrate and vertebrate hosts. Thus their lifestyle is very similar to those of the positive-strand RNA flaviviruses and togaviruses (Chapters 17 and 18). Virus infection has little effect on the invertebrate host even though the virus causes a systemic infection. The vector transmits the virus to a vertebrate when taking a blood meal. Virus can also be transmitted to other individuals of the same invertebrate species by mating or to offspring via the egg. **Transovarial transmission** is an important mechanism for the maintenance of these viruses in nature, as the virus can survive in the egg until conditions are favorable for the arthropod to hatch.

Viruses of a particular genus tend to be transmitted by specific types of invertebrate vectors. Mosquitoes and midges transmit orthobunyaviruses, ticks transmit nairoviruses, and sandflies or ticks transmit phleboviruses. Tospoviruses are plant viruses that are transmitted by **thrips** and are able to replicate both in

TABLE 22.1 Some pathogens in the family *Bunyaviridae*

Genus/Virus	Host: disease	Vector	Distribution
Orthobunyavirus			
Oropouche	Human: fever	Midge	S. America
La Crosse	Human: encephalitis	Mosquito	N. America
Akabane	Cattle: abortion and congenital defects	Midge	Africa, Asia, Australia
Hantavirus			
Hantaan	Human: severe hemorrhagic fever with renal syndrome (HFRS); 5–15% fatality	Field mouse	Eastern Europe, Asia
Puumala	Human: mild HFRS, nephropathica epidemica; 0.1% fatality	Bank vole	Western Europe
Sin nombre	Human: hantavirus pulmonary syndrome; 50% fatality	Deer mouse	N. and S. America
Nairovirus			
Crimean-Congo hemorrhagic fever	Human: hemorrhagic fever; 20–80% fatality	Tick	Eastern Europe, Africa, Asia
Nairobi sheep disease	Sheep, goats: fever, hemorrhagic gastroenteritis, abortion	Tick	E. Africa
Phlebovirus			
Rift Valley fever	Human: encephalitis, hemorrhagic fever, retinitis; 1–10% fatality	Mosquito	Africa
	Domestic ruminants: necrotic hepatitis, hemorrhage, abortion		
Sandfly fever	Human: fever	Sandfly	Europe, Africa, Asia, Naples
Tospovirus			
Tomato spotted wilt	More than 650 plant species; various symptoms	Thrips	Worldwide

vector and in plant cells. Hantaviruses are exceptions; they are not transmitted by arthropod vectors but are maintained in nature as persistent infections of rodents. Although hantaviruses can be present in large amounts in lungs and kidneys of infected rodents, they do not cause obvious disease in the animal host. Humans become infected by inhaling aerosolized excretions from infected rodents.

Some bunyaviruses cause severe hemorrhagic fever, respiratory disease, or encephalitis

Although most bunyaviruses have little impact on humans, some are important pathogens, either directly by causing disease in humans, or indirectly by affecting domestic animals or food crops (Table 22.1). Bunyaviruses were among the earliest animal viruses discovered: Nairobi sheep disease (in 1910) and Rift Valley fever virus (in 1930).

Four types of illness are caused by bunyaviruses in humans: fever, **encephalitis**, **hemorrhagic fever**, and a fatal respiratory syndrome. Some viruses cause a self-limiting febrile illness, often with rash and rarely fatal. In South America, Oropouche virus can cause **epidemics** involving tens or hundreds of thousands of people. Rift Valley fever infection results in a range of disease syndromes, from severe febrile illness to retinal disease to a fatal hemorrhagic disease. La Crosse virus is a significant cause of childhood encephalitis in the Midwestern United States, averaging about 100 cases per year. Although rarely fatal, the acute illness is severe, frequently involving seizures, and can lead to unpleasant aftereffects such as chronic seizures or epilepsy.

Hemorrhagic fevers are caused by some nairoviruses and hantaviruses. Crimean-Congo hemorrhagic fever, a tick-borne disease found in Eastern Europe, Asia and Africa, results in severe illness and high case fatality. Its symptoms resemble those of Lassa fever and Ebola (see Chapter 21) viruses. Hantaviruses are responsible for three forms (mild, moderate, or severe) of **hemorrhagic fever with renal syndrome**. Hundreds of thousands of cases of hemorrhagic fever with renal syndrome are reported throughout Europe and Asia every year, with case fatality rates ranging from less than 0.1% for Puumala-like viruses in Western Europe, to 15% for Hantaan-like viruses in China.

A new type of hantavirus disease called **hantavirus pulmonary syndrome** was described in the southwestern United States in 1993, and was linked to a virus called sin nombre ("no name") virus. Several hundred cases of hantavirus pulmonary syndrome have since occurred in North and South America. This severe respiratory disease, which is a classic example of an emerging viral disease, has an overall fatality rate of 50%.

Bunyaviruses encapsidate a segmented RNA genome in a simple enveloped particle

Like the orthomyxoviruses and the reoviruses (Chapters 23 and 24), the genome of bunyaviruses consists of multiple distinct RNA molecules. All bunyaviruses contain three genome segments, named L, M, and S for large, medium, and small. These genome segments are encapsidated by the nucleocapsid (N) protein to form ribonucleoprotein complexes with helical symmetry. Associated with the nucleocapsids is the RNA polymerase (L) protein.

Virions enclose the nucleocapsids within a lipid envelope derived from the membranes of the Golgi complex. The envelope contains two viral glycoproteins, named Gn and Gc. Thus bunyavirus particles contain only four viral proteins: N, L, Gn, and Gc. Virions examined by cryoelectron microscopy are spherical, with a diameter of about 100 nm (Figure 22.1a). Smaller particles are also seen; these may contain fewer than three nucleocapsids, and therefore would be noninfectious. The mechanisms controlling genome packaging are unknown; as with other viruses that have segmented genomes, the question of how virions package at least one of each genome segment into the virion is not resolved.

The total genome size ranges from just over 12 Kb for orthobunyaviruses to nearly 19 Kb for nairoviruses; much of this difference is accounted for by the very large nairovirus L segment. The terminal 8–11 nucleotides of genome RNAs are highly conserved among all three segments for a given virus, and among all viruses within a genus (Table 22.2). The conserved sequences at the 5′ termini are complementary to the sequences at the 3′ termini, which permits the ends of the RNA to base-pair and form panhandle structures. Circular forms of both the naked RNAs and the helical ribonucleoprotein nucleocapsids (Figure 22.1b) have been observed by electron microscopy.

Bunyavirus protein coding strategies: negative-strand and ambisense RNAs

The genome coding strategies of two representative bunyaviruses are shown in Figure 22.2. Most bunyavirus proteins are coded by messenger RNAs copied from the genome, and therefore the genomes are considered negative-strand RNAs. However, some bunyaviruses have "sense" and "antisense" protein coding regions on the same RNA genome: for an example, see the phlebo virus S RNA segment in Figure 22.2. In this case, the messenger RNA for the N protein is copied from the genome RNA, but the messenger RNA for the NSs protein is of opposite polarity, and must be copied from the antigenome RNA (see below). Such ambisense genomes are relatively rare in the RNA virus world, but arenaviruses

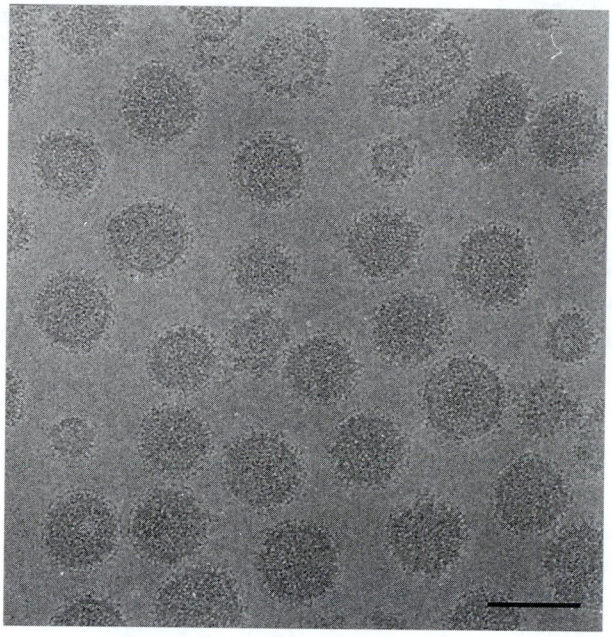

(a)

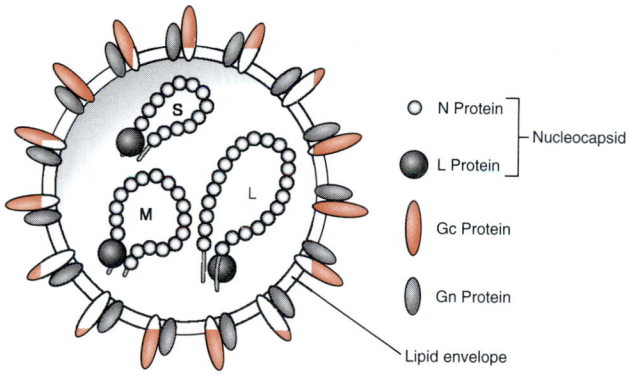

(b)

FIGURE 22.1 **Bunyavirus virions.** (a) Cryoelectron microscopic image of purified virions of La Crosse virus. Bar: 100 nm. (Courtesy of B.V.V. Prasad.) (b) Schematic diagram showing the envelope and three nucleocapsids (L, M, and S) of a bunyavirus virion, with the four structural proteins labeled. Nucleocapsids are shown as circular; they are joined by complementary RNA sequences at the 3′ and 5′ ends.

(not covered in this book) also use this strategy to express their proteins. On the other hand, it is commonplace for many double-stranded DNA viruses to have coding regions for mRNAs and proteins on both DNA strands, in different regions of their genomes.

L RNA codes for viral RNA polymerase

The L genome segments of all bunyaviruses encode a single large protein, the L protein, which is translated from an mRNA complementary to the genome RNA. The L proteins contain motifs characteristic of all RNA polymerases, and can both transcribe viral genome RNA, making viral messenger RNAs, and replicate the genome, making full-length genome copies. L proteins vary in size from 259 to 459 KDa, depending on the genus. Thus, like RNA polymerases of many positive- and negative-stranded RNA viruses, the L proteins are by far the largest viral proteins made by bunyaviruses.

M RNA codes for virion envelope glycoproteins

The M segment encodes the two virion glycoproteins, Gn and Gc, in the form of a precursor polyprotein. These glycoproteins are named for their relative positions in the precursor polyprotein; Gn lies toward the N-terminus and Gc lies toward the C-terminus. This replaces a confusing terminology based on relative sizes of these two proteins.

The polyprotein is cleaved during translation and transport into the **lumen** of the endoplasmic reticulum to yield the two proteins, which are notably rich in cysteine residues. Both Gn and Gc are **type I transmembrane proteins**, with hydophobic membrane anchor regions near their C-termini. These proteins are targeted to the Golgi membranes by a signal contained in Gn. A Gn/Gc dimer probably forms rapidly in the endoplasmic reticulum and is the form transported to the Golgi. Unlike many other viral glycoproteins, Gn and Gc do not migrate to the plasma membrane but are re-

TABLE 22.2 **Consensus 3′ and 5′ terminal nucleotide sequences of *Bunyaviridae* genome RNAs**

Orthobunyavirus	3′ UCAUCACAUGA . . . . . UCGUGUGAUGA 5′
Hantavirus	3′ AUCAUCAUCUG . . . . . AUGAUGAU 5′
Nairovirus	3′ AGAGUUUCU . . . . . AGAAACUCU 5′
Phlebovirus	3′ UGUGUUUC . . . . . GAAACACA 5′
Tospovirus	3′ UCUCGUUAG . . . . . CUAACGAGA 5′

tained in the Golgi, where virus assembly takes place (see below).

These viral glycoproteins have been implicated in many biological attributes of the virus including **hemagglutination**, virulence, tissue **tropism**, neutralization by antibodies, and cell fusion. By making **reassortant viruses** between virulent and nonvirulent bunyaviruses, it was shown that the M segment (and hence its gene products) is associated with the virulent phenotype. Similarly, reassortant viruses have been used to demonstrate that the ability of a bunyavirus to cause a disseminated infection in its mosquito vector maps to the M segment. Monoclonal antibodies to the orthobunyavirus Gc protein prevent infection of vertebrate cells, and an orthobunyavirus Gc mutant is defective in the fusion process.

For orthobunyaviruses and some phleboviruses a small nonstructural protein (NSm) is encoded as part of the polyprotein and is cleaved during translation (Figure 22.2). Tospoviruses also encode an NSm protein, but they use an ambisense strategy. The NSm coding region is contained in the 5′ terminal part of the tospovirus genomic RNA, but is translated from a subgenomic mRNA. The function of NSm is presently unknown.

S RNA codes for nucleocapsid protein and a nonstructural protein

The S RNA segments also show diversity in their coding strategies. The S RNA of orthobunyaviruses (Figure 22.2) encodes two proteins, the nucleocapsid protein (N) and a nonstructural protein (NSs). These proteins are both synthesized from the same complementary-sense mRNA but from different, overlapping reading frames, as the result of alternative initiation at different AUG codons. In contrast, the S RNAs of hantaviruses and nairoviruses contain only a single open reading frame, encoding the nucleocapsid protein.

The S RNAs of phleboviruses (Figure 22.2) and tospoviruses employ an ambisense coding strategy. The N protein is encoded in an mRNA corresponding to the 3′ half of the S RNA; this mRNA is copied from genome RNA. The NSs protein is encoded in an mRNA corresponding to the 5′ half of the S RNA; this mRNA is copied from antigenome RNA (see below). The N protein binds specifically to full-length viral genome and antigenome RNAs to form viral nucleocapsids, and it is in this form that the viral RNA serves as a template for both transcription and replication.

After attachment via the virion glycoproteins, bunyaviruses enter the cell by endocytosis

No cellular receptor has yet been identified for most members of the *Bunyaviridae* family. Antibody neutralization and hemagglutination inhibition sites have been mapped to both virion glycoproteins Gn and Gc encoded by phleboviruses, suggesting that for these viruses both Gn and Gc are involved in attachment to cells. For orthobunyaviruses, Gc is the major attachment protein for vertebrate cells, whereas Gn may contain the major

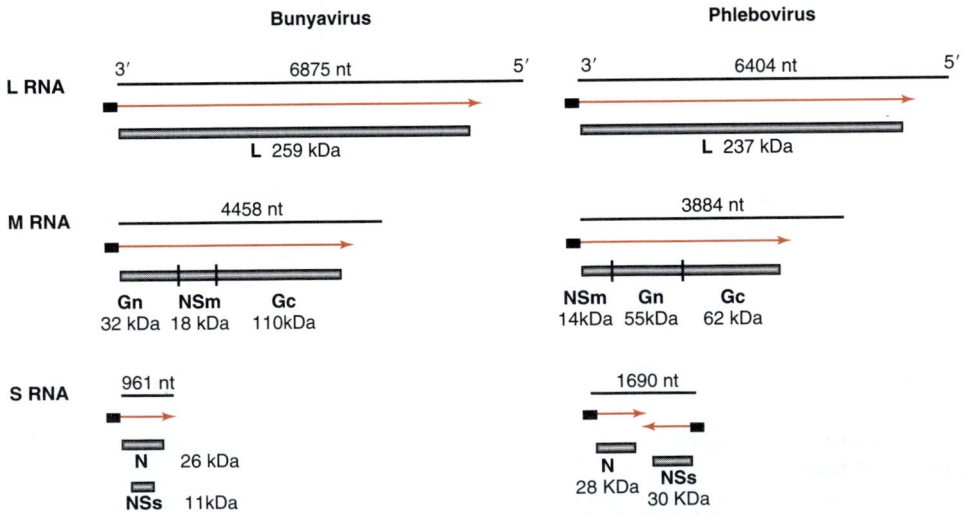

FIGURE 22.2 **Coding strategies of genome segments of orthobunyaviruses and phleboviruses.** Genomic RNAs are represented by black lines (length in nucleotides above each segment) and messenger RNAs as orange arrows; host-derived sequences at the 5′ end of mRNAs are shown as black rectangles. Coding regions for viral proteins, with their molecular weights and cleavage sites, are shown below mRNAs.

determinants for attachment to invertebrate (mosquito) cells. Recently, β_3 **integrins** have been shown to mediate the entry of pathogenic hantaviruses; infection of cells can be blocked by anti-integrin antibodies or the integrin **ligand**, vitronectin. Which of the two hantavirus glycoproteins is involved in the interaction so far remains unknown.

Based on electron microscopic studies of phleboviruses, entry into cells is by endocytosis. It is probable that uncoating occurs when endosomes become acidified, initiating fusion of the viral membrane and endosomal membrane, followed by release of the nucleocapsids into the cytoplasm. In common with many other enveloped viruses, bunyaviruses can fuse with cellular membranes at acidic pH, and this is accompanied by a conformational change in Gn.

Bunyavirus mRNA synthesis is primed by the capped 5′ ends of cellular mRNAs

Like other negative-strand RNA viruses, bunyaviruses begin their replication cycle by transcription of the infecting genome segments into messenger RNAs, using a virion-associated RNA polymerase, the L protein (Figure 22.3). This process, termed primary transcription, can begin as soon as the viral nucleocapsids are released into the cytoplasm. Transcriptase activity can be detected in detergent-disrupted bunyavirus preparations. Surprisingly, mRNA synthesis is stimulated by addition of methylated cap analogues such as mGpppAm, or naturally capped mRNAs such as alfalfa mosaic virus RNA 4. Both cap analogues and capped mRNAs appear to act as primers for transcription. Further support for this notion was provided by sequencing the 5′ ends of bunyavirus mRNAs. They were found to contain an additional 10–18 nucleotides at their 5′ ends whose sequence is unrelated to the viral genome sequence. An endonuclease activity that specifically cleaves capped mRNAs was detected *in vitro*, and this activity has been mapped to the L protein by expression studies using a recombinant vaccinia virus.

Taken together, these data indicate that bunyaviruses, like influenza viruses (see Chapter 23), use a "cap-snatch" mechanism to prime messenger RNA synthesis. This involves binding of the capped 5′ end of a cellular mRNA by the L protein, cleavage at a site 10–18 nt from the 5′ end, and alignment of the 3′ end of the cleaved RNA fragment with the 3′ end of the genome RNA. The L protein then adds nucleotides to this primer RNA, using the genome RNA as a template. Influenza virus transcription takes place in the cell nucleus, and uses newly synthesized cellular transcripts as a source of its capped primers; therefore, it is inhibited by agents that interrupt cellular RNA polymerase II transcription, such as **actinomycin D** or **α-amanitin**. In contrast, bunyavirus transcription takes place in the cytoplasm, and uses preformed cellular mRNAs as primers; thus it is not sensitive to these drugs, because cellular messenger RNAs are abundant and stable. Using capped cellular RNAs as primers for transcription allows these viruses to make their own capped mRNAs even though they do not code for a capping enzyme.

Coupled translation and transcription may prevent premature termination of mRNAs

The presence of a capped primer at the 5′ end of elongating transcripts (Figure 22.3) could allow binding by ribosomes and initation of protein synthesis before synthesis of the mRNA chains is completed, because transcription takes place in the cytoplasm. This would resemble the coordinated transcription and translation known to take place in bacteria, which (unlike eukaryotic cells) do not physically segregate their genomes from the protein synthetic machinery. Interestingly, synthesis of full-length mRNAs of some bunyaviruses has been shown to depend on ongoing protein synthesis in certain cell types. Furthermore, production of full-length mRNAs *in vitro* is stimulated by addition of the protein synthetic machinery and is reduced by inhibitors of protein synthesis. In the absence of protein synthesis, many prematurely terminated mRNAs are made. These results suggest that, at least in some cell types, growing bunyavirus mRNAs can be translated before their synthesis is completed, and that translation is required for the proper elongation of mRNAs. Ribosomes translating or scanning the mRNA in the 5′ to 3′ direction could dislodge interactions between the growing mRNA chain and the template genome RNA that might otherwise slow down or block elongation by the viral RNA polymerase.

Synthesis of most bunyavirus mRNAs terminates 50–100 nucleotides before the end of the genome RNA template. There does not appear to be a universal termination signal for bunyavirus transcription; however, two sequences that have been suggested as termination sites are 3′-G/CUUUUU-5′ and 3′-GGUGGGGG-GUGGGG-5′. Bunyavirus mRNAs are not polyadenylated at their 3′ ends, but many have the potential to form stem-loop structures at their 3′ ends, and these structures may confer stability to the mRNAs.

Genome replication begins once sufficient N protein is made

Viral genome replication requires synthesis of a full-length complementary positive-sense RNA, called the antigenome or replicative intermediate RNA. This molecule is an exact complementary copy of the

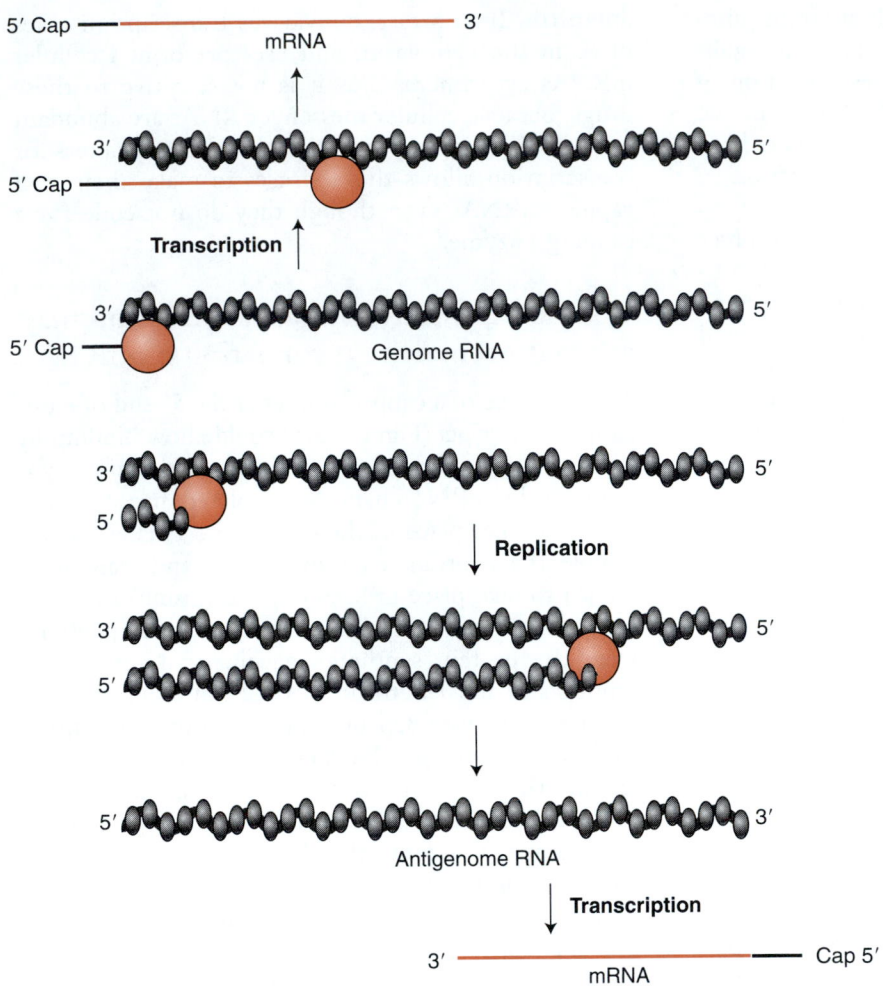

FIGURE 22.3 Transcription and replication of genome segments. Genome RNA is encapsidated by N protein (gray ovals). Transcription: viral L protein (orange sphere) cleaves capped cellular mRNAs and uses 5′ end fragment (black line) as a primer for transcription of genome RNA. A termination signal releases the mRNA. Replication: when sufficient N protein is made, the L protein initiates RNA synthesis directly at the 3′ end of genome RNA and replicates the genome via an antigenome copy which is encapsidated. Antigenome RNA serves as a template for synthesis of progeny genomes (not shown) and in some cases ambisense mRNAs (bottom).

genome, and it does not contain the 5′ primer sequences used to make mRNAs. Therefore the L protein initiates genome replication at the first nucleotide at the 3′ end of the genome. It is not known what controls the switch from transcriptive to replicative mode of the polymerase, though by analogy with other negative-strand viruses it is likely that the level of N protein plays a role. Both genomes and antigenomes are always found complexed with N protein. Encapsidation of the antigenome RNA as it is synthesized may prevent formation of secondary structure in the RNA, thereby overcoming the termination signals recognized during formation of mRNA, and allowing the polymerase to copy to the end of the genome RNA. Once antigenome nucleocapsids are formed, they can be used to initiate synthesis of full-length genome RNAs by the same mechanism. Because the 5′ and 3′ terminal 8–11 nucleotides of bunyavirus genome RNA are exactly complementary, the 3′ end of the antigenome is identical to the 3′ end of the genome, and this is probably the signal for initiation of genome replication by the RNA polymerase.

Bunyaviruses that use an ambisense strategy transcribe some of their mRNAs from antigenome templates (Figures 22.2 and 22.3). Therefore, the synthesis of these mRNAs must await the production of sufficient N protein for genome replication to begin. It is not presently known what signals at the ends of genome and antigenome RNA specify transcription. Neither is it clear why only the S antigenome of phleboviruses, but not the M or L antigenome, is used as a template for mRNA synthesis.

Virus assembly takes place at Golgi membranes

Maturation of the bunyaviruses occurs at the smooth membranes in the Golgi apparatus (Figure 22.4), and is inhibited by the drug **monensin**, a monovalent **ionophore** that disrupts protein transport through Golgi membranes. The viral glycoproteins accumulate in the Golgi complex and cause a progressive vacuolization. However, the morphologically altered Golgi complex retains its ability to glycosylate and transport glycoproteins destined for the plasma membrane. Using vaccinia virus recombinants it has been shown that the targeting of bunyavirus glycoproteins to the Golgi complex is a property of the glycoproteins alone, and does not require other viral proteins or virus assembly.

Electron microscopic studies revealed that viral nucleocapsids condense on the cytoplasmic side of areas of the Golgi vesicles at sites where viral glycoproteins are present. Bunyaviruses lack a matrix protein, which for other viruses serves as a bridge between the nucleocapsid and the envelope glycoproteins. Therefore, direct transmembrane interactions between the nucleocapsid and the glycoproteins may be involed in budding of bunyaviruses. Virions form by budding into the Golgi cisternae. Vesicles containing virions are then transported to the cell surface via **exocytosis**, eventually releasing their contents to the exterior. Although most bunyaviruses mature by this pathway, Rift Valley fever phlebovirus has been observed to bud directly at the surface of infected rat hepatocytes.

Evolutionary potential of bunyaviruses via genome reassortment

Like other viruses with segmented genomes, bunyaviruses evolve by two major pathways: **genetic drift**, or the slow accumulation of point mutations, insertions, or deletions; and **genetic shift**, caused by the exchange of genome segments via reassortment. Viral RNA-dependent RNA polymerases are known to be error-prone because, unlike DNA polymerases, they lack a proofreading ability. This is a significant factor in the accumulation of mutations leading to genetic heterogeneity ("drift") in viral RNA populations. RNA oligonucleotide fingerprinting studies on field isolates of La Crosse virus revealed that no two isolates were identical. However, virus isolates obtained from the brains of children who died from infection by La Crosse virus in 1960, 1978, and 1993 revealed remarkable genetic stability: Only 51 nucleotide changes were detected in the 4526 nt M genome segment over this 33-year period. This suggests that only some genetic variants of the virus are virulent for humans.

Genome segment reassortment results in more dramatic antigenic changes ("shift"), as exemplified by influenza viruses (Chapter 23). Following a mixed infection of a cell with two parental viruses, progeny may be produced containing segments derived from the different parents (Figure 22.5). Bunyavirus reassortment has been studied experimentally both in cells cultured *in*

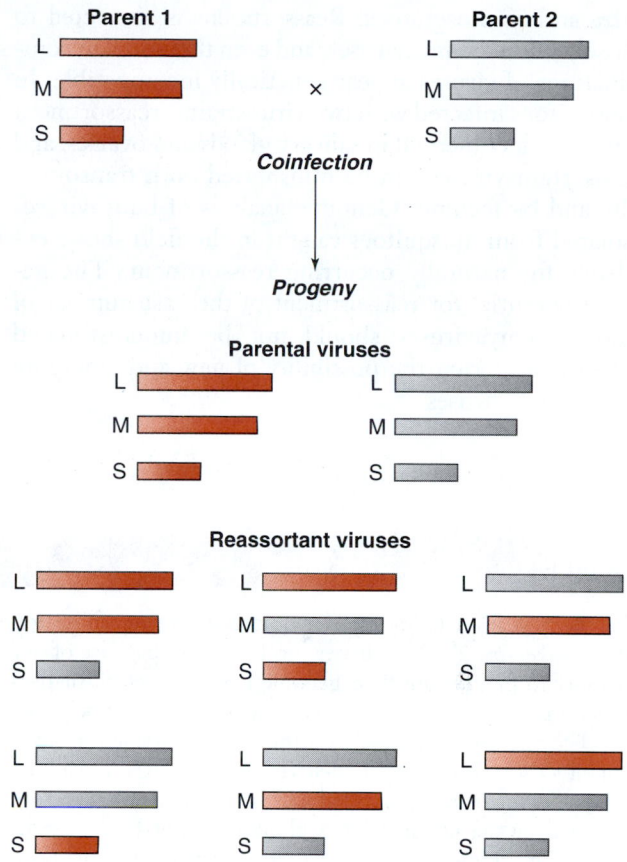

FIGURE 22.5 Reassortment of bunyavirus genome segments. A cell is coinfected with two closely related bunyavirus strains (parent 1 and parent 2). Eight different genotypes can be represented in progeny virus: the two parental genotypes and six possible reassortant genotypes, each with two segments from one parent and one segment from the other parent.

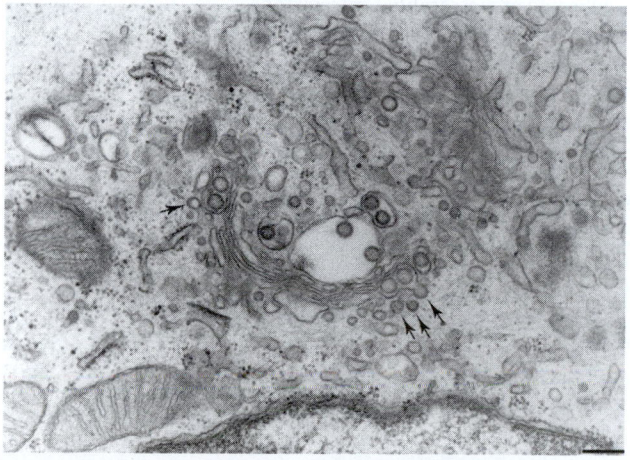

FIGURE 22.4 Electron micrograph of cells infected with Uukuniemi phlebovirus showing virus particles budding into Golgi vacuoles (center) and small vesicles (arrows) budding from Golgi cisternae. Bar: 200 nm. (Micrograph courtesy of Esa Kuismanen.) From *J. Virol.* (1997), 71, 1162–1172.

vitro and in mosquitoes. Reassortment is restricted to closely related bunyaviruses, and even then certain combinations of viruses appear genetically incompatible. In mosquitoes infected with two virus strains, reassortment has been documented in salivary glands and ovaries, and reassortant viruses can be transmitted both transovarially and by feeding. Genomic analysis of bunyaviruses isolated from mosquitoes caught in the field shows evidence for naturally occurring reassortment. The genetic potential for reassortment of the vast number of known bunyaviruses should not be underestimated when considering the possibility of new and emerging pathogenic viruses.

Reverse genetics of bunyaviruses

For many years detailed molecular analysis of the replication of negative-strand RNA viruses was hampered because of our inability to manipulate their genome RNAs directly. Analysis of gene function by site-directed mutagenesis was not possible, because infectious RNA cannot be expressed from a cloned DNA copy. Negative-strand genome RNA must be encapsidated to be transcribed and replicated, and therefore synthesis of the genome RNA alone does not lead to virus production. However, recent dramatic progress has been made in the "reverse genetics" of negative-strand RNA viruses—the mutation of complementary DNA copies of the viral genome followed by the rescue of the mutated cDNA into an RNA genome.

Recently, a simple system for recovery of infectious bunyaviruses from cloned cDNAs was established. This system, outlined in Figure 22.6, involves the transfection of cells with a set of plasmids containing cDNAs of the L, M, and S genome segments flanked by bacteriophage T7 promoters and hepatitis delta virus **ribozyme** sequences (see Chapter 29). A mammalian cell line engineered to express T7 RNA polymerase enables transcription of these three cDNAs into positive-strand antigenome RNAs with the authentic 5′ and 3′ terminal sequences. Surprisingly, these antigenome RNAs can act as messenger RNAs even though they are neither capped nor polyadenylated, and small amounts of viral proteins are made in the transfected cells. Further transcription of the cDNAs provides antigenome RNAs that are encapsidated during transcription by newly synthesized N protein, and these RNAs can then be replicated by the bunyavirus L polymerase. These nucleocapsids are then packaged into infectious particles containing the Gn and Gc envelope proteins.

Individual virus clones can be isolated from the supernatant medium by plaque formation in mammalian cells. A bunyavirus lacking the NSs gene has been created by reverse genetics, leading to the discovery that the NSs protein is an antagonist of the interferon response mechanism. The door is now open for rapid progress on understanding many aspects of bunyavirus replication, and perhaps in the future for the design of modified viruses for use as vaccines.

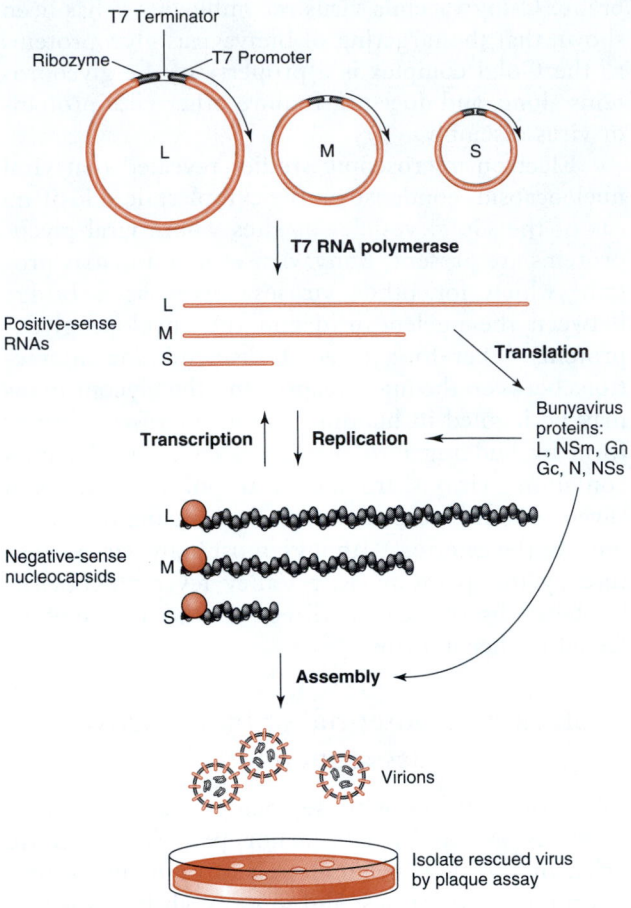

FIGURE 22.6 Rescue of infectious bunyavirus from cloned cDNAs. Mammalian cells constitutively expressing bacteriophage T7 RNA polymerase are transfected with a set of three plasmids containing cDNA copies of each genome segment. The transcribed antigenome (positive-strand) RNAs are directly translated to give bunyaviral replication and structural proteins. The antigenome RNAs are then replicated and assembled into virions.

SUGGESTED READING

1. Bouloy, M. (1991). Bunyaviridae: Genome organization and replication strategies. *Advances in Virus Research* **40,** 235–275.

2. Bridgen, A., Weber, F., Fazakerley, J.K., and Elliott, R.M. (2001). Bunyamwera bunyavirus nonstructural protein NSs is a nonessential gene product that contributes to viral pathogenesis. *Proceedings of the National Academy of Sciences, USA* **98,** 664–669.

3. Elliott, R.M., Ed. (1996). The Bunyaviridae. Plenum Press, New York.

4. Lowen, A.C., Noonan, C., McLees, A., and Elliott, R.M. (2004). Efficient bunyavirus rescue from cloned cDNA. *Virology* **330,** 493–500.

KEY TERMS

α-amanitin
Actinomycin D
Ambisense
Encephalitis
Epidemic
Exocytosis
Genetic drift
Genetic shift
Hantavirus pulmonary syndrome
Hemagglutination
Hemorrhagic fever
Hemorrhagic fever with renal syndrome
Integrins
Ionophore
Ligand
Lumen
Monensin
Reassortant viruses
Ribozyme
Thrips
Transovarial transmission
Tropism
Type I transmembrane protein

Orthomyxoviruses

Orthomyxoviridae

From Greek *ortho* (correct or normal) and *myxa* (mucus), virions bind to mucoproteins

VIRION

Enveloped particles, quasi-spherical or filamentous.
Diameter 80–120 nm.
Envelope derived from host cell plasma membrane by budding.
Compact helical nucleocapsids.

GENOME

Linear ss RNA, negative sense.
Six to eight different segments.
Total genome length 10–15 Kb.

GENES AND PROTEINS

Each genome segment codes for one or two proteins.
Envelope glycoproteins: hemagglutinin (HA) and neuraminidase (NA).
Integral membrane protein (M2) with ion channel activity.
Matrix protein (M1).
Nucleocapsid protein (NP).
Three RNA polymerase proteins (PA, PB1, and PB2).
Nonstructural protein (NS1).
Minor structural protein (NS2).

VIRUSES AND HOSTS

Influenza types A, B, and C.
Hosts include birds, various mammals and humans.
A separate genus, *Thogotovirus*, includes viruses transmitted between vertebrates by ticks.

DISEASES

Symptoms include high fever, sore throat, cough, headache, muscular pain.
Can be fatal in elderly, infants, and chronically ill, often by secondary bacterial infections.
In 1918, an influenza pandemic killed 20 million people worldwide.
Emerging avian influenza virus strains threaten domestic fowl and a possible new human pandemic.

DISTINCTIVE CHARACTERISTICS

Replicate in the nucleus, unlike most RNA viruses.
Viral mRNA synthesis is primed by stealing capped 5′ ends of cellular pre-mRNAs in the nucleus.
Undergo reassortment by exchanging genome segments between related strains.
Reassortment generates new viruses that can cause pandemics because of changed surface antigens.

Influenza viruses cause serious acute disease in humans, and occasional pandemics

Influenza virus was responsible for one of the worst human **epidemics** since those caused by the plague bacterium in the Middle Ages. The so-called "Spanish flu" **pandemic** (worldwide epidemic) of 1918–1920 caused the deaths of 20 million people—more than died in all the battles of World War I. Influenza received its name in eighteenth-century Italy as a reference to the imagined cause of epidemics—the *influence* of the stars. Despite the present availability of vaccines and antiviral drugs, influenza viruses continue to kill hundreds of thousands of humans worldwide every year. A major reason is the ability of influenza viruses to escape immune surveillance by changing the amino acid sequences of their surface glycoproteins.

What the general population calls "flu" is a usually a mild upper respiratory illness (the "common cold") caused not by influenza virus but by a number of upper respiratory viruses including rhinoviruses and coronaviruses. Real influenza is a severe acute disease in all age groups with symptoms of sore throat, cough, high fever, headache, and muscular aches and pains. A rule of thumb for distinguishing influenza from other respiratory virus infections is: if you really can't get out of bed, it's probably influenza! The acute stage lasts about three days, but cough and malaise may last for several weeks.

Epidemics of influenza occur every winter, usually between December and March in the Northern Hemisphere. More than 20,000 deaths per year in North America continue to be caused by influenza virus, most among elderly or otherwise debilitated or immunocompromised individuals. In addition to these regular epidemics, more severe pandemics occur at irregular 10- to 50-year intervals. These have included not only the "Spanish flu" pandemic in 1918–1920, but also the "Asian flu" in 1956 (60,000 deaths in North America) and the "Hong Kong flu" in 1967. In May 1790, a severe influenza epidemic struck Philadelphia, then the capital of the fledgling United States of America, and George Washington, the newly elected President, was close to death but eventually recovered.

Influenza virus infections of the respiratory tract can lead to secondary bacterial infections

The virus initially infects epithelial cells in the upper respiratory tract (Figure 23.1). Virus infection causes loss of the **ciliated epithelium**, leading to loss of the ability of the respiratory tract to clear viruses or bacteria by **mucociliary flow** that normally traps these agents in mucus and disposes of them. Virus replication furthermore induces production of **interferons, cytokines,** and other soluble response factors, leading to local and systemic inflammatory responses. This results in the symptoms that define the "flu" syndrome: fever, headache, chills, malaise, and muscle aches. The loss of ciliated epithelium allows relatively easy entry of bacteria into the lower respiratory tract. Resulting secondary bacterial pneumonias cause many of the deaths attributed to influenza virus infection.

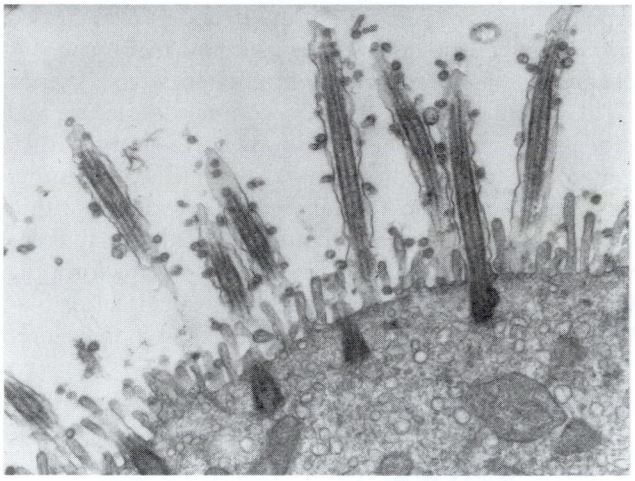

FIGURE 23.1 **Epithelial cells of the upper respiratory tract infected with influenza virus.** Cilia cut tangentially in this electron micrograph of a thin section are shown projecting from an epithelial cell. A number of influenza type A virions are shown attached to the cilia.

Orthomyxoviruses are negative-strand RNA viruses with segmented genomes

Myxoviruses were originally named because virions of influenza and several other enveloped, negative-strand RNA viruses bind to **sialic acid** residues in mucoproteins (*myxo* = mucus). When several of these viruses were understood to have a very different life style from the influenza viruses, the myxoviruses were split into two families, orthomyxoviruses and paramyxoviruses (*ortho* = correct; *para* = alternative). Paramyxoviruses are discussed in Chapter 20.

The *Orthomyxoviridae* family includes influenza virus types A, B, and C, as well as two distinct tick-borne mammalian viruses in the recently named *Thogotovirus* genus. Influenza type A viruses are the best studied orthomyxoviruses; they are responsible for most flu epidemics, and are widespread in many avian and mammalian species. Influenza type B is limited to humans, and type C can infect both pigs and humans. Multiple subtypes are distinguished within each influenza virus type by means of variation in the structure of their two surface glycoproteins. Most of the information in this chapter comes from studies with influenza type A virus.

Orthomyxoviruses share a common structure and mode of replication. Their negative-strand RNA genome is composed of between six and eight distinct gene segments. These are individually wrapped within helical nucleocapsids, which are packaged in a lipid envelope derived from the plasma membrane to form the virion. Virions can be either roughly spherical or filamentous (Figure 23.2). Fresh isolates tend to make predominantly filamentous particles, while virus repeatedly passaged in eggs or cell cultures tend to make smaller, elongated or quasi-spherical particles. It has been speculated that filamentous virions, which can grow as long as the diameter of an epithelial cell, are responsible for cell-to-cell transmission of virus; such virions could infect a neighboring cell while still being extruded. Quasi-spherical virions are much smaller and may be more readily incorporated into tiny droplets in respiratory aerosols; therefore these may be responsible for person-to-person transmission.

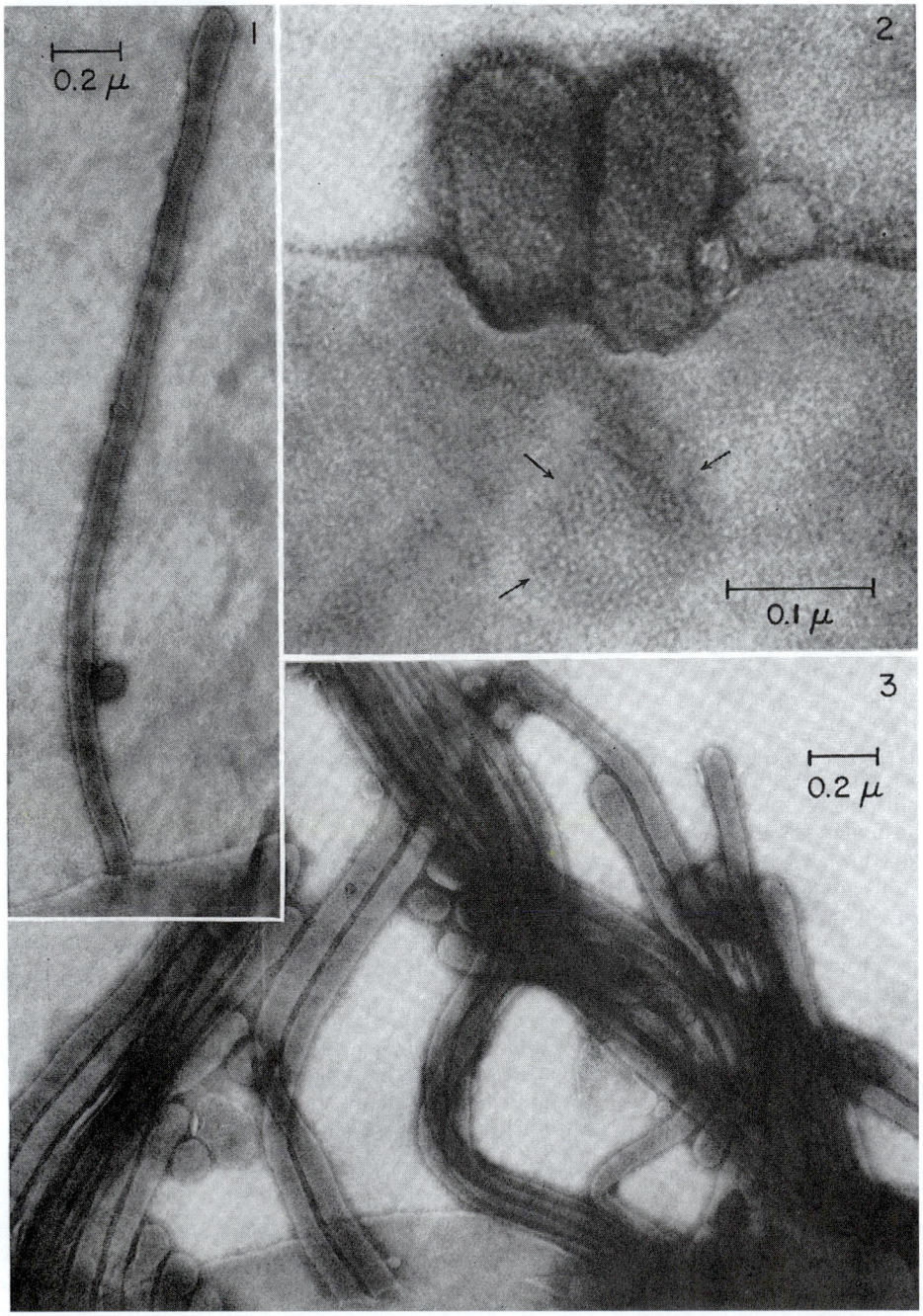

FIGURE 23.2 Electron micrographs of spherical and filamentous influenza virus A particles. Top right: quasi-spherical particles budding from the cell surface. The glycoprotein projections can be readily seen. Top left and bottom: filamentous partles being extruded from the cell surface. These particles can extend for several micrometers and may be an efficient means for cell-to-cell transmission. Negatively stained preparations of infected cells.

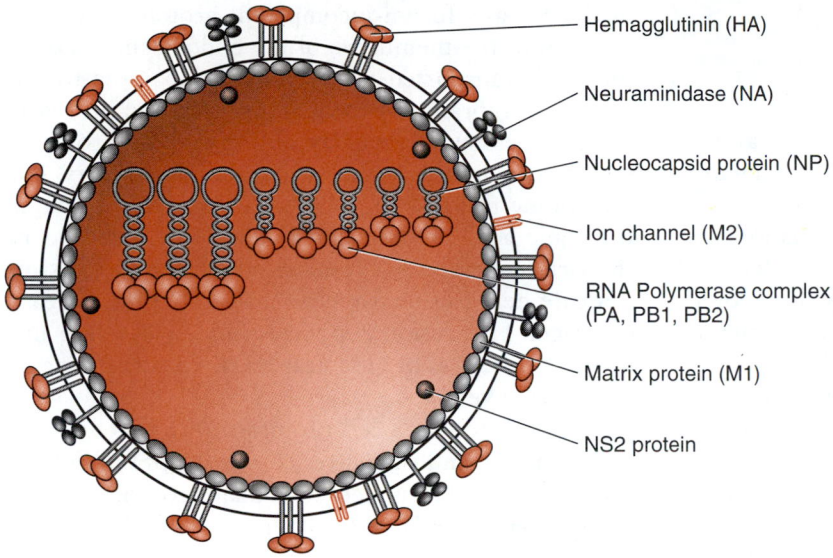

FIGURE 23.3 Schematic diagram of influenza virus virion. Nine virus-coded proteins are shown. Hemagglutinin, neuraminidase, M2 ion channel, and matrix protein are associated with lipid bilayer membrane. Eight different helical nucleocapsids containing NP and the RNA polymerase trimer of PA, PB1, and PB2 are enclosed within the envelope. There is evidence that one copy of each of the eight genome segments is incorporated in a bundled form into each virion. A small amount of NS2 protein is also present in virions.

Eight influenza virus genome segments code for a total of 10 different viral proteins

Of the 10 viral proteins produced by influenza type A viruses, 9 are packaged in virions (Figure 23.3 and Table 23.1). Genome segments 1 to 6 each code for a single protein: three RNA polymerase subunits (PA, PB1, and PB2), the envelope glycoproteins **hemagglutinin** (HA) and **neuraminidase** (NA), and the nucleocapsid protein (NP). Messenger RNAs transcribed from both genome segments 7 and 8 can be spliced, each giving rise to two distinct mRNA species that code for different viral proteins. Genome segment 7 codes for the matrix protein (M1) and a third envelope protein (M2). Genome segment 8 codes for two nonstructural proteins (NS1 and NS2), although NS2 is also found in small amounts in virions. The roles of these 10 viral proteins in the viral replication cycle are described in detail below.

TABLE 23.1 Influenza A virus strain A/PR/8/34 genome RNAs and proteins

RNA segment	Length (nucleotides)	Proteins encoded	Protein size (aa)	Protein function
1	2341	PB_2	759	Binds to cap structures on cellular pre-mRNAs; part of transcriptase complex
2	2341	PB_1	757	Cleaves cellular pre-mRNAs to create primer; RNA polymerase activity for both transcription and replication
3	2233	PA	716	Part of transcription and replication complexes, exact role unknown
4	1778	HA	566	Hemagglutinin; major surface glycoprotein; receptor-binding; mediates membrane fusion at low pH; antigenic determinant
5	1565	NP	498	Nucleocapsid protein; binds to and encapsidates viral RNA; control functions in RNA synthesis
6	1413	NA	454	Neuraminidase; major surface glycoprotein; receptor destruction; dissociation of virus aggregates; antigenic determinant
7	1027	M_1	252	Matrix protein; interacts with envelope, nucleocapsids, and NS_2
		M_2	97	Integral membrane protein; ion channel activity essential for virus uncoating and maturation
8	890	NS_1	230	Nonstructural protein that down-regulates host cell mRNA processing; sequesters dsRNA and reduces interferon response
		NS_2	121	Nonstructural protein that directs nuclear export of viral nucleocapsids; interacts with M_1

Hemagglutinin protein binds to cell receptors and mediates fusion of the envelope with the endosomal membrane

The influenza hemagglutinin protein (HA) was so named because influenza viruses can agglutinate (clump) red blood cells. HA binds to sialic acid–containing receptors, which are found on red blood cells as well as most other cell types. This property is the basis for the rapid and sensitive **hemagglutination assay** that allows detection of many viruses (see Chapter 1). HA forms a trimer in the virus envelope, and is responsible not only for binding of virions to the cell surface but also for entry of the virus genome into the cell via the fusion pathway.

HA is a 549-amino acid **type I transmembrane protein**, with its N-terminus exposed to the outside of the virion, a hydrophobic transmembrane domain near the C-terminus, and a short C-terminal tail projecting inward (Figure 23.4). HA is first synthesized in the cell as a fusion-incompetent protein that is inserted into the membrane of the endoplasmic reticulum and transported via the Golgi apparatus to the plasma membrane, where it is incorporated into virions (see below). Cleavage of HA by cellular proteases into two subunits activates its ability to carry out membrane fusion. HA_1, the surface subunit, contains the sialic acid–binding domain. HA_2 is anchored in the virus envelope and contains a hydrophobic **fusion peptide** near the amino terminus created by protease cleavage. This fusion peptide is hidden within a hydrophobic region of HA, near the lipid membrane of the virus envelope.

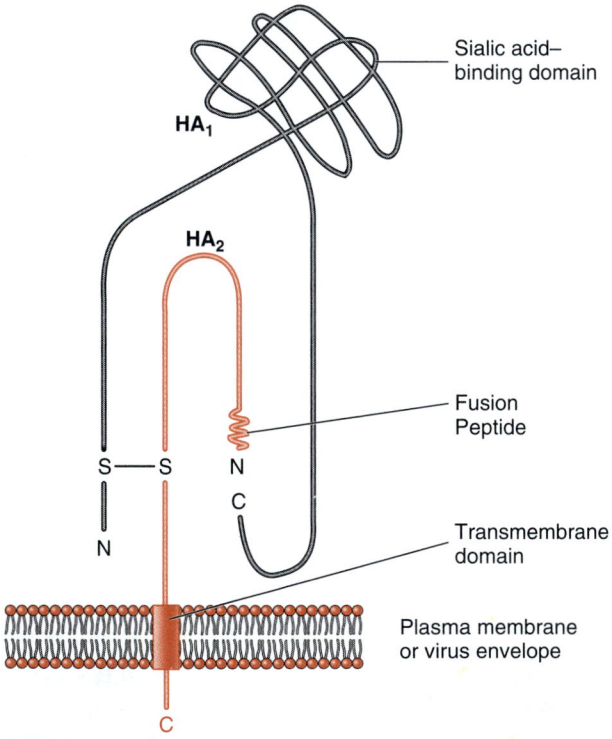

FIGURE 23.4 Influenza A hemagglutinin protein (HA). A schematic diagram showing the HA_1 (black) and HA_2 (orange) fragments created by proteolytic cleavage of the full-length polypeptide (HA_0) shown inserted in the plasma membrane or the virus envelope at neutral pH. N- and C-termini are shown. The sialic acid–binding domain is shown folded at the top. The two polypeptide chains are linked by a S–S bond between cysteines. This diagram does not attempt to describe the exact three-dimensional structure of the protein.

After binding to cell receptors, virions enter the cell within an endosomal vesicle. As the pH of the endosome drops, HA undergoes a major conformational change that is similar to the opening of a coiled spring. The N-terminal part of HA_2 forms a rigid alpha-helical structure, moving the fusion peptide out of the hydrophobic region of HA and toward the endosomal membrane, where it inserts itself into the membrane (see Figure 4.4 in Chapter 4 on virus entry for details). At the same time, the part of HA_2 that is inserted into the virus envelope is brought closer to the extended N-terminal region. One end of HA_1 is now bound to the endosomal membrane, and the other end is bound to the virus envelope; these two membranes are brought into close proximity by this conformational change, resulting in membrane fusion.

M2 is an ion channel that facilitates release of nucleocapsids from the virion

The M2 protein is crucial in the next phase. It is a small 97 aa protein with an N-terminal external domain (residues 1–23), a single transmembrane domain (residues 24–44), and a larger internal domain (residues 45–97). M2 is an essential component of the virus envelope because of its ability to form a highly selective transmembrane ion channel that allows H^+ ions to penetrate the membrane. M2 forms tetramers by lining up four parallel transmembrane α-helices, thus creating a small pore in the viral envelope.

Release of the viral nucleocapsids from the virion is facilitated by the M2 H^+ ion channel. This channel allows protons to enter the interior of the virion during acidification within the endosome (see Figure 32.3 in Chapter 32 on antiviral chemotherapy). The low pH within virions weakens the interaction of the matrix protein M1 with viral nucleocapsids, facilitating their release into the cytoplasm upon membrane fusion. The anti-influenza virus drug **amantadine** is a specific blocker of the M2 H^+ channel. In the presence of amantadine, release of viral nucleocapsids is incomplete, blocking infection (see Chapter 32).

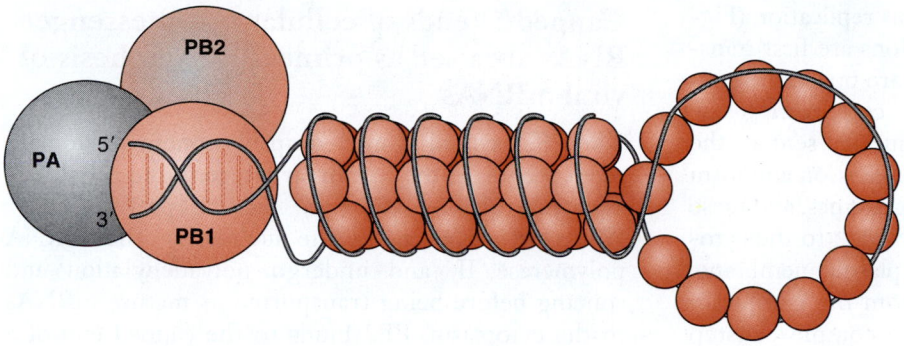

FIGURE 23.5 **Helical nucleocapsid of influenza virus.** The nucleocapsid protein (NP, small orange spheres) binds to negative-strand viral genome RNA (black line) to form a nucleocapsid with a twin helical conformation. A trimer consisting of RNA polymerase proteins PA, PB1, and PB2 is bound via PB1 to the 5′ and 3′ ends of the RNA. Influenza type A has eight distinct genome segments each assembled in a helical nucleocapsid.

Nucleocapsids enter the nucleus, where mRNA synthesis and RNA replication occur

Influenza virus nucleocapsids contain multiple copies of the NP protein that wrap the single-stranded genome RNAs in an unusual twin helical conformation with a central loop (Figure 23.5). Each NP subunit binds to about 20 nucleotides of viral RNA. Nucleocapsids also contain a trimer of RNA polymerase proteins PA, PB1, and PB2. Sequences of the terminal 12–13 nucleotides at both the 5′ and 3′ ends of viral RNAs are highly conserved among all genome segments of a given virus type, and are highly self-complementary. Distinct sites within the PB1 protein bind to the conserved sequences at the 5′ and 3′ ends of the genome RNA (Figure 23.6). The two ends of genome RNAs are therefore maintained in close proximity. In spite of a high degree of self-complementarity, however, they are not believed to be hydrogen bonded to each other.

Unlike most other RNA-containing viruses, orthomyxoviruses replicate in the nucleus. This compli-

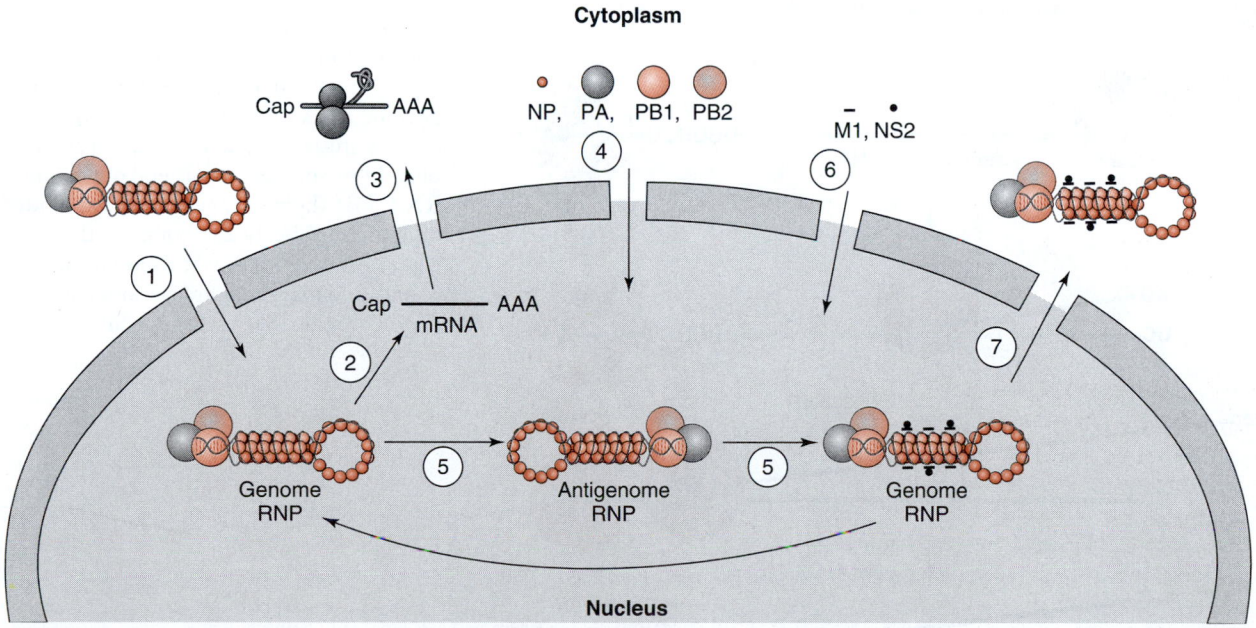

FIGURE 23.6 **Transport of influenza virus RNAs and proteins between nucleus and cytoplasm.** Viral nucleocapsids are transported through the nuclear pores into the nucleus (1), where viral mRNAs are synthesized (2). These mRNAs are transported back to the cytoplasm (3) where they are translated. Viral proteins are transported back into the nucleus (4 and 6), where they mediate genome replication (5) and transport of progeny nucleocapsids back to the cytoplasm (7).

cates the machinery required for virus replication (Figure 23.6). Genomes of infecting virions are first transported into the nucleus, where they are transcribed; viral mRNAs are then sent to the cytoplasm to be translated; viral proteins are subsequently sent to the nucleus where they direct genome replication and form nucleocapsids with newly replicated genomes, and these nucleocapsids are finally transported back to the cytoplasm for assembly of virions at the plasma membrane.

Nucleocapsids are imported from the cytoplasm to the nucleus via the nuclear pore complexes (step 1, Figure 23.6). The NP protein and polymerase proteins contain nuclear localization signals that mediate interactions with cellular **importin-α**. This complex interacts with importin-β, which docks with nuclear pore complexes and promotes entry of its cargo into the nucleus. At this point, the virus must begin making messenger RNAs ("transcription") by copying its negative-sense RNA genome segments (step 2). Surprisingly, the viral transcriptional machinery cannot make mRNAs on its own, but uses cellular RNAs as primers for initiating viral mRNA synthesis (Figure 23.7).

Capped 5′ ends of cellular pre-messenger RNAs are used as primers for synthesis of viral mRNAs

The PB2 protein in nucleocapsids contains a domain that recognizes the 5′ cap structure found on all eukaryotic mRNAs. Capped cellular pre-messenger RNAs are synthesized in the nucleus by cellular RNA polymerase II, and undergo polyadenylation and splicing before being transported as mature mRNAs to the cytoplasm. PB2 binds to the capped end of a cellular pre-messenger RNA, and PB1 then acts as a nuclease, cleaving the bound pre-messenger RNA at an A or G residue 10–13 nt from its 5′ cap (Figure 23.7, top left). This capped RNA fragment, still bound to PB2, is then used as a primer to begin copying the genome RNA. The terminal A or G residue is hydrogen-bonded to the 3′ terminal U residue on genome RNA, and PB1 acts as an RNA polymerase by adding ribonucleotides complementary to the genome RNA sequence, proceeding toward its 5′ end. The role of PA in transcription is not well understood, and it is not shown in Figure 23.7.

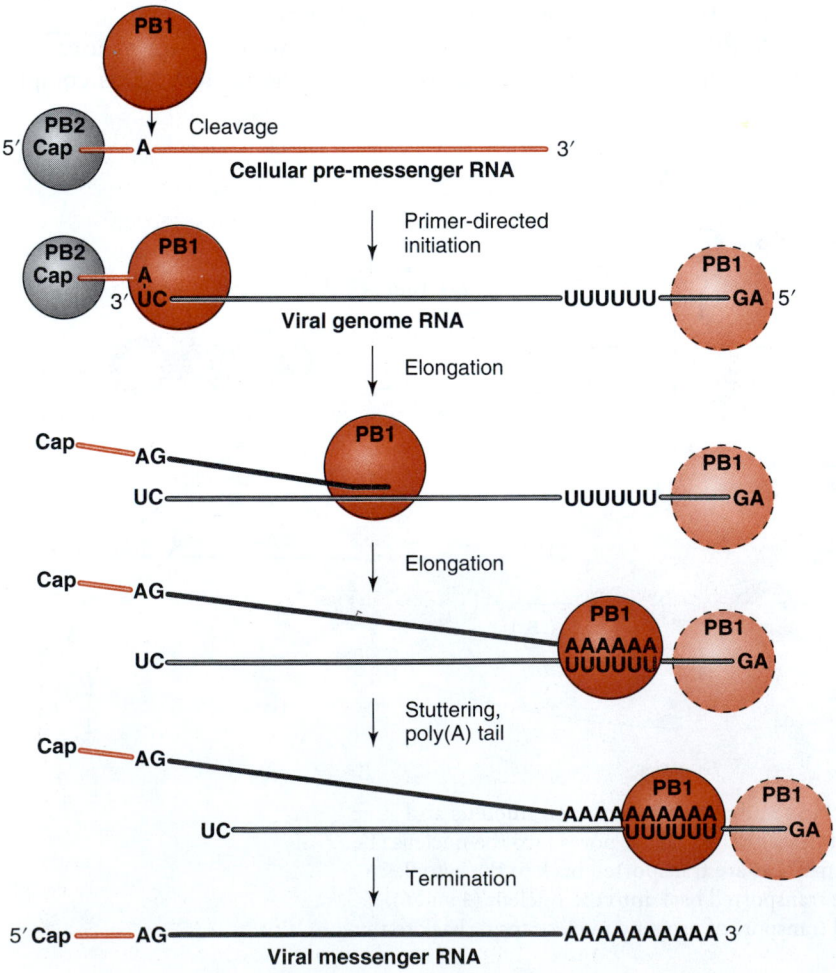

FIGURE 23.7 Production of influenza virus messenger RNAs by cap-stealing. In viral nucleocapsids, PB2 (small gray sphere) binds to the capped 5′ end of a cellular pre-messenger RNA (orange line) and PB1 (orange sphere) cleaves the RNA 10–13 nt from its 5′ end, preferentially after an A or G residue. PB1 then uses this capped fragment as a primer for copying genome RNA (black line). Transcription ends by stuttering and termination at an oligo U region near the 3′ end of genome RNA. PB1 (light orange sphere with dotted outline) is probably also bound to the 5′ end of genome RNA as it copies from the 3′ end, and may block transcription at the oligo U region. See text for details.

The resulting capped viral mRNAs have heterogeneous sequences for the first 10–13 nt at their 5′ ends, derived from whatever cellular pre-mRNAs were used to prime transcription. Orthomyxoviruses therefore can make capped messenger RNAs even though they do not code for a capping enzyme. Bunyaviruses use a similar cap-stealing mechanism to prime viral mRNA synthesis, but they replicate in the cytoplasm and use mature cellular mRNAs as a source of caps (Chapter 22). In contrast, a number of cytoplasmic RNA viruses such as togaviruses, paramyxoviruses, and reoviruses cap their transcripts with virus-coded enzymes.

This intricate mechanism explains why the replication of influenza virus is blocked by treatment of infected cells with inhibitors of cellular RNA synthesis such as **actinomycin D** or **α-amanitin**. Cellular pre-messenger RNAs are rapidly spliced and exported to the cytoplasm, and therefore the pool of pre-messenger RNAs is quickly depleted when their synthesis is inhibited. In the absence of sufficient pre-messenger RNAs to serve as primers, influenza virus cannot make its own mRNAs. Thus virus replication is indirectly dependent on a healthy cellular RNA synthesizing system.

Viral mRNAs terminate in poly(A) tails generated by "stuttering" transcription

Each genome RNA segment contains a short poly(U) stretch 15–22 nt from the 5′ end of the RNA. When the viral RNA polymerase reaches this poly(U) sequence, it pauses and stutters, reading through the poly(U) sequence a number of times and repeatedly adding complementary A residues to the transcript to generate a poly(A) tail (Figure 23.7). The RNA polymerase eventually terminates transcription at that position. This stuttering mechanism recalls a similar mechanism used by paramyxoviruses and rhabdoviruses for polyadenlyation of their mRNAs (Chapter 20).

Stuttering and termination of transcription on influenza genome RNAs is believed to result from blockage of further movement of the RNA polymerase by PB1 itself, because it remains bound to the terminal 12–13 nt at the 5′ end of the same genome RNA it is copying. This implies that the template RNA is actually threaded through the RNA polymerase complex as elongation proceeds, while its 5′ end remains bound to the complex. Figure 23.7 does not attempt to show this spatial relationship between the polymerase complex and the ends of the genome RNA.

Two influenza A mRNAs undergo alternative splicing in the nucleus

Transcription generates a complete set of eight viral mRNAs, one from each genome segment. Six of these mRNAs are exported directly to the cytoplasm where they are translated by ribosomes (step 3, Figure 23.6). However, RNAs generated from genome segments 7 and 8 (those encoding the M1 and NS1 proteins) contain splicing consensus sequences that are recognized by the cellular splicing machinery in the nucleus. Unlike many host cell mRNAs, which are usually spliced to completion, only a fraction of these two viral mRNAs are spliced, and both unspliced (M1 and NS1) mRNAs and spliced (M2 and NS2) mRNAs are exported from the nucleus. The ratio of unspliced to spliced segment 7 and 8 viral mRNAs in the cytoplasm is approximately 9:1. This ensures that M2 and NS2 are made in lesser amounts than the highly abundant M1 and NS1 proteins.

Genome replication begins when newly synthesized NP protein enters the nucleus

Viral proteins that contain nuclear localization signals, including NP and the RNA polymerase subunits, are transported from their site of synthesis in the cytoplasm back into the nucleus (step 4, Figure 23.6). Their presence enables the replication machinery to begin to produce full-length plus-strand copies of the eight viral genome segments (step 5, Figure 23.6; Figure 23.8). In contrast to capped viral mRNAs, these full-length transcripts contain 5′ triphosphate groups and result from unprimed transcription initiation.

The mechanism of switching from primed to unprimed RNA synthesis depends on the presence of free NP protein, but is not fully understood. NP interacts either with the RNA polymerase complex or with the 5′ and 3′ ends of the genome RNAs in nucleocapsids, and directs the RNA polymerase to begin copying the genome RNA directly from its 3′ terminus with no intervening primer. As this plus-strand copy is synthesized it is complexed with NP protein, eventually forming nucleocapsids containing plus-strand antigenome RNAs. In this mode, the RNA polymerase does not stutter and terminate at the poly(U) stretch, but continues to the 5′ end of the genome RNA, generating a full-length plus-strand copy. This presumably occurs because the PB1 subunit has been displaced and is no longer bound to the 5′ end consensus sequence on the RNA genome.

These plus-strand antigenome RNAs are then themselves copied in a similar fashion by the RNA polymerase complex, in the presence of NP, to generate nucleocapsids containing full-length minus-strand genomes (Figure 23.8). Newly synthesized nucleocapsids containing minus-strand genomes can be transcribed to produce more viral mRNAs or can generate more full-length plus-strand RNAs for genome replication (Figure 23.6); the balance between these activities depends on the amount of NP protein present in the nucleus. Because NP is bound to growing RNA chains and incorporated into nucleocapsids, its concen-

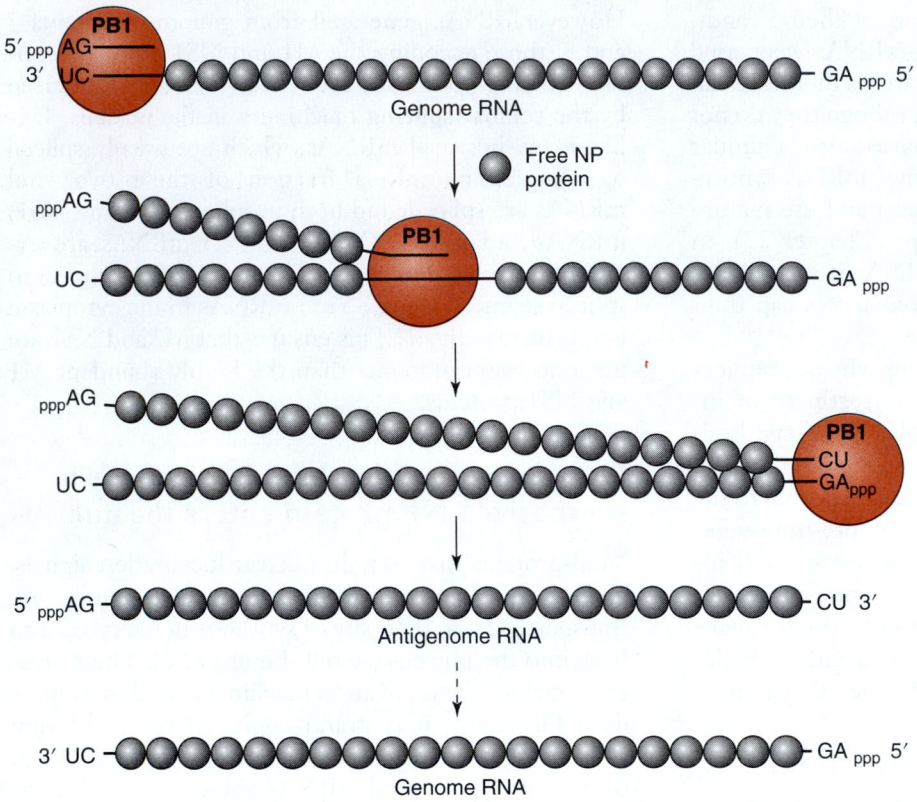

FIGURE 23.8 Replication of influenza virus genome RNA. Genome replication begins when free NP protein enters the nucleus. Unlike transcription, genome replication does not use a primer but begins at the 3′ end of genome RNA and continues to the 5′ end, making a complete copy. PB1 (orange sphere) copies genome RNA into a plus-strand complementary RNA that is coated with NP protein as a viral nucleocapsid. This RNA is subsequently copied into progeny genome RNA by the same mechanism.

tration in the nucleus will drop during genome replication, favoring mRNA synthesis until more NP is imported into the nucleus. NP does not bind to influenza virus mRNAs, presumably because they are capped and contain cellular mRNA sequences, not viral sequences, at their 5′ ends.

Nucleocapsids are exported from the nucleus in a complex with matrix protein and NS2

Remember that during the intial stages of infection, nucleocapsids are dissociated from the matrix protein as a result of low pH within the virion brought about by the activity of the M2 ion channel, and are subsequently transported into the nucleus via nuclear localization signals on the NP and polymerase proteins. Now, nucleocapsids must be transported back out of the nucleus to the cytoplasm, so that they can be incorporated into progeny virions. How does this reverse transport occur?

Current evidence indicates that the matrix protein (M1) is imported into the nucleus and binds to nucleocapsids, forming a complex (step 6, Figure 23.6). At this point the NS2 protein comes into play. NS2 contains a nuclear export signal, and it can also bind to the matrix protein. It is believed that NS2 therefore piggybacks onto nucleocapsid/matrix protein complexes and signals the cellular protein export system to transport the complexes through the nuclear pores to the cytoplasm (step 7, Figure 23.6).

The NS1 protein interferes with polyadenylation of cellular mRNAs

Although the NS1 protein is the most abundant viral protein in virus-infected cells, its roles have only recently begun to be appreciated. NS1 can bind to both single- and double-stranded RNA and to two host cell proteins: (1) cleavage and polyadenylation specificity factor (CPSF) and (2) poly(A)-binding protein II (PABII). These two proteins are crucial players in the formation of polyadenylated cellular messenger RNAs in the cell nucleus. CPSF, as its name implies, is responsible for the cleavage of cellular pre-mRNAs shortly downstream of the consensus AAUAAA polyadenylation signal; PABII facilitates production of full-length poly(A) chains on the 3′ terminus created by cleavage. Binding of NS1 to these proteins blocks their functions, and therefore inhibits the normal 3′ end cleavage and polyadenylation of cellular mRNAs. Because the polyadenylation of viral mRNAs occurs by a distinct mechanism and therefore does not depend on these cellular enzymes, the synthesis of viral mRNAs is not affected.

The cap-binding and endonuclease activities of the influenza PB2 and PB1 proteins lead to the degradation of many cellular pre-mRNAs in virus-infected cells. As a combined result of interference with proper 3′ end processing and cleavage near their 5′ cap structures, most cellular pre-mRNAs are degraded in the nucleus, and cellular mRNA production is therefore suppressed.

NS1 also inhibits activation of PKR, an important antiviral pathway induced by interferon

Perhaps more important to the virus is inhibition of the cellular interferon response by NS1. Because both transcription and replication generate viral RNAs that are complementary to the negative-strand viral genome, some double-stranded RNAs are inevitably formed during the viral replication cycle. Virus infection induces interferon production, and in response, cells produce **double-stranded RNA-dependent protein kinase (PKR)** (see Chapter 31). In the presence of double-stranded RNA, PKR is activated and blocks both cellular and viral protein synthesis, therefore inhibiting production of virus by infected cells. However, NS1 also binds to double-stranded RNA, reducing its concentration in the cell; this blocks activation of PKR. NS1 therefore blocks one of the major mechanisms by which the cell combats virus infection.

The importance of NS1 in facilitating virus replication was shown by poor replication of influenza viruses that have a mutated NS1 gene. These mutant viruses nevertheless replicate well in cells that have a defective interferon response. The NS1 protein therefore can be considered as a potential target for chemotherapeutic agents to suppress influenza virus infection.

Viral envelope proteins assemble in the plasma membrane and direct budding of virions

Viral hemagglutinin, neuraminidase, and M2 ion channel proteins are synthesized and inserted into membranes of the endoplasmic reticulum. The cellular protein-trafficking machinery translocates these proteins via the Golgi network to the cell surface, where virus assembly takes place. These three proteins accumulate in regions of the membrane enriched in cholesterol, named **lipid rafts**, which are believed to be the site of virion formation by budding. The very short cytoplasmic tails of HA (10 to 11 amino acids, Figure 23.4) and NA (6 amino acids) are crucial to virion formation, probably because they interact with matrix protein bound to virus nucleocapsids and transported from the nucleus to the cytoplasm.

Orthomyxovirus virions must contain at least one copy of each genome segment in order to be able to initiate an infection. Recent evidence suggests that most influenza A virus particles contain exactly one copy of each genome segment in an ordered bundle that can be seen in high-resolution electron micrographs. Seven nucleocapsids are arranged parallel to each other as a cylindrical barrel and an eighth nucleocapsid lies in the center of the bundle. The head ends of filamentous particles also apparently contain a single bundle of eight nucleocapsids, and their thinner tails appear to be empty extensions of the envelope. This structural information implies that one copy of each of the eight different nucleocapsids is incorporated into a bundle either before or during virus budding. Experiments using deletion mutants have shown that RNA sequences in the 5′ and 3′ coding and noncoding regions of each genome segment are required to incorporate the segment into virions. This implies that viral proteins recognize and interact with specific RNA sequences in the nucleocapsids to incorporate them, one by one, into the bundles that are packaged into virions during budding.

Neuraminidase cleaves sialic acid, the cellular receptor that binds to HA

The roles of HA and M2 in virus entry were discussed above. Influenza neuraminidase (NA), in contrast to HA, is a type II integral membrane protein with a short cytoplasmic N-terminus, a membrane-spanning domain, and a long C-terminal domain that extends outward from the virus envelope. Neuraminidase was named for its ability to cleave terminal N-acetyl neuraminic acid (sialic acid) from oligosaccharides located on mucoproteins, cell-surface glycoproteins, and glycolipids. When HA binds to sialic acid residues on these macromolecules, NA can reverse this binding by cleaving the bound sialic acid residue from the remainder of the oligosaccharide. The ying-yang relationship between the activities of these two influenza A surface glycoproteins may help virions reach their target cells by releasing them from the surface of virus-producing cells or from mucoproteins that are abundant in the respiratory tract.

Influenza virus strains vary in both transmissibility and pathogenicity

Most influenza A viruses that circulate in wild or domestic animals (particularly in ducks, chickens, and pigs) do not readily infect humans. This host specificity resides in certain viral proteins that have adapted to function optimally in a particular species. These include HA, NA, and the RNA polymerase proteins PA, PB1, and PB2. For example, HA proteins of most avian influenza A viruses bind preferentially to α-2,3 sialic acid, the major form of sialic acid found in the avian enteric tract. In contrast, human influenza A viruses code for HA proteins that bind preferentially to α-2,6 sialic acid, the major form found in the human respiratory tract. Furthermore, some HA proteins are readily cleaved by intracellular **furin** proteases, allowing them to infect many cell types because their fusion mechanism is activated upon cleavage. Other HA proteins are not cleaved intracellularly during virus formation, but require activation by trypsin cleavage after release of virions from an infected cell. Trypsin expression is limited to the upper respiratory tract in humans, and therefore these viruses are less likely to infect lung and other tissues in the body, leading to milder and more self-limiting infections.

Genetic variability generates new virus strains that can cause pandemics

Unlike what is seen with a number of other respiratory viruses, protective immunity after infection with influenza A virus does not last for a lifetime. This results from the ease with which antigenic domains in the virus envelope glycoproteins, HA and NA, can change. Antigenic change occurs in two ways: (1) slowly but continuously ("antigenic drift"), and (2) suddenly but episodically ("antigenic shift").

Antigenic drift results from the accumulation of point mutations within regions of the envelope glycoprotein genes that code for the antigenic domains of these proteins. RNA polymerases can make errors in incorporation of nucleotides into genome RNAs during replication, and do not have the proofreading mechanisms possessed by many DNA polymerases. Therefore, as RNA viruses replicate and spread from cell to cell or animal to animal, variants in antigenic domains of the envelope glycoproteins appear. Because these protein variants are not as well recognized by the immune system as those of the original infecting virus strain, these viruses can selectively replicate and be propagated.

This drift mechanism is common to a number of RNA viruses, including human immunodeficiency virus, and is a major complicating factor in production of vaccines. Antigenic drift has resulted in the generation of numerous influenza A subtypes with antigenically distinct HA and NA surface glycoproteins. So far, 16 different HA subtypes (H1 to H16) and 9 NA subtypes (N1 to N9) have been identified in naturally occurring influenza type A viruses in birds, particularly in waterfowl.

Antigenic shift results from **reassortment** of influenza virus genes during mixed infections with two or more virus subtypes. Cells infected simultaneously with two virus strains can produce reassortant virions that contain some genome segments from one strain and some genome segments from the other strain (see Figure 22.5). If such a reassortant virus can replicate well in the animal host, it is propagated. By this mechanism, viruses with many combinations of the different HA and NA subtypes, as well as other viral genes, are generated in multiply infected animals in the wild.

Reassortment between animal and human strains can create a virus that can replicate in humans but has acquired an HA or NA subtype derived from the animal virus. Such a virus can rapidly spread throughout the human population if antibodies against that subtype are not present, and can cause a pandemic. This seems to be a rare event, as it has occurred only a few times in the past century, but the results can be disastrous.

The 1918 pandemic influenza A virus was probably not a reassortant virus

In an experimental *tour de force*, scientists recently recovered genome fragments of the influenza A virus implicated in the 1918 pandemic from preserved human tissue, and determined the complete genome sequence of this virus. The virus was demonstrated to be subtype H1N1. Surprisingly, analysis of its genome sequence suggests that this pandemic virus strain arose from an avian virus by mutation ("drift"), rather than by reassortment. The polymerase proteins PA, PB1, and PB2 of the 1918 human virus differ by a total of only 10 amino acids from the consensus sequences of the same proteins made by most avain strains. The HA protein of the 1918 virus acquired the ability to bind to the α-2,6 sialic acid present in the human respiratory tract by virtue of a single amino acid change from an HA that binds to the avian α-2,3 sialic acid. The 1918 strain HA lacks a furin cleavage site, but the accompanying NA protein enhances the cleavage of HA in a trypsin-independent fashion that is not yet fully understood, increasing the infectivity and spread of this virus.

An infectious influenza virus with the sequence of the 1918 strain was generated in a high-level biosecurity laboratory by reverse genetic engineering and its properties were tested. This virus showed extremely high pathogenicity for mice and for embryonated chicken eggs, and these properties required the 1918 versions of HA, NA, and the three polymerase proteins.

Genome sequences from some previous influenza A virus strains confirm the antigenic shift hypothesis

The H1N1 subtype that emerged in 1918 was still present in 1933, when the first human influenza virus was isolated in the laboratory. In 1957, a new influenza A virus suddenly appeared in Southeast China. This subtype H2N2 virus caused the "Asian flu" pandemic. At some point during this pandemic, H1N1 strains disappeared completely from the human population. H2N2 strains persisted until 1968, when another new virus suddenly appeared in China. This virus, subtype H3N2, had a new hemagglutinin gene and was the cause of the "Hong Kong" influenza pandemic, during which H2N2 viruses disappeared completely from people. In 1977, the H1N1 virus reappeared in humans in China and then spread around the world. Presently, variants of H3N2 and H1N1 viruses continue to circulate in human populations along with influenza type B strains.

The Asian (H2N2) and the Hong Kong (H3N2) viruses are reassortant viruses. In 1957, the new human virus contained an HA gene, an NA gene, and a PB1 gene that were all apparently derived from an avian influenza virus, while the other genes were the same as those in the original H1N1 human virus. In similar fashion, the 1968 Hong Kong virus contained an HA gene and a PB1 gene derived from an avian virus but other genes derived from the previously circulating human H2N2 virus. Up to the present, influenza viruses that cause widespread human disease have been limited to strains that contain HA types 1, 2, or 3.

Vaccination against influenza is an effective prevention method

Inactivated influenza vaccines are manufactured every year; each year's vaccine reflects the strains of influenza A and B viruses circulating during the end of the preceding season. It is recommended that influenza vaccine be administered yearly to all individuals with significant chronic respiratory disease as well as all individuals over the age of 65. Vaccination in healthy adults is 60–90% effective in preventing clinical illness caused by virus strains similar to those used to manufacture the vaccine. Although this effectiveness is diminished in the elderly and chronically ill, vaccination still reduces illness and death in these populations. Current vaccines are highly purified, and the only common side effect is soreness at the site of vaccination. Because these vaccines are inactivated, they cannot cause influenza.

A new live influenza vaccine has recently been approved for use in humans between 6 and 55 years old. This cold-sensitive strain was developed and tested over a 30-year period. It can replicate in the (cooler) nasal passages but not in the bronchi or lungs, and therefore causes few or no disease symptoms. It is administered by a nose spray, in contrast to inactivated vaccines, which require injection. Furthermore, it protects 90% of recipients against infection by viruses of the same subtype. Further clinical studies are needed before this live vaccine can be used in people who are most susceptible to serious influenza disease, the very young and the very old.

Highly pathogenic influenza A strains in poultry farms could lead to a new pandemic

However, starting in 1997, highly pathogenic avian H5N1 strains spread through commercial poultry farms in Hong Kong, and a small number of humans were infected with this virus, causing six deaths. These strains reappeared in multiple sites in southeast Asia during subsequent years, requiring destruction of millions of chickens and resulting in serious and often fatal disease in a number of people directly exposed to infected birds. Further spread throughout the world is occurring, probably via infected migrating waterfowl.

Fortunately, these strains have not yet shown signs of spreading efficiently among humans. However, the danger of these highly pathogenic strains adapting to humans and causing a serious pandemic is present. The high density of domestic animals living in close proximity to humans in Southeast Asia, as well as the spread of these viruses worldwide by migratory wildfowl, increase the likelihood of generation of mutants or reassortant viruses that will acquire the ability to be transmitted among humans.

Because of increased travel and rapid transportation, the next pandemic could be particularly hard to deal with. These developments have led to constant surveillance of avian viruses circulating in poultry farms worldwide, coupled with attempts to produce a human vaccine that could protect against H5N1 and other highly pathogenic influenza virus strains. National public health officials are also considering stockpiling chemotherapeutic agents active against influenza A, such as the neuraminidase inhibitors **zanamivir** and **oseltamivir** (see Figure 32.10a). These could slow down the spread of influenza A virus during a pandemic, or reduce symptoms in infected people.

SUGGESTED READING

1. Cooper, N.J., Sutton, A.J., Abrams, K.R., Wailoo, A., Turner, D., and Nicholson, K.G. (2003). Effectiveness of neuraminidase inhibitors in treatment and prevention of influenza A and B: Systematic review and meta-analyses of randomised controlled trials. *British Medical Journal* **326,** 1235.

2. Fischer, W.B. and Sansom, M.S. (2002). Viral ion channels: Structure and function. *Biochimica et Biophysica Acta* **1561,** 27–45.

3. Laver, G. and Garman, E. (2002). Pandemic influenza: Its origin and control. *Microbes & Infection* **4,** 1309–1316.

4. Ludwig, S., Pleschka, S., and Wolff, T. (1999). A fatal relationship: Influenza virus interactions with the host cell. *Viral Immunology* **12,** 175–196.

5. Neumann, G. and Kawaoka, Y. (2001). Reverse genetics of influenza virus. *Virology* **287,** 243–250.

6. Portela, A. and Digard, P. (2002). The influenza virus nucleoprotein: A multifunctional RNA-binding protein pivotal to virus replication. *Journal of General Virology* **83,** 723–734.

7. Taubenberger, J.K., Reid, A.H., Jancewski, T.A., and Fanning, T.G. (2001). Integrating historical, clinical and molecular genetic data in order to explain the origin and virulence of the 1918 Spanish influenza virus. *Philosophical Transactions: Biological Sciences* **356,** 1829–1839.

8. Weissenhorn, W., Dessen, A., Calder, L.J., Harrison, S.C., Skehel, J.J., and Wiley, D.C. (1999). Structural basis for membrane fusion by enveloped viruses. *Molecular Membrane Biology* **16,** 3–9.

KEY TERMS

α-amanitin
Actinomycin D
Amantadine
Antigenic drift
Antigenic shift
Ciliated epithelium
Cytokine
Double-stranded RNA-dependent protein kinase (PKR)
Epidemic
Furin
Fusion peptide
Hemagglutination assay
Hemagglutinin
Importin
Interferon
Lipid rafts
Mucociliary flow
Neuraminidase
Oseltamivir
Pandemic
Reassortment
Sialic acid
Type I transmembrane protein
Zanamivir

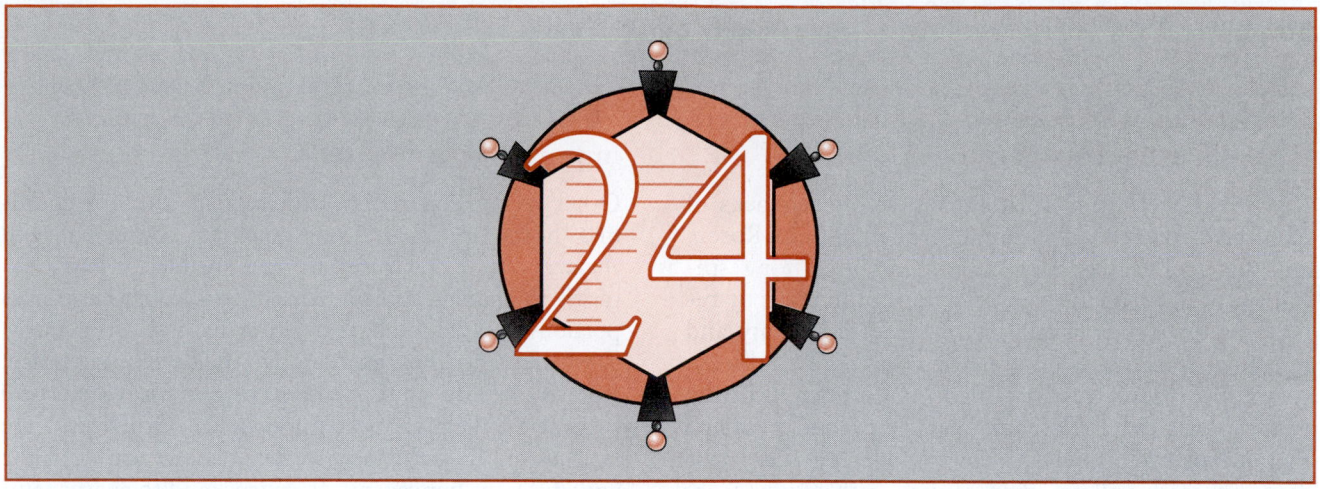

Reoviruses

Reoviridae

From *r*espiratory *e*nteric *o*rphan viruses

VIRION

Naked icosahedral capsid (T = 13).
Diameter 60–85 nm.
Capsid consists of two or three concentric protein shells.
Inner capsid (T = 1), or *core*, contains RNA polymerase and capping enzymes.

GENOME

Linear ds RNA.
10–12 segments.
Total genome length 18–24 Kb.

GENES AND PROTEINS

mRNAs are full-length copies of each genome segment.
Typically one protein encoded per genome segment.
Six–eight capsid proteins.
Three–six nonstructural proteins.

VIRUSES AND HOSTS

Twelve genera, including *Orthoreovirus*, *Rotavirus*, *Orbivirus*.
Infect humans (reoviruses types I–III, rotaviruses, Colorado tick fever virus), mammals, birds, fish, mollusks, plants, insects, and fungi.

DISEASES

Members of genus *Orthoreovirus* cause little or no disease in humans.
Rotavirus gastroenteritis is an important cause of infant disease in the developed world and infant mortality in the developing world.
Viruses spread between hosts by direct transmission, contaminated food or water, or arthropod vectors.

DISTINCTIVE CHARACTERISTICS

Family has members that infect a broad range of hosts from fungi to humans.
mRNAs are synthesized and capped inside intact cores and extruded through channels into the cytosol.
Synthesis of double-stranded genome RNAs occurs within core-like subvirion particles.
A single copy of each gene segment is packaged into each virion by an unknown sorting mechanism.
Gene segments can be reassorted during coinfection of cells by different strains.

Reoviruses were the first double-stranded RNA viruses discovered

Reoviruses were first isolated from stool specimens of children during the 1950s by Albert Sabin, Leon Rosen, and their colleagues. *Reovirus* is an acronym for *r*espiratory *e*nteric *o*rphan virus; infections of humans by reoviruses types 1–3 usually involve the respiratory and intestinal tracts with minimal or no associated disease symptoms (therefore the virus is an "orphan"). In the early 1960s, Peter Gomatos and Igor Tamm noticed that **inclusion bodies** in reovirus-infected cells fluoresce greenish-yellow when stained with acridine orange, whereas single-stranded RNA fluoresces orange. They then demonstrated by chemical and physical studies that the viral genome consists of double-stranded RNA. These were the first double-stranded RNA viruses to be described.

The *Reoviridae* family presently includes 12 genera of double-stranded RNA viruses that infect a wide variety of animals, from mammals to insects, as well as plants and fungi (Table 24.1). The mammalian reoviruses are the type species of the *Orthoreovirus* genus, which also contains viruses that infect birds and reptiles. The term *reovirus* in its common usage refers to the mammalian reoviruses, and will be used as such in this chapter. To date, only three mammalian reovirus serotypes have been identified despite isolation from a broad range of hosts. Three virus strains—type 1 Lang (T1L), type 2 Jones (T2J), and type 3 Dearing (T3D)—are the prototype members of the three reovirus serotypes. Recently, a distinct reovirus strain was isolated from a mouse in the Cameroon; this virus may represent a fourth mammalian reovirus serotype.

TABLE 24.1 Genera within the *Reoviridae*

Genus	Number of gene segments	Hosts
Orthoreovirus	10	Mammals, birds, reptiles
Rotavirus	11	Mammals, birds
Orbivirus	10	Mammals, birds, arthropods
Coltivirus	12	Mammals, arthropods
Aquareovirus	11	Fish, mollusks
Cypovirus	10	Insects
Idnoreovirus	10	Insects
Fijivirus	10	Plants, insects
Oryzavirus	10	Plants, insects
Phytoreovirus	12	Plants, insects
Seadornavirus	12	Insects, humans
Mycoreovirus	11–12	Fungi

Some members of the *Reoviridae* are important pathogens

Other recognized genera of *Reoviridae* that infect animal hosts are *Rotavirus*, *Orbivirus*, *Coltivirus*, and *Aquareovirus*. Rotaviruses are responsible for gastroenteritis in animals and humans; they cause a large proportion of childhood gastroenteritis and are a major cause of infant illness and death in the developing world. Rotaviruses are an important target for vaccination because of their medical importance. Orbiviruses are transmitted by arthropod vectors, as are viruses of the very different bunyavirus, flavivirus, and togavirus families. These virus groups have the interesting ability to replicate both in warm-blooded mammals and in cold-blooded arthropods. The best studied orbivirus is bluetongue virus, an economically significant pathogen of sheep named for a symptom affecting sick animals. Coltiviruses are also transmitted by arthropod vectors, and the prototype member, *Col*orado *ti*ck fever virus, can cause lethal neurologic disease in humans. Aquareoviruses infect fish and mollusks.

Reoviruses have emerged as a model system for studies of viral replication and pathogenesis. Bernard Fields and his co-workers used genetic approaches to study the pathogenesis of reoviruses in rodent models of infection and disease, especially viral neurologic illness. Reoviruses also have been used to study RNA synthesis. Early work by the groups of Wolfgang Joklik and Angus Graham showed that the isolated viral core contains all of the enzymes required to synthesize messenger RNA molecules from the double-stranded genome templates. Aaron Shatkin and colleagues were the first to describe the methylated cap structure that adorns most eukaryotic messenger RNA molecules and is required for normal translation; this work was done with RNAs synthesized in vitro from reovirus cores.

Reoviridae have segmented genomes made of double-stranded RNA

Reoviridae that infect animals have genomes consisting of 10 to 12 discrete, linear segments of double-stranded RNA (Figure 24.1). Reovirus gene segments range from approximately 1200 to 3900 bp in length. They are classified as large (L), medium (M), or small (S) based on their relative mobility in polyacrylamide gels; there are three L genes, three M genes, and four S genes. Mixed infection with different virus strains produces a collection of progeny **reassortant viruses** in which the inheritance of specific parental gene segments can be deduced (Figure 24.2). This feature allows the determination of which viral proteins are involved in particular biologic properties of the virus. Analysis of reassortant viruses to delineate the genetic basis of specific viral phenotypes is termed *reassortant genetics*.

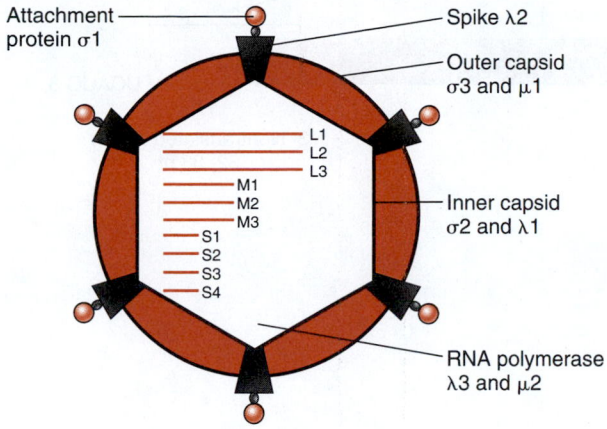

FIGURE 24.1 The reovirus virion. The T = 1 inner capsid encloses the 10 double-stranded RNA gene segments and is covered by a T = 13 outer capsid. Twelve spikes project from the core through the outer capsid. Gene segments are classified as large (L), medium (M), or small (S) based on relative electrophoretic mobilities. Viral capsid proteins are labeled.

Like most viruses, reoviruses demonstrate an extraordinary degree of genetic economy; nontranslated regions on different gene segments range from just 12 to 83 nucleotides in length (Figure 24.3). The extreme 5′-terminal four nucleotides (5′-GCUA-3′) and 3′-terminal five nucleotides (5′-UCAUC-3′) are conserved among all 10 reovirus gene segments. Furthermore, sizable regions of subterminal sequence are conserved among homologous genes of different virus strains. Nucleotide sequence conservation extends inward from both nontranslated regions into the protein-coding region. This pattern indicates that there are selection pressures on reovirus gene segments to maintain certain nucleotide sequences independent of their protein-coding function. Obvious functions for these conserved sequences are recognition by viral RNA polymerase, RNA packaging into capsids, and translational control, but the role of these sequence elements has not been precisely defined.

Nine of the 10 reovirus gene segments code for a single protein; the exception is the S1 gene, which codes for both the σ1 attachment protein and the σ1s nonstructural protein in overlapping reading frames. Viral proteins are named with the Greek letters λ (lambda), μ (mu), and σ (sigma), corresponding to the size classes (L, M, and S) of their respective gene segments; unfortunately, the numbering schemes used for genes and proteins, based on their relative migration rates in agarose or polyacrylamide gels, do not perfectly correspond. Salient features of the eight structural and three replication proteins are provided in Table 24.2.

Reovirus virions contain concentric layers of capsid proteins

Virions of mammalian reoviruses are nonenveloped particles approximately 85 nm in diameter (Figure 24.1). The virion is formed from eight proteins arranged in an unusual two-layered structure of concentric protein shells, each with icosahedral symmetry. The T = 1 inner capsid surrounds the 10 tightly packed double-stranded RNA gene segments, forming the viral core. It is assembled from two major proteins, λ1 and σ2 (120 and 150 copies each, respectively), and minor core proteins μ2 (probably 24 copies) and λ3 (probably 12 copies), which constitute the viral RNA polymerase. The T = 13 outer capsid is constructed from complexes of μ1 and σ3 proteins, present in 600 copies each. A spike, formed by pentamers of the λ2 protein (60 copies), projects from the inner capsid through the outer capsid at the 12 axes of fivefold symmetry. The RNA polymerase complexes are most likely located at the base of each spike in the core. Inserted within a turret-like opening of the spike pentamer is the trimeric attachment protein, σ1 (36 copies).

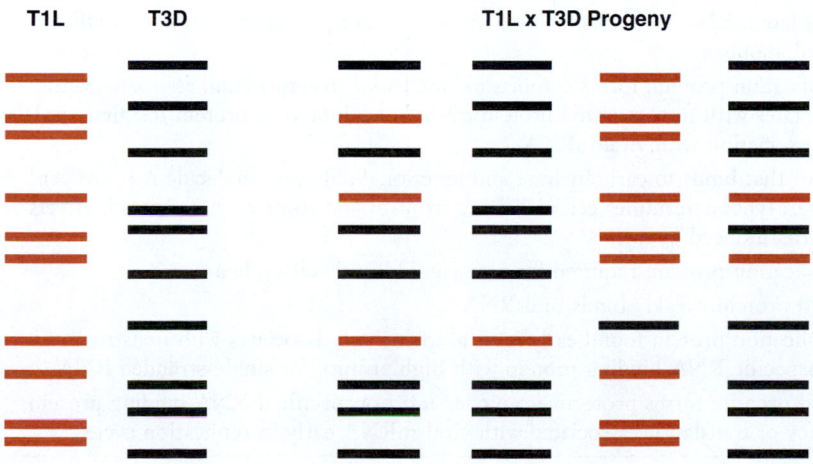

FIGURE 24.2 Generation of reassortant viruses. Reovirus strains are distinguishable by unique electrophoretic profiles of their double-stranded RNA gene segments. Coinfection of cells with different strains produces a collection of progeny reassortant viruses containing various combinations of gene segments from each parent. Shown here are schematized diagrams of electrophoretic profiles of prototype strains type 1 Lang (T1L), type 3 Dearing (T3D), and some reassortant progeny.

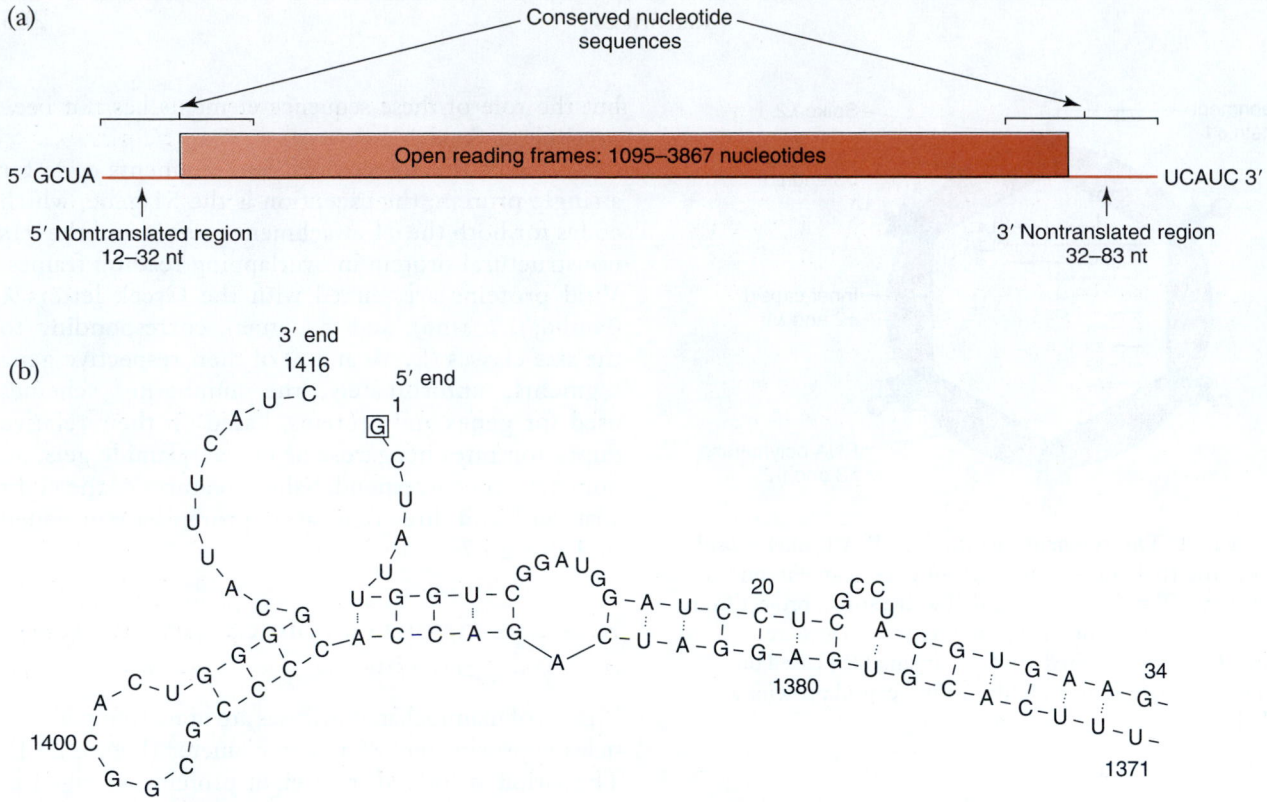

FIGURE 24.3 Features of reovirus gene segments. (a) General characteristics of a mammalian reovirus gene segment (based on strain T3D). Only the plus strand is shown here. Four nucleotides at the 5′ end and five at the 3′ end are invariant among all 10 viral gene segments. (b) Secondary structure prediction for the termini of the plus strand of the T3D S1 gene segment. Long regions of hybridization involving gene-segment termini are a consistent feature of the predicted structures for all reovirus mRNAs.

TABLE 24.2 Reovirus gene segments and functions of their translation products

Gene segment	Protein	Functions
L1	λ3	Minor inner-capsid protein; catalytic subunit of the viral RNA-dependent RNA polymerase
L2	λ2	Core spike protein; forms pentameric turret for insertion of attachment protein σ1; adds a 5′ methylated cap to mRNAs
L3	λ1	Major inner-capsid protein; possible RNA helicase; removes terminal phosphate from newly synthesized mRNA in preparation for capping
M1	μ2	Minor inner-capsid protein; probable RNA polymerase subunit; stabilizes microtubules; nonspecific RNA-binding protein
M2	μ1	Major outer-capsid protein; important for penetration of endosomal membrane by the viral core; probably regulates late mRNA synthesis from secondary transcriptase particles; affects efficiency of reovirus-induced apoptosis
M3	μNS	Nonstructural replication protein; forms cytoplasmic inclusion structures and associates with cytoskeleton; associates with nonstructural protein σNS and minor core protein μ2; first viral protein found in association with viral mRNA
S1	σ1	Attachment protein that binds to carbohydrate and junctional adhesion molecule A (JAM-A); determines virus serotype; determines cell and tissue tropism and route of virus spread; affects efficiency of reovirus-induced apoptosis
S1	σ1s	Nonstructural replication protein; required for reovirus-induced cell-cycle arrest
S2	σ2	Major inner-capsid protein; weakly binds to dsRNA
S3	σNS	Nonstructural replication protein found early in viral inclusions; associates with nonstructural protein μNS; nonspecific RNA-binding protein with high affinity for single-stranded RNA
S4	σ3	Major outer capsid protein; forms protective cap over μ1; nonspecific dsRNA-binding protein; influences efficiency of translation; associated with viral mRNA early in replication cycle

The attachment protein binds to one or two cellular receptors

The reovirus attachment protein $\sigma 1$ is a long, fiber-like molecule with head-and-tail morphology. The $\sigma 1$ protein has the most sequence diversity of any reovirus protein and is the serotype-specific antigen. The N-terminal tails of $\sigma 1$ trimers are inserted into the $\lambda 2$ spikes, while the C-terminal globular heads project away from the virion surface. The structure of this protein closely resembles that of the adenovirus fiber protein, also a trimer (see Chapter 12).

The attachment protein of type 3 reovirus contains two receptor-binding domains: one in the tail, which binds to **sialic acid**, and another in the head, which binds to **junctional adhesion molecule A (JAM-A)**. JAM-A is a transmembrane protein with two immunoglobulin-type domains in the extracellular region. It is located at **tight junctions** of epithelial and endothelial cells and mediates cell-to-cell interactions. Attachment proteins of the two other reovirus serotypes bind to JAM-A but not to sialic acid. Binding of reoviruses to JAM-A alters cell membrane permability to ions, and leads to redistribution of proteins associated with JAM-A from tight junctions to other cellular compartments.

Reovirus serotype 3 virions are likely to bind first to sialic acid residues on cell-surface glycoproteins and glycolipids. Subsequently, virions can diffuse laterally ("walk") on the cell surface until they encounter and bind to JAM-A within a tight junction. This two-stage attachment strategy is similar to those proposed for several other viruses, including bacteriophage φX174, adenoviruses, herpesviruses, and human immunodeficiency virus (Chapters 6, 12, 13, and 16). Reovirus strains that do not recognize sialic acid may bind directly to JAM-A, which is probably necessary for virus entry.

During entry, the outer capsid is stripped from virions and the core is released into the cytoplasm

Following binding to receptors on the cell surface, reovirus virions enter cells by clathrin-mediated endocytosis (Figure 24.4). In endocytic vesicles, virions undergo partial degradation by cellular proteases, resulting in the generation of **infectious subvirion particles (ISVPs)**. Either **cathepsin** B or cathepsin L, proteases located in endocytic vesicles, can carry out this reaction in the acidic environment of these organelles. ISVPs also can be formed extracellularly in the intestine or in the laboratory by *in vitro* treatment with proteases. ISVPs may in fact be the form of virus particle that infects intestinal cells. During conversion of virions to ISVPs, the attachment protein $\sigma 1$ undergoes a conformational change and becomes elongated. The outer capsid protein $\sigma 3$ is degraded and lost from virions, and the exposed $\mu 1$ protein is cleaved to form two fragments, δ and φ, that remain bound to the ISVP. These changes facilitate interactions of the ISVP with endosomal membranes, leading to delivery of the core particle into the cytoplasm and activation of the viral transcriptase.

The disassembly of reovirus virions can be visualized using the technique of **cryoelectron microscopy**

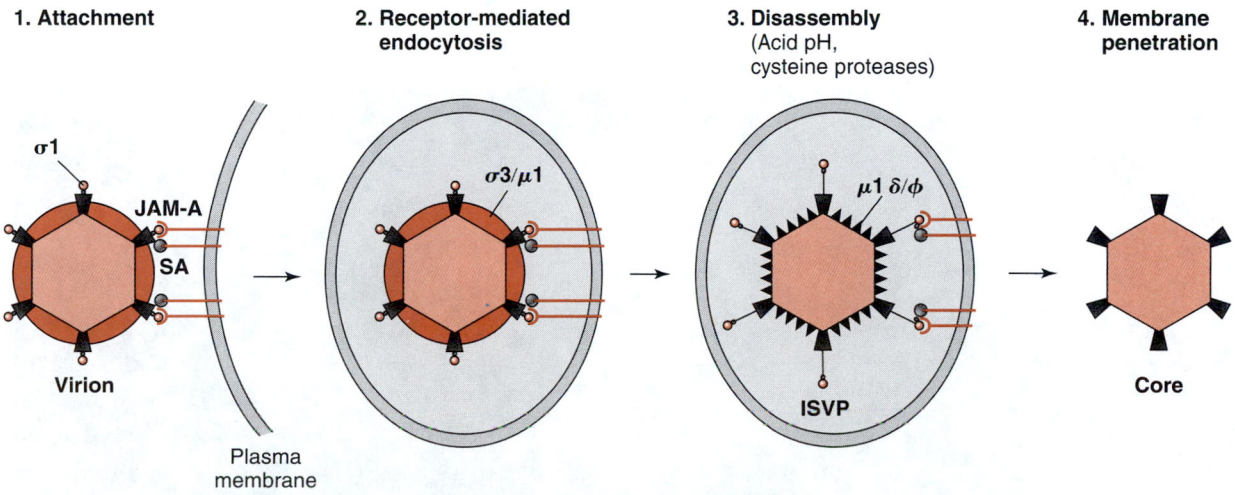

FIGURE 24.4 **Stepwise disassembly of reovirus.** Following attachment to cell-surface carbohydrate (α-linked sialic acid [SA] for type 3 reoviruses) and junctional adhesion molecule A (JAM-A), reovirus virions enter cells by clathrin-mediated endocytosis. Within an endocytic compartment, the viral outer capsid undergoes acid-dependent proteolysis, generating infectious subvirion particles (ISVPs). ISVPs interact with endosomal membranes, leading to delivery of transcriptionally active core particles into the cytoplasm.

and three-dimensional image reconstruction (Figure 24.5). Virions are spherical particles with a star-shaped density corresponding to pentamers of the λ2 protein at the fivefold axes and flower-like hexameric rings corresponding to the σ3 protein covering the remainder of the surface. Conversion to ISVPs involves the loss of σ3 and extension of the attachment protein σ1 outward from the virion. Comparison of image reconstructions of the virion (85 nm in diameter) and ISVP (80 nm in diameter) reveals the loss of 600 finger-like subunits, which correspond to the 600 copies of s3 in the virion. During the transition from ISVP to core (70 nm in diameter), the attachment protein and the μ1 protein fragments are lost, leaving the λ2 spike protein and the viral core proteins. Concurrently, the spike protein in the core undergoes a major conformational change in which the turrets formed by λ2 pentamers open to allow egress of mRNA during transcription.

Enzymes in the viral core synthesize and cap messenger RNAs

The core particles released into the cytoplasm contain all of the machinery needed by the virus to make and cap messenger RNAs. These enzymes are activated on removal of the outer capsid, and ribonucleoside triphosphates required for RNA synthesis diffuse into the core from the cytoplasm.

Messenger RNA synthesis. The reovirus core is a highly structured "mRNA factory" (Figure 24.6). An RNA polymerase complex, composed of one molecule of λ3 and one or two molecules of μ2, is located at the base of each of the λ2 spikes. It is notable that different members of the *Reoviridae* family contain between 10 and 12 gene segments, which allows each segment to be transcribed at one of the 12 spike bases. All 10 reovirus double-stranded RNAs are simultaneously transcribed, probably by translocation through fixed RNA polymerase complexes, generating a full-length plus-strand RNA copy. This has been termed the "moving template" model of transcription, because it postulates that the template RNA moves through the fixed RNA polymerase. It is possible that the double-stranded gene segments are maintained in circular form during transcription, facilitating reinitiation and multiple rounds of transcription at the same locus.

Recent x-ray crystallographic studies of the λ3 protein complexed with oligoribonucleotides suggest that the growing messenger RNA chain passes up into the spike turret and the minus-strand template of the genome RNA passes down into the core interior via independent channels within the RNA polymerase. Core shell protein λ1 appears to possess RNA helicase activity and may locally unwind the double-stranded genome RNA to provide the polymerase access to the minus-strand template. λ1 also possesses an enzymatic activity that removes the γ phosphate from the 5′ terminal nucleotide to yield a 5′-diphosphate end, preparing the RNA for cap addition.

Messenger RNA capping. During the initial stages of synthesis of each mRNA molecule, the λ2 spike protein catalyzes cap addition to the RNA 5′ end. The spike protein has three distinct active sites for the three enzymatic activities required for capping: (1) formation of the cap itself by adding a GMP residue (derived from GTP by pyrophosphate release) to the 5′-diphosphate end of the RNA; (2) methylation of the cap guanine at the N7 position; and (3) methylation of the ribose sugar

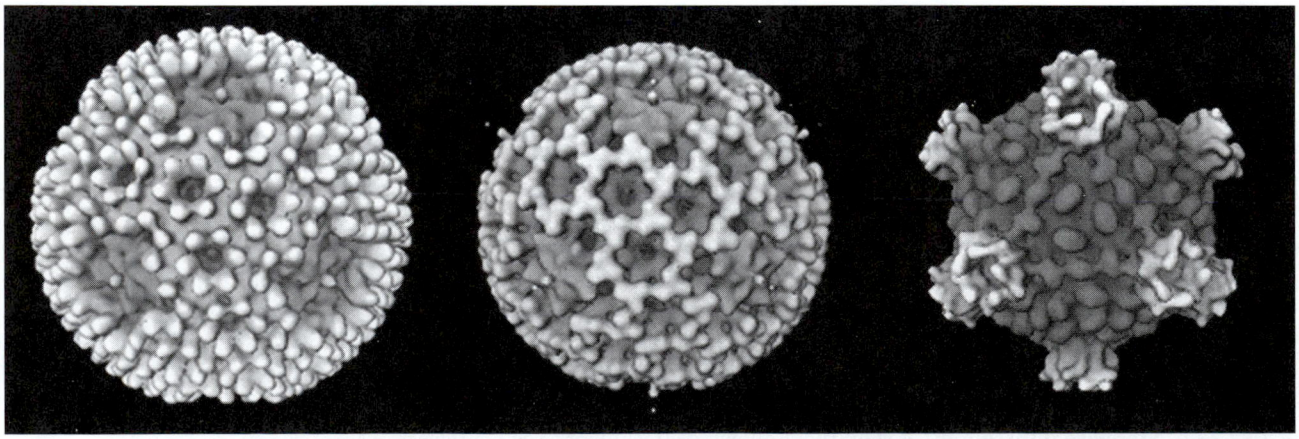

FIGURE 24.5 Structure of reovirus virions and subvirion particles. Surface-shaded views of the reovirus strain T1L virion, infectious subvirion particle (ISVP), and core. The images were obtained using cryoelectron microscopy and three-dimensional image reconstruction. Each particle is viewed along a threefold axis.

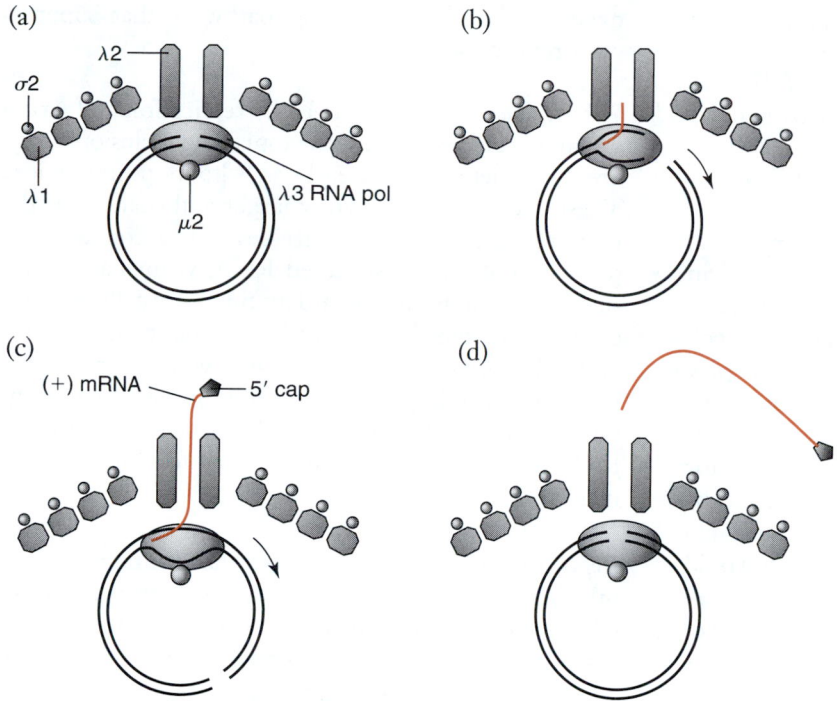

FIGURE 24.6 Synthesis of reovirus messenger RNAs.
(a) Diagram of the viral core transected through one of the 12 spikes. The viral RNA polymerase, λ3, and the polymerase cofactor protein, μ2, are thought to reside at the base of the turret formed by pentamers of the λ2 protein. The viral gene segments may be maintained in circular form as shown here. (b–d) The double-stranded template RNA is locally unwound, possibly by a helicase function of the λ1 protein, and serves as a template for synthesis of full-length plus-strand mRNAs (orange line). These are extruded through the λ2 turret, receiving a 5′ terminal methylated cap (black pentagon) in the process. The template RNA is pulled through the RNA polymerase in the direction shown by the arrows. The circular form of the genome template would allow for multiple rounds of transcription.

of the first template-coded G residue. These reactions probably occur in sequence as the RNA progresses through the λ2 spike during transcription.

Messenger RNA export. The capped RNA transcripts are extruded through channels within the spikes to the exterior, where they are released into the cytoplasm (Figure 24.6). This can be visualized by electron microscopy of actively transcribing cores, which reveals multiple RNA transcripts emerging from core particles. The double-stranded RNA templates remain within the core and can be repeatedly transcribed, giving rise to multiple copies of all mRNAs.

Translation of reovirus mRNAs is regulated

Newly synthesized viral proteins can be detected in infected cells as early as two hours after infection, and as their synthesis increases, there is a concomitant decrease in synthesis of cellular proteins. Not all reovirus proteins are produced in the same amount. One factor affecting protein abundance is the length of the mRNA; the longer the transcript, the longer the time required for its synthesis, and therefore long mRNAs are less abundant than shorter transcripts. Sequences surrounding the initiator AUG codons and differences in the length of 5′ nontranslated regions also account for different levels of synthesis of viral proteins.

Reovirus mRNAs do not contain 3′ poly(A) tails, unlike most cellular mRNAs. Translation efficiency in eukaryotic cells is usually strongly dependent on the poly(A) tail, which is bound by cellular proteins and interacts with the translational machinery. Studies of translational control have identified sequences in the 3′ nontranslated region of the S4 mRNA that interact with cellular proteins and affect translation efficiency. These sequence elements are approximately 20 nucleotides in length and do not appear to affect transcript stability. A reasonable hypothesis is that signals embedded in reovirus mRNA functionally substitute for the poly(A) tail by binding to cellular proteins.

Like reovirus mRNAs, rotavirus transcripts are capped but are not polyadenylated. Efficient translation of rotavirus mRNA requires viral nonstructural protein NSP3, a two-domain protein that binds to the last four nucleotides at the 3′ end of the mRNA and also interacts with the 5′ cap complex. An NSP3 homolog has not been identified among mammalian reovirus proteins, which suggests that a cellular protein (alone or associated with a reovirus protein) functions similarly to NSP3 during translation of reovirus mRNAs.

Interferon and PKR: effects on viral and cellular protein synthesis

Reovirus double-stranded RNA is a potent inducer of type I (α and β) **interferons**. This cellular response to virus infection can block translation of viral mRNAs by induction of **double-stranded RNA-dependent protein kinase R (PKR)** and **ribonuclease L** (see Chapter 31). PKR, a latent protein kinase activated by double-stranded RNA, interferes with translation of mRNAs by phosphorylation (and consequent inactivation) of eukaryotic initiation factor 2α (eIF-2α). Ribonuclease L cleaves mRNAs and is activated by oligomers generated by **2′, 5′-oligo(A) synthetase**, which in turn is activated by double-stranded RNA.

The σ3 protein modulates PKR activation. The translation of cellular mRNAs is strongly inhibited in reovirus-infected cells; by approximately 10 hours after reovirus infection, most proteins synthesized in the cell are viral. The reovirus outer capsid protein $\sigma3$ may play a central role in this selective inhibition of cellular protein synthesis via its effects on PKR activity. $\sigma3$ binds to double-stranded RNA, competing with PKR for binding. In the vicinity of virus factories the concentration of $\sigma3$ may be high, so that any double-stranded RNA would be bound to $\sigma3$ and PKR activation would be correspondingly low, allowing viral protein synthesis to continue. In other areas of the cytoplasm, lower concentrations of $\sigma3$ would allow PKR activation, leading to selective inhibition of cellular protein synthesis.

PKR regulation and cancer. Reoviruses replicate more efficiently in transformed cells cultured *in vitro* than in their normal counterparts. Activated signaling pathways such as **Ras** interfere with PKR function in transformed cells; the resulting low PKR activity may allow higher levels of viral protein synthesis and replication. This suggests a potential use of reoviruses as a treatment for cancer. In several studies, mice were implanted with transformed cells of mouse or human origin, leading to tumor growth. Infection of these mice with reoviruses led to regression of tumors. Because reoviruses replicate in humans but are largely nonpathogenic, they might be useful as novel **oncolytic** agents that could target a wide array of human cancers while sparing healthy, untransformed cells and tissues.

Synthesis of progeny double-stranded genomes occurs within subviral particles

In addition to being used for protein synthesis, reovirus mRNAs have a second function: they serve as templates for the synthesis of progeny double-stranded genome RNAs. Production of genome RNAs (see below) is a two-step process. First, one copy of each of the 10 reovirus mRNA molecules is assembled into a subviral particle; second, each plus-strand mRNA within this particle is copied by viral RNA polymerase to form 10 different double-stranded genome RNAs. This mechanism ensures that one molecule of each gene segment is incorporated into every virion. Furthermore, it keeps most viral double-stranded RNA segregated within subviral particles and away from the cytosol, reducing activation of cellular signaling pathways that suppress virus replication.

Virus factories. Reovirus RNA replication and virion assembly take place in intracytoplasmic inclusions (Figure 24.7), detectable as early as 8 hours postinfection. These "virus factories" grow larger and coalesce as the virus replication cycle progresses. They contain viral proteins and double-stranded RNA, virions at various stages of maturation, and at late times crystalline arrays of mature virions. Viral inclusions do not contain membranes but are intimately associated with the cytoskeleton. Core protein $\mu2$ stabilizes microtubules within viral inclusions and associates with nonstructural protein μNS, forming an organizing center for formation of inclusions.

Replicase particles. Complexes that contain reovirus mRNAs bound to nonstructural viral proteins μNS and σNS and outer-capsid protein $\sigma3$ are postulated to represent the first step in reovirus assembly (Figure 24.8). The next step probably involves formation of "replicase particles" engaged in minus-strand synthesis. These particles contain reduced levels of both core proteins and outer capsid proteins compared with mature virions. The mechanism ensuring that only one copy of each of the 10 unique RNA segments is packaged into each particle is not understood. However, this process most likely involves specific recognition of nucleotide sequences and secondary structures at the termini of each mRNA (Figure 24.3) by viral proteins (for further discussion of this topic, see Chapter 23 on orthomyxoviruses).

Replicase particles can synthesize full-length double-stranded RNAs by copying the packaged plus-strand mRNAs using the viral RNA-dependent RNA polymerase. The same RNA polymerase therefore participates in two very different reactions: (1) synthesis of single-stranded mRNAs in viral cores, using a double-stranded RNA template that remains intact and can be used multiple times, and (2) synthesis of double-stranded genome RNAs within replicase particles, using a single-stranded mRNA template to which the negative-strand copy remains hydrogen-bonded after synthesis.

Secondary transcription and encapsidation. Once double-stranded RNA is made, the replicase particles are able to synthesize more plus-strand viral mRNA. Transcription by these "secondary transcriptase particles" accounts for the bulk of the mRNA made in reovirus-infected cells. Secondary transcriptase particles, like replicase particles, lack the full complement of virion proteins, but the two particle forms are physicochemically distinct. When sufficient capsid proteins

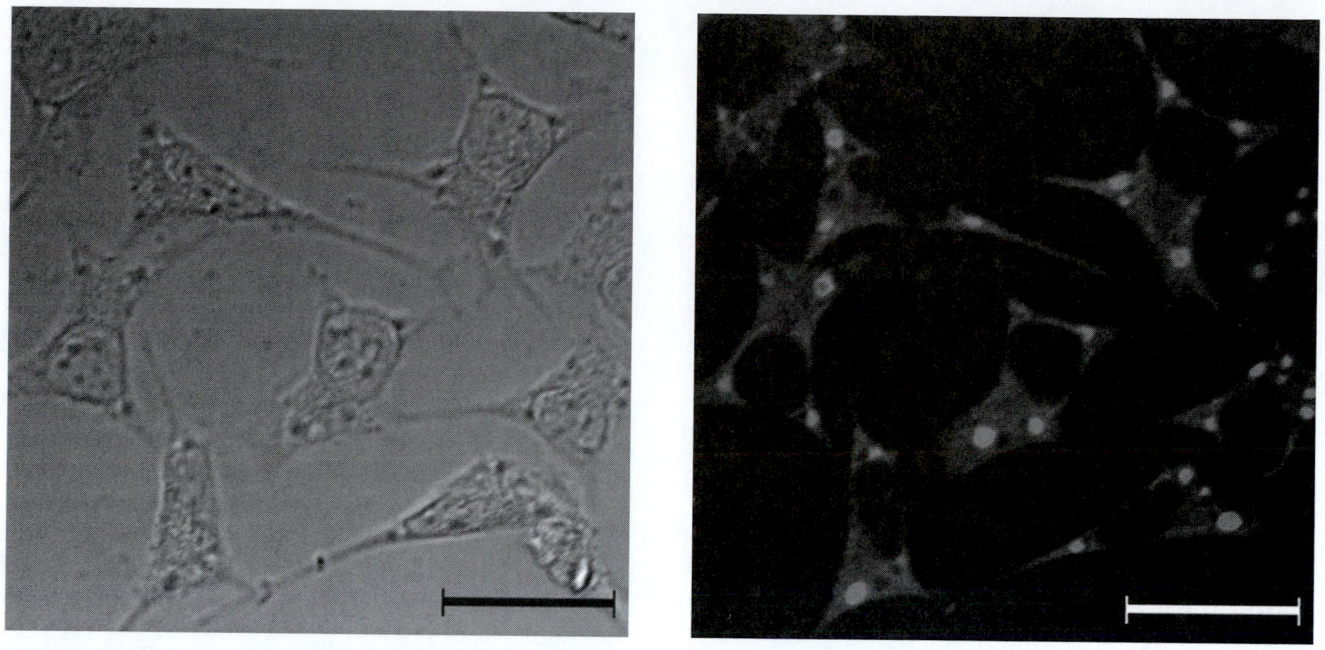

(a) (b)

FIGURE 24.7 Formation of inclusions in reovirus-infected cells. Mouse L fibroblast cells were infected with reovirus strain T3D and examined 24 hours after infection using a confocal microscope. (a) Differential interference contrast image. (b) Cells were stained with fluorescent antibodies that detect reovirus proteins, which are revealed as bright patches within cells. Note the large inclusion structures in the cytoplasm that contain reovirus proteins and the absence of reovirus proteins in the nucleus. Bar: 25 μm.

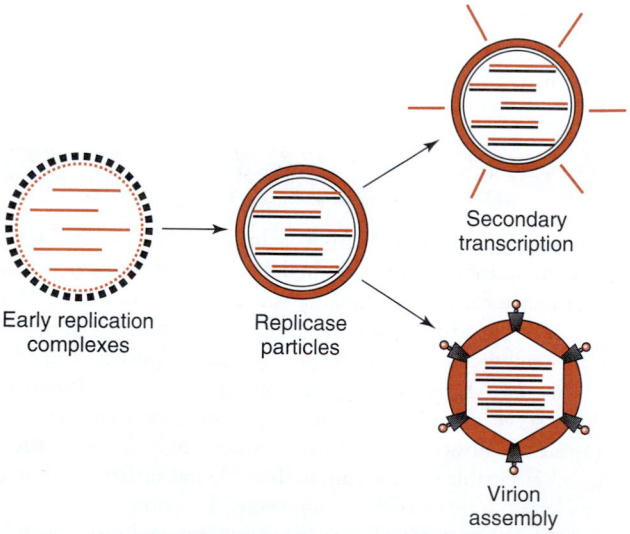

FIGURE 24.8 Assortment of gene segments and virion assembly. Reovirus plus-strand RNAs are assembled into particles called early replication complexes, which contain only one copy of each of the 10 different transcripts. Viral RNA polymerase then synthesizes double-stranded genome RNAs by copying the plus-strand mRNAs. These replicase particles can proceed to synthesize new mRNAs (upper arrow) or they can be assembled into virions by the addition of outer capsid proteins (lower arrow).

have been synthesized, assembly is completed by the addition of a layer of outer capsid proteins to replicase particles, resulting in the formation of infectious double-shelled virions (Figure 24.8).

Reoviruses induce apoptosis via activation of transcription factor NF-κB

Reoviruses can induce **apoptosis** (programmed cell death) in cultured cells and in intact animals. Reovirus infection leads to activation of **nuclear factor kappa B (NF-κB)**, a family of transcription factors known to play important roles in regulating cellular stress responses, including apoptosis. Reovirus is capable of activating NF-κB in a variety of cultured cell lines including mouse L cells, dog MDCK cells, and human HeLa cells. NF-κB complexes activated by reovirus are comprised of NF-κB subunits p50 and p65.

Apoptosis induced by reovirus is significantly reduced in cells expressing an inhibitor of NF-κB. Furthermore, apoptosis is blocked in reovirus-infected cells deficient in the expression of the p50 or p65 NF-κB subunits. These findings demonstrate that reovirus-induced apoptosis requires functional NF-κB, which presumably enhances the expression of genes that induce this cellular response to reovirus infection. Some NF-κB-responsive genes function in host innate immune defense mechanisms; these include interferon-stimulated genes and the gene encoding interferon-β (see Chapter 31). Treatment of reovirus-infected cells with antibodies specific for interferon-β attenuates apoptosis. This points to a cellular response axis linking NF-κB, interferon-β, and reovirus-induced apoptosis. Reovirus infection also activates **caspases**, which are cellular proteases that mediate apoptosis. Caspases linked to both the intrinsic (mitochondrial) and extrinsic (death receptor) pathways of apoptosis induction are activated by reovirus.

Receptor binding and apoptosis. Binding of viral attachment protein σ1 to different types of cell-surface receptors regulates the efficiency of the apoptotic response. Binding of virions to both sialic acid and JAM-A is required to induce maximal levels of apoptosis as well as maximal NF-κB activation. One model to explain these observations asserts that multivalent interactions of σ1 with sialic acid and JAM-A surpass a critical threshold required for NF-κB activation and apoptosis induction.

However, there is evidence that steps beyond virion binding also induce apoptosis. Inhibitors of virion uncoating in endocytic vesicles block reovirus-induced apoptosis, but inhibitors of viral mRNA synthesis do not. Therefore, virus entry seems to be a requirement for induction of apoptosis, but not viral protein synthesis.

Cell-cycle arrest. In addition to apoptosis, reovirus also causes cell-cycle arrest in infected cells. Arrest occurs at the transition between the G2 stage of the cell cycle and mitosis. It is mediated by nonstructural protein s1s, and occurs independently of the apoptotic response to infection. Cell-cycle arrest may be a result of hyperphosphorylation and consequent inactivation of the **cyclin-dependent kinase (CDK)**, $p34^{cdc2}$/cdk1, by σ1s.

Studies of reovirus pathogenesis in mice

Central nervous system. Although reoviruses are mostly nonpathogenic in humans, these viruses have served as a very productive experimental system for studies of viral pathogenesis. In newborn mice, reoviruses cause patterns of disease that vary with the virus serotype. The best-characterized model is reovirus pathogenesis in the mouse central nervous system. After primary replication in the intestine, type 1 reovirus strains spread through the bloodstream to the brain where they infect **ependymal cells**, leading to nonlethal **hydrocephalus**. In contrast, type 3 reovirus strains spread primarily by neural routes to the central nervous system and infect neurons, causing fatal **encephalitis**.

The use of reassortant viruses showed that these distinct patterns of central nervous system disease depend on the source of the S1 gene, which encodes the attachment protein, σ1. This suggested that type 1 and type 3 reoviruses may interact with different receptors expressed on ependymal cells and neurons. However, both type 1 and type 3 reovirus strains are known to use JAM-A as a receptor. Because type 3 reovirus also binds to sialic acid residues, the distribution or density of cell-surface carbohydrates may determine the types of cells infected by a given reovirus strain. It is also possible that cells in the central nervous system have alternative reovirus receptors that have not yet been discovered.

Liver and heart. Reovirus infection of mice causes pathology in a wide range of other organs and tissues, including the liver and bile ducts, heart muscle, lungs, exocrine pancreas, and endocrine tissues. Reovirus liver and bile duct disease and **myocarditis** have become particularly well-established experimental models of viral injury at these sites.

Some type 3 reovirus strains infect the epithelial lining of bile ducts within the liver, leading to ductal obliteration and obstructive **jaundice**. The capacity of attachment protein s1 to bind to cell-surface sialic acid is a determinant of viral **tropism** for bile duct epithelium and inflammatory damage to bile ducts. There are indications that injury to bile ducts leading to jaundice in newborn humans may also be caused by reovirus infection.

A new cellular site for viral receptors

The reovirus receptor, JAM-A, is localized to tight junctions between epithelial and endothelial cells, which seems an unusual locale for a virus docking station. It is noteworthy that several other viruses bind to receptors located in regions of cell-to-cell contact. The receptor for coxsackievirus and adenoviruses, CAR, is also expressed at tight junctions. **Nectins**, which serve as receptors for herpes simplex virus, are expressed at **adherens junctions**. Interestingly, each of these viruses is capable of infecting both epithelial surfaces and neurons in some types of host organisms. Junctional regions are sites of enhanced membrane recycling, uptake into endocytic vesicles, and intracellular signaling. By targeting junction-associated proteins as receptors, these viruses may enhance their uptake and prepare the cell for optimal virus replication by stimulating selected signaling pathways. Understanding the molecular mechanisms by which viruses exploit such vulnerabilities in the host may eventually lead to the design of inhibitors that could modify virus uptake and improve the outcome of virus infections.

Unlike myocarditis caused by several other viruses, reovirus-induced myocarditis is not mediated by immune responses. Rather, reovirus-induced cytopathic effects directly cause injury of heart muscle cells. These cytopathic effects result from a complex interplay of the interferon and apoptosis pathways of host cellular defenses. The efficiency of viral RNA synthesis is a key factor in determining the extent of myocardial injury.

Reovirus infection of mice with a defect in NF-κB activation due to absence of the p50 subunit results in massively increased levels of myocardial apoptosis and marked cardiac dysfunction compared to wildtype animals. This cell damage is due to decreased production of type I interferon in response to viral infection. Treatment of p50-negative mice with interferon protects heart cells from apoptosis following reovirus infection. Interestingly, p50-negative mice exhibit greatly reduced levels of apoptosis in the central nervous system in comparison to normal animals. Thus, NF-κB appears to play organ-specific roles in reovirus pathogenesis.

SUGGESTED READING

1. Barton, E.S., Forrest, J.C., Connolly, J.L., Chappell, J.D., Liu, Y., Schnell, F.J., Nusrat, A., Parkos, C.A., and Dermody, T.S. (2001). Junction adhesion molecule is a receptor for reovirus. *Cell* **104,** 441–451.

2. Chandran, K., Zhang, X., Olson, N.H., Walker, S.B., Chappell, J.D., Dermody, T.S., Baker, T.S., and Nibert, M.L. (2001). Complete in vitro assembly of the reovirus outer capsid produces highly infectious particles suitable for genetic studies of the receptor-binding protein. *Journal of Virology* **75,** 5335–5342.

3. Chappell, J.D., Prota, A.E., Dermody, T.S., and Stehle, T. (2002). Crystal structure of reovirus attachment protein s1 reveals evolutionary relationship to adenovirus fiber. *EMBO Journal* **21,** 1–11.

4. Coffey, M.C., Strong, J.E., Forsyth, P.A., and Lee, P.W. (1998). Reovirus therapy of tumors with activated Ras pathway. *Science* **282,** 1332–1334.

5. Connolly, J.L., Rodgers, S.E., Clarke, P., Ballard, D.W., Kerr, L.D., Tyler, K.L., and Dermody, T.S. (2000). Reovirus-induced apoptosis requires activation of transcription factor NF-κB. *Journal of Virology* **74,** 2981–2989.

6. Dryden, K.A., Wang, G., Yeager, M., Nibert, M.L., Coombs, K.M., Furlong, D.B., Fields, B.N., and Baker, T.S. (1993). Early steps in reovirus infection are associated with dramatic changes in supramolecular structure and protein conformation: Analysis of virions and subviral particles by cryoelectron microscopy and image reconstruction. *Journal of Cell Biology* **122,** 1023–1041.

7. Ebert, D.H., Deussing, J., Peters, C., and Dermody, T.S. (2002). Cathepsin L and cathepsin B mediate reovirus disassembly in murine fibroblast cells. *Journal of Biological Chemistry* **277,** 24609–24617.

8. O'Donnell, S.M., Connolly, J.L., Hansberger, M.W., Chappell, J.D., Watson, M.J., Han, W., Barton, E.S., Forrest, J.C., Valyi-Nagy, T., Pierce, J.M., Yull, F.E., Blackwell, T.S., Rottman, J.N., Sherry, B.J., and Dermody, T.S. (2005). Organ-specific roles for transcription factor NF-κB in reovirus-induced apoptosis and disease. *Journal of Clinical Investigation.* **115,** 2341–2350.

9. Reinisch, K.M., Nibert, M.L., and Harrison, S.C. (2000). Structure of the reovirus core at 3.6 Å resolution. *Nature* **404,** 960–967.

10. Tao, Y., Farsetta, D.L., Nibert, M.L., and Harrison, S.C. (2002). RNA synthesis in a cage: Structural studies of reovirus polymerase λ3. *Cell* **111,** 733–745.

KEY TERMS

Adherens junction
Apoptosis
Caspases
Cathepsins
Cryoelectron microscopy
Cyclin-dependent kinase (CDK)
Double-stranded RNA-dependent protein kinase (PKR)
Encephalitis
Ependymal cells
Hydrocephalus
Inclusion bodies
Infectious subvirion particles (ISVPs)
Interferons
Jaundice
Junctional adhesion molecule A (JAM-A)
Myocarditis
Nectins
NF-κB (nuclear factor-κB)
2′, 5′-oligo(A) synthetase
Oncolytic
Ras
Reassortant viruses
Ribonuclease L
Sialic acid
Tight junction
Tropism

Retroviruses

Retroviridae

From *re*verse *tr*anscription, characteristic of these viruses

VIRION

Spherical enveloped particle.
Diameter 100 nm.
Envelope assembled at plasma membrane.
Icosahedral or conical capsid.
Contains reverse transcriptase.

GENOME

Linear ss RNA, positive sense.
Two identical molecules, 7–10 Kb, packaged in virions.
5′ cap, 3′ poly(A) tail.

GENES AND PROTEINS

mRNAs transcribed from integrated provirus DNA by RNA polymerase II.
Three major genes, each translated as a polyprotein and cleaved into mature proteins:

- *gag: nucleocapsid proteins*
- *pol: reverse transcriptase, protease, and integrase*
- *env: envelope proteins*

Gag and Pol proteins made on unspliced mRNA.
Env proteins made on singly spliced mRNA.
Lentiviruses and human T-cell leukemia viruses have additional regulatory genes expressed on multiply spliced mRNAs.

VIRUSES AND HOSTS

Seven genera are recognized: *Alpha-* to *Epsilonretrovirus*, *Lentivirus*, *Spumavirus*.
Simple retroviruses: Rous sarcoma virus (chickens), mouse mammary tumor virus (mice), Jaagsietke sheep retrovirus.
Lentiviruses: human, feline, bovine, and simian immunodeficiency viruses
Spumaviruses: simian foamy virus.

DISEASES

Humans: AIDS, leukemia.
A variety of cancers in monkeys, mice, cats, sheep, birds, etc.

DISTINCTIVE CHARACTERISTICS

Two identical copies of genome per virion.
Genome RNA converted to DNA by reverse transcription.
A cellular tRNA used as primer to initiate reverse transcription.

A DNA copy of the genome (provirus) is integrated into host cell DNA.
Retrovirus populations exists as quasispecies in their hosts because of frequent errors made by reverse transcriptase.
Can cause cancer by expressing mutated cellular regulatory genes or by integrating into or nearby cellular regulatory genes.

Retroviruses have a unique replication cycle based on reverse transcription and integration of their genomes

The first members of the retrovirus family were described nearly 100 years ago, when a virus responsible for equine infectious anemia was isolated. Shortly afterward, in 1911, Peyton Rous showed that the Rous sarcoma virus induces tumor formation in chickens. In neither case was the nature of these viruses understood until much more recently. The discovery of **reverse transcriptase** independently in the laboratories of Howard Temin and David Baltimore in 1971 demonstrated that retroviruses integrate a DNA copy of their RNA genome into the chromosomes of infected cells. Shortly thereafter the laboratory of Michael Bishop and Howard Vamus showed that the ability of Rous sarcoma virus to induce tumors results from the acquisition of an altered cellular gene (an **oncogene**) by the virus. These discoveries were all awarded Nobel prizes, although Peyton Rous had to wait 55 years between his discovery and his prize!

Retroviruses have been isolated from numerous animal species, including chickens, mice, cattle, horses, cats, monkeys, and humans. Retroviruses are presently grouped into seven genera (Table 25.1), based on differences in morphology and genome organization. A previous but related classification was based on the pathology associated with infection: *oncoviruses* (many viruses in the *Alpha-* to *Epsilonretrovirus* genera) are tumor-inducing viruses; *lentiviruses* induce slowly progressing, wasting disease; and *spumaviruses* ("foamy" viruses) induce persistent infection without any associated pathology. The discovery of human immunodeficiency virus (HIV), a lentivirus, in the mid-1980s (see Chapter 26) led to an explosion of work on this newly emerged human pathogen, responsible for a worldwide epidemic of acquired immunodeficiency syndrome (AIDS).

Viral proteins derived from the gag, pol, and env genes are incorporated in virions

Retroviruses are roughly spherical and approximately 100 nm in diameter (Figure 25.1). The envelope contains the external surface protein (SU), bound by noncovalent interactions to the transmembrane protein (TM), which traverses the lipid bilayer. Coating the inner surface of the membrane is the viral matrix protein (MA). The capsid protein (CA) forms an icosahedral or conical core, depending on the virus strain. Three virus-coded enzymes, a protease (PR), an integrase (IN), and a reverse transcriptase (RT), are associated with the virus core. The viral strucural proteins are often identified by their glycosylation status and their molecular weights; for example, the human immunodeficiency virus SU protein is called gp120, TM is gp41, and CA is p24.

The virus genome (Figure 25.2) is a positive-strand RNA 7 to 10 Kb long, complexed with the nucleocapsid protein (NC). Genome RNAs are capped at their 5′ termini and polyadenylated at their 3′ termini, as are eukaryotic mRNAs. A 150–200 nt repeated sequence (R) is present at both the 5′ and 3′

TABLE 25.1 Retrovirus genera

Genus	Examples of virus	Host
Alpharetrovirus	Rous sarcoma virus	Chickens
Betaretrovirus	Mouse mammary tumor virus	Mice
Gammaretrovirus	Murine leukemia virus	Mice
Deltaretrovirus	Human T cell leukemia virus type 1	Humans
Epsilonretrovirus	Walleye dermal sarcoma virus	Fish
Lentivirus	Human immunodeficiency virus type 1	Humans
	Simian immunodeficiency virus	Monkeys
	Feline immunodeficiency virus	Cats
Spumavirus	Simian foamy virus	Monkeys

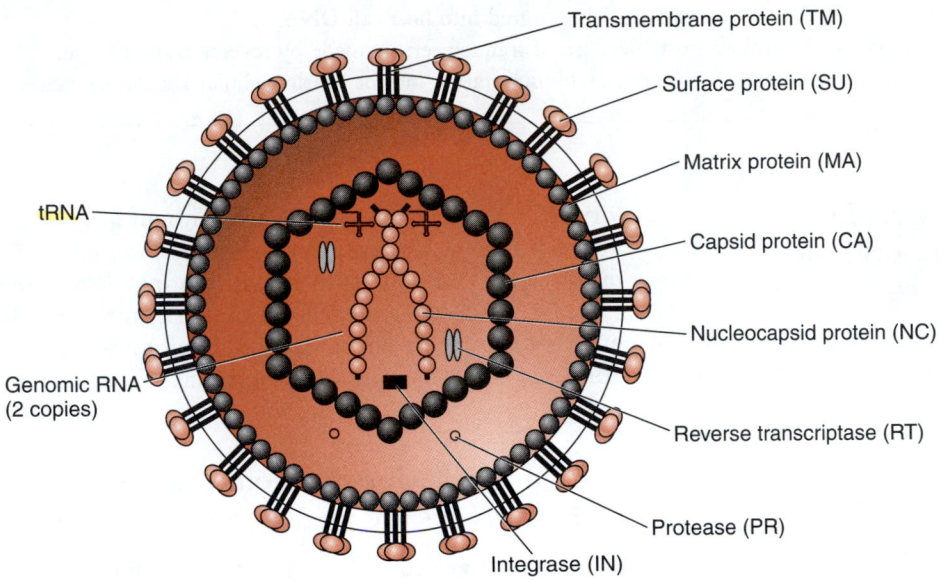

FIGURE 25.1 A typical retrovirus virion. Schematic diagram showing relative positions of structural proteins and RNAs in the virion.

ends of the genome RNA. Adjacent to the repeated sequences at either end are unique regions designated U5 (80–200 nt) and U3 (240–1200 nt). Within the virion the two RNAs exist as a dimer, held together in a head-to-head configuration by interaction of sequences known as a "kissing loop" located in the U5 region. A specific cellular transfer RNA is bound to the genome RNA by base-pairing between the primer binding sequence (PBS), located just downstream of U5, and 18 nucleotides at the 3′ end of the tRNA. Different virus strains bind different cellular tRNAs. Three genes encoding viral proteins are designated gag (for "group-specific antigen"), pol (for DNA polymerase), and env (for envelope proteins).

Retroviruses enter cells by the fusion pathway

The retrovirus life cycle takes place in two phases. During the early phase (Figure 25.3) the virus enters the cell, makes a DNA copy of its RNA genome, and inserts this copy into the host cell genome. Retroviruses can only infect certain species and types of cells, a property based mainly on the interaction of SU with receptor proteins on the cell surface. Examples of virus receptors are CD4 and the **chemokine** receptors CCR5 and CXCR4 for HIV-1 (see Chapter 26); the amino acid transporter EcoR/Rec-1 for murine leukemia virus; the phosphate transporters Pit-1/Glvr-1 and Pit-2/Glvr-2 for the gibbon ape leukemia virus; and Tva, a homologue of the low-density lipoprotein receptor, for avian sarcoma leukosis virus type A. Following interaction of SU with its receptor, the viral envelope either fuses directly with the plasma membrane, or the virus undergoes receptor-mediated endocytosis and fusion is initiated by a decrease in pH within the endosome. Fusion is induced by a conformational shift in the SU protein that exposes the hydrophobic amino terminus of the TM protein, which is then inserted into the cell membrane. Fusion releases the virion core into the cytoplasm.

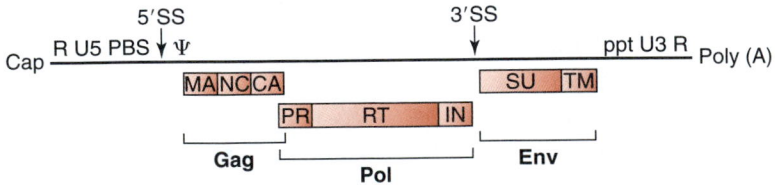

FIGURE 25.2 Structure of retrovirus RNA. Beginning at the left end, features are: methylated cap; repeat region (R); untranslated 5′ sequence (U5); primer binding sequence (PBS); 5′ splice site (5′ ss); psi (*ψ*) packaging sequence; gag, pol, and env reading frames for viral structural genes; 3′ splice site (3′ ss); polypurine tract (ppt) used during reverse transcription; untranslated 3′ sequence (U3); repeat region (R), poly(A) tail.

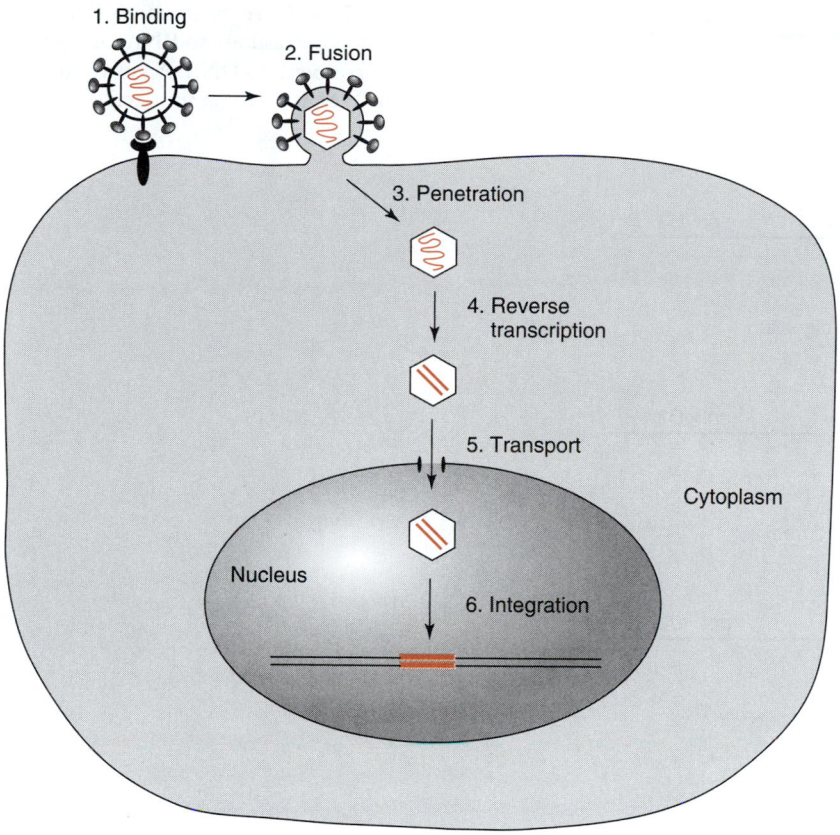

FIGURE 25.3 **Early phase of retrovirus life cycle.** RNA genomes are shown as curved orange lines; they are converted to double-stranded DNBA (provirus), shown as parallel straight orange lines, by reverse transcription.

Viral RNA is converted into a double-stranded DNA copy by reverse transcription

In the cytoplasm, deoxynucleoside triphosphates penetrate into the core, and the process of conversion of viral RNA to double-stranded DNA begins. Reverse transcriptase, the enzyme responsible for this conversion, exists as a dimer that has two distinct activities: an RNA/DNA-dependent DNA polymerase activity, and a **ribonuclease H** activity. Ribonuclease H is an enzyme that selectively hydrolyses the RNA part of an RNA-DNA hybrid, but does not degrade unhybridized RNA. The cellular tRNA molecule that is bound to the primer-binding site (PBS) on the viral RNA serves as a primer for the initiation of synthesis of DNA complementary to the genome RNA. Conversion of RNA to DNA is a complex process that proceeds as follows (Figure 25.4):

1. *Synthesis of minus-strand strong-stop DNA.* DNA synthesis initially extends from the tRNA primer up to the 5′ end of the viral RNA, resulting in a short (100–400 nt) DNA molecule that is a copy of the U5 and R regions. This is called "minus-strand strong-stop DNA" because it is a complementary copy of part of the positive-strand genome RNA, and DNA synthesis stops abruptly at the end of the RNA template during *in vitro* reactions.

2. *Removal of template RNA.* RNase H digests the RNA part of the newly made DNA-RNA hybrid (but not the RNA-RNA hybrid at the primer binding site), removing the RNA template and exposing the DNA copy of the U5 and R sequences.

3. *Strand transfer.* The DNA copy of the R sequence hybridizes with the R sequence at the 3′ end of the genome RNA. This is called the "first strand transfer," since the strong stop DNA strand is transferred from the 5′ end to the 3′ end of the genome RNA. The actual geometry of how this transfer takes place within the virus core during reverse transcription is not clear. Although the resulting complex is drawn as a linear molecule in Figure 25.4, it could remain circular.

4. *Copying of full-length genome.* The strong stop DNA is extended by reverse transcriptase to copy the entire remaining part of the genome RNA up through the primer binding site, making a full-length minus-strand copy of the genome. Because of the strand transfer, this DNA molecule contains a copy of the R sequence sandwiched in between the U3 sequence, originally at the right end of the viral RNA, and the U5 sequence, originally at the left end. The U3-R-U5 sequence is called a "long terminal repeat" (LTR).

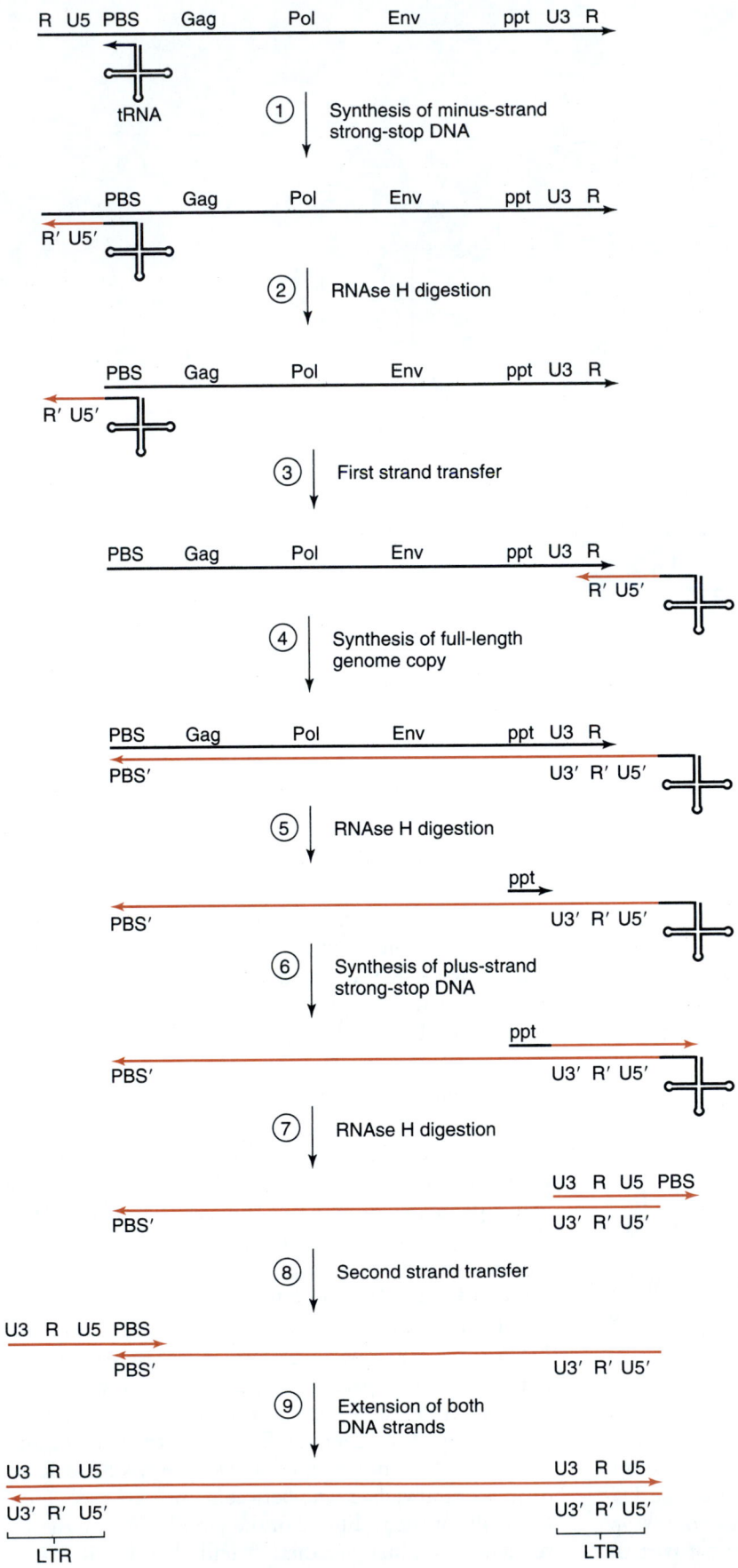

FIGURE 25.4 Reverse transcription. Black lines correspond to RNA, orange lines correspond to DNA. Arrowheads denote 3′ ends. See text for details.

5. *Removal of template RNA.* As it is copied into DNA, most of the genome RNA is digested by RNAse H. However, an RNA sequence just to the left of U3, consisting of a polypurine tract (ppt), is resistant to digestion, and remains hybridized to the newly synthesized minus-strand DNA. This residual RNA serves as a primer for the subsequent synthesis of plus-strand DNA by reverse transcriptase.

6. *Synthesis of plus-strand strong-stop DNA.* Synthesis is initiated by the ppt RNA primer and extends through the U3-R-U5 long terminal repeat just formed, and on through the 18 nucleotides of the tRNA that were initially hybridized to the primer binding site on genome RNA. The short DNA intermediate made is designated "plus-strand strong-stop DNA."

7. *Removal of tRNA and ppt primer.* RNase H digestion of the 3′ end of the tRNA, now part of an RNA-DNA hybrid, removes the tRNA from the minus-strand DNA copy and exposes the primer binding site on the plus-strand strong stop DNA. The ppt primer is also removed.

8. *Second strand transfer.* This exposed primer binding site (PBS) can hybridize with its complementary PBS′ sequence at the other end of the newly synthesized minus strand DNA. This is called the "second strand transfer" since, like the first strand transfer, the plus-strand strong-stop DNA is transferred from one end of the template to the other end. Again, this is drawn as a linear molecule in Figure 25.4 but could involve a circular intermediate, with the plus-strand strong-stop DNA acting as a bridge between the two ends of the minus-strand DNA.

9. *Extension of both DNA strands.* Both the minus and plus DNA strands are then extended by reverse transcriptase to the ends of their respective template strands. This results in a linear, double-stranded DNA with long terminal repeats at both ends. This DNA is called **proviral DNA**.

The reverse transcriptase of retroviruses lacks the 3′-to-5′ exonuclease activity that cellular DNA polymerases use for proofreading. This exonuclease activity is used by cellular DNA polymerases to remove regions that contain mispaired bases, allowing their repair. Because this activity is missing, between 1 and 10 nucleotide errors are introduced during synthesis of each proviral DNA molecule. As a result, retrovirus populations cannot be considered uniform but rather are a collection of variants or **quasispecies**.

The linear double-stranded viral DNA resulting from reverse transcription remains associated with components of the virus core in a **preintegration complex**. The large size of this complex prevents its entry through the nuclear pores, and most retroviruses must await the breakdown of the nuclear envelope that occurs during cell division. For this reason, these retroviruses can only productively infect cells that undergo mitosis. However, lentiviruses code for proteins that mediate the active transport of the viral DNA complex through the nuclear pore (shown in Figure 25.3; see also Chapter 26).

A copy of proviral DNA is integrated into the cellular genome at a random site

Insertion of the proviral DNA into the host cell chromosome is catalyzed by the viral integrase, present in the core of the infecting virion. The integrase binds to the two ends of linear viral DNA and brings them together and in close proximity to cellular DNA. The reaction proceeds as follows (Figure 25.5): (1) The integrase removes the two 3′-terminal nucleotides of each strand of the linear viral DNA. (2) Viral DNA is inserted into host DNA by a concerted cleavage and ligation reaction. The integrase brings the two 3′-OH ends of viral DNA close to two phosphodiester linkages 4–6 nt apart on the host DNA, and joins viral to cellular DNA strands. This leaves a 4–6 nt single-stranded gap on the target host DNA, and a 2 nt unpaired region on the viral DNA. (3) Host enzymes carry out repair synthesis of the gap, simultaneously removing the terminal 2 unpaired nucleotides of the viral DNA. This generates a direct repeat of host DNA (4–6 bp depending on the virus) and results in the loss of the terminal 2 bp of viral DNA. Note that the loss of viral sequences at the ends of viral DNA has no effect on progeny virus because the ends are not used in the synthesis of viral RNA (see below).

Integration sites appear to be distributed randomly over the host genome. Once integrated, the proviral DNA becomes part of a host cell chromosome and is replicated along with host DNA, just like any cellular gene. Consequently, spread of the infection within an animal can be achieved either by infection of new cells with progeny virus or by multiplication of cells already containing proviral DNA. Furthermore, virus infections can be transmitted from parent to offspring if an egg or sperm cell becomes infected and contains integrated proviral DNA. The stable integration of the viral DNA also renders "curing" of a retrovirus infection difficult, requiring the destruction of all cells containing proviral DNA whether or not it is actively expressed.

Sequence elements in the long terminal repeats direct transcription and polyadenylation by host cell enzymes

With integration, the early phase of retrovirus replication is completed. The late phase involves expression of viral RNA, synthesis of viral proteins, and assembly of virions. Proviral DNA can remain unexpressed in the host cell for long periods of time, like any other cellu-

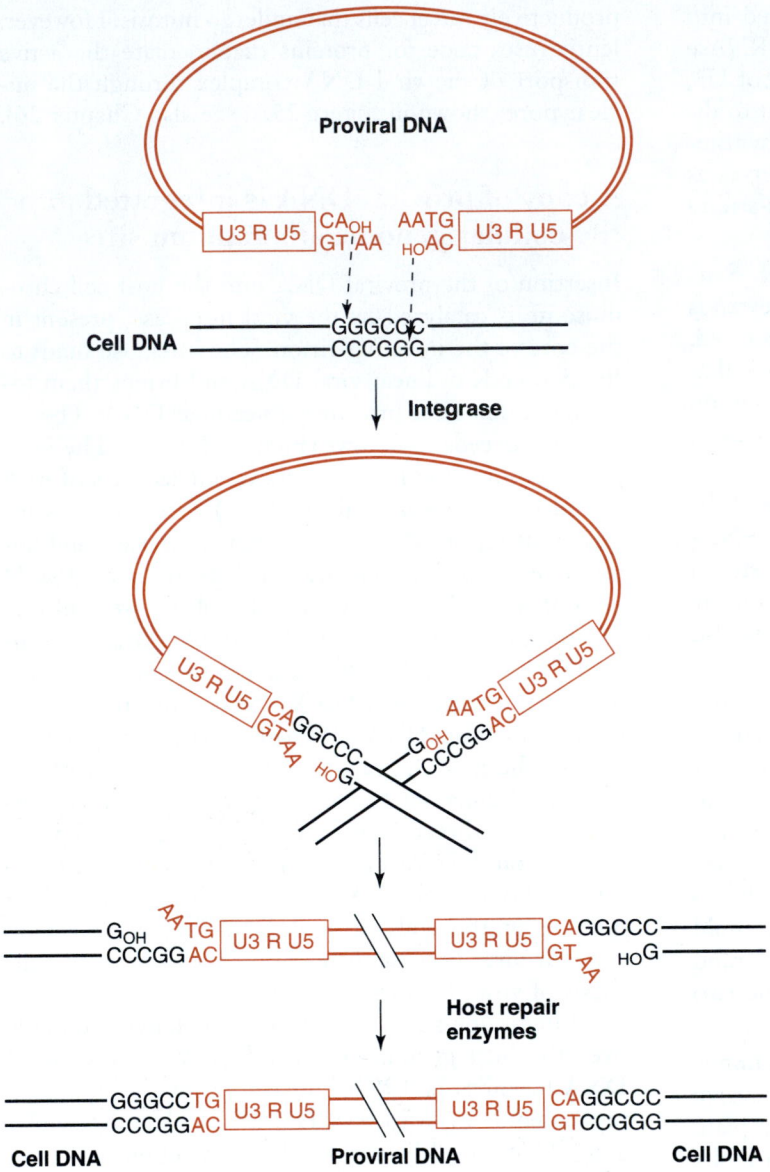

FIGURE 25.5 Integration of proviral DNA into host cell DNA. Viral integrase cleaves 2 nucleotides from the 3′ ends of the viral DNA, then joins these 3′ ends to host DNA in a concerted cleavage-ligation reaction at sites 4–6 nt apart on the host DNA (dashed arrows). Host cell enzymes repair the resulting gapped structure. Proviral DNA is orange, host DNA is black.

lar gene, if the cell does not contain the appropriate set of transcription factors. The U3 region contains transcriptional enhancers that interact with cellular transcription factors and determine in what cell type and to what extent transcription takes place. Transcription of the provirus DNA may also be affected by external stimuli, which can change the mix of active transcription factors present. For example, the U3 sequence of mouse mammary tumor virus contains elements that confer responsiveness to glucocorticoids.

A TATA box just upstream of the U3/R junction directs the initiation of transcription by cellular RNA polymerase II (Figure 25.6). Transcription begins precisely at the junction of the U3 and R regions within the left-hand LTR and proceeds through the entire genome and the right-hand LTR. A polyadenylation signal in the right-hand LTR directs cleavage of the transcript precisely at the R/U5 boundary, and the RNA 3′ end is polyadenylated by host cell enzymes. This gives rise to full-length RNA identical to the genome RNA of the infecting virus.

There are two identical LTRs, one at each end of the proviral DNA; why is only the left-hand LTR used for signaling initiation of transcription? RNA polymerases that initiate transcription within the left-hand LTR and progress through the proviral DNA can dislodge transcription factors bound to the right-hand LTR as they pass through that region, inactivating its ability to initiate transcription. This is called **promoter occlusion**. This concept is supported by the observation that mutation or deletion of the left-hand LTR can lead to initiation of transcription in the right-hand LTR, resulting in transcription of cellular genes downstream of the proviral DNA (see below).

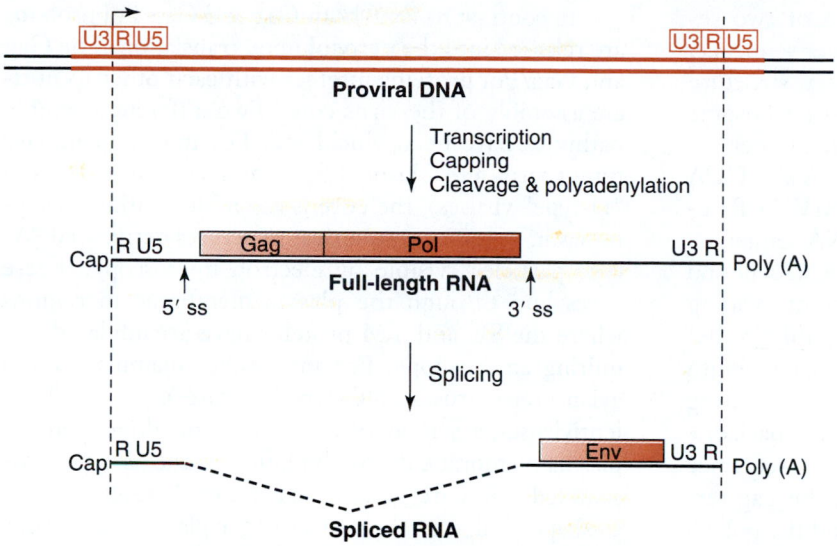

FIGURE 25.6 Production of retrovirus RNAs. Transcription initiates at junction of U3 and R, and RNA is cleaved and polyadenylated at junction of R and U5. Full-length RNA serves as mRNA for Gag and Gag/Pol and is packaged in virions. A fraction of genome-length RNA is spliced, removing the Gag/Pol reading frames, and serves as mRNA for Env.

Similarly, why is only the right-hand LTR used for signaling cleavage and polyadenylation of the transcript? The cellular cleavage and polyadenylation machinery recognizes the highly conserved AAUAAA polyadenylation signal in RNA; some retroviruses position this signal within the U3 region so that it is transcribed only once, from the right-hand LTR. However, the AAUAAA is located within the R region of other retroviruses, and therefore it is transcribed and present at both the 5′ and 3′ ends of viral RNA. In these cases, the U3 region contains sequences that enhance recognition of the polyadenylation signal on viral RNAs, ensuring that cleavage and polyadenylation take place only at the distal end of the RNA.

Differential splicing generates multiple mRNAs

Transcription produces genome-length RNA. How are the numerous viral proteins, some of which are encoded in different reading frames, made from this RNA? This problem is solved in part by splicing of the primary transcript (Figure 25.6). All retroviruses make at least two mRNAs: unspliced RNA is used for synthesis of the Gag and Gag/Pol proteins, and a singly spliced form, from which the Gag/Pol reading frames have been removed, is used for synthesis of the Env proteins. In some retroviruses, more complex splicing patterns generate additional mRNAs. For example, Rous sarcoma virus (see Figure 25.8) has an alternative 3′ splice site downstream of the env gene that permits the expression of the Src protein. For human T-cell leukemia virus (Chapter 27) and members of the lentivirus subfamily (Chapter 26), viral RNA can undergo multiple splicing events to generate mRNAs encoding regulatory proteins.

The Gag/Pol polyprotein is made by suppression of termination and use of alternative reading frames

The unspliced, full-length RNA is used to synthesize Gag and Gag/Pol **polyproteins**, the precursors of the MA, CA, NC, PR, RT, and IN proteins (Figure 25.2). The generation of these two polyproteins from a single mRNA requires modification of the normal translation process. For viruses of the *gamma retrovirus* genus, suppression of translation termination is used. The termination codon UAG separating the gag and pol reading frames is correctly recognized 19 times out of 20, resulting in synthesis of only the Gag polyprotein. However, occasionally a glutamine-tRNA misreads the UAG codon as CAG and inserts glutamine at that site. This allows translational readthrough, permitting the generation of the Gag/Pol polyprotein. The misreading process is stimulated by a particular secondary structure called a **pseudoknot**, located just beyond the termination codon in the mRNA.

Avian retroviruses and human immunodeficiency virus type 1 (HIV-1) employ a different strategy. Not only is there a termination codon between the gag and pol regions, but the two proteins are also in different reading frames. This barrier is overcome by inducing the ribosome to shift its reading frame at a precise position within the RNA prior to the termination codon.

Frameshifting is induced by the presence of two sequence elements within the RNA: a heptamer sequence, where the ribosome stalls, and a secondary structure downstream of the heptamer that induces ribosome stalling (Figure 25.7). Examples of such heptamer sequences are A AAA AAC (HTLV-1), A AAU UUA (Rous sarcoma virus), and U UUU UUA (HIV-1). Base-pairing between the codons on these mRNA sequences and the anticodon sequences of the two tRNAs bound to the ribosome is similar in two different reading frames. The stalled ribosome is able to equilibrate between these two states because of the similar strength of base-pairing interactions. Therefore the translating ribosome occasionally shifts the reading frame back one nucleotide, and resumes translation of the sequence (Figure 25.7). In the new reading frame, the gag termination codon is no longer recognized and the pol region is translated, resulting in the generation of the Gag/Pol protein.

Why have these viruses evolved such a complex scheme for the generation of the Gag/Pol polyprotein? This process ensures the synthesis of Gag and Gag/Pol in a particular ratio, so that the required quantities of each protein necessary for proper virus assembly are present. Only a few molecules of the enzymes reverse transcriptase, protease, and integrase are needed, but many molecules of the structural proteins encoded by Gag are required to form a single virion.

Virions mature into infectious particles after budding from the plasma membrane

The Env protein is translocated directly into the **lumen** of the endoplasmic reticulum as it is being synthesized. It is subsequently transported through the Golgi apparatus and the endosome compartment and finally arrives at the plasma membrane. In the course of these events, the protein undergoes glycosylation and cleavage by host enzymes to generate the mature forms of SU and TM.

In contrast to Env, both Gag and Gag/Pol proteins are released into the **cytosol** upon translation. The Gag and Gag/Pol proteins interact with each other to initiate assembly of the virus core. Two different assembly pathways have been elucidated. For mouse mammary tumor virus and Mason-Pfizer monkey virus ("B-" and "D-type" viruses), the core is assembled within the cytoplasm, generating spherical structures designated "A-type particles" visible by electron microscopy. These cores bud through the plasma membrane in regions where the SU and TM proteins have accumulated, acquiring an envelope. For most other mammalian and avian retroviruses ("C-type" viruses), as well as lentiviruses, assembly of the core occurs directly at the plasma membrane during budding. As the particle is assembled, the core is extruded outward, resulting in a C-shaped bulge on the surface of the plasma membrane, whence the designation "C-type."

The Gag protein is targeted to the plasma membrane by the fatty acid **myristate**, linked post-translationally to its N-terminal amino acid. The NC portion of Gag binds to a **packaging signal** designated psi (ψ), located near the 5′ end of viral genome RNA (Figure 25.2). Selective encapsidation of full-length viral RNA is achieved by positioning the psi sequence downstream of the 5′ splice site, so that only unspliced viral RNA contains the packaging signal. The Gag/Pol polyprotein is incorporated into the assembling virus core by interaction between its CA region and the corresponding region of Gag. Finally, the tRNA primer is incorporated into the assembling core via binding to the RT and NC portions of the Gag/Pol protein.

As virions are assembled and extruded from the cell, the viral protease becomes activated. The protease then cleaves the Gag and Gag/Pol polyproteins into the individual structural (MA, CA, and NC) and enzymatic (PR, RT, and IN) proteins, and they rearrange to form the electron-dense core characteristic of mature virions. It is only at this stage that virions become infectious; therefore, the protease is an important target of antiviral chemotherapy directed against retroviruses (see Chapter 32).

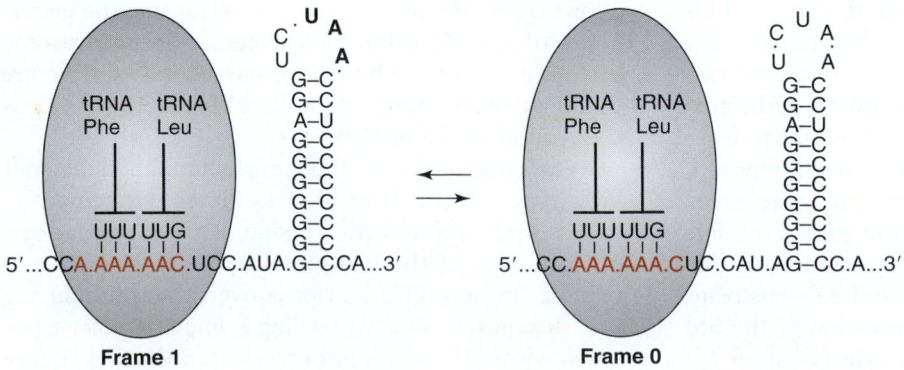

FIGURE 25.7 Ribosome frameshifting. A base-paired RNA stem induces stalling upstream by the translating ribosome (orange ellipse) at the heptamer sequence. While stalled, the ribosome can shift from reading frame 1 to frame 0. Translation in frame 0 does not recognize the UAA termination codon (bold letters).

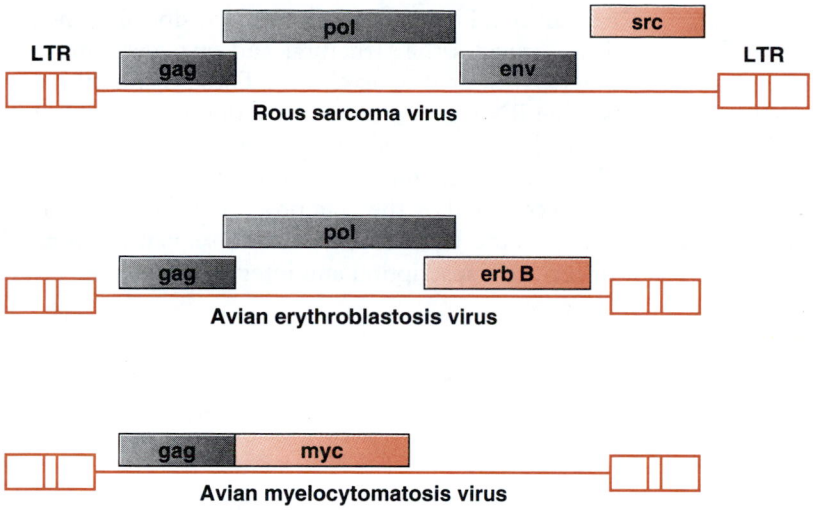

FIGURE 25.8 Genomes of acute transforming retroviruses. Oncogenes are indicated in color. Rous sarcoma virus is replication-competent because it contains all structural genes; however, most transducing retroviruses have deletions of structural genes and often express their oncogenes as fusion proteins.

Acute transforming retroviruses express mutated forms of cellular growth signaling proteins

Study of the mechanisms by which retroviruses transform cells in culture and induce tumors in animals has provided important insights into the causes of cancer. There are two classes of oncogenic retroviruses: the *acute transforming* viruses, which transform cultured cells rapidly and induce cancer within a short time after inoculation; and the *nontransforming* viruses, which induce cancer only after a long latency period.

Tumor formation by acute transforming retroviruses is directed by proteins that are not required for virus replication. An example is one of the first retroviruses discovered, Rous sarcoma virus (Figure 25.8). A viral gene was identified by mapping mutations that inactivated tumor formation by Rous sarcoma virus in chickens, but that did not affect virus replication. This gene was designated src, for sarcoma. Src was one of the first oncogenes identified, and its presence in a virus lent credence to the theory that certain cancers were caused by oncogenes of viral origin. However, the viral src gene was subsequently shown to be nearly identical to a cellular gene, called c-src, that codes for a protein tyrosine kinase involved in cellular signaling pathways. The previous assumption that src was an oncogene of viral origin was proved incorrect. It became clear that Rous sarcoma virus had acquired a mutated copy of the c-src gene, probably by integrating its proviral DNA nearby the c-src gene and incorporating that gene into the viral genome during subsequent transcription from the LTR.

Numerous other transforming retroviruses contain a variety of oncogenes that are derived from cellular genes. The cellular genes (**proto-oncogenes**) code for growth factors, receptor proteins, protein kinases, or transcription factors involved in pathways that regulate cell growth and division. Comparison of viral oncogenes with their cellular counterparts revealed that the viral forms have mutations or deletions that alter the activity of the proteins. Consequently, cells expressing these viral forms of the protein behave as if they are receiving a growth stimulus independent of the external environment of the cell, and show uncontrolled growth and division.

Rous sarcoma virus turns out to be a unique example of a transforming retrovirus that is both tumor-inducing and able to replicate on its own. Most transforming retroviruses that have picked up cellular genes

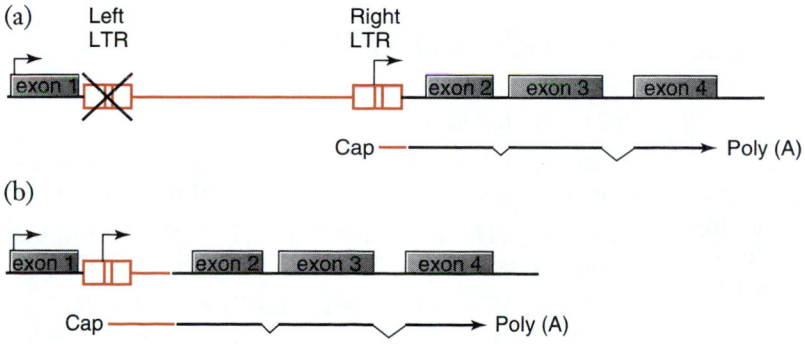

FIGURE 25.9 Promoter insertion near proto-oncogenes can lead to cell transformation and tumor formation. Transcription of the proto-oncogene (exons shown as filled boxes) initiates within the viral LTR, either as a result of inactivation of the left-hand LTR (a) or deletion of the right-hand LTR (b). Result is a mixed viral/cellular mRNA that expresses the proto-oncogene under control of viral promoter/enhancer.

have also lost large segments of viral genes by deletion, rendering them unable to replicate on their own; they depend on coinfection with a replication-competent retrovirus. Two examples of such deleted viral genomes are shown in Figure 25.8. The nontransforming "helper virus" provides the structural and enzymatic proteins necessary for producing virions. These virions incorporate the RNA genomes of either the helper virus itself or the defective transforming virus. The defective viral RNA must at the minimum contain a packaging signal (psi) recognized by the Gag protein of the helper virus, as well as the U3, R, U5, and PBS sequences required for reverse transcription and integration into the cellular genome.

Retrovirus-based gene therapy

The ability of retroviruses to stably introduce foreign genes into cells suggested that they could be used for gene therapy by reintroducing into the cell valid copies of missing or malfunctioning genes. However, effective therapy requires minimizing risks associated with virus infection, such as tumor induction. Defective gene therapy vectors (see Chapter 33) retain only the 5′ and 3′ end regions, the primer-binding site, and the packaging signal, as these are necessary for reverse transcription of the RNA, integration of the proviral DNA into the target cell, and packaging of genome RNA in virions. The removal of the viral structural genes (gag, pol, and env) ensures that the virus cannot replicate, and provides room for insertion of the desired gene.

The structural proteins required for virus assembly are supplied by "packaging" cell lines. These cells express the gag, pol, and env genes on mRNAs that lack the RNA packaging signals and therefore cannot be incorporated into virus. Vector RNA can be expressed in these cells and packaged in virions. Target cells are then infected with the recombinant virus. To permit selection of cells that integrate provirus DNA, some vectors incorporate genes that confer resistance to toxic drugs such as G418 (neomycin), hygromycin B, or puromycin. Such vectors are designed with two promoters; one gene is driven by the 5′ LTR, while the other uses an internal promoter. Selection of cells for drug resistance is done *in vitro*, and these cells are then reintroduced into the patient.

Several problems limit the utility of this approach for gene therapy. Transgene expression can be variable and unstable in the absence of drug selection. Integration of the proviral DNA only takes place in cells undergoing mitosis, rare among terminally differentiated cells. Recent developments may provide a solution to this problem. Packaging cell lines that express lentivirus proteins have been developed; lentiviruses can infect cells that are not undergoing division. Another limitation is the size of the gene which can be expressed; only RNAs less than approximately 10 Kb can be packaged. Consequently, vectors containing selection genes and the gene of interest must fall within this size limit.

Despite these restrictions, retroviral gene transduction has been used successfully in a limited number of situations. The most common use is to mark cells that are being introduced into patients; this does not require expression of the gene, because integrated proviral DNA can be detected by PCR or other methods. Marking of cells with retroviruses has been used in cancer therapies that destroy both tumor and stem cells and replace them with transplanted bone marrow. Bone marrow-derived cells are infected with a defective retrovirus prior to reintroduction into the patient. If any tumor containing the viral sequence develops, the cells responsible for the tumor must have been present within the bone marrow used. Such procedures allow for the refinement of treatment strategies to maximize the benefit and minimize the risk to the patient.

Retroviruses lacking oncogenes can transform cells by insertion of proviral DNA near a proto-oncogene

In contrast to acute transforming viruses, nontransforming retroviruses lack oncogenes. However, the replication of a retrovirus can result in mutagenesis of the host cell because it involves integration of proviral DNA into the host genome. Since integration takes place at random sites within the cellular genome, it can occasionally disrupt crucial genes or insert a strong promoter and enhancer near certain genes, changing their normal pattern of regulation. Should insertion occur adjacent to or within a cellular proto-oncogene, the altered pattern of expression of the gene can lead to tumor formation if the infected cell is not killed by the resident retrovirus.

The most common mechanism of insertional activation of cellular proto-oncogenes is called "promoter insertion." A viral LTR is inserted just upstream of a cellular gene, and transcription that initiates within the LTR progresses through the adjacent cellular gene. This can happen if the left-hand LTR is inactivated by mutation or deletion, allowing the right-hand LTR to initiate transcription (Figure 25.9a). Alternatively, the integrated proviral DNA may undergo a deletion that removes most of the genome, leaving the left-hand LTR intact and adjacent to a cellular proto-oncogene (Figure 25.9b). In both cases, transcription of the adjacent cellular gene is now regulated by the viral LTR rather than its own control elements, and overexpression of the cellular gene may result in inappropriate cell growth and division.

SUGGESTED READING

1. Blankenstein, T. (1999). Gene therapy: Principles and applications. Birkhauser Verlag, Basel, Switzerland.

2. Blesch, A. (2004). Lentiviral and MLV based retroviral vectors for ex vivo and in vivo gene transfer. *Methods* **33,** 164–172.

3. Coffin. J., Hughes, S., and Varmus, H. (1997). Retroviruses. Cold Spring Harbor Laboratory Press, Plainview, NY.

4. Gotte, M. and Wainberg, M.A. (1999). HIV-1 reverse transcriptase: A brief overview focused on structure-function

relationships among molecules involved in initiation of the reaction. *Archives of Biochemistry & Biophysics* **365,** 199–210.

5. Huber, B. and Magrath, I. (1998). Gene therapy in the treatment of cancer: Progress and prospects. Cambridge University Press, Cambridge, UK.

6. Thomas, M. and Brady, L. (1997). HIV Integrase: A target for AIDS therapeutics. *Trends in Biotechnology* **15,** 167–172

KEY TERMS

Chemokine
Cytosol
Lumen
Myristate
Oncogene
Polyprotein
Preintegration complex
Promoter occlusion
Proto-oncogene
Proviral DNA
Pseudoknot
Quasispecies
Reverse transcriptase
Ribonuclease H

Human Immunodefiency Virus Type 1

Lentivirus

From Latin *lentis* (slow), for slow progression of disease

VIRION

Spherical enveloped particle.
Diameter 100 nm.
Conical capsid.

GENOME

Linear ss RNA, positive sense.
Two identical segments, each 9.3 Kb.
Cellular $tRNA^{lys3}$ molecules packaged in virions used as primers for reverse transcription.

GENES AND PROTEINS

Four capsid proteins: MA (p17), CA (p24), NC (p7), p6.
Three enzymes: PR (p10), RT (p51/66), IN (p32).
Two envelope proteins: SU (gp120) and TM (gp41).
Six regulatory proteins: Vif, Vpu, Vpr, Tat, Rev, Nef.

VIRUSES AND HOSTS

Human immunodeficiency virus types 1 and 2 (HIV-1, HIV-2).
Simian immunodeficiency virus.
Equine, bovine, feline immunodeficiency viruses.

DISEASES

Acquired immune deficiency syndrome (AIDS) first described in 1981.
A major global pandemic today (more than 30 million people infected).
HIV replicates in and kills lymphocytes and macrophages.
Results in depletion of CD4+ T cells to render host immune-incompetent.
As a result, opportunistic infections by other pathogens are often fatal.
HIV is transmitted through sexual contact and blood exchange.

DISTINCTIVE CHARACTERISTICS

Proviral DNA can enter nucleus without requirement for cell division.
Lentiviruses make a complex set of singly and doubly spliced mRNAs.
Six regulatory proteins control virus production and pathogenesis.

Human immunodeficiency virus type 1 (HIV-1) and acquired immunodeficiency syndrome

In 1981, the term *acquired immunodeficiency syndrome (AIDS)* was coined to describe a condition in a group of previously healthly young males within the Los Angeles/ San Francisco area who displayed a marked depletion of their immune **CD4-positive T lymphocytes**, rendering them immune-incompetent. As a consequence, they suffered from a number of **opportunistic infections** (the most prevalent being pneumonia) that were often fatal. Subsequent epidemiological studies suggested that the syndrome was due to a transmissible agent that was acquired through sexual contact or blood exchange; hemophiliacs, recipients of blood transfusions, and intravenous drug users were also affected.

In 1983, a retrovirus was isolated from the blood of individuals with AIDS by groups led by Luc Montagnier in Paris and Robert Gallo in Maryland. This virus, subsequently named human immunodeficiency virus type 1 (HIV-1), was demonstrated (despite much controversy and debate) to be the causative agent of AIDS. HIV-1 defined a new subfamily of retroviruses (see Chapter 25), the lentiviruses (Table 26.1), so-named because they often cause disease slowly, up to many years after primary infection.

Infection by HIV-1 is a major global pandemic that affects males and females equally. The World Health Organization estimated in 2005 that more than 44 million people were infected worldwide, most of these being inhabitants of Africa and South-East Asia. In the absence of effective vaccines and affordable drug treatments, the number of infected people continues to grow as the pandemic expands into other developing countries. Despite intensive information programs and the availability of treatment, the number of infected people also continues to increase in many developed countries, where AIDS can be a major cause of mortality in people between the ages of 16 and 40. In response to this health crisis considerable resources have been committed to understanding HIV-1 biology and pathology, rendering it perhaps the most intensively studied disease agent of our time.

TABLE 26.1 Lentiviruses

Virus	Host
Human immunodeficiency virus type 1	Humans
Human immunodeficiency virus type 2	Humans
Simian immunodeficiency virus	Monkeys
Feline immunodeficiency virus	Cats
Equine infectious anemia virus	Horses
Caprine arthritis-encephalitis virus	Goats
Visna-maedi virus	Sheep

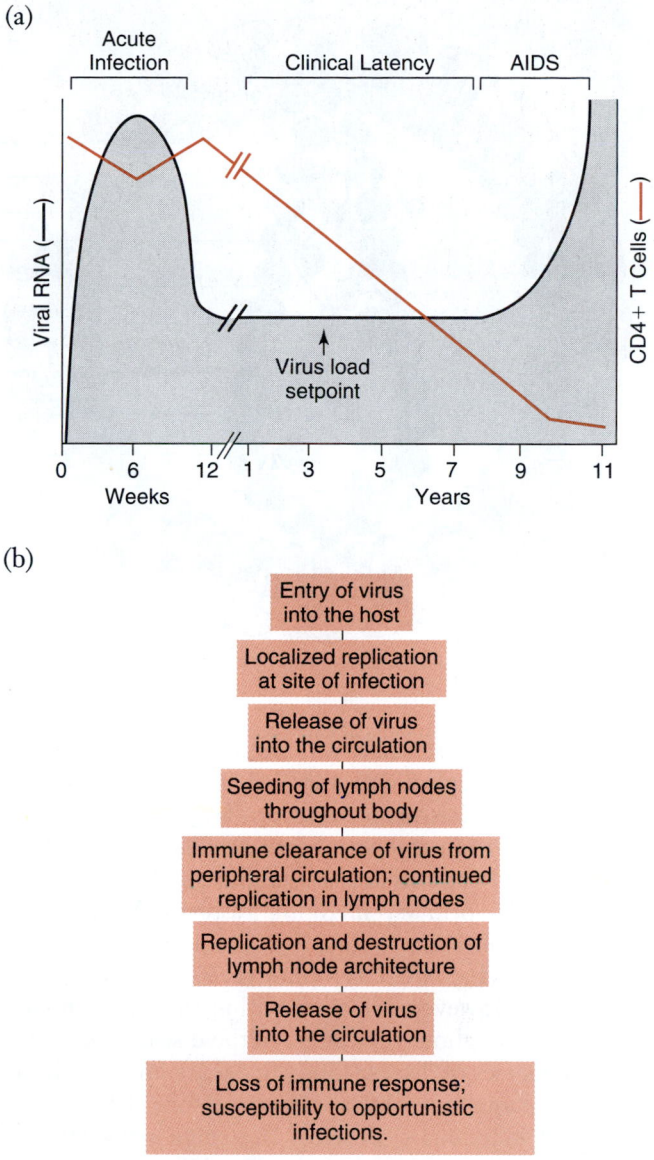

FIGURE 26.1 Events associated with progession to AIDS. (a) Changes in viral RNA and CD4-positive T cell levels in the peripheral circulation from acute infection to end-stage disease. (b) Stages of HIV-1 infection.

HIV-1 infection leads to a progressive loss of cellular immunity and increased susceptibility to opportunistic infections

The course of HIV-1 infection can be roughly divided into three phases: acute infection; a subsequent clinically latent phase of variable duration; and finally AIDS (Figure 26.1). Two to six weeks following exposure to the virus, individuals can develop a mononucleosis or influenza-like syndrome (fever, malaise,

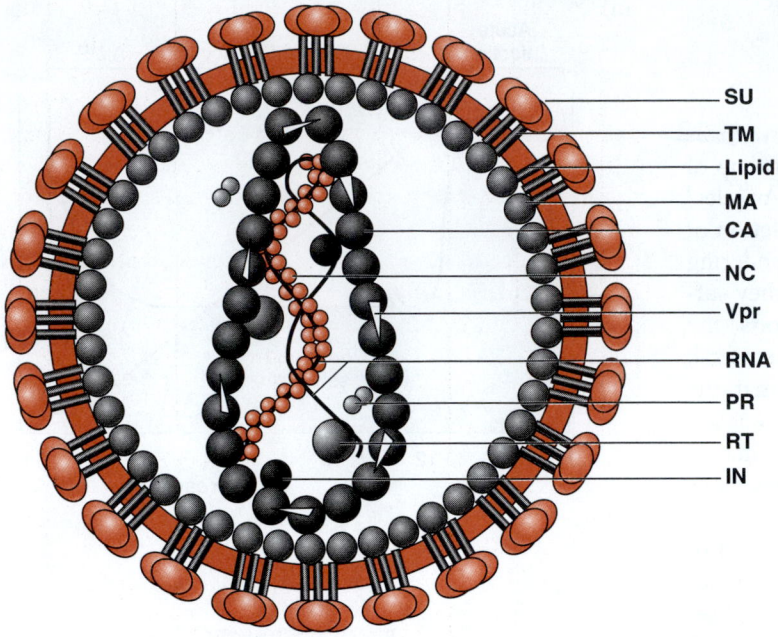

FIGURE 26.2 Diagram of structure of HIV-1 virion. Components of the virion are labeled; see Table 26.2. For the sake of clarity, only one of the two genome RNA molecules is shown covered by NC protein.

lethargy, nausea, diarrhea, headaches, stiff neck, swelling of lymph nodes) that only in a minority of individuals requires hospitalization. High levels of virus are found within the peripheral circulation and infection of the lymph nodes is established. Within 1 to 2 months, the primary infection resolves. Virus titers drop as the immune system (see Chapter 31) responds with both **cytotoxic T lymphocytes** and humoral **antibodies.** However, infection within the lymph nodes persists throughout the course of the disease. Four different courses of disease following acute infection have been documented: (1) rapid progressors (10–15% of infected individuals) develop late stage symptoms in 2–3 years; (2) slow progressors (70–80%) develop late stage symptoms in 8–10 years; (3) a small percentage of individuals designated long-term survivors end up with low CD4-positive "helper" T lymphocyte levels but do not develop late stage complications; (4) long-term nonprogressors (5%) show no decline in CD4-positive T cell levels.

Several factors help predict clinical outcome. After resolution of the acute infection, varying levels of viral RNA genomes ("virus load setpoint," Figure 26.1) can be detected in the blood of different individuals. The higher the value of this basal level of RNA, the more rapidly patients progress to full-blown AIDS. Also important is the nature of the virus itself. For example, patients infected with a strain of virus that lacks the viral nef gene progress to AIDS very slowly if ever. The nature of the immune response is also crucial. Patients who develop a predominantly cytotoxic T cell-based immune response have a better chance of long-term survival. The diversity of the **epitopes** recognized by the immune system also appears to play a role; recog-

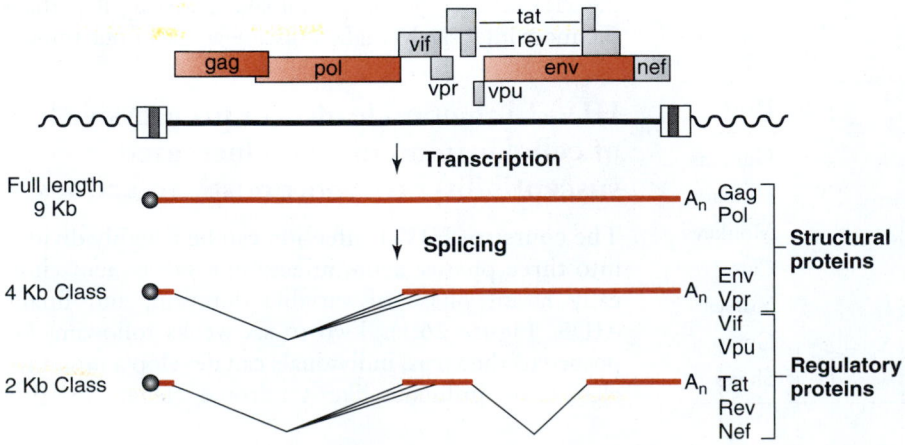

FIGURE 26.3 Genome structure and RNA splicing pattern of HIV-1. Above: a map of the integrated proviral DNA framed by two long terminal repeats, and the reading frames for viral proteins. Below: the pathway by which the primary transcript is converted into more than 25 different mRNAs in three size classes by alternative splicing. Right: viral proteins produced from the different classes of viral mRNA.

TABLE 26.2 HIV-1 structural proteins

Name	Abbreviation	M. Wt.
Matrix	MA	p17
Capsid	CA	p24
Nucleocapsid	NC	p7
Protease	PR	p14
Reverse transcriptase	RT	p66/51
Integrase	IN	p32
Surface protein	SU	gp120
Transmembrane protein	TM	gp41
Virion protein R	Vpr	p15

nition of a limited number of epitopes is associated with a poor prognosis.

Over the course of clinical latency, virus replication persists in the lymph nodes, resulting in a gradual depletion in the level of circulating CD4-positive T cells and destruction of the lymph node architecture. Individuals progressing toward end-stage disease have high levels of virus in the blood, indicative of the failure of the immune system to contain the infection. Patients exhibit chronic fever, night sweats, diarrhea, a number of infections such as cytomegalovirus, pneumonia, oral thrush, herpes simplex, neoplasms such as Kaposi's sarcoma and lymphomas, and neurological syndromes including dementia and neuromuscular disorders. Neurological symptoms correlate with virus replication in the central nervous system. While current therapies delay or prevent disease progression, they do not eradicate the infection. Consequently, withdrawal from therapy results in reemergence of the virus and continued disease progression.

HIV-1 is a complex retrovirus

The basic replication pattern of lentiviruses is identical to that of other retroviruses. However, sequence analysis of the genome of HIV-1 revealed it to be considerably more complex than many other retroviruses. In addition to the standard gag, pol, and env genes, six additional reading frames were identified: vif, vpr, vpu, tat, rev, and nef (Figures 26.2 and 26.3, Tables 26.2 and 26.3). Since all of these additional reading frames are conserved among various clinical isolates of the virus, they must play important roles in the virus life cycle. The following discussion of the replication of HIV-1 will emphasize the unique features of this virus, including the roles of these additional viral proteins.

In contrast to simpler retroviruses, which make only two mRNAs (unspliced and singly spliced), splicing of the HIV-1 primary transcript generates more than 25 mRNAs that fall into three size classes (Figure 26.3): (1) the unspliced 9 Kb full-length RNA, used to produce Gag and Gag-Pol proteins; (2) the singly spliced 4 Kb class of RNAs, which encode Vif, Vpr, Vpu, and Env; and (3) the doubly spliced 2 Kb class of RNAs, which encode Tat, Rev, and Nef. In each class of spliced RNA there are multiple species, generated by the presence of several different 3′ and 5′ splice sites.

HIV-1 targets cells of the immune system by recognizing CD4 antigen and chemokine receptors

The initial interaction of HIV-1 virions with cells is mediated by the **CD4 antigen**, found on the surface of T lymphocytes. Thus HIV-1 attacks at the very heart of the immune system, as CD4-positive T cells are vital for development of both humoral antibody and cell-mediated immunity (see Chapter 31). However, the presence of CD4 alone on the cell surface is not sufficient for infection; for example, mouse cells engineered to express human CD4 antigen cannot be infected by HIV-1. In addition, HIV-1 strains display differences in their ability to infect T cells or monocyte/**macrophage** cells even though both cell types express CD4. This resulted in viruses being classified as "T cell-tropic" or "macrophage-tropic." These strain differences were found to depend on variations in the SU protein (gp 120), particularly in a region that undergoes a high rate of evolution, designated variable domain 3. The **chemokine** receptors CCR5 and CXCR4 were then shown to be obligatory co-receptors for the virus.

TABLE 26.3 HIV-1 nonstructural proteins

Name	Abbreviation	M. Wt.
Viral infectivity factor	Vif	p23
Virion protein unique to HIV-1	Vpu	p16
Transactivator of transcription	Tat	p15
Regulator of expression of virion proteins	Rev	p19
Negative effector	Nef	p27

Chemokines and their receptors are normally involved in regulating trafficking of cells of the immune system to sites of infection. Cell tropism of HIV-1 strains depends on the ability of gp120 to interact with these two receptors; CXCR4 is expressed on T cells, and CCR5 is expressed on macrophage/monocyte cells.

Viruses that can infect cells expressing only chemokine receptors and not CD4 have been isolated. Binding to CD4 is therefore not obligatory, but may serve either to enhance interaction of gp120 with the chemokine receptor or to induce a conformational change in gp120 that exposes determinants for interaction with the chemokine receptors. Once interaction with the chemokine receptors occurs, the envelope protein TM (gp41) is triggered to undergo a conformational change that induces the fusion of the viral envelope with the plasma membrane (Figure 26.4).

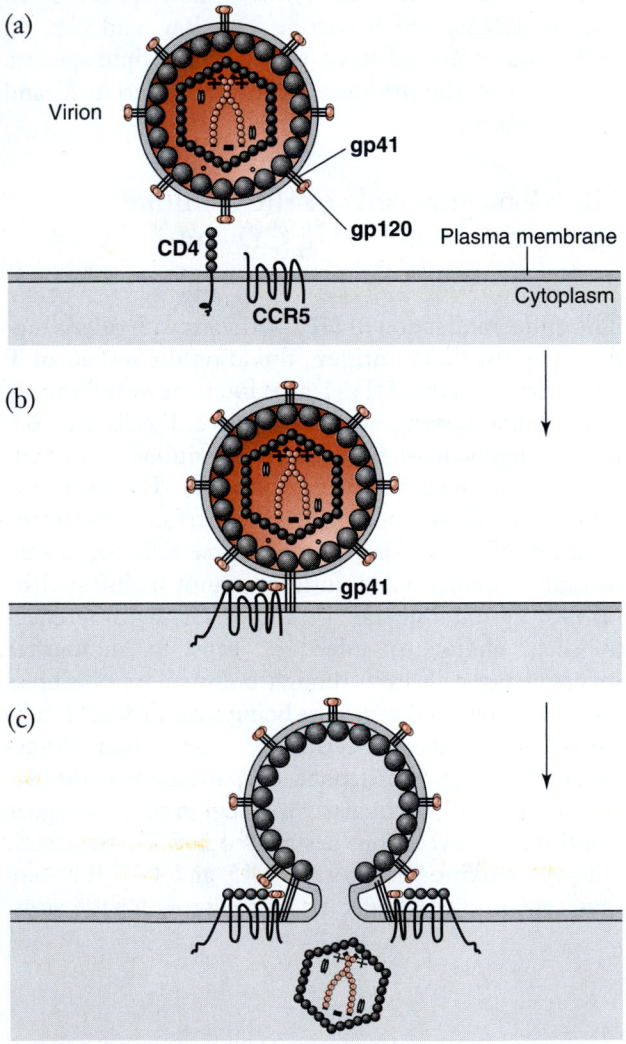

FIGURE 26.4 **Model of HIV-1 entry.** (a) gp120 on the virion binds to CD4 on the cell surface. (b) gp120 undergoes a conformational change that exposes regions that bind to chemokine receptors (CCR5 or CXCR4). Changes in gp120 conformation expose the fusion domain of gp41, resulting in its insertion into the cell membrane. (c) The close proximity of the viral and cellular membranes results in their fusion, releasing the virion core into the cytoplasm.

Virus mutants arise rapidly because of errors generated during reverse transcription

The error rate associated with reverse transcription (1–10 mutations per proviral DNA molecule synthesized) renders HIV-1 infections a moving target for both the immune system and treatment strategies. Approximately 2×10^9 virus particles are made and destroyed every day in an untreated infected individual; only a small fraction of these particles successfully integrates proviral DNA into infected cells, but this still translates into approximately 10^4 to 10^5 virus variants being generated every day. Thus HIV-1 has the ability to evolve rapidly to overcome barriers to its replication, be they an immune response or drug treatment. Within six months of initiating treatment with reverse transcription inhibitors such as **azidothymidine (AZT)**, dideoxyinosine (ddI), or dideoxycytosine (ddC) (see Chapter 32), viruses displaying significant resistance to the drug were isolated.

Current therapy calls for the use of multiple drugs affecting several processes in an attempt to generate a higher barrier for the virus to overcome. Although this approach has met with great success, the emergence of viruses resistant to the combination of three drugs suggests that even this barrier may not be sufficient to control the infection. The same issues apply to vaccine design. Immunization with one or a small group of viral antigens will generate an immune response to only a limited number of the possible variants of the virus, something that the virus can easily overcome.

Unlike other retroviruses, HIV-1 directs transport of proviral DNA into the cell nucleus

Most retroviruses cannot productively infect nondividing cells because the **preintegration complex**, containing proviral DNA, is unable to enter the intact nucleus (see Chapter 25). Integration of proviral DNA must therefore await the disintegration of the nuclear membrane when the cell divides. In contrast, HIV-1 is able to transport the preintegration complex into the nucleus via the nuclear pores. Two virus-encoded proteins in the preintegration complex, MA and Vpr, facilitate this transport. Therefore HIV-1 can infect cells that are not actively dividing. This probably affects virus spread within infected individuals, and is of interest in the construction of gene therapy vectors targeted to terminally differentiated cells (see Chapter 33).

However, in resting T cells the preintegration complex remains in the cytoplasm. Only after T-cell activation can the preintegration complex be transported into the nucleus and viral DNA be integrated. This situation presents a problem for the immune system. To generate an immune response, T cells are activated by presentation of antigen fragments during contact with **dendritic cells** within lymph nodes. However, dendritic cells collect HIV-1 virions on their surface. Consequently, T cells that would support an immune response against HIV-1 are exposed to virus while interacting with dendritic cells; these conditions enhance the ability of the virus to replicate by facilitating entry of proviral DNA into the nucleus, and eventually lead to death of the T cells.

Latent infection complicates the elimination of HIV-1

After integration of the HIV-1 proviral DNA into the host cell genome, there are two possible outcomes: **latent infection**, in which viral RNA is not expressed, or active infection, leading to production of progeny virus. Up to 1% of the cells within the lymph nodes and circulation in recently infected individuals contain integrated HIV-1 provirus, but only about 1 in 100 of those infected cells actively express viral RNA. Although latently infected cells do not directly contribute to disease, they do represent a significant hurdle to treatment and virus eradication. Because they do not express viral proteins, these cells cannot be readily distinguished from uninfected cells by the immune system. Removal of patients from multiple antiviral drug therapy invariably results in reemergence of the infection. The virus that appears is not drug-resistant, and was probably carried in an unreplicated form in latently infected cells. Many of these cells are memory T cells that have a very long lifetime in the body. Therefore, unless latently infected cells can be activated in some fashion to render them recognizable to the immune system, treatment for HIV-1 must be continued for life if the disease is to be held in check.

Viral latency is regulated by transcriptional control elements present in the HIV-1 LTR. Binding sites for transcription factors **NFκB** and NFAT, whose activities are sensitive to external stimuli, are present in the U3 region of the LTR. NFκB is held in the cytoplasm in an inactive state by association with the inhibitor IκB (see Figure 27.4). A variety of stimuli induce the degradation of IκB, resulting in the transport of NFκB into the nucleus where it can activate transcription. One such stimulus is activation of T cells via the **T-cell receptor**; others are virus infections or treatment with various cytokines. The U3 region contains binding sites for other cellular transcription factors such as Sp1, LEF, Ets-1, and USF, which also play a role in modulating HIV-1 transcription.

The Tat protein increases HIV-1 transcription by stimulating elongation by RNA polymerase II

Tat is a highly basic 86–amino acid protein produced by doubly spliced HIV-1 mRNA; it is localized in the cell nucleus. Expression of this protein dramatically increases the amount of viral RNA produced in infected cells, thus its name, "*trans*activator of *t*ranscription," abbreviated to "Tat." Key to the function of Tat is the presence of a sequence element located just beyond the HIV-1 transcription start site designated the *Tat*-*r*esponsive element (TAR) (Figure 26.5). Tat binds to the RNA sequence on nascent (growing) RNAs corresponding to TAR, but it does not bind to the same sequence on proviral DNA. This was the first demonstration that transcription by RNA polymerase II could be modulated by an RNA element on the nascent transcript. TAR forms a stem-loop structure in viral RNA, and two regions are essential for its function: (1) a bulge in the stem, which forms the recognition element for Tat binding; and (2) the nucleotide sequence within the

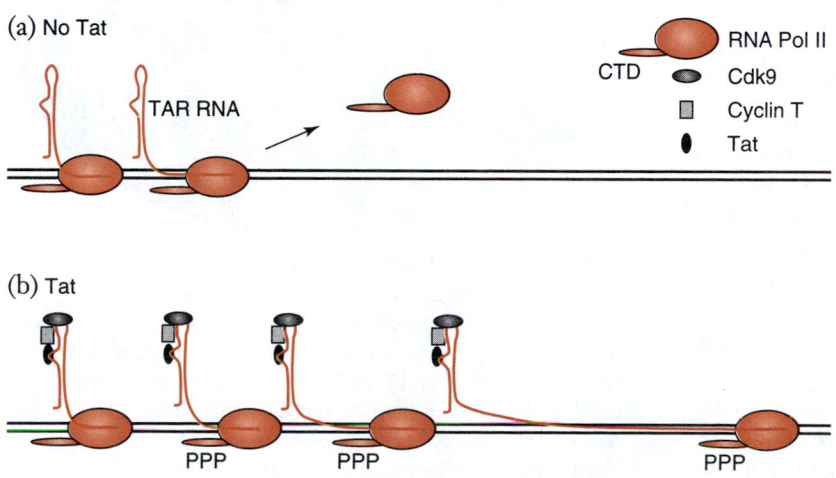

FIGURE 26.5 Mechanism of Tat function. (a) In the absence of Tat, RNA polymerase II molecules that initiate transcription at the HIV-1 promoter lack processivity and most are released from the template before reaching the polyadenylation site. (b) Tat recruits cyclin T and cyclin-dependent kinase-9 (Cdk9) to the transcription complex shortly after initiation, resulting in hyperphosphorylation (P) of the carboxy-terminal domain of RNA polymerase II. This increases processivity of the polymerase and generates increased levels of full-length transcripts. CTD: carboxy terminal domain of RNA polymerase II.

loop. Tat increases viral RNA abundance not by increasing the rate of initiation of transcription by RNA polymerase II but rather by increasing the elongation efficiency of RNA polymerase molecules (Figure 26.5).

Elongation by RNA polymerase II is stimulated by increasing the extent of phosphorylation of the carboxy terminal domain (CTD) of RNA polymerase II. This domain contains 52 repeats of a seven–amino acid sequence that is rich in serines and threonines, amino acids that can be phosphorylated by specialized protein kinases. The form of RNA polymerase II that participates in initiation contains few phosphate groups within its carboxy-terminal domain. Movement of the polymerase away from the point of initiation is facilitated by phosphorylation of the carboxy-terminal domain.

Tat increases phosphorylation of the carboxy-terminal domain by recruitment of a protein kinase to the transcription complex (Figure 26.5). Two kinases have been implicated in this process: **cyclin-dependent kinase (cdk)-7/cyclin** H and cyclin-dependent kinase-9/cyclin T. The cdk 7/cyclin H complex (not shown in Figure 26.5) is a part of the transcription initiation factor TFIIH, and binds to the domain of Tat involved in transactivation. The cdk 9/cyclin T complex interacts with the same domain of Tat but also binds to the loop sequences within TAR and stabilizes the interaction of Tat with TAR. These observations led to the suggestion that Tat acts via a two-stage mechanism: (1) Tat initially activates the cdk7/cyclin H kinase complex within TFIIH to increase phosphorylation of the CTD as the polymerase clears the promoter; (2) recruitment of cdk9/cyclin T further increases the phosphorylation of the CTD and the elongation efficiency of RNA polymerase II.

The Rev protein mediates cytoplasmic transport of viral mRNAs that code for HIV-1 structural proteins

Mutations in the rev gene of HIV-1 result in the loss of synthesis of all proteins encoded by the 9 Kb (unspliced) and 4 Kb (singly spliced) classes of viral RNA; therefore this gene was named *r*egulator of *e*xpression of *v*irion proteins. In the absence of Rev, these mRNAs are retained in the nucleus and therefore are not translated, because ribosomes are confined to the cytoplasm. However, export and translation of the 2 Kb (doubly spliced) class of viral mRNAs is not dependent on Rev. Expression of Rev was found to induce accumulation of the 9 Kb and 4 Kb viral RNAs in the cytoplasm, demonstrating that Rev is required for transport of the unspliced and singly spliced viral RNAs from the nucleus to the cytoplasm.

Regulation of mRNA transport by Rev requires two classes of sequences on viral RNA. (1) *Cis*-acting repressive sequences (CRS) present in the gag, pol, and env regions inhibit the transport of RNAs to the cytoplasm. (2) A 240 nt sequence within env, termed the *Rev response element (RRE)*, is a target for Rev binding, which is required for Rev action (Figure 26.6). The 110–amino acid Rev protein contains both a nuclear localization signal and a nuclear export signal, indicating that this protein can shuttle continuously between the nucleus and cytoplasm. Transport to the cytoplasm involves binding of the nuclear export signal on Rev to the cellular protein exportin 1, which mediates the docking of Rev to the nuclear pore. If Rev is simultaneously bound to an mRNA via its RRE, the export of Rev results in the export of the mRNA.

Messenger RNAs that contain *cis*-acting repressive sequences cannot leave the nucleus unless they contain an RRE, in which case Rev supplies a pathway for their export. Since the doubly spliced 2 Kb RNAs (and most cellular mRNAs) do not contain a CRS, their export does not require Rev. Once HIV-1 mRNAs are in the cytoplasm, Rev can remain bound and can enhance their translation by a mechanism that is not yet understood. Transport of Rev back into the nucleus requires dissociation of the Rev/RNA complex. Subsequently, Rev binds to **importin** β, which docks Rev to the cytoplasmic face of the nuclear pore

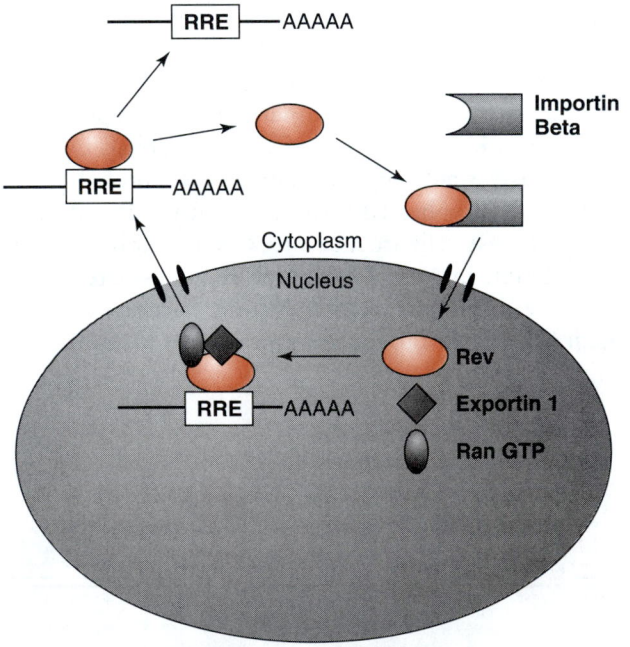

FIGURE 26.6 **Mechanism of Rev function.** Rev is transported into the nucleus via nuclear pores by direct interaction with importin beta. Within the nucleus, Rev binds to the Rev-responsive element (RRE) in viral RNAs, and recruits exportin 1 and Ran GTP to generate a complex capable of exporting the RNA to the cytoplasm. The complex dissociates in the cytoplasm, releasing the mRNA and allowing Rev to return to the nucleus and repeat the cycle.

(Figure 26.6). Rev is returned to the nucleus, where the cycle is repeated.

Together, the Tat and Rev proteins strongly upregulate viral protein expression

Tat and Rev are essential for virus replication because they regulate HIV-1 transcription and transport of mRNAs that code for viral structural proteins. Their action results in the expression of viral proteins in two stages. Following integration of proviral DNA and its transcription at a basal level, only the doubly spliced 2 Kb RNAs (Figure 26.3) are transported to the cytoplasm. This permits the synthesis of Tat, Rev, and Nef. Both Tat and Rev are then transported to the nucleus where they act to augment transcription of provirus DNA (Tat) and the transport of all viral mRNAs to the cytoplasm (Rev). Expression of proteins encoded by the 9 Kb and 4 Kb classes of mRNAs (Gag, Gag/pol, Env, Vif, Vpr, and Vpu) can then take place.

Mutational studies of viral genes determined that vif, vpr, vpu, and nef are not essential for production of infectious HIV-1 in many cell culture systems *in vitro*. However, the conservation of the reading frames for these proteins in most clinical isolates of the virus suggests that they play important roles during infection within the host. These roles are described below.

The Vif protein increases virion infectivity by counteracting a cellular deoxcytidine deaminase

Vif (*v*iral *i*nfectivity *f*actor) is a 193–amino acid protein found in the cytoplasm of infected cells and present at low levels in virions. Its deletion reduces infectivity of HIV-1 both in cell cultures and in animals. Vif acts by binding to a cellular protein, CEM15/APOBEC3G, and inducing **ubiquitination** and degradation of this protein by **proteasomes**. CEM15 is a member of a family of cellular proteins that deaminate cytidine or deoxycytidine residues (in RNA or DNA) to either uridine or deoxyuridine. The original member of this family, APOBEC, was so-named because it deaminates a specific cytidine residue in the mRNA coding for *Apo*lipoprotein *B*, allowing two different proteins to be made by either the native or the edited messenger RNA.

CEM15 is incorporated into HIV-1 virions during their assembly in infected cells. CEM15 within virions can induce deamination of multiple deoxycytidine residues in the DNA product of reverse transcription made during a subsequent infection. This leads to mutations in viral structural and regulatory proteins, with a corresponding reduction in infectivity. CEM15 therefore functions to defend the cell against infection by mutating the incoming viral genome, or in this case, the DNA copy of the HIV-1 genome. By degrading CEM15 in infected cells, Vif prevents its incorporation into virions, and therefore counteracts this cellular antiviral defense mechanism.

The Vpr protein enables the preintegration complex to be transported to the nucleus

Vpr (*v*irion *p*rotein *R*) is a 100–amino acid protein that is recruited into virions (10–100 molecules per virion) by virtue of its interaction with the carboxy-terminal region of Gag. It permits the infection of nondividing cells by signaling the transport of the preintegration complex into the cell nucleus (see above). In addition, Vpr retains cells in the G2 stage of the cell cycle, possibly as a consequence of its interaction with the protein complex p34-cdc2/cyclin B, which is involved in the transition from G2 to mitosis. Arresting cells in the G2 stage may be beneficial for virus replication since transcription directed by the HIV-1 LTR is more active at this stage of the cell cycle.

The Vpu protein enhances release of progeny virions from infected cells

Vpu (*v*irion *p*rotein *u*nique to HIV-1) is an 81–amino acid protein that is inserted into membranes via its amino terminal domain. This protein accumulates in the Golgi apparatus and the endosome compartment of the cell. No homologs have been identified in related lentiviruses such as HIV-2 and simian immunodeficiency virus. Vpu has two known activities within the cell.

Degradation of CD4. The cellular CD4 protein is a receptor for HIV that interacts with gp120 at the cell surface (see above). However, both CD4 and gp160, the precursor of gp120, are made in the endoplasmic reticulum, and they can bind to each other at that intracellular site. This retains gp160 inside the cell and therefore reduces virus production (Figure 26.7a). Vpu acts by binding to CD4 and to the cellular protein β-TrCP. This induces the ubiquitination of CD4 and its subsequent degradation by proteasomes, thus releasing gp160 and increasing surface expression of its cleavage products, gp 41 and gp 120.

Enhancement of virus release from the plasma membrane. In the absence of Vpu, virions accumulate on the cell surface in a partially budded state, and are not released. Therefore, Vpu somehow enhances budding and release of virions. This activity is dependent on the transmembrane portion of Vpu. Remarkably, this effect

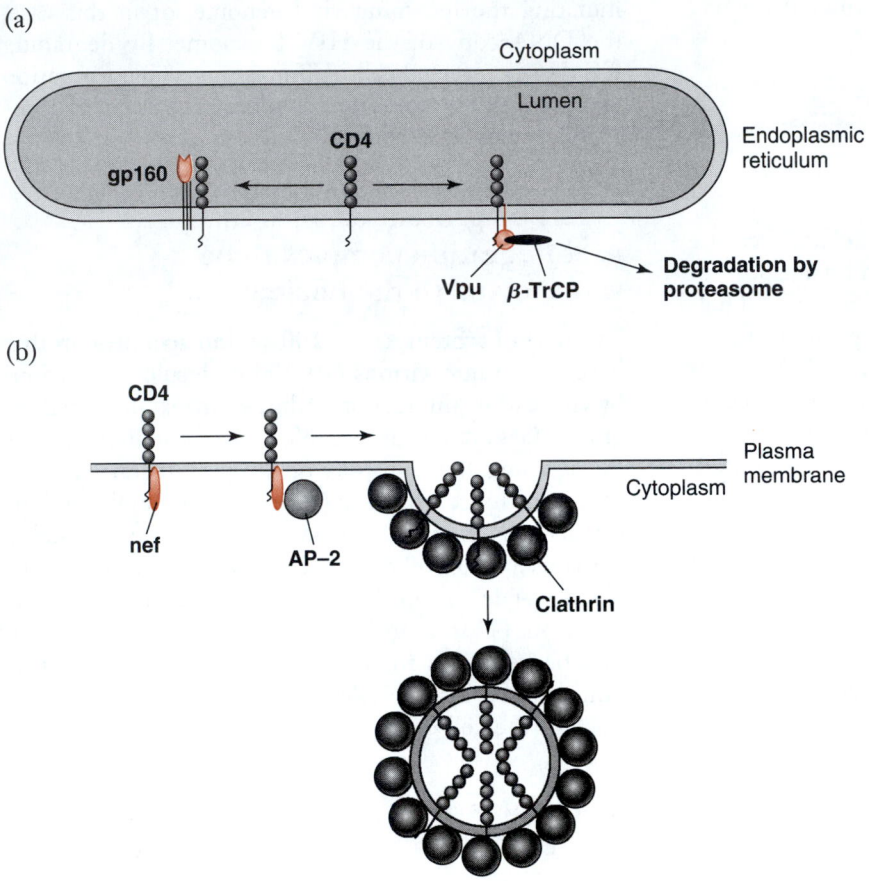

FIGURE 26.7 Down-regulation of CD4 expression. (a) In the absence of Vpu, CD4 interacts with gp160 in the endoplasmic reticulum, resulting in formation of a complex that blocks transport to the cell surface. Vpu binds to the cytoplasmic domain of CD4 and through interaction with β-TrCP directs CD4 to the proteasome degradation pathway. (b) Nef binds to the cytoplasmic tail of CD4 at the inner surface of the plasma membrane, and via interaction with AP-2 increases uptake of CD4 into clathrin-coated pits, lowering its concentration on the cell surface.

is not restricted to HIV-1; Vpu also enhances the release of other, unrelated viruses. The exact mechanism by which Vpu does this is not known. However, Vpu is a membrane protein that enhances membrane fluidity and is a cation-selective pore channel. Vpu may allow Gag to associate efficiently with the plasma membrane, therefore enhancing budding of virions.

The Nef protein is an important mediator of pathogenesis

Nef (*n*egative *ef*fector) is a 210–amino acid protein that is localized at the inner face of the plasma membrane via a residue of **myristate** added to its N-terminal amino acid. Infection of monkeys with simian immunodeficiency virus containing point mutations in the nef gene resulted in the rapid generation of revertant viruses with a functional nef gene, showing that there is considerable selection pressure for an active Nef protein. Simian immunodeficiency virus mutants with deletions of the nef gene are viable, but have lower titers and fail to induce disease in infected monkeys. Expression of Nef as a **transgene** in T cells and macrophages in mice causes a disease that resembles the late stages of AIDS. These findings implicate Nef as an important determinant of pathology in infected animals, and therefore much attention is being paid to this protein. Three activities have been attributed to Nef, but it is unclear which of these activities is important for disease.

Decrease in the surface expression of CD4 and MHC 1. Nef expression decreases the levels of CD4 and the **major histocompatibility complex (MHC) protein** MHC 1 on the cell surface. Because these are important mediators of immune responses (see Chapter 31), their absence can be important in disease progression. Decreased levels of these proteins are caused by a selective increase in their cycling between the cell surface and the endosome compartment. Protein cycling is carried out by interaction between Nef and a CD4/MHC-1 complex. This complex is recruited into clathrin-coated pits by interaction between Nef and the adaptor proteins AP1 and AP2 (Figure 26.7b).

Enhancement of virus infectivity. HIV nef mutants make virus particles that have a low level of infectivity. Virion infectivity cannot be rescued by expressing Nef in cells being infected. Nef may therefore modify virion structure in cells producing the virus. Nef has been found only at very low levels in virions, and its physical presence is therefore unlikely to play an important role in virion structure. It has been suggested that Nef

HIV-1 vaccines?

The high cost and complex regimen of current multiple drug therapies for HIV-1 render them inaccessible to most people in the developing world, where the bulk of infections occur. This has spurred efforts to develop vaccines capable of inducing protective immunity. Preliminary attempts focused on the use of gp120 as the immunizing antigen. However, the immune responses generated could not neutralize clinical strains of HIV-1, raising questions about the efficacy of this approach. Since gp120 is highly variable in the infected population, it is not clear which type of gp120 should be used in vaccines. More recent vaccine trials employ recombinant poxviruses expressing multiple HIV-1 antigens in an effort to selectively induce cytotoxic lymphocyte responses in the host.

Several alternative strategies for vaccination have been suggested. The demonstration that simian immunodeficiency virus strains lacking nef are unable to induce disease generated much interest. Monkeys inoculated with the nef-deleted strain were also found to be resistant to superinfection by pathogenic strains of the virus. This suggested the use of replication-competent, mutated forms of HIV-1 as a vaccine in humans. To reduce the possibility of reversion to wildtype, viruses were constructed containing deletions in multiple reading frames. While inoculation of monkeys with multiply deleted simian immunodeficiency virus prevents disease upon challenge with wildtype virus, some concerns still exist. First, these deleted viruses can induce disease in young animals. Second, infection by any replication-competent retrovirus could induce tumor formation in the host (see Chapter 25). At present, this approach remains in the development phase.

Another strategy attempts to make use of differences between the proteins coded by the major histocompatibility complex (MHC) to induce immunity. Inactivated simian immunodeficiency virus grown in human cells is able to protect monkeys against infection by live virus grown in human cells, but not against infection by virus grown in monkey cells. Surprisingly, immunization with human cells alone is able to provide the same degree of protection to monkeys. In epidemiological studies of prostitutes in Africa, a subgroup of women displays resistance to HIV-1 infection despite repeated exposure. Resistance is correlated with a rare MHC subtype. These findings suggest that resistance to infection is based on the recognition of foreign MHC protein subtypes that are incorporated into the virus envelope in the course of budding from the cell. It has been postulated that immunity to HIV-1 infection could be induced by immunization with proteins encoded by several different MHC alleles.

may induce the phosphorylation of the matrix protein, which could increase virus infectivity.

Modification of cell signaling. Nef has been implicated in alterations of signaling in T cells, resulting in general activation of the cell, which promotes virus replication. However, unlike T cells activated by antigens, T cells activated by Nef cannot effectively mount an immune response. Nef contains a conserved proline motif similar to the consensus binding sequence for the SH3 domain of the src family of protein tyrosine kinases, important players in cell signaling. Interaction of Nef with the **src-related kinases** lyn and hck has been shown to lead to their activation; in contrast, interaction with lyc and fyn leads to their inhibition. Activated hck results in increased expression by T cells of a number of **cytokines** and chemokines that enhance HIV replication and recruit more T cells to the sites of virus infection.

SUGGESTED READING

1. Cullen, B.R. (1998). HIV-1 auxiliary proteins: making connections in a dying cell. *Cell* **93,** 685–692.

2. Doms, R. and Peiper, S. (1997). Unwelcome guests with master keys: How HIV uses chemokine receptors for cell entry. *Virology* **235,** 179–190.

3. Fackler, O. and Baur, A. (2002). Live and let die: Nef functions beyond HIV replication. *Immunity* **16,** 493–497.

4. Finzi, D. and Siliciano, R.F. (1998). Viral dynamics in HIV-1 infection. *Cell* **93,** 665–671.

5. Goff, S. (2004). Retrovirus restriction factors. *Molecular Cell* **16,** 849–869.

6. Hanna, Z., Kay, D.G., Rebai, N., Guimond, A., Jothy, S., and Jolicoeur, P. (1998). Nef harbors a major determinant of pathogenicity for an AIDS-like disease induced by HIV-1 in transgenic mice. *Cell* **95,** 163–175.

7. Pope, M. and Haase, A. (2003). Transmission, acute HIV-1 infection and the quest for strategies to prevent infection. *Nature Medicine* **9,** 847–852.

8. Stevenson, M. (2003) HIV-1 pathogenesis. *Nature Medicine* **9,** 853–860.

9. Tang, H., Kuhlen, K.L., and Wong-Stahl, F. (1999). Lentivirus replication and regulation. *Annual Review of Genetics* **33,** 133–170.

KEY TERMS

Azidothymidine
CD4 antigen
CD4-positive T lymphocytes
Chemokines
Cytokines
Cytotoxic T lymphocytes
Dendritic cells
Epitopes
Importin
Latent infection
Macrophage
Major histocompatibility complex (MHC) protein
Myristate
NF-κB (nuclear factor-κB)
Opportunistic infections
Preintegration complex
Proteasomes
Src-related kinases
T-cell receptor
Transgene
Ubiquitination

Human T-Cell Leukemia Virus Type 1

Deltaretrovirus
From Greek letter *delta*

VIRION
Spherical enveloped particle.
Diameter 100 nm.
Icosahedral capsid.

GENOME
Linear ssRNA, positive sense.
Two identical segments, each 8.7 Kb.
Cellular $tRNA^{pro}$ molecules packaged in virions used as primers for reverse transcription.

GENES AND PROTEINS
Gag, Pol, and Env proteins packaged in virion.
Also codes for regulatory proteins Rex, Tax, $p12^{I}$, p13, and p30.

VIRUSES AND HOSTS
A human retrovirus, member of genus *Deltaretrovirus*.
Related viruses: bovine leukemia virus, simian T-lymphotropic viruses.

DISEASES
Adult T-cell leukemia, which affects CD4+ T cells.
Tropical spastic paraparesis, a neurological demyelinating disease.

DISTINCTIVE CHARACTERISTICS
Infects human T lymphocytes.
Tax protein regulates gene expression and leads to oncogenic transformation.
Causes T-cell proliferation and transformation by multiple pathways.
A human tumor virus.

Discovery of the first human retrovirus

Retroviruses infecting horses and chickens were first isolated at the beginning of the 20th century. Numerous avian and mouse retroviruses were described from 1950 onward, and spectacular advances in understanding the biology of retroviruses were made in the 1960s and 1970s (see Chapter 25). However, it was not until 1980 that Robert Gallo and colleagues isolated the first human retrovirus. They and others had been searching long and hard for human retroviruses by examining human cells for the presence of reverse transcriptase activity. They learned how to propagate human **T lymphocytes** *in vitro*, and detected reverse transcriptase in a cell line derived from the blood of a patient suffering from a cutaneous **T-cell lymphoma**. Therefore, the virus was named human T-cell leukemia (or lymphotropic) virus type 1 (HTLV-1). HTLV-1 is now considered to be the cause of **adult T-cell leukemia (ATL)**, an aggressive and fatal leukemia involving CD4-positive T lymphocytes.

In 1985, a number of West Indian patients with a neurological demyelinating disease known as **tropical spastic paraparesis (TSP)** were found to have antibodies directed against HTLV-1, implicating the virus in this disease as well. A similar disease prevalent in Japan is called **HTLV-1-associated myelopathy (HAM)**. Both ATL and HAM/TSP are localized in regions of the world where HTLV infection is endemic, notably southern Japan, sub-Saharan Africa, the Caribbean basin and the southwestern United States. At present, it is estimated that between 10 and 20 million people around the world are infected with HTLV-1.

A closely related virus named human T-cell leukemia virus type 2 (HTLV-2) was identified in a T-cell line established from a patient with **hairy-cell leukemia**. However, the role of HTLV-2 in malignancy is still unclear. HTLV-2 is found clustered in specific groups of Native American populations and in intravenous drug users.

Like lentiviruses, HTLV-1 codes for regulatory proteins by producing doubly spliced mRNAs

The overall genetic structure of HTLV-1 is similar to that of other retroviruses. The genome contains gag, pol, and env genes that code for the viral capsid proteins, protease, reverse transcriptase and integrase, and envelope proteins (Figure 27.1). In HTLV-1, the protease is encoded by a reading frame that overlaps part of the gag region and part of the pol region. Synthesis of the protease as part of the gag polyprotein precursor is accomplished by **ribosomal frameshifting**, a process in which the ribosome slips to the −1 position and changes its translational reading frame (see Figure 25.7 in the chapter on retroviruses). A second ribosomal frameshift event is necessary to express the pol gene as a part of a Gag/Pro/Pol polyprotein that is eventually cleaved by the protease. The use of ribosomal frameshifting to express these genes leads to lower amounts of these enzymatic proteins (protease, reverse transcriptase, integrase) being made compared to viral structural proteins (Gag) that are required in multiple copies in each virion.

In addition to viral structural genes, HTLV-1 RNA contains a number of open reading frames between the envelope gene and the U3 region. Three of these code for important regulatory proteins: Tax, Rex, and $p12^I$. To produce viral proteins, three different species of mRNA are generated. Full-length genome RNA encodes the Gag and Gag/Pol fusion proteins, singly spliced mRNA encodes the Env proteins and $p12^I$, and doubly spliced mRNA encodes the HTLV-1 regulatory proteins Tax and Rex (Figure 27.1).

The major steps of virus replication, including reverse transcription, integration, gene expression, and assembly of virions, are covered in detail in the retrovirus chapter (Chapter 25). Here we will discuss the roles of the regulatory proteins Rex, Tax, and $p12^I$ in the virus life cycle, and theories of how this virus causes leukemia and degenerative central nervous system disease.

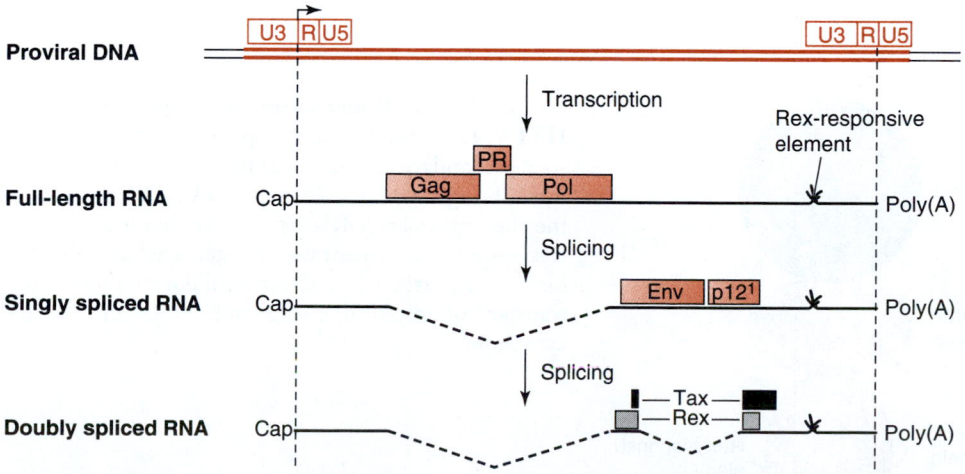

FIGURE 27.1 Structure and expression of HTLV-1 proviral DNA. Integrated proviral DNA is shown in orange. Boxes above RNAs show open reading frames coding for HTLV-1 structural and regulatory proteins. Introns are shown by diagonal dashed lines. The Rex-responsive element is shown in the U3 region.

1 line short

HTLV-1 Rex regulates polyadenylation, splicing, and nuclear export of viral RNAs

Rex is a 245–amino acid protein that is localized mainly in the nucleolus of infected cells. Rex induces the export of singly-spliced or unspliced mRNAs, which encode viral structural proteins, to the cytoplasm. It is generally accepted that Rex escorts viral RNAs from the nucleus to the cytoplasm in a fashion very similar to the way in which the Rev protein of human immunodeficiency virus (HIV-1) acts (refer to Figure 26.5).

Rex binds to a sequence called the Rex-responsive element, located in the U3 region of viral RNAs. The Rex-responsive element is folded into a highly stable 255-nucleotide stem-loop RNA structure (Figure 27.2). In addition to its role in mediating Rex responsiveness, this element also plays a role in the processing of the 3′ end of viral transcripts. The viral polyadenylation signal (AAUAAA) is found at a position far upstream from the sequence where viral RNA is cleaved and polyadenylated by cellular enzymes. For most cellular genes, this signal is only 15–20 nucleotides from the cleavage site, and host cell proteins that bind to the AAUAAA direct cleavage by a cellular endoribonuclease shortly downstream. In the case of HTLV RNA, formation of the correct secondary structure in the Rex-responsive element brings the cleavage site close to the AAUAAA (Figure 27.2), permitting cleavage and polyadenylation at the correct site.

Rex possesses two important structural domains. A region in the amino terminus of the protein (aa 1–19), rich in basic amino acids, mediates binding to the Rex-responsive element (Figure 27.2). This region also functions as a nuclear and nucleolar targeting signal for the Rex protein. A second essential region, located between aa 79 and aa 99, contains a leucine-rich activation domain that functions as a nuclear export signal by binding to the cellular cofactor **exportin** 1.

In the absence of Rex, HTLV transcripts in the nucleus are either spliced to completion or are degraded. As a result, only doubly spliced mRNAs (coding for Tax and Rex) are exported to the cytoplasm. When Rex is present, it binds to unspliced or singly spliced RNAs and prevents further splicing. These transcripts are then exported from the nucleus to the cytoplasm, where they are either translated or (in the case of unspliced RNA) packaged as genomes into progeny virions.

The analogous HIV-1 Rev protein binds to the Rev-responsive element, located within the coding region of the HIV-1 env gene instead of the U3 region, where the Rex-responsive element is located. Interestingly, the HTLV-I Rex protein is capable of binding to the HIV-1 Rev-responsive element, and expression of Rex can therefore rescue HIV-1 mutants lacking Rev. This indicates some functional conservation between HTLV Rex and HIV Rev, although there is no sequence homology between the two proteins.

Import of cellular proteins into the nucleus is generally mediated by basic nuclear localization signals that bind to **importin** α. Importin α in turn binds to importin β, which targets the resultant protein complex to the nucleus via the nuclear pores. Interestingly, neither HIV-1 Rev nor HTLV-1 Rex bind to importin α; instead, they bypass it and bind directly to importin β for import to the nucleus (see Figure 26.6 in HIV chapter). Both Rex and Rev therefore shuttle between the cytoplasm and the nucleus, transporting viral RNAs out of the nucleus and then returning to the nucleus to pick up further viral RNA cargo.

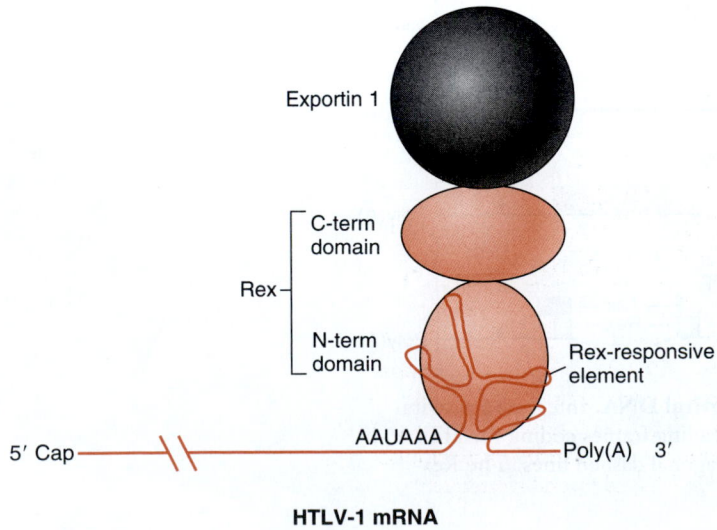

FIGURE 27.2 Polyadenylation and export of HTLV-1 RNAs. The Rex-responsive element forms a stable secondary strucure near the 3′ end of HTLV-1 RNA, bringing the poly(A) signal AAUAAA close to the cleavage/polyadenylation site. Rex binds to the Rex-responsive element via an N-terminal domain and binds to exportin 1 via a C-terminal domain, mediating transport of unspliced or singly spliced RNA to the cytoplasm.

HTLV-1 Tax regulates transcription of viral and cellular genes

The oncogenic potential of HTLV-1 resides in the 353–amino acid Tax protein. Tax is a positive regulator of viral gene transcription, whose activity is mediated by cyclic AMP-responsive sequence elements on DNA known as **Tax-responsive elements (TRE)**. Several of these 21 bp elements lie within the U3 region of the HTLV-1 long terminal repeat (Figure 27.3a). Tax also stimulates the expression of numerous cellular genes involved in early events leading to transformation of T lymphocytes, including cell cycle regulatory proteins, **cytokines**/growth factors, and transcription factors (Table 27.1). Since Tax is not itself a DNA-binding protein, transactivation of these genes occurs indirectly via physical interaction between Tax and host transcription factors. There are three major classes of transcription factors whose activity is stimulated by Tax.

CREB/ATF. Each of the three 21 bp repeat Tax-responsive elements contains a central domain sharing homology with *c*yclic AMP *r*esponse *e*lements (CRE). Members of the CRE-binding CREB/ATF family of cellular transcription factors bind both to the 21 bp repeats and to Tax. The Tax-CREB/ATF interaction is mediated through a **leucine zipper domain** found in these cellular transcription factors. The formation of the TRE-CREB-Tax complex leads to the recruitment of the transcriptional coactivator, **CREB-binding protein (CBP)** (Figure 27.3b). CBP is a 256 KDa **histone acetylase**, which is part of the **RNA polymerase II holoenzyme** complex. The recruitment of CBP promotes strong transcriptional activation through **chromatin remodeling** induced by the histone acetylase activity of CBP. Local acetylation of histones within **nucleosomes** reduces their affinity for DNA, allowing proteins such as TFIID and RNA polymerase to bind to the promoter and initiate transcription. In summary, Tax acts by increasing the binding of CREB transcription factors to the viral Tax-responsive elements on the viral genome, thus recruiting the coactivator CBP to the viral promoter, leading to an increase in transcription of viral genes.

NF-κB. One of the important targets of the Tax protein is the ***NF-κB/IκB*** transcription complex. The NF-κB/Rel transcription factors are a family of dimer-forming proteins that bind to the consensus DNA sequence 5′-GGGANNPyPyCC-3′, found in the promoter regions of cellular genes implicated in immunoregulation, cell growth control, lymphoid development, and apoptosis (Table 27.1). NF-κB is composed of homodimers and heterodimers of the Rel family of transcription factors, which include p65 or RelA, c-Rel, RelB, p50, and p52. NF-κB/Rel proteins are coupled to inhibitory molecules such as IκBα that are responsible for cytoplasmic retention of the NF-κB complex, rendering it unable to activate transcription. Multiple inducers (cytokines, virus infection, mitogens) lead to the phosphorylation of IκBα at serine residues 32 and 36; phosphorylation at these sites is a signal for subsequent **ubiquitination** and degradation of IkBα (Figure 27.4). Loss of IkB permits translocation of NF-κB Rel dimers to the nucleus, resulting in binding to

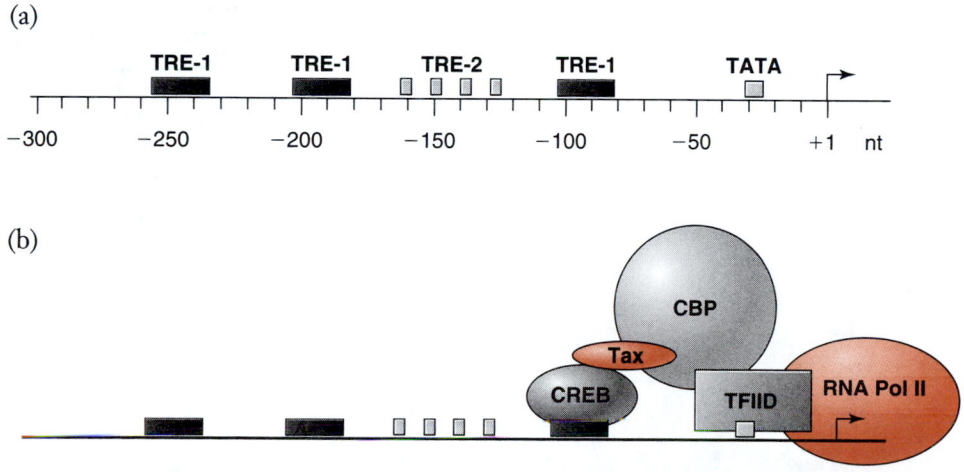

FIGURE 27.3 **Tax activates transcription of HTLV-1 RNA.** (a) Part of the U3 region of the HTLV-1 long terminal repeat. Tax-responsive elements (TRE) are shown as black and gray rectangles. A TATA box is present at nt −30; transcription starts at nt +1. (b) Cyclic AMP-responsive element binding protein (CREB) binds to TRE; Tax recruits the transcriptional coactivator, CREB-binding protein (CBP), leading to enhanced transcription of proviral DNA by enhancing interaction of TFIID with the TATA box and subsequent binding of RNA polymerase II.

TABLE 27.1 Activation of gene expression by HTLV-1 Tax

Genes induced by Tax	Examples	Transcription factor used[a]
Viral genes	HTLV-1	CREB/ATF
	HIV-1	NF-κB
	Cytomegalovirus	NF-κB
Cytokines and receptors	Interleukin-1, 2, 6, 8, 15	
	Tumor necrosis factor-α, β	
	Interferon-β	NF-κB
	Interleukin-2 receptor α	
	T-cell receptor	
	MHC class I	
Transcription factors and proto-oncogenes	c-fos, erg-1	SRF
	fra-1	NF-κB
	c-myc, c-rel, c-sis	EGR-1
Cell adhesion molecules	Endothelial leukocyte adhesion molecule-1 (ELAM-1)	NF-κB
	Intracellular cell adhesion molecule-1 (ICAM-1)	NF-κB
	Vimentin	?
Acute phase proteins	Angiotensinogen	
	Serum amyloid A precursor	NF-κB
	Complement factors	

[a]CREB/ATF: cAMP responsive element binding/activation transcription factor; NF-κB: nuclear factor-κB; SRF: serum response factor; EGR-1: early growth response factor-1.

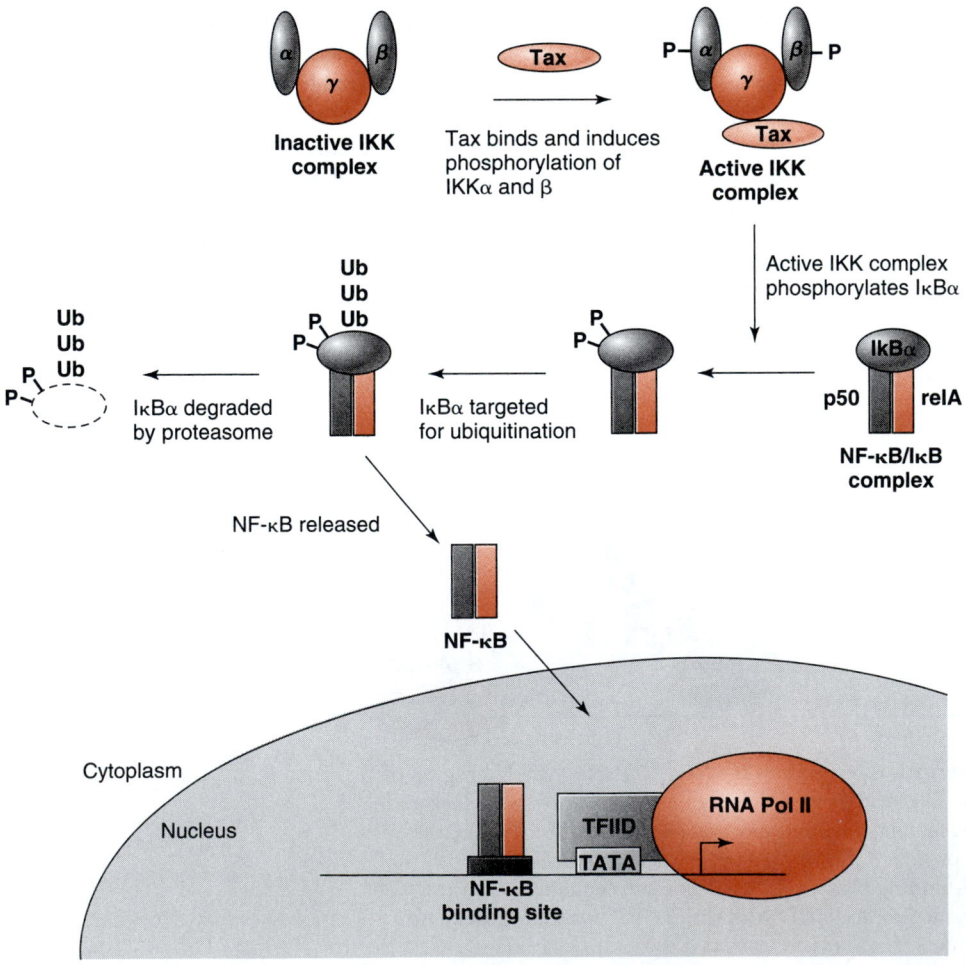

FIGURE 27.4 NF-κB signaling cascade. The Iκ kinase (IKK) complex, composed of IKKα, IKKβ, and IKKγ, is activated by phosphorylation induced by Tax binding. The active complex phosphorylates IκBα at Ser-32 and Ser-36. The phosphorylated IκBa is subsequently ubiquitinated by the E3 ubiquitin ligase and targeted for degradation by the proteasome. NF-κB dimers are then free to translocate to the nucleus, bind to DNA, and activate target gene transcription.

target genes and transcriptional activation. Recently a novel protein kinase, the **IκB kinase complex (IKK)**, was shown to be involved in phosphorylation of IκBα at serines 32 and 36 (Figure 27.4). This kinase complex is itself activated by phosphorylation through multiple upstream activators.

Tax expression in HTLV-1 infected cells targets the NF-κB/IκB complex at multiple regulatory levels via protein-protein interactions. Tax stimulates IκBα phosphorylation by interacting with and activating the IKK complex. Tax does this by binding to the IKKα subunit, thereby leading to the continuous activity of the catalytic IKKα and IKKβ subunits (Figure 27.4). Tax also binds directly to IκB and directs it to the proteasome for degradation. Finally, Tax indirectly mediates the transcriptional activation of c-Rel, NF-κB2, and IκBα genes, leading to constitutive NF-κB activity and aberrant gene expression (Table 27.1).

Serum response factors. Serum response factors are also involved in Tax-mediated transactivation. These factors bind to "serum responsive elements" (SRE) within promoters of the c-fos, egr-1, and fra-1 genes. Tax binds specifically to one of the serum response factors, enhancing its activity and therefore the transcription of these genes, which are involved in cell growth and proliferation.

Cell transformation by HTLV-1 is mediated by Tax

HTLV-1 can infect a variety of cell types in the body, including T lymphocytes, **B lymphocytes**, **macrophages**, immature bone marrow cells and neurons. However, only T lymphocytes are transformed in infected humans, leading to adult T-cell leukemia. The mechanisms by which this retrovirus transforms T cells in the body into leukemic cells are not fully understood. An important clue to the mechanism of transformation was revealed recently with the discovery that the cellular receptor used by the virus to gain entry into T cells is the glucose transporter receptor known as GLUT-1; this discovery will undoubtedly contribute to new information concerning the early events of viral entry and infection.

There are two pathways by which retroviruses are known to transform cells (Chapter 25): (1) **acute transforming retroviruses** express an activated **oncogene** that they have picked up from the cellular genome. These viruses can rapidly and efficiently transform cells *in vitro*, and infection of animals can lead to rapid tumorigenesis. (2) Retroviruses that do not contain oncogenes transform cells and cause tumors by integrating nearby a cellular **proto-oncogene** and activating its expression, usually under the control of the long terminal repeat, which contains transcriptional enhancer and promoter sequences (Figure 25.9). Such viruses transform cells only rarely *in vitro*, and animal tumors appear infrequently and only a long time after infection. These tumors are clonal, in that all tumor cells contain proviral DNA integrated at the same position, meaning that all cells in the tumor are derived from a single transformed cell caused by a rare event.

Unlike acute transforming retroviruses, HTLV-1 does not contain a cellular oncogene. However, neither do HTLV-1 transformed cells or tumor cells have proviral DNA integrated nearby a specific proto-oncogene as do cells transformed by other nonacute transforming retroviruses. How, then, does HTLV cause tumors and transform T cells *in vitro*?

The most widely accepted explanation of HTLV-1 tumorigenesis is that infected T cells are stimulated to divide and, as they proliferate, they accumulate certain mutations that allow them to grow in the absence of the normal stimuli required for T-cell growth. These cells can therefore further proliferate and accumulate a variety of mutations that eventually lead to the emergence of a T-cell population growing in an uncontrolled manner, causing leukemia. A number of different pathways have been implicated in the **polyclonal** T-cell proliferation caused by HTLV-1. Tax appears to play a critical role in many of these pathways. Tax expression leads to T-cell activation and proliferation by generating an interleukin-2/interleukin-2 receptor **autocrine loop**. Tax also induces the G1 to S phase transition in the cell cycle. Finally, Tax physically associates with cellular transcription factors to activate transcription of a number of cell cycle and growth-controlling genes. Several of these pathways are briefly described below.

The interleukin-2 autocrine loop stimulates T-cell proliferation

CD4-positive T cells are stimulated to divide when they encounter **interleukin-2** (originally called "T-cell growth factor"). Increased proliferation of HTLV-1 infected T cells in the early stages of infection is dependent on the production of interleukin-2 (Figure 27.5a). Up-regulation of the interleukin-2 receptor α subunit (Table 27.1) leads to an increase in sensitivity of infected T cells to this cytokine, and increased production of interleukin-2 stimulates cell division. This results in an autocrine stimulatory loop that drives the early stages of T-cell proliferation.

Activation of the Jak-Stat pathway by p12^{I} mimics interleukin-2 stimulation

Over time, T-cell proliferation becomes interleukin-2 independent by activation of the **Jak-Stat pathway**, which is normally activated by binding of interleukin 2

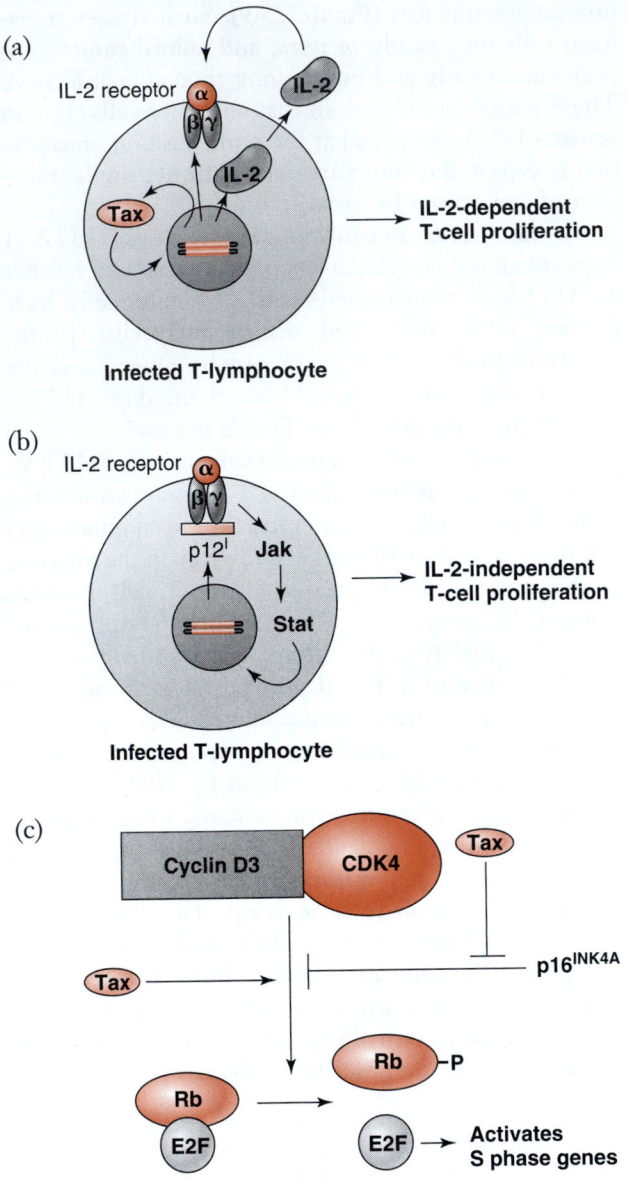

FIGURE 27.5 Possible mechanisms of HTLV-1 induced T-cell proliferation and leukemogenesis. (a) HTLV-1 infected T cells that express Tax from an integrated proviral DNA (in orange) secrete interleukin 2 (IL-2) and up-regulate its receptor, causing polyclonal IL-2 dependent proliferation. (b) $p12^I$ can bind to and dimerize α_c and β_c subunits of the IL-2 receptor, leading to activation of the Jak-Stat pathway in absence of IL-2 and resulting in IL-2 independent proliferation. (c) Tax blocks $p16^{INK4A}$ inhibition of cyclin-dependent kinase activity, and binds to cyclin D3 to increase cyclin-dependent kinase activity, resulting in increased phosphorylation of Rb protein and subsequent activation of S phase genes by free E2F transcription factor.

(and other cytokines) to its receptor (see Figure 31.3, chapter on interferons). One way this can happen is via the $p12^I$ protein. $p12^I$ is a small hydrophobic protein encoded by the an open reading frame near the 3′ end of the HTLV-1 genome (Figure 27.1). $p12^I$ binds to and dimerizes the β_c and γ_c chains of the interleukin-2 receptor, mimicking the action of interleukin-2 on the receptor (Figure 27.5b). The dimerized β_c and γ_c chains activate the Jak-Stat pathway, leading to gene activation and cell proliferation.

Cell cycle progression: $p16^{INK4A}$ and cyclin-dependent kinases

Transition from one phase of the cell cycle to the next is regulated at "checkpoints," which are in part governed by **cyclin-dependent kinases (CDKs)** assembled with their partner **cyclins**. Active CDK-cyclin complexes are further regulated by phosphorylation and dephosphorylation. CDKs can also be inactivated through physical interaction with inhibitory proteins. D- and E-type cyclins are important in cell cycle transition from G1 to S. The cyclin D-CDK complex phosphorylates retinoblastoma (Rb) protein, resulting in activation of transcription factor E2F (see Figure 12.6, chapter on adenoviruses). E2F transactivates genes that facilitate the transition from G1 to S and induce cellular proliferation.

Tax plays a role in cell cycle progression by direct interference with the cell cycle inhibitor, $p16^{INK4A}$. The $p16^{INK4A}$ gene is a **tumor suppressor** gene that is often deleted in human cancers. $p16^{INK4A}$ binds to the cyclin-dependent kinases CDK4 and CDK6, and inhibits their kinase activity, stopping the progression of the cell from G1 into the S phase. The binding of Tax to $p16^{INK4A}$ reverses its inhibition of CDK4 activity, allowing CDK4 kinase to induce E2F-stimulated gene expression (Figure 27.5c). By inhibiting $p16^{INK4A}$ activity, Tax therefore causes unregulated cell cycling and cellular proliferation. Tax also binds to cyclin D3 and mediates an increase in cyclin D3-dependent CDK4 and CDK6 activity. This increase in kinase activity is accompanied by Rb phosphorylation and E2F activation. Tax therefore has at least two effects on the cell cycle control pathways: (1) inactivation of $p16^{INK4A}$ (an inhibitor of the kinases), and (2) direct activation of cyclin D-CDKs.

The mitotic spindle checkpoint and MAD1

The failure to maintain proper cell cycle control is recognized as a contributing factor to genetic instability, a hallmark of human and experimental cancer. The **mitotic spindle checkpoint** prevents the onset of **anaphase** and subsequent commitment to cellular division until chromosomes are aligned properly on a bipolar spindle. Experiments in yeast have shown that the disruption of genes required for spindle checkpoint during mitosis dramatically increases the rate of chromosome loss. Several proteins in yeast and in humans have been identified as factors required for pre-anaphase arrest or delay in response to microtubule depolymerization. These factors are called **mitotic arrest-defective (MAD)** proteins. Recent studies have shown that Tax interacts with human MAD1 protein and compromises its functions by preventing dimerization, thus leading to the loss of the mitotic checkpoint. T lymphocytes from adult T-cell leukemia patients have abnormal chromosome complements and are frequently present as **multinucleated giant cells**. The expression of Tax correlates with the progressive accumulation of damaged DNA in cells. This Tax-MAD1 interaction provides a molecular explanation for HTLV-1-induced chromosome abnormalities in ATL cells.

Downregulation of p53 activity by Tax allows T-cell proliferation

The long latency period in the development of adult T-cell leukemia suggests that increased HTLV-1 mediated T-cell proliferation permits the accumulation of genetic changes that result in leukemogenesis. The cellular tumor suppressor protein p53 is a target for alterations in HTLV-1 induced transformation. The levels of p53 are normal or even slightly elevated in T cells transformed by HTLV-1 or from patients with adult T-cell leukemia, but p53 transactivation activity is reduced. This tumor suppressor protein induces the expression of many cellular genes: (1) MDM2, which is involved in DNA repair; (2) Gadd45, which serves as a marker for DNA damage; (3) p21, which arrests cells at the G1/S border to allow repair of DNA damage; and (4) Bax, which can induce apoptosis in cells that have damaged DNA. p53 also down-regulates Bcl-2, an anti-apoptotic protein. By inhibiting p53 activity, Tax inhibits DNA repair and apoptosis, thus increasing the chance that T cells with mutations can survive and proliferate. Tax was also shown to repress the expression of DNA polymerase β, which is involved in DNA repair. A decrease in DNA repair could lead to an increase in the accumulation of DNA mutations and ultimately contribute to cellular transformation.

Diseases caused by HTLV-1 develop slowly and can be severe

Adult T-cell leukemia. ATL develops in a small proportion (1/1000) of HTLV-1 infected people after a long latency period; the mean age at disease onset is 43 years. ATL is classified into four clinical subtypes, based on the number of abnormal CD4-positive T cells, tumor lesions, and the clinical course: chronic, smoldering, lymphoma, and acute ATL (Table 27.2). The acute phase of ATL is the most severe. Patients with acute ATL have elevated numbers of abnormal T cells, high levels of blood calcium, prominent skin lesions, and enlarged spleen and liver. These patients also suffer from serious bacterial, viral, fungal, and protozoan infections because of reduced immune responses. The prognosis of acute ATL is poor, with a median survival time for acute disease of 6 months.

Almost all ATL patients suffer from the complications of high levels of blood calcium. A parathyroid-related protein (PTHrP), whose level is increased by Tax-mediated transactivation, is responsible for high calcium levels in patients suffering from ATL and trop-

TABLE 27.2 Clinical phases of adult T-cell leukemia

Disease phase	Clinical signs	Survival
Pre-ATL	• No clinical signs for decades	
Smoldering	• Less aggressive • Normal white blood cell count • Skin lesions • 5% abnormal T lymphocytes	More than 10 months
Chronic	• Less aggressive • Normal white blood cell count • Prominent skin lesions • 5% abnormal T lymphocytes • Modest bone marrow and visceral involvement	More than 10 months
Lymphoma	• Skin lesions • 5% abnormal T lymphocytes • Liver and spleen pathology	10 months
Acute	• Very aggressive • Elevated white blood cell count • Prominent skin lesions • >5% abnormal T cells • Liver and spleen pathology	Less than 6 months

ical spastic paraparesis, as well as in people who are infected with HTLV-1 but have no symptoms.

HTLV-1 associated myelopathies (HAM)/tropical spastic paraparesis (TSP). The prevalence of HAM/TSP among people infected with HTLV-1 is approximately 1 per 1500 individuals. Compared with acute ATL, the latency period is shorter (1–3 years) and disease onset occurs earlier in life (mean age of 25 years). Clinical manifestations include progressive **demyelination** of motor neurons in the spinal cord, **spasticity**, disturbances in bladder control and lower-extremity muscle weakness. HTLV-1 proviral DNA has been found in spinal cord lesions that contain infiltrating CD4-positive T lymphocytes.

Three possible mechanisms have been suggested for the pathogenesis of HAM/TSP. The "autoimmune" model proposes that HTLV-1 activates **autoreactive T lymphocytes**, which then migrate to the central nervous system, where they recognize target antigens presented on the surface of normal cells and proceed to destroy those cells. The "direct infection" model proposes that tissue destruction is caused by persistent infection of central nervous system cells by HTLV-1, leading to attack and destruction of these cells by the cellular immune system. The "bystander damage" model proposes that the supply of infected CD4-positive T lymphocytes from blood flowing to the central nervous system is crucial for the development of lesions. These lymphocytes proliferate as a result of HTLV-1 replication and then respond to viral antigens presented on central nervous system cells and destroy them.

Coinfection by HIV-1 and HTLV-1 is an emerging problem

An increasing number of individuals are found to be coinfected with HIV and HTLV. This trend is particularly common among intravenous drug users. Evidence suggests that coinfection with both viruses may accelerate the progression to AIDS and worsen the consequences of HTLV infection. The transmission modes for HTLV-1 are identical to those of HIV: through sexual contact, blood transfusion, pregnancy, and breast milk. However, transmission of HTLV-1 infection requires transfer of cells infected with HTLV-1 into the recipient individual, whereas transfer of a small number of free HIV-1 virions can be sufficient to infect an individual.

Antiviral therapy of disease caused by HTLV-1 has not met with great success

Standard chemotherapeutic regimens that have proven to be highly successful in the treatment of malignant leukemia and lymphomas have failed miserably in the treatment of adult T-cell leukemia. These conventional chemotherapy regimens (even in combination) have produced a 4-year survival rate of less than 5%. A combination of **azidothymidine** (zidovudine) and interferon-α was shown to produce a synergistic response and induce rapid remission within 1 month, with a maximum response reached after several months. Despite initial successes, this combination therapy does not appear to be a cure, since relapse occurred when patients in remission were taken off treatment. The use of immunosuppressants such as cyclosporin A and FK506 had only minimal effects on the progression of HTLV-1 induced HAM/TSP.

SUGGESTED READING

1. Edlich, R., Arnette, J., and Williams, F. (2000). Global epidemic of human T cell lymphotropic virus type I (HTLV-1). *Journal of Emergency Medicine* **18,** 109–119.

2. Goon, P.K. and Bangham, C. (2004). Interference with immune function by HTLV-1. *Clinical and Experimental Immunology* **137,** 234–236.

3. Jeang, K.T. (2004). Life, death and Tax: Role of HTLV-1 oncoprotein in genetic instability and cellular transformation. *Journal of Biological Chemistry* **279,** 1991–1994.

4. Mandel, N., Kim, F.J., Kinet, S., Taylor, N., Sitbon, M., and Battini, J.L. (2003). The ubiquitous glucose transporter is a receptor for HTLV-1. *Cell* **115,** 449–459.

5. Nicot, C. (2005). Current views in HTLV-1 associated adult T cell leukemia/lymphoma. *American Journal of Hematology* **78,** 232–239.

KEY TERMS

Acute transforming retroviruses
Adult T-cell leukemia (ATL)
Anaphase
Autocrine loop
Autoreactive T lymphocytes
Azidothymidine
B lymphocytes
Chromatin remodeling
CREB-binding protein (CBP)
Cyclin-dependent kinases (CDKs)
Cyclins
Cytokines
Demyelination
Exportins
Hairy-cell leukemia
Histone acetylases
HTLV-1 associated myelopathy (HAM)
IκB
IκB kinase complex (IKK)
Importin
Interleukin-2
Jak-Stat pathway
Leucine zipper domain
Macrophages
Mitotic arrest-defective (MAD)
Mitotic spindle checkpoint
Multinucleated giant cells
NF-κB (nuclear factor-κB)
Nucleosomes
Oncogene
Polyclonal
Proto-oncogene
Ribosomal frameshifting
RNA polymerase II holoenzyme
Spasticity
Tax-responsive elements (TRE)
T-cell lymphoma
T lymphocytes
Tropical spastic paraparesis (TSP)
Tumor suppressor
Ubiquitination

Hepadnaviruses

Hepadnaviridae

From *hepa*titis and *DNA*

VIRION

Spherical enveloped particle.
Diameter 42 nm.
Icosahedral capsid, T = 4.
Abundant smaller spherical and filamentous forms lack nucleocapsid and are not infectious.

GENOME

Circular dsDNA with a ss gap on one strand, 3.2 Kb.

GENES AND PROTEINS

mRNAs transcribed by cellular RNA polymerase II from five promoters.
Four overlapping reading frames, seven viral proteins.
Three surface proteins: LS, MS, SS.
Core proteins: C, E.
Polymerase (reverse transcriptase) protein: P.
Regulatory protein: X.

VIRUSES AND HOSTS

Hepatitis B viruses of human, chimpanzee, duck, gibbon, gorilla, ground squirrel, heron, orangutan, snow goose, woodchuck, woolly monkey.

DISEASES

Hepatitis: incubation period 30–180 days.
Transmission by blood and sexual contact.
Acute disease can be mild or severe.
Chronic or associated diseases: cirrhosis, liver cancer, serum sickness.
Treatment: interferon, lamivudine.
Prevention: vaccination with recombinant hepatitis B surface antigens; injection with anti-hepatitis B immunoglobulin after exposure to virus.

DISTINCTIVE CHARACTERISTICS

Unusual partly ss, partly ds circular genome results from incomplete replication in cell.
Pregenome RNA is both an mRNA and a template for synthesis of genome DNA.
Reverse transcriptase (polymerase protein) generates genome DNA from pregenome RNA within capsid during virus assembly.
Makes large amounts of noninfectious spherical and filamentous particles.
Has a satellite virus: hepatitis delta virus (see Chapter 29, viroids).

At least seven distinct viruses cause human hepatitis

Hepatitis (infection of the liver) consitutes a major worldwide public health problem. Among the many agents that cause hepatitis are a number of unrelated viruses. Before the viruses causing hepatitis were identified, these pathogens were classified by modes of transmission and epidemiology. The terms "hepatitis A" and "hepatitis B" were first introduced by MacCallum in 1947 to differentiate between infectious (**enteric**) and serum hepatitis. Type A hepatitis was predominantly transmitted via the fecal–oral route while type B hepatitis was transmitted primarily through blood contact. Eventually these terms were adopted by the World Health Organization Committee on Viral Hepatitis. Hepatitis A virus is now known to be a member of the *Picornaviridae* (Chapter 16). A new bloodborne hepatitis virus, distinct from A and B, was named C, and is a member of the *Flaviviridae* (Chapter 17). At least four other hepatitis viruses, named D (or delta) to G, are presently known.

The discovery of hepatitis B virus

In 1963, Baruch Blumberg discovered a previously unknown protein, which he denoted the Australia antigen, in the blood of an Australian aborigine. It soon became apparent that this protein was related to type B hepatitis, and the Australia antigen is now known as the hepatitis B surface antigen. In 1973, David Dane found virus-like particles in the serum of patients with type B hepatitis; these particles subsequently became known as Dane particles and were shown to be virions of hepatitis B virus (HBV: Figure 28.1). Paul Kaplan detected endogenous RNA/DNA-dependent DNA polymerase (reverse transcriptase) within the core of hepatitis B virions. The hepatitis B virus DNA genome was eventually cloned and sequenced by Pierre Tiollais, Patrick Charny, Pablo Valenzuela, and William Rutter in 1979.

Hepatitis B virus is distinct in the world of viruses with its very compact 3.2 Kb circular DNA genome, extensive use of overlapping reading frames, production of a pregenome RNA, and dependence on a reverse transcription step for replication. Because of its use of reverse transcriptase, hepatitis B virus is compared here with the retroviruses, which were the first viruses known to use reverse transcriptase for replication. Although they share reverse transcription as an obligatory step in their replication cycles, these viruses differ in an important way: retroviruses package an RNA genome that is reverse transcribed into DNA once it enters the host cell, whereas hepadnaviruses package a DNA genome that is reverse transcribed from a pregenomic RNA in the cell during virus assembly (see below).

Human hepatitis B virus became the archetype of the *Hepadnaviridae* family. Similar viruses are found in a number of animals (see chapter opener), and some of these viruses have served as models for the study of hepadnavirus replication and pathogenesis. Although the hepatitis B surface antigen has been found in other primates, humans remain the primary reservoir for this virus. Cauliflower mosaic virus, a plant virus, also has circular DNA, has overlapping reading frames, and uses a reverse transcriptase during its replication. It is believed to be a distant cousin of hepadnaviruses (see Table 3.6).

The World Health Organization estimates that over two billion people worldwide have been exposed to hepatitis B virus, and around 500 million of these individuals are chronic carriers. Transmission is primarily through blood (transfusions, needle sharing) and sexual contact. A satellite virus known as hepatitis delta virus (see Chapter 29) is associated with hepatitis B virus infection. Hepatitis delta virus can replicate only in cells infected with hepatitis B virus, since it borrows the hepatitis B surface proteins to package and form an envelope around its own capsid proteins and genomic RNA. However, hepatitis delta virus resembles viroids and is quite different from the hepadnaviruses.

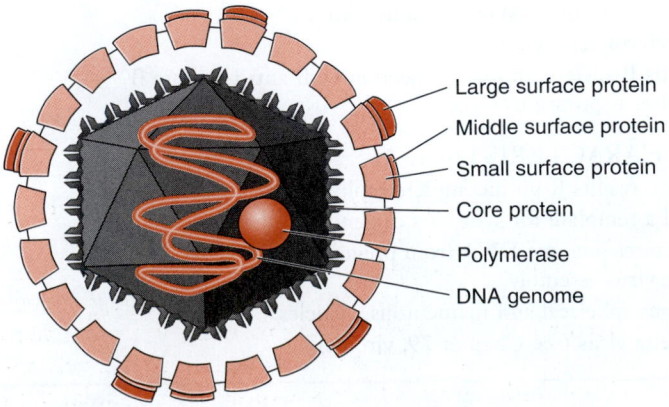

FIGURE 28.1 Structure of virions of hepatitis B virus. Genome is shown as a orange wavy line bound to polymerase (orange sphere). Capsid, shown as an icosahedron, is constructed from core protein. Envelope, containing three forms of surface protein, is shown as orange subunits surrounding the capsid.

Dane particles are infectious virions; abundant noninfectious particles lack nucleocapsids

The hepatitis B virion, or Dane particle (Figure 28.1), has a diameter of 42 nm. Its envelope contains hepatitis B surface (S) proteins and lipids; however, electron microscopy does not reveal a typical unit membrane structure, and the envelope contains a lower proportion of lipids than most membranes and viral envelopes. The envelope surrounds the nucleocapsid or core, which is composed of 180 molecules of hepatitis B core (C) protein arranged with T = 4 icosahedral symmetry. The core contains at least one molecule of reverse transcriptase (P), as well as the viral genome.

High concentrations of noninfectious particles can be found in the serum together with Dane particles during the acute phase of infection. These particles are composed primarily of lipids and hepatitis B surface proteins. Both filamentous and spherical noninfectious particles, each having a diameter of 22 nm, are found in the serum. The spheres contain only small and middle hepatitis B surface proteins (SS and MS), whereas the filaments also contain large hepatitis B surface proteins (LS). These particles contain no core proteins, reverse transcriptase, or viral genome DNA. Since noninfectious particles contain the viral surface antigens, they can induce an immune response in the infected host. Production of massive amounts of noninfectious particles may provide a "stealth" mechanism for evading the immune response; they can divert the immune attack by binding to antibodies and **complement**, allowing the less abundant virions to traverse the bloodstream undetected.

Eight major strains with different genomic sequence variations (genotypes) and four major antigenic strains (serotypes) of hepatitis B virus are found worldwide. The genotypes, denoted Group A through Group H, are distributed in specific regions of the world. For example, genotypes A and D are found in Europe, D and E in Africa, B, C, and D in Asia, B in Alaska, C in Australia, and F, G, and H in Central and South America. Genotypes vary in the severity of disease symptoms, in the ability to induce chronic liver disease, and in their response to **interferon** or chemotherapy.

Serotypes are defined by antibody recognition sites on the small surface protein. A common **epitope** present on the surface protein of all known hepatitis B strains is called determinant "a". The four distinct serotypes contain two pairs of mutually exclusive epitopes named "d" or "y" and "w" or "r". Determinant d has a lysine at position 122 on the small surface protein, while y has an arginine at that position; w has a lysine at position 160 while r has an arginine at that position. Thus, a common strain of hepatitis B virus might be classified as adw or adr.

The viral genome is a circular, partly single-stranded DNA with overlapping reading frames

Hepatitis B virus has one of the smallest viral genomes known: the circular DNA is 3.2 Kb long. This DNA has the unusual property of being partly double-stranded, with a single-stranded region of variable length (Figure 28.2a). Furthermore, unlike other circular DNA viral genomes, neither DNA strand forms a covalently closed circle. The (full-length) minus DNA strand is joined at its 5′ end to a tyrosine residue in the P protein, a virus-coded DNA polymerase with reverse transcriptase activity. The complementary plus DNA strand is joined to a short, capped RNA at its 5′ end. The plus strand is shorter than the minus strand, because its synthesis is interrupted before completion, giving rise to the single-stranded region. The ends of the DNA are held together as a circle by hydrogen bonding between short repeated sequences called DR2, near the 5′ ends of both strands. These peculiar features arise from the mechanism of generation of the genome DNA via reverse transcription (see below). *In vitro* incubation of disrupted virions with deoxyribonucleoside triphosphates allows the virion-associated DNA polymerase to extend the plus DNA strand, filling in the single-stranded gap (Figure 28.2b).

Coding regions in the genome of hepatitis B virus are organized in a highly efficient fashion (Figure 28.3). Four partly overlapping open reading frames, designated C (for capsid protein), P (for polymerase), S (for surface protein), and X (for a regulatory protein), are translated to yield seven different viral proteins. Every base-pair in the genome of hepatitis B virus is involved in coding for at least one viral protein! The genome also contains two enhancer elements (Enh 1 and 2) that regulate levels of transcription, a polyadenylation signal, a packaging signal (ε), and direct repeats (DR1 and DR2) that are involved in reverse transcription. These sequence elements necessarily overlap with one or more coding regions (see inset in Figure 28.3). Few other viruses have so many overlapping functions within such a compact genome.

Nucleocapsids enter the cytoplasm via fusion and are transported to the nucleus

Replication of hepatitis B virus is most efficient in liver cells (hepatocytes), although other cell types, including monocytes, epithelial cells, and endothelial cells, have been found to support minimal viral replication. Hepatocytes freshly explanted from duck liver support duck hepatitis B virus infection, but there are currently no *in vitro* cultivated cell lines that can be productively infected with hepatitis B virus. Consequently, initial steps

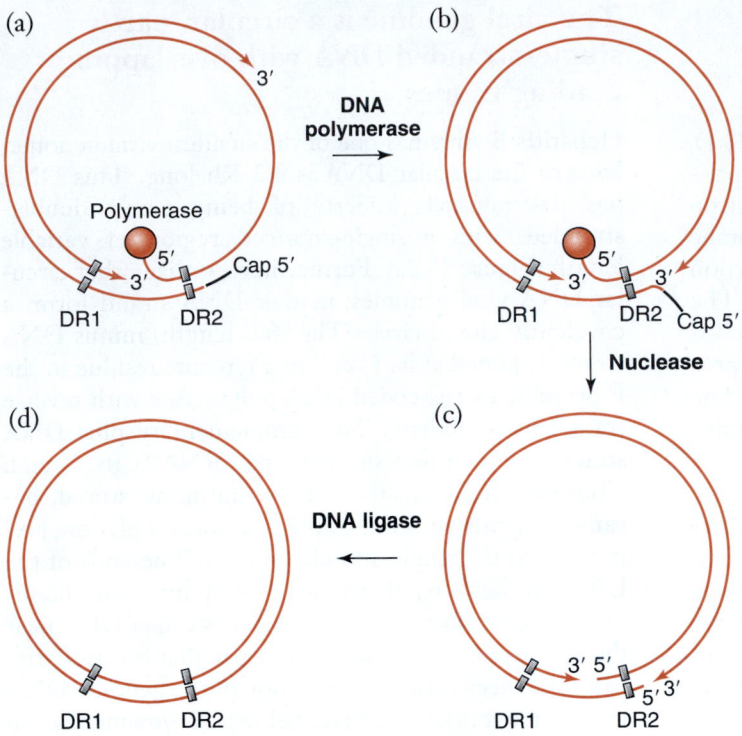

FIGURE 28.2 Structure of hepatitis B virus genome, and its conversion to a covalently closed, circular DNA.
(a) The 5′ ends of minus-strand (inside circle) and plus-strand (outside circle) genomic DNA are joined by short complementary regions named DR2 (gray boxes). The 5′ end of the minus strand is attached to the viral polymerase protein (orange sphere) and the 5′ end of the plus strand is attached to a short, capped RNA (black line). The plus strand is incomplete. (b) DNA polymerase extends the plus strand, filling in the ss gap. (c) Nucleases remove the 5′ ends of both strands. (d) DNA ligase joins the 3′ and 5′ ends of each strand, forming a fully covalently closed double-stranded circular DNA.

of virus entry are poorly understood. However, several cell lines are capable of supporting hepatitis B virus DNA synthesis upon transfection with cloned viral DNA. These cell systems have helped to elucidate much of the hepadnavirus life cycle.

Attempts to define a receptor for hepatitis B virus have yielded various candidates, including apolipoprotein-H, fibronectin, interleukin-6, annexin V, and an unknown 80 KDa protein. Which of these proteins are functional cellular receptors remains to be established.

It is believed that the large hepatitis B surface protein undergoes proteolytic cleavage as it interacts with the cellular membrane, exposing a fusion peptide. This results in the fusion of the virion envelope and the host

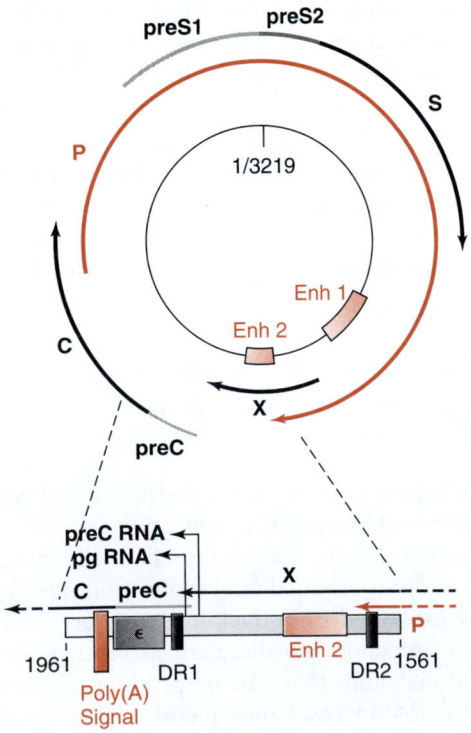

HBV Genome (adw subtype) 3219 bp

Some Regions of Interest

Enh 1	1060-1260
DR2	1590-1600
Enh 2	1635-1714
preC RNA start	1785-1791
pg RNA start	1818,1819
DR1	1824-1835
ε-Stem Loop	1847-1907
Poly (A) Signal	1916-1921

FIGURE 28.3 Coding and signaling regions on the hepatitis B genome. The four partially overlapping open reading frames (C, P, S, and X) are shown with arrowheads pointing in the direction of transcription and translation. Translation of the C and S open reading frames starts at two or three alternative sites (preC or C and preS1, preS2, or S). The overlap of control sequences and coding sequences in a 400 nt stretch of the genome is shown in the inset.

cell membrane and the release of the nucleocapsid into the cytoplasm. The nucleocapsid is then transported to the nuclear membrane. Studies with duck hepatitis B virus have suggested that release of the viral genome into the nucleus occurs by interaction of the core with the nuclear pore complex.

The viral genome is converted within the nucleus into a covalently closed, circular form (Figure 28.2). This process involves extension of the plus-strand DNA across the single-stranded gap region to form a fully double-stranded DNA, removal of the terminal primers (the P protein on the minus strand, and the short, capped RNA primer on the plus strand), and covalent ligation of the resulting 3′ and 5′ ends of each DNA strand (Figure 28.2b, c, and d). The P protein may take part in the extension of the plus strand, but inhibitor studies have shown that all of these steps (strand extension, primer removal, and ligation) can be carried out by host cell enzymes in the nucleus.

Transcription of viral DNA gives rise to several mRNAs and a pregenome RNA

In contrast to retroviruses, hepadnavirus DNA is not integrated into the host cell genome, but remains in a free, circular form. This circular viral DNA uses cellular RNA polymerase II to direct the synthesis of a set of five messenger RNAs, one of which also serves as a pregenome RNA (pg RNA). Transcription is influenced by two enhancer elements (Enh I and Enh II, Figure 28.3), which are binding sites for several cellular transcription factors. Five different promoter elements (preC, pregenome, preS1, S, and X) direct initation of transcription by RNA polymerase II (Figure 28.4). The preC and pregenome promoters are closely spaced, only 10–15 nucleotides from each other (see insert, Figure 28.3). A single polyadenylation signal specifies the 3′ ends of all viral RNAs. This gives rise to five classes of RNA transcripts, ranging in size from 900 to 3500 nucleotides (Figure 28.4).

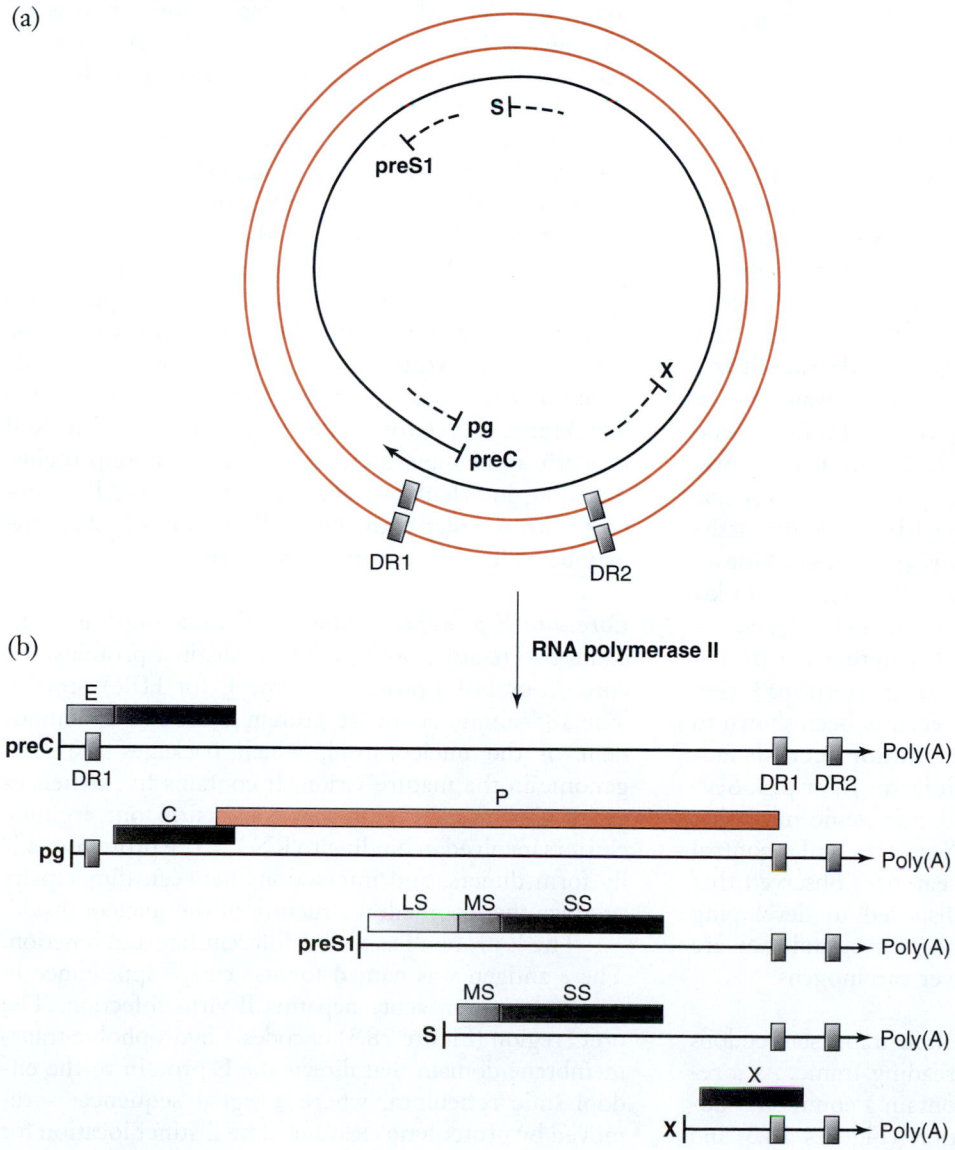

FIGURE 28.4 Transcription of hepatitis B virus DNA. **a.** Circular DNA is transcribed by host RNA polymerase II, which recognizes five promoters (preC, pg, preS1, S, and X) to generate five classes of capped mRNAs that share a common 3′ polyadenylated end. b. Open reading frames and translation products are shown above each RNA. pg RNA also serves as a template for virus replication.

The smallest mRNA encodes the X protein. Transcripts initiated at the S promoter encode both the small (SS) and medium (MS) surface proteins, while those initiated at the preS1 promoter encode the large surface protein (LS). Transcripts initiated at the pregenome promoter have a dual role. They can serve as messenger RNAs for the C and P proteins, and can also serve as pregenome RNA (pgRNA), an intermediate in the synthesis of progeny viral genome DNA. Transcripts initiated at the nearby preC promoter encode the E protein.

The 3.5 Kb pg and preC transcripts are exceptional in that they extend more than once around the circular DNA genome. Their promoters lie upstream of the polyadenylation signal, but RNAs initiated at these promoters are not cleaved on first passage of the polyadenylation signal, which lies very close to the promoter region. This may be due in part to the sequence of the polyadenylation signal, UAUAAA, which differs from the consensus AAUAAA found in most genes in vertebrate cells.

The roles of hepatitis B virus proteins

X protein. The smallest open reading frame encodes the 154–amino acid X protein. This protein is aptly named, because its function remains elusive; many contradictory findings have been made over the years. X has been implicated in the development of **hepatocellular carcinoma** (liver cancer). X is not found in mature virions or nucleocapsids, but it stimulates viral gene transcription during the early phases of infection. It was originally called the "promiscuous transactivator" since it increases the synthesis of many viral and cellular gene products. However, the bulk of X protein has recently been shown to reside in the cytoplasm, and its major function may be to regulate signal transduction pathways. X has been reported to up-regulate c-**src kinase**, **Ras**/raf/MAP kinase, stress kinase (SAPK), protein kinase C, the **Jak/Stat pathway** (see Chapter 31), and the transcription factor **NF-κB**, and to interact with and sequester the **tumor suppressor protein** p53 (see Chapters 10, 11, and 12). It has recently been shown to inhibit **apoptosis** mediated by fas, **tumor necrosis factor (TNF)**, transforming growth factor β, or p53. Several laboratories have generated transgenic mice that synthesize the hepatitis B virus X protein under control of its own promoter. These researchers observed that the transgenic mice were predisposed to developing liver tumors within a year of their birth, and they are more sensitive to exposure to liver carcinogens.

Surface proteins. Three different in-frame start codons are utilized within the S open reading frame. As a result, all three surface proteins contain a common S domain in their C-terminal portion (Figures 28.3 and 28.4). These proteins are involved in envelope formation, as well as the formation of noninfectious particles found in serum.

The small hepatitis B surface protein (SS, also known as SHBs), originally designated as Australia antigen, is the smallest and most abundant surface protein. SS has four α-helical hydrophobic regions. It contains 14 cross-linked cysteine residues, providing for high stability, and it can be glycosylated at asparagine-146. SS has at least two transmembrane regions. Prior to DNA sequence analysis, different strains of hepatitis B virus were classified based on the antigenic epitopes contained within SS (see above).

The second most abundant membrane protein produced is the middle hepatitis B surface protein (MS, or MHBs). The MS protein contains an additional 55 N-terminal amino acids encoded by the preS2 domain (Figure 28.3). These additional amino acids are primarily hydrophilic and are believed to reside on the external face of hepatitis B virions. The preS2 domain also contains a glycosylation site at asparagine-4 that is always glycosylated in the MS protein. However, viruses that lack the ability to express MS protein still retain the ability to produce infectious virions.

The largest and least abundant envelope protein is the large hepatitis B surface protein (LS, or LHBs). It contains, besides the preS2 and S domains, an additional 119 N-terminal amino acids in the preS1 domain. The sequence of preS1 varies among different strains and it is likely involved in host cell attachment. The preS1 domain does not contain any additional glycosylation sites, but it has a **myristate** residue at its N-terminus, a modification that serves to anchor the N-terminus in the membrane. **Myristoylation** may also help LS fold properly at the membrane surface. A variety of proteins, including interleukin-6 and annexin V, have been reported to interact with the preS1 region of LS; these may be cellular receptors for the virus.

Core and E proteins. The C open reading frame encodes two related, yet functionally distinct proteins: the core (C or HBc) protein and the E (or HBe) protein. The 185–amino acid core protein is the major component of the nucleocapsid, which packages the viral genome in the mature virion. It contains five α-helices and a C-terminal region that consists of four arginine clusters involved in binding to RNA. Core proteins readily form dimers, and interactions between dimer pairs produce the icosahedral structure of the nucleocapsid.

The E protein has a very different fate and function. The *e* antigen was named for its "early" appearance in serum during an acute hepatitis B virus infection. The preC region (Figure 28.3) encodes a hydrophobic transmembrane domain that directs the E protein to the endoplasmic reticulum, where a signal sequence is removed by proteolytic cleavage. The distinct location for

E alters its folding and antigenicity when compared to the nearly identical C protein. Among other differences, the basic C-terminal region is removed by cleavage from E. A function for E has not been established. One theory is that high levels of E protein may suppress the host immune system and prevent it from eliminating cells that contain hepatitis B virus. In support of this hypothesis, woodchuck hepatitis virus variants that lack E are incapable of establishing persistent viral infections.

Polymerase protein. The largest open reading frame in the genome encodes the hepatitis B DNA polymerase protein (P, or HBp). This 90 KDa protein functions as an RNA/DNA-dependent DNA polymerase (reverse transcriptase). The polymerase plays a critical role in hepadnavirus genome replication and pregenome RNA encapsidation. P has four characteristic domains: an N-terminal domain for priming minus-strand DNA synthesis; a spacer domain; a reverse transcriptase domain, which occupies roughly 40% of the protein; and a **ribonuclease H** domain. The ribonuclease H activity degrades the pregenome RNA during the process of genome replication.

The pregenome RNA is packaged by interaction with polymerase and core proteins

The pregenome RNA (pgRNA), in addition to serving as mRNA for C and P proteins, is used as a template to produce the DNA genome of hepatitis B virus. This occurs in a "packaged" RNA-protein complex including both the C and P proteins. The viral replication machinery targets the pregenome RNA for packaging by recognition of a region known as the epsilon (ε) stem-loop (Figures 28.3 and 28.5). All viral transcripts have a copy of ε at their 3′ ends, just upstream of the polyadenylation site; however, only pregenome and preC RNAs have a second copy of the ε stem-loop near their 5′ ends. It appears that only the 5′ ε stem-loop is functional in packaging, and only in the pregenome RNA. The slightly longer preC RNA is not packaged, apparently because translation of the preC open reading frame (which overlaps with the ε stem-loop) interferes with packaging. The pregenome RNA starts just downstream of the beginning of the preC open reading frame and therefore does not translate this region (Figure 28.3).

Analysis of the ε region sequence predicts a series of inverted repeats that fold into a three-dimensional stem-loop structure (Figure 28.5). This secondary structure is conserved among all hepadnaviruses despite many differences in the primary sequence. The polymerase protein is believed to recognize and interact directly with the ε stem-loop, initiating both encapsidation and reverse transcription of the pregenome RNA. The C-terminus of the polymerase protein interacts with the core protein to begin encapsidation only when the polymerase is bound to the ε stem-loop.

Genome replication occurs via reverse transcription of pregenome RNA

The generation of the hepatitis B virus DNA genome from pregenome RNA is formally similar to the synthesis of retrovirus proviral DNA from genome RNA (Chapter 25). In both cases, a reverse transcriptase initiates DNA synthesis near the 5′ end of the template RNA, then jumps to a repeated sequence at the 3′ end, and copies the remainder of the RNA, degrading the template as it progresses. However, many details of the process are different, as described below (Figure 28.5 and Table 28.1).

Step 1. Initiation of the reverse transcription process begins when P binds to the 5′ ε stem-loop. Retrovirus reverse transcriptases use a cellular tRNA bound to their genome RNA as a primer to initiate synthesis of a DNA copy (Chapter 25). Hepatitis B virus, in contrast, initiates DNA synthesis on the pregenome RNA template by covalently linking the first nucleotide of the DNA chain to the hydroxyl group of a tyrosine residue in the polymerase protein itself. This has similarities to the mechanism by which adenoviruses initiate replication of their linear, double-stranded DNA genomes (see Chapter 12), but in the case of hepadnaviruses the polymerase plays the roles of both the adenovirus terminal protein and the DNA polymerase. The P protein remains linked to DNA throughout the process, so presumably the active site extends the 3′ end of the growing DNA molecule, while the 5′ end remains attached to another part of the same molecule of P, forming a circular complex.

Steps 2 and 3. Reverse transcription progresses for only a few base-pairs beyond the 5′ ε stem-loop, after which the enzyme and the newly synthesized DNA are translocated onto an identical sequence in the direct repeat region 1 (DR1) near the 3′ end of the pregenome RNA. This step is reminiscent of the transfer of the retrovirus "minus strong stop" DNA to the repeat region at the 3′ end of the retrovirus genome RNA.

Steps 4 and 5. The P protein then makes a complete (minus strand) copy of the remainder of the pregenome RNA by extending the DNA chain to the 5′ end of the RNA template. As reverse transcription proceeds, the RNA template is degraded by the ribonuclease H activity of P. A short RNA fragment at the capped 5′ end, including the DR1 region, remains resistant to ribonuclease H degradation. This fragment is translocated to the identical DR2 region near the 5′ end of the newly

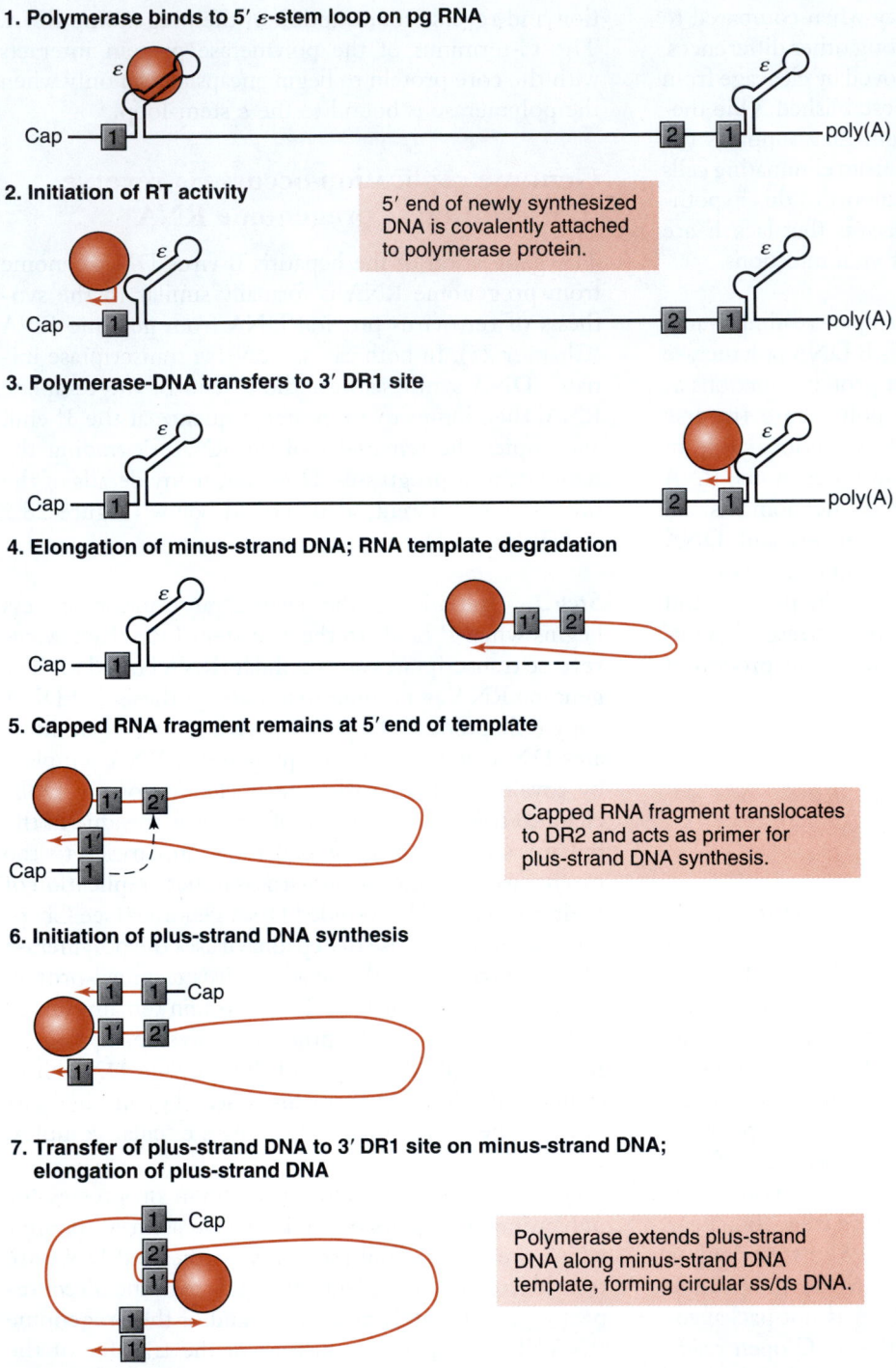

FIGURE 28.5 Reverse transcription of pregenome RNA to form hepatitis B virus genome DNA. Boxes labeled 1 and 2 represent DR1 and DR2 sequence elements. ε stem-loops are shown at both ends of pregenome RNA (pg RNA). Orange sphere represents reverse transcriptase (P protein). 3′ ends of growing DNA chains (orange) are shown as arrows. RNA (black) degraded by RNAse H is shown as dashed line. See text for details.

synthesized minus-strand DNA, where it forms a short RNA/DNA hybrid.

Steps 6 and 7. The translocated RNA fragment serves as a primer for subsequent synthesis of plus-strand DNA. As in step 2, the polymerase begins copying the template minus-strand DNA near its 5′ end, making a short DNA product with the capped RNA fragment at its 5′ end. This nascent DNA chain then jumps to the 3′ end of the same minus strand DNA molecule by annealing to the DR1 sequence, and the polymerase proceeds to copy the minus strand. The newly synthesized plus-strand DNA remains hydrogen-bonded to the minus strand, forming a double-stranded DNA. However, plus-strand DNA synthesis is usually not completed by the time the virion is released from the host cell, and stalls when the dNTP pools within the virion are exhausted. This explains the region of single-stranded

TABLE 28.1 Comparison of reverse transcription by retroviruses and hepadnaviruses

Step	Retroviruses	Hepadnaviruses
Template for DNA synthesis	Plus-strand virion RNA	Plus-strand pregenome RNA
Site of DNA synthesis	Residual capsid after virus entry (cytoplasm)	Newly assembled capsid before virion formation (cytoplasm)
Primer for initiation of minus-strand DNA synthesis	Cellular tRNA bound near 5′ end of genome RNA	Polymerase protein bound near 5′ end of pregenome RNA
First switch: to 3′ end of RNA template	Via terminal repeat (R) sequence	Via subterminal DR1 sequence
RNA template degradation	RNAse H activity in RT	RNAse H activity in Polymerase
Initiation of plus strand-DNA synthesis	RNAse H-resistant polypyrimidine tract in genome RNA	RNAse H-resistant capped 5′ RNA fragment bound to DR2 sequence
Second switch: to 3′ end of minus-strand DNA	Via primer-binding site	Via DR1 sequence
Final product	ds linear DNA with long terminal repeats	Partly ss/ds circular DNA
Fate of DNA	Integrated into cellular chromosomal DNA	Incorporated into virions

DNA found within the genomes of purified virions. The resulting molecule, shown in step 7 of Figure 28.5, is identical to the molecule shown in Figure 28.2a. Table 28.1 and Figure 28.6 summarize in schematic form the similarities and differences between the reverse transcription mechanisms and the replication cycles of retroviruses and hepadnaviruses.

Virions are formed by budding in the endoplasmic reticulum

Because more S mRNA is made than preS1 mRNA, larger amounts of the SS and MS proteins are made than LS protein. These proteins are inserted into the endoplasmic reticulum as they are translated (Figure

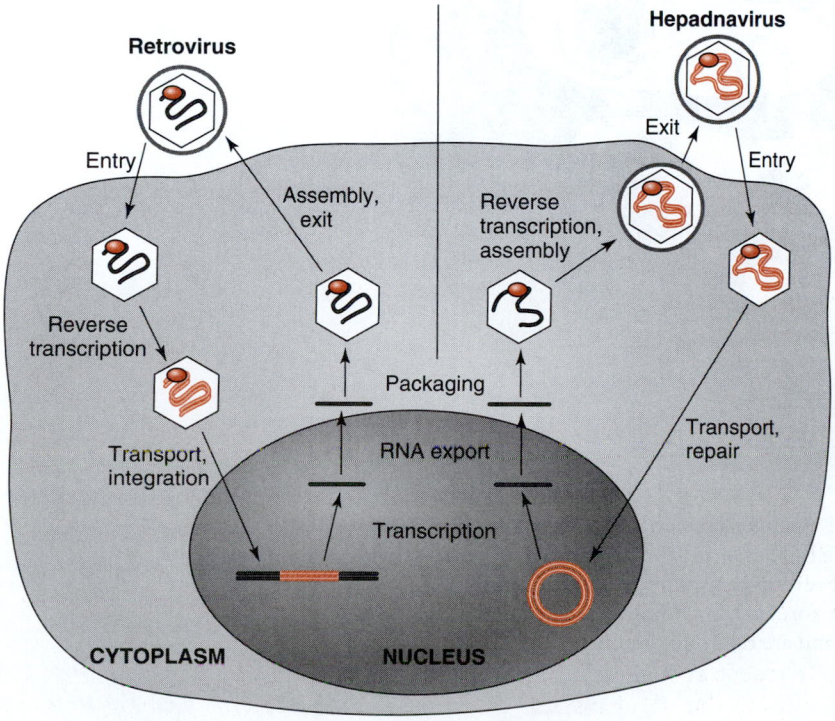

FIGURE 28.6 Schematic comparison of replication cycles of retroviruses and hepadnaviruses. RNAs are shown in black and DNAs in orange. Capsids are shown as hexagons, and the virus envelope is shown as a gray circle. Reverse transcriptase is shown as orange sphere. Numerous details in the replication cycles are omitted.

28.7). By interaction between their transmembrane regions, S proteins aggregate in specific regions of endoplasmic reticulum membranes and exclude host membrane proteins. Regions rich in S proteins bud into the lumen of the endoplasmic reticulum in the absence of cores to produce the abundantly secreted 22 nm diameter spheres and filaments (Figure 28.7). These noninfectious particles contain little or no LS protein, since its presence favors the retention of surface proteins within the endoplasmic reticulum.

The assembled core associates with areas of the Golgi membranes rich in these surface proteins. LS is believed to interact with the core protein at the cytoplasmic face of the Golgi membranes, pulling the nucleocapsid into a vesicle that forms a 42 nm enveloped Dane particle within the lumen of the Golgi vesicle (Figure 28.7). The mature virions are then secreted into the extracellular environment via **exocytosis**.

Hepatitis B virus can cause chronic or acute hepatitis, cirrhosis, and liver cancer

Hepatitis B virus does not immediately kill infected cells but instead leads to chronic, persistent infections. It appears that the majority of tissue damage in the host is the result of a host immune response directed against viral antigens on the surface of infected cells. In the initial stages of infection, patients suffer from flu-like symptoms, **jaundice**, and **serum sickness**, a hypersensitivity response to large amounts of viral antigen in the serum. Fortunately, about 90% of infected adults resolve their infection within six months and show no further symptoms. About 1% of those infected suffer from a severe disease known as **fulminant hepatitis**, which results in massive levels of liver damage and often is fatal. The remaining 9% of infected adults become chronic carriers, almost half of whom may eventually

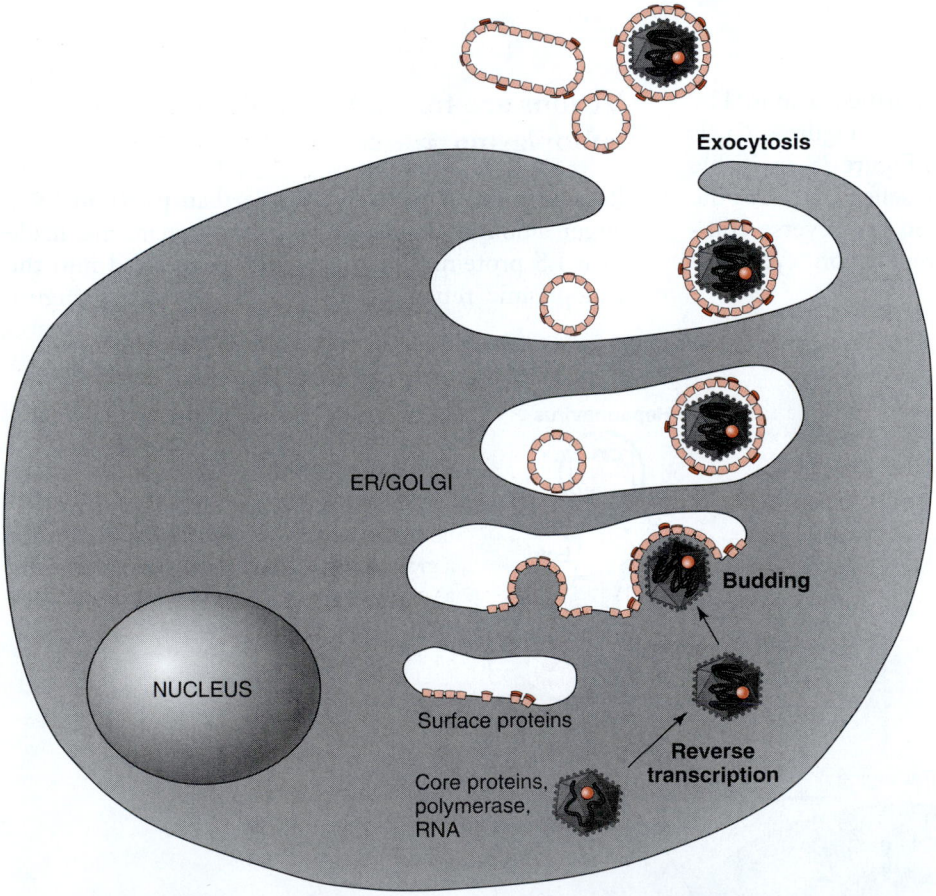

FIGURE 28.7 **Assembly of nucleocapsids and virions of hepatitis B virus.** Surface proteins are inserted into membranes of the endoplasmic reticulum and Golgi apparatus. Nucleocapsids are assembled in the cytosol and reverse transcription generates the DNA genome. Virions and noninfectious particles are formed by budding into the lumen of endoplasmic reticulum and Golgi membranes, and are released upon fusion of vesicles with the plasma membrane.

develop liver cancer. Other chronically infected individuals progress to **cirrhosis**, which is characterized by fibrin deposition in the liver, the production of **ascites** fluid, infiltration of lymphocytes, flu-like symptoms, and hemorrhaging. Infected children have different disease outcomes; only about 50% manage to clear the virus and 40% become chronic carriers, perhaps due to the underdeveloped state of a child's immune system.

Hepatitis B virus is transmitted by blood transfusions, contaminated needles, and unprotected sex

Transmission of hepatitis B virus is primarily through blood or sexual contact. Hepatitis B virus can also be passed on to babies at birth. There are presently no drugs or treatments that ensure clearance of hepatitis B virus from infected individuals. However, there are several ways to minimize the likelihood of contracting the virus. Risky activities to avoid include intravenous drug use and unprotected sexual intercourse with multiple partners. In professions where the virus might be expected to be in the environment, care when handling potentially contaminated surfaces, fluids, and sharp objects will minimize accidental exposure. Members of these professions are encouraged to be vaccinated. Infection with hepatitis B virus nevertheless remains a problem for certain occupational groups such as morticians, doctors, dentists, dialysis workers, hemophiliacs, mental care facility workers, tattoo-parlor workers, firefighters, police officers, daycare workers, and barbers.

A recombinant vaccine is available

In 1986, Pablo Valenzuela and William Rutter of Chiron Corporation produced the first recombinant hepatitis B vaccine in yeast, and this is the basis of the vaccines still used today. In a new approach, Charles Arntzen of the Boyce Thompson Institute at Cornell University is currently testing edible hepatitis B vaccines engineered into bananas! Another method of prevention is through **immunoprophylaxis**, used primarily on newborns of mothers who are carriers of hepatitis B virus. In this procedure, anti-hepatitis B immunoglobulins are injected into the exposed individual to neutralize circulating virus and prevent it from entering target cells. In spite of the availability of a vaccine, millions of individuals still are chronically infected with the virus.

Antiviral drug treatment has real but limited success

Currently, only two antiviral drugs are effectively used to treat chronic carriers. One is interferon-α (see Chapter 31), an immune system modulator that works by stimulating cytokine synthesis and by enhancing host antiviral responses. The second is lamivudine (3TC), a nucleoside analogue that was originally developed for the treatment of HIV (see Chapter 32). 3TC interferes with the DNA polymerase activity of the P protein. Both of these treatments can reduce the viral load. However, this reduction is often only temporary and virus titers can begin to increase after prolonged therapy, due to the emergence of drug-resistant mutants. The drugs also have some secondary effects, such as nausea and fever, which make treatment uncomfortable. Clearly there is a need for the further development of antiviral therapies.

There is still no cure for individuals infected with hepatitis B virus, and infection remains a major worldwide health problem, particularly because it can be a persistent, chronic infection that can eventually lead to serious liver disease. Availability of the vaccine worldwide, especially in poorer countries, is still a major problem. More effective vaccines and antiviral agents are currently being developed by a number of laboratories and pharmaceutical companies.

Some interesting facts about hepatitis B virus infection

- You should never use someone else's toothbrush. It might transmit hepatitis B virus to you!
- The first hepatitis B virus vaccine was prepared from the blood of paid chronic carriers.
- Hepatitis B virus can linger on drying surfaces like door handles for over a week and still remain infectious.
- Most people who are acutely infected with hepatitis B virus do not even realize they have it.
- The hepatitis B virus vaccine is only effective in about 80–90% of those who are vaccinated with the three-injection regimen.
- Two billion people worldwide have been exposed to hepatitis B virus.
- Some laboratories have been working on generating an edible hepatitis B virus vaccine by expressing HBs proteins in potatoes, carrots, and bananas.

SUGGESTED READING

1. Bouchard, M.J. and Schneider, R.J. (2004). The enigmatic X gene of hepatitis B virus. *Journal of Virology* **78,** 12725–12744.

2. Gerlich, W.H. (1995). Hepatitis B virus. *Intervirology*, 38(1–4), 1–154. Karger Press, Zurich (collection of reviews).

3. Koshy, R. and Caselmann, W.H. (1998). Hepatitis B virus: Molecular mechanisms in disease and novel strategies for therapy. Imperial College Press, River Edge, New Jersey.

4. Locarini, S. (2004). Molecular virology of hepatitis B virus. *Seminars in Liver Disease* **24,** 3–10.

5. Nassal, M. (1999). Hepatitis B virus replication: Novel roles for virus–host interactions. *Intervirology* **42**(2–3), 117–126 (special section).

6. Tiollais, P. and Buendia, M.-A. (1991). Hepatitis B virus. *Scientific American* **264,** 116–123.

7. Zoulim, F. and Trepo, C. (1999). New antiviral agents for the therapy of chronic hepatitis B virus infection. *Intervirology* **42**(2–3), 125–144 (special section).

KEY TERMS

Apoptosis
Ascites
Cirrhosis
Complement
Enteric
Epitope
Exocytosis
Fulminant hepatitis
Hepatitis
Hepatocellular carcinoma
Immunoprophylaxis
Interferons
Jak-Stat pathway
Jaundice
Myristate
Myristoylation
NF-κB (nuclear factor-κB)
Ras
Ribonuclease H
Serum sickness
Src kinases
Tumor necrosis factor (TNF)
Tumor suppressor protein

Viroids and Hepatitis Delta Virus

Viroids

Name means "virus-like" particles

VIROIDS

Autonomously replicating pathogens.
Unencapsidated, circular single-stranded RNA, 250–400 nt.
Extensive internal basepairing.
Do not encode proteins.
Some possess catalytic activity (hammerhead self-cleaving structure).

VIROID-LIKE PLANT SATELLITE RNAS

Require helper virus for replication
Circular single-stranded RNA, 200–500 nt.
Highly base-paired structure like viroids.
Do not encode proteins.
Unlike viroids, are encapsidated by helper virus proteins.
All possess catalytic activity (hairpin or hammerhead self-cleaving structure).

HEPATITIS DELTA VIRUS (HDV)

The only viroid-like satellite RNA known to infect humans and animals.
Depends on hepatitis B virus for its replication.
Circular ss RNA, 1682 nt.
Extensive internal base-pairing.
Encodes two proteins from one reading frame by RNA editing.
Packaged by hepatitis B virus S protein.
RNA is capable of self-cleavage (delta ribozyme).

DISEASES

Viroids are transmitted between plants via mechanical breaks.
Symptoms of viroid infections include leaf decoloration, dwarfing, enhanced fruit production, orange spotting, decline in fruit yield, stunting, etc.
Hepatitis delta virus is a bloodborne pathogen that replicates in the liver.
Hepatitis delta virus frequently causes fulminant hepatitis in both primate and nonprimate hosts.
Hepatitis delta virus is only detected in the presence of hepatitis B virus.

DISTINCTIVE CHARACTERISTICS

Viroids are the only known infectious, pathogenic RNAs that are not packaged and do not code for any protein product.
They use cellular RNA polymerases to replicate RNA.
They replicate via rolling circle forms, RNA cleavage, and ligation.
Some viroids have enzymatic activity ("ribozymes": self-cleavage).
They may be relics of the ancient RNA world.

Viroids are small, circular RNAs that do not encode proteins

Until the early 1970s, viruses were believed to be the smallest agents of infectious diseases. The discovery of small, self-replicating RNA pathogens changed this view. With genomes ranging from 250 to 1700 nucleotides, these infectious RNAs cause various diseases in plants and in animals. They are divided into four major groups: viroids, satellite RNAs, viroid-like plant satellite RNAs, and hepatitis delta virus.

Viroids were first discovered as the causative agent of potato spindle tuber disease. This novel pathogen consists solely of a small, unencapsidated single-stranded RNA, for which the term "viroid" ("virus-like") was proposed. Since then, different viroids have been identified as the causative agents of a number of diseases that affect plants of economic importance. Viroids are composed only of a circular, single-stranded RNA molecule 240–400 nucleotides in length, a size significantly smaller than the genomes of the smallest viruses. The differences between viruses and viroids are not only restricted to size: viroids, unlike viruses, do not code for any proteins and therefore must rely on host enzymes for their replication. It is believed that they cause disease by direct interaction between the viroid RNA, or other viroid-specific RNAs generated in the course of the infection, and cellular targets. However, the specific molecular mechanisms of pathogenesis are presently unknown.

Group A and group B viroids have distinct properties

Viroids can be classified into two groups based on comparative sequence analyses and whether they possess a central conserved region (Table 29.1). There are currently only three members in the avocado sunblotch viroid (ASBVd) group, also known as group A or the *Avsunviroidae* family. In this group, both plus- and minus-strand viroid RNAs have the ability to self-cleave their RNA multimers (see below). All other viroids identified to date fall into the potato spindle tuber viroid (PSTVd) group, also known as group B or the *Pospiviroidae* family. The RNAs of group B viroids possess a highly conserved central region (C), and are not capable of self-cleavage.

Sequence analysis indicates that group B viroid RNAs are characterized by five domains: T_L (terminal left), P (pathogenic), C (central), V (variable), and T_R (terminal right) (Figure 29.1b). The boundaries between these domains are defined by sharp changes in degree of sequence homology among different viroids. Initially, these structural domains were presumed to have specific functional roles; for example, the P domain was associated with pathogenicity. However, the situation is more complex, and expression of disease symptoms is now thought to be controlled by discrete determinants located within different domains. Group B viroids are divided into five subgroups (Table 29.1) based mainly on the presence of highly conserved sequences within the C domain, and in part on the presence of sequence homologies in the other domains.

Viroid RNAs have a high degree of base-pairing throughout the molecule. Members of group B fold into a rod-like secondary structure that includes short double-stranded regions separated by small single-stranded loops (Figure 29.1b). Although avocado sunblotch viroid is also believed to adopt a rod-like secondary structure, the other two members of group A, peach latent mosaic viroid and chrysanthemum chlorotic mottle viroid, have complex secondary structures (Figure 29.1a), in which several stem-loops (hairpins) emerge from a central core. The actual structure of viroids inside

TABLE 29.1 Classification of viroids

Group/Subgroup	Characteristics	Examples
A (ASBVd) (*Avsunviroidae*)	Self-cleaving RNA; replicate in chloroplasts	Avocado sunblotch viroid Peach latent mosaic viroid Chrysanthemum chlorotic mottle viroid
B (PSTVd) (*Pospoviroidae*)	RNAs do not self-cleave; replicate in nucleus	
PSTVd		Potato spindle tuber viroid (8 others)
CCCVd		Coconut cadang-cadang viroid (3 others)
HpSVd		Hop stunt viroid
ASSVd		Apple scar skin viroid (8 others)
CbVd		Coleus blumei-1 viroid (2 others)

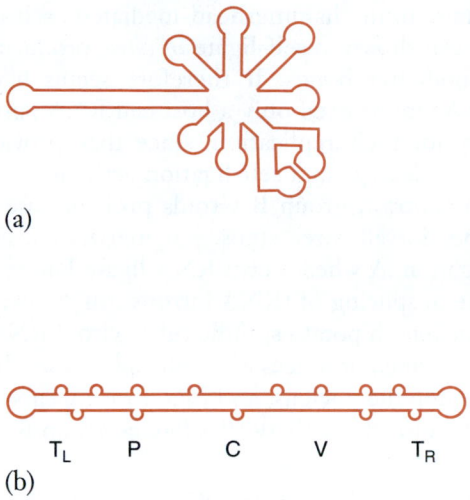

FIGURE 29.1 **Structure of viroid RNA.** (a) Branched structure proposed for peach latent mosaic viroid, a member of group A. (b) Rod-like structure proposed for group B viroids, showing the five domains (T_L: terminal left; P: pathogenic; C: central; V: variable; T_R: terminal right).

the cell remains undetermined, and it is possible that these RNAs adopt several alternative conformations during the different steps of their life cycles.

Viroids replicate via linear multimeric RNA intermediates

By convention, the plus polarity is assigned to the most prominent viroid RNA species, and the minus polarity to its complementary strand; since viroids do not code for proteins, these designations are arbitrary. Group A viroids replicate by a **rolling-circle replication** mechanism that generates linear multimers from a circular RNA template (Figure 29.2). A host RNA polymerase copies plus-strand circular RNA (top of Figure 29.2) into single-stranded linear, complementary minus strands (step 1). Because there is no strong termination signal for the RNA polymerase on the template RNA, transcription can continue around the circle for several cycles, producing a multimeric linear RNA (step 2). These RNAs are subsequently cleaved to unit-length minus strands (step 3), which circularize (step 4). The circular minus-strand RNAs then serve in the same fashion as templates for the synthesis of multimeric, linear plus-strand RNAs, which are cleaved and circularized (steps 5–8).

This cycle can generate several new circular plus-strand RNA daughter molecules from each plus-strand template, because transcription continues for multiple rounds at both steps 1 and 5. This strategy is called "symmetric" replication, since both plus- and minus-strand circles are replicated in the same fashion. Group B viroids follow a similar scheme but directly copy the linear minus-strand multimers made in step 2 into linear plus-strand RNA multimers, therefore skipping steps 3–6. These multimers are subsequently cleaved and circularize into plus-strand progeny circles. This is called "asymmetric" replication.

Three enzymatic activities are needed for viroid replication

Rolling-circle replication as depicted in Figure 29.2 requires three distinct enzymatic activities for completion: (1) an RNA-dependent RNA polymerase, to generate the multimeric linear strands; (2) an RNA-cleaving activity (endoribonuclease), to process them to unit-length linear molecules; and (3) an RNA ligase, to circularize the linear monomers.

Polymerization. Cellular DNA-dependent RNA polymerase II appears to be involved in the replication of group B viroids, since their replication can be inhib-

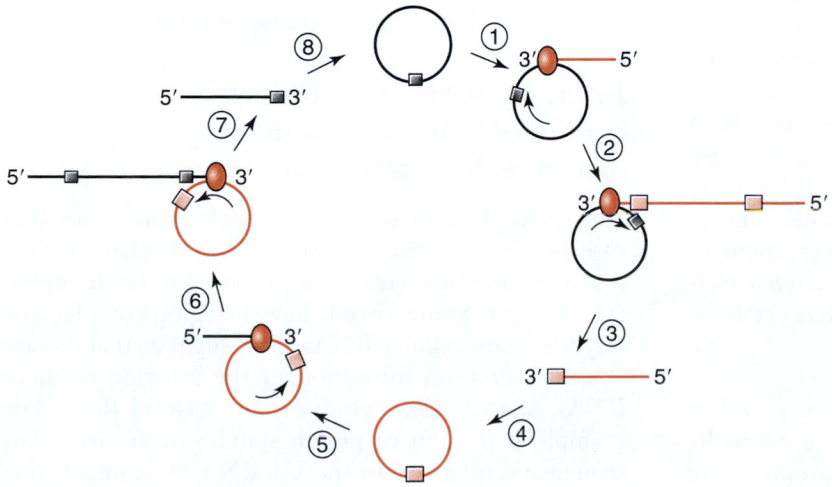

FIGURE 29.2 **Rolling-circle replication of circular RNAs.** Shown is symmetric replication, in which both plus- and minus-strand circles are formed and used as templates for RNA synthesis. Plus strands are in black, minus strands in orange. RNA polymerase is shown as an orange oval. Arrows inside circles show direction of rotation of the circular RNA through the RNA polymerase, which is shown here as stationary. Rectangles on the RNAs show cleavage/ligation sites. 3′ and 5′ ends of linear RNAs are labeled. See text for details.

ited by low concentrations of **α-amanitin,** a known inhibitor of this enzyme. The polymerization step for group A viroids is clearly different since it is unaffected by high levels of α-amanitin. This is probably related to the distinct sites where viroids in the two groups are localized. Group B viroids are found in the cell nucleus, whereas group A viroids are found within chloroplasts in the cytoplasm of plant cells. Chloroplast RNA polymerase resembles bacterial RNA polymerase in its resistance to α-amanitin. Therefore group A viroids probably replicate in chloroplasts using the chloroplast RNA polymerase.

Cellular RNA polymerases normally transcribe only double-stranded DNA templates into RNA. It is not clear how viroids recruit these polymerases and induce them to use their single-stranded RNA genomes as templates to be copied into complementary RNAs. Furthermore, the sequences that signal initiation and termination of RNA synthesis on the viroid RNAs are not presently known.

Cleavage. Group A viroids can cleave their multimeric linear RNAs *in vitro* in the absence of any proteins, yielding monomeric linear RNA molecules. *Self-cleavage* is a term used for the intrinsic property of an RNA to cleave itself; such RNAs are **ribozymes**, because they are RNA molecules that have enzymatic activity in the absence of proteins. Self-cleavage of group A viroid RNAs is carried out by a **hammerhead structure** that directs cleavage at a specific phosphodiester bond in the RNA (Figure 29.3a). This structure is so named because it resembles the head of a ball-pean hammer. Other self-cleaving RNA structures have also been identified in viroid-like satellite RNAs (**hairpin structure**, Figure 29.3b) and hepatitis delta virus RNA (**delta structure**, Figure 29.3c). All of these structures align the target cleavage site by base-pairing with other regions of the RNA molecule. These structures have been shown to be responsible for the production of monomers both *in vivo* and *in vitro*. Self-cleavage occurs by nucleophilic attack by the 2′-hydroxyl of the nucleotide at the cleavage site, generating a 2′, 3′-cyclic phosphate at one side of the cleavage site, and a 5′-hydroxyl terminus at the other side (Figure 29.3d).

It is unclear how the multimeric plus strands of group B viroids are cleaved to give monomers, because a self-cleavage activity has not been found. It has been proposed that host cell enzymes carry out cleavage of these RNA molecules. Some effort has been spent in searching for a plant cell enzyme that can catalyze cleavage of viroid RNA multimers, but none has yet been found.

Ligation. Ligation of peach latent mosaic viroid (a group A viroid found in the chloroplast) can occur without the participation of host proteins. Linear monomers resulting from hammerhead-mediated self-cleavage have been shown to self-ligate *in vitro*, producing 2′-5′ phosphodiester bonds. It therefore seems likely that group A viroids need only a host cell RNA polymerase activity for their replication, since they provide their own self-cleavage and self-ligation activities.

In contrast, group B viroids probably use cellular enzymes for all three steps: polymerization, cleavage, and ligation. A wheat germ RNA ligase known to participate in splicing of tRNA introns can circularize linear unit-length potato spindle tuber viroid RNAs. This enzyme, which produces 3′-5′ phosphodiester bonds, is localized in the nucleus, and therefore appears a logical candidate for the activity that ligates group B viroids.

How do viroids cause disease?

Viroids are most probably transmitted between plants via mechanical breaks. This mechanism of transmission is widespread among plant pathogens. The infectious agent is transmitted following contact between a wounded uninfected plant and an infected plant, a contaminated agricultural tool, or a carrier insect. Following infection, viroids replicate and propagate in the plant, most likely using cellular junctions to pass from cell to cell. Viroid infections give rise to a wide range of symptoms such as leaf decoloration, dwarfing, enhanced fruit production, orange spotting, decline in fruit yield, stunting, and so forth.

Despite their extreme simplicity, viroids cause plant diseases as varied as those caused by plant viruses. Because viroids do not code for any proteins, their effects on plants must be a consequence of direct interactions between the viroid RNA and host constituents. Analysis of molecular **chimeras** between group B viroids of different pathogenicity levels has revealed that the severity of disease symptoms is the result of complex interactions among three out of the five viroid domains. Sequence changes in the pathogenicity (P) domain can affect infectivity and the severity of symptoms. For example, some mutations that increase base-pairing interactions in this domain decrease severity of symptoms.

Interaction of viroid RNA with cellular RNAs or proteins may disrupt cell metabolism

The molecular mechanisms by which viroids cause disease in plants remain mysterious. Both plant nucleic acids and proteins have been proposed to be the initial viroid target. Some viroids have a region complementary to certain cellular RNAs. This suggests that disease could result from inhibition of the function of these RNAs, or their cleavage directed by viroid RNA. For example, a portion of potato spindle tuber viroid has sequence similarities to the U1 RNA of mammals (in-

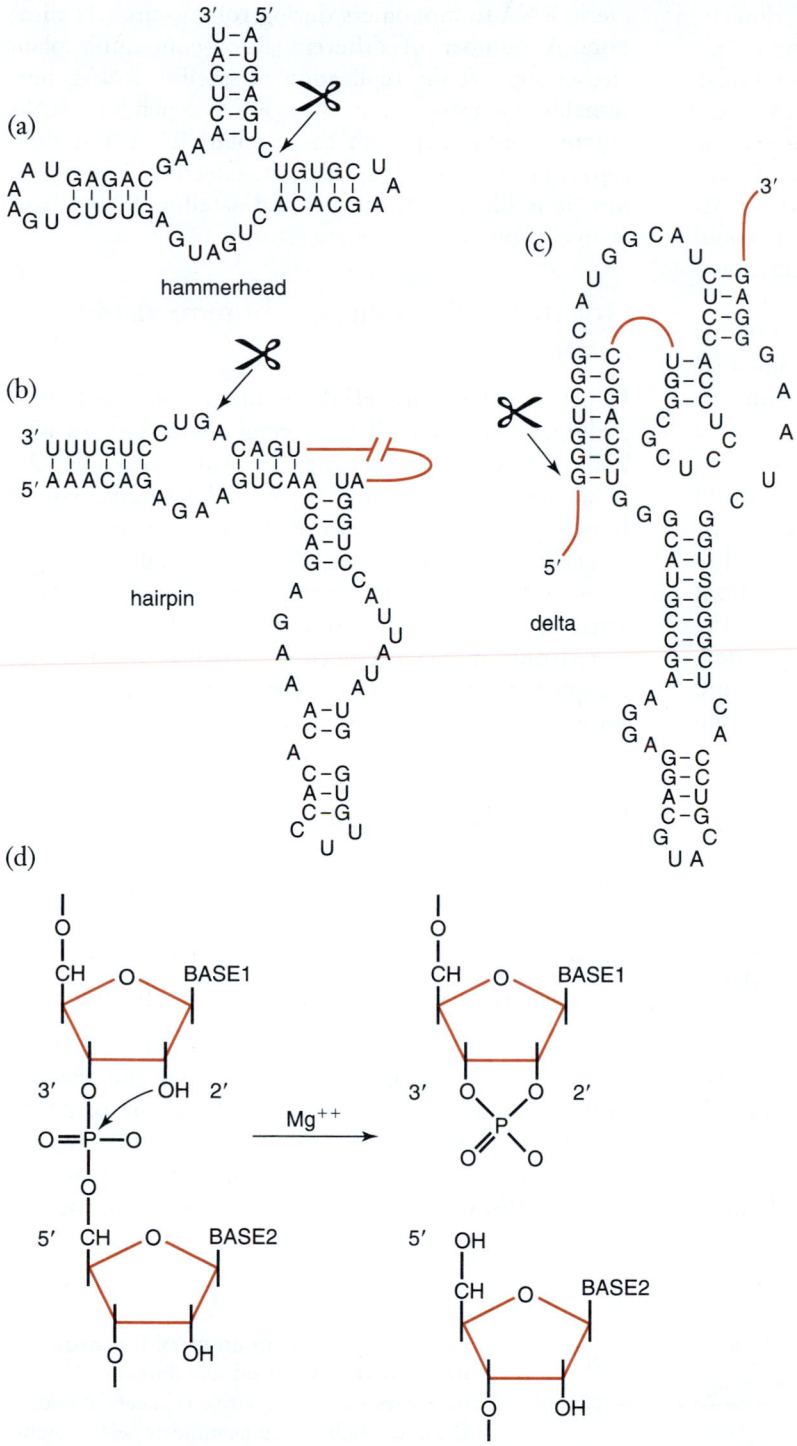

FIGURE 29.3 Self-cleaving RNA structures and the cleavage reaction. (a) Hammerhead self-cleaving structure of peach latent mosaic viroid. (b) Hairpin self-cleaving structure of satellite tobacco ringspot virus satellite RNA. (c) Delta self-cleaving structure of hepatitis delta virus. (d) The self-cleavage reaction: this involves a Mg^{2+}-catalyzed phosphoryl transfer reaction that cleaves the RNA, producing 2′-3′ cyclic phosphate and 5′-hydroxyl termini. The arrows and scissors indicate the cleavage sites in (a)–(c).

volved in RNA splicing), and some viroids contain sequences that could base-pair with 7S ribosomal RNA. However, it is difficult to explain how changing a few nucleotides can convert a severe variant into a mild one, since these changes generally are not located in the regions proposed to base-pair.

Alternatively, there may be host proteins that recognize and interact with different viroid structures. Mammalian **PKR (double-stranded RNA-dependent protein kinase)** is activated (phosphorylated) by potato spindle tuber viroid, as its secondary structure resembles that of double-stranded RNA. The level of activation has been correlated to the severity of plant disease symptoms. Activated PKR phosphorylates the alpha subunit of eukaryotic protein synthesis initiation factor 2 (eIF2), resulting in inhibition of cellular protein synthesis. Activation of a PKR plant homologue may be the triggering event in viroid pathogenesis, as PKR is induced by **interferons** and activated by double-stranded RNAs in mammalian cells (see Chapter 31).

Pathogenicity could also be the result of molecular mimicry. Viroid RNAs could, by virtue of their sequence or structure, act in the place of certain cellular RNAs. Homologies have been observed between potato spindle tuber viroid and a group I intron, as well as U3B RNA (involved in RNA splicing). Viroid RNAs could therefore interfere with splicing by replacing functional RNAs in splicing complexes.

Cross-protection is known to occur between strains of the same viroid, and between different viroids that are very similar in sequence. Plants infected with one viroid will not allow replication and pathogenesis by a related viroid. This is reminiscent of viral interference, in which the presence of one virus inhibits replication by a superinfecting virus. The mechanism of cross-protection in viroids is not understood. One hypothesis is that a limiting host factor is required for viroid replication, transport, or accumulation. Differential affinity of the viroid RNA for this factor could determine which of two co-inoculated viroids prevails, and the interaction of viroid RNA with the host factor could determine pathogenicity.

Plant satellite RNAs resemble viroids but are encapsidated

Plant satellite RNAs are small single-stranded RNAs that depend on a **helper virus** for replication and encapsidation, but show little or no sequence similarity with their helper virus. Like viroids, they have a circular, highly base-paired structure and do not encode any protein. Unlike viroids, they are encapsidated and can therefore be transmitted more efficiently from plant to plant. Furthermore, like group A viroids, all viroid-like plant satellite RNAs studied so far have a hairpin or hammerhead structure that catalyzes cleavage of multimeric RNA to monomers during rolling-circle replication. A number of different RNA-containing plant viruses support the replication of satellite RNAs, presumably by using their own RNA-dependent RNA polymerases to replicate the satellite RNA and their capsid proteins to encapsidate it. Given their similarities, it is likely that viroids and satellite RNAs have evolved from a common ancestor.

Hepatitis delta virus is a human viroid-like satellite virus

Hepatitis delta virus (HDV) is unique among human pathogens in that it shares several characteristics with both plant viroids and viroid-like plant satellite RNAs. It has been proposed that hepatitis delta virus evolved from a primitive viroid-like RNA through the capture of a cellular transcript. This bloodborne pathogen replicates in the liver and frequently causes **fulminant hepatitis** in both primate and nonprimate hosts. It is detected only in the presence of hepatitis B virus (see Chapter 28), and it uses the hepatitis B virus envelope protein (S antigen) to package its RNA genome. Because hepatitis delta virus needs a hepatitis B virus protein for packaging, it is considered an RNA satellite of hepatitis B virus. With greater than 15 million people affected worldwide, hepatitis caused by hepatitis delta virus is a serious public health problem.

Hepatitis delta virus may use two different cellular RNA polymerases to replicate

The genome of hepatitis delta virus (Figure 29.4) is a 1682 nt single-stranded circular RNA molecule that can be folded into a rod-like structure by virtue of extensive internal base-pairing. It possesses a single protein-coding open reading frame, complementary to the genome RNA. An 800 nt polyadenylated messenger

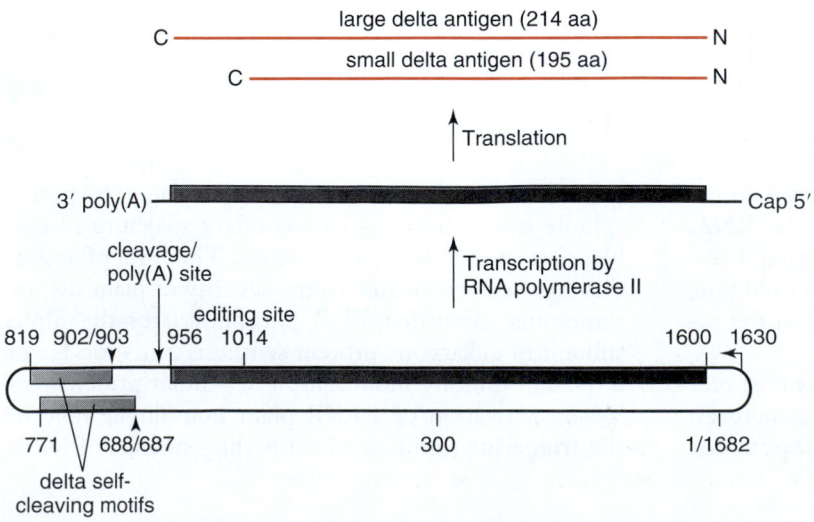

FIGURE 29.4 The genome of hepatitis delta virus. The viroid-like domain necessary for rolling-circle replication is at the left. Shaded regions indicate self-cleaving motifs on the genome RNA (nt 668–771) and antigenome RNA (nt 819–902); short arrows indicate the cleavage sites. The complement of the single open reading frame encoding both forms of delta antigen is shown in black. The larger form results from editing (A to G) at position 1014. Messenger RNA is shown above the genome map in black; the transcriptional start site at nt 1630 and the cleavage/polyadenylation sites on the genome RNA are noted. The two delta antigens are shown above in orange (N = amino terminus, C = carboxy terminus).

RNA is synthesized by RNA polymerase II and is translated into proteins called hepatitis delta antigens. The 3′ end of this mRNA is specified by the classical polyadenylation signal, AAUAAA, which directs RNA cleavage by cellular proteins and the addition of poly A to the cleaved 3′ end of the RNA.

The left-terminal domain (about 360 nt) of the genome RNA of hepatitis delta virus is reminiscent of viroids, and includes self-cleaving motifs used on both genome and complementary antigenome RNAs (delta motif; see Figure 29.3c). Hepatitis delta virus replicates in the nucleus via a symmetric rolling-circle mechanism as shown for viroids in Figure 29.2. The production of antigenome RNA from genome RNA templates (steps 1 and 2) may use cellular DNA-dependent RNA polymerase I, because it is resistant to α-amanitin, which inhibits RNA polymerase II. This could explain two differences between transcription (mRNA synthesis) and replication of genome RNA. First, there is evidence that transcription and replication initiate at different sites, implying the use of different polymerases. Second, the machinery responsible for genome replication does not recognize the cleavage/polyadenylation signal that leads to mRNA 3′ end formation. It is known that RNA polymerase I, used by the cell to transcribe ribosomal RNA genes, does not interact with the cellular factors used for cleavage and polyadenylation of RNA polymerase II transcripts. This would explain why multimeric antigenome RNAs are generated by rolling-circle replication using circular genome RNA as template, but are not cleaved at the polyadenylation signal.

An alternative interpretation of experimental results suggests that RNA polymerase II is used for both transcription and replication. This model proposes that the polyadenylation signal is not always recognized by the cellular cleavage factors, allowing production of multimeric antigenome templates as well as 800 nt mRNAs

Engineering ribozymes for novel uses

The plant pathogenic RNAs, and the more complex hepatitis delta virus, have provided us with most of the naturally occurring self-cleaving RNAs known to date. The hammerhead, hairpin, and delta self-cleaving RNA motifs (Figure 29.3a–c) have been studied extensively. These motifs can be formed by annealing two different RNA fragments *in vitro*, such that one RNA fragment acts as the ribozyme and the other as the substrate (Figure 29.5). If the complementary region on the two RNAs is short enough, the cleavage products will dissociate from the ribozyme, thereby permitting the binding of more substrate molecules for cleavage. Via successive rounds of binding and cleavage, a single ribozyme molecule can therefore cleave many substrate molecules, establishing a classic enzyme/substrate relationship. Furthermore, by changing the binding sequence between the ribozyme and its substrate, it is possible to create a ribozyme with a new substrate specificity.

Novel ribozymes with catalytic profiles different from those of the naturally occurring molecules have been made by designing oligonucleotides synthesized *in vitro* or by mutagenesis of existing ribozymes. A wide variety of RNAs can be targeted for cleavage by such engineered ribozymes. Because of their ability to interact directly with RNA, ribozymes are currently being developed as potential therapeutic agents for a wide range of diseases, including virus infections (human immunodeficiency virus, viral hepatitis, etc.), cancer, diabetes, cardiovascular disease, and osteoporosis.

Hammerhead and hairpin ribozymes derived from plant RNA pathogens can cleave many RNAs and inhibit viral replication. However, the capacities of these two ribozymes are limited by their requirement for high magnesium concentrations; mammalian cells contain one-tenth the concentration of magnesium needed for optimal activity of these ribozymes. The delta ribozyme is therefore extremely interesting, because it can function in the environment of human and animal cells.

Viroids and related satellite RNAs have been a rich source of ribozymes to date, and there is considerable promise of more to be found. In the coming years, altered forms of these versatile molecules will surely emerge as a new class of drugs.

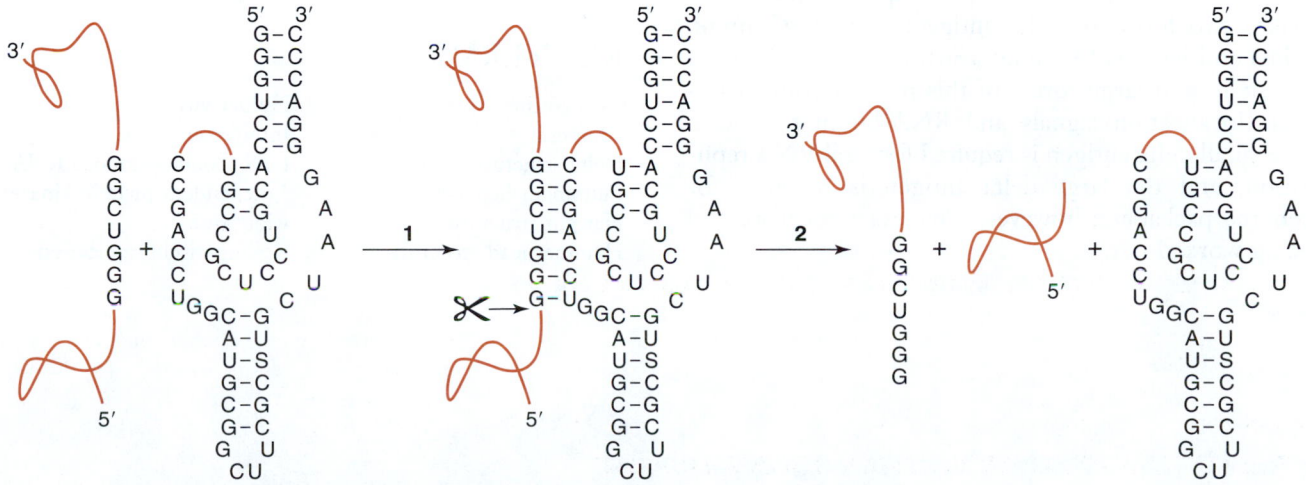

FIGURE 29.5 Proposed cleavage of an RNA substrate by a trans-acting delta ribozyme. The target RNA substrate (left) anneals to the ribozyme (step 1) and is cleaved (scissors), producing two products that are released (step 2). The ribozyme (right) is then ready to process a new molecule of substrate RNA.

by the same cellular RNA polymerase. However, insensitivity of antigenome synthesis to α-amanitin would remain unexplained by this model.

The second stage of replication, formation of genome RNA from antigenome templates (steps 5 and 6), is sensitive to α-amanitin, and therefore presumably uses RNA polymerase II. There is no polyadenylation signal on the antigenome RNA, allowing multigenome-length transcripts to accumulate by rolling-circle replication. Monomeric circular antigenome and genome RNAs are formed by self-cleavage and ligation of multimeric linear RNAs (steps 3, 4 and 7, 8). The small delta antigen and several other cellular factors are also involved in genome replication.

RNA editing generates two forms of hepatitis delta antigen

A fraction of hepatitis delta virus antigenome RNAs are edited during replication, resulting in the alteration of a specific A residue (at nucleotide 1014) to a G. This editing event is carried out by a cellular adenosine deaminase that specifically recognizes regions of double-stranded RNA. However, it is not clear how only a single A residue on the extensively base-paired 1682 nt antigenome RNA is selected for editing.

As a result of editing, the UAG termination codon that normally ends the open reading frame at amino acid 195 is changed to a UGG codon, which codes for tryptophan. Edited antigenome RNAs give rise to genome RNAs by replication; these genome RNAs are then transcribed by RNA polymerase II to form modified messenger RNAs. Up to 30% of hepatitis delta virus messenger RNA molecules contain this modified termination codon. On these altered messenger RNAs, translation progresses to amino acid 214 before encountering a stop codon. Thus hepatitis delta virus makes two forms of delta antigen: small (195 amino acids) and large (214 amino acids) (Figure 29.4). Both the small and large forms of this protein contain nuclear localization signals and RNA-binding regions. The small delta antigen is required for viral RNA replication, and the large delta antigen is required for genome packaging; however, their exact roles are still being worked out.

Conclusion: viroids may be a link to the ancient RNA world

The various small self-replicating RNA pathogens share a considerable number of characteristics and may well have evolved from a common ancestor. To date, viroids have only been found in the plant kingdom, although hepatitis delta virus, a viroid-like RNA, replicates in humans. Because viroids can replicate and possess enzymatic activities (RNA cleavage and ligation), they are considered as a model living fossil from a precellular RNA world. They are used to study the evolutionary connection between RNA and DNA genomes. They are also ideal biological molecules with which to study RNA structure/function relationships.

SUGGESTED READING

1. Bagheri, S. and Kashani-Sabet, M. (2004). Ribozymes in the age of molecular therapeutics. *Current Molecular Medecine* **4,** 489–506.

2. Bergeron, L. Jr., Ouellet, J., and Perreault, J.P. (2003). Ribozyme-based gene-inactivation systems require a fine comprehension of their substrate specificities. The case of delta ribozyme. *Current Medicinal Chemistry* **10,** 2589–2597.

3. Bussiere, F., Lehoux, J., Thompson, D.A., Skrzeczkowski, L.J., and Perreault, J.P. (1999). Subcellular localization and rolling circle replication of peach latent mosaic viroid: Hallmarks of group A viroids. *Journal of Virology* **73,** 6353–6360.

4. Diener, T.O. (1999). Viroids and the nature of viroid diseases. *Archives of Virology*, Supplementum **15,** 203–220.

5. Hadidi, A., Flores, R., Randles, J.W., and Semanik, J.S. Eds. (2003). Viroids. CSIRO Publishing, Collingwood, Australia.

6. Taylor, J.M. (1999). Human hepatitis delta virus: An agent with similarities to certain satellite RNAs of plants. *Current Topics in Microbiology and Immunology* **239,** 107–122.

KEY TERMS

α-amanitin
Chimera
Delta structure
Fulminant hepatitis
Hairpin structure
Hammerhead structure
Helper virus
Interferon
PKR (double-stranded RNA-dependent protein kinase)
Ribozyme
Rolling-circle replication

Prions

Prions

From *p*roteinaceous *i*nfectious particles

PROPERTIES

Small, filterable infectious particles that contain protein but no detectable nucleic acid.
Prion proteins (PrP^C) are encoded by the host genome.
PrP^C is found in neuronal synapses, binds copper, has unknown function.
Prion proteins become infectious and pathogenic (PrP^{Sc}) as a result of protein conformational changes.
PrP^{Sc} can catalyze its own formation from PrP^C in animals.
PrP^{Sc} aggregates and accumulates in diseased brain.

DISEASES

Chronic, progressive, invariably fatal central nervous system degeneration.
Brain pathology is spongiform encephalopathy—large vacuoles in cortex and cerebellum give brain a sponge-like appearance.
Affected areas contain microscopic insoluble amyloid fibrils and macrocrystalline arrays known as amyloid plaques.
No signs of a host immune response.
Can arise spontaneously or by ingestion of infected tissue.
Affects wild and domesticated ruminants (sheep and goats: scrapie; cattle: mad cow disease); mink, cats.
Experimentally transferred to mice, hamsters, chimpanzees.
Human diseases: Creutzfeldt-Jakob disease (CJD), kuru.

DISTINCTIVE CHARACTERISTICS

Proteinaceous infectious agent that contains no nucleic acid and consists only of a single species of protein called PrP.
A new kind of infectious agent that can transmit a disease and replicate itself without the intervention of informational nucleic acids.

Prions are proteins that cause fatal brain diseases

The concept that a protein can transmit an infectious disease from one individual to another without the intervention of a nucleic acid genome has caused considerable surprise and controversy within the scientific community, but has now become widely (if not universally) accepted. A proteinaceous infectious agent called a **prion** ("pree-on" or "pry-on") appears to contain no nucleic acid, and only a single species of protein called PrP (prion-associated protein).

All viruses use nucleic acid as a repository for the information necessary for their replication. A number of viruses package in their virions proteins that are necessary for replication once the genome enters the host cell. These include virion-associated RNA or DNA polymerases (retroviruses, negative-strand viruses, reoviruses), and tegument proteins of herpesviruses. However, these proteins are encoded by information in the viral nucleic acid genome, and their activities serve to express and replicate the genome. Prions appear to represent a new kind of infectious agent that can transmit a disease (**prion disease**) and replicate itself without the intervention of informational nucleic acids.

Prion diseases were first detected in domestic ruminants

Sheep and goat herders in Britain have for hundreds of years observed a disease in their animals that they termed **scrapie**. Affected animals develop behavioral changes, tremor (especially of head and neck), and uncoordinated movements that progress to prostration and death without associated fever. Scrapie got its name from the habit of affected animals to itch, rub, and scrape their skin. In severe cases, the sheep's wool is rubbed off their sides and back. Central nervous system tissue from such animals reveals large vacuoles in the cortex and cerebellum, giving the brain the appearance of a sponge. This condition is therefore termed **spongiform encephalopathy**. There are no signs of brain inflammation, suggesting that there is no infectious organism present. However, sheep and goat herders have long recognized that scrapie behaves like an infectious disease. Isolated, unaffected flocks appeared never to develop the disease unless they were introduced to animals from affected flocks. Under such circumstances previously healthy animals would develop scrapie, albeit with an incubation period of from 2 to 5 years.

Cattle have increasingly been fed meat and bone meal prepared from animal sources as the beef industry has evolved. This started in the United Kingdom as early as 1926, but the volume of recycled animal protein increased rapidly over the twentieth century. In the 1970s and early 1980s, a number of changes were made in the rendering process for transforming diseased and otherwise unusable animal carcasses into meat, bone meal, and animal feed. Continuous instead of batch processing led to a greater risk of cross-contamination between feeds from different species. In addition, the falling price of tallow (animal fat) led to a decrease in the amount of expensive processing using organic solvents.

A disease similar to scrapie, called bovine spongiform encephalopathy (**BSE**) or **mad cow disease**, spread throughout cattle herds in Britain during the 1980s, presumably via an infectious agent present in their feed. It is presently unclear whether this disease crossed the species barrier from sheep and goats to cattle, or whether it is a rare, preexisting disease of cattle that was propagated by feeding remains of diseased cattle to livestock. During the subsequent 20 years, millions of cows in the United Kingdom and parts of continental Europe were slaughtered in attempts to control this disease (Figure 30.1). BSE has since apparently itself crossed the species barrier and has infected humans. The first human case of an apparent BSE-derived disease was diagnosed in the United Kingdom in 1994. Since then, more than 100 individuals have been affected by this disease, called **new variant Creutzfeldt-Jakob disease (nvCJD)**, with an incidence in the United Kingdom of 10 to 15 cases per year.

FIGURE 30.1 Incineration of British cattle infected with mad cow disease.

Human prion diseases can be either inherited or transmitted

During the 1950s, an apparently transmissible spongiform encephalopathy of humans was identified among the Fore people in the mountainous interior of Papua New Guinea. This disease was called **kuru** ("shivering" or "trembling" in the Fore language). Kuru was characterized by cerebellar movement disorders and a shivering-like tremor progressing to total muscular incapacity, dementia, and death within one year of the onset of symptoms (Figure 30.2). The disease had a very high yearly incidence rate of 1%. Early investigators were able to explain the high frequency of transmission of the disease by ritual funeral preparation of victims, including ritual cannibalism of their brains. D. Carleton Gajdusek proved the transmissibility of this human spongiform encephalopathy, transferring it by direct inoculation of diseased human brain tissue into chimpanzee brains. Gajdusek was awarded the Nobel Prize in Medicine in 1976 for his work on spongiform encephalopathies.

A number of central nervous system disorders in humans were shown to have similar symptoms and progression, and to share the spongiform pathology of the brain; these include Creutzfeldt-Jakob disease (CJD), Gerstmann-Sträussler-Scheinker syndrome, and fatal familial insomnia (Table 30.1). Many of these cases were shown to have a familial distribution, and are therefore linked to inherited genes. However, other cases were labeled "sporadic" and have no obvious genetic component. In some cases, Creutzfeldt-Jakob disease was shown to be transmitted from human to human by (1) corneal allografts, (2) reusable intracerebral electrodes used for electroencephalograms, (3) grafts of **dura mater**

FIGURE 30.2 **Fore victims of kuru.** Five women and one girl, all victims of kuru, who were still ambulatory, assembled in 1957 in the South Fore village of Pa'iti. The girl shows the spastic strabismus, often transitory, which most children with kuru developed early in the course of the disease. Every patient required support from the others in order to stand without the aid of the sticks they had been asked to discard for the photograph.

(covering of the brain) from cadavers, and (4) human growth hormone purified from pituitary glands of cadavers. Similarities between symptoms and brain pathology in these diseases and kuru and scrapie led to extensive efforts to identify a causative infectious agent.

The infectious agent of prion diseases contains protein but no detectable nucleic acid

Initial efforts to identify a conventional infectious agent in affected nervous tissue were fruitless. However, direct inoculation of affected brain tissue from kuru and Creutzfeldt-Jakob disease victims into the brains of chimpanzees, and scrapie brain tissue into mice and hamsters, eventually led to transmission of the disease. Despite incubation periods of months before symptoms appeared, it was now possible to detect and to attempt to identify the infectious agent. It was quickly found that prion infectivity is unusually resistant to inactivation by ionizing or ultraviolet irradiation, which damage nucleic acids. The target size for inactivation is several orders of magnitude lower than had ever previously been found for a virus, meaning that if there is a nucleic acid genome, it is unusually small. Infectious fractions of scrapie-infected brain contain no demonstrable nucleic acid even when analyzed by highly sophisticated techniques. Thus the ratio of nucleic acid to infectious particles would have to be much lower than ever previously encountered in viruses.

Scrapie infectivity was found to be resistant to treatment with agents that were usually effective inactivators of nucleic acids (psoralen cross-linking agents, nucleases, high concentrations of zinc ions, ammonium hydroxide) while sensitive to agents that were usually effective inactivators of proteins (strong alkali, phenol, sodium dodecyl sulfate). Purification of infectious material resulted in preparations highly enriched for a 27–30 kilodalton protein not detected in similarly purified fractions of uninfected brain tissue. This ultimately led Stanley Prusiner in 1982 to propose that scrapie was caused by a novel *pro*teinaceous *i*nfectious particle for which he coined the term "prion" (Pr); the scrapie prion protein was named PrP^{Sc}. His proposal was very controversial at the time, but has gained acceptance and eventually led to his being awarded the Nobel Prize in 1997.

PrP^{Sc} is encoded by a host cell gene

Soon after the purification of PrP^{Sc}, its N-terminal amino acid sequence was determined. This led to the demon-

TABLE 30.1 **Species affected by prion diseases**

Species	Disease	Abbreviation
Sheep	Scrapie	
Elk, mule deer	Chronic wasting disease	CWD
Mink	Transmissible mink encephalopathy	TME
Cattle	Bovine spongiform encephalopathy	BSE
Cats	Feline spongiform encephalopathy (probable cross-species spread from BSE)	FSE
Zoo ruminants	Transmissible spongiform encephalopathy of zoo ruminants	TSE
Humans	Creutzfeldt-Jakob disease	CJD
	(New) variant Creutzfeldt-Jakob disease	vCJD
	Gerstmann-Sträussler-Scheinker syndrome	GSS
	Fatal familial insomnia	FFI
	Kuru	

stration that PrP^{Sc} is encoded by a cellular gene. Uninfected animals express a cellular isoform (PrP^{C}) of the prion protein. PrP^{C} (Figure 30.3) is a 208–amino acid glycoprotein normally expressed on the surface of neurons, lymphocytes, follicular **dendritic cells**, and other tissues. During synthesis and transport to the plasma membrane via the endoplasmic reticulum and the Golgi membranes, a 22–amino acid N-terminal signal peptide and 24 C-terminal amino acids are cleaved from the precursor 253–amino acid polypeptide. Oligosaccharide chains are added at one or two asparagine residues toward the carboxy terminal end of the protein. A carboxy-terminal phosphatidylinositol glycolipid anchors PrP^{C} in the plasma membrane. A number of octapeptide amino acid repeats in the N-terminal region bind to copper ions via histidine residues.

Although PrP^{C} is highly conserved among mammalian species, attempts to identify its normal function by genetic means have proven frustrating. Knockout mice lacking PrP^{C} remain healthy throughout development, suffering only minor electrophysiological defects. PrP^{C} is presumed to function in transmembrane transport or signaling. A current possibility involves shuttling or sequestration of copper ions at nerve synapses via binding to the N-terminal octapeptide residues.

Differences between PrP^{C} and PrP^{Sc}

Proteinase K treatment of PrP^{C} present in brains of normal animals results in complete digestion of the protein, while similar treatment of PrP^{Sc} from diseased animals gives a proteinase-resistant 27–30 kilodalton core fragment (PrP 27–30) that retains infectivity. This fragment is missing N-terminal amino acids 23 to 89, but is otherwise identical in amino acid sequence and glycosylation pattern to PrP^{C}. Although PrP^{C} is a soluble protein, PrP^{Sc} tends to form oligomers and aggregates that can be detected as fibrils in purified preparations or in infected brains (Figure 30.4).

The structure of PrP^{C} has been determined by nuclear magnetic resonance and x-ray crystallography (Figure 30-5). PrP^{C} contains three α-helices in the C-terminal half of the molecule, while the N-terminal region, containing octapeptide repeats that bind to copper ions, is relatively unstructured. Because of its tendency to aggregate, PrP^{Sc} has not been crystallized, but physical measurements show that it contains less α-helical content and substantial amounts of β strands, and therefore has undergone a significant conformational change. Two-dimensional crystals of PrP^{Sc} have been analyzed by electron microscopy (Figure 30.4) and a model of the structure of this protein has been proposed (Figure 30.6). In this model, the N-terminal region of PrP^{Sc} is folded into a highly stable, parallel β-helix. Therefore the conformational change from PrP^{C} to PrP^{Sc} could involve formation of this β-helix structure.

The prion hypothesis: formation of infectious and pathogenic prions from normal PrP^{C}

Prusiner's original statement of the prion hypothesis proposed that a misfolded form of a protein (PrP^{Sc}) can catalyze the refolding of properly folded native protein molecules (PrP^{C}) into a similar misfolded conformation. In one model, formation of PrP^{Sc} is a nucleation-dependent oligomerization process (Figure 30.7). This model states that the conformational change leading to conversion of PrP^{C} into PrP^{Sc} is a reversible reaction, but PrP^{Sc} monomers are less stable than PrP^{C}, resulting in little or no accumulation of PrP^{Sc}. Aggregation of PrP^{Sc} would promote the conversion of PrP^{C} by binding to and stabilizing the otherwise unfavored PrP^{Sc} conformation. Thus massive conversion of PrP^{C} to PrP^{Sc} would require an initial nucleation process. Infection with exogenous PrP^{Sc} would introduce a "seed" that initiates aggregation. Genetically inherited prion diseases would result from an increased stability of mutated PrP^{Sc} or an increased tendency of the protein to undergo aggregation and thus stabilize the altered conformation.

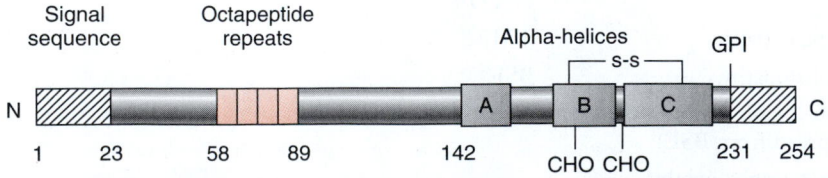

FIGURE 30.3 Domains and modifications of the prion protein. Amino acid numbers are shown below the map. N is terminus at left, C-terminus at right; the mature protein begins at amino acid 23 and ends at amino acid 231. CHO: carbohydrate residues added to two aspararagine residues; GPI: glycophosphatidyl inositol membrane anchor; S–S: putative disulfide bond; A,B,C: α-helical regions in PrP^{C}.

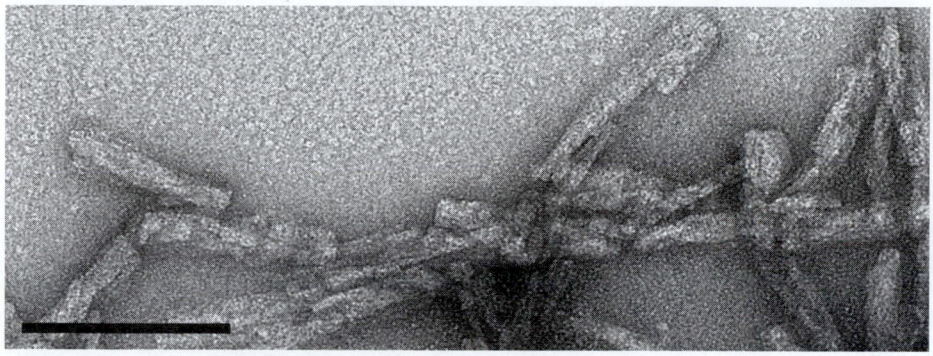

(a)

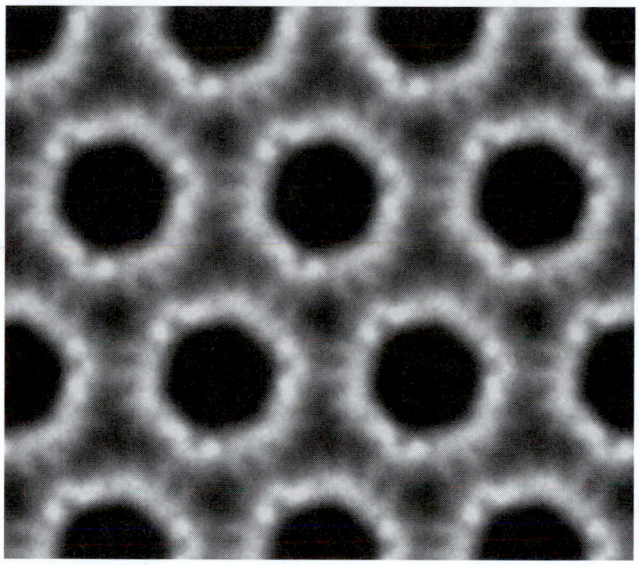

(b)

FIGURE 30.4 Prion fibrils and two-dimensional crystals. Negatively stained preparation of PrP 27–30 were observed by electron microscopy. (a) a typical prion rod or fibril. Black bar = 100 um. (b) Crystallographic averaging of a two-dimensional crystal shows threefold symmetry. Each hexagon is believed to be formed from trimers of PrP 27–30 dimers. Further analysis led to the model shown in Figure 30.6.

In an alternative proposed mechanism, the PrP^{Sc} form is postulated to be inherently more stable than PrP^{C}, but its formation is highly improbable because PrP^{C} folds in a relatively stable conformation upon its synthesis in the cell. In this case, exogenous PrP^{Sc} would catalyze the rearrangement of a molecule of PrP^{C}, or of a partially destabilized intermediate, to the more stable PrP^{Sc} conformation. The presence of the product PrP^{Sc} would then catalyze more transformation of PrP^{C} to PrP^{Sc}. If this model is corect, genetically inherited prion diseases would result from mutations in PrP^{C} that increase the population of unstable intermediates, or the rate of spontaneous conversion of PrP^{C} to PrP^{Sc}.

The nucleation-polymerization model resembles the well-studied processes of crystal growth, tubulin polymerization, viral capsid assembly, and bacterial flagellar polymerization. In these processes, proteins bind to each other and form specific aggregates that can become very large and insoluble. The template-assisted, catalyzed conversion model is similar to what is seen with a number of proteases (subtilisin and alpha-lytic protease) and the serpin family of protease inhibitors. These molecules are folded in a stable state as originally synthesized, even though a distinct, more stable folding state exists. Transition from the original folded state to the more stable state is a highly improbable event, because a large energy barrier separates the two states. Nevertheless, once the process begins, it is autocatalytic, as the product catalyzes formation of more of itself from the originally folded protein.

Is the prion hypothesis correct?

Although generally accepted, the prion hypothesis has not convinced all scientists in the field. It remains possible that a small virus might transmit prion disease by infecting and replicating in brain neurons, leading by some change in the infected cell to the conformational change in PrP^{C} that causes disease. The most highly purified preparations of PrP^{Sc} contain 100,000 protein molecules per infectious unit. This could reflect inefficient infection of the host, as with some viruses. However, it could also mean that an unknown virus remains hidden among the protein in the prion preparations. If

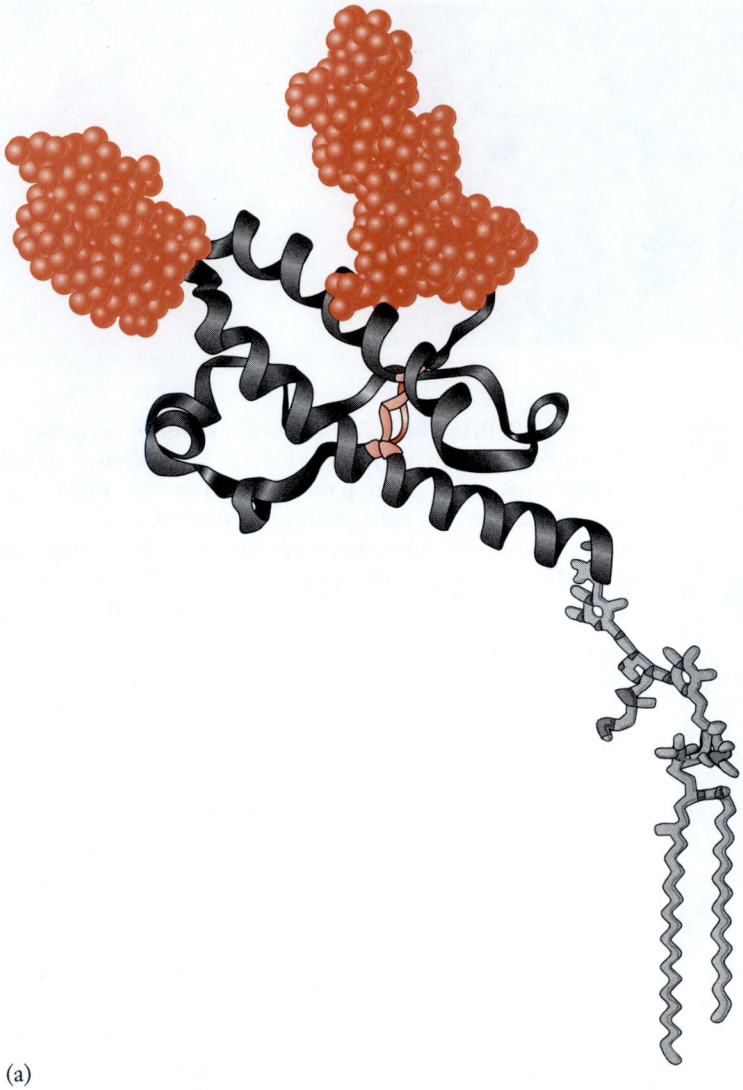

(a)

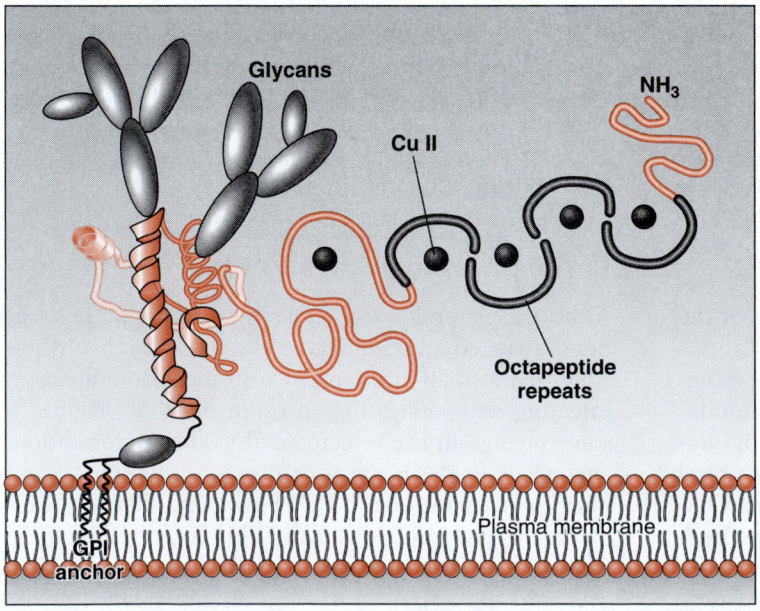

(b)

FIGURE 30.5 The structure of the prion protein. (a) The conformation of the backbone of the structured region of mouse PrP^C (residues 124–231), determined by nuclear magnetic resonance studies, is shown as a gray ribbon. The carbohydrate groups are shown as orange space-filling structures and the glycoinositol phospholipid anchor is shown below. The disulfide bridge is shown below in light gray.
(b) The prion protein is shown in a different orientation inserted into the plasma membrane via the glycoinositol phospholipid anchor. The three α-helical regions in the C-terminal region are at the left, and at the right is the flexible N-terminal region containing the octapeptide repeats (gray half-circles) with bound copper ions (gray spheres).

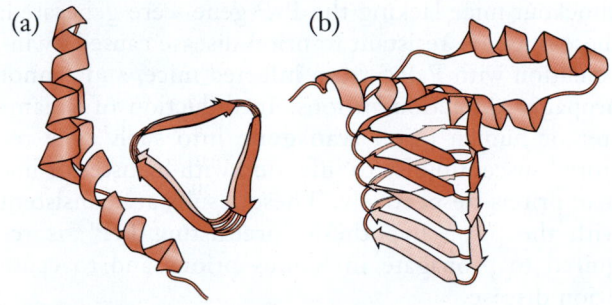

FIGURE 30.6 β-helical model of PrP 27–30. (a and b) Top and side views, respectively, of PrP 27–30 modeled with the N-terminal portion of the protein as a left-handed β-helix (ribbon arrows displayed in a triangular barrel). The structure of the α-helical part (ribbon helices) was derived from the known strucure of the C-terminal region of PrP^C.

this were the case, it is not clear why the virus should purify along with the prion protein, nor why the virus should remain resistant to treatments that normally destroy nucleic acids. Also, a virus hypothesis would not easily explain genetically inherited prion diseases.

Prion proteins can be produced by recombinant DNA technology in bacteria; these proteins can be folded into states that resemble the native PrP^C and the altered PrP^{SC} conformations. However, it has not yet been possible to generate infectivity from purified recombinant prion proteins. This could mean that other cellular proteins are required to help catalyze the conformational switch. Success in this kind of experiment might constitute definite proof of the prion hypothesis.

Pathology and diagnosis of prion diseases

A number of human diseases are caused by accumulation of a specific protein that occurs in insoluble fibers known as **amyloid**. These fibers can be detected by their characteristic binding of the dye, Congo red. Amyloid diseases include Alzheimer's disease, in which neurodegeneration is associated with deposition of the amyloid β protein. All of the proteins involved in these diseases undergo a conformational alteration to a common structure in the amyloid fibril, a "cross β" repeat structure in which β strands are aligned perpendicular to the axis of the amyloid fiber. As with prion diseases, these amyloidoses can be initiated by inherited mutations in the gene coding for the responsible protein.

All tissue abnormalities of prion diseases are confined to the central nervous system. Affected areas contain microscopic insoluble amyloid fibrils (Figure 30.4) and macrocrystalline arrays known as amyloid plaques. The prion protein accumulates selectively and abnormally in neurons in the central nervous system during the course of the disease. Until recently, diagnosis of prion disease in individuals with suspicious clinical presentations was made by histological examination of brain biopsies or, more commonly, autopsy specimens. Recent advances have been made in concentrating and detecting abnormal PrP in blood and lymphoid tissues such as tonsils. Immunodetection using western blots or direct fluorescent detection before and after proteinase digestion have allowed detection of as little as 10^{-8} μg (10^6 molecules) of abnormal PrP.

Genetics of prion diseases

PrP^C is encoded by a single exon of a unique gene (two copies per diploid genome). There are at least 20 known mutations in the PrP^C gene that are associated with prion disease, presumably becuase of a high rate of spontaneous PrP^{Sc} formation. A number of families that suffer from genetic Gerstmann-Sträussler-Scheinker syndrome have a common double mutation in PrP^C at amino acids 178 and 200 (D178N; E200K), while other such families have a mutation of leucine to proline at amino acid 102 (P102L). Mice with a transgene containing the P102L mutation spontaneously develop a lethal scrapie-like disease. On the other hand, sheep homozygous for a PrP gene encoding substitution of arginine for glutamine at codon 171 (R171Q) are entirely resistant to scrapie.

Humans who are homozygous for valine at position 129 are predisposed to both acquired and sporadic Creutzfeldt-Jakob disease. All victims of new variant Creutzfeldt-Jakob disease in the United Kingdom up to the present have been found to be homozygous for methionine at position 129. Homozygosity at this locus has recently been determined to be present among most victims of kuru. Taken together, this information suggests that heterozygosity at position 129 protects against both inherited and infectious prion disease. In fact, heterozygosity in the PrP gene, particularly at position 129, is present at unexpectedly high levels in humans worldwide. This implies that there has been selective pressure for maintenance of heterozygosity in humans, suggesting that prion diseases similar to kuru may have been commonplace over the past 500,000 years of human evolution. It has been speculated that this may be the result of widespread cannibalism among humans in the past; many of our ancestors may have succumbed to prion disease as a result of consumption of human tissues.

Prion diseases are not usually transmitted among different species

To a large extent, prion diseases respect species barriers; prions from one animal cannot usually infect

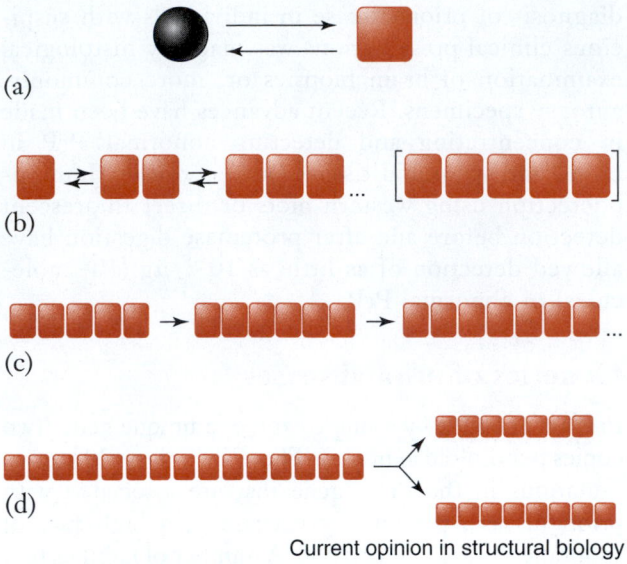

FIGURE 30.7 A mechanism for prion propagation. The reaction is broken into four stages. In (a), there is an equilibrium between the PrP^{C} state (black disc) and a rare monomeric precursor of PrP^{Sc} (orange square). (b) Monomeric precursors can interact with each other, weakly and reversibly, until a stable infective seed is formed (shown as a bracketed structure). Once this critical size has been acquired, then monomeric precursors of PrP^{Sc} can add irreversibly to allow growth of the PrP^{Sc} particle (c). The infective particle can then "reproduce" by breakage to smaller, stable particles (d).

animals of a different species. This suggests that prion infectivity depends on sequence similarity between donor and recipient prion proteins. This requirement may mean that interactions of PrP^{Sc} that lead to its stabilization or aggregation are stronger between molecules of the same amino acid sequence and therefore the same three-dimensional conformation. However, the species barrier is not absolute. Sheep scrapie can be used to infect mice and hamsters, a discovery that eventually led to the purification and identification of prions. Sheep are susceptible to infection with bovine spongiform encephalitis at the same frequency and incubation time as scrapie, although sheep PrP differs from bovine PrP at amino acids 98, 100, 143, 146, 155, 168, 198, and 218. In contrast, cattle are very resistant to infection with sheep scrapie but are susceptible to BSE. This raises the question of whether BSE was acquired during the epidemic in cattle by ingestion of remains not of scrapie-infected sheep, but of cattle that had been infected by a rare, spontaneous case of BSE sometime in the past.

Mice are resistant to infection with hamster prions, which have 16 differences in amino acid sequence from mouse PrP. A striking finding was made when knockout mice lacking the PrP gene were generated: these mice are resistant to prion disease caused by inoculation with PrP^{Sc} from infected mice, and cannot propagate infectious prions. Introduction of a hamster or human prion **transgene** into such mice restored susceptibililty to infection with hamster or human prions, respectively. These results are consistent with the prion hypothesis: preexisting PrP^{C} is required to propagate infectious prions and to cause prion disease.

Strain variation and crossing of the species barrier

Distinct strains of prions can produce distinct and reproducible patterns of disease, incubation time, and glycosylation and protein cleavage patterns of PrP^{Sc}. In some cases, these strains are attributed to different conformational structures of PrP^{Sc}. For example, two strains of transmissible mink encephalopathy exist: hyper (HY) and drowsy (DY). These two strains reproducibly transmit different behavioral changes, with different neuropathology and incubation periods. Proteinase K treatment of the PrP^{Sc} species corresponding to HY and DY gives rise to digested proteins with distinct amino termini, although the parent PrP^{C} proteins have the same amino acid sequence. This shows that amino acid sequence differences are not necessary to produce different strains.

Human cases of new variant Creutzfeldt-Jakob disease (nvCJD) appear to be linked to the BSE epidemic in the United Kingdom. New variant CJD is distinct from sporadic CJD; it has a distinct pathology, has a far younger age of onset than sporadic CJD, and presents with very different initial symptoms. Intracerebral inoculation of BSE-infected brain extract into macaque monkeys produced disease and pathology very similar to those of new variant CJD patients. It seems possible, therefore, that new variant CJD is caused by a different prion strain that is less restricted by the species barrier than that of scrapie. CJD and new variant CJD show different proteinase K-resistant PrP^{Sc} species and different patterns of di-, mono-, and nonglycosylated PrP^{Sc} are revealed by electrophoresis. The protein pattern of new variant CJD is similar to that of BSE-infected animals, but quite different from that of sporadic or acquired CJD. A proteinase K-resistant diglycosylated species is particularly prominent among the new variant CJD cases.

The nature of the prion infectious agent

Prions challenge our traditional concepts of infectious agents. The prion hypothesis proposes that pri-

Possibilities for the future

At the moment, no proven effective treatments exist for the prevention, cure, or even delay of progression of prion diseases. All are uniformly fatal. However, initial hopes have been raised for two tricyclic drugs that are already approved for other uses: the antimalarial agent **quinacrine** and the antipsychotic drug **chlorpromazine**. These have been confirmed to inhibit prion infection in cells and are being tried in patients. Both drugs can cross from bloodstream to brain. Chlorpromazine crosses the blood–brain barrier more readily than quinacrine, but quinacrine is 10 times more potent against prion production. Other work has been conducted on antibodies and antibody fragments that bind to prion protein and, in some cases, block prion formation in cells in a dose-dependent fashion. Since some variants of prion proteins seem to protect against prion disease, vaccination or gene therapy are other possible solutions to curing or preventing prion diseases.

ons transform a preexisting normal cellular protein into an infectious, pathogenic agent. We are used to the concept that information contained in nucleic acids can be transferred from cell to cell or from organism to organism, as with viruses. Virus genomes code for proteins that enable their replication and transmission, and thus viruses are self-propagating. Similarly, computer viruses transfer information from computer to computer, propagating themselves, although their "program" is carried by electronic bytes rather than by a nucleic acid. Nevertheless, computer viruses are analogous to real viruses, because their sequence of bytes (analogous to nucleic acid) codes for programs (analogous to enzymes that direct nucleic acid replication) that enable them to be replicated, de novo, using elements provided by a computer (analogous to the cell). Both viruses and computer viruses can cause disease or damage to their host cell or computer.

The prion hypothesis proposes that information coded in the three-dimensional structure of a protein, and therefore ultimately in its amino acid sequence, can be transferred from organism to organism and propagate itself, without the intervention of a nucleic acid. To be sure, the amino acid sequence of prions is encoded in nucleic acid genes of the host organism, but only the protein, not the prion gene, is transmitted and infectious. It may prove useful to modify the concept of an "infectious agent" to include any transfer of information between cells or organisms that results in repetitive self-propagation.

Richard Dawkins, in his best-selling 1976 book *The Selfish Gene*, postulated that evolution occurs whenever information is copied over and over again with introduction of variations and with selection of some variants over others. The most familiar unit of evolving information is the gene, but evolution can occur based on any mode of information transfer. Human ideas, for example, can be transmitted by human communication from generation to generation. The copying of ideas from one person to another is imperfect, just as the copying of genes from parent to child is sometimes inaccurate. Agriculture may be thought of as an idea that first arose in the ancient Middle East and then spread and evolved among humans for thousands of years.

Prions are transmissible, replicable, and variable disease-causing agents that are distinct from viruses. Whether we define them as "living" or "nonliving" or as an "infectious enzyme," we do know the following about them: (1) they have arisen in organisms during evolution; (2) they are able to propagate themselves and the diseases they cause; and (3) they appear to be able to evolve and to adapt themselves to different hosts.

SUGGESTED READING

1. Gajdusek, D.C. (1977). Unconventional viruses and the origin and disappearance of kuru. *Science* **197,** 943–960.

2. Jackson, G.S. and Clarke, A.R. (2000). Mammalian prion proteins. *Current Opinion in Structural Biology* **10,** 69–74.

3. Korth, C., May, B.C., Cohen, F.E., and Prusiner, S.B. (2001). Acridine and phenothiazine derivatives as pharmacotherapeutics for prion disease. *Proceedings of the National Academy of Sciences USA* **98,** 9836–9841.

4. Mallucci G. and Collinge J. (2004). Update on Creutzfeldt-Jakob disease. *Current Opinion in Neurology* **17,** 641–647.

5. Mead, S., Stumpf, M.P.F., Whitfield, J., Beck, J.A., Poulter, M., Campbell, T., Uphill, J.B., Goldstein, D., Alpers, M., Fisher, E.M.C., and Collinge, J. (2003). Balancing selection at the prion protein gene consistent with prehistoric kurulike epidemics. *Science* **300,** 640–643.

6. Prusiner, S.B. (1982). Novel proteinaceous infectious particles cause scrapie. *Science* **216,** 136–144.

7. Silveira, J.R., Caughey, B., and Baron, G S. (2004). Prion protein and the molecular features of transmissible spongiform encephalopathy agents. *Current Topics in Microbiology and Immunology* **284,** 1–50.

8. Telling, G.C. (2000). Prion protein genes and prion diseases: Studies in transgenic mice. *Neuropathology and Applied Neurobiology* **26,** 209–220.

KEY TERMS

Amyloid
BSE
Chlorpromazine
Creutzfeldt-Jakob disease (CJD)
Dendritic cells
Dura mater
Kuru
Mad cow disease
New variant Creutzfeldt-Jakob disease (nvCJD or vCJD)
Prion
Prion disease
Quinacrine
Scrapie
Spongiform encephalopathy
Transgene

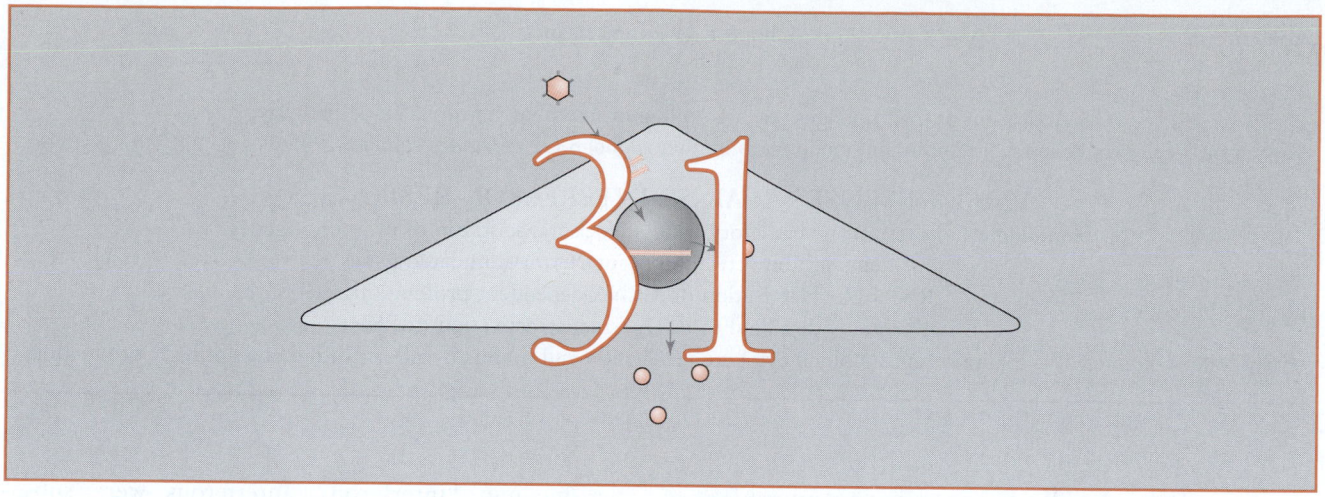

Interferons

Interferon

From viral *interference*

INTERFERONS

Type I interferon:

- *165–amino acid glycoproteins with antiviral activity*
- *Stable at pH 2*
- *Interferon-α: 13 genes in humans, made in leukocytes and other cells*
- *Interferon-β: 1 gene in humans, made in fibroblasts and epithelial cells*

Type II interferon:

- *146–amino acid glycoprotein, immune modulator*
- *Labile at pH 2*
- *Interferon-γ: 1 gene in humans, made in lymphocytes and macrophages*

INDUCTION OF SYNTHESIS

Induced by virus infection and double-stranded RNA.
Interferon regulatory factors (IRF) activated by phosphorylation.
IRF bind to interferon genes, activating transcription.
Interferon mRNAs are unstable; short burst of interferon synthesis.
Interferons are secreted into the extracellular medium.

INDUCTION OF ANTIVIRAL ACTIVITY

Interferons interact with specific receptors on the surface of target cells.
Binding leads to dimerization of receptors, phosphorylation and activation of Jak kinases.
Jak kinases phosphorylate Stat proteins, which dimerize and transit to nucleus.
Stat proteins bind to interferon-stimulated genes in nucleus and activate transcription.
More than 100 cellular genes are activated by the Jak-Stat pathway.

ANTIVIRAL ACTIVITIES INDUCED BY INTERFERONS-α and -β

Mx proteins suppress replication of negative-strand RNA viruses.
2′, 5′-oligoadenylate synthetase activates ribonuclease L, which degrades host and viral mRNAs.
Double-stranded RNA-dependent protein kinase (PKR) inactivates protein synthesis by phosphorylating initiation factor 2α (eIF-2α).

ANTIVIRAL ACTIVITIES INDUCED BY INTERFERON-γ

Expression of major histocompatibility protein (MHC) class I promotes development of cytotoxic T cells leading to killing of virus-infected cells.
Expression of MHC class II promotes antibody responses against viral proteins.
Activation of proteasomes increases immune display of viral peptides.
Synthesis of interleukin-2 promotes development of Th1 cells, leading to cytotoxic response and killing of virus-infected cells.

Activation of macrophages leads to killing of microbes via nitric oxide pathway.
Switch of B-cell production from immunoglobulin M to immunoglobulin G2 generates soluble antibodies against virus proteins.

VIRAL DEFENSES AGAINST INTERFERON RESPONSES

Homologues of interferon-regulatory factors block transcription of interferon genes.
Viral proteins inhibit activation of interferon-stimulated genes.
Small viral dsRNAs block activation of dsRNA-dependent protein kinase (PKR).
Viral proteins that bind to dsRNA reduce activation of PKR.
Soluble homologues of cytokine receptors block cytokine production and inhibit B-cell activation and antibody production.

Virus-infected cells secrete interferons, which protect nearby cells against virus infection

One of the first responses of an organism to viral infection is the secretion of **cytokines** and **chemokines**, proteins that stimulate protective reactions in the infected host. Among the cytokines are the **interferons** (Table 31.1). Interferons released from virus-infected cells bind to receptors on nearby uninfected cells and stimulate transcription of a set of genes encoding proteins with antiviral activities, resulting in protection of those cells against viral infection. Interferon was discovered in 1957 by Alick Isaacs and Jean Lindenmann. They were studying the phenomenon called **viral interference**, in which infection of cultured mammalian cells by one virus can interfere with subsequent superinfection by the same or another virus. Isaacs and Lindenmann found that the culture medium from cells infected with influenza virus could transmit antiviral protection to cells that had not previously come in contact with the virus; they named the substance present in the medium "interferon." Interferons were subsequently shown to be proteins that could be purified and used to protect cells against virus infection.

Their discovery stimulated an enormous amount of research whose underlying hope was that interferons could be used to prevent and cure virus infections in humans. It was soon found that interferon induced by one kind of virus can inhibit replication of many different virus types. However, the antiviral activity of interferon is generally limited to the species in which the interferon is produced. Three major classes of interferon, designated α, β, and γ, were described. Besides their antiviral effects, interferons were found to inhibit cell growth and division and to stimulate the immune system.

Interferons are a first line of host defense against viruses but therapeutic use has been limited

A gene coding for interferon-α was cloned by Charles Weissmann and collaborators in 1980, allowing the protein to be produced in large amounts and opening up

TABLE 31.1 Properties of interferons

	Interferon-α	Interferon-β	Interferon-γ
Alternative name	Leukocyte interferon	Fibroblast interferon	Immune interferon
Type	I	I	II
Stability	Stable at pH 2	Stable at pH 2	Labile at pH2
Length, amino acids	165	165	146
Molecular weight, KDa	16 to 27	28 to 35	20 to 25
Introns	None	None	Three
Number of genes	Human: 13 Mouse: 10	Human: 1 Mouse: 1	Human: 1 Mouse: 1
Chromosomal location	Human: #9 Mouse: #4	Human: #9 Mouse: #4	Human: #12 Mouse: #10
Principal producer cells	Leukocytes, fibroblasts, macrophages, epithelial cells	Leukocytes, fibroblasts, macrophages, epithelial cells	T lymphocytes, macrophages, natural killer cells
Inducing agents	Viruses, double-stranded RNA	Viruses, double-stranded RNA	Mitogens, antigens, interleukin-2
Major activity	Antiviral action	Antiviral action	Immune stimulation

the possibility of therapeutic use. Recombinant α- and β-interferons are presently used to treat chronic infections by hepatitis B and C viruses, cancers such as melanoma, multiple myeloma, renal cell carcinoma and Kaposi's sarcoma, and the neurodegenerative disease, multiple sclerosis. However, their original promise as major antiviral drugs has not been fulfilled. Therapeutically administered interferon can lead to unpleasant side effects and its antiviral effects are short-lived. Interferons are most effective in reducing viral spread during the initial stages of virus infection, before most symptoms are detected, but are less effective if administered late during an acute infection. Nonetheless, we now know a lot about the induction of interferon synthesis in virus-infected cells, the signaling pathways by which interferons stimulate the expression of antiviral genes, and the mechanisms by which antiviral proteins inhibit virus replication. Interferons play a major role in the "rapid reaction" first line of defense against viral infections.

Interferons α, β, and γ are made by different cells and have distinct functions

The interferons are classified into two groups based on amino acid sequence homology and stability at acidic pH (Table 31.1). Type I interferons, stable at pH 2, include interferon-α and interferon-β. Thirteen different interferon-α genes have been identified and are clustered on the short arm of human chromosome 9. Similarly, 10 different mouse interferon-α genes are grouped in a region close to the centromere of the mouse chromosome 4. Human interferon-α genes, like those of the mouse, lack introns and share a high degree of homology at the nucleotide sequence level, suggesting that the gene cluster was derived from a common ancestor gene by successive duplications. A single gene encodes interferon-β; it also lacks introns and is more distantly related to the interferon-α family.

Interferon-α, previously called leukocyte interferon, is produced by peripheral blood leukocytes and many other cell types in response to viral infection or treatment with double-stranded RNA. The different interferon-α subtypes all have the same number of amino acids (165) and share a high degree of amino acid sequence homology, but have molecular weights ranging between 16 and 27 KDa due to differential glycosylation. Interferon-β, also called fibroblast interferon, is a 165–amino acid glycosylated protein of 28 to 35 KDa produced by fibroblasts or epithelial cells.

There is only one type II interferon, called interferon-γ or "immune interferon." Interferon-γ is labile at low pH and is only distantly related to the type I interferons. It is encoded by a unique gene containing three introns, located in the long arm of human chromosome 12. It is produced exclusively by lymphocytes and **macrophages**. The secreted protein is a mixture of two 146-aa polypeptides of molecular weights 20 and 25 KDa, differing by levels of glycosylation. Secretion of interferon-γ is induced by mitogenic stimuli, antigens, and cytokines such as **interleukin-1**, interleukin-2, or interferon-γ itself. Although interferon-γ possesses antiviral activity, it is primarily an immune modulator rather than an antiviral agent.

Transcription of interferon genes is activated by virus infection or double-stranded RNA

Virus infection is the best natural inducer of interferon expression, although other infectious agents can also induce interferon production. Most RNA- and DNA-containing viruses induce interferon synthesis; however, double-stranded RNA viruses, such as reovirus and blue tongue virus, are the best inducers. Synthetic and natural double-stranded RNAs are also potent inducers of interferon synthesis, and double-stranded RNA may represent the natural signaling molecule by which virus infection leads to interferon production (Figure 31.1). Many DNA viruses transcribe both strands of certain regions of their DNA genomes, giving rise to self-complementary RNAs that can anneal to form double-stranded RNA in infected cells. RNA viruses replicate by synthesizing full-length complementary RNA strands that can form double-stranded RNA by hybridizing to the genome RNA. Thus the presence of double-stranded RNA in a cell is an indicator of active virus replication, and it makes sense that the cell recognizes and responds to this molecule.

Toll-like receptors are cell surface molecules that recognize distinct conserved structures in pathogens, such as bacterial lipopolysaccharide and viral double-stranded RNA. Binding of these molecules to the appropriate toll-like receptor activates signaling pathways that result in the generation of immune responses against microbial and viral infection. This proceeds via a series of intracellular associations between the toll-like receptors and adaptor molecules that recruit kinases, ultimately leading to activation of inflammatory genes as well as genes involved in antiviral defense.

Transcriptional activation occurs by binding of transcription factors to interferon gene enhancers

Induction of interferon synthesis requires transcriptional activation of interferon genes, which are strongly repressed in uninfected cells. However, induction of interferon mRNA synthesis can take place in the presence of inhibitors of protein synthesis. This indicates that the transcription factors required for activation of inter-

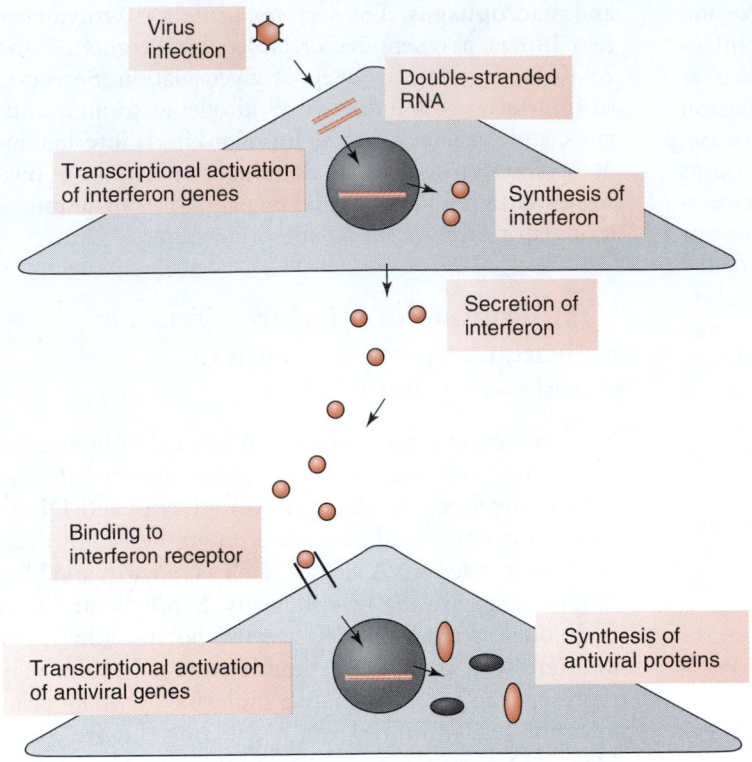

FIGURE 31.1 The interferon system. Virus infection leads to production of double-stranded RNA, which activates transcription of interferon genes. Interferons are synthesized and secreted, then bind to receptors on nearby cells, inducing expression of genes that make antiviral proteins.

feron gene transcription preexist in the cell and are activated upon virus infection. These factors bind to regulatory sequences called positive regulatory domains (PRD), which lie within 200 nucleotides upstream of the transcription start sites on both interferon-α and interferon-β genes (Figure 31.2). The binding of transcription factors to the interferon-β gene enhancer has been carefully studied and serves as a paradigm for the interaction between transcription factors and the general transcriptional machinery.

Four different transcription factors are known to bind to the interferon-β enhancer (Figure 31.2). The ATF-2/c-Jun heterodimer is activated by phosphorylation by a cellular protein kinase called MEKK1 kinase. In its active state, ATF-2/c-Jun can bind to positive regulatory domain IV within the enhancer. The NF-κB heterodimer is bound in the cytoplasm to an inhibitor, IκBα; when this inhibitor is phosphorylated, it releases NF-κB, which then translocates to the nucleus where it binds to positive regulatory domain II. Two *i*nterferon *r*egulatory *f*actors, IRF-3 and IRF-7, are activated by phosphorylation and bind to positive regulatory domains I and III. TBK1 and IKKε, two kinases that are related to the NF-κB-IKK pathway, are directly responsible for the phosphorylation and activation of IRF-3 and IRF-7. Finally, the high mobility group protein HMG-1, a small histone-like protein associated with chromatin, binds to positive regulatory domains IV and II. Together with general transcription factors including TFIID and RNA polymerase II, this collection of DNA-bound proteins assembles to form an active transcription complex called the "enhanceosome." Although not identical, the interferon-α enhancer also contains multiple positive regulatory domains and is activated by binding to similar transcription factors.

There are presently 9 known members of the IRF family of transcription factors. They function in a number of biological processes such as pathogen response,

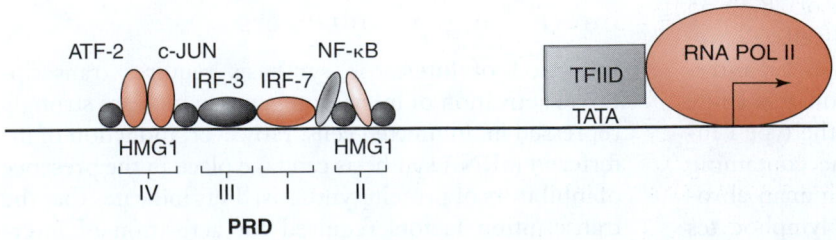

FIGURE 31.2 Transcriptional activation of the interferon β gene. Virus infection leads to activation of cellular transcription factors that bind to positive regulatory domains (PRD) upstream of the TATA box and start site on the interferon-β gene, forming an "enhanceosome."

cytokine signaling, cell growth regulation, and hematopoiesis. All interferon regulatory factors share a high degree of homology in the N-terminal DNA-binding domain. They bind to the consensus DNA sequence GAAANNGAAANN; the C-terminal portion is unique to each specific IRF member. Recently, viral versions of IRFs were discovered in the genome of human herpesvirus 8, the virus responsible for Kaposi's sarcoma. These viral IRFs block the interferon response, allowing human herpesvirus 8 to replicate more readily.

Interferon mRNAs are transported to the cytoplasm and interferon is synthesized and subsequently secreted into the extracellular medium (Figure 31.1). Interferon mRNAs contain AU-rich elements that render them unstable, and the transcriptional activation of interferon genes is short-lived. As a result, infected cells give rise to a burst of interferon that is sent out to neighboring cells, but synthesis is soon halted.

Interferon signal transduction is carried out via the Jak-Stat pathway

Cytokines and interferons are intercellular messengers that induce many important biological responses in target cells. Signaling by interferons is mediated by a pathway that includes the *J*anus tyrosine *k*inases (Jak kinases) and the *s*ignal *t*ransducers and *a*ctivators of *t*ranscription (Stat proteins). These proteins also mediate cellular responses to many other cytokines (e.g., interleukins) and growth factors such as growth hormone, prolactin, and erythropoietin. This pathway is activated by a cascade of phosphorylation events on tyrosines located in the receptor molecules, in the intermediary Jak kinases, and in the Stat proteins that activate transcription of a variety of target genes (Figure 31.3).

Signaling begins by the binding of the cytokine to a cell surface receptor. The interferon-α and -β receptors are transmembrane proteins with a binding domain on the extracellular side and an intracellular domain that binds to one of the Jak kinases, either Jak1 or Tyk2. Binding to interferon brings two or more receptor molecules together, and their associated Jak kinases phosphorylate tyrosine residues on each other as well as on the cytoplasmic domains of the receptor molecules. The phosphorylated Jak kinases are now activated. Subsequently, Stat1 and Stat2 proteins are recruited by binding to the phosphorylated receptor, and they in turn are tyrosine-phosphorylated by the activated Jak kinases, leading to the dimerization of the Stat proteins. These dimers are released and translocate to the nucleus, where they bind to sequences on a variety of interferon-stimulated genes to activate their transcription.

The **Jak-Stat pathway** represents an extremely rapid plasma membrane–to–nucleus signaling system

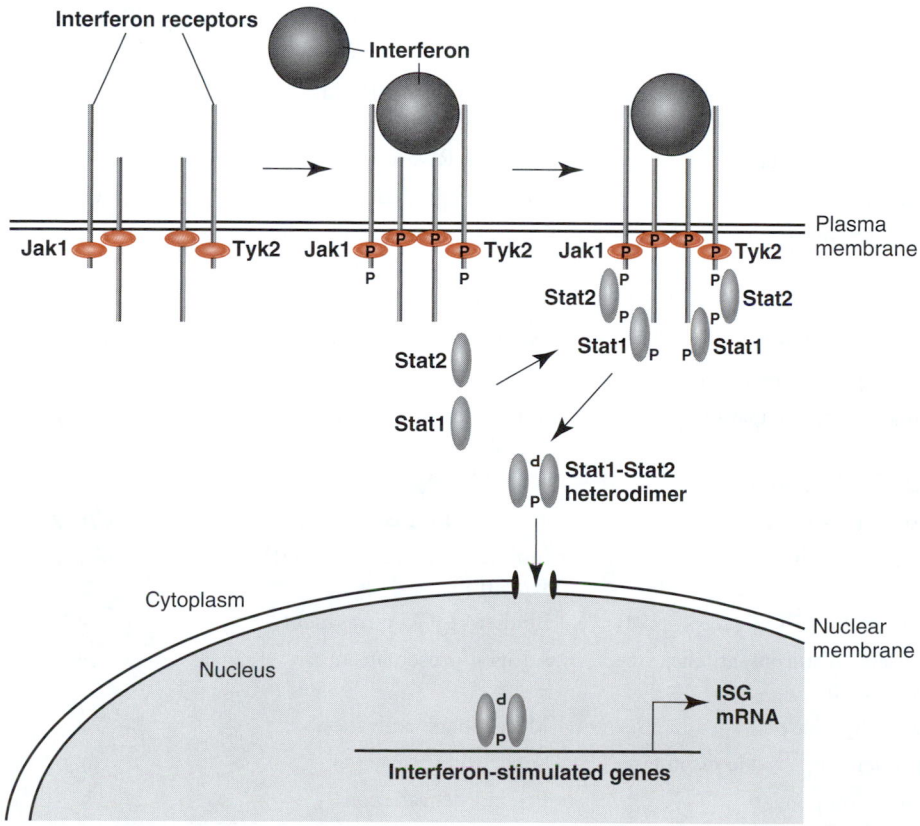

FIGURE 31.3 **Interferon signal transduction.** Binding of interferon to receptors at the surface of target cells activates the Jak-Stat pathway and leads to transcriptional activation of interferon-stimulated genes (ISG). See text for details.

and explains at least part of the basis for the specificity of signals that are induced by different cytokines. A remarkable feature of this system is that newly induced Stat DNA binding activity can be detected in the nucleus within minutes of cytokine binding. This timing reflects the rapidity of the activation of these emergency response genes. Multiple cellular responses affecting cell growth and differentiation, resistance to a wide range of pathogens, and modulation of immune functions are induced by interferon via the Jak-Stat pathway. These responses are carried out by proteins expressed from the *interferon-stimulated genes* (ISG). More than 100 such genes have been identified to date; a partial list is presented in Table 31.2.

Antiviral activities induced by interferon

The ability of interferons to confer an antiviral state to uninfected cells is their defining activity. Interferons provide an early line of defense against viral infections, and are part of the host's innate immune response to infection. Adaptive immunity, provided by humoral antibodies and specific B- and T-lymphocyte responses, is much slower to respond (days to weeks). The vital role of interferon in antiviral defense has been demonstrated by studies in which animals that cannot mount an interferon response succumb to viral infections more rapidly and at a lower infectious dose than normal animals.

Although there is evidence that the earliest stages of virus infection (virus entry and uncoating) and the final stages (assembly and release) can be inhibited by interferon treatment of cells, the best-characterized antiviral effects are directed against viral mRNA synthesis and stability, and viral protein synthesis. Three interferon-induced systems that are beginning to be understood in some detail are (1) the Mx proteins; (2) **2′, 5′-oligo(A) synthetase** and **ribonuclease L**; and (3) **PKR (double-stranded RNA-dependent protein kinase)**.

1. The Mx proteins. Mx (for "myxovirus") proteins are interferon-inducible, 70- to 80-KDa proteins with the ability to hydrolyze GTP. Expression of Mx affects virus replication by interfering with the transcription of influenza virus and other negative-strand RNA viruses. In mice, Mx1 suppresses the growth of influenza viruses; in human cells, MxA inhibits influenza, vesicular stomatitis virus, measles, and certain other negative-strand viruses. Mx proteins inhibit the activity of the viral RNA polymerase, thus blocking viral mRNA production.

TABLE 31.2 Genes activated by interferon treatment

Gene	Protein	Function	Induced by interferon type
PKR	dsRNA-dependent ser/thr protein kinase	Phosphorylates eIF-2α, inhibits protein synthesis	α/β
2′, 5′-oligo(A) synthetase	2′, 5′-oligo(A) synthetase	Activates RNase L	α/β
RNase L	Ribonuclease L	mRNA degradation	α/β
MxA	Myxovirus-inhibiting protein	GTPase; inhibits replication of negative-strand RNA viruses	α/β
IRF-2	Interferon regulatory factor-2	Transcription factor	α/β
ISG-15	Interferon-stimulated gene protein-15	Cytokine immunomodulator	α/β
CRG-2	Cytokine-responsive gene protein-2	Chemokine	α/β, γ
MHC-I	Class I major histocompatibility complex antigen	Antigen presentation	α/β, γ
GBP	Guanylate-binding protein	GTP binding	α/β, γ
IRF-1	Interferon regulatory factor-1	Transcription factor	α/β, γ
Rbp-27	Rev-binding protein-27	Inhibits Rev-dependent HIV activation	α/β, γ
FcγRI	Immunoglobulin G-Fcγ receptor	Binds to IgG Fc fragment	γ
MHC-II	Class II major histocompatibility complex antigen	Antigen presentation	γ
iNOS	Nitric oxide synthase	Macrophage activation	γ
MnSOD	Mn-dependent superoxide dismutase	Superoxide scavenger	γ
RING-12	Proteasome component	Protein degradation	γ

They also interfere with the transport of influenza virus ribonucleoprotein complexes from the cytoplasm to the nucleus, where they are transcribed and replicated. It is not clear exactly how the Mx proteins block these steps, and the role of the GTPase activity of Mx proteins in their function is not understood.

2. 2′, 5′-oligo(A) synthetase and ribonuclease L. The enzyme 2′, 5′-oligo(A) synthetase is produced in interferon-treated cells, and is activated by binding to double-stranded RNA in virus-infected cells. This enzyme has the unique ability to produce oligomers of ATP via a 2′, 5′ linkage, in contrast to the normal 3′, 5′ linkage found in natural RNAs (Figure 31.4). These oligomers contain up to 10 adenylate residues. They in turn bind to another interferon-induced protein, ribonuclease L, and activate its ribonuclease activity. Ribonuclease L then degrades both host and viral mRNAs, leading to inhibition of virus replication by suppression of viral protein synthesis. Since host mRNAs are also cleaved, the cell may die if sufficient ribonuclease L activity is present.

3. Double-stranded RNA-dependent protein kinase. PKR is a serine-threonine protein kinase that is present in normal cells at low levels; its expression is increased by interferon treatment. Like 2′, 5′-oligo(A) synthetase, PKR is activated by double-stranded RNA. When two molecules of PKR bind to the same double-stranded RNA molecule, they phosphorylate each

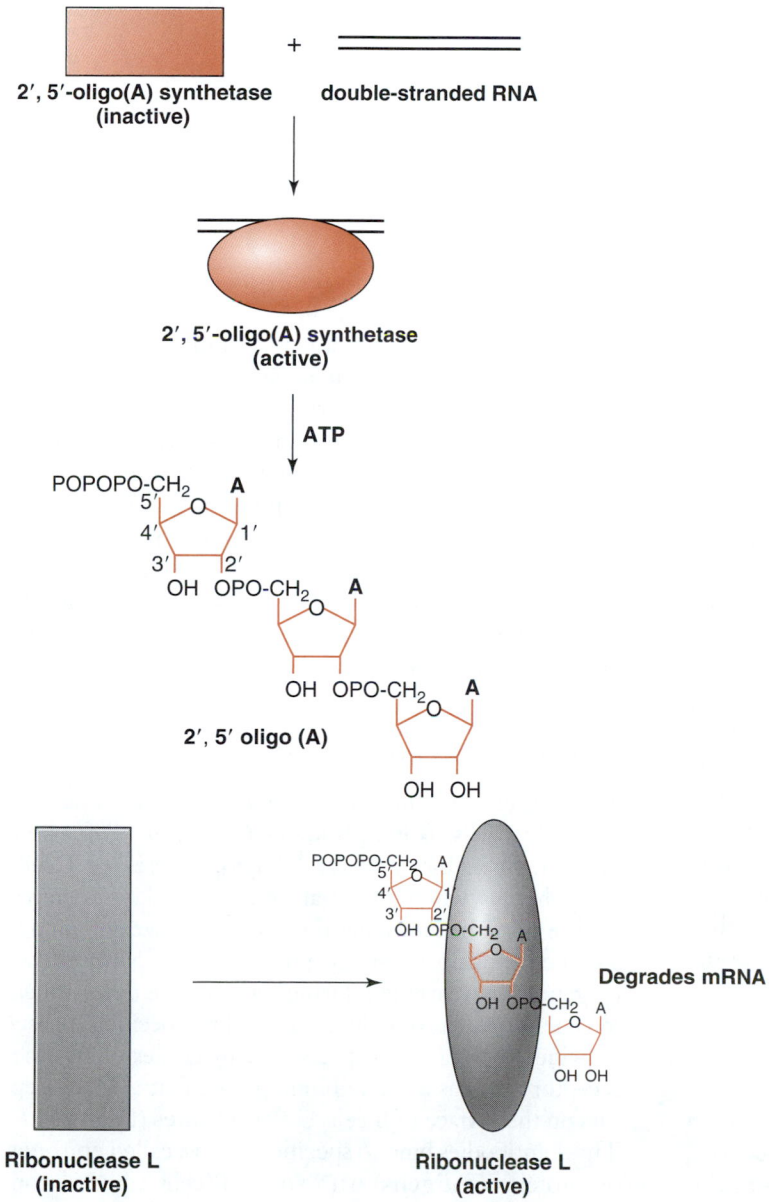

FIGURE 31.4 Antiviral activity directed by 2′, 5′-oligo(A) synthetase and ribonuclease L. Double-stranded RNA activates 2′, 5′-oligo(A) synthetase, which produces oligomers of ATP joined by 2′, 5′ linkages. These oligomers bind to and activate ribonuclease L, which degrades viral and cellular mRNAs.

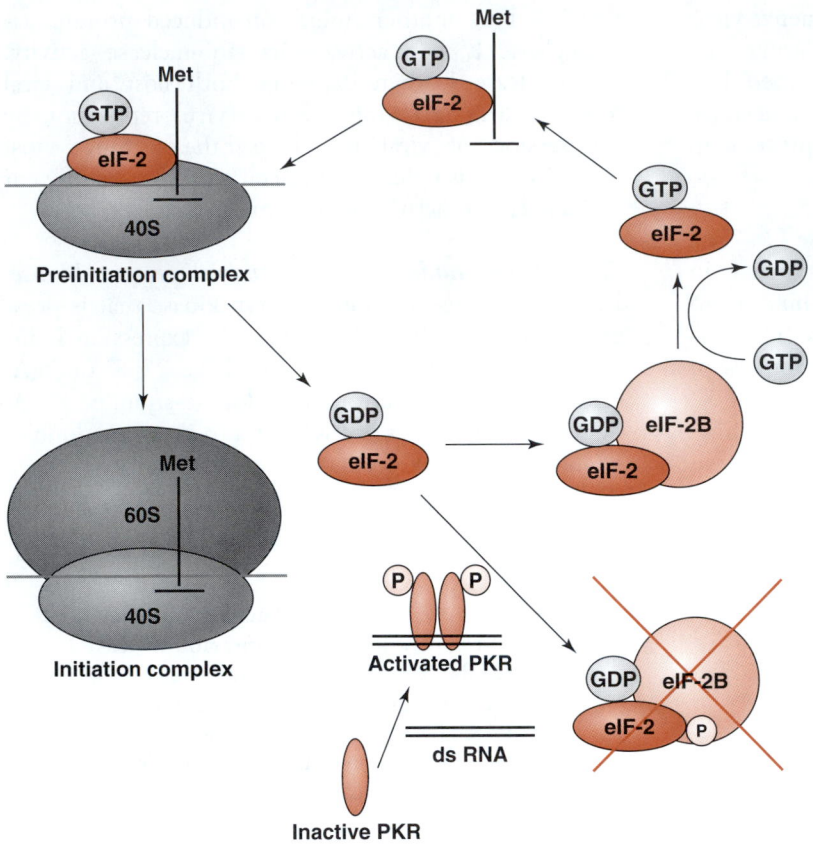

FIGURE 31.5 Mechanism of antiviral activity directed by PKR. When activated by double-stranded RNA, PKR phosphorylates the eIF-2α subunit of translation initiation factor eIF-2 (bottom right), blocking the recycling of GDP with GTP normally carried out by eIF-2B, and ultimately blocking initiation of protein synthesis. 40s and 60s: small and large ribosomal subunits.

other, inducing a conformational change that greatly increases their serine-threonine kinase activity (Figure 31.5). Activated PKR then phosphorylates a variety of cellular proteins. The best-known substrate is the alpha subunit of translation initiation factor eIF-2, which is phosphorylated by PKR at serine-51. Phosphorylation results in the formation of an inactive complex that involves eIF-2 bound to guanosine diphosphate (GDP) and the recycling factor eIF-2B. Normally, eIF-2B catalyzes the exchange of GDP for GTP, and the regenerated eIF-2-GTP complex binds to initiator methionine tRNA and forms a translation intiation complex with mRNA and a 40S ribosomal subunit. Since the GDP-GTP exchange reaction is blocked, insufficient met-tRNA-eIF-2-GTP complex is formed, and translation initiation can no longer take place. Apoptosis may also play a role in the antiviral effect of PKR, since its overexpression results in programmed cell death. PKR is believed to play an important role in regulation of cellular protein synthesis in the absence of virus infection.

Interferons have diverse effects on the immune system

Interferon-γ, or immune interferon, is produced by a restricted set of immune cells (**T lymphocytes**, macrophages, and **natural killer cells**) in response to immune and/or inflammatory stimuli, and its major function is to stimulate the development and actions of immune effector cells. Interferons-α and -β play a more limited role in modulating the adaptive immune response to viral infection. Detailed discussion of the immune system is beyond the scope of this book, but since interferons affect the immune response in such a dramatic fashion, we here give a brief introduction to the cells and molecules involved in the adaptive immune response. Immunology textbooks should be consulted for more detailed information.

The adaptive immune system

The immune system is built upon two major classes of circulating cells: **B lymphocytes** (or B cells: B for bone marrow, where they arise) and T lymphocytes (or T cells: T for thymus, where they mature). Both cell types express surface receptor molecules that can recognize and bind to specific peptides. These receptor molecules are produced by genes that rearrange during lymphocyte development to generate millions of different binding specificities; each individual B or T cell expresses many copies of a surface receptor that has a single binding specificity. The receptors on the surface of B cells are **antibodies** (Figure 31.6). These antibodies bind to specific regions, called **epitopes**, on proteins (**antigens**) with which B cells come in con-

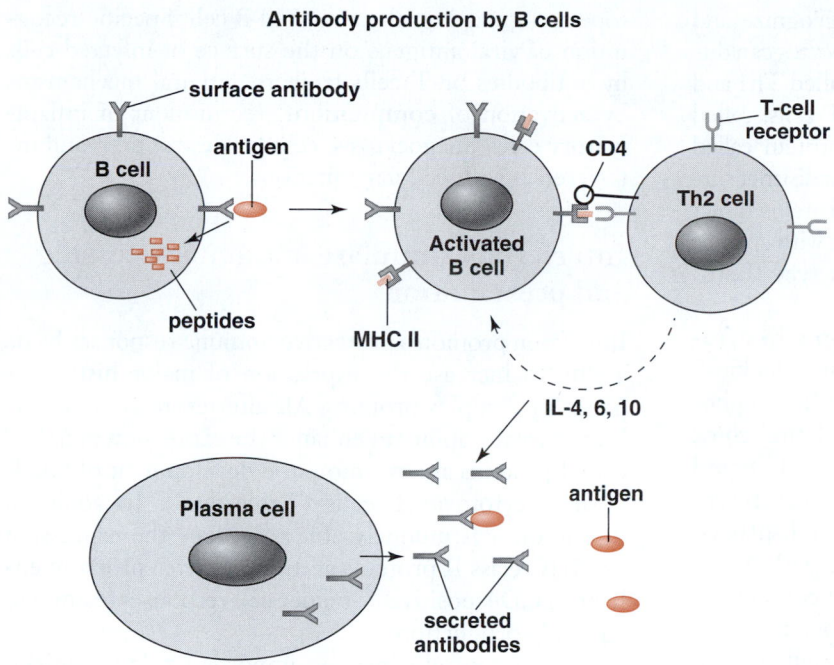

FIGURE 31.6 Antibody production by B cells. Proteins that bind to antibodies on the surface of immature B cells are taken up in the cell and digested; peptide fragments are presented on the surface of the B cell in association with MHC type II protein. These cells are stimulated to differentiate into antibody-producing plasma cells when they come in contact with T-helper cells whose T-cell receptor shares specificity for the displayed peptide.

tact in the circulation. The receptors on the surface of T cells (Figure 31.7) are called **T-cell receptors**. T-cell receptors bind not to intact proteins but to peptides produced by intracellular digestion of proteins. These peptides are complexed with specialized presenting molecules called **major histocompatibility complex (MHC) proteins**, which are located on the surface of many different cell types in the body.

When a particular B lymphocyte in the circulation comes in contact with a protein that its receptor antibody can recognize, the B cell is stimulated to divide and to differentiate into **plasma cells** that produce antibodies with the same specificity as the original receptor (Figure 31.6). Similarly, when the receptor on a T lymphocyte comes in contact with a peptide bound to a MHC protein on the surface of another cell in the body, the T cell is stim-

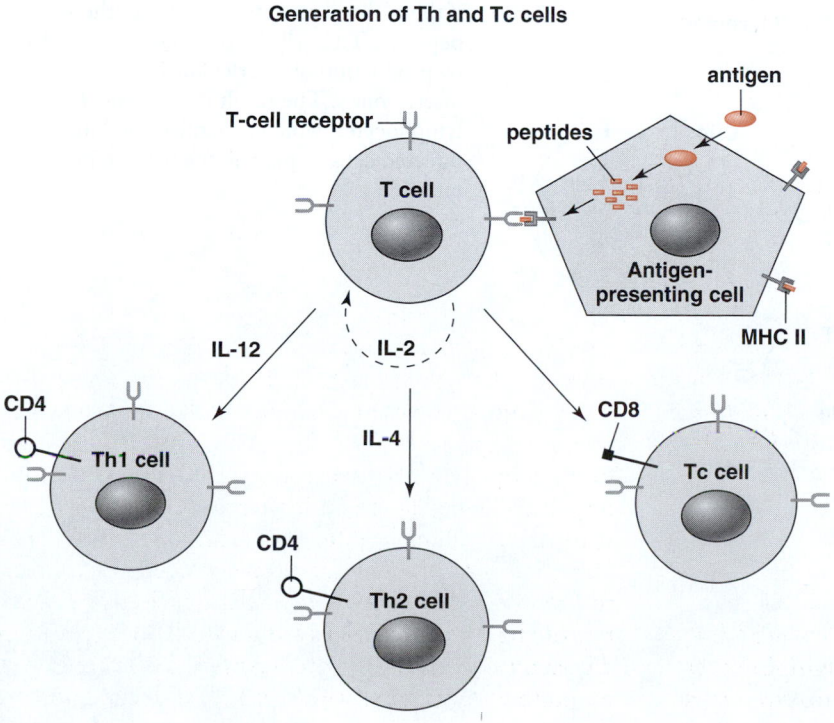

FIGURE 31.7 Generation of helper (Th) and cytotoxic (Tc) T lymphocytes. Antigen-presenting cells ingest proteins and display peptide fragments on their surface in association with MHC type II protein. Immature T cells whose T-cell receptor recognizes these peptides are activated and differentiate into Th1, Th2, and Tc cells, expressing CD4 or CD8 surface proteins.

ulated to differentiate if its receptor can recognize and bind to that peptide (Figure 31.7). T lymphocytes can differentiate into T-helper cells of two kinds, called Th1 and Th2 cells, or into precursors of cytotoxic T cells, called Tc cells. T-helper cells usually express a protein called CD4 on their surface, while Tc cells express a distinct surface protein, called CD8. CD4 and CD8 proteins direct these different T-cell populations to interact with specific target cells displaying either MHC type I or type II proteins on their surface.

Th1 cells stimulate Tc cells to develop into **cytotoxic T lymphocytes** by production of interleukin-2 and interferon-γ (Figure 31.8). Cytotoxic T lymphocytes have the capacity to kill cells with which they come into contact if those cells express surface MHC type I bound to the specific peptides that their T-cell receptor recognizes. Thus virus-infected cells that display viral peptides on their surface in a complex with MHC type I proteins will be killed when they encounter cytotoxic T lymphocytes that recognize those peptides. Similarly, Th2 cells stimulate activated B cells to mature into plasma cells, by expression of a number of other interleukins including interleukins-4, -6, and -10 (Figure 31.6). Plasma cells synthesize antibodies that specifically bind to proteins containing the peptide epitope recognized on the activated B cell. Specific recognition of viral antigens on the surface of infected cells by antibodies or T cells triggers antiviral mechanisms by activation of **complement**, recruitment of inflammatory cells, phagocytosis, cellular destruction, and interference with cellular functions.

Interferons stimulate antigen processing and presentation

Interferon promotes protective immune responses by its ability to increase the expression of major histocompatibility complex proteins. All interferon family members share the ability to enhance the expression of MHC class I proteins and promote the development of CD8-positive cytotoxic T cells (Table 31.2). In addition, interferon-γ is uniquely able to induce the expression of MHC class II proteins and therefore to promote enhanced CD4-positive T-helper cell responses leading to antibody production.

Interferons also play an important role in antigen processing by regulating the expression of proteins required to generate antigenic peptides. Interferon modifies the activity of several components of **proteasomes**, which digest viral proteins and make peptides available

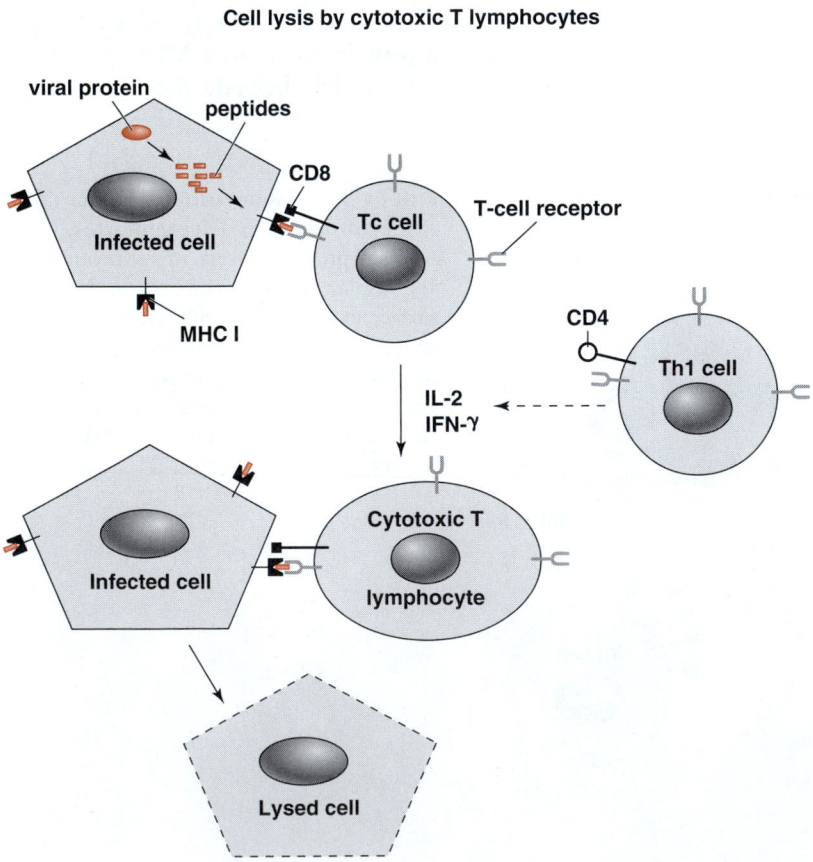

FIGURE 31.8 Cell lysis by cytotoxic T lymphocytes. Virus-infected cells display peptides derived from digested viral proteins on their surface in association with MHC type I protein. When Tc cells come in contact with infected cells, they are activated if their T-cell receptor recognizes these peptides. Th1 cells help to activate Tc cells by production of interleukin-2 and interferon-γ. The result is cytotoxic T lymphocytes, which recognize and kill infected cells displaying the peptide in question.

to MHC class I proteins. This action enhances the immunogenicity of viral proteins by increasing the quantity and repertoire of peptides displayed in association with MHC class I proteins.

Interferon and the development of CD4-positive helper T-cells

The distinct T-helper cell populations are defined by the cytokines they produce after stimulation. Th1 cells synthesize interferon-γ and interleukin-2, and promote cell-mediated immunity and the cytotoxic response. Th2 cells synthesize interleukins -4, -5, -6, and -10, and thereby facilitate antibody production by B cells. Interferon has an important effect on Th1 cell development and plays a dual role in this process: (1) it facilitates Th1 production by enhancing the synthesis of interleukin-12 in antigen-presenting cells; and (2) it blocks the development of Th2 cells by inhibiting interleukin-4 production, which is required for Th2 cell formation. Thus interferon directs the immune system to generate a cytotoxic response, which is most effective in suppressing viral infections, as infected cells are killed and eliminated.

The role of interferon in macrophage activation and cellular immunity

Macrophages function as key effector cells in innate and adaptive immune responses. To carry out these functions, macrophages must first become activated, a process that enhances cytocidal activities. Interferon-γ alone is able to stimulate macrophage activation. Activated macrophages use a variety of interferon-induced mechanisms to kill microbial targets. Two of the most important killing mechanisms involve the production of reactive oxygen and reactive nitrogen intermediates. For example, nitric oxide is generated in macrophages as a result of the interferon-dependent induction of the inducible form of nitric oxide synthase (iNOS).

Effects of interferons on antibody production

Interferons play complex roles in regulating antibody production since they exert their effects either indirectly, by regulating the development of specific T-helper cell subsets (as described above), or directly at the level of B cells. The best-characterized action of interferons on B-cell activation is their ability to influence immunoglobulin (Ig) type switching. For example, interferon-γ can switch production from IgM to IgG2a in stimulated B cells. In this fashion, interferons modulate the immune response and increase host defenses against certain bacteria and viruses.

Interferons regulate cell growth and apoptosis

Interferons inhibit cell growth and induce apoptosis, activities that affect the suppression of cancer and infection. Different cells exhibit varying degrees of sensitivity to the antiproliferative activity of interferons. In some cases, growth arrest may be due to differentiation, particularly when interferons are used in combination with agents such as retinoids (vitamin A derivatives). Specific interferon-induced gene products have not been linked directly to antiproliferative activity; however, interferon targets components of the cell-cycle control apparatus, including c-myc, pRB, cyclin D3, and cdc25A.

TABLE 31.3 Inhibition of interferon system by viruses

Virus	Gene product	Effect on interferon system
Adenovirus	VA I RNA	Blocks PKR activation
	E1A protein	Inhibits activation of interferon-stimulated genes
Epstein-Barr virus	EBV small RNA	Blocks PKR activation
	Epstein-Barr nuclear antigen-2	Inhibits activation of Stat proteins
Reovirus	Sigma 3 protein	Blocks PKR activation by binding to dsRNA
Vaccinia virus	SK-1 protein	Blocks PKR activation by binding to dsRNA
	E3-L protein	Substrate for PKR
Human immunodeficiency virus type I	Tat protein	Blocks PKR activation
Human herpesvirus-8	Viral interferon response factor (v-IRF)	Homologue of IRF family; blocks gene activation

Viruses have developed numerous strategies to evade the interferon response

Viruses fight back against the host interferon response by making proteins specifically designed to interfere with these host defense systems. Table 31.3 gives some examples of a growing list of viral gene products that interfere with the interferon response.

An important target for many viruses is the double-stranded RNA-activated protein kinase, PKR. Adenovirus produces high levels of VA I RNA, which is a small (160 nt) RNA transcribed from the adenovirus genome by RNA polymerase III. By virtue of self-complementary sequences, it takes on a structure similar to double-stranded RNA. VA I RNA binds to PKR as do other double-stranded RNAs, but fails to activate the protein kinase, perhaps because it is too small to bind to two PKR molecules at the same time (see above). Therefore it occupies the ds RNA binding site and inhibits PKR activation, allowing viral protein synthesis to take place even in cells that express high levels of PKR induced by interferon. Epstein-Barr virus (a herpesvirus) encodes a small RNA that performs a similar function.

Both reovirus $\sigma 3$ capsid protein and vaccinia virus SK-1 protein block PKR activation by binding strongly to double-stranded RNA in the infected cell. These proteins therefore sequester the double-stranded RNA and do not allow it to come in contact with PKR, so that activation cannot occur. The HIV transcriptional transactivator Tat and the hepatitis C virus NS5A protein appear to inhibit PKR antiviral activity by binding directly to PKR. In poliovirus-infected cells, PKR is proteolytically cleaved.

Adenovirus E1A protein inhibits the production and action of interferon at the level of transcription. E1A sequesters p300/CBP, a co-activator required for transcription of the interferon genes. Human herpesvirus 8 produces viral homologues of the interferon regulatory factors, called v-IRFs, that block interferon gene activation by inhibiting, rather than stimulating, transcription when bound to interferon genes.

Some viruses evade apoptosis or the inflammatory response by synthesizing proteins that interfere with signaling pathways. For example, poxviruses produce CrmA, CrmB, and CrmC proteins that bind to specific proteases called **caspases**, involved in apoptosis, resulting in a delay in cell death and allowing the virus to replicate. Poxviruses also produce soluble forms of cytokine receptors that perform the dual functions of blocking the production of antiviral cytokines and inhibiting B cell activation, thus delaying the onset of the humoral immune response.

Conclusion: interferons are a first line of defense against virus infection

The objective of all viruses is to infect target cells, produce large numbers of progeny virions, and release these progeny to initiate new rounds of infection in other cells. Viruses encode highly efficient proteins to optimize their replication; however, the host organism also possesses defenses to limit virus infection. Interferons constitute an early line of defense against virus infections. The antiviral, antiproliferative, and immune-regulatory functions of interferons justify their clinical use in the treatment of several viral and neoplastic diseases. Knowledge of the mechanisms that lead to interferon and cytokine induction upon virus infection, and the strategies developed by viruses to escape the host response, provides a means to develop new antiviral agents and a window on critical molecular events occurring during the struggle between the virus and the host cell.

SUGGESTED READING

1. Fitzgerald, K.A., McWhirter, S.M., Faia, K.L., Rowe D.C., Latz, E., Golenbock D.T., Coyle, A.J., Liao, S.M., and Maniatis, T. (2003). IKK epsilon and TBK1 are essential components of the IRF3 signaling pathway. *Nature Immunology* **4,** 491–496.

2. Leonard, W.J. and O'Shea, J.J. (1998). Jaks and STATs: Biological implications. *Annual Review of Immunology* **16,** 293–322.

3. LePage, C., Genin, P., Baines, M., and Hiscott, J. (2000). Interferon activation and innate immunity. *Reviews in immunogenetics* **2,** 374–386.

4. Mamane, Y., Heylbroeck, C., Genin, P., Algarte, M., Servant, M.J., LePage, C., DeLuca, C., Kwon, H., Lin, R., and Hiscott, J. (1999). Interferon regulatory factors: The next generation. *Gene* **237,** 1–14.

5. Roulston, A., Marcellus, R.C., and Branton, P.E. (1999). Viruses and apoptosis. *Annual Review of Microbiology* **53,** 577–628.

6. Sen, G. (2000). Novel functions of interferon-induced proteins. *Cancer Biology* **10,** 93–101.

7. Sharma, S., tenOever, B.R., Grandvaux, N., Zhou, G.P., Lin, R., and Hiscott, J. (2003). Triggering the interferon antiviral response through an IKK-related pathway. *Science* **300,** 1148–1151.

8. Stark, G.R., Kerr, I.M., Williams, B.R., Silverman, R.H., and Schreiber, R.D. (1998). How cells respond to interferon. *Annual Review of Biochemistry* **67,** 227–264.

9. Takeda, K. and Akira, S. (2004). TLR signaling pathways. *Seminars in Immunology* **16,** 3–9.

KEY TERMS

Antibodies
Antigens
B lymphocytes
Caspases
Chemokines
Complement
Cytokines
Cytotoxic T lymphocytes
Epitopes
Interferons
Interleukin-1
Jak-Stat pathway
Macrophages
Major histocompatibility complex (MHC) proteins
Natural killer cells
2′, 5′-oligo(A) synthetase
PKR (double-stranded RNA-dependent protein kinase)
Plasma cells
Proteasomes
Ribonuclease L
T-cell receptors
T lymphocytes
Toll-like receptors
Viral interference

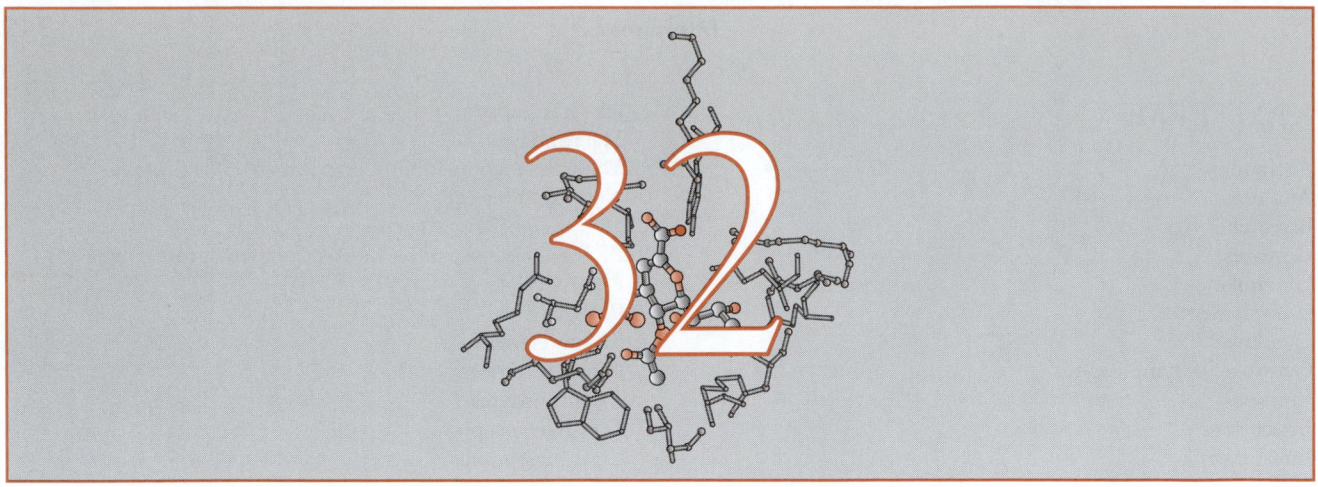

Antiviral Chemotherapy

DISCOVERY OF ANTIVIRAL DRUGS

"Serendipity"—trying out compounds used for other purposes (amantadine and acyclovir).
Chemical modification of known active compounds (ganciclovir and azidothymidine).
High throughput screening assays of many compounds (nevirapine).
Rational design, often with the aid of three-dimensional structures of viral proteins (ritonavir and zanamivir).

TARGETS OF ANTIVIRAL DRUGS

Capsid-binding drugs: picornavirus capsid proteins (attachment/entry).
Uncoating inhibitors: influenza M2 ion channel (uncoating).
Nucleoside analogues: viral DNA and RNA polymerases (genome replication):
- *May be selectively phosphorylated by virus-encoded kinases*
- *Selectively inhibit viral DNA polymerases*

Protease inhibitors: HIV-1 protease (virion maturation).
Neuraminidase inhibitors: influenza neuraminidase (virion release).

The discovery and widespread use of antiviral compounds began only recently

Both bacteria and viruses (filterable infectious agents) were first shown to cause disease in the late nineteenth century. Antibiotics active against bacterial diseases began to be widely used in the 1940s. However, the clinical use of drugs to treat viral diseases in humans lagged behind, and the first antiviral drugs were introduced only in the 1960s. A major factor accounting for this delay is that viruses replicate within host cells, using many elements of the host cell machinery. Therefore, there are fewer obvious targets for chemotherapy against viruses than for chemotherapy against bacteria. For example, many antibiotics are directed against bacterial cell wall synthesis or bacterial protein synthesis, which are clearly distinct from their human counterparts. The development of compounds active against a broad spectrum of viruses is unlikely, because different viruses do not share common features that distinguish them from their host cells. Each virus has its own array of specific proteins required for its replication, and therefore different antiviral drugs may have to be developed for each virus.

Because of these difficulties, pharmaceutical companies have not always been enthusiastic about investing resources in the development of antiviral drugs. Individual virus diseases often represent smaller potential markets than the wide variety of diseases caused by bacteria. Moreover, vaccination has been a very successful approach to controlling viral diseases, further dampening the interest of pharmaceutical companies to develop drugs against these diseases. Efforts to develop antiviral drugs have focused mainly on those viruses for which effective vaccines or public health measures are not available, and that cause widespread and easily diagnosed diseases whose treatment would earn substantial profits.

However, since 1980 many effective antiviral agents have been developed. This began with the success of **acyclovir**, a drug directed against certain herpesviruses. The AIDS epidemic led to a pressing need for drugs active against human immunodeficiency virus (HIV) and opportunistic pathogens, and coincided with the development of expanding knowledge about virus genetics, molecular biology, enzymology, and protein structure. These factors have led to a rapid increase in the number and kinds of antiviral drugs available to treat human diseases. Indeed, many advances in **rational drug discovery** have come from efforts to develop antiviral drugs.

Although protein drugs such as **interferons** (see Chapter 31) and nucleic acid–based drugs such as oligonucleotides are of considerable interest for medicine and virology, for purposes of brevity only small-molecule drugs will be covered here. DISCLAIMER: The inclusion of certain drugs in this chapter should not be construed as a comment on their clinical merits compared with drugs that are not included. The spectrum of drugs available is changing rapidly, and the drugs described here are used only as examples of the mechanisms of a much larger group of compounds.

Importance of antiviral drugs for basic science

Aside from their importance in treating viral diseases, antiviral drugs make excellent laboratory tools. As will be illustrated in detail below, drugs can block the functions of specific viral proteins and thus specific stages in viral infection. This is crucial for understanding the timing of particular events in the virus life cycle and which proteins are needed for each step. Antiviral drugs, in conjunction with viral mutations that confer drug resistance, can help to identify the roles of viral proteins and dissect the details of how these proteins do their jobs. Drug resistance mutations can provide selectable markers for engineering interesting mutant viruses and performing viral genetics.

How are antiviral drugs obtained?

The first antiviral drugs were discovered by testing chemicals for their ability to inhibit virus replication in labor- and time-intensive assays of viral replication, such as assays of cytopathic effect or plaque assays. These chemicals had usually been synthesized for some other purpose, and discovery of their antiviral activities was often serendipitous. Examples of drugs that were discovered this way are **amantadine** and acyclovir. When particular chemicals were known to have antiviral activity, a variety of other chemicals with similar structures were tested for activity. By use of this somewhat more rational procedure, compounds such as the nucleoside analogues **ganciclovir** and **azidothymidine** were discovered.

More modern and rational approaches have been used to discover some newer drugs. One approach is to find a way to assay virus replication that can be automated, using robots, and performed relatively rapidly (i.e., within one to a few days). An example of such an assay would be the use of a dye that measures viability and proliferation of cells following virus infection. This assay is then used to screen large libraries of chemicals (hundreds of thousands of compounds) for their ability to inhibit virus replication. This approach is called a **cell-based high throughput screen**, because it involves applying chemicals to cells infected with the virus and assaying the effect on virus replication. Such an assay has been used to discover novel inhibitors of herpesvirus replication.

A related approach, called a **target-based high throughput screen**, starts with the identification of a viral gene product that would make a good drug target. In principle, the best targets are essential viral gene products (almost always proteins) that are sufficiently different from their cellular counterparts that the activity of the viral protein can be inhibited without affecting the function of cellular proteins. Nonessential viral proteins can also be targeted if they activate a drug, transforming it into an inhibitor. For high throughput screening, it is important that the target protein can be expressed at high levels and purified and assayed rapidly (i.e., within hours or even minutes) using robots. This assay is then used to screen large libraries of chemicals. For example, HIV-1 reverse transcriptase can be expressed and purified, and its DNA polymerase activity can be easily assayed *in vitro*. This approach was used to discover nonnucleoside reverse transcriptase inhibitors such as **nevirapine**.

A still more rational approach is to identify a viral protein that is crucial for virus replication and then to exploit detailed knowledge about this protein to design drugs that will inhibit its activity. Many proteins interact with other viral or cellular proteins in functional complexes, and these interactions can be targeted by drugs. Furthermore, knowledge of the three-dimensional structure of the protein can enable visualization of its active site and understanding of its mechanism of action. This information can then be used to design small molecules that bind to the active site and block activity of the protein. This approach has been used to discover anti-HIV protease inhibitors such as **ritonavir** and anti-influenza neuraminidase inhibitors such as **zanamivir**.

Targeting drugs to specific steps of virus infection

As outlined in Chapter 1 and detailed in most chapters in this book, the infection of cells by viruses can be broken down into a common set of steps or stages (Figure 32.1). In principle, any step of viral infection can be tar-

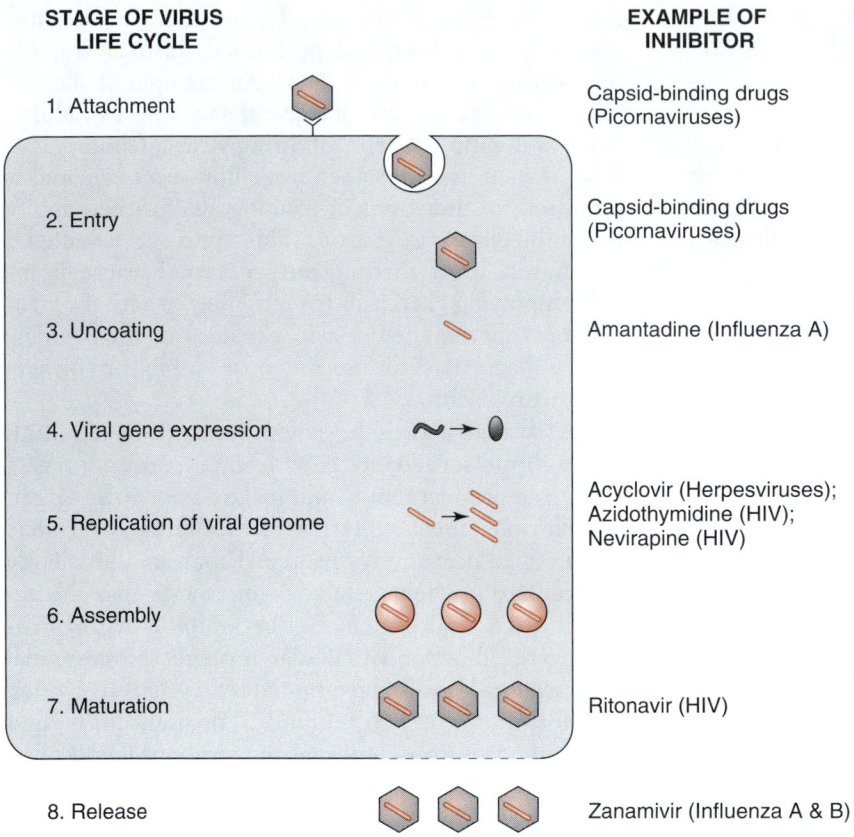

FIGURE 32.1 Inhibitors useful at different stages in virus replication cycle. Mature virion or capsid is shown as a hexagon, genome as a thick diagonal line, viral mRNA as a curved black line, and viral protein, as an oval. Immature virion is shown as a circle. For some viruses (e.g., HIV), maturation takes place after release of virion from the cell.

geted for inhibition. There are advantages to very early or late steps such as attachment, entry, and release, because inhibitors of these steps do not have to enter cells to exert activity. Steps such as genome replication, assembly, and maturation require specific viral proteins, often acting as enzymes. Enzymes make good drug targets because they are usually at low concentrations inside cells (due to being catalysts), they are relatively well-understood mechanistically, and they normally interact with small-molecule substrates. Furthermore, pharmaceutical companies have a great deal of experience in developing enzyme inhibitors as drugs. It is therefore not surprising that most clinically useful antiviral drugs are enzyme inhibitors that target genome replication, assembly, and maturation. Nevertheless, there are antiviral drugs that act at nearly every step of viral infection.

Capsid-binding drugs prevent attachment and entry of virions

Inhibition of attachment and entry prevents all subsequent steps in virus infection and permits the virion to be cleared by immune and other mechanisms. Members of one class of drugs called **capsid-binding drugs** block attachment and/or entry of picornaviruses, especially rhinoviruses, a major cause of the common cold. There are many different serotypes of rhinoviruses, making vaccination impractical. There is already a large market for drugs that mitigate cold symptoms, and some think that market could be tapped by an effective antiviral agent, if it is very safe.

Capsid-binding drugs act by direct binding to virions. These drugs inhibit attachment and/or entry or even later steps depending on the drug-virus combination. Our understanding of these drugs comes largely from x-ray crystallographic studies of complexes between picornaviruses and such drugs. In Figure 32.2 is a model showing how picornaviruses bind to their cellular receptor via the so-called "canyon" in the surface of the capsid. Beneath the canyon floor is the drug-binding site—a pocket. This pocket is lined with hydrophobic amino acid side chains, and is ordinarily occupied by a lipid molecule. Binding of the cellular receptor molecule to the canyon is thought to release the lipid molecule from the pocket. This allows the capsid to undergo a conformational change that results in the eventual entry of the genome RNA into the cytoplasm. Binding of a capsid-binding drug to the pocket displaces the lipid molecule. The presence of the drug can alter the structure of the canyon floor, thereby preventing proper interaction with the cellular receptor and preventing attachment. Alternatively, the drug can stabilize the capsid and prevent the conformational

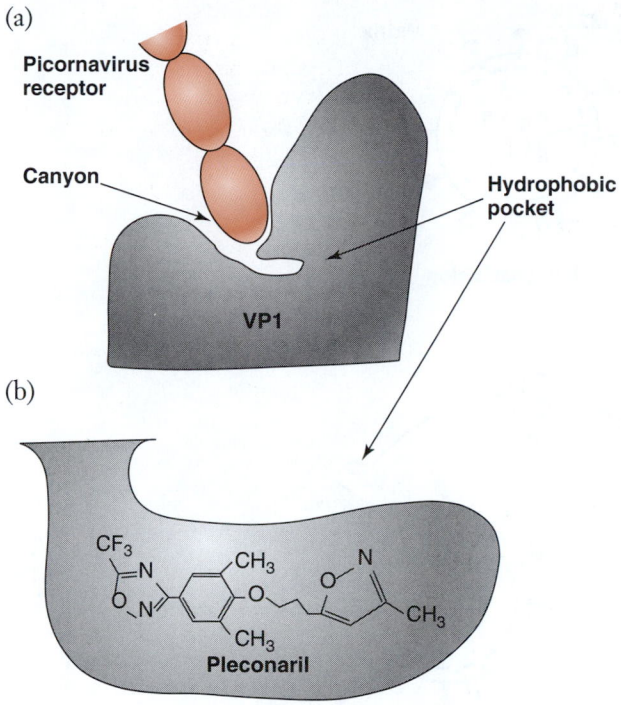

FIGURE 32.2 **Receptor and drug binding to the picornavirus capsid.** (a) Binding of the N-terminal domain of a picornavirus receptor to the "canyon" on the surface of the capsid. The hydrophobic pocket lies just below the canyon floor. (b) Binding of pleconaril, a capsid-binding drug, to the hydrophobic pocket displaces the bound lipid and interferes with attachment of virion to receptor and/or entry of viral genome into the cytoplasm.

change that releases the RNA, thereby preventing uncoating and genome entry. An example of a capsid-binding drug is **pleconaril**, which is active against many picornaviruses, including rhinoviruses.

Although anti-picornavirus capsid-binding drugs have not yet been licensed for clinical use, an anti-HIV drug, **enfuvirtide**, was approved by the U.S. Food and Drug Administration in 2003. This peptide drug blocks a conformational change in the viral envelope protein gp41 that is required for fusion between the viral envelope and the cell membrane.

Amantadine blocks ion channels and inhibits uncoating of influenza virions

Amantadine (Figure 32.3) was discovered by random screening in the 1960s. It has been most useful in certain settings, such as nursing homes, for prevention of infection among contacts of flu victims. In most influenza strains, amantadine inhibits virus uncoating. At first amantadine, which is a weak base, was thought to inhibit the acidification of endosomes, the cellular compartment in which influenza uncoating occurs. However, the concentrations of drug required to inhibit endosomal acidification are much higher than those that have a therapeutic effect.

The elucidation of the mechanism of action of amantadine represents a triumph of molecular genetic approaches using drug-resistant viruses. Alan Hay and collaborators at the National Institute for Medical Research in London (U.K.) isolated a set of influenza virus mutants resistant to amantadine. They then used influenza genetics to map the mutations to the M segment of the virus. Upon sequencing, all of the mutations were found to lie within the M2 gene (the other gene in the M segment is the M1 gene). The product of the M2 gene is a small protein that spans the virion envelope. It was noticed that the M2 protein resembles known channel-forming transmembrane proteins. Subsequent biochemical and electrophysiological experiments showed that the M2 protein is a channel for H^+ ions and that amantadine blocks this channel.

How would blocking an ion channel lead to inhibition of uncoating? A reasonable but not completely verified model is diagrammed in Figure 32.3. This starts with the fact that influenza virus enters cells by receptor-mediated endocytosis. As a result, the virions find themselves in endosomes whose pH is lowered. This permits fusion of the influenza envelope with the endosome membrane, mediated by the low pH conformation of the viral hemagglutinin (see Chapter 4). It is thought that M2 in the virion envelope functions to let H^+ ions enter the virion, and the low pH changes the conformation of the matrix protein. This results in liberation of the viral nucleocapsids from within the envelope, so that they can find their way to the nucleus for transcription. Amantadine would block the entry of H^+ ions, thereby preventing uncoating.

A key point about amantadine is that this drug and the drug-resistant mutants opened up a whole new area of virus biology. No one had imagined that a virus might encode an ion channel. An exciting possibility is that other viruses encode similar proteins that might serve as good drug targets.

Nucleoside analogues target viral DNA polymerases

Virus genome replication is often performed by virus-coded enzymes that are essential for viral replication and differ substantially from host proteins. Therefore these viral proteins make good drug targets. Accordingly, more antiviral drugs target genome replication than any other stage of virus infection.

Many clinically useful antiviral drugs are nucleoside analogues that target viral DNA polymerases. Figure 32.4 shows three examples: acyclovir and ganciclovir, both active against herpesviruses, and azidothymidine (AZT), active against HIV-1. Upon entering cells, nucleoside analogues must first be activated by phospho-

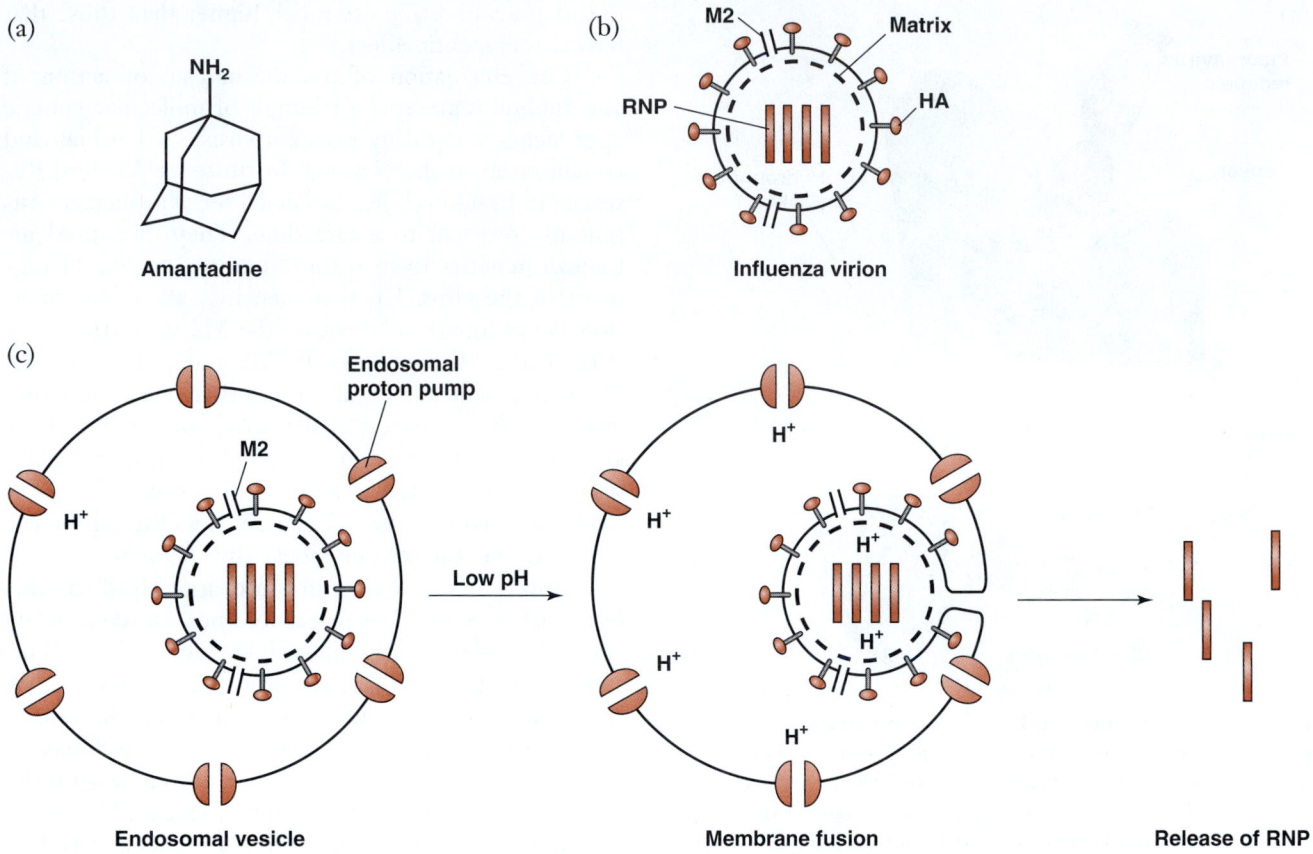

FIGURE 32.3 Proposed mechanism of action of amantadine on influenza virus uncoating. (a) Stucture of amantadine. (b) Structure of influenza virus particle, showing HA and M2 proteins in envelope, matrix protein, and viral ribonucleoprotein (RNP). (c) Influenza virions in endosomes undergo fusion with endosomal membrane upon drop in pH induced by an endosomal proton pump. The M2 protein allows hydrogen ions to enter virion, releasing the RNP from the matrix protein. Amantadine blocks the M2 channel, inhibiting this process.

rylation so that they mimic the natural substrates of DNA polymerases, deoxynucleoside 5′-triphosphates. These phosphorylated analogues can inhibit viral DNA polymerases in two ways. They can act as a competitive inhibitor, blocking binding of the natural nucleoside triphosphate to the active site of the enzyme. Alternatively, analogues can be incorporated into the growing DNA chain and prematurely terminate its elongation. Both of these events can be important for antiviral activity and for selectivity.

If host cell enzymes can phosphorylate a nucleoside analogue and the phosphorylated forms can inhibit or be used as a substrate by host enzymes such as cellular DNA polymerases, the nucleoside analogue will usually be toxic. Selectivity of these drugs therefore depends on two factors: (1) whether viral nucleoside kinases can phosphorylate them more efficiently than their cellular counterparts; and (2) whether viral DNA polymerases are preferentially inhibited by the phosphorylated analogues compared with cellular enzymes.

Acyclovir is selectively phosphorylated by herpesvirus thymidine kinases

Acyclovir exhibits a very high **therapeutic index** (toxic dose divided by effective antiviral dose). Accordingly, it has been a very successful drug in treating both common herpes simplex virus infections such as genital herpes (many millions of doses of the drug have been administered) and less common but more serious effects of herpesvirus infections such as encephalitis.

The drug consists of a guanine base attached to what is essentially a broken sugar ring (Figure 32.4). It can be thought of as deoxyguanosine missing the CHOH moiety at the 3′ position and the CH_2 moiety at the 2′ position of ribose. This **acyclic** sugar-like molecule accounts for the name of the compound and for aspects of acyclovir action. Figure 32.5 shows what happens to acyclovir in cells infected with herpes simplex virus or varicella zoster virus. Each of these viruses encodes a **thymidine kinase** that is able to phosphorylate

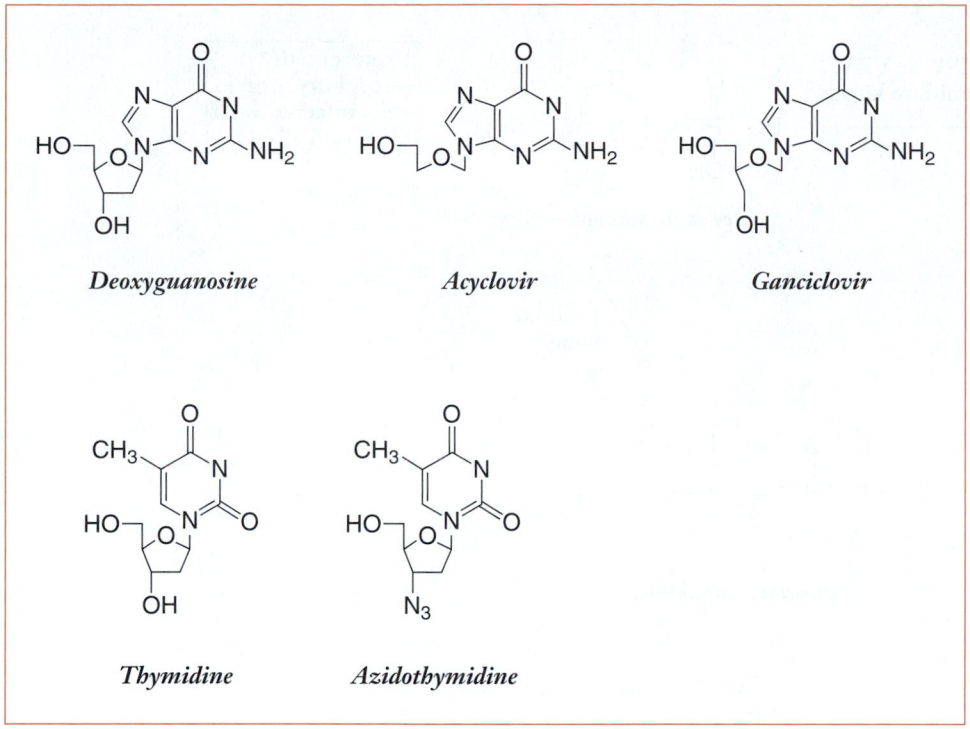

FIGURE 32.4 **Structures of selected nucleosides and nucleoside analogues.**

not only thymidine but also deoxyuridine, deoxycytidine, thymidine monophosphate (dTMP), and a variety of nucleoside analogues, including some like acyclovir, which contains a purine instead of a pyrimidine base. No mammalian enzyme phosphorylates acyclovir as efficiently as do the thymidine kinases of these two herpesviruses, although some phophorylation does occur in uninfected cells (1–3% of that found in cells infected with herpes simplex virus). Therefore, herpesvirus-infected cells contain a much higher concentration of phosphorylated acyclovir than do uninfected cells. This accounts for much of the antiviral selectivity of acyclovir.

Acyclovir is preferentially incorporated by herpesvirus DNA polymerases

Acyclovir monophosphate is then converted to acyclovir diphosphate and acyclovir triphosphate, probably exclusively by cellular enzymes. Acyclovir triphosphate is an inhibitor of herpesvirus DNA polymerases. Moreover, it inhibits viral DNA polymerases more potently than cellular DNA polymerases. The details of the inhibition of herpes simplex virus DNA polymerase by acyclovir triphosphate have been worked out with the purified enzyme *in vitro*. This fascinating mechanism, termed "induced substrate inhibition," involves three steps (Figure 32.6).

In the first step, acyclovir triphosphate acts as a competitive inhibitor of the incorporation of deoxyguanosine triphosphate (dGTP) into DNA, by competing with dGTP for binding to the DNA polymerase; thus, high concentrations of dGTP can reverse inhibition at this stage. In the second step, acyclovir triphosphate serves as a substrate and is incorporated by the viral DNA polymerase into the growing DNA chain opposite a C residue on the template strand.

In the third step, the DNA polymerase translocates to the next position on the template (G in the figure). However, there is no 3′-hydroxyl group on acyclovir on which to add the next deoxynucleoside triphosphate. Moreover, if the next deoxynucleoside triphosphate (dCTP in the figure) is present, this induces the polymerase to freeze in a "dead-end complex," leading to apparent inactivation of the enzyme.

Selectivity of inhibition by acyclovir triphosphate can occur at any of these steps. For example, compared with herpes simplex virus DNA polymerase, cellular DNA polymerase alpha uses acyclovir triphosphate less efficiently as a substrate and it is less potently inhibited by the next deoxynucleoside triphosphate. The precise differences between herpesvirus and cellular polymerases that contribute to selectivity of acyclovir remain poorly undersood.

Cytomegalovirus encodes a protein kinase that phosphorylates ganciclovir

Ganciclovir was the first drug licensed for use against human cytomegalovirus, another herpesvirus. Cy-

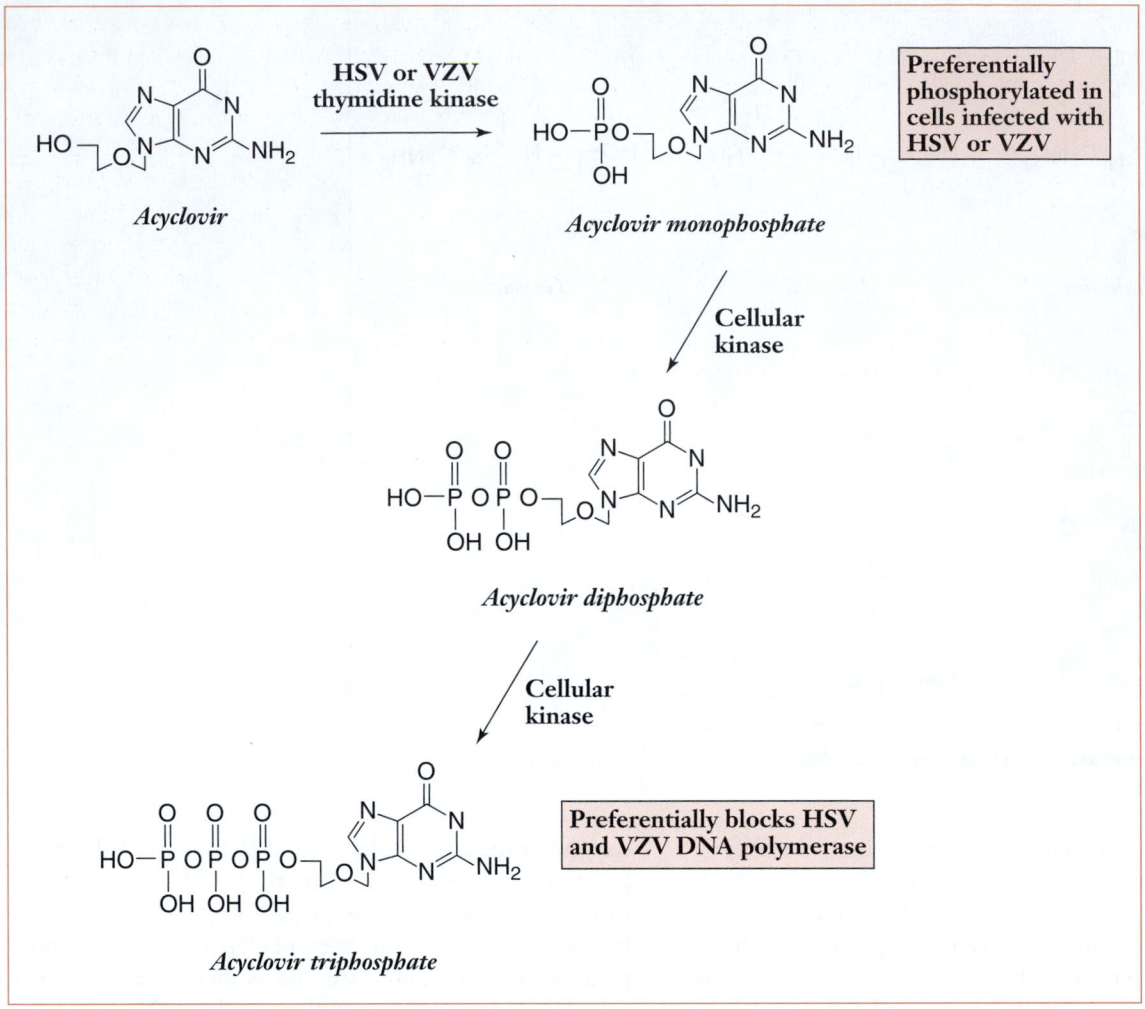

FIGURE 32.5 **Phosphorylation of acyclovir.** Shown are the enzymatic steps that result in the phosphorylation of acyclovir to the mono-, di-, and triphosphate forms, the last of which blocks viral replication. HSV: herpes simplex virus; VZV: varicella zoster virus.

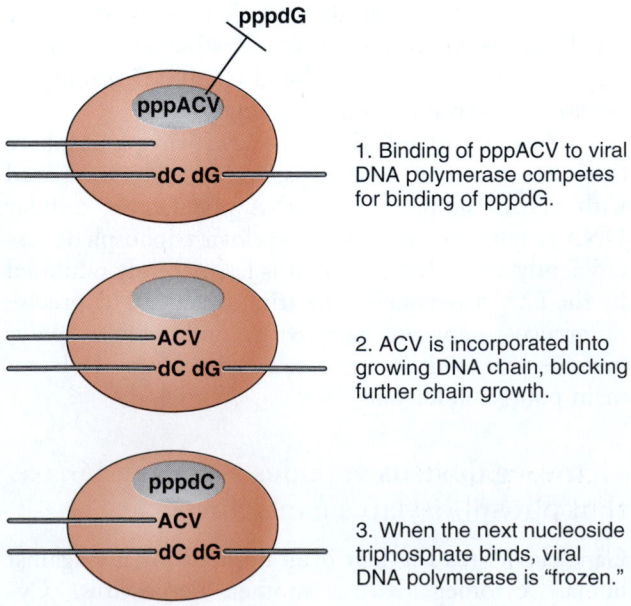

tomegalovirus infections are usually inapparent in normal adults, but can be life-threatening or sight-threatening in the immunocompromised. Ganciclovir is relatively efficacious in this setting. Its structure is shown in Figure 32.4. Like acyclovir, ganciclovir is a guanine linked to an acyclic sugar-like molecule, and lacks a 2′ CH_2 moiety. However, it contains the 3′ CHOH group that is missing in acyclovir. Thus, it more closely resembles the natural compound, deoxyguanosine, which may account for its greater toxicity.

FIGURE 32.6 **Mechanism of inhibition of herpes simplex virus DNA polymerase by acyclovir triphosphate.** Large orange oval is viral DNA polymerase, small gray oval is dNTP binding site, horizontal lines are template and growing daughter DNA strands. pppACV: acyclovir triphosphate; pppdG: deoxyguanosine triphosphate; pppdC: deoxycytidine triphosphate.

Ganciclovir must be phosphorylated to act as an inhibitor, in a similar fashion to acyclovir (Figure 32.5). However, cytomegalovirus does *not* encode a homologue of herpes simplex virus thymidine kinase (which phosphorylates ganciclovir very efficiently). Despite this, cytomegalovirus infection induces a roughly 30-fold increase in intracellular phosphorylated forms of ganciclovir. What enzyme is responsible for this? Biochemical experiments failed to uncover the correct answer, but, as was true for amantadine, molecular genetics succeeded in defining the drug target. A ganciclovir-resistance mutation that prevents cytomegalovirus from inducing phosphorylation of ganciclovir was mapped to a gene encoding a viral protein called UL97. It was subsequently shown that this protein can directly phosphorylate ganciclovir to the monophosphate form; cell kinases then generate the di- and triphosphates. Interestingly, UL97 does not phosphorylate natural nucleosides and it is not homologous to nucleoside kinases. Rather, it is a serine/threonine protein kinase, although it lacks some motifs common to protein kinases. How a protein kinase can phosphorylate a nucleoside analogue is an interesting problem in enzymology.

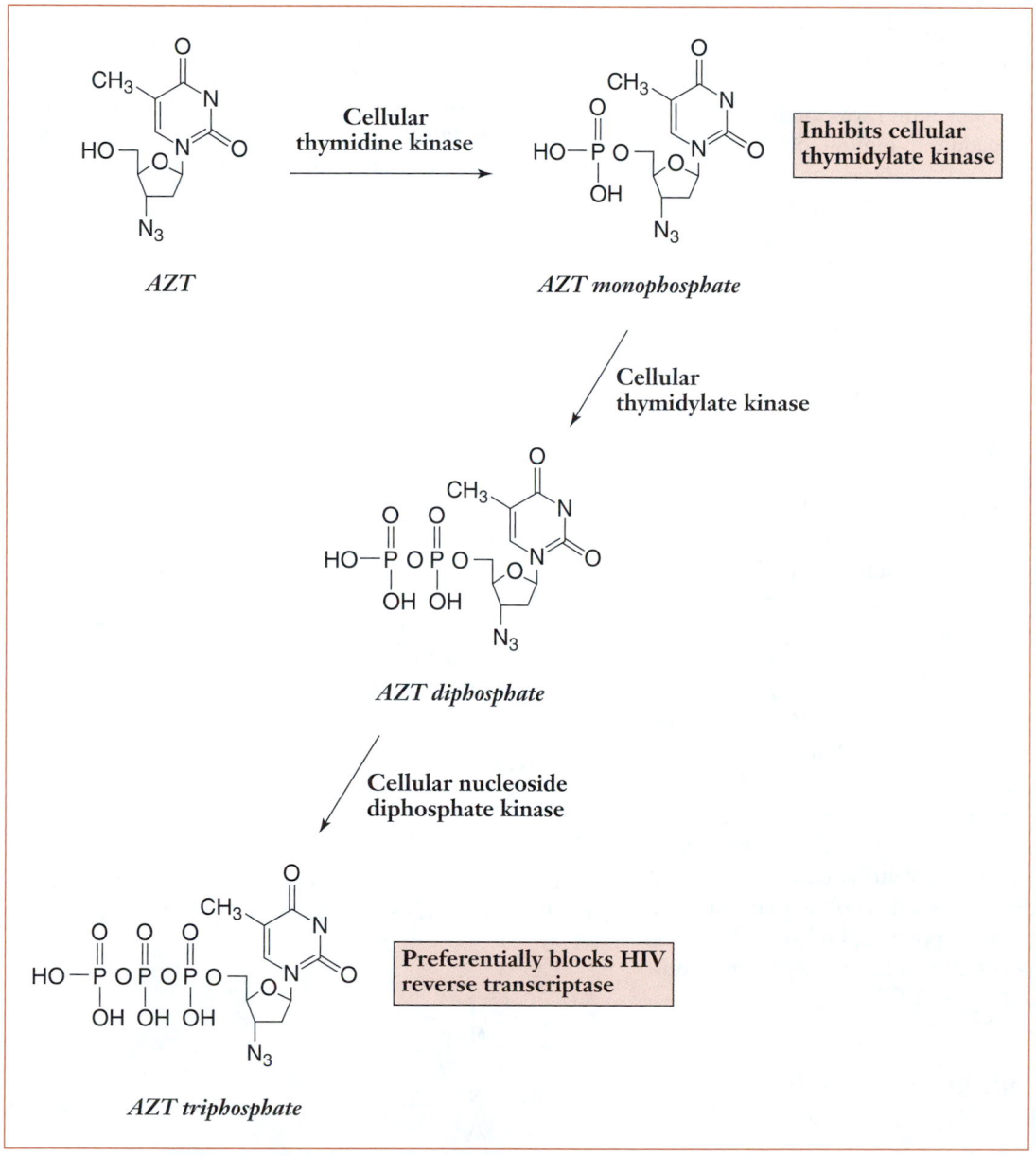

FIGURE 32.7 Phosphorylation of azidothymidine. Unlike acyclovir (Figure 32.5), AZT is not selectively phosphorylated by a viral kinase, but follows the phosphorylation pathway of thymidine in the cell. As is the case with acyclovir, the triphosphate species exerts antiviral activity. However, the monophosphate species of AZT blocks the cellular enzyme, thymidylate kinase, contributing to toxicity.

Ganciclovir triphosphate is a more potent inhibitor of cytomegalovirus DNA polymerase than it is of cellular DNA polymerases. Cytomegalovirus DNA polymerase mutations that confer resistance to ganciclovir have been identified. Thus, as is the case for acyclovir and herpes simplex virus, ganciclovir has selective activity against cytomegalovirus at two steps—activation by phosphorylation, and inhibition of DNA polymerase. However, selectivity at these two steps is not as great for ganciclovir as it is for acyclovir. Accordingly, ganciclovir is more toxic and is used only for very serious infections.

HIV-1 reverse transcriptase preferentially incorporates azidothymidine into DNA, leading to chain termination

Azidothymidine (AZT; zidovudine) was the first drug licensed for treatment of acquired immune deficiency syndrome (AIDS) caused by HIV-1, based on clinical studies showing that it could prolong life of patients with AIDS. However, when used alone its benefits are limited, and it is now used mainly in combination with other drugs. AZT, like acyclovir and ganciclovir, is a nucleoside analogue with an altered sugar moiety. In this case, a thymine base is attached to a ribose in which the normal 3′ OH has been converted to an azido (N_3) group (Figure 32.4). Thus, like acyclovir, it leads to chain termination when incorporated into a growing DNA molecule.

The intracellular metabolism of AZT is very similar to that of thymidine (Figure 32.7). HIV-1 does not encode an enzyme that activates AZT by phosphorylation. Rather, AZT is a substrate for cellular thymidine kinase. Thus, unlike acyclovir and ganciclovir, activation of this drug is not selective; it occurs in uninfected as well as in infected cells. AZT monophosphate is then converted to AZT diphosphate by cellular **thymidylate kinase**. However, AZT monophosphate is also an inhibitor of thymidylate kinase, which is essential for synthesis of thymidine triphosphate and therefore DNA. This likely accounts for at least some of the toxicity of this drug, which most often manifests itself clinically as bone marrow suppression. AZT diphosphate is converted to the triphosphate by cellular **nucleoside diphosphate kinase**. AZT triphosphate is a more potent inhibitor of HIV reverse transcriptase than of human DNA polymerases; this accounts for the selectivity of AZT.

Nonnucleoside inhibitors selectively target viral replication enzymes

Despite problems with AZT, it helped to validate the concept developed from viral genetics and biochemistry that HIV-1 reverse transcriptase is a useful drug target. Subsequently, several pharmaceutical companies developed target-based high throughput screens against HIV reverse transcriptase, leading to the identification of a number of promising compounds that are not nucleoside analogues. Interestingly, the various compounds are similar to each other in structure (Figure 32.8 shows one, nevirapine), and they act in a similar fashion. These **nonnucleoside inhibitors** bind to reverse transcriptase at a site close to the active site for DNA polymerase activity, and they drastically slow the rate of DNA polymerization. They are very selective—they don't even inhibit the reverse transcriptase of the closely related virus HIV-2—and their discovery makes a strong case for the validity of the target-based high throughput approach. Due to the ability of the virus to rapidly develop resistance to these compounds, they are mainly used in combination with other drugs. Using several different drugs simultaneously reduces the likelihood that a virus will become resistant to all of the drugs, and certain mutations that provide resistance against one drug may make the same protein more sensitive to a second drug.

Protease inhibitors can interfere with virus assembly and maturation

Many viruses encode proteases that are essential for viral replication, often at the stage of assembly and maturation. A great deal of effort has been made to discover drugs active against these enzymes. Much of the impetus for this effort came from the successes and the lessons learned from the development of anti-HIV-1 protease inhibitors. These drugs were developed rationally rather than by random screening, high throughput or otherwise. The use of protease inhibitors in combination with reverse transcriptase inhibitors led to strong reductions in the amount of HIV circulating in patients and to dramatic improvements in the health and longevity of the patients.

Ritonavir: a successful protease inhibitor of HIV-1 that was developed by rational methods

One example of these drugs is ritonavir (structure in Figure 32.9). HIV-1 protease is an excellent target for

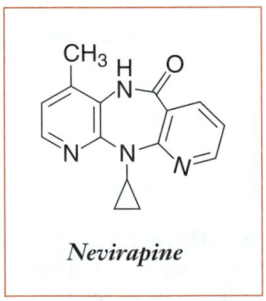

FIGURE 32.8 **Nevirapine, a nonnucleoside inhibitor of HIV-1 reverse transcriptase.**

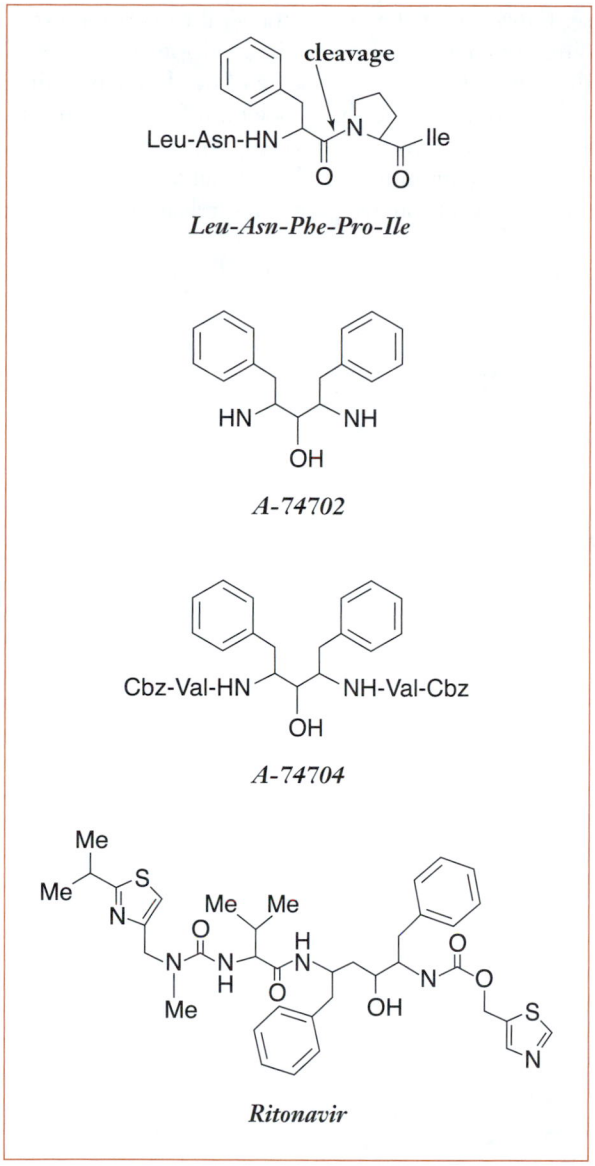

FIGURE 32.9 **Steps in the development of ritonavir.** Top: a natural substrate of HIV protease; the peptide bond between the phenylalanine and proline (arrow) is cleaved by the protease. Middle: two early symmetric inhibitors in which the peptide bond is replaced by a hydroxyethylamine. Bottom: the end result, ritonavir. Cbz = carboyxbenzyl; Me = methyl.

antiviral chemotherapy because it is an enzyme that is essential for HIV replication; it cleaves the Gag precursor of viral capsid proteins within the assembled virion. Without these cleavages, the progeny virions are unable to infect cells and therefore the infection does not spread. HIV-1 protease is also a good target because it has an unusual substrate specificity; it cleaves between phenylalanine and proline, a site that is rarely if ever cleaved by human proteases.

Ritonavir is a **peptidomimetic** drug (mimics a peptide). Its design began with one of the natural substrates of HIV-1 protease (Leu_{165}-Asn-Phe-Pro-Ile), a region of the Gag-Pol polyprotein that is cleaved to release reverse transcriptase. A key insight was provided by the crystal structure of HIV-1 protease. The enzyme was found to be a symmetrical dimer of two identical subunits. The most closely related cellular proteases are not symmetrical dimers. The group that developed ritonavir therefore made a correspondingly symmetrical inhibitor in which the proline was replaced with phenylalanine and the peptide bond between the two phenylalanines was modified by converting the normal C=O to CHOH, creating a hydroxyethylamine (A-74702 in Figure 32.9). Hydroxyethylamine compounds not only mimic the original peptide, they also mimic the transition state of protease catalysis (i.e., the catalytic intermediate that binds most tightly to the enzyme). However, unlike the original peptide and the actual transition state, they cannot be cleaved. These features make them good inhibitors.

A-74702 was a very weak inhibitor, but adding symmetrical valine-carboxybenzyl groups at both ends (A-74704 in Figure 32.9) resulted in a 40,000-fold increase in potency. Further modifications increased potency by more than 100-fold, resulting in a compound that can inhibit the protease when present at picomolar concentrations. The modifications also vastly improved the uptake and stability of the drug in the body (**pharmacokinetics**). These improvements were facilitated by an iterative process of synthesizing an inhibitor and solving the structure of the protease bound to the inhibitor by x-ray crystallography. Then by looking at these structures, scientists can make informed guesses about what groups to add or subtract to generate a new inhibitor. After many cycles of this process, the result was ritonavir.

The HIV-1 protease inhibitors work as expected. In cell culture, these compounds are potent inhibitors of virus replication, although the concentrations needed are higher than required to inhibit the protease *in vitro*. As expected, HIV-infected cells continue to make viral proteins, but the Gag polyprotein is not cleaved when the protease inhibitor is present. Therefore, virus particles bud from the infected cells, but they are immature and do not successfully infect other cells. The proof that these drugs act as expected is that drug-resistance mutations are located in the HIV-1 gene encoding the protease. These protease inhibitors are now used along with other drugs in the very successful triple therapy drug combinations that efficiently suppress HIV-1 replication and virus levels, and lead to healthier and longer lives for people with HIV-1.

Neuraminidase inhibitors inhibit release and spread of influenza virus

The inhibitors of influenza virus **neuraminidase** are perhaps the best examples of the success of a purely rational approach to drug design. These drugs inhibit virus release from infected cells. Most viruses do not require special mechanisms to be released from the cell surface, but influenza virus does. It initially attaches to cells by interactions between the virion hemagglutinin protein and cellular **sialic acid** (Figure 32.10a), also known as N-acetyl neuraminic acid, which is present on many membrane glycoproteins and **gangliosides.** This greatly facilitates virus entry. However, at the end of the viral replication cycle, binding of hemagglutinin protein on virions to sialic acid on the cell surface inhibits release of newly formed virus particles. To overcome this problem, influenza virus encodes another envelope protein, the enzyme neuraminidase. Neuraminidase cleaves sialic acid from the membrane glycoproteins, permitting release of the virus;

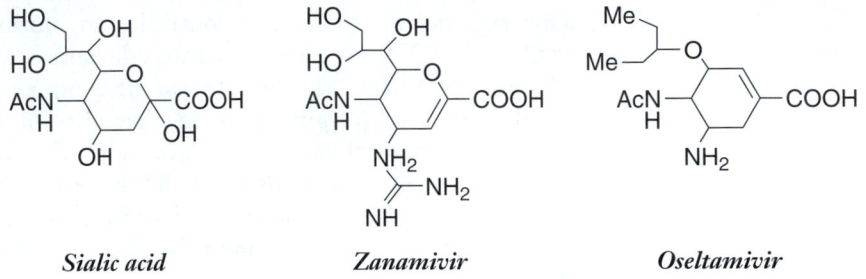

Sialic acid *Zanamivir* *Oseltamivir*

(a)

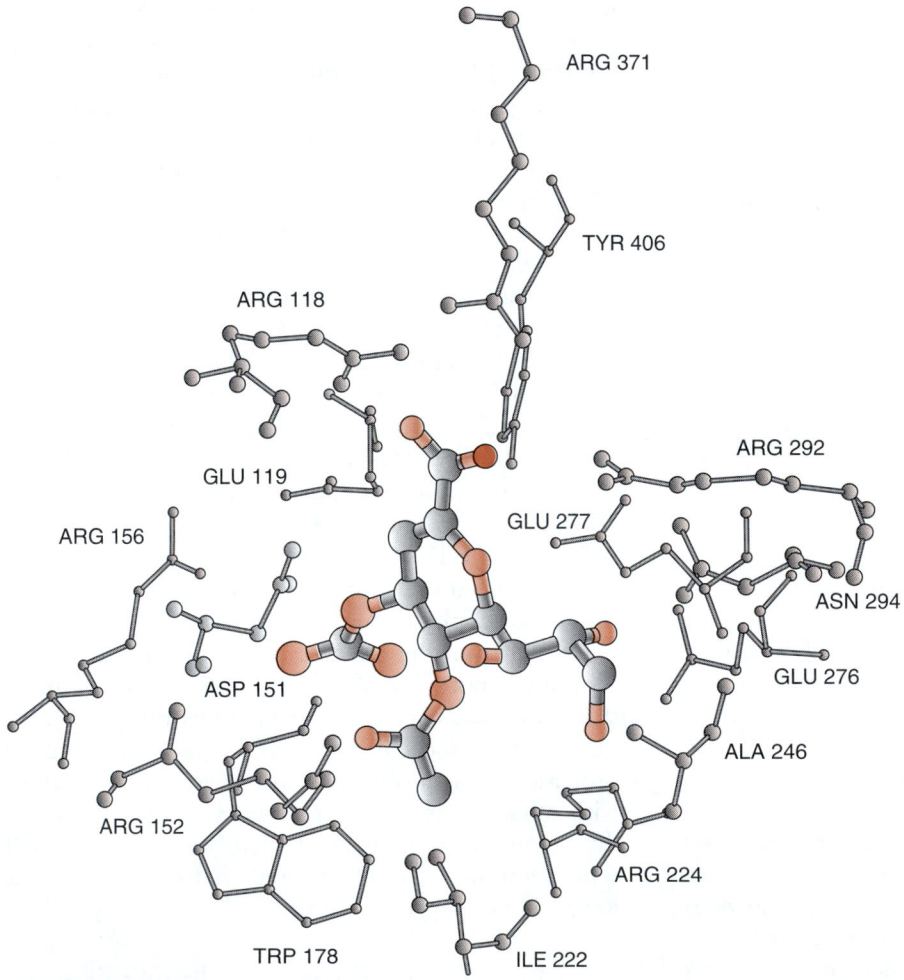

(b)

FIGURE 32.10 Rational design of neuraminidase inhibitors. (a) Structures of sialic acid, zanamivir, and oseltamivir. Ac = acetyl; Me = methyl. (b) Model of part of the active site of influenza neuraminidase bound to zanamivir, shown as ball-and-stick model in the center. Amino acids near the binding site are labeled.

without this enzyme, the virus remains stuck and cannot spread to other cells. Influenza neuraminidase differs sufficiently from cellular enzymes to make it an excellent drug target.

The design of the two neuraminidase inhibitors presently available, zanamivir and **oseltamivir** (Figure 32.10a), began when the three-dimensional atomic structure of the neuraminidase was determined by x-ray crystallography in 1983. Subsequently, the structure of the neuraminidase complexed with sialic acid was solved, showing that sialic acid occupies two of three well-formed pockets on the enzyme (Figure 32.10b). Zanamivir and oseltamivir were designed based both on this structure and that of neuraminidase bound to a transition state inhibitor. The idea was to maximize energetically favorable interactions with specific amino acid residues of the enzyme in the three pockets. The idea worked and highly potent inhibitors were derived.

Although these compounds are potent *in vitro*, clinical trials show that they do not rapidly alleviate flu symptoms. They can reduce the duration of fever by one or two days, but only if administered during the first 48 hours. They have been marketed aggressively to patients, using, for example, actors from sitcoms to play the role of influenza virus. Regardless of these clinical and marketing aspects, the neuraminidase inhibitors are one of the best examples of rational drug design starting from the structure of a protein and a natural ligand.

Antiviral chemotherapy shows promise for the future

In summary, the field of antiviral chemotherapy is a very active and fertile area. Part of the excitement is that we are taking the knowledge we have gained over many years about viruses and are putting it to work to treat viral diseases. The successes with influenza virus, HIV-1, and herpesviruses lead to optimism that other viral diseases will be susceptible to antiviral therapies in the future. Furthermore, antiviral drugs provide a terrific avenue for learning more about viruses.

SUGGESTED READING

1. Balfour, H.J. (1999). Antiviral drugs. *New England Journal of Medicine* **340,** 1255–1268.

2. Challand, R. and Young, R.J. (1998). Antiviral chemotherapy. Biochemical and Medicinal Chemistry Series, Oxford University Press, Oxford, UK.

3. Flexner, C.F. (1998). HIV protease inhibitors. *New England Journal of Medicine* **338,** 1281–1292.

4. Hay, A.J., Wolstenholme, A.J., Skehel, J.J., and Smith, M.H. (1985). The molecular basis of the specific anti-influenza action of amantadine. *EMBO Journal* **4,** 3021–3024.

5. LaBranche, C., Galasso, G., Moore, J.P., et al. (2001). HIV fusion and its inhibition. *Antiviral Research* **50,** 95–115.

6. Richman, D.D. Ed. (1996). Antiviral drug resistance. John Wiley & Sons, Chichester, UK.

7. von Itzstein, M., et al. (1993). Rational design of potent sialidase-based inhibitors of influenza virus replication. *Nature* **363,** 418–423.

8. Yeh, R. and Coen, D.M. (2004). Pharmacology of viral replication. In Golan, D.E., Tashjian, A.H. Jr., Armstrong, E.J., Galanter, J.M., Armstrong, A.W., Arnaout, R.A., and Rose, H.S., Eds., Principles of pharmacology: The pathophysiologic basis of drug therapy, pp. 545–564. Baltimore: Lippincott, Williams & Wilkins.

KEY TERMS

Acyclic
Acyclovir
Amantadine
Azidothymidine
Capsid-binding drugs
Cell-based high throughput screen
Enfuvirtide
Ganciclovir
Gangliosides
Interferon
Neuraminidase
Nevirapine
Nonnucleoside inhibitors
Nucleoside diphosphate kinase
Oseltamivir
Peptidomimetic
Pharmacokinetics
Pleconaril
Rational drug discovery
Ritonavir
Sialic acid
Target-based high throughput screen
Therapeutic index
Thymidine kinase
Thymidylate kinase
Zanamivir

Eukaryotic Virus Vectors

USES OF VIRUS VECTORS
As recombinant vaccines.
For gene therapy.
As anticancer agents.

CONSTRUCTION OF VECTORS
Deletion of some or all virus genes.
Retention of *cis*-acting replication and packaging sequences in the vector genome.
Insertion of gene(s) of interest into the vector genome.

PRODUCTION OF VECTORS
Transfection of genome into cell lines engineered to express virus genes deleted in the vector, or
Transfection of genome into cells that are coinfected with helper viruses that express missing genes.

ADENOVIRUS VECTORS
Are easy to produce at high titers.
Have been widely used in human trials.
Can induce immune responses that reduce their utility.

RETROVIRUS VECTORS
Can permanently introduce a foreign gene into target tissues by integration.
Are less immunogenic than first-generation adenovirus vectors.
Can be packaged with the of vesicular G envelope protein stomatitis virus to allow infection of many cell types.

ADENO-ASSOCIATED VIRUS VECTORS
Are considered safe because the parent virus is nonpathogenic for humans.
Are less immunogenic than other vectors.
Have limited capacity for insertion of large foreign genes.

Many viruses can be engineered to deliver and express specific genes

When viruses are genetically modified to deliver and express either their own or foreign genes, they are called **virus vectors**. Although DNA can be delivered to cells using a variety of physical and chemical methods, virus vectors are capable of much higher gene transfer efficiency. Virions are highly efficient gene transfer machines that interact with specific receptors on the cell surface and deliver their nucleic acid genomes to the appropriate compartment for optimal expression and replication (Chapters 2 and 4). This gives virus vectors an enormous advantage over other methods for gene transfer. The genomes of many different viruses of eukaryotes have been cloned and can be manipulated in bacterial plasmids and subsequently reintroduced into their natural host cells to produce recombinant virions.

Virus vectors have been designed for four major objectives: (1) the production of specific proteins in cultured cells; (2) vaccination of animals or humans against specific diseases; (3) gene therapy in animals or humans, usually to replace deficient genes; and (4) anticancer agents that specifically kill tumor cells. As Table 33.1 shows, vectors have been developed from a large variety of viruses whose genomes can be positive- or negative-strand RNA, single- or double-stranded DNA, ranging from 5 to 200 Kb in size. Tens of thousands of research papers have been published on the subject in the past 20 years.

Virus vectors are used to produce high levels of specific proteins in cultured cells

The first virus used as a vector for protein production in mammalian cells was simian virus 40 (SV40), a polyomavirus. Richard Mulligan and Paul Berg expressed rabbit β-globin in monkey cells in 1979 using the SV40 genome as a vector. The 5.3 Kb circular, double-stranded DNA SV40 genome was one of the first eykaryotic virus genomes to be cloned in a bacterial plasmid, and its complete nucleotide sequence was determined in the late 1970s. However, the small genome size was a handicap, because it limits the size of foreign DNA that can be packaged in the virion. Both adenoviruses (35 Kb genome) and baculoviruses (135 Kb genome) proved to be more effective vectors for protein production because of their larger genome sizes and the ease of producing recombinant virions. They are still used for such purposes, and a number of vectors are available commercially as kits, along with selection systems for recombinant viruses and susceptible cell lines for propagation. More recently, high-level protein production has been achieved with other vector systems, par-

TABLE 33.1 Some eukaryotic virus vectors and their applications

Application	Virus family	Virus species
Protein production	Adenovirus	Adenovirus type 5
	Baculovirus	*Autographa californica* nucleopolyhedrovirus
	Togavirus	Semliki Forest virus, Sindbis virus
Recombinant vaccines	Adenovirus	Adenovirus types 2 and 5
	Coronavirus	Transmissible gastroenteritis virus, mouse hepatitis virus, infectious bronchitis virus
	Flavivirus	Japanese encephalitis virus, yellow fever virus
	Herpesvirus	Bovine herpesvirus
	Orthomyxovirus	Influenza A and B viruses
	Paramyxovirus	Sendai virus, simian virus 5
	Parvovirus	Adeno-associated virus
	Picornavirus	Poliovirus
	Poxvirus	Vaccinia virus, canarypox virus, fowlpox virus
	Rhabdovirus	Rabies virus, vesicular stomatitis virus
	Togavirus	Semliki Forest virus, Sindbis virus, western equine encephalitis virus, Venezuelan equine encephalitis virus
Gene replacement therapy	Adenovirus	Adenovirus types 2 and 5
	Baculovirus	*Autographa californica* nucleopolyhedrosis virus
	Flavivirus	Kunjin virus, tick-borne encephalitis virus
	Herpesvirus	Cytomegalovirus
	Retrovirus	Human immunodeficiency virus type 1, simian immunodeficiency virus type 1, murine leukemia virus
	Paramyxovirus	Sendai virus, human parainfluenza virus
	Parvovirus	Adeno-associated virus, B19 virus
	Polyomavirus	Simian virus 40
	Togavirus	Semliki Forest virus, Sindbis virus
Antitumor agents	Adenovirus	Adenovirus type 5
	Herpesvirus	Herpes simplex virus
	Paramyxovirus	Newcastle disease virus, simian virus 5
	Parvovirus	Minute virus of mice, H-1 parvovirus, LuIII virus
	Poxvirus	Vaccinia virus, avian poxviruses
	Reovirus	Reovirus
	Rhabdovirus	Vesicular stomatitis virus

ticularly those derived from togaviruses such as Semliki Forest virus. Large-scale transfection of mammalian cells with plasmids has begun to supplant the use of recombinant viruses for the purposes of protein production.

Gene therapy is an expanding application of virus vectors

Vaccination and gene therapy are the most widely publicized uses of virus vectors. The idea of gene therapy arose in the late 1950s, when it was realized that viruses are able to transport their own genetic material into cells. Furthermore, tumor viruses such as polyomaviruses and retroviruses were shown in the late 1960s to integrate copies of the virus genome into the host cell during transformation (Chapters 10 and 25). This led to the realization that viruses could be engineered to bring foreign genes (**transgenes**) into cells, and that those genes could in some cases be stably integrated into the cellular genome, replacing defective host cell genes.

Gene replacement therapy was initiated in the mid-1980s with the successful transfer of the adenosine deaminase gene to mouse bone marrow stem cells to correct a severe immune deficiency, using a retrovirus vector. But it was not until 1990 that the equivalent experiments were carried out in humans, with limited success. Developments in the past 15 years have led to more than 600 clinical trials of gene therapy, most using virus vectors. However, gene therapy is still largely an experimental procedure, and much work remains to be done to make it available for general use in humans.

Gene transfer can be carried out *in vitro* by infecting cells isolated from an individual and then reintroducing cells expressing the desired gene into the host organism. Alternatively, genes can be transferred directly into cells in the body by injection of virus vectors. For gene therapy applications such as vaccination and anticancer therapy, transient but high-level expression of gene products in the target cells may be sufficient. Other applications such as gene replacement therapy require long-term expression, ideally for the lifetime of the individual.

With the advent of gene transfer technologies, new horizons are opening for the development of biopharmaceuticals based on recombinant DNA technology. Presently, administration of therapeutic proteins is restricted to secreted and circulating proteins such as **antibodies**, serum proteins, **cytokines**, and hormones. In the future, it will be possible to synthesize therapeutic proteins directly in the cells of an individual. With this new drug delivery system, the possibilities are great: the delivery of genes encoding receptors for signaling molecules, membrane transport proteins, structural proteins, enzymes, transcription factors, and so on, can be used in myriad therapeutic applications.

Virus vectors are produced by transfection of cells with plasmids containing deleted genomes

For gene replacement, an ideal vector should be capable of delivering a transgene specifically to the appropriate target cells and expressing that gene sufficiently strongly to obtain a therapeutic effect. In most cases, this will require (1) efficient, targeted delivery, (2) control of both the level and duration of transgene expression, and (3) minimal toxicity and/or immune reaction to either the vector components or the transgene.

A typical virus vector is unable to replicate because of deletion of some or all essential viral genes (Figure 33.1). The vector retains the *cis*-acting sequence elements of the virus genome required for replication (ori) and packaging (Ψ), and the deleted viral genes are replaced with cassettes controlling the expression of the foreign genes. Vectors are propagated in cell lines that express the deleted viral genes and can therefore provide the essential replication and packaging components normally provided by the intact virus. Development of an efficient producer cell line is often the most challenging requirement in the construction of a vector system. In some cases, viral proteins are provided by infection with a **helper virus**; however, this complicates production of recombinant vector stocks, because the helper virus must be removed.

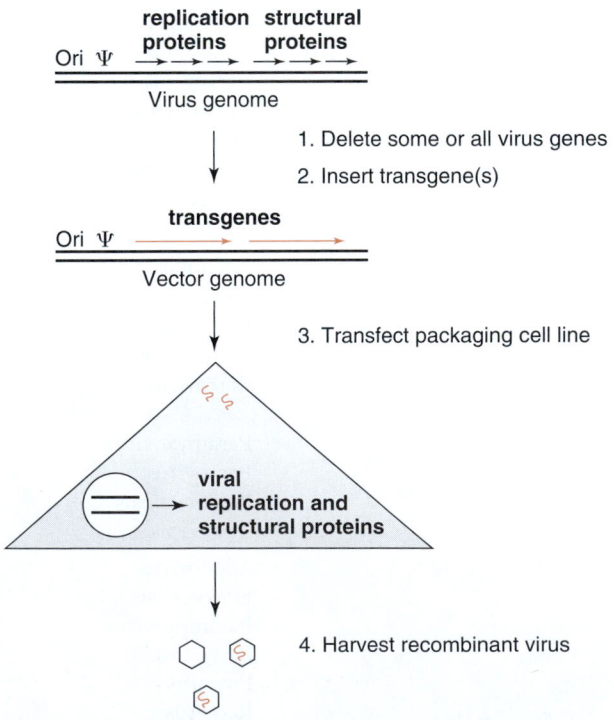

FIGURE 33.1 Construction and production of a typical virus vector. Ori = sequences required for genome replication; Ψ = *cis*-acting packaging sequences.

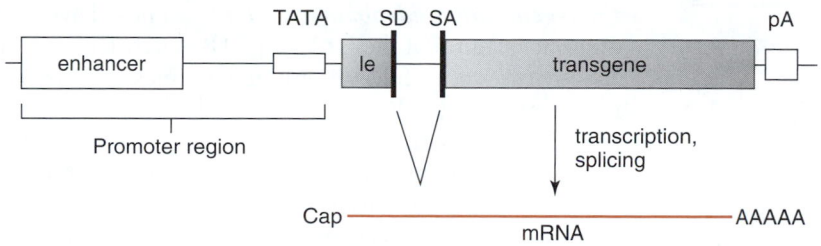

FIGURE 33.2 **Typical mammalian expression cassette.** Shown is a typical transgene with two exons. The first exon contains leader sequence (le) for efficient translation. Strong transcriptional enhancer and TATA box ensure high-level transcription from a defined site. Strong splice donor (SD) and splice acceptor (SA) sequences as well as a strong polyadenylation signal (pA) ensure efficient processing and export of mRNA.

Vectors based on viruses with RNA genomes are constructed in DNA plasmids containing copies of the virus genome made by reverse transcription. These plasmids are introduced into cells where they are transcribed either by endogenous cellular RNA polymerases or by bacteriophage T7 RNA polymerase, which is provided by specialized cell lines or by a vaccinia virus vector. In most cases, both the 5′ and the 3′ ends of the genome RNA of the vector must be precisely reproduced by transcription from the plasmid DNA to enable replication of the transcripts by viral RNA polymerases. Various genetic engineering tricks are used to produce RNAs with precise 3′ ends, including the use of T7 transcriptional terminator signals (see Chapter 7) and self-cleaving **ribozymes** (Chapter 29). The resulting virus genome RNAs are replicated by the viral RNA polymerase, which is usually provided by coinfection of the cells with an expression plasmid.

Virus vectors are engineered to produce optimal levels of gene products

The expression cassette (Figure 33.2) is one of the chief elements of a virus vector. It is a recombinant DNA molecule containing sequence information for efficient transcription of the desired gene and translation of its messenger RNA. A typical recombinant DNA used for gene expression in mammalian cells includes a strong **transcriptional enhancer** and promoter, appropriate sequences at the 5′ end of the mRNA for efficient initiation of translation, RNA splicing signals, and finally a strong **polyadenylation signal**. If two transgenes must be expressed coordinately, they can be cloned as two independent transcription units with their own enhancers, promoters, splice sites, and polyadenylation signals. Alternatively, they can be expressed on the same mRNA molecule but be controlled by independent translation initiation cassettes (internal ribosome entry sites or IRES; see Chapter 16). Expression cassettes for RNA virus vectors do not usually contain transcriptional promoters/enhancers, RNA splicing sites, or polyadenylation signals because they are expressed as RNA by viral RNA polymerases in the cytoplasm.

Although many different gene therapy vectors have been developed, this chapter will focus on three of the most widely used vectors: those derived from adenoviruses, retroviruses, and parvoviruses (Table 33.2). Descriptions of vectors derived from bacteriophage T7, poxviruses, baculoviruses, flaviviruses, togaviruses, and bunyaviruses can be found in their respective chapters.

TABLE 33.2 **Properties of selected virus vectors**

Vector	Transgene capacity	Transduced cells: Dividing	Transduced cells: Nondividing	Integrates transgene	Antigenicity
First-generation adenovirus	8 Kb	Yes	Yes	No	Strong
Third-generation adenovirus	35 Kb	Yes	Yes	No	Weak
Oncoretrovirus	8 Kb	Yes	No	Yes	Weak
Lentivirus	10 Kb	Yes	Yes	Yes	Weak
Adeno-associated virus	4.5 Kb	Yes	Yes	Yes/No	Weak

ADENOVIRUS VECTORS

Adenovirus vectors are widely used in studies of gene transfer and antitumor therapy

Adenoviruses (Chapter 12) have a double-stranded linear DNA genome of 25–45 Kbp, depending on the serotype. Adenoviruses do not integrate their genome during productive infection, but the adenovirus genome can persist as a stable episome in some cell types for months or even years. Adenovirus vectors have been extensively used for gene transfer into mammalian cells during the past 20 years. They are the best characterized vectors, and the most common vectors used in preclinical and clinical trials. High titer stocks can be produced easily, and high levels of transgene expression can be achieved with these vectors.

Replication-defective adenovirus vectors are propagated in complementing cell lines

The development of adenovirus vectors has involved the following genetic modifications: (1) eliminating the capacity of the vector to replicate on its own; (2) increasing its capacity to package exogenous genes; (3) reducing the host immune response to viral proteins; and (4) retargeting the vector to specific cell receptors. The various adenovirus vectors engineered up to the present are shown in Figure 33.3.

First-generation adenovirus vectors. These have undergone deletion of the E1A and E1B genes. E1 deleted adenoviruses can replicate only in cell lines such as human embryonic kidney 293 cells, which express integrated copies of the E1 region and therefore complement these functions. The E3 region is not required for virus production in cultured cells *in vitro* and its deletion increases the space available (total 8 Kbp) for insertion of exogenous DNA.

Construction of first-generation vectors is routinely performed by transfecting 293 cells with two DNA fragments. One contains the wildtype adenovirus genome cleaved within the E1 region to remove the left-hand **inverted terminal repeat** and **packaging signals**. The second is a transfer plasmid that contains the left-hand end of the genome, the transgene cassette, and a homologous sequence to the right of the transgene cassette that allows for recombination to occur. This procedure, favored for its simplicity, is nevertheless inefficient. Recombination is a rare event and incompletely digested adenovirus genomes may give rise to nonrecombinant virus.

Other strategies have been developed to minimize contamination with nonrecombinant virus and to reduce the time and effort required for vector production. These strategies utilize a virus genome that is carried in bacteria as a plasmid or cosmid. Recombinant adenoviruses can be produced by either: (1) subcloning the transgene directly into the plasmid containing the virus genome; (2) using recombination within bacteria to insert the transgene into the virus genome; or (3) using Cre recombinase (see below) to insert the transgene.

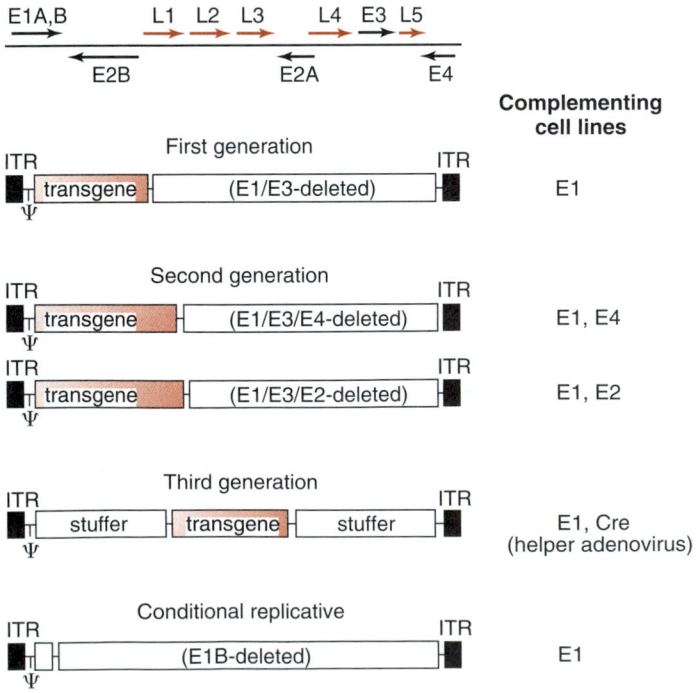

FIGURE 33.3 Production of adenovirus vectors. The adenovirus genome map showing major early (E) and late (L) transcription units is shown at top. Below are schematic structures of the genomes of five adenovirus vectors, and at the right, viral genes expressed in complementing cell lines. See text for details. ITR: inverted terminal repeat; Ψ: packaging signal; Cre: recombinase.

Second-generation adenovirus vectors. First-generation adenovirus vectors can sustain transgene expression for years in postmitotic cells of immunodeficient animals. However, expression persists for only a few weeks in immunocompetent animals, because the host immune system destroys transduced cells expressing viral proteins. To reduce immune rejection, adenovirus vectors were further crippled by deletion of genes in the E2 or E4 regions (Figure 33.3). The production of these second-generation vectors required the development of producer cell lines expressing the corresponding deleted genes, which are required for virus replication.

Deletions in E2A (single-stranded DNA binding protein), E2B (viral DNA polymerase/preterminal protein), and E4 were engineered. In most cases, *in vivo* studies showed significant reduction of the immune response, lower liver toxicity, and prolonged vector stability. However, adenovirus vectors lacking E4 are retained less stably in transduced cells and produce lower amounts of transgene protein. It appears that E4 is required for efficient expression of transgenes under the control of several promoters used in these vectors, particularly the cytomegalovirus and Rous sarcoma virus promoters.

Third-generation adenovirus vectors. Because second-generation adenovirus vectors do not fully escape the host immune response, and because there was a need for larger cargo space for some applications, the more drastic approach of eliminating all viral genes was undertaken. Fully deleted vectors have been given several names: third generation, helper-dependent, high capacity, or "gutless." Those vectors offer two main advantages: (1) no viral genes are present, eliminating problems of immune response to viral proteins; (2) up to 35 Kb are available for insertion of transgenes for expression. However, these vectors are more difficult to produce; it is almost impossible to generate a stable cell line that complements all of the essential adenovirus genes, due to the toxicity associated with the overexpression of most of these proteins. The functions required for complementation of third-generation adenovirus vectors are most often provided by coinfection with a helper virus, usually a first-generation adenovirus vector. However, this leads to a further problem: the helper virus replicates at the same time as the vector, and it must be removed from the vector stock.

The most efficient strategy to eliminate the helper virus is to remove the packaging signal on the helper virus genomes made during propagation of the vector. This can be conveniently done by the use of the Cre recombinase, derived from bacteriophage P1. The helper adenovirus is engineered to contain *lox*P sequences flanking its packaging signal. In 293 cells expressing Cre recombinase, the packaging signal of the helper virus is deleted, resulting in DNA that still can produce viral proteins, but cannot be packaged into virus particles. Although cleavage by Cre recombinase is not complete and some helper viruses are still packaged, contamination of the vector with helper virus can be reduced to less than 0.1%.

Replication-competent adenovirus vectors are useful tools in antitumor therapy

The disadvantages of some adenovirus vectors, namely cytotoxicity and induction of an immune response, are assets in applications such as vaccination or cancer gene therapy. Adenoviruses that do not express E1B genes but are otherwise intact replicate poorly in normal cells but well in many tumor cells. This is a result of the complex interplay between adenovirus early gene products and cellular signaling pathways leading to **apoptosis** (see Chapter 12). The selective replication of E1B-deleted adenoviruses in tumor cells leads to death of tumor cells, and enables spread of virus progeny to nearby tumor cells, increasing the antitumor effect of the virus. Other strategies rely on tumor-specific promoters to drive E1A gene expression, or expression of cytokine or suicide genes to further improve the antitumor effects of replicative adenovirus vectors.

Onyx-015 is a conditionally replicative adenovirus that lacks the E1B region and preferentially replicates in cancer cells. In recent clinical trials it was observed that injections of Onyx-015 directly into solid tumors, or its delivery through the intra-arterial or intravenous routes of administration, was safe but was ineffective at killing cancer cells. On the other hand, in conjunction with chemotherapy, the use of Onyx-015 resulted in complete or partial regression of tumors and to a significant extension of the life of the patients. This success is even more impressive considering that the treated individuals were affected by recurrent cancer of the head and neck, or by gastrointestinal cancer that had spread to the liver, two types of cancer with particularly poor prognosis.

Advantages and limitations of adenovirus vectors

Several features make first-generation adenovirus vectors attractive vehicles for gene transfer and protein production: (1) they can be easily produced at high titers, (2) they can transduce many cell types, (3) they do not require cell division for transgene expression, (4) their genome remains episomal, and therefore the target cell does not undergo permanent mutation. The high levels of recombinant proteins synthesized by first-generation adenovirus vectors in mammalian cells have not been achieved with any other vector system. However, because adenoviruses do not integrate their genome into cellular chromosomes, cell division leads to dilution of the transgene and eventually to loss of transgene expression. Furthermore, first-generation adenovirus vectors are strongly immunogenic, causing the loss of transgene expression in immunocompetent hosts.

The use of second- and third-generation adenovirus vectors greatly reduces problems caused by host immune responses. However, vectors that lack the E4 region usually express lower levels of transgene proteins expressed from viral promoters. Furthermore, expression of these transgenes is often prematurely silenced. The good news is that long-term transgene expression without evidence of toxicity can be achieved with third-generation adenovirus vectors if nonviral promoters are employed. Because of their lack of toxicity, their weak antigenicity, their demonstrated long-term transgene expression, and their very large insert capacity, third-generation adenovirus vectors are very promising for gene therapy applications. The main hurdle limiting their use is the production of clinical-grade vectors due to the complexity of the production process and to the low level of helper virus contamination required.

RETROVIRUS VECTORS

Retrovirus vectors incorporate transgenes into the cell chromosome

Retroviruses (Chapter 25) use reverse transcriptase to generate a double-stranded DNA copy of their RNA genome and subsequently integrate this DNA into the cellular chromosome. They are therefore ideal vectors for gene replacement therapy, as transgenes intoduced into target cells remain stably integrated. In the case of retroviruses such as murine leukemia virus, integration of the proviral DNA occurs only in dividing cells. The integration complex of these retroviruses is unable to pass through the nuclear pores, and therefore has access to the cellular chromosomes only after the nuclear membrane dissassembles during mitosis. Lentiviruses such as human immunodeficiency virus are capable of transporting their integration complex into intact nuclei, and vectors based on lentiviruses are therefore useful for expressing transgenes in nondividing cells.

Packaging cell lines express retrovirus enzymatic and structural proteins

Retrovirus vectors are produced by packaging cell lines that express all essential viral proteins: Gag (capsid proteins), Pol (reverse transcriptase and integrase), and Env (envelope proteins). The plasmids used to produce these cell lines contain neither the long terminal repeats (LTR) nor viral packaging signals (Ψ), ensuring that no virus genomes are made or packaged. Currently the best packaging cell lines are generated from split genome constructs (Figure 33.4a). These lines further ensure safety by reducing the possibility of generating replication-competent virus by recombination.

To produce a retrovirus vector, a plasmid called the transfer construct is introduced into the packaging cell line (Figure 33.4b). This plasmid includes the gene of interest (up to 8 Kb) flanked by two long terminal repeats, and contains the primer binding site (PBS) for reverse transcription and the packaging signal, both near the left-hand LTR. Transcription by cellular RNA polymerase II begins in the left-hand LTR, using the enhancer and promoter sequences located in the U3 region, and ends in the right-hand LTR where the polyadenylation signal defines the RNA 3′ end. The enzymes necessary for reverse transcription and integration are produced by the packaging cell line and are incorporated along with the RNA into virions. In most cases, the cell is not lysed and continuously releases virions by budding into the extracellular medium.

The ability of a retrovirus to infect specific cell types (**tropism**) is determined by the envelope protein, which mediates attachment of the virion to the cell surface as well as entry by fusion or endocytosis. It is possible to modify the tropism of a retrovirus by replacing the envelope gene of the producer cell line with a gene derived from a different virus. This strategy is known as **pseudotyping**. The G glycoprotein of vesicular stomatitis virus (VSV-G) is often used for pseudotyping retroviruses. Retroviruses pseudotyped with VSV-G can infect many different cell types and are very stable. Unfortunately, VSV-G is toxic to cells in which it is made. For this reason, retroviruses pseudotyped with VSV-G are produced either by transient transfection of the producer cells with a plasmid encoding the VSV-G gene, or by using a producer cell line containing the VSV-G gene tightly regulated by an inducible promoter.

Strategies for controlling transgene transcription

Transgene expression in the target cells can be controlled by the enhancer/promoter elements of the LTR or by using a second promoter that directly controls the transgene (Figure 33.4c). A common problem that arises when the transgene is controlled by the LTR is reduction or loss of transgene expression. This phenomenon, known as **silencing**, is most likely caused by methylation of DNA sequences located in the LTR, leading to loss of transcriptional activity. Some of the sequences within the LTR responsible for promoter silencing have been characterized and can be deleted to correct the problem.

An alternative approach is to insert a second promoter to control transgene expression (Figure 33.4c).

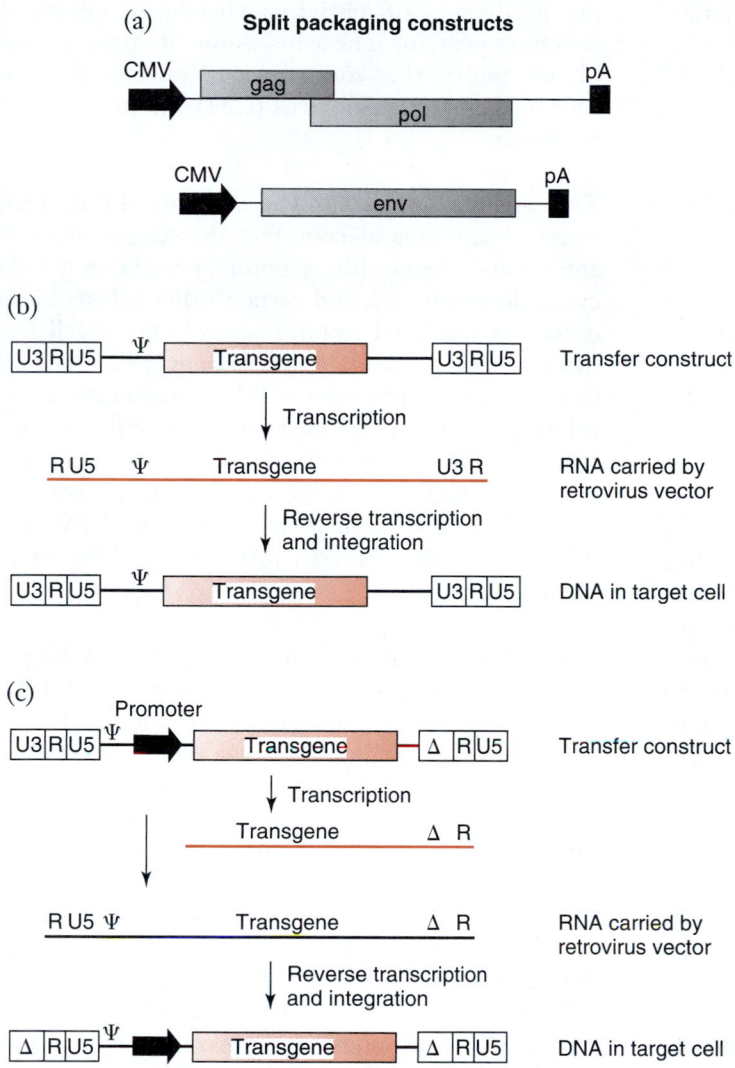

FIGURE 33.4 Production of retrovirus vectors. (a) Plasmid constructs used to establish a packaging cell line expressing gag, pol, and env gene products. (b) Production of a simple retrovirus vector by transfection of a transfer construct into the packaging cell line. The vector DNA integrated in the target cell is identical to the DNA in the transfer construct. (c) Production of a self-inactivating retrovirus vector. See text for details. CMV: cytomegalovirus promoter; U3, U5, and R, unique 3′, unique 5′ and repeat regions of the retrovirus long terminal repeat; Ψ: packaging signal; pA: polyadenylation signal.

However, interference between the LTR promoter and the second promoter (**promoter occlusion**) often occurs. This interference is thought to be caused by dislodging of transcription factors from the downstream promoter by RNA polymerases that initate at the upstream promoter. For this reason an elegant scheme has been designed to inactivate the transcriptional activity of the LTR after integration. Such vectors, referred to as self-inactivating retrovirus vectors, are safer because they cannot produce viral RNA after integration into their target cells. Furthermore, they cannot use the right-hand LTR to drive transcription of adjacent cellular genes, which could be problematic if the proviral DNA integrated next to an oncogene. A deletion is made within the U3 region of the right-hand LTR of the transfer construct, thus eliminating its transcriptional activity (Figure 33.4c). The left-hand LTR is normal, and drives transcription of the full-length RNA. After reverse transcription and integration, the U3 regions of both the left and right LTRs will be deleted and both LTRs will therefore be transcriptionally inactive (see the mechanism of reverse transcription in Chapter 25).

Lentivirus vectors are used for gene delivery to nondividing cells

Lentiviruses have a major advantage over other retroviruses when used as vectors: the integration complex, containing the proviral DNA made by reverse transcriptase, is transported into the nucleus via the nuclear pores. This means that lentivirus vectors can transduce stationary cells as well as dividing cells. The best characterized lentivirus is human immunodeficiency virus type 1, the causative agent of AIDS; most lentivirus vector systems used today are derived from this virus. The first experiment using a lentivirus as a vector was accomplished in 1996 by the group of Inder Verma at the Salk Institute in California. These researchers demonstrated that a replication-defective lentivirus could be used to stably express a reporter gene in mouse brain neurons, which are nondividing cells.

Production of lentivirus vectors requires additional *cis*-acting sequences

Lentiviruses are complex retroviruses that encode six accessory proteins (Vif, Nef, Tat, Rev, Vpr, and Vpu) in addition to the proteins coded by the gag, pol, and env genes (Chapter 26). Of these proteins, only Tat and Rev are essential for virus production in many cell lines. Tat is a powerful transcription factor that stimulates transcription from the LTR. Consequently, the presence of Tat is essential only if the wildtype LTR is used to control expression of the transgene or the RNA of the vector genome. The Rev protein promotes the efficient transport of unspliced or singly spliced viral RNA from the nucleus to the cytoplasm. Rev interacts with a sequence on viral RNA that is named the Rev-responsive element, or RRE. Hence, efficient production of lentivirus vectors requires the presence of both Rev and the RRE. The viral RNA must also contain the central polypurine tract (cPPT). This is a sequence of 118 nucleotides that significantly increases the transduction efficiency of lentivirus vectors in both dividing and nondividing cells.

Because some essential lentivirus proteins are toxic for cells, it has not been possible to establish good producer cell lines for lentivirus vectors without tight regulation of expression of the packaging proteins. To circumvent this problem, vectors are produced by simultaneously transfecting susceptible cells with three different plasmid constructs, referred to as packaging, envelope, and transfer constructs (Figure 33.5).

The packaging construct. This plasmid encodes all viral enzymes and structural proteins necessary for the production of lentivirus virions except for the envelope protein. The expression of these genes is placed under the control of a strong promoter such as the cytomegalovirus (CMV) immediate early promoter.

The envelope construct. The env gene of HIV-1 produces an envelope glycoprotein that targets the CD4 antigen and **chemokine** receptors, present on lymphocytes, **macrophages**, and certain other cell types. Because this restricted tropism severely limits cell types that can be transduced, the HIV-1 env gene is not used to generate lentivirus vectors. The use of a different envelope protein also improves the safety of the vector by reducing the likelihood of producing replication-competent HIV-1. The envelope protein most often used is that from vesicular stomatitis virus (VSV-G), as this enables the vector to infect many different cell types.

The transfer construct. The simplest form of this plasmid contains the transgene flanked by the two LTRs of HIV in addition to the other essential *cis*-acting elements, which are the packaging signal, the RRE, and the cPPT. The transgene carried by lentivirus vectors can be expressed from a strong promoter such as the CMV promoter, or from an inducible promoter that allows regulation of protein levels. In more recent constructs, the U3 region of the left-hand LTR, which contains the promoter and enhancer elements, has been replaced with a strong constitutive promoter such as CMV, or with the U3 region of another retrovirus such as Rous sarcoma virus (RSV). This modification permits transcription of viral RNA without the need of the Tat protein. An additional improvement is the deletion of the enhancer/promoter elements of the right-hand LTR, which generates self-inactivating vectors as described above.

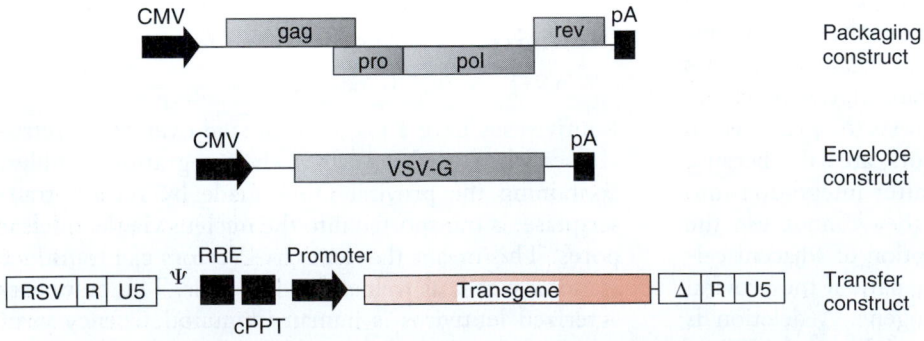

FIGURE 33.5 **Production of lentivirus vectors.** Cells are transfected with three different plasmids. The packaging construct provides the structural proteins and enzymes necessary for the assembly of the virion and processing of the RNA. The envelope construct encodes the gene for the envelope protein, in this example VSV-G. The transfer construct contains the transgene expression cassette and all the necessary *cis*-acting DNA sequences. RRE: Rev-responsive element; cPPT: central polypyrimidine tract.

Applications of retrovirus vectors: treatment of blood disorders

One of the most promising applications of retrovirus vectors is the treatment of blood disorders. Hematopoietic stem cells isolated from the affected individual are infected with a retrovirus containing a normal version of the faulty gene causing the disease. Cells expressing the corrected gene are then reintroduced into the body where they can differentiate and proliferate. Such a strategy was successfully employed to treat children suffering from severe combined immunodeficiency disease (SCID-XI). This disease is caused by mutations in a cellular receptor (the γc cytokine receptor subunit) whose function is required for differentiation of T lymphocytes. Children affected with SCID-XI die at an early age because of the absence of T cell–mediated and humoral immune responses. Children treated by retrovirus vector-mediated therapy demonstrated a complete recovery of their immune system; they were able to leave their protective isolation in the hospital and return home with their parents to enjoy a normal life. The result of this clinical trial made news in 2000 and is regarded as the first example of a successful treatment of a human disease by gene therapy.

Advantages and limitations of retrovirus vectors

The main advantage of retrovirus vectors is their capacity to integrate their genome efficiently into the cell chromosome, which ensures the permanent presence of the transgene in transduced cells and their progeny. In addition, they are far less immunogenic than first-generation adenovirus vectors, because they do not encode any viral proteins. The major advantage of lentivirus vectors over other retrovirus vectors is their ability to transduce nondividing cells, which include most cells in the body. Interestingly, lentivirus vectors are also less prone to silencing. As a consequence, transgene expression with lentivirus vectors can be proportionally increased by increasing the dose of infecting virus. At present, the main problem with lentivirus vectors is the absence of efficient packaging cell lines, which greatly impairs large-scale vector production.

An additional problem with retrovirus vectors is the possibility of generating replication-competent virus during large-scale vector production. Even though retrovirus vectors do not encode any viral genes, possible recombination events between sequences of the transfer construct and viral genes integrated in the packaging cells may generate a replication-competent retrovirus. Because such retroviruses could cause cancer, their presence in preparations of retrovirus vectors intended for gene therapy applications is intensively scrutinized. Another safety issue concerns the fact that the integration of proviral DNAs of retroviruses can happen at numerous sites in the cell genome. This raises the possibility that the vector might integrate near an oncogene and cause cancer by disturbing the function or expression of this oncogene. Unfortunately, this seems to have happened recently in some children who have undergone retrovirus gene therapy to cure severe immunodeficiencies.

ADENO-ASSOCIATED VIRUS VECTORS

Adeno-associated virus vectors can insert transgenes into a specific chromosomal locus

Adeno-associated virus (AAV) has a small, nonenveloped virion 25 nm in diameter. Its 4.7 Kb single-stranded DNA genome contains only two genes: rep, which encodes replication and integration enzymes, and cap, which produces the capsid structural proteins (Chapter 9). Replication of AAV requires the presence of a helper virus, either an adenovirus (thus the name) or a herpesvirus. These viruses provide functions needed for efficient replication and transcription of AAV genomes and abundant synthesis of capsid proteins. Although the use of AAV as a gene transfer vehicle was first documented in 1984, its utility as vector for gene therapy applications has been exploited only during the past decade.

AAV can infect dividing as well as nondividing cells, and its life cycle involves integration of its genome into the host chromosomes. In contrast to retroviruses, integration of AAV is site-specific. In humans, integration occurs preferentially on chromosome 19 at a site called *AAVS1*. Recombinant AAV vectors are deleted of all viral genes and consequently do not express the Rep proteins. In the absence of Rep proteins, integration of AAV is less efficient and is no longer restricted to the *AAVS1* site.

Production of AAV vectors usually requires a helper virus

The essential genes that must be provided by helper adenoviruses include E1A and E1B, open reading frame 6 of early region 4 (E4orf6), E2A, and the small VA RNAs. One of the first strategies employed to produce AAV vectors was to cotransfect 293 cells, which express adenovirus E1A and E1B genes, with a transfer plasmid and a packaging plasmid (Figure 33.6). The transfer plasmid contains the transgene expression cassette flanked by the AAV inverted terminal repeats (ITRs), which include the viral origin of DNA replication and the DNA packaging signal. The packaging plasmid con-

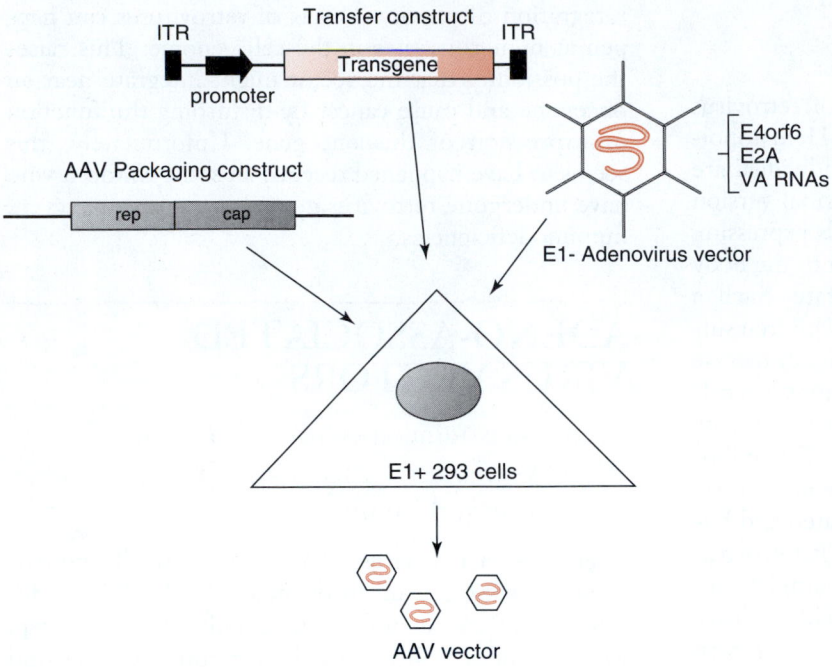

FIGURE 33.6 Production of adeno-associated virus (AAV) vectors. A cell line expressing the E1 region of adenovirus (usually 293 cells) is transfected with two plasmids, the transfer construct and the packaging construct, followed by infection with a first-generation adenovirus vector to provide helper functions (listed on right). The packaging construct provides AAV Rep and Cap proteins. The transfer construct provides the transgene expression cassette, which is replicated and packaged in AAV capsids because of the inverted terminal repeats (ITR).

tains the rep and cap genes, regulated by their own endogenous promoters, but no ITRs. The remaining adenovirus gene functions are provided by infecting the cells with a first-generation adenovirus vector. Although relatively efficient, this method has the disadvantage that vector preparations are contaminated by adenovirus virions. Fortunately, the adenovirus can be partly removed by heat denaturation and by density gradient centrifugation. Another disadvantage of this method is that the AAV vector stock can be contaminated by wild-type AAV generated by nonhomologous recombination between the two plasmids used. To eliminate the adenovirus contamination from the AAV vector preparations, all the essential adenovirus functions can be expressed from a helper plasmid instead of intact adenovirus.

Producing AAV vectors by transfection is cumbersome and difficult to scale up. For this reason, attempts were made to produce AAV vectors using cell lines that express all the necessary gene functions. Because expression of these proteins can be toxic, cell lines expressing the rep and cap genes under the control of their own endogenous promoters were generated. These promoters are inactive in the absence of the adenovirus E1A protein. AAV vectors can be produced by infecting these cells with two recombinant adenoviruses. One lacks the E1 region but contains the transgene expression cassette flanked by the two AAV ITRs. The second contains the E1 region but is deleted in E2B (viral DNA polymerase) to avoid production of adenovirus DNA. E1A proteins induce synthesis of Rep and Cap proteins, which replicate the ITR-flanked expression cassette and package the resulting single-stranded DNA in AAV capsids.

Applications of adeno-associated virus vectors: treatment of hemophilia

Transduction of the liver and skeletal muscle with AAV vectors is nontoxic, relatively efficient, and provides long-term transgene expression. For these reasons, a promising application of AAV vectors is to use these two tissues as a platform for systemic delivery of therapeutic proteins. Hemophilia type B is a disease caused by a reduced level of coagulation factor IX in the blood. The current available treatment of this disease is infusion of factor IX concentrates. In a clinical trial mainly designed to test the safety of AAV vector administration, hemophilia patients were injected several times in their thigh muscles with AAV encoding human factor IX. No toxic effects or inflammation were observed, indicating the vector was safe at the dose employed (2×10^{11} genomes/kg). Low level but stable expression of factor IX was detected in the blood. More importantly, mild clinical improvements were observed. The low level of factor IX produced was most likely due to the low dose of vector injected, which was about 50 times lower than that previously used in laboratory animals. This trial is currently ongoing using a higher dose of AAV.

Advantages and limitations of AAV vectors

Because AAV does not cause disease, AAV vectors are considered to be very safe for administration to humans. AAV virions are very stable; they are resistant to heat inactivation as well as large pH variations. Infection with AAV vectors does not trigger a strong cellular immune response, which allows for long-term transgene

Recent progress in vector development

Retargeting virus vectors. The majority of gene therapy strategies using virus vectors target specific tissues or organs within the body. Many of the vectors used either have too broad a tissue tropism or are not able to infect the targeted tissue. Therefore significant research efforts are being made to target virus vectors by manipulating viral proteins responsible for attachment to receptors and uptake.

Adenovirus binding to cells is mediated by the fiber protein, which projects from the capsid. Modification of domains on the fiber protein involved in receptor binding can lead to the production of differently targetted vectors. An alternative is conjugation of the fiber protein with small-molecular-weight **ligands** that will bind to specific cellular receptors. Most retrovirus and lentivirus vectors are pseudotyped with the vesicular stomatitis virus G glycoprotein, a protein conferring a very wide tropism. Methods have been developed to target retrovirus vectors to a specific cell type by modifying the gene encoding the viral envelope protein. The envelope proteins derived from viruses such as Ebola, Molokai, rabies, influenza, and respiratory syncytial virus have been successfully employed to alter the tropism of retrovirus vectors.

Chimeric viruses. The goal in designing chimeric viruses (also known as hybrid vectors) is to combine into one vector the advantageous properties of different viruses. An adenovirus/retrovirus **chimera** utilizes the ease of production, high titer, and wide tropism of adenoviruses in conjunction with the ability of retroviruses to integrate their proviral DNA. In this system, the transgene expression cassette flanked by retrovirus LTRs is delivered to cells using a first-generation adenovirus vector. Recombinant retroviruses are produced by coinfecting the cells with a second adenovirus carrying the retrovirus env, gag, and pol genes. The recombinant retrovirus vectors can infect nearby cells and integrate and express the transgene they carry.

Long-term maintenance of virus genomes. Two elements derived from the herpesvirus, Epstein-Barr virus, have been very useful for allowing long-term maintenance of vector DNA in dividing cells. Expression of the DNA-binding protein EBNA-1 permits the replication of circular DNA containing the latent origin of DNA replication, oriP, in dividing cells, thus maintaining the DNA in daughter cells. A circular template can be made from an adenovirus vector by inserting two recombination sites, such as *lox*P sites, in the vector backbone. Upon expression of the Cre recombinase, recombination between the two *lox*P sites will generate a circular DNA.

expression. This is partly due to the lack of expression of any viral genes by these gutless vectors. Furthermore, AAV vectors poorly infect professional antigen presenting cells, whose function is to induce the cascade of events leading to the activation of the immune system (see Chapter 31). The prolonged transgene expression observed after gene transfer *in vivo* using AAV vectors was first thought to be a consequence of their integration into the host chromosome. However, AAV DNA has been found in the form of stable, unintegrated concatamers in the nucleus. It is not clear if the prolonged transgene expression observed with AAV vectors is due to the transcription of the integrated vector or of the episomal AAV DNA.

The major limitation of AAV vectors is their small insert size capacity of roughly 4.5 Kb, which reflects the limited packaging capacity of the AAV capsid. This is sufficient to carry cDNAs from small genes such as that coding for blood clotting factor IX, but not larger genes affected by several common genetic diseases. Additionally, current production methods of vector stocks are cumbersome and difficult to scale up.

SUGGESTED READING

1. Barquinero, J., Eixarch, H., and Perez-Melgosa, M. (2004). Retroviral vectors: New applications for an old tool. *Gene Therapy* **11** Suppl 1, S3–S9.

2. Flotte, T.R. (2004). Gene therapy progress and prospects: Recombinant adeno-associated virus (rAAV) vectors. *Gene Therapy* **11,** 805–810.

3. Kanerva, A. and Hemminki, A. (2004). Modified adenoviruses for cancer gene therapy. *International Journal of Cancer* **110,** 475–480.

4. Monahan, P.E. and Samulski, R.J. (2000). Adeno-associated virus vectors for gene therapy: More pros than cons? *Molecular Medicine Today* **6,** 433–440.

5. Oualikene, W. and Massie, B. (2000). Adenovirus vectors in functional genomics. In: Al-Rubeai, M., Ed., Cell engineering, Vol. 2, pp. 80–154. Kluwer Academic Publishers.

6. Thomas, C.E., Ehrhardt, A., and Kay, M.A. (2003). Progress and problems with the use of viral vectors for gene therapy. *Nature Reviews Genetics* **4,** 346–358.

7. Vigna, E. and Naldini, L. (2000). Lentiviral vectors: Excellent tools for experimental gene transfer and promising candidates for gene therapy. *Journal of Gene Medicine* **2,** 308–316.

8. Volpers, C. and Kochanek, S. (2004). Adenoviral vectors for gene transfer and therapy. *Journal of Gene Medicine* **6,** Suppl **1,** S164–S171.

KEY TERMS

Antibody
Apoptosis
Chimera
Chemokine
Cytokine
Helper virus
Inverted terminal repeat
Ligand
Macrophages
Packaging signal
Polyadenylation signal
Promoter occlusion
Pseudotyping
Ribozyme
Silencing
Transcriptional enhancer
Transgene
Tropism
Virus vectors

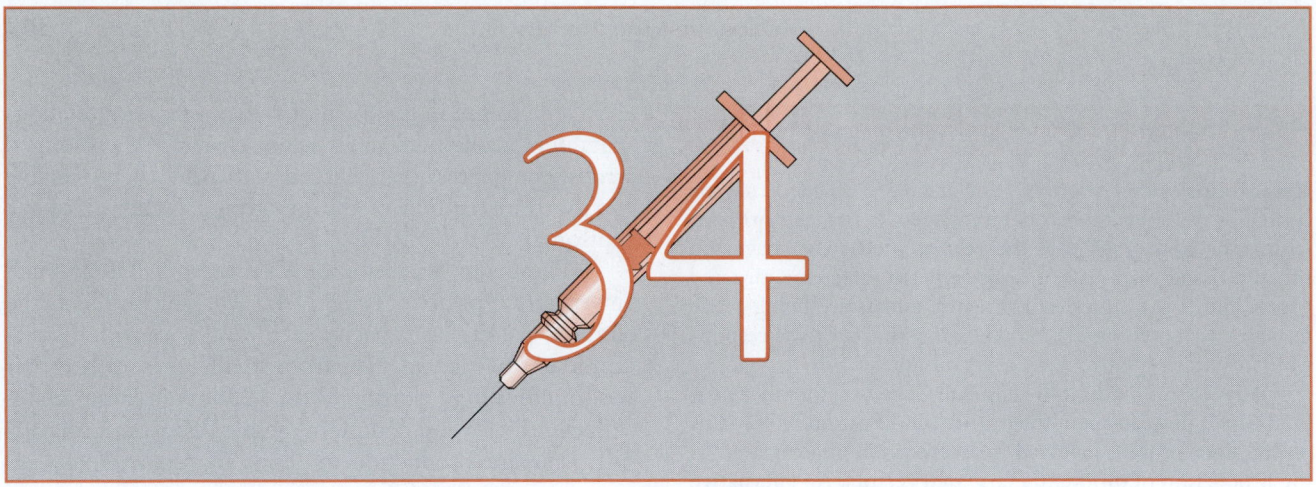

Antiviral Vaccines

Vaccine

From Latin *vacca* (cow), because vaccinia virus was thought to be cowpox virus

Vaccination is the most effective means available for prevention of viral infections.
Smallpox and rabies vaccines were groundbreaking developments in medical history.
Adverse effects of vaccination are rare but sometimes important.

TYPES OF VIRAL VACCINES

Live, wildtype viruses that infect other animals and are attenuated in humans.
Live, attenuated human viruses with mutations that reduce virulence.
Inactivated viruses: retain immunogenicity but are no longer infective.
Component/subunit vaccines: contain one or more viral proteins but no viral genome.
DNA vaccines: express immunogenic viral proteins but cannot generate virus.
Chimeric vaccines: benign or attenuated microbes (viruses or bacteria) that express specific immunogenic viral proteins.

MODERN VACCINE PRODUCTION

Virus production in cultured cells or embryonated eggs.
Sophisticated virus purification and detection methods.
Genetic engineering to introduce specific mutations that attenuate virus infectivity.
Production of viral proteins in bacteria, yeast, or eukaryotic cell expression systems.

LIVE, ATTENUATED VIRUS VACCINES

Can stimulate both cell-mediated and humoral immunity.
Are among the most successful virus vaccines.
Can give long-lived immunity after even a single dose.
Can occasionally cause serious illness in certain individuals.
Can be expensive to make and store.

INACTIVATED VIRUS VACCINES

Usually stimulate only humoral immunity.
Can induce immunity to several different antigens.
Usually require multiple doses to achieve protection.
Are incapable of replication and therefore do not cause disease if properly made.
Are expensive to manufacture, but are less sensitive to storage conditions.
Are safer for immunocompromised individuals.

COMPONENT OR SUBUNIT VACCINES

Usually stimulate only humoral immunity.
Usually require multiple doses to achieve protection.
Present no danger of inadvertent infection with live virus.
Are cheaper to manufacture and test.

Can be very stable.
Avoid delivery of "irrelevant" and potentially pathologic viral antigens.
Are safer for immunocompromised individuals.

DNA VACCINES

Can stimulate both cell-mediated and humoral immunity.
Share other advantages of component or subunit vaccines.
May involve risks because foreign genes are expressed in the body.

A brief history of viral vaccines

Viral pathogens have been serious threats to health and society throughout the evolution of *Homo sapiens*. The horror of watching an individual die from disease caused by rabies or Ebola virus is surely the same today as it was for our hunter-gatherer ancestors. However, these deadly **zoonotic** (animal-borne) virus infections have had relatively little impact on human civilization due to their local distribution and their low capacity to spread from person to person.

In contrast, widespread **epidemics** of viral infections that are spread directly from person to person and are highly contagious, such as smallpox, influenza, and measles, have been limited only by population size and the speed of transportation. The incubation periods of these pathogens are all relatively short (7–10 days) and the diseases run their course to recovery or death quite rapidly. When transportation of humans was restricted to foot, hoof, and oar, and population centers were small and dispersed, travelers infected in one city were either dead, or had recovered and were immune by the time they reached the next city. With population growth, the great trading cities of the Middle Ages became large enough to maintain foci of these diseases year round. Also, faster sailing vessels ensured periodic delivery of unwanted viral passengers to all parts of the known world.

From the middle of the thirteenth century until recent times, human population growth has been slowed and even reversed by waves of destruction wrought by viruses such as measles, yellow fever, and smallpox. As much as 50% of the population of Europe is estimated to have been killed by these agents between 1300 and 1700. The Americas were essentially depopulated by viral diseases in the 100 years following the arrival of Europeans in the mid-sixteenth century. The terrible reality of these viruses and the impact of vaccines are readily apparent from the data shown in Table 34.1.

Edward Jenner and his European counterparts are widely credited with discovering vaccination in the late 1700s (see Chapter 15 on poxviruses). However, attempts to confer active immunity using microbial products have been documented in many cultures extending back at least one to two centuries before Jenner's birth. For example, Chinese physicians of the sixteenth century attempted active immunization against smallpox using dried scabs in various forms, soiled undergarments, and even powdered scabs or fleas fed on scabs! Many cultures in the Middle East have long recognized the durability of immunity to leishmania parasites after natural infection. Children were intentionally inoculated with fully virulent organisms in inconspicuous body locations to prevent subsequent, and less controlled, naturally occuring leishmania infections. Jenner

TABLE 34.1 Impact of antiviral vaccines on disease frequency: pre-vaccination era vs. 1998 (USA)

		Annual reported cases:		
Disease	**Dates**	**Pre-vaccination**	**Post-vaccination (1998)**	**Decrease (%)**
Smallpox	1900–04	48,164	0	100
Poliomyelitis (paralysis)	1951–54	1,314	1*	100
Measles	1958–62	503,282	89	100
Mumps	1968	152,209	606	99.6
Rubella	1966–68	47,745	345	99.3
Congenital rubella syndrome	1958–62	823	5	99.4

*Single case of vaccine-associated paralytic poliomyelitis.
From *Morbidity and Mortality Weekly Report* 1999, **48**, 243–248.

TABLE 34.2 Milestones in viral vaccine development

Pre-1700s	Chinese doctors use powdered smallpox scabs to "immunize" intranasally. Mediterranean-area doctors use directed leishmania-infected sandfly bites to induce long-term protection from reinfection.
1721	Lady Montagu brings concept of variolation (inoculation with pus from recovering smallpox victim) from Turkey to England.
1798	Jenner publishes *Variolae Vaccinae*, the use of cowpox inoculation to protect against smallpox.
1885	Pasteur and collaborators introduce air-dried rabbit spinal cord as rabies vaccine.
1900	Walter Reed demonstrates that yellow fever is caused by a filterable virus.
1930–45	Introduction of vaccines for Japanese B encephalitis (1930), yellow fever (1935), and influenza (1936).
1946–75	Introduction of vaccines for polioviruses types 1–3 (Sabin attenuated strains and Salk inactivated virus); measles, mumps, and rubella viruses; tick-borne encephalitis virus; mouse brain, duck embryo, and tissue culture vaccines for rabies virus; inactivated influenza A and B viruses; and adenoviruses.
1975–present	Introduction of vaccines for hepatitis B virus, hepatitis A virus, varicella zoster virus (chickenpox), and live, cold-adapted influenza virus.

and his contemporaries were nevertheless the first to recognize the full potential of active immunization and to systematically study vaccines and vaccination. The fruits of their labors have fundamentally changed human history. Some milestones in vaccine development are presented in Table 34.2.

Early vaccine technology was crude but effective

Like all medical interventions, the development of new vaccines has been limited by the availability of appropriate technologies (Table 34.3). Nowadays we may shudder at the idea of spraying powdered scabs up the nose, but the inevitability and destructiveness of smallpox went a long way toward easing any squeamishness about the purity or the origins of early vaccines. It would be hard to imagine people lining up today to be inoculated with crudely purified pus from cowpox lesions or dried spinal cord from rabbits infected with rabies.

Early in the history of vaccine development there were occasional setbacks due to contaminated vaccine lots (infections from bacterial contamination of pus) and serious vaccine-associated side effects (encephalomyelitis following rabies vaccine). Many technologically crude vaccines made from natural products actually worked, and their popularity spread. However, public acceptance was far from universal. Modern vaccine developers could learn a great deal from the showmanship of vaccine pioneers like Louis Pasteur, who arranged public demonstrations of his anthrax and rabies vaccines in the early 1880s. Pasteur's most potent ally in convincing Parisians to try vaccines was fear. Injection of spinal cord material from a rabid rabbit would certainly be less

TABLE 34.3 Technological advances supporting viral vaccine development

1850	Germans add glycerin to "cow lymph" to stabilize and decontaminate smallpox vaccine.
1879	Galtier succeeds in passaging rabies from rabbit to rabbit.
1884	Metchnikoff descibes role of phagocytes in immunity.
1885	Pasteur and colleagues fix biological characteristics of rabies by passage in rabbit brains and discover that virulence in spinal cord is lost with drying.
1887	Petri describes overlapping glass plates to use for sterile cultures.
1897	Loeffler and Frosch describe "filterable viruses" (foot-and-mouth disease virus).
1931	Goodpasture first uses embryonated eggs to culture viruses.
1936	Stanley isolates nucleic acids from tobacco mosaic virus.
1950	Enders, Weller, and Robbins launch systematic study of cell culture.
1953	Description of first immortalized cell line (HeLa).
1960	Production of reassortant influenza A viruses in eggs.
1972	First DNA cloning performed.
1974	First recombinant proteins expressed in bacteria.
1985	Hepatitis B surface antigen gene cloned and expressed.

intimidating to someone who had just been bitten by a rabid wolf. Dried pus from a smallpox or cowpox lesion would likewise seem quite benign in the setting of a smallpox epidemic with 20–25% mortality.

Embryonated chicken eggs and cell culture played major roles in recent vaccine development

The technological leaps that had the greatest impact on viral vaccine development in the twentieth century were (1) serial passage of human viruses in animals; (2) the adaptation of viruses for growth in embryonated chicken eggs; and (3) the growth of viruses in vertebrate cells cultured *in vitro*. As late as the 1880s, pioneering microbiologists were restricted to growing "germs" either in whole animals or on potato slices and in open jars containing broth. In 1881, Robert Koch introduced gelatin and agar-based culture systems and, in 1887, R. Petri invented the overlapping glass plates that have immortalized his name. These innovations set the stage (in 1931) for the first use of eggs to culture viruses; think of an egg as a sterile tissue culture flask with a calcium shell. Early vaccines made possible by growth in embryonated eggs include the yellow fever virus strain 17D (1932) and influenza virus (1936).

In the early 1950s, John Enders, Thomas Weller, and Frederick Robbins began the first systematic study of cell culture, using tissues from children who had died at birth. The use of such **primary cell cultures**, and the subsequent establishment of immortalized cell lines such as HeLa, dramatically accelerated the development of antiviral vaccines in the late 1950s and early 1960s. The achievements of this era include improved vaccines for smallpox and rabies as well as poliovirus, measles, mumps, and rubella vaccines.

Nevertheless, the following story illustrates how technological limitations persist to modern times. Beginning in the last years of the 1990s and up to the time of writing, many parts of the world have experienced a major **panzootic** (widespread epidemic in several animal species) of H5N1 avian influenza. These viruses are variably pathogenic in domestic and wild birds and have led to billions of bird deaths in South and East Asia either naturally or through culling of infected stock. Although relatively few cases of H5N1 disease have been reported in humans, mortality rates have been very high (33–70%).

There is increasing concern that continued mutation of these viruses could lead to human-to-human spread and a major influenza **pandemic** (widespread epidemic). At the time of writing, the manufacturing process for influenza A vaccine still depends on the creation of reassortant vaccine-strain virus (see Chapter 23) in eggs. Unfortunately, the H5N1 viruses are highly pathogenic for avian cells and can rapidly kill the embryos before sufficient reassortant viruses can be produced. As a result of this technological glitch, it was initially quite difficult to make candidate vaccines against these dangerous viruses. Since the current influenza vaccine production cycle typically takes ~12 months to produce the world's annual capacity of ~250 million doses, any such technical delays in vaccine production could have devastating consequences in the setting of an actual pandemic.

Major categories of viral vaccines

Viral vaccines can be grouped into several categories (Table 34.4).

Live wild-type viruses. These are the simplest vaccines to make and are among the most effective. A virus that infects one animal species can sometimes be used to infect another species to protect against a closely related pathogen. Viruses are often adapted to replicate well only in their natural host; they can infect other species, but replicate poorly and have limited or no pathogenic effects. However, these viruses may share **immunogenic** determinants with the target virus, resulting in an effective immune response against that virus. In addition to the wildtype viruses used as human vaccines listed in Table 34.4, vaccine-strain measles virus can protect dogs from canine distemper.

Live attenuated viruses. Blind serial passage of pathogenic viruses in animals, eggs, or tissue culture can lead

TABLE 34.4 Classification of currently available human antiviral vaccines

Live wildtype viruses	Vaccinia (cowpox)[a]
	Rotavirus (rhesus monkey)[b]
Live attenuated viruses	Adenovirus[c]
	Influenza A (cold adapted)
	Measles
	Mumps
	Polio (Sabin)
	Rubella (German measles)
	Varicella (Chickenpox)
	Yellow fever
Whole inactivated viruses	Hepatitis A
	Influenza A
	Influenza B
	Polio (Salk)
	Rabies
	Tick-borne encephalitis
Component/subunit vaccine	Hepatitis B surface antigen

[a]Smallpox eradicated (supply of vaccine tenuous).
[b]Introduced in 1998–99, but rapidly withdrawn due to association with **intussusception**.
[c]Military use only.

to acquisition of a variety of mutations that reduce the pathogenic properties of the virus in its natural host. Although there is often little knowledge of the mechanisms of attenuation, these are among the most successful viral vaccines currently used. These attenuated viruses replicate in humans and therefore can often induce immune responses as effectively as the parent viruses, but they cause little or no disease. Table 34.5 illustrates how two such vaccines were produced.

The recently introduced cold-adapted influenza A virus vaccine is a live attenuated virus designed with specific, molecularly defined characteristics. Cold-adapted influenza vaccines are produced by combining the H and N genes of the targeted virus strain with six other gene segments from mutant viruses known to have restricted growth at 37°C. The H and N genes code for the envelope glycoproteins hemagglutinin and neuraminidase, which are the most immunogenic influenza virus proteins. The reassortant viruses produced cannot replicate in the lung at core body temperature, but grow well in the cooler nasal mucosa, where they elicit an excellent immune response without endangering the vaccinated individual.

Inactivated virus. This is a common route for production of vaccines where a live virus vaccine is not available. Virulent viruses are grown to high concentrations in eggs or in cultured cells and purified. The purified virus is then subjected to inactivation by treatment with formaldehyde, organic solvents, detergents, or high temperature. Careful control of the treatment can completely eliminate infectivity of the virus, while retaining the antigenicity of the viral proteins. Such inactivated viruses can retain their ability to induce an immune response but are incapable of replication and therefore do not cause disease.

Component or subunit vaccines. These consist of purified viral proteins that are known to be immunogenic. These vaccines do not contain other components of the virus, in particular the nucleic acid genome. Their major advantage is that they present no danger of inadvertent infection with live virulent virus, which can be a problem with the other vaccine categories. They may also be significantly cheaper to manufacture, because they do not require expensive virus growth in cell culture systems and testing of the virus produced. At the time of writing, there is only one licensed subunit viral vaccine, the hepatitis B virus surface antigen. This protein was originally purified from infected human serum, but is now produced as a recombinant protein by expressing the cloned surface antigen gene in yeast or mammalian cells. Subunit vaccines against tumorigenic strains of human papillomaviruses have recently been shown to be effective and may soon be licensed for general use.

Advantages and drawbacks of vaccine types

The immune response elicited by live attenuated vaccines comes closest to that observed in natural infection. Immunity after even a single dose of these vaccines is typically long-lived (10 years to life). However, these vaccines are expensive to manufacture and are relatively unstable, requiring careful attention to refrigeration via the **cold chain**, a problem in developing countries. Live vaccines are also subject to inactivation by preformed maternal antibodies when administered in early childhood. For example, maternal antibodies can inactivate measles vaccine-strain virus in some children before 12 months of age. Live virus strains can occasionally cause serious illness and even death in subjects with abnormal immune systems.

TABLE 34.5 Passage histories for two vaccine-strain viruses

Edmonston-B and Moraten measles vaccine strains
- Isolated by Enders & Peebles in 1954 from patient (D. Edmonston)
- Twenty-four passages (35°–36°C) in primary human kidney cells
- Twenty-eight passages in primary human amnion cells
- Six passages in chicken embryos
- Marketed as Edmonston B vaccine in 1963
- Seven passages (36°C) in chicken embro fibroblasts
- Forty passages (32°C) in chicken embryo fibroblasts
- Marketed as "further attenuated" measles vaccine (Moraten strain) in 1968

Oka-strain varicella virus
- Isolated by Takahashi et al. in 1974 from three-year-old Japanese boy (family name Oka)
- Initial growth in human lung fibroblast cells
- Twelve passages in primary guinea pig fibroblasts
- Two passages in WI-38 cells (human diploid fibroblasts)
- Three to six passages in MRC-5 cells (human diploid lung cells)
- Licensed for use in Japan and Korea in late 1980s, and in USA in 1995

Inactivated whole virus vaccines, like live virus, can generate immune responses to a wide range of viral antigens. These vaccines are less sensitive to storage conditions and are safer for immunocompromised individuals. However, they are also relatively expensive to produce. Furthermore, multiple doses are usually required to achieve protection, and immunity often fades with time, requiring booster doses.

If recombinant DNA technologies can be used, component or subunit vaccines are potentially the simplest and least expensive vaccines to make. These vaccines can be remarkably stable and their use avoids the delivery of irrelevant and potentially pathologic viral antigens responsible for vaccine-associate adverse events (see below). These vaccines can also be used without risk in immunocompromised subjects. However, like inactivated virus vaccines, multiple doses of component vaccines are normally required for protection, and periodic boosting may be necessary to ensure long-term immunity.

How do viral vaccines work?

In the current environment of molecular sophistication, it is sometimes hard to believe that products developed 40 to 50 years ago are still in use. Until very recently, the process of attenuation was essentially black-box science. Pathogenic viruses were repeatedly passaged through eggs or tissue culture cells under varying conditions until they acquired an attenuated phenotype. Astonishingly by today's ethical standards, the appearance of the attenuated phenotype was typically demonstrated by inoculation of children! Several of our most successful vaccines, including yellow fever, measles, mumps, rubella, and Sabin polio, were generated in this way.

We have learned a great deal about the molecular characteristics of some of these products in recent years. The Sabin attenuated poliovirus vaccine strains very occasionally revert to neurovirulence, and we now know the sites on the viral genome of many of the mutations that affect virulence (Table 34.6). These data help to explain the observation that types 2 and 3 Sabin vaccine-strain viruses are most often isolated from subjects with vaccine-associated paralytic polio. Other vaccines continue to be used effectively even though we remain ignorant of the precise mechanisms of attenuation and protection against infection. Smallpox has already been eradicated by systemic vaccination, and it is likely that poliovirus and measles virus will follow smallpox into extinction before we fully understand how these vaccines have actually been attenuated at a molecular level.

The role of the immune system in fighting viral infections

Cellular immune responses (see Chapter 31) mediate recovery from the vast majority of viral illnesses. Individuals who lack the ability to make soluble antibodies via the **humoral immune response** recover from most viral infections in spite of their defect, showing that antibody formation is not required for recovery. However, individuals with defects in cell-mediated immunity are highly susceptible to severe, protracted, and frequently fatal infections with a wide range of viral pathogens. Thus killing of infected cells by activated **cytotoxic lymphocytes** appears to be the major mechanism by which the immune system clears viral infections from the body. This makes sense, as infected cells can be recognized and killed before they produce large amounts of virus, which could then spread throughout the body. The only exception to this rule appears to be the enterovirus family, including poliovirus and the echo and Coxsackie viruses. Control of these pathogens requires an intact humoral response.

However, antibodies do have a role in protecting against viral infections. High titers of preformed **neutralizing antibodies** can confer immunity against most viral pathogens. Antibodies acquired from the mother near the end of pregnancy are critically important for the protection of infants between birth and 6–9 months of age. Antiviral antibodies are also important for limiting reinfection with the same virus and can sometimes provide partial or complete protection against closely related viruses. Therefore, the humoral response can be thought of as sufficient but not always necessary for protection against viral pathogens.

TABLE 34.6 Attenuated oral polioviruses—reversion to neurovirulence

Virus strain	Number of mutations	Number of amino acid changes	Number of mutations required for return to neurovirulence
Sabin strain 1	57	23	>10
Sabin strain 2	23	5	5–6
Sabin strain 3	~6	3	1–2

From Minor, P.D. (1992). The molecular biology of poliovaccines. *Journal of general Virology* 73, 3065–3077.

Some vaccines target mainly antibody production; others target the cellular immune response

Because they are able to replicate in the body, live attenuated virus vaccines induce both cellular and humoral immunity as do most natural virus infections. However, inactivated whole virus or subunit vaccines cannot replicate and therefore are less efficient in inducing immune responses. In order to stimulate the immune response, these vaccines are mixed with aluminum hydroxide or aluminum phosphate salts, called **adjuvants**. These are the only adjuvants presently licenced for use in humans. Aluminum salts preferentially stimulate **type 2 T helper lymphocytes** (Th-2 cells), which secrete **cytokines** that lead to the production of antibodies (see Chapter 31). However, these adjuvants do not efficiently induce cellular immune responses. Early vaccine developers showed that levels of neutralizing antibodies correlate well with protection against infection, and therefore this was one of the primary goals of vaccine development. The viruses targeted by most vaccines developed up to the present cause acute infections by one prevalent serotype (measles, rubella, smallpox) or can be neutralized by antibodies before reaching their target tissues (poliovirus, hepatitis B, rabies).

It is now clear to vaccine developers that focusing primarily on the generation of antibodies is not always sufficient for an effective antiviral vaccine. Viruses with the capacity to establish latent infections (herpesviruses), viruses with multiple serotypes and limited cross-neutralization (respiratory syncytial virus, dengue), and those that rapidly mutate their structural proteins (influenza A, human immunodeficiency virus, hepatitis C) present much more difficult targets for vaccines that generate predominantly antibody responses. Antibodies are not useful against viruses that have already established a chronic infection (human immunodeficiency virus, hepatitis B virus, papillomaviruses). Furthermore, inappropriate or insufficient neutralizing antibody responses generated by vaccines can lead to immunopathology (respiratory syncytial virus, atypical measles) or can even facilitate virus entry into target cells (dengue). Presently, a major focus of vaccine designers is the harnessing of cell-mediated immunity to create vaccines for these elusive and complicated viral targets that remain serious threats to human health.

New approaches to vaccine development show great promise

In the last 50 years, spectacular advances in molecular biology, immunology, biotechnology, and protein chemistry have been the driving forces behind viral vaccine development. The modern era of vaccine development was foreshadowed by the introduction of the serum-based hepatitis B surface antigen (HBsAg) vaccine in 1981, and was fully inaugurated with the cloning and expression of the HBsAg gene in yeast and mammalian cells and the licensing of a recombinant vaccine in 1986. Several of the most promising advances in vaccine development (Table 34.7) are discussed here.

New adjuvants and delivery systems. As noted above, only aluminum salts are currently used as adjuvants in human vaccines. Although aluminum-based adjuvants help the immune system to respond to many viral antigens, they do not stimulate **type 1 T helper lymphocytes** (Th-1 cells), involved in the cellular immune response. A variety of novel adjuvants are being considered for use in human vaccines. CpG elements are short, unmethylated DNA motifs not commonly found in the human genome but often present in bacterial genes. Injection of DNA containing these motifs can elicit potent, innate cellular (Th-1-like) responses and therefore increase the host response to antigens. **Chimeric** molecules that contain cytokines, **lectins**, or **monoclonal antibodies** fused to viral antigens can help to direct the antigen to an appropriate target cell or enhance immune responses by interacting with cells in the immune system. **Dendritic cells**, which present antigens to the immune system, can be isolated from an individual, "pulsed" with a viral antigen, and reintroduced into the body. This can lead to enhanced antigen presentation, which may result in clearance of chronic viral infections.

A variety of auto-assembling vesicles within which viral antigens can be trapped are being tested as delivery systems for subunit vaccines. These include **proteosomes**, vesicles made from the outer membrane of *Neisseria* bacterial species. Viral antigens can be trapped in these vesicles, which promote both humoral and cellular immune responses. An alternative to introduction of antigens is the insertion of their genes into expression plasmids, which are then introduced into cells in the body. These **DNA vaccines** can produce the protein of interest intracellularly, leading to establishment of both humoral and cellular immunity, without the dangers inherent in live or inactivated virus vaccines.

Historically, the options for mechanical delivery of vaccines into the human body have been remarkably limited, based for the most part on inoculation by needle, contributing to many people's fear of vaccines. Aerosols, eye drops, and suppositories are being developed to direct vaccines to specific body locations. Genetic engineering of plants may soon allow specific viral antigens to be produced by natural foods such as bananas. Consumption of the antigen-containing food would result in acquisition of immunity to the targeted viruses. Finally, **gene guns** deliver tiny particles coated with plasmid DNA at high velocity into cells. This can

TABLE 34.7 New technologies for human virus vaccines

A. Adjuvant systems	
Cholera toxin B subunit	Elicits strong humoral response without toxicity.
Detoxified bacterial lipopolysaccharide	Elicits strong humoral and cellular immune responses.
CpG motifs	Short, unmethylated DNA motifs that elicit potent, innate cellular (Th-1-like) responses.
Pulsed dendritic cells	Presentation of antigens to dendritic cells outside of the body and their reintroduction can lead to enhanced response to antigens.
Conjugates	A variety of molecules such as cytokines, monoclonal antibodies and lectins can be conjugated to antigens to enhance immune reactions.
B. Delivery systems	
Liposomes	Antigens are trapped inside vesicles made from synthetic lipids, enhancing immune reactions. Stability and consistency have been major problems.
Immune stimulating complexes (ISCOM)	More complex liposomal preparations that retain the capacity to elicit strong immune responses, but are more stable.
Proteosomes	Auto-assembling vesicles containing bacterial outer membrane proteins from Neisseria that can trap antigens and promote strong humoral and cellular responses.
Plasmid DNA	An expression plasmid containing the gene that encodes a viral antigen is introduced into the body. Expression of this gene *in vivo* can lead to potent induction of both humoral and cellular responses.
Viral or bacterial vectors	Genes that encode viral antigens of interest can be inserted into nonpathogenic viral vectors (avian poxviruses, Semliki Forest virus) or bacterial vectors (attenuated Salmonella typhi) that retain some useful characteristics of the vector organisms (binding and entry, tissue tropism).
C. Delivery technologies	
Aerosol	A variety of ingenious mechanical devices are being studied for aerosolization of viral vaccine preparations for delivery either to the nasal or the lower respiratory mucosa.
Eye drops, troches, suppositories	Formulations that deliver vaccine antigens to conjunctival, vaginal, or gastrointestinal mucosae.
Foods	Genetic manipulation of plants has made it possible to consider the delivery of viral and other antigens in foods such as bananas.
Gene gun	Accelerates gold particles carrying plasmid DNA to high velocities for penetration of skin to the desired depth, targeting dermal layers or muscle.

be a quick and efficient system for localized delivery of DNA vaccines.

Vaccination with defined proteins. With the exception of the hepatitis B vaccine, which contains only the viral surface antigen, all other viral vaccines in use at the time of writing are based on either live or inactivated whole viruses. As more is learned about the biology of viral pathogens, it should be possible to create a new generation of "minimalist" vaccines that contain only the antigens or peptides that elicit protection while avoiding those that contribute to vaccine-associated adverse events (see below). Many such products are currently under development or in field trials for a wide range of viral pathogens including measles, respiratory syncytial virus, influenza, HIV, dengue, papilloma, and several herpes viruses. Other advantages of these simpler vaccines include ease of manufacture and the possibility of combining several antigens in a multivalent vaccine to protect against a range of pathogens.

While attractive in many ways, the utility of some of these narrowly targeted vaccines may eventually prove to be limited, because viruses can undergo mutations in the targeted gene and therefore escape from the immune response. Because of this, vaccine developers are focusing on highly conserved, essential viral proteins and are considering vaccination with mixtures or "cocktails" of defined proteins. Plasmid-based DNA vaccination may be particularly well suited for the delivery of either single or small groups of viral proteins to the immune system.

Use of live viruses with defined attenuation characteristics. The intentional exposure of healthy subjects to a live, vaccine-strain virus can make both vaccinator and vaccinee quite nervous, particularly when the vaccine has been produced by random passage until it lost its original virulence. Such black-box vaccines continue to be introduced (e.g., the Oka strain varicella [chickenpox] vaccine in the late 1990s). However, by using genetic engineering methods, vaccine developers now are able to create molecularly defined attenuated vaccine-strain viruses. The best example of this approach is the cold-adapted influenza A vaccine described above.

Use of live vectors and chimeric viruses. Because we have accumulated many years of experience with hundreds of millions of vaccinated people, the advantages and dangers of live virus vaccines such as vaccinia and measles are now well known. We have simultaneously gained considerable experience with live bacteria used as vaccines such as Bacille Calmette-Guerin (BCG, against tuberculosis) and attenuated *Salmonella typhi* (against typhoid fever). Most importantly, advances in molecular biology have given us the capacity to modify these organisms to carry foreign genetic material. The safe track records of these viral and bacterial agents make them very attractive vectors for the delivery of antigens from unrelated organisms for which vaccines are not yet available. Examples of such chimeric products include genetically modified vaccine-strain measles virus carrying the structural proteins of respiratory syncytial virus, and attenuated *Salmonella typhi* expressing rotavirus glycoproteins.

A chimeric vaccinia virus expressing the major glycoprotein protein of rabies virus is currently in use in the northeastern part of North America to control a major rabies **epizootic** in raccoons. Other viruses are also the subject of intense investigation as possible vectors for foreign genetic material. Examples include bovine or simian rotaviruses engineered to express human rotavirus glycoproteins, canarypox viruses that express HIV sequences, and alphaviruses such as Semliki Forest virus that express paramyxovirus proteins. Several of these chimeric virus vaccines are currently in clinical trials. All seek to exploit the capacity of the vector organism to express the targeted proteins within host cells, eliciting a cytotoxic T cell response.

Vaccines that can break tolerance. Until very recently, vaccine development efforts have focused almost exclusively on products capable of preventing or modifying acute disease. Many hundreds of millions of people around the globe are already infected with viruses such as hepatitis B and C, papillomaviruses, and human immunodeficiency virus. In many cases, these viruses persist in the body because of the establishment of some degree of **tolerance** by the human immune system. Several of the new approaches outlined in Table 34.7 have the proven capacity to break tolerance in animal models. Among the most promising include DNA vaccination, pulsed dendritic cells, and the use of potent Th1-biasing adjuvants such as ligands for **toll-like receptors**. Several of these strategies are already in advanced-stage clinical trials.

The changing vaccine paradigm

Just as there can be no one-size-fits-all for the development of antiviral vaccines, there is no single optimal vaccination strategy for all viruses. There is still an urgent need for vaccines that can protect against viruses that cause significant illness and death, such as HIV, dengue virus, hepatitis C virus, papillomaviruses, and Ebola virus. However, development efforts are not limited to these viruses. Viruses with more limited capacity to do harm, such as agents of mild childhood diarrhea and the common cold, are currently being targeted. Vaccine developers are also targeting ubiquitous and persisting viruses such as the herpesviruses Epstein-Barr virus and cytomegalovirus. However, the consequences for human biology of vaccinating against these types of viral agents are wholly unknown.

Finally, many of the scientific advances outlined above have to some extent democratized the process of vaccine development, which was once the exclusive purview of the major pharmaceutical companies. Presently, thousands of small academic and biotechnology company laboratories around the world are actively pursuing vaccine development programs. The range of viruses against which vaccines are currently under development is impressive (Table 34.8). These developments could lead to both favorable and unfavorable outcomes. On the one hand, it is likely that in the near future more humans throughout the world will enjoy protection against more viral diseases than ever before. On the other hand, the unbridled use of vaccines for everything from tooth decay to cancer has the potential to tarnish the relatively good name that vaccines have enjoyed for almost half a century.

Vaccine-associated adverse events

Although the currently available viral vaccines are among the safest pharmaceutical products ever introduced, none is 100% safe. The history of vaccine development contains a small number of unfortunate incidents in which vaccinated individuals have suffered severe adverse events (Table 34.9). In some cases, these incidents can be explained by the biological properties of the product. For example, vaccine-associated paralytic polio due to reversion of Sabin polio vaccine-strain virus to neurovirulence is understood in terms of the mutations required to maintain attenuation (see above).

TABLE 34.8 Some antiviral vaccines currently under development

Virus family	Virus species	Most promising strategies
Arenavirus	Junin Lassa fever	Live attenuated Vaccinia vectored
Calicivirus	Norwalk	Recombinant capsid "ghosts"
Filovirus	Ebola/Marburg	DNA Vesicular stomatitis virus vectored
Flavivirus	Hepatitis C Dengue	Recombinant envelope glycoprotein Tetravalent attenuated, DNA
Hepadnavirus	Hepatitis B	DNA, recombinant surface antigen including pre-S1/S2 regions; curative hepatitis B vaccine: DNA with CpG motifs
Herpesvirus	Epstein-Barr Herpes simplex 1, 2	Whole inactivated, component (gp350) Recombinant polypeptide, vectored
Orthomyxovirus	Influenza A	Vectored, DNA
Papillomavirus	Human strains	Recombinant virus-like particles
Paramyxovirus	Measles Parainfluenza Respiratory syncytial	Vectored, DNA Live, attenuated Subunit G and fusion protein, DNA, vectored, peptide
Parvovirus	Parvovirus B19	Empty capsid with VP1 protein
Picornavirus	Enterovirus 71	Whole inactivated
Reovirus	Rotavirus	Live reassortants of simian or human origin
Retrovirus	HIV-1, 2	Recombinant polypetide, DNA, vectored
Unclassified	Hepatitis E	Recombinant protein ORF2

However, due to incomplete knowledge about how vaccines work, most vaccine-associated adverse events are not fully understood and therefore are difficult to foresee.

A great deal of energy is being expended internationally to refine the tools available for the surveillance of problems due to vaccination. These surveillance activities are essential not only to ensure the safety of new vaccines but also to protect the enormous societal investment already made in national and international vaccination programs. Even transient losses of public confidence in vaccination programs can have devastating effects. Examples inclulde the pertussis vaccine in Japan in the mid-1990s, measles vaccine in the United Kingdom, and poliovirus vaccine in Northern Nigeria at the time of writing. It is important to understand that

TABLE 34.9 Severe adverse events associated with virus vaccine administration

Vaccine implicated	Event
Smallpox ("lymph" vaccines)	Sepsis due to bacterial contamination
Yellow fever (1942)	Vaccine lot contaminated with hepatitis B virus, leading to 28,000 cases of hepatitis B
Inactivated polio (1955)	Incomplete inactivation of virus, leading to 204 cases of paralytic disease (Cutter incident)
Inactivated measles (1960s)	Atypical (severe) disease upon exposure to natural measles infection
Attenuated polio (Sabin)	Vaccine-associated paralysis (~1 per 2.4 million persons vaccinated) due to reversion to more pathogenic strain
Live vaccines (vaccinia, polio, measles)	Dissemination and death in immunocompromised individuals
Attenuated live measles	Unexplained mortality in girls who received high titer formulations (relative risk of death doubled to age 5); more than 20 million doses of vaccine distributed
Inactivated influenza A	Apparent risk of Guillain Barré Syndrome (<1 per 1 million persons vaccinated), only in some years
Rotavirus (1999)	Intussusception—bowel folding and obstruction (vaccine withdrawn)

The story of the rabies vaccine

Human rabies is a devastating disease with periods of relative lucidity and apparent madness alternating over a period of days to weeks in a downward spiral to inevitable death. Beginning with Zinke's observation in 1804 that rabies could be transmitted by the injection of saliva from a rabid animal, several investigators worked through the nineteenth century to develop a successful rabies vaccine. By 1879, Victor Galtier had succeeded in passaging the virus in rabbits and had demonstrated limited efficacy with a rabbit tissue vaccine in herbivores. Simultaneously, Émile Roux and colleagues in Pasteur's group had been able to "fix" rabies virus (i.e., establish consistent biological properties) by serial passage in rabbit brains. They had also made the key discovery that viral virulence in spinal cord disappeared over a 15–20 day period of drying. Armed with this information, they were able to demonstrate that a series of injections with spinal cord material dried for variable periods of time could protect dogs.

In today's terms, it is stunning that preexposure prophylaxis had been demonstrated in only 50 to 60 dogs before the first human, a preteen boy named Joseph Meister, received this vaccine in 1885. Desperation and fear go some way toward explaining why Meister's parents and the investigators themselves conducted this experiment. To everyone's relief (including Joseph's), the vaccine appeared to work and its popularity spread. However, with more widespread use it became apparent that this vaccine induced serious **autoimmune demyelination** of the central nervous system in as many as four per thousand vaccinees. The associated mortality was significant (30–50%) and many of those who recovered were left with permanent neurological sequelae. This disease is now called postinfectious encephalomyelitis, and is known to be mediated by autoreactive T cells. The myelin component of the vaccine was eventually implicated.

Subsequent generations of rabies vaccines produced in suckling mouse brain and duck embryos had progressively less myelin and dramatically lower incidences of such complications (1/8000 and 1/32,000 respectively). The successful adaptation of rabies virus to the human diploid fibroblast cell line WI-38 in the early 1960s finally permitted the development of a rabies vaccine that was both effective and safe. To date, more than 10 million people have been inoculated with rabies vaccines grown in cultured cells, with seroconversion and protection rates approaching 100% and without a single proven case of vaccine-associated neurological damage.

the safety demanded of a vaccine can change in different situations. For example, an occasional serious complication with a rotavirus vaccine would be tolerable in the developing world, where this virus kills more than a million children every year. Such a problem would, however, be unacceptable in North America, where only a small number of children die from rotavirus infection annually. As we approach worldwide eradication of measles and poliovirus, there are increasing concerns about the safety of the currently used vaccines. In the 1990s, most developed countries replaced the live Sabin (oral) polio vaccines with inactivated vaccines in their national programs to avoid the small but measurable risk of reversion (~1 per 2 million doses). At the time of writing, there is no inactivated measles virus vaccine.

Ethical issues in the use of viral vaccines

Many of the viruses targeted by vaccine programs are among the most highly transmissible infectious agents known. As a result, public health authorities have great interest in maintaining high levels of vaccine coverage to protect the small number of individuals who are missed by the program, who have failed to respond to vaccination, who have a valid contraindication to vaccination, or who refuse vaccination. The protection afforded these individuals by high levels of vaccine coverage in the community is termed **herd immunity**.

The desire to establish herd immunity has led public health authorities to enforce vaccination or to make avoidance of vaccination difficult. In the past, compulsory or heavy-handed vaccination programs for smallpox occasionally resulted in street riots. The debates surrounding some childhood vaccination programs are only slightly less vigorous today. Some individuals at each end of the spectrum of arguments are quite prepared to label those at the opposite extremes as "abusers" of children. Ironically, the great successes of the quasi-compulsory vaccination programs that have eliminated smallpox and nearly eliminated polio and measles have also contributed to the growing wave of anti-vaccination sentiment, because people feel less at risk.

Another ethical problem arises from the concept of "altruistic" vaccination. For example, annual influenza vaccination of health care providers can reduce the seasonal impact of influenza on hospitalized and institutionalized individuals, although the danger of influenza to the (younger and healthier) vaccinated individuals is not as great. Although many institutions have strong promotional programs, so far there has been no attempt to legislate or mandate annual influenza vaccination for all health care workers.

SUGGESTED READING

1. Anonymous. (2004). Canadian immunization guide, 6th Ed. Laboratory Centers for Disease Control—Division of Immunization. Minister of Public Works and Government Services: Ottawa.

2. Centers for Disease Control and Prevention (CDC). (2000). Recommended childhood immunization schedule—United States, 2000. *Morbidity and Mortality Weekly Reports* **49,** 35–38, 47.

3. Estcourt, M.J., Ramshaw, I.A., and Ramsay, A.J. (1998). Cytokine responses in virus infections: Effects on pathogenesis,

recovery and persistence. *Current Opinion in Microbiology* **1,** 411–418.

4. Horwitz, M.S. and Sarvetnick, N. (1999). Viruses, host responses, and autoimmunity. *Immunological Reviews* **169,** 241–253.

5. Plotkin, S.A. and Orenstein, W.A. Eds. (2004). Vaccines, 4th Ed. WB Saunders, Philadelphia.

6. Stratton, K.R., Howe C.J., Johnson R.B. Eds. (1994). Adverse events associated with childhood vaccines: Evidence bearing on causality. National Academy Press, Washington, D.C.

7. Ward, B.J. (2000). Vaccine adverse events in the new millennium: Is there reason for concern? *Bulletin of the World Health Organization* **78,** 112–123.

KEY TERMS

Adjuvant
Autoimmune demyelination
Cellular immune responses
Chimera (adj: chimeric)
Cold chain
Cytokines
Cytotoxic T lymphocytes
Dendritic cells
DNA vaccines
Epidemic
Epizootic
Gene gun
Herd immunity
Humoral immune response
Immunogenic
Intussusception
Lectins
Monoclonal antibodies
Neutralizing antibodies
Pandemic
Panzootic
Primary cell cultures
Proteosomes
Tolerance
Toll-like receptors
Type 1 and type 2 T helper lymphocytes
Zoonotic

Glossary

α-amanitin: a cyclic octapeptide, produced by the fungus *Amanita phalloides*, that binds to and inhibits eukaryotic RNA polymerase II.

Abortive: an infection that does not lead to a complete virus replication cycle.

Actinomycin D: a tricyclic unsaturated hydrocarbon linked to two cyclic pentapeptides, made by a streptomyces species, that binds tightly to dG residues in DNA and blocks transcription by DNA-dependent RNA polymerases.

Acute transforming retroviruses: retroviruses that transform cells rapidly and efficiently by means of cellular oncogenes incorporated into the viral genome.

Acyclic: a version of a compound in which a group that is usually cyclic is broken or removed.

Acyclovir: nucleoside analogue used for treatment of herpes simplex and varicella zoster virus infections. It is selectively phosphorylated by viral thymidine kinase and, in its triphosphate form, selectively inhibits viral DNA polymerase.

Adenoid: growth of lymphoid tissue in the upper part of the throat, behind the nose.

Adherens junction: intercellular junction of epithelial cells, containing microfilaments: membranes are separated by 15–25 nm.

Adjuvant: a substance that enhances immune responses when administered along with antigens.

Adult T-cell leukemia: a malignant blood disease caused by T cell proliferation induced by HTLV-1 infection.

Amantadine: an antiviral agent that inhibits the activity of the M2 ion channel of influenza A viruses and blocks uncoating.

Ambisense: coding strategy used by arenaviruses and some members of the bunyavirus family, in which an RNA genome segment contains both positive- and negative-sense genes (*ambi* = both). Therefore some viral proteins are made from messenger RNAs copied from the genome RNA, while other proteins are made from messenger RNAs copied from the antigenome RNA.

Amyloid: insoluble proteins that accumulate in microscopic fibers and macrocrystalline arrays called plaques, usually resulting from a conformational change in a normal cellular protein.

Anaphase: phase of mitosis during which the chrosomes begin to migrate to opposite ends of the cell.

Antibodies: immunoglobulin proteins that recognize and bind to specific regions (epitopes) on a variety of types of molecules including proteins, carbohydrates, lipids, and nucleic acids.

Antibody-dependent enhancement: enhanced entry of certain viruses when the virions are complexed with antibody, resulting in an increased variety of host cells that can be infected.

Antigens: molecules or parts of a molecule that can induce antibody formation by binding to a specific antibody or T-cell receptor.

Antigenic drift: a slow change in the antigenic specificity of a protein due to the accumulation of point mutations in the gene coding for the protein. RNA viruses tend to mutate rapidly because their RNA polymerases do not have proofreading mechanisms to eliminate mutations that occur during genome replication.

Antigenic shift: a rapid change in the antigenic specificity of a protein. For RNA viruses with multiple genome segments, this is usually due to reassortment of genome segments between two virus strains that infect the same cell.

Antigenome: a full-length complementary copy of a single-stranded (usually RNA) genome.

Aplastic crisis: a severe disease condition in which no red blood cells are produced.

Apoptosis: programmed cell death, involving a number of signaling pathways that lead to activation of caspases (proteases), fragmentation of chromosomal DNA, condensation of the nucleus, and blebbing of the cytoplasm.

Arthropod-borne: transmitted (usually to mammals and birds) by arthropod vectors (e.g., mites, ticks, fleas, mosquitoes). Arthropod-borne viruses ("arboviruses") generally replicate both in the arthropod vector and in the vertebrate host.

Ascites: fluid that accumulates in the abdominal cavity as a result of impaired liver function or other pathologic conditions.

Asialoglycoprotein receptor: a liver-specific protein that binds to and internalizes glycoproteins that lack terminal sialic acid.

Autocrine loop: self-stimulation of growth by a cell population that produces a growth-promoting molecule and responds to the same molecule.

Autoimmune demyelination: loss of myelin in the central nervous system due to an immune response against myelin introduced with an impure vaccine.

Autophagy: a cellular process responsible for the degradation and recycling of cellular components by ingestion into vesicles called autophagosomes, which fuse with lysosomes containing digestive enzymes.

Autoreactive T lymphocytes: T lymphocytes that recognize antigens on normal cells in the body, and destroy them.

Azidothymidine: nucleoside analogue used for treatment of human immunodeficiency virus infections. It is activated by cellular nucleoside kinases and, in its triphosphate form, selectively inhibits viral reverse transcriptase.

β-galactosidase: an enzyme that cleaves lactose to galactose plus glucose; often used as a reporter enzyme, because artifical substrates such as Xgal produce colored cleavage products that are easy to detect and measure.

Bafilomycin A1: inhibitor of the vacuolar proton pump that acidifies endosomes.

Basal cells: cells in the edpidermis of mammalian skin that divide and differentiate into the outermost squamous cells of the epithelium.

B lymphocytes: white blood cells that originate in bone marrow and are involved in the adaptive immune response, especially in steps leading to the production of humoral antibodies.

Bronchitis: inflammation of the bronchi that results in coughing and excess sputum production.

BSE: bovine spongiform encephalopathy; prion disease in cattle.

Buckminsterfullerene: C_{60} molecule with icosahedral symmetry.

Budded virions: for baculoviruses, virions that are produced by budding from the plasma membrane (see **Occluded virions**).

Budding: the process of forming an enveloped virus particle by the folding of a cellular membrane around the virus capsid as it is extruded through the membrane.

Calyx: a carbohydrate-rich layer on the surface of occlusion bodies, sometimes called the outer or occlusion body envelope.

Capsid: the protein coat that encloses a viral genome.

Capsid-binding drugs: a class of drugs that bind to hydrophobic pockets on picornavirus capsids and inhibit viral attachment and/or entry.

Carboxylic ionophors: compounds that bind monovalent cations (including protons) and increase their membrane permeability.

Caspases: proteases that are activated during the events leading to apoptosis.

Cathepsins: a family of intracellular proteases with a variety of specificities.

Caveolae: small flask-shaped invaginations of the plasma membrane that are enriched in cholesterol, glycosphingolipids, and the protein caveolin.

Caveolin: major membrane protein component of caveolae.

CD4 antigen: a protein on the surface of helper T lymphocytes that is involved in interactions with other cells in the immune system.

CD4-positive T lymphocytes: T lymphocytes that carry CD4 antigen on their surface.

Cell-based high throughput screen: method that uses automated assays for screening large numbers of drugs for their ability to inhibit virus replication or other viral processes in infected cells.

Cellular immune responses: adaptive immune responses mediated by T lymphocytes. During virus infection, these responses are mediated primarily by cytotoxic $CD8^+$ T cells but also include $CD4^+$ T cell responses.

Chemokines: secreted proteins that attract leukocytes and are active in inflammatory reactions.

Chimera: an organism or genome resulting from the artificial combination (usually by genetic engineering) of genomes of two or more distinct species. In Greek mythology, a chimera was a fire-breathing monster that had a lion's head, a goat's body, and a snake's tail. (adj.: **chimeric**.)

Chloramphenicol acetyltransferase: an enzyme that transfers an acetyl group from acetyl-coenzyme A to chloramphenicol. This reaction can be measured by separating the acetylated from unacetylated chloramphenicol, and is used as a reporter assay to measure activity of genes.

Chlorpromazine: a tricyclic compound used as an antipsychotic drug that acts as an antagonist of certain neurotransmitters; inhibits accumulation of infectious prion protein PrP^{Sc} in cultured cells.

Chromatin remodeling: alteration of chromatin by modification (acetylation and methylation) of nucleosomal histones, changing their binding affinities to DNA and the access of chromatin to transcription factors.

Ciliated epithelium: epithelial cells that line body surfaces such as the respiratory tract and which are covered with cilia. Cilia are thin, elongated extensions of the plasma membrane containing microtubules; cilia move rhythmically to sweep up foreign particles and mucus.

Cirrhosis: severe damage to the liver, involving loss of normal strucure, caused by environmental factors or infection by pathogens.

***Cis*-acting replication elements (cre):** sequence elements located in viral RNA that are required for initiation of viral RNA replication. Cre probably interact with specific viral proteins.

CJD: Creutzfeldt-Jakob disease; prion disease in humans. The new variant CJD, suspected to be transmitted to humans by consumption of BSE-contaminated beef, strikes mostly young people.

Class I fusion proteins: envelope glycoproteins that form long alpha-helical coiled-coils, which undergo major conformational changes during fusion to bring the virus envelope and the target cell membrane in close proximity. Examples: orthomyxoviruses, paramyxoviruses, coronaviruses, filoviruses, retroviruses.

Class II fusion proteins: envelope glycoproteins that achieve fusion by a mechanism distinct from that of class I fusion proteins. Fusion involves rearrangement of multimers of protein subunits but only minor conformational changes in protein tertiary structure. Examples: flaviviruses, togaviruses, rhabdoviruses, hepadnaviruses.

Clathrin: a protein involved in the generation of endocytic vesicles; lines the cytoplasmic side of coated pits.

Coated pits: invaginations of the plasma membrane involved in endocytosis, and covered on their cytosolic surface by a clathrin protein coat.

Co-chaperone: a protein that associates with chaperones and stimulates their protein-folding and unfolding activities.

Coiled-coil: a protein structural element consisting of several alpha-helical chains coiled about each other, forming a rod-shaped domain.

Cold chain: a system to maintain the storage temperature for a temperature-sensitive product such as live vaccines all the way from the manufacturer to the site of administration. Many vaccines are susceptible to both excessive heat and freezing.

Complement: a large set of proteins involved in cascades of reactions stimulated by bacteria or antigen/antibody complexes, and leading to immune cytolysis and other protective mechanisms.

Concatemer: a linear or circular oligomer of nucleic acid genome sequences produced by rolling circle replication and/or recombination. Monomers can be arranged as direct repeats (head to tail) or inverted repeats (head to head; tail to tail) depending on how they are formed.

Condyloma: growth resembling a wart on the skin or a mucous membrane, usually of the genitals or anus, caused by infection with a human papillomavirus.

Confocal microscope: a microscope that uses focused lasers and computer imaging to visualize thin segments in living or fixed cells.

Core proteins: proteins associated with the viral genome inside the capsid.

CREB-binding protein (CBP): a histone acetylase that binds to a cellular transcription factor called cyclic AMP response element-binding protein (CREB). CREB binds to cyclic AMP response elements on DNA and activates transcription.

Cryoelectron microscopy: a method for visualizing biological structures, including viruses, at high resolution (to about 1 nm) by suspending unstained objects in a thin sheet of noncrystalline ice at liquid nitrogen temperatures. Computer-assisted averaging is utilized to reconstruct three-dimensional images of the object.

Cyclin-dependent kinases (CDKs): protein kinases that, in combination with cyclins, regulate the cell cycle by phosphorylating other proteins.

Cyclins: cellular proteins that regulate the cell cycle by associating with and activating cyclin-dependent kinases.

Cytokines: secreted proteins that act on cells of the innate and adaptive immune systems to stimulate or inhibit various immune functions.

Cytopathic: causes cell injury and death.

Cytoplasm: the portion of a euykaryotic cell enclosed by the plasma membrane, excluding the nucleus.

Cytosol: the soluble portion of the cytoplasm, excluding membrane-bound organelles and their contents.

Cytotoxic T lymphocyte: a class of lymphocytes, usually bearing the CD8 antigen, that attack and kill cells expressing foreign antigens presented by MHC class I molecules. In the context of virus infection, cytotoxic T lymphocytes are the primary effectors for killing virus-infected cells.

DC-SIGN (*d*endritic *c*ell-*s*pecific *i*ntercellular adhesion molecule-grabbing *n*onintegrin): a type II transmembrane mannose-binding protein found on dendritic cells, involved in their interaction with T cells during antigen presentation.

Delta structure: structure of a ribozyme derived from hepatitis delta virus.

Demyelination: loss of the myelin sheaths covering axons in the central nervous system, leading to dysfunction and loss of muscle control.

Dendritic cells: cells that accumulate and present foreign antigen to lymphocytes in lymph nodes and elsewhere in the body.

Discontinuous transcription: mechanism that copies and joins together nonadjacent sequences on an RNA or DNA template by displacement of the RNA polymerase along with its nascent RNA chain from one site to another.

Disseminated intravascular coagulation: systemic clotting of the blood; can be caused by activated neutrophils.

Divergent transcription: transcription of adjacent genes that progresses in opposite and diverging directions from a central point. RNAs are therefore transcribed from complementary genome strands.

DNA gyrase: a bacterial enzyme that overwinds double-stranded DNA by cleaving a phosphodiester bond and rotating one strand around the complementary strand using energy provided by ATP, to generate supercoiled DNA.

DNA vaccine: introduction into the body of plasmids that express specific proteins, usually of a selected pathogen, resulting in an immune response against those proteins.

DNA-dependent RNA polymerase: An enzyme that synthesizes RNA using a DNA template.

Dorsal root neurons: sensory neurons whose ends are in peripheral organs and whose cell bodies are grouped in dorsal root ganglia near the spinal cord.

Double-stranded RNA-dependent protein kinase: see **PKR**.

Dura mater: the tough outer membrane covering the brain and spinal cord (from Latin words for "hard" and "mother").

Dynactin: an adaptor protein complex that allows dynein to bind to its cargo.

Dynein: motor complex that typically carries cargo in the minus end direction (towards the center of the cell) along microtubules.

Early endosome: intermediate vesicle to which molecules bound to receptors during receptor-mediated endocytosis are first delivered.

Ectodomain: the part of a membrane protein that projects into the extracellular space.

Encephalitis: disease involving inflammation of the brain.

Endemic: natural to a specific animal or human population or place; indigenous; native. Refers to a disease or pathogen that is present and circulates among hosts in a specific location, but does not necessarily infect a large segment of the population.

Endoplasmic reticulum–Golgi intermediate compartment (ERGIC): a membranous cytoplasmic organelle that is continuous with the endoplasmic reticulum and the Golgi apparatus, and is involved in transport, processing, and modification of proteins.

Endosome: intermediate membrane vesicles in the receptor-mediated endocytosis pathway.

Enfuvirtide: a peptide that blocks a conformational change in HIV-1 gp41 necessary for fusion of the viral envelope with the cell membrane.

Enteric: intestinal.

Envelope: a lipid bilayer membrane, usually derived from a cellular membrane, into which viral glycoproteins are inserted. Envelopes are used by many viruses as an external layer in which their nucleocapsids are wrapped.

Ependymal cells: epithelial cells that line the brain and spinal cord.

Epidemic: increase in infectious disease incidence over time and/or geographic range. Epidemics are larger than "outbreaks" (which are more geographically limited) and smaller than "pandemics."

Epitopes: short regions on antigens such as proteins (usually 10 amino acids or less) or carbohydrates, to which specific antibodies bind.

Epizootic: an epidemic in one or more animal species.

Erythema infectiosum: Latin for "infectious rash." A childhood viral disease with typical rash and fever, also known as "fifth disease."

Erythroid precursors: cells on the pathway of differentiation from hemopoietic stem cells to mature red blood cells.

Ethidium bromide: a fluorescent polycyclic aromatic molecule that intercalates between bases in DNA or RNA. Used to detect and quantitate nucleic acids.

Exocytosis: the process of delivery of material within vesicles derived from the endoplasmic reticulum to the outside of the cell by their passage through the Golgi membranes and fusion with the plasma membrane.

Exportins: proteins that mediate the export from the nucleus to the cytoplasm of proteins or RNAs.

Extracellular enveloped virus (EEV): for poxviruses, a fraction of intracellular mature virions (IMV) are wrapped in a double membrane derived from the Golgi apparatus. These particles are exported by fusion of the outermost membrane with the plasma membrane, resulting in a virion with two envelopes called extracellular enveloped virus. Important for spread of virus within host.

Family: in virology, a taxonomic grouping of genera whose members share common virion structure, genome organization and size, and replication strategy, and for which there is evidence of a common evolutionary origin by comparison of nucleotide and amino acid sequences.

Filterable virus: the name given to viruses when they were first distinguished from other microorganisms on the basis of their small size and their ability to pass through fine-pore filters.

Folate receptor-α: a cellular membrane protein that binds to folic acid, a soluble B vitamin.

F-pili: filamentous protrusions from the surface of *E. coli*, coded by the F plasmid, that mediate bacterial conjugation.

Fulminant hepatitis: severe and rapidly progressive hepatitis often leading to hepatic failure and death.

Furin: a host cell endopeptidase, localized in the *trans*-Golgi compartment of the cell's secretory pathway, that cleaves proteins at arginyl residues.

Fusion peptide: a short hydrophobic region within a viral envelope protein that is inserted into the target cellular membrane during virus-induced membrane fusion.

Ganciclovir: nucleoside analogue used for treatment of human cytomegalovirus infections. It is selectively phosphorylated by the viral UL97 protein kinase and, in its triphosphate form, selectively inhibits viral DNA polymerase.

Gangliosides: sialic acid–containing glycosphingolipids found in cellular membranes.

Gastroenteritis: inflammation of the gastrointestinal tract resulting in diarrhea, vomiting, and/or cramps.

Gene gun: apparatus that projects tiny particles (gold or other inert material) coated with DNA into cells at high speeds, allowing penetration of cell membrane and intracellular expression of proteins encoded by the DNA.

Genetic drift: see **Antigenic drift**.

Genetic economy: principle that organisms with genomes of limited size use their protein-coding capacity economically, for example, by making capsids with multiple copies of a few different proteins.

Genetic shift: see **Antigenic shift**.

Genus: in virology, a grouping of closely related virus species. (Plural: **genera**.)

Glomerular nephritis: inflammation of glomeruli in the kidney.

Glycolipid: lipid molecule with a polar head group that contains carbohydrate.

Glycoprotein: protein that has covalently bound carbohydrate residues.

Granulin: protein that forms ovoid occlusion bodies of granuloviruses.

Green fluorescent protein: a jellyfish protein that emits green light when exposed to ultraviolet light. Widely used in cell biology for intracellular localization of specific cellular proteins fused to green fluorescent protein by genetic engineering.

Guanylyltransferase: an enzyme that transfers a guanosine residue to the phosphorylated 5′ end of an RNA, creating an RNA "cap" structure.

Hairpin structure: a stem structure formed by base-pairing in a nucleic acid that is internally self-complementary.

Hairy-cell leukemia: a malignant blood disease caused by proliferation of a type of B cell.

Hammerhead structure: secondary structure found in ribozymes of group A viroids. It can be drawn in a shape that resembles a hammerhead.

Hantavirus pulmonary syndrome: an often-fatal lung disease resulting from infection by hantaviruses such as sin nombre virus, transmitted by rodents.

Helical symmetry: symmetry defined by rotational and translational displacement about a single helix axis.

Helicase: an enzyme that unwinds DNA or RNA helices.

Hemagglutination: the agglutination or clumping of red blood cells.

Hemagglutination assay: a method of titering viruses that agglutinate red blood cells.

Hemagglutinin: a protein that binds to receptor molecules on the surface of red blood cells, leading to their agglutination.

Hemifusion: intermediate in the membrane fusion reaction where only the outer bilayer halves have fused.

Hemocoel: in arthropods, an open body cavity in which blood flows and bathes the tissues and organs.

Hemolytic anemia: a disease caused by shortage of red blood cells due to their fragility.

Hemorrhagic fevers: diseases caused by a number of viruses, involving high fever, vascular damage leading to internal bleeding, and resulting shock, which if untreated can lead to organ failure and death.

Hemorrhagic fever with renal syndrome: a disease caused by various hantaviruses involving internal bleeding and kidney pathology, which can be mild or severe with a high fatality rate.

Hepadnavirus: a family of DNA-containing animal viruses that replicate via an RNA intermediate, using reverse transcriptase.

Heparan sulfate: a sulfated polysaccharide (glycosaminoglycan) made from repeating disaccharides. These are covalently attached to core proteins to form proteoglycans, found on the cell surface.

Hepatitis: disease involving damage to and inflammation of the liver.

Hepatocellular carcinoma: malignant tumor of the liver.

Herd immunity: sufficiently high vaccine coverage in a population such that the capacity of viral pathogens to spread is limited, thus protecting the small number of unvaccinated individuals in the community.

Hexon: structural subunit of adenovirus virions, made from trimers of the hexon protein.

Hippocampus: a section of the brain that is located inside the temporal lobe and is part of the limbic system. The major functions are involved in memory and navigation.

Histone acetylases: enzymes that acetylate nucleosomal histones and other proteins, reducing their binding affinity for DNA and resulting in a more relaxed chromatin structure that is more accessible to specific transcription factors and is more easily transcribed.

Holin: a small, bacteriophage-coded protein that self-associates in the bacterial membrane under certain conditions of pH and ionic concentrations to open holes in the membrane, allowing phage endolysin to enter the periplasmic space and attack the bacterial cell wall.

HTLV-1 associated myelopathies (HAM): a disease similar to **tropical spastic paraparesis.**

Humoral immune response: all aspects of the response to foreign antigens by the adaptive immune system that result in production of antibodies.

Hydrocephalus: swelling of the cerebral ventricles.

Hypoxemia: low oxygen level in the blood, leading to lack of oxygen supply to tissues and organs.

Icosahedral symmetry: symmetry of a regular icosahedron, with 12 axes of fivefold symmetry, 20 axes of threefold symmetry, and 30 axes of twofold symmetry.

IκB: an inhibitor that binds to the transcription factor NF-κB, maintaining it in an inactive state in the cytoplasm.

IκB kinase complex (IKK): a serine-threonine protein kinase that phosphorylates IκB, resulting in the dissociation and activation of NF-κB.

Immunogenic: able to induce an immune response.

Immunoglobulin family: a family of membrane proteins that have characteristic repeated domains usually stabilized by disulfide bonds, first described in immunoglobulins.

Immunoprophylaxis: treatment of an infectious disease by injection of antibodies directed against the disease agent.

Importin (also called karyopherin): receptor protein that interacts with nuclear localization signals in proteins.

Inclusion bodies: aggregates of viral proteins or nucleic acid visible by light microscopy in infected cells by their staining characteristics.

Infectious subvirion particles (ISVPs): in reoviruses, particles that have lost part of the external capsid as a result of protease digestion in endosomal vesicles or the intestine.

Intasome: an assembly of the integrase and the host protein IHF on the phage lambda *attp* site; the intasome captures the bacterial *attb* site and carries out recombination between the two *att* sites.

Integrins: a family of cell surface proteins (usually heterodimers) that interact with extracellular adhesion proteins and initiate intracellular signaling pathways.

Interferons: a family of cytokines that induce antiviral activity and protect cells against virus infection.

Intergenic region: for polyomaviruses, a 400 nt region located on the viral genome between the early and late genes. Contains important transcription and replication control sequences.

Interleukin-1: a cytokine, secreted early in the immune response, that activates both T and B lymphocytes.

Interleukin-2: a cytokine that stimulates T-cell growth and proliferation.

Internal ribosome entry sites (IRES): regions on viral RNAs (particularly picornaviruses and flaviviruses) that bind to ribosomes and lead to cap-independent initiation of protein synthesis at internal sites distant from the 5′ end of the RNA. IRES contain extensive secondary and tertiary structures that interact with a variety of cellular proteins. Some cellular genes have been shown to contain IRES.

Intracellular mature virus (IMV): infectious poxvirus particles made in the cytoplasm. Some of these can be wrapped in an additional envelope to form extracellular enveloped virus (EEV). IMV are more stable than EEV, and are important for spead of virus from host to host.

Intussusception: a pathological folding of the intestine into itself leading to obstruction, ischaemia, and sometimes death.

Inverted terminal repeat: a nucleotide sequence present at both ends of a nucleic acid molecule but in the opposite orientation; the two ends of a single-stranded DNA with such a repeat are complementary and therefore can anneal to each other.

Ionophores: small lipid-soluble molecules that bind specifically to certain ions and transport them across membranes.

Jak-Stat pathway: signaling by interferons, other cytokines, and growth factors is mediated by a pathway that includes the Janus tyrosine kinases (Jak kinases) and the signal transducers and activators of transcription (Stat proteins), via a cascade of tyrosine phosphorylation events leading to the activation of transcription of a variety of target genes.

Jaundice: yellowing of the skin due to the presence of bile pigments in the blood, often caused by poor liver function.

Jelly-roll β-barrel: folding of a polypeptide chain into a number of parallel β strands, usually as two sheets, found in many virion capsid proteins.

Junctional adhesion molecule A (JAM-A): a cellular membrane protein found in tight junctions; member of the immunoglobulin superfamily of cell-adhesion molecules (CAMs).

Karyopherin: see **Importin**.

Keratinocytes: squamous epithelial cells that comprise skin and mucosa and provide a barrier between the host and the environment.

Kuru: prion disease of the Fore people in Papua New Guinea, caused by ingestion of prions during ritual cannibalism. Means "shivering" or "trembling" in the Fore language.

Laminin receptor: a cell surface protein that binds to laminin. Laminins are a family of proteins, each containing three polypeptides of about 820 KD, that are a component of the basal lamina, which lies underneath epithelial cells.

Latency: a state of inactivity in which the viral genome is present in the cell in a dormant state; little or no viral transcription or replication take place.

Latent infection: an infection in which a pathogenic agent remains present in the organism but replicates slowly or not at all and produces few or no symptoms.

Lectins: plant proteins that bind to specific oligosaccharides in cell membranes.

Leucine zipper domain: in proteins, an alpha-helical region with leucine residues at each 7th position, which enables dimerization between proteins.

Ligand: a specific molecule that is bound by a protein.

Lineage: a collection of individual organisms linked by close genetic relationships.

Lipid rafts: microdomains in plasma membranes of eukaryotic cells that are enriched in cholesterol, glycolipids, and sphingolipids and in certain membrane proteins. Lipid rafts are believed to be targets for binding of certain viruses during virus entry, as well as for assembly of certain enveloped viruses.

Lipopolysaccharide: A complex molecule containing fatty acids and oligosaccharides, found in the outer membrane of gram-negative bacteria.

Lumen: the inner cavity of cellular vesicles, such as the endoplasmic reticulum.

Lysis: breaking open of a virus-infected cell, leading to its death and the release of virus particles within the cell.

Lysogen: a bacterial cell containing a prophage.

Lysosome: subcellular organelle in eukaryotic cells responsible for degradation of substances taken up by endocytosis.

Lysosomotropic agent: weakly basic compound that elevates the pH in acidic organelles of the cell.

Macrophages: cells present in a variety of tissues and organs that engulf and digest extracellular debris, cell fragments, and pathogens, and are involved in antigen presentation to cells of the adaptive immune system.

Mad cow disease: popular name for bovine spongiform encephalopathy.

Major histocompatibility complex (MHC) proteins: cell surface proteins that bind to antigens or antigen fragments and present them to cells involved in the immune response.

Matrix protein: in virology, virus-coded proteins that lie just underneath the viral envelope and form a connecting bridge between the envelope and the viral nucleocapsid.

Membrane fission: the process by which one membrane divides or "buds" into two.

Membrane fusion: the process by which two membranes combine to form one.

Metalloproteinases: enzymes that degrade proteins by cleaving peptide bonds, and that use metal ions such as zinc in the catalytic mechanism.

Methyltransferase: an enzyme that transfers a methyl group from S-adenosyl-methionine to an acceptor molecule (e.g., a nucleotide base or ribose during RNA cap formation)

Micelle: small, water-soluble droplets of polar lipids.

Microvilli: finger-like projections of the plasma membrane, containing a core of actin filaments, on the surface of various cell types such as those of the intestinal epithelium. Microvilli greatly increase the cell surface and play an important role in absorption of substances from the extracellular medium. (Singular: **microvillus**.)

Mitotic arrest-defective (MAD): proteins that control entry into anaphase, preventing separation of chromosomes on the mitotic spindle in dividing cells.

Mitotic spindle checkpoint: a control mechanism that prevents the onset of anaphase until chromosomes are aligned properly on a bipolar spindle; controlled by mitotic arrest-defective (MAD) proteins.

Monensin: monovalent **ionophore** that disrupts the passage of proteins through the Golgi complex, and has multiple effects on glycosylation.

Monoclonal antibodies: antibodies produced by cloned antibody-producing cells. Each monoclonal antibody therefore has a unique amino acid sequence and binds to a highly specific epitope.

Mucociliary flow: flow of mucus and debris along the epithelial lining of the respiratory tract modulated by the movement of cilia on the epithelial cells.

Mucosal epithelium: epithelial cells of the mucosal membrane.

Multinucleated giant cells: see **Syncytium**.

Multiplicity of infection (m.o.i.): the number of infectious virus particles inoculated per cell.

Multivesicular bodies: cytoplasmic vacuoles containing many small membrane vesicles and lysosomal degradative enzymes.

Myocarditis: disease involving inflammation of the muscular walls of the heart.

Myocardium: heart muscle.

Myristate: a C_{14} linear, saturated fatty acid that can be joined via its carboxyl group to N-terminal glycine in a number of proteins; the fatty acid becomes embedded in a cellular membrane, linking the protein to the membrane.

Myristoylation: covalent binding of the 14-carbon fatty acid myristate to N-terminal glycine residues on proteins, leading to their association with lipid membranes.

Natural killer (NK) cells: lymphocytes that attack and kill abnormal cells, particularly cells infected with pathogens and cancer cells; part of the innate immune response.

Nectins: a family of cell surface proteins, with an immunoglobulin-like structure, that function in cellular adhesion.

Negative (or minus) strand: for RNA viruses, the strand complementary to the RNA that contains protein-coding regions. Some viruses specifically package the negative RNA strand in virions.

Negative staining: staining of specimens for electron microscopy by use of heavy-metal salts that absorb electrons but do not bind to the specimen. The image is formed by exclusion of the stain by the specimen, therefore the image is "negative."

Nested set: a set of similar objects, each of which fits into the next larger object.

Neuraminidase: enzyme that cleaves terminal sialic acid (N-acetyl neuraminic acid) from glycolipids or glycoproteins.

Neutralizing antibodies: antibodies that inactivate or "neutralize" virus infectivity when they come in contact with the virus particle.

Nevirapine: nonnucleoside reverse transcriptase inhibitor.

New variant Creutzfeldt-Jakob disease (nvCJD or vCJD): prion disease in humans believed to be caused by transmission of prions from diseased cattle.

NF-κB (nuclear factor-κB): a heterodimeric transcription factor that regulates many aspects of immune responses, apoptosis, and a variety of signaling pathways. It is sequestered in an inhibitory complex in the cytoplasm, from which it is released upon receipt of a variety of extracellular signals; it travels to the nucleus and activates transcription of numerous target genes.

Nonnucleoside inhibitors: inhibitors of DNA polymerases that are not nucleoside analogs.

Nonpermissive: describes a cell type that cannot support productive infection by a particular virus. Nonpermissive cells may block virus infection at many possible stages, including virion binding, uptake, gene expression, genome replication, and virion assembly.

Nucleocapsid: viral capsid containing the viral nucleic acid genome.

Nucleoside diphosphate kinase: an enzyme that phosphorylates nucleoside diphosphates (e.g., deoxyguanosine diphosphate) to nucleoside triphosphates.

Nucleosomes: complexes assembled from two copies each of the four cellular histones H2A, H2B, H3, and H4. Chromosomal DNA is condensed by wrapping around the exterior of cylindrical nucleosomes.

Obligatory intracellular parasite: an infectious element or organism that can replicate only within other cells.

Occlusion body: a crystalline protein matrix formed in the nucleus of cells infected with baculoviruses, consisting of enveloped viral nucleocapsids embedded in polyhedrin or granulin.

Occlusion-derived virions: virions embedded in occlusion bodies, formed in the nucleus.

2′, 5′-oligo(A) synthetase: an enzyme that produces oligomers of up to 10 ATP residues linked via a 2′, 5′ phosphodiesterase bond. Part of the interferon response, this enzyme is activated by double-stranded RNA. 2′, 5′-oligoadenylate binds to and activates ribonuclease L, which degrades mRNAs.

Oncogene: a viral or cellular gene whose expression can lead to cell transformation or tumorigenesis.

Oncolytic: able to kill or lyse tumor cells.

Oncotropism: the ability of a virus to replicate more efficiently in cancer cells than in normal cells.

Operator: a DNA or RNA binding site for a repressor protein that negatively regulates gene expression.

Opportunistic infections: infections that flourish in an immune-deficient individual but that are usually absent from healthy individuals.

Order: in virology, a rarely used taxonomic grouping of virus families whose members share distinctive characteristics of replication strategy, but which may diverge widely in host organism and genome size, and which may share little or no nucleotide sequence homology.

Oseltamivir: neuraminidase inhibitor used for treatment of influenza A and B virus infections.

Packaging signal: sequence on a DNA or RNA genome that is recognized by viral proteins involved in packaging the genome in the capsid.

Palmitoylation: covalent attachment of palmitic acid to cysteine sulfhydryl groups in a protein; increases affinity of the protein for lipid membranes.

Pandemic: increase in infectious disease incidence over time and a large geographic range spanning many countries and continents (e.g., the human influenza pandemic that spread worldwide in 1918–1920).

Panleukopenia: infectious disease (of cats) leading to shortage of white blood cells, among other symptoms.

Panzootic: increase in infectious disease incidence in animal species over time and large geographic range (e.g., panzootic of rabies in raccoons extending up eastern North America through the 1970s to the present).

Papain-like cysteine proteinases: enzymes that have structural homology to the proteinase papain, and that use cysteine as the nucleophile for hydrolysis of peptide bonds.

Papanicolaou test: a diagnostic test for cervical cancer at an early stage, by examination of exfoliated cells stained with Papanicolaou stain. It is the most successful test invented for cervical cancer prevention.

Penton base: structural subunit of adenovirus virions, made from pentamers of the penton protein and located at the 12 vertices of the icosahedral virion.

Peptidomimetic: drug that mimics a peptide, but is not itself a peptide.

Peritonitis: inflammation of the peritoneum (the lining surrounding the abdominal cavity).

Peritrophic membrane: chitinous protective sheath secreted around food in insect gut.

Permissive: a cell type is considered permissive for infection by a particular virus if it can be productively infected and give rise to progeny virus.

Persistent infection: an infection (in this case, by a virus) of cultured cells or of an animal during which the virus is not eliminated, but continues to replicate for a prolonged time at a low level, without killing the host.

Phagocytosis: the uptake of particulate material from the extracellular medium by vesicles in phagocytic cells.

Pharmacokinetics: study of the fate of a drug when administered to an organism, including its uptake, its distribution, its degradation, and its clearance.

Picornavirus 3C-like proteinases: enzymes that have structural homology to the picornavirus 3C proteinase, which is involved in cleavage of picornavirus polyproteins into mature proteins.

PKR (double-stranded RNA-dependent protein kinase): a cellular serine-threonine protein kinase that inactivates protein synthesis by phosphorylating eukaryotic initiation factor 2B (eIF-2B), in response to virus infection or the presence of double-stranded RNA.

Plaque assay: a method for enumerating infectious virus particles by generation and counting of plaques, or focal areas of cell death, on a monolayer of cells. Each plaque usually arises from a single infectious virus particle.

Plaque-forming units (PFU): infectious virus particles as detected by a plaque assay. Usually expressed as PFU/ml.

Plasma cells: terminally differentiated B cells specialized in producing and secreting antibodies.

Plasmodesmata: membrane channels connecting neighboring plant cells through the cell wall.

Pleconaril: a capsid-binding drug active against rhinoviruses and other picornaviruses.

Pock: a localized blistering lesion on skin or mucosae caused by poxviruses or other agents.

Poliomyelitis: disease caused by poliovirus, involving central nervous system lesions that can lead to paralysis.

Polyadenylation signal: in eukaryotes, a consensus AAUAAA sequence in messenger RNAs that is recognized by proteins that bind to the RNA, cleave it downstream of the AAUAAA to create a new 3′ end, and add a poly(A) tail to the 3′ end.

Polyclonal: derived from a number of individual clones. Tumor cells or cells of the immune system are considered polyclonal if they have numerous different origins or antigenic specificities.

Polyhedrin: protein that forms the polyhedral occlusion bodies of nucleopolyhedroviruses.

Polyprotein: a precursor protein that is cleaved by proteolysis during or after synthesis to generate several mature functional proteins.

Positive (or plus) strand: the RNA strand that contains coding regions for proteins.

Positive staining: staining of specimens for electron microscopy by use of heavy-metal salts that bind to the specimen and impart electron density. Usually used for observation of thin sections of cells.

Preintegration complex: a complex of retrovirus capsid proteins and newly synthesized proviral DNA in a state ready to be integrated into the cellular genome.

Primary cell cultures: cells grown *in vitro* that are derived directly from an animal tissue. Cells from embryos or newborn animals are often used because of their ease of growth.

Primosome: a protein complex that synthesizes RNA primers on which DNA replication can be initiated.

Prion: pathogen that replicates and causes prion diseases, presumably a conformational variant $PrPp^{sc}$ of a normal cell protein, PrP^{c}. The term *prion* stands for "*pr*oteinaceous *i*nfectious" particle. [Note: in biology, the name "prion" also refers to a group of 6 or 7 different species of small petrels (oceanic birds) restricted to the Southern Ocean surrounding Antarctica.]

Prion disease: neurodegenerative disease caused by the accumulation of prions in the brain, leading to spongiform encephalopathy, and characterized by accumulation of amyloid fibrils and plaques. Invariably fatal. Can be caused by inherited mutations in PrP gene or transmitted from infected animals, usually of same species.

Procapsid: a precursor to a viral capsid that usually does not contain the viral genome. Procapsids often change in structure or protein composition when they incorporate the viral genome to form a mature capsid.

Processivity: the ability of a polymerase to extend a growing chain without detaching from the template.

Progressive multifocal leukoencephalopathy: a rare, fatal demyelinating disease of humans caused by the polyomavirus JC.

Prohead: term used frequently to designate procapsids of bacteriophages. See **Procapsid.**

Promoter occlusion: silencing the use of a transcriptional promoter, usually as a result of the presence of an active upstream promoter. Promoter occlusion may occur by dislodging of transcription factors from the downstream promoter by transcribing RNA polymerases that initiate upstream.

Prophage: a repressed bacteriophage chromosome that is integrated into the bacterial chromosome or replicates as a free plasmid.

Proteasomes: large specialized complexes that degrade proteins in the cell and make peptides available to MHC class I proteins for antigen presentation.

Protein tyrosine kinase: an enzyme that phosphorylates proteins at specific tyrosine residues.

Proteoglycan: extracellular macromolecule composed of large negatively charged polysaccharide chains and protein.

Proteosomes: vesicles made from the outer membrane of *Neisseria* bacterial species, into which exogenous antigens are introduced. Injection of proteosomes into animals can induce immunity to these antigens.

Protomers: subunits used to build a larger structure.

Proto-oncogene: a cellular regulatory gene that, when mutated or overexpressed, can cause cellular transformation and oncogenesis.

Proviral DNA: a DNA copy of a retrovirus genome that is inserted into the chromosome of an infected host cell.

Pseudoknot: an RNA secondary structure in which bases in the loop at the end of a stem form hydrogen bonds with bases at a distant site on the same RNA molecule.

Pseudotype: see **Pseudotyping**.

Pseudotyping: production of a recombinant virus (pseudotype) that carries a foreign receptor-binding protein, allowing the virus to bind to and infect different cell types or species.

Pseudovirion: virions of polyomaviruses that encapsidate fragments of cellular chromatin instead of the viral genome.

Quasi-equivalent: in virus capsid structure, describes the symmetrical distribution of protein subunits on the surface of a capsid in such a way that they form similar but not identical bonds with each other.

Quasispecies: virus variants that differ by a small number of mutations.

Quinacrine: a tricyclic compound previously used as a drug against malaria; a fluorescent dye, used to detect DNA; inhibits accumulation of PrP^{Sc} in cultured cells.

Raft culture: in vitro culture system used to differentiate epithelial cells by growth at an air–liquid medium interface.

Ras: a cellular signaling protein that is activated by receptor tyrosine kinases and in turn activates the MAP kinase pathway, which itself induces transcription of cell cycle and other genes.

Rational drug discovery: approaches for discovering new drugs that use basic knowledge from genetics, molecular biology, enzymology, and protein structure to choose drug targets and molecules to act on these targets.

Reassortant viruses: viruses that have packaged different genome segments from two (or more) parent viruses and therefore express some viral proteins from one strain and some viral proteins from the other strain. Reassortment can occur at high frequency with viruses that have segmented genomes.

Reassortment: see Reassortant viruses.

Receptor tyrosine kinases: cell surface receptor proteins that respond to binding of signaling molecules by phosphorylating specific tyrosine residues on intracellular proteins, activating a variety of signaling pathways leading to activation or inactivation of numerous cellular genes.

Receptor-mediated endocytosis: endocytic uptake of molecules from the extracellular space involving their specific binding to plasma membrane receptors.

Replication complex: for RNA viruses, a factory for viral RNA synthesis consisting of viral and cellular proteins associated with membranes in the cytoplasm of an infected cell.

Replicative intermediate: for RNA viruses, an RNA molecule on which one or several growing complementary RNA strands are being synthesized. The growing strands are usually base-paired to the template RNA only near their growing 3′ ends.

Replicon: a nucleic acid that, when transfected into cells, is capable of replication directed by proteins encoded in the nucleic acid. RNA virus replicons may lack coding sequences for viral structural proteins.

Reporter gene: a gene coding for a protein that can be detected within cells easily and quantitatively. Examples include green fluorescent protein, beta-galactosidase, and chloramphenicol acetyltransferase. The insertion of a reporter gene into a virus genome allows for the detection of virus infection and replication by assaying for the appearance of the reporter protein.

Reservoir: a population of organisms in which a pathogenic agent circulates in nature, not necessarily causing overt disease. Humans or domestic animals can be infected by coming in contact with members of this population, and they can suffer from disease, but do not necessarily transmit the agent efficiently to other members of their species.

Resolvase: an enzyme that makes staggered single-strand breaks at the ends of a concatemer junction, allowing formation of single-length DNA molecules with hairpin ends.

Retinoblastoma: a tumor of the retina. Can be caused by absence of retinoblastoma protein (pRb), a tumor suppressor protein.

Retrograde transport: transport of macromolecules within nerve axons in the direction from the nerve periphery to the nucleus, or cell body.

Retrovirus: a family of viruses with positive-strand RNA genomes, which replicate via a DNA intermediate, synthesized by a virus-coded RNA-dependent DNA polymerase (reverse transcriptase).

Reverse genetics: The process of engineering specific genetic changes in the genome of an organism and then looking for resulting phenotypic changes.

Reverse transcriptase: an enzyme that synthesizes DNA using an RNA template: also known as an RNA-dependent DNA polymerase.

Ribonuclease H: a ribonuclease that specifically degrades the RNA part of RNA/DNA hybrids.

Ribonuclease L: an enzyme that, when activated by binding of 2′, 5′-oligoadenylate, degrades cellular and viral mRNAs. Part of the interferon response.

Ribose methyltransferase: an enzyme that transfers a methyl group to the 2′ position of ribose during generation of a methlylated cap structure at the 5′ end of an RNA.

Ribosomal frameshifting: a process by which ribosomes pause during translation of a coding region on a messenger RNA, move back one or two nucleotides on the messenger RNA, and recommence reading the messenger RNA in a different translational reading frame.

Ribosome shunting: a mechanism by which ribosomes are able to locate initiation codons for translation of messenger RNAs without "scanning" the RNA from its capped 5′ end by the classical pathway using initiation factor eIF4F.

Ribozyme: a *ribo*nucleic acid that acts as an en*zyme*. The RNA catalyzes a specific biochemical reaction in a manner that can transform several substrate molecules per ribozyme molecule. Some known ribozymes can cleave or ligate RNAs.

Ritonavir: protease inhibitor used for treatment of human immunodeficiency virus infections.

RNA helicase: an enzyme that unwinds helical structures (stems) in single-stranded RNA molecules, or unwinds the strands of double-stranded RNA molecules. This enzyme is usually required for transcription and replication of RNA.

RNA polymerase II holoenzyme: a complex consisting of eukaryotic RNA polymerase II plus accessory proteins, which is active in transcription.

RNA world: a hypothetical prebiotic world in which self-replicating RNAs developed and evolved, preceding and perhaps leading to development of the DNA world that is universal in present-day life.

RNA-dependent DNA polymerase: an enzyme that synthesizes DNA using an RNA template; reverse transcriptase.

RNA-dependent RNA polymerase: an enzyme that synthesizes RNA using an RNA template. Viruses with RNA genomes must code for such an enzyme, which is not available in the host cell.

Rolling-circle replication: synthesis of a linear RNA or DNA from a circular template by an RNA or DNA polymerase that cycles around the template numerous times without detaching, resulting in an extended linear molecule that contains tandem repeats of the complement of the template sequence. These are usually cleaved into unit-length molecules.

Satellite nucleic acid: a nucleic acid genome that is able to replicate only in the presence of a helper virus, and which is packaged as a virion by capsid proteins coded by the helper virus. Genomes are usually less than 2 Kb long and in some cases code for no proteins.

Satellite virus: a virus whose genome is able to replicate only in the presence of a helper virus, but which codes for its own capsid protein (and usually no other proteins).

Scaffolding proteins: proteins that are involved in the assembly of complex biological structures such as viral capsids, but are not retained in the final structure.

Scrapie: prion disease in sheep, characterized by rubbing or scraping of the skin.

Secondary receptor: cell-surface receptors that interact with viral capsid or envelope proteins to direct uptake, but only after the virus particle has bound to a primary receptor on the cell surface.

Self-assembly: ability of a structure such as a viral capsid to assemble spontaneously from its protein subunits, without the aid of other machinery.

Seroconversion: the appearance in serum of antibodies specific to a pathogen, indicating exposure to the pathogen but not necessarily overt disease.

Serpin: *se*rine *p*rotease *in*hibitor. Can block cleavage of cytokines and therefore block their normal biological functions.

Serum sickness: a hypersensitivity response to large amounts of antigen in the circulatory system.

Shine-Dalgarno sequence: a consensus sequence located just upstream of translational start sites in bacterial mRNAs that facilitates initiation of protein synthesis by binding to 16S ribosomal RNA.

Shock: inability of circulatory system to maintain adequate blood supply to vital organs; most often caused by low blood volume due to loss of blood.

Sialic acid: a 9-carbon acidic sugar derived from mannose, found on many glycolipids and glycoproteins. Also called N-acetyl neuraminic acid.

Sickle cell disease: a hereditary form of anemia caused by an abnormal type of hemoglobin and characterized by unusual crescent-shaped red blood cells—"sickle cells."

Signal peptidase: a host protease that cleaves the signal sequence from membrane or secretory proteins during or shortly after their transport into the lumen of the endoplasmic reticulum.

Signal sequence: amino acid sequence that directs insertion of the N-terminus of proteins into the endoplasmic reticulum during protein synthesis, and is cleaved from the protein in the process.

Silencing: reduction or elimination of the activity of a transcriptional promoter, often because sequences in the promoter region are methylated.

Sodium dodecyl sulfate: a 12-carbon ionic detergent that is used to denature proteins.

SOS genes: a collection of LexA-regulated genes that function to repair damaged DNA.

Spasticity: continuous contraction of certain muscles, usually resulting from damage to brain centers that control voluntary movement.

Species: in virology, a taxonomic grouping of virus isolates that share a high degree of nucleic acid and amino acid sequence homology in addition to having a common genome structure and size, common virion structure and replication strategy, and (often but not always) a limited host range.

Spliceosome: a complex made up of a number of small ribonucleoproteins bound to pre-messenger RNAs. Carries out cleavage and joining reactions leading to production of spliced messenger RNAs.

Spongiform encephalopathy: disease accompanied by large vacuoles in the cortex and cerebellum, giving the brain the appearance of a sponge.

Squamous cell carcinoma: a type of skin cancer arising from the upper layers of skin epidermis.

Src-related kinases: enzymes like c-src that phosphorylate tyrosines on specific target proteins and regulate numerous cellular signaling pathways.

Subacute sclerosing panencephalitis (SSPE): a progressive, fatal neurological disease that is a rare complication of measles virus infection. Only one in 300,000 measles victims are affected by SSPE, which seems to result from a slow, persistent infection and strikes up to 10 years after the acute illness.

Subgenomic mRNA: a messenger RNA that includes only part of the sequence of the genome of an RNA virus.

Supercoiled: double-stranded (usually circular) DNA molecules that are overwound and as a result are twisted into a coiled structure.

Surface lattice: a set of lines drawn on the surface of an object that connect points defining axes of symmetry.

Synchrotron: device used to accelerate electrons, which can then be used to create high-energy x-rays.

Syncytium: a group of cells that share a common cytoplasm with more than one nucleus. Can be formed by cell-cell fusion. (Plural: syncytia.)

Target-based high throughput screen: method for screening large numbers of drugs for their ability to affect biological activity of a purified protein by use of automated *in vitro* assays.

TATA box: a consensus sequence, often TATAAA, to which the TATA-binding protein (TBP), a part of the general transcription factor TFIID, binds. Usually located 25–30 nt upstream of eukaryotic transcription start sites.

Taxonomy: the science of classification of living organisms on the basis of genetic and evolutionary relatedness.

Tax-responsive elements (TRE): 21 bp repeated sequences in HTLV-1 proviral DNA that help form a complex with Tax and other proteins, recruiting histone acetylases to the DNA and increasing viral gene transcription.

T-cell lymphoma: a T-cell malignancy in which the proliferating T cells are localized mostly in lymphoid tissues rather than in the blood.

T-cell receptor: a receptor on the surface of T lymphocytes that binds to specific antigens in association with major histocompatibility proteins, leading to activation of T cells and production of antibodies or cytotoxic T lymphocytes.

Tegument: region between the envelope and the nucleocapsid in virions of certain viruses, notably baculoviruses and herpesviruses. Contains a number of virus proteins.

Telomerase: an enzyme that maintains and extends ends of linear chromosomes in eukaryotic cells by using a small template RNA to make repeated DNA sequences.

Temperate phage: a bacterial virus that can either replicate to produce progeny virions and cell lysis, or enter a repressed prophage state that can be harbored in a cell for multiple generations.

Terminal protein: an adenovirus-coded protein that is covalently bound to the 5′ ends of linear viral DNA. Directs and primes initiation of viral DNA replication.

Therapeutic index: the ratio of the dose of a drug that exerts a half-maximal toxic effect, to the dose that exerts a half-maximal therapeutic effect. Effective, safe drugs have a high therapeutic index.

Thrips: a minute insect of the family Thripidae (order Thysanoptera) commonly found on plants, some of which are vectors for tospoviruses. ("Thrips" is both singular and plural.)

Thymidine kinase: an enzyme that phosphorylates thymidine to thymidine monophosphate.

Thymidylate kinase: an enzyme that phosphorylates thymidine monophosphate (thymidylate) to thymidine diphosphate.

Tight junction: intercellular junction that brings plasma membranes of neighboring cells into very close proximity (1–2 nm); found in endothelial and epithelial cells.

T lymphocytes: white blood cells that mature in the thymus and are involved in the adaptive immune response, regulating development of B lymphocytes for antibody production, and generating cytotoxic T lymphocytes, leading to killing of target cells.

Tolerance: a state in which the immune system accepts as "self" organisms or cells of foreign origin. In the context of viral illness, tolerance can occur when infection occurs at a very young age or can be induced as a part of the biology of the infecting virus (HIV or human papillomaviruses).

Toll-like receptors: a set of cell-surface proteins that recognize molecules specific to invading pathogens (e.g., lipopolysaccharide, CpG-containing DNA sequences, double-stranded RNA) and signal activation of innate and adaptive immune responses via a protein kinase cascade.

Topoisomerase I: enzyme that cleaves a phosphodiester bond in a double-stranded DNA, forming a temporary covalent bond with the DNA, and allows one strand to rotate around the complementary strand, subsequently resealing the phosphodiester bond and leading to relaxation of overwound (supercoiled) DNA.

Topoisomerases: enzymes that relieve torsional stress in supercoiled DNA molecules by cleavage and resealing of phosphodiester bonds.

Toroidal coil: a spring-like coil that consists of a cylindrical form bent into a circle.

Transcription factors: proteins that bind to specific DNA sequences in transcriptional enhancers or promoters and alter the transcriptional activity of the gene by interacting with other transcription factors and RNA polymerases.

Transcriptional enhancer: a region that contains consensus binding sequences for transcription factors and that enhances the transcription of specific genes. Often located immediately upstream of a gene, but can be found within the transcribed sequence or thousands of base pairs from the transcription start sequence.

Transgene: a foreign gene that is introduced artificially into a cell and expressed, sometimes as part of a virus vector.

Transmembrane anchor: a domain of 20 hydrophobic amino

acids that is embedded in a lipid bilayer and therefore anchors a protein in the membrane.

Transovarial transmission: the transmission of a pathogen from mother to offspring via the egg.

Triangulation number: $T = h^2 + hk + k^2$, where h and k are integers; the number of protein subunits on the surface of viral capsids with icosahedral symmetry is $60 \times T$.

Tropical spastic paraparesis (TSP): a neurological disease, caused by HTLV-1 infection, that leads to demyelination of motor neurons.

Tropism: the ability of a virus (or other pathogen) to infect and replicate in a particular tissue, organ, or species. Often depends on the initial interaction of the pathogen with cell surface receptors.

Tumor necrosis factor (TNF): a cytokine that activates the inflammatory response and can cause apoptosis.

Tumor suppressor protein: a protein, such as pRb or p53, whose absence or mutation can facilitate or cause tumor formation; therefore the wildtype protein suppresses tumor development.

Twofold symmetry axis: an axis around which an object rotated 180° appears identical to the original object.

Type 1 and type 2 T helper lymphocytes: T lymphocytes that secrete cytokines that contribute to the development of cellular immune responses (Th1 cells) or humoral immune responses (Th2 cells)

Type I transmembrane protein: an integral membrane protein oriented with its amino terminus facing the lumen of the endoplasmic reticulum or the extracellular space and its carboxy terminus facing the cytosol. Its single hydrophobic membrane-spanning anchor is usually near the carboxy terminus.

Type II transmembrane protein: an integral membrane protein oriented with its carboxy terminus facing the lumen of the endoplasmic reticulum or the extracellular space and its amino terminus facing the cytosol. Its single hydrophobic membrane-spanning anchor is usually near the amino terminus.

Ubiquitin: a 76-amino acid polypeptide that is covalently joined to lysine residues on proteins, which are then degraded in proteasomes.

Ubiquitination: covalent linkage of ubiquitin to a protein, marking it for degradation by proteasomes.

Ubiquitin ligase: an enzyme that catalyzes covalent binding of ubiquitin to proteins, marking them for degradation in proteasomes.

Uncoating: the removal of the protein coat, or capsid, from a viral genome within a cell, in preparation for its expression and replication.

Vaccination: a term coined by Edward Jenner and initially used to describe immunization against smallpox using cowpox. The term was generalized by Louis Pasteur in 1881, in honor of Jenner, to include artificial immunization against any disease.

Variolation: artificial inoculation with variola (smallpox) virus, designed to produce immunity to natural smallpox infection. Practiced in various forms from the tenth through the eighteenth centuries in several countries worldwide, variolation produced a disease of diminished severity while conferring lifelong immunity to smallpox. The practice was replaced by vaccination in the nineteenth century.

Vesicle transport: the transfer of vesicular membranes and their contents between two membrane compartments by coupled membrane fission and fusion.

Viral interference: the state of resistance of a cell to virus infection brought about by previous exposure of the cell to a similar or different virus.

Virion: a complete, infectious virus particle.

Virogenic stroma: electron-dense material that accumulates in baculovirus-infected nuclei, believed to be the site of viral replication and nucleocapsid assembly.

Viroid: a small (250–400 nt), circular RNA that can replicate in the absence of a helper virus by reprogramming cellular DNA-dependent RNA polymerases to copy its genome RNA. Viroids do not code for proteins and are not packaged into virions, but are probably transferred as naked RNAs from one host to another via insects or mechanical abrasion.

Viroplasm: name given to localized regions in the cytoplasm of infected cells where poxviral DNA accumulates and where viral assembly takes place.

Virus vectors: viruses that are genetically modified to express foreign genes and/or to infect cell types or species not originally targeted by the virus.

X-ray diffraction: the diffraction of x-rays by oriented molecular assemblies, usually in crystals, used to determine the atomic structure of the object.

Zanamivir: neuraminidase inhibitor used for treatment of influenza A and B virus infections.

Zoonosis: infection of an animal population with a pathogenic agent that can be transmitted to human beings.

Zoonotic disease: a disease that can be transmitted to human beings from animals.

Index

Acid-activated penetration, 47
Acqua, C., 148
Acquired immunodeficiency syndrome (AIDS), 285, 286. *See also* HIV-1
Actinomycin D, 243, 255
Acute transforming viruses, 281
Acyclovir, 350–351, 352
Adeno-associated virus (AAV), 90, 367
Adeno-associated virus vectors, 367–369
Adenoviridae, 35
Adenovirus, 19, 25, 26, 123–133
  cell killing, 132
  cell transformation/oncogenesis, 132–133
  DNA replication, 129–131
  E1A, 127–128, 133
  EIB, 128–129, 133
  gene products, 126
  inverted terminal repeat, 130
  late mRNAs, 131–132
  late promoter, 131
  terminal protein, 124, 125
  therapeutic applications, 133
  transcription units, 126–127
  transcriptional map, 126
  tripartite leader, 132
  tumors, 124
  virion, 124, 125
  virus release, 132
Adenovirus vectors, 362–364
Adherens junctions, 270
Adjuvant systems, 376, 377
AIDS, 4, 285, 286. *See also* HIV-1
Allolevivirus, 54
α-amanitin, 243, 255
Alpharetrovirus, 273
Alphavirus, 191–200. *See also* Togavirus
Alternative splicing, 255
Altruistic vaccination, 380
Amantadine, 51, 252, 349, 350
Ambisense coding, 239
Aminopeptidase-N family, 205
Amyloid, 330
Anti-apoptosis genes, 145
Antibody-dependent enhancement, 185
Antigenic drift, 258
Antigenic shift, 258
Antigenome RNAs, 194
Antitumor therapy, 359, 363
Antiviral chemotherapy, 346–357
  acyclovir, 350–351, 352
  amantadine, 349, 350
  AZT, 353, 354
  capsid-binding drugs, 348–349
  future directions, 357
  ganciclovir, 351–354
  how drugs obtained, 347
  importance, 347
  neuraminidase inhibitors, 355–357
  nonnucleoside inhibitors, 354
  nucleoside analogues, 349–350, 351
  protease inhibitors, 354
  ritonavir, 354–355
  stages in virus replication cycle, 347–348
Antiviral vaccines, 370–381. *See also* Vaccination
Aphthovirus, 171
Aplastic crisis, 96
Apoptosis
  adenovirus, 128
  baculovirus, 153
  coronavirus, 207
  herpes simplex virus, 145
  interferon, 343
  reovirus, 269
  togavirus, 198
Aquareovirus, 262
Arbovirus, 192
Arenaviridae, 37
Arntzen, Charles, 313
Arteriviridae, 37
Arthropod-borne virus, 192
Asfarviridae, 35
Asian flu, 249
Astroviridae, 37
Asymmetric replication, 317
Autographa californica MNPV, 147–156
Autophagy, 207
Avery, Oswald, 7
Avian erythroblastosis virus, 281
Avian infectious bronchitis virus, 204
Avian myelocytomatosis virus, 281
Avsunviroidae, 316
Azidothymidine (AZT), 353, 354

B lymphocytes, 340
B19, 96–97
Bacteriophage fd, 23
Bacteriophage f1, 53, 54
Bacteriophage f2, 53, 54
Bacteriophage lambda, 79–88
  chaperone proteins, 87
  CI repressor, 82–83
  Cro, 83
  DNA packaging/virion assembly, 87
  DNA replication, 87
  gene products, 85
  genetic map, 81
  integration/excision of λ DNA, 85–86
  liberation of λ DNA, 86
  lytic or lysogenize?, 83–85
  lytic transcription program, 81
  outline of λ biology, 80
  promoters, 84
  recombineering, 88
  retroregulation, 86
  scaffolding proteins, 87
  virion, 81
Bacteriophage MS2, 54, 55
Bacteriophage ϕX174, 63–70
  atomic structure, 65
  binding site, 65
  cell lysis, 69
  delivery of genome into cell, 66
  gene products, 65
  genome, 64
  icosahedral ssDNA virus families, 69–70
  J protein, 69
  morphogenetic pathway, 68
  mutation detector, as, 70
  procapsid, 68, 69
  promoters, 66–67
  proteins, 65
  rolling circle replication, 67
  scaffolding proteins, 67–68
  stage I DNA replication, 66
  stage II DNA replication, 67
  stage III replication, 68
  transcriptional terminators, 66–67
  virion, 69
Bacteriophage Q_β, 54–55
Bacteriophage T7, 71–78.
  class II genes, 74
  class III gene expression, 74–75
  classes, 72–73
  DNA concatemers, 76
  DNA replication, 75
  F-restriction, 77
  gene products, 73–75
  genetic engineering, 76–77
  genetic map, 72
  host restriction/modification, 77
  novel T7-coded RNA polymerase, 73–74
  special features, 76
  transcription, 73
  transcriptional map, 72
Baculoviridae, 35
Baculovirus, 147–156
  acquired cellular genes, 153
  apoptosis, 153
  budded virion, 149
  DNA replication, 153
  expression of foreign proteins, 155, 156
  expression of viral proteins, 152
  genetic map, 151
  genome, 150–151
  human gene therapy, 156
  immediate early gene products, 152–153
  infection in insects, 151–152
  LEFs, 153–154
  occluded-derived virion, 150
  pest control, 148
  recombinant virus, 155, 156
  virus-coded polymerase, 153–156
Bafilomycin, 51
Baltimore, David, 6, 13, 170, 273
Baltimore classification of viruses, 13–14
Barnaviridae, 37
Bat coronavirus, 204
Bax, 120, 129, 301
Bcl-2, 129
Bdellomicrovirus, 64
Beijerinck, Martinus, 5, 6
Berg, Paul, 6, 359
Bertani, Giuseppe, 77
Betaretrovirus, 273

Birnaviridae, 38
Bishop, Michael, 6, 273
BK virus, 99
Blood disorders, 367
Blumberg, Baruch, 6, 304
BmNPV, 148
Border disease virus, 182
Bornaviridae, 37
Bovine coronavirus, 204
Bovine papillomavirus, 114
Bovine spongiform encephalopathy (BSE), 324
Bovine viral diarrhea virus, 182
Buckminsterfullerene, 23, 24
Budded baculovirus virion, 149
Budding, 15, 27–29
Bunyaviridae, 37
Bunyavirus, 238–247
  arthropod vectors, 239–240
  cap snatch mechanism, 243
  coding strategies, 240–241
  entry, 243
  evolutionary potential, 246–247
  genome, 240
  genome replication, 244–245
  hemorrhagic fever, 240
  L protein, 241–242
  M RNA, 242–243
  pathogens, 239
  reassortment of genome segments, 245–247
  reverse genetics, 246
  S RNA, 242, 243
  transcription and replication, 244
  virion, 240, 241
  virus assembly, 245–246
Burkhart, James, 70

c-Src, 106
C-type viruses, 280
C_{60}, 23, 24
Caliciviridae, 37
Campbell, Allan, 80
Cancer
  antiviral chemotherapy, 346–357. *See also* Antiviral chemotherapy
  cervical, 121–122
  herpesvirus, 135
  reovirus, 268
Canine coronavirus, 204
Cap stealing, 254
Caprine arthritis-encephalitis virus, 285
Capsid-binding drugs, 348–349
Carboxylic ionophors, 51
Cardiovirus, 171
Caspases, 270, 344
Caulimoviridae, 38
Cavalier-Smith, Thomas, 93
Caveolae, 47, 103
Caveolin, 47
CD4, 287–288
cdk 7/cyclin H complex, 290
CDKs, 300
Cell-associated enveloped virus (CEV), 166
Cell-based high throughput screen, 347
Cell-cycle arrest, 269
Cervical cancer, 121–122
Cervical carcinoma, 4
Chanock, Robert, 202
Chaperone proteins, 87
Chapman, J. W., 148
Charny, Patrick, 304
Chase, Martha, 6, 7
Chemotherapy, 346–357. *See also* Antiviral chemotherapy
Chickenpox, 4
Chimeric molecules, 376
Chimeric viruses, 369, 378
Chlamydiamicrovirus, 64
Chlorpromazine, 331, 344
Cholera toxin B subunit, 377
Chordopoxvirinae, 159
Chronic wasting disease (CWD), 325
Chrysoviridae, 38
Circoviridae, 33, 69
Circovirus, 35
Circular single-stranded RNA satellites, 39
Cis-acting replication elements (cre), 178
Class I fusion proteins, 49, 185, 206
Class II fusion proteins, 49, 185, 206
Classical swine fever virus, 182
Classification of viruses, 30–32. *See also* Virus groups
Clathrin-coated pits, 46
Clear lysis, 8
Co-chaperone, 107
Coiled-coil, 204
Cold sores, 4, 135
Coltivirus, 262
Common cold, 4
Complement, 166, 167
Component/subunit vaccines, 374
Concatemer, 76, 87, 164
Concatemer junction, 164
Condylomas, 115
Confocal microscopes, 12
Core proteins, 26
Cornalia, Emilio, 148
Coronaviridae, 37
Coronavirus, 201–213
  adaptability, 212
  binding sites, 205–206
  common cold, 202
  discontinuous transcription, 208–211
  fusion, 206
  genome, 203
  group II proteins, 207
  groups, 203
  membrane protein, 204–205
  polyprotein, 206
  replicase gene, 207
  replication complexes, 207
  replicative intermediate, 208
  reverse genetics, 211
  SARS, 202
  small envelope protein, 205
  spike protein, 204
  subgenomic mRNAs, 208–211
  virion, 204, 205
  vision assembly/release, 210, 211–212
  in vitro assembly, 211, 212
Corticoviridae, 35
Court, Dan, 88
Cowpox, 159
Coxsackievirus, 172
CpG motifs, 377
Creutzfeldt-Jakob disease (CJD), 324, 331
Crick, Francis, 7
Crimean-Congo hemorrhagic fever, 239, 240
Cryoelectron microscopy, 18
Cyclin-dependent kinases (CDKs), 300
Cypovirus, 262
Cystoviridae, 38
Cytokine signalling, 167
Cytotoxic lymphocytes, 342

Dane, David, 304
Dane particles, 305
Dawkins, Richard, 332
de la Condamine, Comte, 158
Delbrück, Max, 6, 7, 71
Delta self-cleaving structure, 318, 319, 321
Deltaretrovirus, 273
Deltavirus, 39
Dendritic cells, 376, 377
Dengue virus, 182–185
Densovirus, 91
Deoxyguanosine, 351
Dependovirus, 91
Detection/titration of viruses, 8–10
Detoxified bacterial lipopolysaccharide, 377
d'Herelle, Felix, 6, 7, 63
Diarrhea, 4
Discontinuous transcription, 208–211
Divergent transcription, 102
DNA concatemers, 76
DNA-ethidium bromide complexes, 101
DNA gyrases, 101
DNA J domain, 107
DNA polymerase delta, 111
DNA vaccines, 376
Double-stranded DNA genomes, 34, 35
Double-stranded RNA-dependent protein kinase (PKR). *See* PKR
Double-stranded RNA satellites, 39
Dulbecco, Renato, 6
Dynactin, 50
Dynein, 50

Early endosomes, 46–47
Early viral proteins, 14
Ebola virus, 226–237. *See also* Filovirus
Ectodomain, 27
Eddy, Bernice, 98
Edmonston, David, 216
Edmonston-B and Moraten measles vaccine strains, 374
Ellis, Emory, 7
Encephalomyocarditis virus, 172
Enders, John, 6, 170, 215, 373
Endocytosis, 44, 46
Endoplasmic reticulum-Golgi intermediate compartment (ERGIC), 211
Enfuvirtide, 349
Enhanceosome, 336
Enterovirus, 171
Entomopox virus, 159
Entomopoxvirinae, 159
Envelope, 2, 17, 27

Epitopes, 340
Epsilonretrovirus, 273
Epstein-Barr virus, 135
Equine encephalitis viruses, 192
Equine infectious anemia virus, 210, 285
Erythema infectiosum, 97
Erythrovirus, 91
Ethidium bromide, 101
Eukaryotic virus vectors, 358–369
  AAV vectors, 367–369
  adenovirus vectors, 362–364
  antitumor therapy, 359, 363
  applications, 359
  blood disorders, 367
  chimeric viruses, 369
  expression cassette, 361
  gene replacement therapy, 360
  how produced, 360–361
  lentivirus, 365–366
  long-term maintenance of virus genomes, 369
  protein production, 359–360
  retargeting virus vectors, 369
  retrovirus vectors, 364–367
Evolutionary origin of viruses, 39–41
Exocytosis, 44
Expression of foreign proteins, 155, 156
Expression vectors, 198, 199
Extracellular enveloped virus (EEV), 160–161, 166

F-pili, 55
Fane, Bentley, 64
Fatal familial insomnia (FFI), 325
Feline immunodeficiency virus, 273, 285
Feline infectious peritonitis virus, 204
Feline spongiform encephalopathy (FSE), 325
Fijivirus, 262
Filoviridae, 37
Filovirus, 226–237
  cloned CDNA copies, 230
  editing, 232
  equatorial Africa, 234–235
  genome, 228–229
  GP, 231–232
  hemorrhagic fever, 228
  infection, 235–237
  multiplasmid transfection system, 230–231
  proteins, 229
  sGP, 232–233
  structure, 227
  transcription and replication, 230
  vaccine, 236
  virion, 228
  VP30, 233–234
  VP40, 234
Filterable viruses, 5
Finlay, Carlos, 6
First-generation adenovirus vectors, 362
Flaviviridae, 37
Flavivirus, 181–190, 182
  E protein, 184–185
  genome, 185–186
  hepatitis C virus, 183
  NS1, 187
  NS2A, 187
  NS2B, 188
  NS3, 188
  NS4A, 188
  NS4B, 188
  NS5, 188
  pH-dependent fusion, 185
  polyprotein, 186–187
  replication pathway, 188
  RNA synthesis, 188–189
  virion, 183–184
  virus assembly, 189
  West Nile virus, 182, 189
  yellow fever, 182–183, 185, 186
Flu, 248–260. *See also* Orthomyxovirus
Flu pandemic, 249, 258, 259
Flu vaccine, 259
Foot-and-mouth disease virus, 171, 172
Fowlpox virus, 159
F-pili, 55
Franklin, Richard, 170
F-restriction, 77
Frosch, Paul, 6
Fuller, Buckminster, 23
Furin, 186, 197, 231, 258
Fusion, 274
Fusion peptide, 48, 252
Fusion proteins, 185, 206

Gajdusek, D. Carleton, 6, 324
Gallo, Robert, 285, 294
Galtier, Victor, 380
Gammaretrovirus, 273
Ganciclovir, 351–354
Gangliosides, 218
Geminiviridae, 33, 69
Gene gun, 376, 377
Gene replacement therapy, 360
Genetic economy, 18
Genetic engineering, 76–77
Genital herpes, 4
Genital herpes lesions, 135
Geodesic dome, 24
Gerstmann-Sträussler-Scheinker syndrome (GSS), 325
Glaser, Rudolf, 148
Global smallpox eradication program, 159
Glycosaminoglycans, 161
Gokushovirus, 64, 70
Gomatos, Peter, 262
Graham, Angus, 262
Granulin, 149
Granulovirus, 149
Gross, Ludwig, 98
Guanylyltransferase, 163

Hamre, Dorothy, 202
Hantaan virus, 239
Hantavirus, 239, 240
Hantavirus pulmonary syndrome, 240
Hausmann, Rudolf, 72
Hayashi, Masaki, 64
Helical nucleocapsid, 218, 219
Helicase, 67
Helper virus, 38
Hemagglutination, 217, 242
Hemagglutination assay, 9–10
Hemagglutinin-esterase (HE), 204, 205, 207
Hemagglutinin protein (HA), 252
Hemifusion, 48
Hemolytic anemia, 96
Hemorrhagic fever, 4, 228, 240
Hendra virus, 216
Hendrix, Roger, 88
Hepacivirus, 182
Hepadnaviridae, 38
Hepadnavirus, 303–314
  antiviral treatment, 313
  conversion of viral genome, 305–307
  core proteins, 308
  Dane particles, 305
  diseases, 312–313
  E protein, 308–309
  genome, 305
  noninfectious particles, 305
  pgRNA, 309
  reverse transcription, 309–311
  surface proteins, 308
  transcription, 307
  transmission of hepatitis B virus, 313
  vaccine, 313
  virion assembly, 311–312
  X protein, 308
Heparan sulfate, 193
Hepatitis, 4
Hepatitis A virus, 19, 170
Hepatitis B virus, 303–314. *See also* Hepadnavirus
Hepatitis B vaccine, 377
Hepatitis C virus, 183
Hepatitis delta virus (HDV), 315, 320–322
Hepatovirus, 171
Herd immunity, 380
Herpes simplex virus, 134–146
  α genes, 140
  anti-apoptosis genes, 145
  β genes, 140–141
  diseases, 135
  DNA replication, 141–143
  entry, 138
  envelopment and egress, 143, 144
  fusion, 138
  γ genes, 138
  genome, 135–137
  ICP0, 145
  ICP27, 143
  ICP34.5, 143–145
  ICP47, 143
  latent infections, 135
  LATs, 145–146
  nomenclature, 137
  replication cycle, 138
  rolling circle replication, 143
  tegument proteins, 137, 139–140
  uncoating, 138
  Vhs, 143
  viral nucleocapsids, 143
  virion, 137
  virus latency, 145
Herpesviridae, 35
Hershey, Alfred, 6, 7, 80
Hexon, 25
Hilleman, Maurice, 124
Histone acetylases, 128

Historical overview, 5–8, 371–373
HIV-1, 284–293
AIDS, 285, 286
CD4 antigen/chemokine receptors, 287–288
down regulation of CD4 expression, 292
entry, 288
genome, 285
HTLV-1, and, 302
latent infection, 289
Net, 292–293
nonstructural proteins, 287
Rev, 290, 291
structural proteins, 287
Tat, 289–290, 291
transport of proviral DNA, 288–289
vaccine, 293
Vif, 291
virus mutants, 288
Vpr, 291
Vpu, 291–292
Holin, 76
Hong Kong flu, 249
HTLV-1, 294–302
adult T-cell leukemia, 301–302
antiviral therapy, 302
CDKs, 300
cell transformation, 299
doubly spliced mRNAs, 295
HIV-1, and, 302
interleukin-2, 299
Jak-Stat pathway, 299–300
MAD1, 301
mitotic spindle checkpoint, 300
$p16^{INK4A}$, 300
p53, 301
Rex, 296
Tax, 297–299
Human, Mary, 77
Human adenovirus, 123–133. *See also* Adenovirus
Human gene therapy, 156
Human herpesviruses, 135. *See also* Herpes simplex virus
Human immunodeficiency virus, 19, 284–293. *See also* HIV-1
Human papillomavirus, 19, 114–122. *See also* Papillomavirus
Human papillomavirus vaccine, 122
Human parainfluenza virus, 19
Human T-cell leukemia virus type 1. *See* HTLV-1
Hybrid vectors, 369

Icosahedral symmetry, 19–21
Icosahedron, 20
Idnoreovirus, 262
IKK, 299
Immune stimulating complexes (ISCOM), 377
Importin-α, 254, 296
Incardona, Nino, 64
Inclusion bodies, 230, 262
Infectious subvirion particles (ISVPs), 265–266
Influenza virus, 4, 248–260. *See also* Orthomyxovirus
Influenza vaccine, 259
Influenza virus hemagglutinin-mediated membrane fusion, 48
iNOS, 338
Inoviridae, 33
Intasome, 86
Interferon, 267–268, 333–345
antibody production, 343
antigen processing and presentation, 342
antiviral activities, 338–340
apoptosis, 343
CD4-positive helper T cells, 342–343
cell growth, 343
cellular immunity, 343
first line of defense, as, 344
future directions, 344
genes, 338
groups, 335
immune system, 340–342
Jak-Stat pathway, 337
macrophage activation, 343
properties, 334
transcription, 335
transcriptional activation, 335–337
viruses, and, 343–344
Interferon-α, 313, 334, 335
Interferon-γ, 334, 335
Interferon-stimulated genes (ISG), 338
Interferon system, 336
Interleukin-2, 299
Internal ribosome entry site (IRES), 173–176, 185
International Committee on Virus Taxonomy (ICTV), 31
Intracellular enveloped virus (IEV), 166
Intracellular mature virus (IMV), 160–161
Introns, 104–105
Inverted terminal repeat, 124, 130
IRES, 173–176, 185
Iridoviridae, 35
Isaacs, Alick, 6, 334
Ivanovski, Dimitrii, 5, 6

Jak-Stat pathway, 299–300, 337
Japanese encephalitis, 182
JC virus, 99
Jelly-roll β-barrel, 21
Jenner, Edward, 5, 6, 158–159, 371
Joklik, Wolfgang, 262

Kaiser, Dale, 80
Kaplan, Paul, 304
Kaposi's sarcoma, 135
Karyopherins, 50
Katsura, Isao, 88
Keratinocytes, 117
Kinetic trap, 60, 61
Koch, Robert, 5, 373
Kornberg, Arthur, 64
Kuru, 324

La Crosse virus, 239, 246
Lamivudine, 313
Large single-stranded RNA satellites, 39
Large T antigen (LTAg), 107–111
LAT, 145–146
Late expression factor (LEF), 153–154
Late messenger RNAs, 15
Late viral proteins, 15
Latency, 145
Latency-associated transcripts (LATs), 145–146
Latent infection, 289
Latent phase, 11
Lederberg, Esther, 80
Lentivirus, 273, 285, 365–366
Leukemia, 294–302. *See also* HTLV-1
Leviviridae, 37, 53
Levivirus, 54
LexA repressor, 83
Lindenmann, Jean, 6, 334
Lipid rafts, 257
Lipopolysaccharide, 65
Liposomes, 377
Live attenuated viruses, 373–374
Live viruses with defined attenuation characteristics, 378
Live wildtype viruses, 373
Loeb, Tim, 53
Loeffler, Friedrich, 6
Long control region (LCR), 118
Long-term maintenance of virus genomes, 369
Long terminal repeat (LTR), 275, 278–279, 282, 364
LTR, 275, 278–279, 282, 364
Luria, Salvador, 6, 7, 77
Lwoff, Andre, 6
Lysis hairpin, 56, 57
Lysogen, 80
Lysosomotropic agents, 51

MacLeod, Colin, 7
Mad cow disease, 324
Maestra, A., 148
Maizel, Jacob, 171
Major histocompatibility complex (MHC), 293
Malling, Heinrich, 70
MAP kinase, 105
MAP kinase pathway, 106
Marburg virus, 226–237. *See also* Filovirus
Marnaviridae, 37
Martini, Gustav-Adolf, 227
Mason-Pfizer monkey virus, 280
Matrix protein, 28
McCarty, Maclyn, 7
Measles virus, 4, 215–216. *See also* Paramyxovirus and rhabdovirus
Meister, Joseph, 380
MEKK1 kinase, 336
Membrane protein (M), 204–205, 207
Messenger RNA capping, 266–267
Metalloproteinases, 205
Methyltransferase, 196
MHC proteins, 342
Micelles, 166
Microviridae, 33, 63, 69
Microvirus, 64
Milestones in virology research, 6, 372
Miller, Lois, 156
Mimivirus, 34, 35
Minichromosomes, 27, 100
Minimalist vaccines, 377

Minus-strand strong-stop DNA, 275
Minute virus of mice (MVM), 89–97
Mitotic arrest-defective (MAD), 301
Mitotic spindle checkpoint, 300
MnSOD, 338
m.o.i., 11
Molluscum contagiosum virus, 159
Monensin, 245
Monkeypox virus, 159
Mononegavirales, 215
Mononucleosis, 4
Montagnier, Luc, 285
Mouse mammary tumor virus, 280
Mouse polyomavirus, 11, 98–113
Movement proteins, 45
Mulligan, Richard, 359
Multinucleated giant cells, 301
Multiplasmid transfection system, 230–231
Multiplicity of infection (m.o.i.), 11
Mumps, 4, 216
Murine hepatitis virus, 202, 204, 208
Murine leukemia virus, 273
Mx proteins, 338
Mycoreovirus, 262
Myoviridae, 35
Myristate, 292, 308
Myxoma, 159

N-acetyl neuraminic acid, 356
Nairobi sheep disease, 239, 240
Nairovirus, 239
Naming conventions, 32
Nanoviridae, 33
Natural killer cells, 340
Nectins, 270
Negative staining, 10, 18
Negative-strand RNA genomes, 13–14, 34, 36, 37
Nelmes, Sarah, 159
Nested set, 208, 223
Neuraminidase, 217, 257
Neuraminidase inhibitors, 355–357
Neuraminidases, 45
Nevirapine, 354
New variant Creutzfeldt-Jakob disease (nvCJD), 324, 331
Newcastle disease virus, 216
NF-κB, 269, 270, 297–299
Nipah virus, 216
Nodaviridae, 37
Nonnucleoside inhibitors, 354
Northrup, John, 6
Norwalk virus, 22
Nucleocapsid, 17, 204, 218, 219, 253–254
Nucleocapsid export, 256
Nucleocapsid protein (N), 205, 207
Nucleopolyhedrovirus, 149
Nucleoside analogues, 349–350, 351
Nucleosome, 100
nvCJD, 324, 331

Obligatory intracellular parasites, 2
Occluded baculoviruses, 150
Occlusion bodies, 149–150
Oct-1, 145
Oka-strain varicella virus, 374
Oncogene, 8
Oncotropism, 90
ONYX-015, 133, 363
Ophiovirus, 37
Orbivirus, 262
Orf, 159
Oropouche virus, 239, 240
Orthobunyavirus, 239
Orthomyxoviridae, 37
Orthomyxovirus, 248–260
  alternative splicing, 255
  Asian flu, 249
  cap stealing, 254
  drift, 258
  genome, 250
  genome replication, 255–256
  HA, 252
  Hong Kong flu, 249
  M2, 252
  NA, 257
  NS1, 256–257
  nucleocapsid, 253–254
  nucleocapsid export, 256
  pandemics, 249, 258, 259
  proteins, 251
  secondary bacterial infections, 249
  Spanish flu pandemic, 249, 258
  stuttering transcription, 255
  synthesis of viral mRNAs, 254–255
  vaccine, 259
  viral envelope export, 257
  virion, 250, 251
  virus strains, 257–259
Orthoreovirus, 262
Oryzavirus, 262
Oseltamivir, 356, 357

p53, 108, 120, 128, 301
Packaging sequence, 25
Packaging signals, 25, 196, 280
Palmitoylation, 197
Panleukopenia, 96
Panum, Peter, 215
Pap test, 121
Papain-like proteinases (PLPs), 206
Papillomaviridae, 35
Papillomavirus, 25, 114–122
  cancer, 121–122
  diagnosis of diseases, 121–122
  genes, 117
  genetic map, 117
  genome, 116
  infectious cycle, 116–117
  p53, 120
  Rb, 119
  synergism between E6 and E7, 121
  transcriptional map, 117
  types, 115
  vaccine, 122
  viral E1/E2 proteins, 118
  viral E6 protein, 120–121
  viral E7 protein, 118–120
  viral mRNAs, 117–118
  virion, 116
Paramyxoviridae, 37
Paramyxovirus and rhabdovirus, 214–225
  C protein, 224
  editing, 223
  envelope proteins, 218
  fusion, 218–219
  gene order, 219–220
  genome, 218, 219
  helical nucleocapsid, 218, 219
  measles, 215–216
  N protein, 224
  P/C/V gene, 223
  P protein, 224
  promoter, 220–222
  rabies, 215
  rule of six, 225
  transcription and replication, 222–223
  V protein, 224
  viral messenger RNAs, 220
  viral RNA polymerase, 220
  virion, 216–217
  virion assembly, 224–225
Parvovirus, 89–97
  adenovirus functions, 96
  antitumor effects, 97
  B19, 96–97
  capsid proteins, 91
  DNA replication, 93–95
  end replication problem, 93
  genes and transcripts, 91
  integration, 96
  nonstructural proteins, 95–96
  subfamilies, 91
  terminal structures, 92
  uncoating, 92–93
  virion, 90, 91
Partitiviridae, 38
Parvoviridae, 33, 69
Parvovirus, 21
Pasteur, Louis, 5, 6, 215
PCNA, 111, 153
P/C/V gene, 223
Penicillin, 69
Peptidomimetic drug, 355
Pest control, 148
Pestivirus, 182
pET-vectors, 76, 77
Petri, R., 373
PFU, 9
pgRNA, 309
pH-dependent fusion, 185
pH-independent penetration, 47
pH-induced fusion, 193
Phipps, James, 159
Phlebovirus, 239, 242
Phosphatase 2A (PP2A), 105
Phospholipase C, 106
Phosphotyrosines, 106
Proto-oncogenes, 281
Phycodnaviridae, 35
Phytoreovirus, 262
PI3 kinase, 106
Picornaviridae, 37
Picornavirus, 169–180
  cell receptors/co-receptors, 172
  diseases, 170, 179
  entry, 172
  genome, 173
  inhibition of host cell functions, 179
  initiation of protein synthesis, 174
  IRES, 173–176

Picornavirus (*continued*)
poliovirus, 170–171, 172
polyprotein, 176
proteinases, 176
replication of RNAs, 176–177
RNA synthesis, 177
virion, 171
virion assembly, 178–179
VPg, 177
Picornavirus 3C proteinases, 206
Picornavirus capsids, 21
PKR
herpes simplex virus, 144–145
interferon, 338, 339–340, 343–344
orthomyxovirus, 257
poxvirus, 167
reovirus, 267–268
viroids, 319
Plant satellite RNAs, 320
Plant-to-plant dissemination, 15
Plaque assay, 8–9
Plaque-forming units (PFU), 9
Plasmaviridae, 35
Plasmid-based DNA vaccination, 377
Plasmodesmata, 15, 44
Pleconaril, 349
Podoviridae, 35
Poliomyelitis, 4, 170
Poliovirus, 21, 170–171, 172, 375
Polyadenylation signal, 102
Polyhedrin, 149
Polyomaviridae, 35
Polyomavirus, 98–113
alternative splicing, 104, 112
capsid, 99–100
control regions, 103
divergent transcription, 102
genome, 100, 102
introns, 104–105
late mRNAs, 112
late transcripts, 111
LTAg, 107–111
MTAg, 105–107
STAg, 105
supercoiled DNA, 100, 101
T antigens, 105–111
transcription factors, 103–104
tumors, 112
virion, 100
in vitro cell transformation, 112
Polyprotein, 176, 186–187, 193, 206
Porcine transmissible gastroenteritis virus, 204
Positive regulatory domain (PRD), 336
Positive staining, 18
Pospoviroidae, 316
Poxviridae, 35
Poxvirus, 157–168
complement, 167
cytokine signalling, 167
early genes, 162
EEV, 160–161, 166
enzymes, 163–164
extracellular virions, 166
genome, 159–160, 161
IMV, 160–161
interferon response, 167
large concatemeric DNA molecules, 164
mature virions, 165
postreplicative mRNAs, 164–165
smallpox, 158–159
vaccinia-T7 system, 167
viral transcription factors, 162
virion, 161
virus-coded enzymes, 162–163
Pregenome RNA (pgRNA), 309
Preintegration complex, 277
Primer binding site (PBS), 277
Primosome, 66
Prion, 323–332
crossing the species barrier, 331
diseases, 324–325, 330–331
fibrils, 327
future research, 331
hypothesis, 326–330
infectious agent, 325, 331–332
mechanism for propagation, 330
protein, structure of, 328
PrP^{Sc}, 325–326
PrP^{c}, 326
strain variation, 331
two-dimensional crystals, 327
Prion diseases, 324–325, 330–331
Prion hypothesis, 326–330
Prion protein, 328
Procapsid, 68, 69, 178
Promoter insertion, 282
Promoter occlusion, 278, 365
Prophage, 80
Protease inhibitors, 354
Proteosomes, 376, 377
Proto-oncogene, 282
Proviral DNA, 277–279, 288–289, 295
Provirion, 69
PrP 27-30, 329
PrP^{C}, 326
PrP^{Sc}, 325–326
Prusiner, Stanley, 6
Pseudoknot, 206, 279
Pseudotyping, 218, 364
Pseudovirions, 27, 112
Pulsed dendritic cells, 377
Puumala, 239

Q_{β} polymerase, 61
Quasi equivalence, 21–22
Quinacrine, 331, 344

Rabies encephalitis, 4
Rabies vaccine, 380
Raft cultures, 115
Ratio of physical particles to infectious particles, 10
Rb protein, 119, 128
Rbp-27, 338
Reassortant genetics, 262
Reassortant viruses, 242, 262, 263
Reassortment, 258
Receptor-mediated endocytosis, 44, 46, 193
Recombinant vaccines, 359
Reed, Walter, 6, 183, 192
Relaxed DNA, 101
Reoviridae, 38
Reovirus, 261–271
apoptosis, 269
attachment protein, 265
cancer, 268
capsid proteins, 263
cell-cycle arrest, 269
genera, 262
interferon, 267–268
ISVPs, 265–266
JAM-A, 265
messenger RNA capping, 266–267
messenger RNA export, 267
messenger RNA synthesis, 266, 267
mice, 270–271
PKR, 267–268
replicase particles, 268
σ^3, 268
secondary transcription, 268–269
segmented genomes, 262–263, 264
translation, 267
virion, 263
virion assembly, 269
virus factories, 268
Replication-competent adenovirus vectors, 363
Replication complexes, 207
Replication cycle, 10–15
Replication protein C (RPC), 110, 111
Replicative form (RF) DNA, 66
Replicative intermediate, 208
Resolvase, 164
Retargeting virus vectors, 369
Retroregulation, 86
Retroviridae, 38
Retrovirus, 14, 272–283
conversion of RNA to DNA, 275–277
fusion, 274
Gag, 279, 280
Gag/Pol, 279, 280
gene therapy, 282
genera, 273
genome, 273–274
infectious particles, 280
insertional activation of proto-oncogenes, 282
integration, 277, 278
life cycle, 274, 275
LTRs, 278–279
oncogenesis, 281–282
replication cycle, 273
reverse transcription, 275–277, 311
ribosome frameshifting, 280
RNA production, 279
virion, 273, 274
Retrovirus vectors, 364–367
Rev, 287, 290, 291
Rev-responsive element (RRE), 290
Reverse genetics, 211, 246
Reverse transcriptase, 8, 14, 36, 38, 273
Reverse transcription, 275–277, 309–311, 311
Rex, 296
RF DNA, 66
Rhabdoviridae, 37
Rhabdovirus, 214–225. *See also* Paramyxovirus and rhabdovirus

Rhinovirus, 170–172
Ribonuclease L, 339
Ribose-methyltransferase, 163
Ribosome frameshifting, 280, 295
Ribozyme, 321
Rift Valley fever, 239, 240
Ris, Hans, 80
Ritonavir, 354–355
RNA editing, 223, 232, 322
RNA helicase, 196
RNA phages. *See* Single-stranded RNA bacteriophages
RNA world, 40
RNAi, 199
RNase L, 338
Robbins, Frederick, 6, 373
Roberts, Richard, 6, 124
Roeder, Robert, 124
Rolling circle replication, 67, 87, 143
Roniviridae, 37
Roosevelt, Franklin D., 170
Rosen, Leon, 262
Ross River virus, 192, 196
Rossmann, Michael, 64
Rotavirus, 262
Rous, Peyton, 6, 273
Rous sarcoma virus, 279–281
Roux, Émile, 380
Rowe, Wallace, 124
Rubella virus, 192, 196
Rubivirus, 192
Rule of six, 225
Rutter, William, 304, 313

Sabin, Albert, 6, 170, 262
Salk, Jonas, 6, 170
Sandfly fever Naples, 239
Sanger, Fred, 64, 80
Satellite nucleic acids, 38–39
Satellite viruses, 38–39
Scaffolding proteins, 15, 27, 67–68
Scrapie, 324
Seadornavirus, 262
Second-generation adenovirus vectors, 363
Secondary receptors, 126
Segmented genomes, 262–263, 264
Self-assembly, 18
Self-cleaving RNA structures, 319
Selfish Gene, The (Dawkins), 332
Semliki Forest virus, 19, 46, 191–200. *See also* Togavirus
Sendai virus, 214–225. *See also* Paramyxovirus and rhabdovirus
Serpin, 167
Serum responsive elements (SRE), 299
Severe acute respiratory syndrome (SARS), 4
Sharp, Phillip, 6, 124
Shatkin, Aaron, 262
Shine-Dalgarno sequence, 60, 61
Shope, Richard, 114
Sialic acid, 102, 206, 218, 249, 356
Sickle cell disease, 96
Siegert, Rudolf, 227
Signal peptidase, 186
Silencing, 364
Simian foamy virus, 273
Simian immunodeficiency virus, 273, 285
Simian virus 5, 216
Simian virus 40 (SV40), 19, 26, 99, 359. *See also Polyomaviridae*
Sin nombre virus, 239
Sindbis virus, 191–200. *See also* Togavirus
Single-cycle virus replication experiment, 10–11
Single-stranded DNA phage. *See* ϕX174
Single-stranded DNA satellites, 39
Single-stranded RNA bacteriophages, 53–62
  assembly/release of virions, 62
  binding to F-pili, 55
  genera, 54–55
  genetic maps, 54
  host ribosomal protein, 59
  proteins, 58–59
  reinitiation, 57–58
  replicase translation, 57
  replication/translation, 55–56, 58
  RNA secondary structure, 56
  synthesis of maturation protein, 60–61
  synthesis of plus strands, 59–60
  terminal A - minus strand, 59
Sinsheimer, Robert, 64
Siphoviridae, 35
Slenczka, Werner, 227
Small linear single-stranded RNA satellites, 39
Small T antigen (STAg), 105
Smallpox, 4
SOD, 153
Sodium dodecyl sulfate, 100
Sonenberg, Nahum, 171
SOS genes, 83
Spanish flu pandemic, 249, 258
Spike protein (S), 204, 205, 207
Spiromicrovirus, 64
Spliceosome, 104
Spongiform encephalopathy, 324
Spumavirus, 273
Src, 281
Stanley, Wendell, 5, 6, 148
Stewart, S., 98
Studier, F. William, 72
Stuttering transcription, 255
Subacute sclerosing panencephalitis (SSPE), 216
Subgenomic RNAs, 194
Subunit vaccines, 374
Summers, Donald, 171
Supercoiled DNA, 100, 101
SV40, 19, 26, 99, 359. *See also Polyomavirus*
Symmetric replication, 317
Syncytia, 44, 206

T antigens, 105–111
T-cell leukemia, 294–302. *See also* HTLV-1
T-cell receptors, 340
T-helper cells, 341
T lymphocytes, 340
T7 RNA polymerase, 167
Tamm, Igor, 262
TAR, 289
Target-based high throughput screen, 347
Targeted recombination, 211
Tat, 287, 289–290, 291
Tax, 297–299
Tax-responsive elements (TRE), 297
Taxonomy, 30
Tc cells, 341
Tectiviridae, 35
Tegument, 137, 149
Tegument proteins, 137, 139–140
Telomerase, 86
Temin, Howard, 6, 273
Temperate phages, 80
Tenuivirus, 37
Terminal protein, 124, 125
TFIID, 104
TFIIH, 290
Th1 cells, 341, 342, 376
Th2 cells, 341, 342
Theiler, Max, 6, 183
Theiler's murine encephalomyelitis virus, 179
Thioredoxin, 75
Third-generation adenovirus vectors, 363
3C proteinase, 176
Threefold axis, 18
Thymidine, 351
Tick-borne encephalitis virus, 182
Tight junctions, 265, 270
Tiny T antigen (TTAg), 105–107
Tiollais, Pierre, 304
Tobacco mosaic virus, 7, 23, 25
Togaviridae, 37
Togavirus, 191–200
  arthropod borne, 192
  expression vectors, 198, 199
  genome RNA, 193
  mutations in viral proteins, 198
  nonstructural proteins, 193–196
  pH-induced fusion, 193
  polyprotein, 193
  replication and transcription, 196
  RNA synthesis, 195
  structural proteins, 195, 196–197
  translation, 196
  treatment of neurological diseases, 199
  uses (vectors), 199–200
  virion, 192–193
  virion assembly/egress, 197–198
Tolerance, 378
Toll-like receptors, 335
Tomato bushy stunt virus, 22
Tomato spotted wilt virus, 239
Topoisomerase I, 101
Toroidal coil, 100
Tospovirus, 239
Totiviridae, 38
Transcription factors, 103–104
Transcription-regulating sequences (TRS), 208–211
Transcriptional enhancer, 103
Transmissible mink encephalopathy (TME), 325
Transmissible spongiform encephalopathy (TSE), 325
Triangulation number, 22
Tropical spastic paraparesis (TSP), 295
Tropism, 13, 212, 242, 270, 364

2′, 5′-oligo (A) synthetase, 338, 339
Twofold axis, 18
Twort, Frederick, 6
Type I transmembrane proteins, 242, 252
Tyrrell, David, 202

Uncoating, 2, 51
Upstream complementary sequence, 60, 61
Uukuniemi virus, 245

Vaccination, 370–381
  adjuvant systems, 376, 377
  adverse events, 378–380
  changing vaccine paradigm, 378
  chimeric viruses, 378
  cocktails, 377
  component/subunit vaccines, 374
  delivery systems, 376–377
  efficacy, 375
  ethical issues, 380
  filovirus, 236
  hepatitis B, 313
  historical overview, 371–373
  HIV-1, 293
  how vaccines work, 376
  human adenovirus, 133
  human papillomavirus, 122
  immune system, 375
  inactivated virus, 374
  influenza, 259
  live attenuated viruses, 373–374
  live viruses with defined attenuation characteristics, 378
  live wildtype viruses, 373
  narrowly targeted vaccines, 377
  poliovirus, 170
  rabies, 215, 380
  smallpox, 158–159
  tolerance, 378
  vaccines currently under development, 379
Vaccine-associated adverse events, 378–380
Vaccinia virus, 157–168. *See also* Poxvirus
Vaccinia-T7 system, 167
Valentine, Carrie, 70
Valenzuela, Pablo, 304, 313
Variola virus, 159
Variolation, 158
Varmus, Harold, 6
Vectors, 8. *See also* Eukaryotic virus vectors
Vegetative replication, 117
Venezuelan equine encephalitis virus, 192
Vesicle transport, 44
Vesicular stomatitis virus, 218
Vida, Marco, 148
Vif, 287, 291
Viral envelopes, 17, 27
Viral interference, 334
Virion, 2, 17
Viroids, 315–322
  ancient RNA world, 322
  cleavage, 318
  discovery of, 316
  disease, and, 318
  disruption of cell metabolism, 318–320
  groups, 316
  HDV, 320–322
  ligation, 318
  plant satellite RNAs, 320
  polymerization, 317–318
  replication, 317
  RNA structure, 316–317
Viroplasm, 165
Virus
  Baltimore classification, 13–14
  classification, 30–32. *See also* Virus groups
  detection/titration, 8–10
  entry into host cell, 43–52
  evolutionary origin, 39–41
  naming conventions, 32
  nature of, 2–3
  size, 3, 34–36
  structure, 17–29
  why study them, 3–5
Virus classification, 30–32
Virus entry, 43–52
  acid-activated penetration, 47
  binding sites, 45–46
  capsids, 49–50
  cell surface proteins used as receptors, 45
  direct passage from cell to cell, 44–45
  early endosomes, 47
  enveloped viruses, 44
  fusion proteins, 47–49
  nonenveloped viruses, 44, 49
  pH-independent penetration, 47
  prevention of, 51
  receptor-mediated endocytosis, 46
  uncoating, 51
  viral genomes, 50–51
  virions, 49–50
Virus factories, 268
Virus groups, 32–39. *See also* Virus types
  double-stranded RNA genomes, 36, 38
  double-stranded DNA genomes, 34, 35
  negative-strand RNA genomes, 34, 36, 37
  positive-strand RNA genomes, 34, 37
  reverse transcriptase, 36, 38
  satellite nucleic acids, 38–39
  satellite viruses, 38–39
  single-stranded DNA genomes, 33
Virus latency, 145
Virus replication cycle, 10–15
Virus structure, 17–29
Virus types. *See also* individual subentries
  adenoviruses, 123–133
  baculoviruses, 147–156
  bunyaviruses, 238–247
  coronaviruses, 201–213
  filoviruses, 226–237
  flaviviruses, 181–190
  HDV, 315–322
  hepadnaviruses, 303–314
  herpes simplex virus, 134–146
  HIV-1, 284–293
  HTLV-1, 294–302
  orthomyxoviruses, 248–260
  papillomaviruses, 114–122
  paramyxoviruses, 214–225
  parvoviruses, 89–97
  picornaviruses, 169–180
  polymaviridae, 98–113
  poxviruses, 157–168
  reoviruses, 261–271
  retroviruses, 272–283
  rhabdoviruses, 214–225
  togaviruses, 191–200
Virus vectors, 8, 358. *See also* Eukaryotic virus vectors
Visna-maedi virus, 285
von Bergold, Gernot, 148
Von Prowzek, Stanislaus, 148
Vpr, 287, 291
Vpu, 287, 291–292

Walleye dermal sarcoma virus, 273
Wanner, Barry, 88
Warts, 4
Washington, George, 182, 249
Watson, James, 7
Weigle, Jean, 77
Weissmann, Charles, 334
Weller, Thomas, 6, 373
West Nile virus, 182, 189
Whole inactivated viruses, 373, 374
Wimmer, Eckard, 171
WIN compounds, 51

X-ray diffraction, 18
Xis, 86

Yellow fever virus, 4, 182–183, 185, 186
Yellow Fever Commission, 183

Zanamivir, 356, 357
Zidovudine, 354
Zinder, Norton, 53

Credits

Chapter 1

Fig. 1.2a: Courtesy Robley Williams, University of California at Berkeley and Harold Fisher, University of Rhode Island, **Fig. 1.2b:** Courtesy John Finch, Cambridge University, **Fig. 1.2c:** Courtesy Dr. Frank Fenner, The Australian National University.

Chapter 2

Fig. 2.1a: Electron micrograph courtesy of Dr. Carla W. Gray, University of Texas, Dallas: Gray, C.W., R.S. Brown, and D.A. Marvin, reprinted from "Adsorption complex of filamentous fd virus," *Journal of Molecular Biology* (1981) 146: 621–627, **Fig. 2.1b, 2.1e–g:** Courtesy Robley Williams, University of California at Berkeley and Harold Fisher, University of Rhode Island, **Fig. 2.1c–d:** Linda Stannard/Photo Researchers, **Fig. 2.1h:** Courtesy Richard Kuhn, Purdue University, **Fig. 2.1i:** Courtesy Jean-Yves Sgro, University of Wisconsin, Madison, **Fig. 2.1j:** Courtesy Jan Orenstein, George Washington University Medical Center; **Fig. 2.2a–d:** Reprinted from Ch. 3, S.C. Harrison, "Principles of Virus Structure", *Fundamental Virology* 4th ed., D.M. Knipe and P.M. Howley (eds), Lippincott Williams & Wilkins (Baltimore), (2001), Figs. 2–4, **Fig. 2.2e:** Reprinted with permission from Tsao, J., et al., "The three-dimensional structure of canine parvovirus and its functional implications *Science* 251: 1456–1464, copyright 1991, AAAS; **Fig. 2.3a, b:** Reprinted from Ch. 18, V. Racaniello, "Picornaviruses", *Fundamental Virology*, D.M. Knipe and P.M. Howley (eds), Lippincott Williams & Wilkins (Baltimore), 2001, Figs. 1b, 1h; **Fig. 2.3c:** Courtesy Norman H. Olson, University of California, San Diego; **Fig. 2.4a, c:** Reprinted from Ch. 3, S.C. Harrison, "Principles of Virus Structure," *Fundamental Virology*, D.M. Knipe and P.M. Howley (eds), Lippincott Williams & Wilkins (Baltimore), 2001, Figs. 6, 8, **Fig. 2.4b:** Reprinted with permission from Prasad, B.V. et al., "X-ray crystallographic structure of Norwalk virus capsid," *Science* 286: 287–290, Fig. 2, copyright 1999 AAAS; **Fig. 2.5a–c:** Courtesy John E. Johnson, Department of Molecular Biology, The Scripps Research Institute, **Fig. 2.5d:** Courtesy Wes Sundquist, University of Utah, **Fig. 2.5d–e:** Reprinted from Ganser, B.K. et al., "Assembly and analysis of conical models for HIV-1 core," *Science* 283: 80–83, **Fig. 3a, d;** copyright 1999 AAAS; **Fig. 2.6a, c:** Courtesy N. Acheson, **Fig. 2.6b:** PhotoDisc, Inc.; **Fig. 2.7:** Reprinted from Ch. 3, S.C. Harrison, "Principles of Virus Structure", *Fundamental Virology*, D.M. Knipe and P.M. Howley (eds), Lippincott Williams & Wilkins (Baltimore), 2001, Fig. 19; **Fig. 2.8a–b:** Reprinted from Liddington, R.C. et al., "Structure of Simian Virus 40 at 3.8 A Resolution," *Nature* (1991) 354: 278–284, **Fig. 2.8c:** Reprinted from Burnett R.M., et al. (eds) *Biological Macromolecules and Assemblies*, John Wiley and Sons, (1984) **Fig. 2.9a: Fig. 24.2,** p. 902 from *Molecular Biology of the Gene*, 4th ed. by James D. Watson, Nancy H. Hopkins, Jeffrey W. Roberts, Joan Argetsinger Steitz, and Alan M. Weiner. The Benjamin/Cummings Publishing Company, Inc. Copyright 1965, 1970, 1976, 1987 Reprinted by permission of Pearson Education, Inc (Atlanta).

Chapter 4

Fig. 4.3a–c: Courtesy Jurgen Kartenbeck, German Cancer Research Center; **Fig. 4.5:** Courtesy of Ari Helenius, modified from a figure from Whittaker, G.R.et al., "Viral entry into the nucleus," *Annual Review of Cell and Developmental Biology* (2000) 16: 627–651.

Chapter 5

Fig. 5.3: Courtesy of A.B. Jacobson, Reprinted from Fauquet, C. et al. (eds.), *Virus Taxonomy*, VIIIth Report of the International Committee on Taxonomy of Viruses, (2004) p. 741, with permission from Elsevier Press (San Diego).

Chatper 7

Fig. 7.1: Courtesy of N. Acheson. Redrawn from F.W. Studier and J.J. Dunn, Cold Spring Harbor Symposia on Quantitative Biology (1983) 47: 999, Fig. 3, with permission from Cold Spring Harbor Laboratory Press (Cold Spring Harbor); **Fig. 7.2:** Courtesy of N. Acheson. Redrawn from Molineux, I.J., "No Springs Please, Ejection of Phage T7 DNA from the Viron is Enzyme Driven," *Molecular Microbiology* (2001) 40: 1–8, Fig. 1, with permission of Blackwell Publishing Co.

Chapter 9

Fig. 9.2: Reproduced by permission of Edward Arnold from Siegl, G. and Cassinotti, P., "Parvoviruses," T*opley & Wilson's Microbiology and Microbial Infections*, Vol. 1, 9th ed., Mahy, B.W.J., and Collier, L. (eds.). London: Edward Arnold Publishers (1998) 261–279, Fig. 14.1; **Fig. 9.4:** Courtesy of N. Acheson. Adapted from Fig. 4 in Berns, K.I. and Linden, R.M., "The Cryptic Lifestyle of Adeno-Associated Virus," *Bioessays* (1995) 17: 237–245, with permission from John Wiley and Sons; **Fig. 9.5:** Reprinted from Ch.32, N. Muzyczka and K.I. Berns, "Parvoviruses," *Fundamental Virology*, D.M. Knipe and P.M. Howley (eds), Lippincott Williams & Wilkins (Baltimore), 2001, Fig. 5; **Fig. 9.6:** Reprinted with permission from Saudan, P., Ph.D. Thesis, University of Lausanne (2000);

Chapter 10

Fig. 10.10: Reprinted from Fig. 6a–c in *Cell* 119, Gai, D., et al., "Mechanisms of conformational change for a replicative hexameric helicase of SV40 large tumor antigen," pp 47–60, copyright 2004, with permission from Elsevier ;

Chapter 11

Fig. 11.1a–b: Courtesy Norman H. Olson, University of California, San Diego; **Fig. 11.3:** Modified from Fig. 2 in Longworth, M.S. and Laimins, L.A., "Pathogenisis of human papilomaviruses in differentiating epithelia," *Microbiology and Molecular Biology Reviews* (2004) 68: 362–372, with permission from the American Society for Microbiology.

Chapter 12

Fig. 12.1: Linda Stannard/Photo Researchers; **Fig. 12.2:** Reprinted from Rux, J.J, and Burnett, R.M., "Review: Adenovirus Structure," *Human Gene Therapy* (2004) 15: 1167–1176, Fig. 1a, with permission from Mary Ann Liebert Publishers; **Fig. 12.9:** Fig. 24.23, p. 932 from *Molecular Biology of the Gene, 4 th ed.* By James D. Watson, Nancy H. Hopkins, Jeffrey W. Roberts, Joan Argetsinger Steitz, and Alan M. Weiner. Copyright 1965, 1970, 1976, 1987 by The Benjamin/Cummings Publishing Company, Inc. Reprinted by permission of Pearson Education, Inc. (Atlanta).

Chapter 13

Fig. 13.3b: Courtesy Wah Chiu, Baylor College of Medicine. Zhou, A.H., et al., "Visualization of tegument/capsid interactions and DNA in intact herpes simplex virus type 1 virions," *Journal of Virology* 73: 3210–3218 (1999); **Fig. 13.4:** Courtesy Beate Sodeik, Hannover Medical School. Reprinted from Sodeik, B. et al., "Microtubule-mediated transport of incoming herpes simplex virus 1 capsids to the nucleus," *The Journal of Cell Biology* (1997) 136: 1007–1021, Fig. 1a–h with permission from Rockefeller University Press; **Fig. 13.6:** Reprinted from Ch. 33, B. Roizman, "Herpesviruses", *Fundamental Virology*, D.M. Knipe and P.M. Howley (eds), Lippincott Williams & Wilkins (Baltimore), 2001, Fig. 6; **Fig. 13.9:** Copyright Bernard Roizman.

Chatper 14

Fig. 14.1–14.2: Courtesy Dr. Eric B. Carstens, Queen's University, Kingston, Ontario; **Fig. 14.4:** Reprinted from A.L. Passarelli and L.K. Miller, "Identification and characterization of Lef-1, a baculovirus gene involved in late and very late gene expression," *Journal of Virology* (1993) 67: 3481–3488, with permission from the American Society for Microbiology;

Chapter 15

Fig. 15.1a–b: Reprinted from Moss, B., et al., in *Mechanics of DNA Replication and Recombination*, Cozarelli, N. (ed.). New York: A. Liss (1983), pp. 449–461; **Fig. 15.1c:** Reprinted from *Cell* 28, Baroudy, B.M., et al., "Incompletely base paired flip flop terminal loops link the two . . .", pp. 315–324, copyright 1982, with permission from Elsevier; **Fig. 15.2:** Courtesy of Richard Condit, modified from Fig. 3 in Condit, R.C. and Niles, E.G., "Orthopoxvirus genetics," *Current Topics in Microbiology and Immunology* (1990) 163: 1–39, with kind permission of Springer Science and Business Media (Heidelberg); **Fig. 15.3a:** Courtesy Dr. Frank Fenner, The Australian National University. Reprinted from *The Journal of Cell Biology* (1963) 18(1): 51–72, Fig. 4 with permission from Rockefeller University Press, **Fig. 15.3b:** Courtesy Samuel Dales, Rockefeller University; , **Fig. 15.4:** Courtesy of Richard Condit. Reprinted from Moyer, R.W. and Condit, R.C., "Poxviruses," *Topley & Wilson's Microbiology and Microbial Infections*, London: Edward Arnold Publishers (2005), reproduced by permission of Edward Arnold; **Fig. 15.6:** Courtesy Mariano Esteban, National Center of Biotechnology, Madrid, Spain.

Chapter 16

Fig. 16.1: Courtesy Ny Carlsberg Glyptotek; **Fig. 16.2:** Courtesy Norman H. Olson, University of California, San Diego, **Fig. 16.3:** Adapted from J.M. Hogle and V.R. Racaniello, "Poliovirus receptors and cell entry," *Molecular Biology of Picornaviruses*, Semler, B.L. and Wimmer, E. (eds.) (2002), with permission of ASM Press, Washington, D.C.; **Fig. 16.4:** Adapted from Leong, L.E., Cornell, C.T., and Semler, B.L., "Processing determinants and functions of the cleavage products of picornavirus polyproteins," *Molecular Biology of Picornaviruses*, Semler, B.L. and Wimmer, E. (eds.) (2002), with permission of ASM Press, Washington, D.C.; **Fig. 16.6:** Reprinted from *Seminars in Virology 8*, Stewart, S.R. and Semler, B.L., "RNA determinants of picornavirus and cap-dependent translation initiation," pp. 242–255, copyright 1997, with permission from Elsevier; **Fig. 16.8:** Reprinted from Cornell, C.T. and Semler, B.L., "Gene expression and replication of the picornaviruses," *Encyclopedia of Molecular Cell Biology & Molecular Medicine*, 2nd ed., Meyers, R.A. (ed.). Weinheim: Wiley-VCH (2004) 93–117, Fig. 2. **Figure 17.1, 17.2:** Courtesy Richard Kuhn, Purdue University.

Chapter 18

Fig. 18.1: Courtesy Richard Kuhn, Purdue University; **Fig. 18.3:** Modified from Fig. 7 in J.A. Lemm et al., "Polypeptide requirements for assembly of functional Sindbis virus replication complexes: a model for temporal regulation of minus and plus strand RNA synthesis" *European Molecular Biology Organization Journal* (1994) 13: 2925–2934, by permission of Oxford University Press; **Fig. 18.4:** Adapted from Figs. 6 and 7 in Schlesinger, S. and Schlesinger, M.J., "Togaviridae," *Fundamental Virology, 3rd ed.*, Fields, B.N., Knipe, D.M., and Howley, P.M., (eds.). Philadelphia: Lippincott-Raven (1996) pp. 523–540;

Chapter 19

Fig. 19.2a: Courtesy Dr. Guy Schoehn, CNBS, France; photo provided courtesy of Mark Denison.

Chapter 20

Fig. 20.1a,b: Courtesy Professor Rob Ruigrok, Universite Joseph Fournier, Grenoble, France; **Fig. 20.2a:** Courtesy Guy Schoehn, CNBS, France, **Fig. 20.2b:** Courtesy Professor Rob Ruigrok, Université Joseph Fournier, Grenoble,

France; **Fig. 20.5a–b:** Reprinted from Kolakofsky et al, (*Journal of General Virology* 86, 1869–1877, 2005).

Chapter 21

Fig. 21.1a–c: Courtesy Dr. Yoshihiro Kawaoka and Dr. Takeshi Noda. Reprinted from Watanabe, S., et al., "Production of novel Ebola virus-like particles from cDNAs: an alternative to Ebola virus generation by reverse genetics." *Journal of Virology* (2004) 78: 999–1005.

Chapter 22

Fig. 22.1a: Courtesy B. V. Venkataram Prasad, Baylor College of Medicine; Fig. 22.4: Courtesy of Esa Kuismanen.Reprinted from Jantti, J. et al., *Journal of Virology* (1997) 71: 1162–1172, Fig. 1A.

Chapter 23

Table 23-1: Reprinted from Ch. 24, Lamb, R.A. and Krug, R.M., "Orthomyxoviridae: the viruses and their replication," *Fundamental Virology*, D.M. Knipe and P.M. Howley (eds), Lippincott Williams & Wilkins, 2001,; **Fig. 23.1:** Courtesy Robert Dourmashkin, London Metropolitan University, **Fig. 23.2:** Courtesy of P.W. Choppin. Reprinted from Choppin, P.W., "On the emergence of influenza virus filaments from host cells." *Virology* (1963) 21: 278–281, Figs. 1–3: **Fig. 23.5:** Redrawn from Portela, A. and Digard, P., "The influenza virus nucleoprotein: a multifunctional RNA-binding protein pivotal to virus replication," *Journal of General Virology* (2002) 83: 723–734, with permission from Society for General Microbiology; adapted from Compans, R.W., Content, J., and Duesberg, P.H., "Structure of the ribonucleoprotein of influenza virus," *Journal of Virology* (1972) 10: 795–800, with permission from American Society for Microbiology.

Chapter 24

Fig. 24.5a–c: Courtesy Max L. Nibert, Harvard Medical School. Reprinted from Centonze, V.E., et al., "Structure of reovirus virions and subvirion particles." Journal of Structural Biology (1995) 115: 215–225; **Fig. 24.7a–b:** Reprinted from Becker, M.M., et al., "Reovirus sigma NS protein is required for nucleation of viral assembly complexes and formation of viral inclusions." *Journal of Virology* (2001) 75: 1459–1475, Figs. 4a, c.

Chapter 26

Fig. 26.1: Courtesy Liwie Rong, McGill University; Fig. 26.2a–b: Reprinted from Panteleo, G. and Fauci, A.S., "New concepts in the immunopathogenesis of HIV infection," *Annual Review of Immunology* (1995) 13: 487–512.

Chapter 30

Fig. 30.1: Peter Morrison/AP/Wide World Photos; Fig. 30.2: Nobel Lecture, December 13, 1976 by D.C. Gajdusek. Copyright The Nobel Foundation, 1976; Fig. 30.3: Adapted from Fig. 6 in *Journal of Molecular Biology* vol. 292, R. Moore et al., "Ataxia in prion protein (PrP)-deficient mice is associated with upregulation of the novel PrP-like protein doppel," pp. 797–817, copyright 1999, with permission from Elsevier; **Fig. 30.4a:** Courtesy of Holger Wille. **Fig. 30.4b** and **Fig. 30.6a–b:** Reprinted from Holger Wille, et al., "Structural studies of the scrapie prion protein by electron crystallography." Proceedings of the National Academy of Sciences, USA (2001) 99: 3563–3568, Figs. 2d–e; **Fig. 30.5a and 30.7:** Reprinted from Figs. 1 and 2 in *Current Opinion in Structural Biology* vol. 10, Jackson, G.S. and Clarke, A.R., "Mammalian Prion Proteins," pp. 69–74, copyright 2000, with permission from Elsevier; **Fig. 30.5b:** Courtesy of Katherine Sutliffe and P. Morrighan, Science. Reprinted with permission from Priola, S.A., Chesebro, B., and Caughey, B., "A view from the top: Prion diseases from 10,000 feet," Science 300: 917–919, copyright 2003, AAAS.

Chapter 32

Fig. 32.2a: Adapted from Smith, T., et al., "Neutralizing antibody to human rhinovirus 14 penetrates the receptor-binding canyon," *Nature* (1996) 383: 350–354, Fig. 3b, **Fig. 32.2b:** Reprinted with permission from Badger, J., et al., "Structural analysis of a series of antiviral agents complexed with human rhinovirus 14," PNAS (1988) 85: 3304–3308, Fig. 2; **Fig. 32.10b:** Courtesy of Elspeth Garman, Reprinted with permission from Graeme, L. and Garman, E., "The origin and control of pandemic influenza," *Science* 293: 1776–1777, copyright 2001 AAAS.

SCHEMATIC DIAGRAMS OF VIRUSES COVERED IN THIS BOOK

The diagrams on the following two pages illustrate most of the viruses covered in detail in this book, and can be used as a study guide and a source of information about these viruses. Nucleocapsids with icosahedral symmetry are shown schematically as icosahedrons; nucleocapsids with helical symmetry are shown as coils or helices. Envelopes are shown as bilayers with integral membrane proteins shown as clubbed spikes inserted with the envelope. The topology (linear or circular) number of segments, and strandedness of the viral genomes are indicated by the black lines within the virions. Relative sizes of virions are shown. The smallest virion illustrated, a single-stranded RNA bacteriophage (MS2 or Q,), has a diameter of 26 nm; the largest virion illustrated, a poxvirus (vaccinia virus), has dimensions of 150 × 300 nm. Filovirus virions are approximately 1000 nm long; only a part of length of the filovirus virion is shown.

ssRNA phage

Parvo

Picorna

ϕX 174

Polyoma

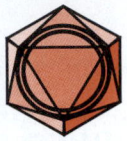

Papilloma

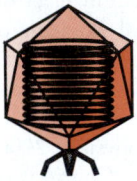

T7 phage

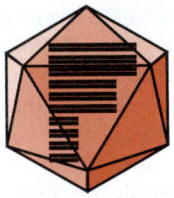

Reovirus

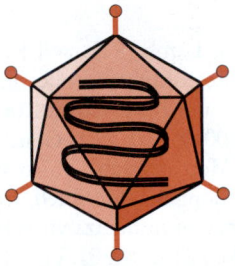

Adenovirus

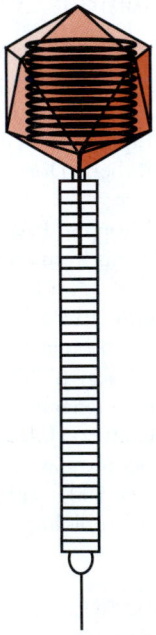

Lambda phage

Baculovirus

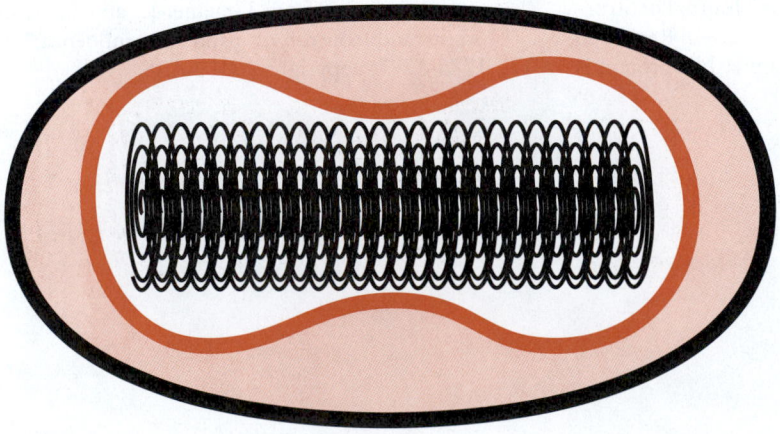

Poxvirus

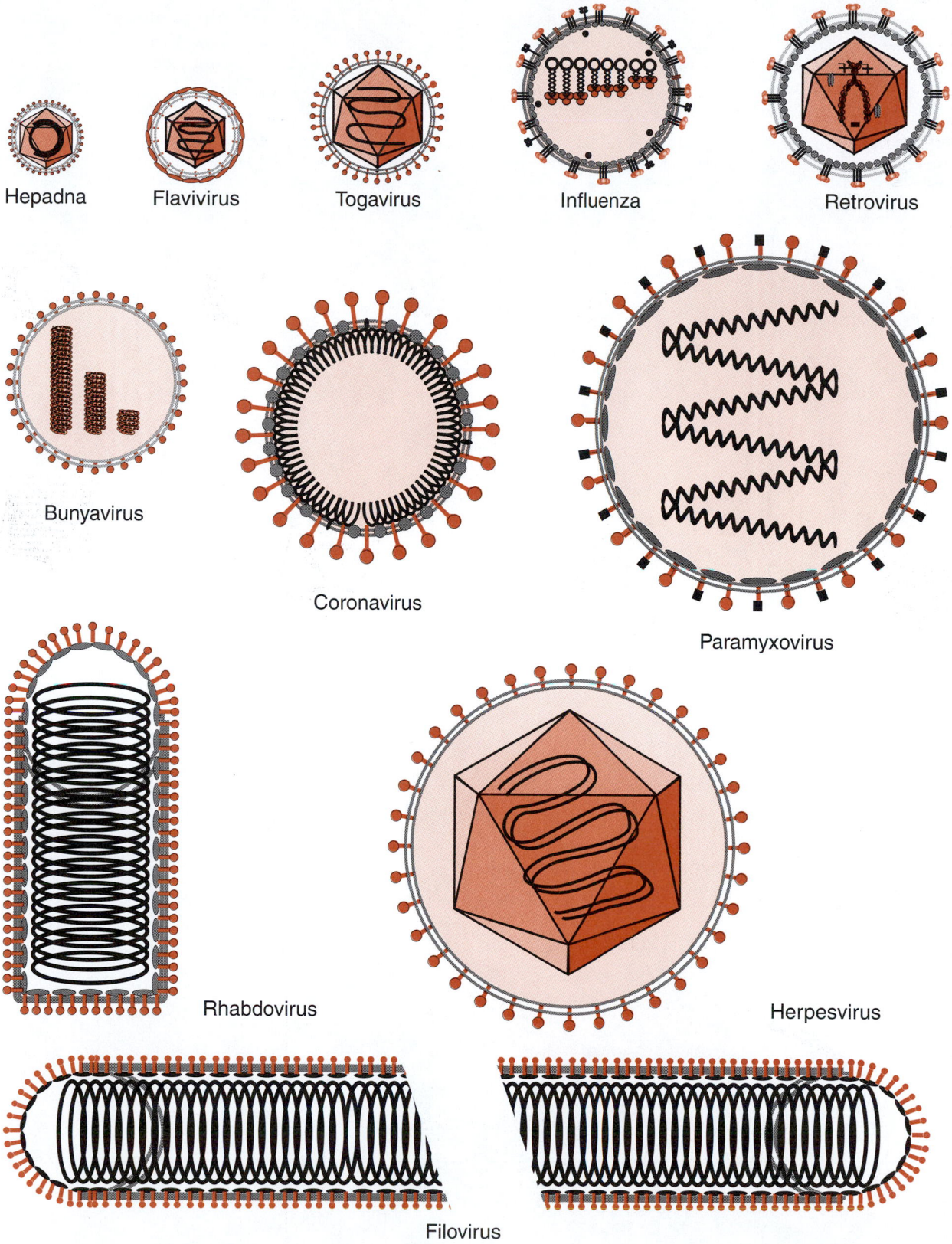
Hepadna
Flavivirus
Togavirus
Influenza
Retrovirus
Bunyavirus
Coronavirus
Paramyxovirus
Rhabdovirus
Herpesvirus
Filovirus